T0203541

ELECTROCATALYSIS
COMPUTATIONAL, EXPERIMENTAL, AND INDUSTRIAL ASPECTS

ELECTROCATALYSIS
COMPUTATIONAL, EXPERIMENTAL, AND INDUSTRIAL ASPECTS

Edited by

Carlos Fernando Zinola
Universidad de la Republica
Montevideo, Uruguay

CRC Press
Taylor & Francis Group
Boca Raton London New York

CRC Press is an imprint of the
Taylor & Francis Group, an **informa** business

CRC Press
Taylor & Francis Group
6000 Broken Sound Parkway NW, Suite 300
Boca Raton, FL 33487-2742

First issued in paperback 2019

ISBN-13: 978-1-4200-4544-4 (hbk)
ISBN-13: 978-0-367-38436-4 (pbk)

Library of Congress Cataloging-in-Publication Data

Zinola, Carlos Fernando.
   Electrocatalysis : computational, experimental, and industrial aspects / Carlos Fernando Zinola.
     p. cm. -- (Surfactant science)
   Includes bibliographical references and index.
   ISBN 978-1-4200-4544-4 (alk. paper)
   1. Electrocatalysis. I. Title.

QD569.Z56 2010
541'.395--dc22
                                              2009050553

Obituary

In memory of my mentor and colleague

During the final stages of preparation of Chapter 19 of this book, Professor Algirdas Vaškelis passed away on February 5, 2009, in Vilnius, Lithuania. Professor Vaškelis was a widely recognized scientist who worked in the field of electrochemistry and electrocatalysis, electroless deposition of metals and alloys, and related topics. This chapter summarizes some of the important aspects of these electrocatalytic processes and their specific features, and provides a general background and a basic understanding based on the original contributions from Vaškelis' school.

Zenonas Jusys

Contents

Preface

Electrocatalysis is an important field for a potentially wide variety of applications, including inorganic and organic electrosyntheses, conversion processes, auxiliary power devices, new transportation technologies, distributed generation applications, and metallic electrodepositions. These applications are employed in a large number of industries worldwide both in old technologies such as galvanoplasty and in new ones such as ultracapacitors. This book attempts to provide a more comprehensive approach to electrocatalysis than many others and includes industrial applications that use newer technologies.

First, a basic overview is provided as to the current uses of electrocatalysis, and this is followed by future perspectives on this field. Some revised introductory concepts are also presented to provide a comprehensive picture. The rate-determining step and what happens when one rate-determining step is not dominant are key points that are clearly discussed here. One of the first plain approaches, *otrora*, reported by J. Appleby, is also presented: the distinction between electrochemical and electrocatalytic reactions, that is, the use of both variable electrode potentials and the presence of adsorbates. The three classic electrocatalytic reactions are discussed—hydrogen evolution, methanol oxidation, and oxygen electroreduction—along with the approaches formerly studied by the author. In the same chapter, Bokris showed some examples of important applications, such as the electrochemical removal of wastes and the catalytic removal of electro-toxic organic wastes, are briefly discussed. Electrocatalytic sewage disposal and the influence of electrocatalysis on global warming are also dealt with.

Second, the quantum nature of the electron transfer is depicted from a different viewpoint. The fundamentals of quantum electrochemistry are briefly presented, and Pauli's, Heisenberg's, Wigner and Von Neumann's, Heitler and London's, and Weyl's historical first approaches to quantum physics are also included. Some aspects of solid-state physics, especially phonon dispersion in the absence and in the presence of a periodic electric potential, are described. Exact and semiempirical methods, such as CNDO INDO and NDDO, are presented with special emphasis on the simple extended Hueckel method. The fundamentals of density functional theory are explained along with the Hohenberg–Kohn theorems, Kohn–Sham equations, and exchange correlation functional energy with the local density approximation. Some examples of computational electrochemistry, such as simple organic molecules and some inorganic species, are studied here through their electroadsorption and decomposition on iron and platinum single crystal clusters. Finally, enzyme chemical catalysis and some aspects of bioelectrocatalysis by an overall structure and active site metallocenter of Ni–Fe hydrogenases are depicted in Chapter 7.

Third, one of the most significant and classical views of electrocatalysis is described with various set points, that is, a global theoretical background covering the Müller–Calandra, Srinivasan–Gileadi, and instantaneous nucleation–growth overlap models. The electrochemical consequences of the three global models are presented for tin in buffered solutions. As a consequence of new advances in the surface properties of single crystals and the characterization of metallic topologies, the physical properties of monocrystalline electrodes are explained with examples of atomic rearrangement. The changes in the surface structure of platinum-type electrodes induced by metal underpotential deposition and anion adsorbates are also included. Besides, the new surface chemistry at bimetallic catalysts is presented with some examples employing electrochemical techniques and ultra high vacuum and spectroscopic *in situ* and *ex situ* methodologies. Finally, a comparison between non-noble and noble metal dissolution is presented along with some experimental aspects of simple organic adsorption inhibition of metal corrosion.

The importance of electrocatalysts as the core of an electrochemical reactor gave us a whole new comprehension using introductive concepts and personal experiences. The main success of this discipline lies in selecting the right electrocatalyst, and is presented here using the influence of electrode potential and current distribution on smooth and rough electrocatalytic surfaces. The influence of the metal roughness at regular and periodic surfaces: were taken as examples. Electrocatalysis in fuel cells is presented considering the electrochemical fuel cell reactor again with the current and potential distributions on the cylindrical pores. In this sense, a diffusion-convective system in the electrochemical reactor using the classical boundary layer condition is fully described. The effects of temperature, double layer, and pressure drop in a packed bed reactor are discussed for an electrocatalytic reactor design. The study of multi-electron charge transfer reactions is a subject of much interest to many electrochemists. Thus, the electrode surface design for electrocatalysis seems to be more important and complex. The use of metal adatoms to modify platinum and gold electrodes for electrosynthesis in an aqueous medium is presented considering the electro-oxidation of carbohydrates and alcohols and the electro-reduction of carbonyl compounds.

Even when working with semi-noble metals in an electrochemical reactor, corrosion processes take place. The basics of corrosion science and technology are analyzed in detail by first considering band theory with electronic and ionic levels defining Fermi energies of electrode reactions. The use of semiconductor electrodes is also worth noticing: the anodic dissolution of covalent semiconductors and anodic photo-dissolution of semiconductors is also important in the new technologies available in engineering. In certain cases, the passivation of electrocatalysts produces disadvantages in the faradaic efficiencies of the reactor, and thus the passivity of metals and semiconductors is studied in the case of localized corrosion.

When working in galvanoplasty, erosion and corrosion can change the deposition processes together with changes in the current distribution due to macroscopic variations in the substrate morphology. The use of microradiology to follow fine electrodeposition with coherent x-rays is also described as an effective real-time instrument for electrodeposition studies. One of the most important problems in electrocatalysis is the shielding of metal deposition by the formation of bubbles; the building on bubbles on zinc and copper-ramified structures or nickel overlayers will therefore be a practical application desirable for any specialized reader. In the case of studying electroplating some slim or fine deposits can be achieved by electroless plating. The theoretical considerations by the action of hydrogen-containing reducing agents with metal ion complexes are worth noticing in this case. However, with the arrival of new experimental techniques, it is possible to verify the electroless plating mechanism. For example, the reduction of Ni(II) ions by hypophosphite using online mass spectrometry and the reduction of Cu(II) ions by formaldehyde using both mass spectrometry and *in situ* electrochemical quartz crystal microgravimetry to measure deposition rates.

Finally, the production, storage, use, and delivery of hydrogen in industrial electrochemistry are briefly presented. The production of hydrogen via industrial electrolysis and gas-derived or petroleum-derived reforming has been specifically emphasized. The steam-reforming reactions have been analyzed together with the water–gas shift reaction. A brief explanation of coal gasification, biomass gasification, thermochemical decomposition, and photo-electrochemical water splitting has also been provided. Since we still believe that the cleanest and simplest method is water electrolysis, we emphasize the possibilities of making water electrolysis an efficient method: alkaline water and polymer electrolyte electrolyzers. However, we have to be realistic and the energy losses within a hydrogen economy have to be studied in more detail and solved; the technical and economic overviews of electrolytic hydrogen production systems are also provided along with the calculation of some cost results.

C. Fernando Zinola
Fundamental Electrochemistry Department
Universidad de la Republica
Montevideo, Uruguay

Overview

A major concern with electrochemistry is that the scientific community does not understand the theoretical concepts since they still believes that electrochemistry is only a part of classical physical chemistry. All these people are not able to generate fundamental knowledge and huge mistakes arise in important reports of energy efficiency, or energy conversion, etc., such as saying that alkaline metal deposition occurs before hydrogen evolution. There is an enormous misunderstanding between the actual requirements in today's undergraduate courses in basic sciences and engineering and what the "politicians" dictate to be good, that is, which science to teach and which not to. These politicians should not be making a decision in this regard.

When we teach electrochemistry, we understand that there are almost no converging issues with all the necessary theoretical concepts with rapid devolution to experimental situations and technological applications. Some books cover only fundamental aspects wherein concepts are discussed based either solely on equations or descriptive explanations, while other books cover only the latest techniques from a practical and experimental viewpoint without providing any real insight into their potential applications in the industry. Moreover, in the field of industrial electrochemistry, there are no new wide-ranging approaches to the classical industries from a fundamental viewpoint either for manufacturing, syntheses, conversion processes, or machinery applications. The principal books on this subject take either a chemical or an electrical engineering viewpoint, and besides five books published in the 1980s, there are almost no books written from an electrochemical engineering viewpoint. I have encountered both situations, and only some books published in the mid-1960s helped me to discover what really happens in an electrochemical reactor. Thus, as a consequence of unpleasant professional experiences and after discussions with colleagues, I decided to write this book with the support of Prof. A. Hubbard, to whom I will always be grateful.

While planning this book, we realized that it would be impossible to include the fundamental aspects, experimental techniques, and technologies of every electrochemical industry in a single volume. Therefore, we decided to concentrate only on the most important developments in the science of electrocatalysis, which are the most relevant to new technologies. Therefore, this book provides only a brief summary of the fundamental technological progresses in this field. Some of the most important concepts in electrocatalysis are presented here keeping in mind the three major industrial applications: metal electrodeposition, fuel cells, and hydrogen production. Despite the *boom* in the petroleum industry and loads of conflicting speculation, hydrogen technology seems to be one of the most promising alternatives for energy conversion.

Chapter 1 is written by one of the most renowned electrochemists, Prof. J. O'M. Bockris, to whom I am greatly indebted, as he provides strong encouragement to young scientists from Third World countries. Prof. Bockris' books have provided me an in-depth understanding of electroplating while I was still working in the industry. Bockris views in galvanoplasty opened my mind to explain the new advances to the "old-fashioned engineers."

I have always been interested in electrochemistry from my early days in the industry. Since it was not possible to do research simultaneously in that direction in an electrochemical plant, I decided to quit and became a full research teacher. Against all odds I have found success, or rather what is called "success" in my country. I was initially inspired by Prof. E.Y. Spangenberg from whom I learned many aspects of academic life that I still practice. As there was no possibility of further specialization in electrochemistry in my country, Prof. A. Arvia invited me to his research institute in Argentina, where more than 50 scientists were already conducting research. At the institute, a whole new world opened up to me and all that was obscure appeared in a new light. My dearest friend, Prof. M. Martins, showed me the intricacies of electrochemical techniques and taught me to be patient with results. The grounding in electrocatalysis is very academic and formative and it is highly recommended for all young scientists who intend to work in the field of electrochemistry. Prof. G. Estiú, one of the most renowned theoretical electrochemists to have emerged from this institute, gave us an impetus for this book: theoretical bio(electro)catalysis. I have added some important points to the fundamentals of quantum electrochemistry and solid-state physics, which are required for the understanding of theoretical electrocatalysis.

Since the last editions of the books on electroplating in the late 1970s and 1980s, only a few books on the subject have appeared in the market, but electrochemical deposition has evolved from what I have called art into an exact science. This development can be attributed to the augmentation and diversification of *in situ* experimental techniques. Therefore, I have asked my friends Prof. B. Gianetti, S. Bonilla, and C. Almeida, from the Universidade Paulista, São Paulo, Brazil, to present, from a global theoretical viewpoint, a detailed description of the deposition processes with the corresponding voltammetric consequences associated to each process. Since all of the equations derived in every case have to be proven, Prof. G. Margaritondo with P. Hsu, C. Lin, Y. Hwu, and J. Je elaborated the possibility of making coherent x-ray microradiology an experimental blueprint for the deposition process. On the other hand, the use of electroless plating is still more applicable than expected. I have learned this in my work with galvanoplasty when preparing the silver electroless film between the first copper electrodeposition on iron and the final bright chromium electrodeposition. Thus, my friends Prof. Z. Jusys and Prof. A. Vaškelis, put special emphasis on the fundamental aspects of copper and nickel electroless deposition, as two main examples, using a wide variety of *online* experimental techniques, such as electrochemical mass spectrometry and quartz crystal microgravimetry. We are especially grateful to Prof. Vaškelis's family for their support during his last days, as well as to his daughter and Dr. Jusys for their constant support.

Only those who possess an in-depth understanding of the fundamental aspects of electrocatalysis can conduct noteworthy research in electrochemical science and technology. I have presented a complementary theory of this field focusing mainly on the properties found for platinum-type and gold-type metals. This essentially makes this book a completely new contribution and not just a revised version of the earlier theoretical and fundamental approaches to electrocatalysis. Thus, for the sake of illustration, this book includes some chapters devoted to the fundamental aspects of electrochemistry, and theoretical concepts related to electrochemical reactors and current and potential distributions under various laws. Prof. R. White along with S. Santhanagopalan have shared their extensive experience, and have covered subjects ranging from rotating ring–disk electrodes to bubble formation and corrosion modeling. They have presented all the effects, such as pressure drop, double layer, temperature, etc., in the design of an electrochemical reactor with concepts and results that will be very useful to the reader.

When I first began working in electrochemistry 25 years ago, my first love was organic electrochemistry, and with the encouragement and support of my friends, Prof. M. González and Prof. H. Cerecetto, my dreams were realized. Thus, my new friend, Prof. N. Alonso-Vante (from Poitiers) along with B. Kokoh offered us the electrocatalytic aspects of fine organic electrocatalysis of large organic molecules such as carbohydrates and alcohols on platinum and gold electrodes modified by metal adatoms with the help of new *in situ* hybrid techniques.

Electrochemists around the world clearly agree on one point: corrosion of structures either in atmospheric or submarine environments is a topic that deserves special attention. This is especially true in the case of "developing countries" as it involves ∼20% of annual economic losses. The electrochemical processes in large reactors, such as those of chlor-alkali and high-temperature electrolysis, are also subjected to this undesirable process. Electrocatalytic processes are not an exception and who better than Prof. N. Sato to guide us through this subject. I will always be grateful to Prof. Sato for being one of the main contributors. His chapter (Chapter 22) covers a wide range of subjects from the fundamentals of metal and semiconductor solid-state physics, corrosion thermodynamics, and kinetics to the influence of light on corrosion and photo-corrosion, which is also approached to both metals and semiconductors. Some concepts of the chemistry of anticorrosive rust are also presented with an emphasis on localized corrosion in weathering steels.

I hope that this book succeeds in generating more interest in the study of electrocatalysis and in providing updated information in this field. The chapters have been written by different authors including myself, and distinct forms in style and approach will, therefore, be evident. However, in my dual role as editor and author, I have tried to smooth those diversities, without making any changes whatsoever to the basic content of each chapter. Therefore, every section is almost complete by itself and can be read and consulted separately. I would like to thank many members of the Electrochemical World Society who helped me in this project, especially Arthur Hubbard. I would also like to thank my entire family and friends for their patience and understanding during the long, hectic years spent in the preparation of this book.

C. Fernando Zinola
Montevideo, Uruguay

Contributors

C. M. V. B. Almeida
Laboratory of Production
 and Environment
Paulista University
São Paulo, Brazil

N. Alonso-Vante
Laboratory of Electrocatalysis
University of Poitiers
Poitiers, France

J. O'M. Bockris
Haile Plantation
Gainsville, Florida

S. H. Bonilla
Laboratory of Production
 and Environment
Paulista University
São Paulo, Brazil

Guillermina L. Estiu
Walther Cancer Research Center
University of Notre Dame
Notre Dame, Indiana

and

Department of Chemistry and Biochemistry
University of Notre Dame
Notre Dame, Indiana

B. F. Giannetti
Laboratory of Production
 and Environment
Paulista University
São Paulo, Brazil

P. C. Hsu
Institute of Physics
Academia Sinica
Taipei, Taiwan

Y. Hwu
Institute of Physics
Academia Sinica
Taipei, Taiwan

J. H. Je
X-Ray Imaging Center
Department of Materials Science
 and Engineering
Pohang University of Science and Technology
Pohang, South Korea

Z. Jusys
Department of Catalysis
Center of Renewable Energies
Institute of Chemistry
Vilnius, Lithuania

and

Institute of Surface Chemistry and Catalysis
Ulm University
Ulm, Germany

K. B. Kokoh
Laboratory of Electrocatalysis
University of Poitiers
Poitiers, France

C. S. Lin
Institute of Physics
Academia Sinica
Taipei, Taiwan

G. Margaritondo
Ecole Polytechnique Fédérale de Lausanne
Lausanne, Switzerland

Shriram Santhanagopalan
Celgard LLC
Charlotte, North Carolina

N. Sato
Graduate School of Engineering
Hokkaido University
Sapporo, Japan

A. Vaškelis
Department of Catalysis
Institute of Chemistry
Vilnius, Lithuania

Ralph E. White
Department of Chemical Engineering
University of South Carolina
Columbia, South Carolina

C. Fernando Zinola
Fundamental Electrochemistry Laboratory
School of Sciences
Universidad de la República
Montevideo, Uruguay

1 About Electrocatalysis Until 2000

J. O'M. Bockris

CONTENTS

1.1   GENERAL VIEW

1.1.1   TAFEL (1905)

I begin my chapter by reminding the reader about the beginning of electrode kinetics. This was in a paper presenting an empirical law (the first law of electrode kinetics) that is the basis for our discussion. Of course, there is much else to say apart from this equation, but it plays the same part in electrochemistry as the famous Arrhenius equation, $r = Ae^{-E/RT}$, plays in physical chemistry. Julius Tafel made a very direct contribution to the subject of kinetics at electrodes. In fact, he founded it as a science with equations. Insofar as electrocatalysis concerns the acceleration of electrode reactions by means of the change of substrate, Tafel's equation is relevant, indeed [1], and, written in the original form, was

$$\eta = a \pm b \log j \tag{1.1}$$

where η is the overpotential, the potential of the working electrode, diminished by the reversible potential, for the electrode reaction concerned. η might well be then renamed the excitation potential, because when it is zero no reaction occurs and when finite, the reaction goes either forward or backward, according to the sign of η, positive for backward, negative for forward.

The law is limited in its applicability to the lengthy middle part of the current versus the potential relation of electrochemical reaction. At sufficiently low values of η, there is a linear region, in which the current density, j, is proportional to η (see Figure 1.1a and b).

At sufficiently high j values, the availability of material for the reaction at the desired rate becomes deficient to meet the demands of the electrode at this relatively high value of η. One comes

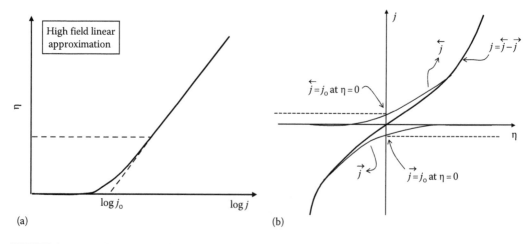

FIGURE 1.1 (a) The typical Tafel line for a single electron-transfer process. The exponential relationship under the high field approximation tends to a linear overpotential (η) versus the logarithm current density (j) expression. (b) The current density (j) versus overpotential (η) curve has a single point of interest, the point of zero overpotential and null current density, that is, the equilibrium situation.

to a limited current region, $j = j_L = \text{constant}$, and Tafel's law ($\eta = a \pm b \log j$) is no longer applicable. Tafel's law can be reexpressed as

$$j = j_o \exp(\pm \alpha F \eta / RT) \tag{1.2}$$

where

j_o is the so-called exchange current density (see [7, p. 24]), a characteristic of a given electrode reaction on a given surface (at a constant temperature)

α is a parameter known as the electronic transfer coefficient

F is the Faraday constant (96,500 C mol^{-1})

R and T are symbols, the meanings of which are well known in physical chemistry

It turns out that α is also dependent on the mechanism of the reaction. Electrode materials on which a given electrode reaction takes place at a relatively high j, for a minimum η, are good electrocatalysts.

1.1.2 BOWDEN AND RIDEAL (1928)

Bowden and Rideal became known in the field of "electrode processes" because of their 1928 publication from Cambridge University in the United Kingdom. They started an electrochemical school at Cambridge that included such names as Agar and also Evans and Hoar who were outstanding in applied work (1936), evolving the mechanism behind corrosion and then applying it to practical problems faced by the Royal Navy during World War II. They were the first to use transient techniques in electrochemistry. Cathode-ray oscillographs were far from being readily available in 1928, but Bowden and Rideal introduced a substitute, the string galvanometer, for electrode kinetic measurements. This allowed measurements in the millisecond range [2,3]. Now, if one plots the electrode potential and its variation with time at a constant current density, the result can be seen as in Figure 1.2. Bowden and Rideal assumed that the linear section of current versus

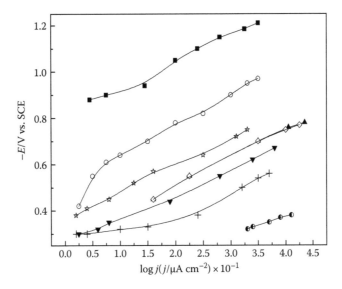

FIGURE 1.2 Tafel plots as electrode potential, E, versus the logarithm of the current density, $\log j$, for mercury (■-■), platinized mercury to 1% (o-o), platinized mercury as thin film (▲-▲), fresh etched silver (▼-▼), aged etched silver (◇-◇), polished silver (☆-☆), bright platinum (+-+), and spongy platinum electrodes (●-●).

potential relation represented the charging of "the double layer,"* a process in which nearly all the electron flow goes into changing the electrode potential up to a value sufficient to cause a reaction to occur. (See the part of Figure 1.2 in which the electrode potential bends over and becomes constant.) Charging is now complete, and all the potential is causing all the electrons to crossover, thus causing a reaction to occur "at a steady state."

For example

$$H_3O^+{}_{(solution)} + e_{(metal)} \rightarrow HM_{(catalyst)} + H_2O_{(solution)} \qquad (1.3)$$

where M is the metal catalyst on which H atoms are being adsorbed.

Bowden and Rideal achieved two things with the new equipment. The double-layer charging part of the graph in Figure 1.1 can be represented by

$$j = C\frac{dE}{dt} \qquad (1.4)$$

where
  C is the electrical capacity of the double layer
  E is the varying electrode potential with time t

Using liquid mercury, the Cambridge workers assumed that the area of the electrocatalyst was what it seemed to be: no invisible roughness. The capacity (around 20 μF cm^{-2}) was then taken by Bowden and Rideal to be the real capacity of the double layer. (By "real," they meant "unaffected by invisible micro-roughness.")

Other materials (e.g., tin or platinum) were found to show bigger values of capacity, C, for an apparent 1 cm^2 area of the electrode (as measured externally neglecting micro-roughness), and the authors took these higher values of C to reflect the micro-roughness of the surfaces of tin or platinum, which a liquid would not have. Assuming, then, that a difference in real area was the only reason for the differences in C of various solid metals, they took it that measuring C from the slope of the straight-line section of the double-layer region (Figure 1.3) gave a measure of the "real" surface area. If a metal catalyst had a value of three times greater than that of the same area of a mercury electrode, its real surface area, including all the invariable roughness of the surface, made the "real" area 3 cm^2 instead of 1.†

This technique, being able to measure the real surface area, was relevant to the as yet unrealized "catalytic properties of the surface," for Bowden and Rideal claimed that to compare the electro-catalytic properties of various metals one had to first eliminate any difference in real areas for the comparison of reaction rates (measured later when the potential no longer varied with time [Figure 1.2] on the same area).

Bowden and Rideal's work was decidedly electrocatalytic in nature. They measured the rate of hydrogen evolution on several electrode materials. Taking into account the real area, that is, the geometric (apparent) area, decided by the ratio of C_M/C_{Hg}, they found a true, substantial, difference in rates of the hydrogen evolution on various different metals of the same real area.

It is remarkable that the erudite Rideal (a specialist in catalysis) did not introduce the term electrocatalysis in 1928. The reason, I think, was that electrochemistry in 1928 was generally focused on the electrode potential at a constant current and not the current (= rate) at a constant

---

* Electrochemists frequently refer to the interfacial region between a metal and the first layer of ions adjacent to the electrode—quite lengthy—as a double layer.

† The Bowden and Rideal method of dealing with apparent and real surface areas was accepted by electrochemists for about 50 years. Gallium is liquid at an early acceptable temperature, but does not have the same capacity as mercury! So the Bowden and Rideal method must now be regarded as inaccurate at best and inapplicable at worst. However, it is still used as a "first approximation."

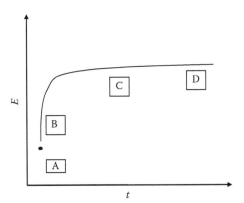

FIGURE 1.3 The basic galvanostatic experience, represented as electrode potential (E) versus time (t). The starting point is A, and the A–B section denotes the ohmic drop. This figure is blank since the experiments in the 1960s and 1970s were performed using a cathode-ray oscilloscope. The linear section in the B–C segment represents the charging of the double layer. After a while, the lines begin to bend and near C it gets almost horizontal. In this situation, some of the ions in solution approach the electrons at the metal due to the E effect. Under the C–D segment (*plateau situation*) all the ions quickly reach the double layer to discharge on the metal. Thus, the value of E does not vary with time anymore, especially for solid electrodes.

potential. They held the current density constant and found the potential at the same current, thus, suppressing the "rate" aspect of the phenomenon.*

Thus, if you know the current density, the current per unit area, for example, per square centimeter, when the square centimeter is real, one is also measuring the rate of the electrode reaction in moles per square centimeter (mol cm^{-2}) of the product. It is 79 years since Bowden and Rideal, but electrochemists still talk in terms of amperes (A) or milliamperes per square centimeter (mA cm^{-2}) and in a sense diminished the chance of a linkup with the general field of catalysis in chemistry, where, of course, results are expressed in gram moles per square centimeter per second (g mol cm^{-2} s^{-1}).

1.1.3 GURNEY (1931)

Ronald Gurney had already become well known, because of his book with Mott, often regarded as the foundation of solid-state physics, and also because of his work on the basic theory of radioactivity. One recalls that the Schrödinger equation was published in 1926 so that Gurney's paper on charge transfer through an energy barrier at the electrode–solution interface was the first application of tunneling in chemistry and was published in 1931, only five years after the beginning of "wave mechanics" (1926). It bore, conceptually, much resemblance to the first paper on radioactive decay by Condon and Gurney. In this subject, it is electrons that tunnel out of the atom and in the case of the electrode, it is the electrons that tunnel through the "energy barrier" across the electrode–solution interface.

Although we can venerate Gurney as the first quantum electrochemist, his model had a flaw in it, and this flaw is relevant to this chapter because its presence prevented Gurney's theory being applied to give a basis to electrocatalysis. Gurney had the electrons leaving the metal and "neutralizing" H_3O^+ in the double layer.

What resulted were a H atom and a water molecule, neither attached to the electrode, that is, there was no chemical bonding between the electrode and the resulting H atom. The atom, in fact, in

---

* In fact, $v = j/nF$, where v is the rate in moles per square centimeter per second, n is the number of electrons used up in achieving one act of the overall reaction.

Gurney's model, ended up unbonded to anything, although a few angstroms away from the catalyst surface. One might think that the work function of the metal would play a part here, and it does indeed enter Gurney's equations. But the problem is that measurements in electrochemistry involve a metal–metal junction between the material of the working electrode and that of the reference electrode, so that the work function of the catalyst (which contributes to the metal–metal contact potential) cancels out. Thus, from the electrocatalytic point of view, Gurney's famous 1931 paper contributed nothing, although the physicists praised the paper because it brought in quantum mechanical tunneling at that time, a rarely used concept. There was, in fact, no metal–hydrogen bonding from which was to be the basis of the first catalysis theory introduced by Butler, independently by Horiuti and Polanyi [4].

1.1.4 BUTLER (1936)

Gurney was a solid-state physicist connected in his earlier days with Neville Mott, the most recognized solid-state physicist of the twentieth century. J.A.V. Butler was a physical chemist, less recognized at the time of the publication I am about to describe, but nevertheless without doubt one of only four or five founders of physical electrochemistry.*

The contribution I want to outline was basically along the same lines as the 1931 Gurney paper described earlier. In 1936, it was still rare to find a paper on the basic mechanism of electrode reactions with quantum mechanical content; this would nevertheless be a way Butler's 1936 paper can be seen [6]. Its main value historically was that it hit on the mistake in Gurney's paper. Being a physicist, Gurney was less sensitive to the need for chemical bonding and that is perhaps why he had left his H atom, in the reaction Gurney treated, neutralized but unattached. Butler's paper was similar to that of Gurney but his H atom was chemically bonded to the metal after the receipt of the electron by the proton in H_3O^+. Thus, in Butler's contribution, the proton would have to receive a lesser degree of overpotential (about 1 eV) than the unbonded one in the prior Gurney paper. Depending on the nature of the metal, the $H_3O^+ + e_{metal} \rightarrow MH + H_2O$ reaction would be easier for a metal that encouraged strong bonding with H. A higher degree of overpotential would be required to stimulate a certain reaction on a metal bonded weakly to H.

Now, any reader of this chapter will realize that, to use a phrase, this strikes a bell. We are drawing close to the birth of theoretical electrocatalysis and Butler's 1936 paper can be seen as a birthing of the explanation that in fact has gone into most parts of electrochemistry and affected particularly those that have financial consequences in industrial processes for a lesser overpotential translates into a fall in price of the product on sale. But it is most interesting to find out that even then, in 1936, there was still no explicit recognition of the connection in the phenomenon of the metal–hydrogen bonding in electrochemistry to catalysis in chemistry.

The slight mathematical operation, which placed the overpotential in the exponential position and the current density on the line, was not yet normal. It was introduced by Agar, also of Cambridge University, in 1938 [7]. One had to wait 23 years before the connection to the well-known phenomenon of catalysis in chemistry became so obvious that, at last, the right name, "electrocatalysis," was used in fuel cell work by Grubb of the General Electric Company in a 1959 publication [8].

---

* Butler was a man utterly wrapped up in his work illustrating that research is principally done at the desk. His absentmindedness is the origin of several storied incidents. Thus, he was wanted to ponder a great deal, bent over pads of white paper on which gradually appeared equations and the order of magnitude calculations, occasionally a diagram. It was characteristic of Butler's mode of thinking that he would whistle rather tunelessly, as the equations and calculations appeared. However, from time to time this unconscious whistling would get too much for him and he would turn on Brian Conway—who occupied the same office—and say, petulantly, "Cannot you stop that whistling, Conway? You are disturbing my thinking."

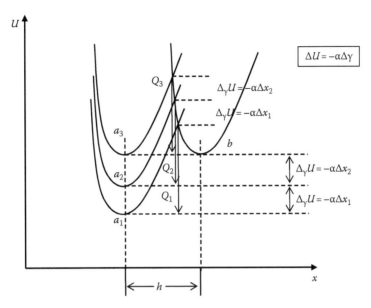

FIGURE 1.4 First explicit expression of a simple electrocatalytic mechanism showing the potential energy versus position diagram of the H^+–OH_2 system. The three curves on the left depict the potential energy curves for the H-adatom interaction with three different metal catalysts. The height of the first intersection point (the heat of activation) with each of the three H-adatom/metal minus the lowest vibration energy level of the H^+–OH_2 bond is the height of the activation energy for the proton discharge reaction. As the strength between the H-adatom and the metal increases, the minimum of the curve gets lower than the lowest vibrational level of the H^+–OH_2 bond. Then, the reaction has a lesser activation energy, and the reaction goes faster for the highest H-adatom/metal interaction.

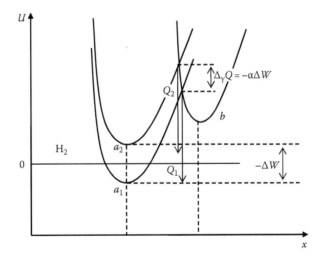

FIGURE 1.5 The right-hand side potential energy versus position diagram is that of the H^+–OH_2 system, whereas the two energy-shaped curves at the left side are the ones representing the H-adatom/metal bond of the adsorbed state. The difference in the energy state arises from being, first, a part of the solution state and, later, to the adsorbed state. The intersection point of the three curves defines the heat of activation $\Delta_\gamma Q$ that produces a faster reaction rate for the formation of H-adatom/metal bond and a stronger bond, that is, a larger heat of adsorption, $-\Delta W$. The good adsorber with the highest hydrogen/metal bonds produces a lower heat of activation and a faster reaction: electrocatalysis! It is clear that the heat of activation is a fraction (often $1/2$) of the heat of adsorption.

1.1.5 HORIUTI AND POLANYI (1937)

Juro Horiuti later became the most well-known physical chemist in Japan and held the chair of that subject at the University of Hokkaido in Sapporo. His fame was due not only to his huge ability to concentrate and produce highly charged mathematical treatments while working in a noisy laboratory but also to his eminence in sculling.

Polanyi was what is called a polymath; he ended his career having held positions in physical chemistry, medicine, and, finally, philosophy. Horiuti was a student when he collaborated with Polanyi in the paper cited. This work, published in English, shows no signs of having been influenced by the prior publication of Butler.

Horiuti and Polanyi's paper was much more explicit in pointing out how, assuming that the rate-determining step (r.d.s.) in the hydrogen evolution reaction is the proton discharge, the rate of the reaction on various electrode catalysts would depend primarily on the strength of the bonding between the adsorbed H atom resulting from the discharge of the proton onto an empty site on the metal catalyst [9]. The appropriate accompanying figure shows what they mean. This was a more explicit demonstration of one root to electrocatalysis, in addition to the rather implicit contribution by Butler.

It can be seen from the figure that, with the r.d.s. quoted, the stronger the H–M bonding, the smaller the heat of activation and the faster the reaction. If one makes a plot of the logarithm of the exchange current density, j_o, for hydrogen evolution, as a function of the bond strength of H to the metal, one can see that log i_o increases with an increase of M–H. This, then, is electrocatalysis in a very explicit way [10].

Now, here is an interesting point. As the value of M–H increases with changes of the metal, the increase of the rate of the "pseudo-equilibrium" velocities* at the reversible potential goes on getting faster. Of course, an increase of M–H bond strength means that the steady-state coverage (θ) of the metal with H tends to increase. Thus, the initial model of protons discharging onto bare metal sites becomes more difficult to hold to (less free sites) and another possibility becomes worth considering. It is that, as the empty sites on the electrode fill up, the direct discharge of H^+ from H_3O^+ onto one of them will be replaced by the discharge of H^+ on top of an absorbed H site (high value of θ) to produce molecular hydrogen in one step.

Let us suppose (there is independent evidence for it) that $MH + H^+ + e \rightarrow H_2$ now (with the increase of H) becomes the r.d.s., the H filling tends to cover most of the surface ($\theta_H \rightarrow 1$). One can see that it will now be more difficult for the reaction of the hydrogen evolution to occur; on increase of M–H, the release of molecular hydrogen begins to slow down and the maximum in the rate of the hydrogen evolution reaction will occur. This is exactly what occurs (see Figure 1.6).

These descriptions about the rate of the mechanism of the hydrogen evolution reaction and how the mechanism changes with increase of M–H are rather simple, but they are consistent with the observations and the proposed change of r.d.s. can be regarded as probable. In fact, over the 78 years that have passed, it is still current thinking.

1.1.6 AGAR (1939)

John Agar was a Cambridge man and for many years a stalwart of the university's Department of Chemistry, where he made contributions to physical electrochemistry. Why he is mentioned among those who founded the subject of electrocatalysis is because it was he who introduced the term "exchange current density" [15], and it is in this term that academics tend to define and measure electrocatalysis.

---

* There is no net velocity in one direction at equilibrium; but there may be two reactions, equal in speed and opposite in direction.

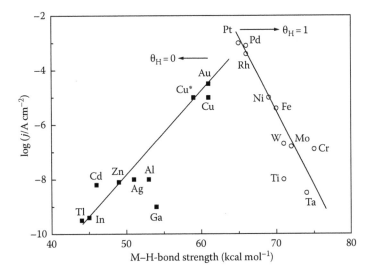

FIGURE 1.6 Logarithm of exchange current density for the hydrogen evolution reaction as a function of the M–H bonding strength.

This means that we have to explain the meaning of the term, which at first is a little puzzling because it introduces doubtful conceptual phrases such as "the velocity of the forward reaction at equilibrium," the apparent oxymoron being that those uneducated in Agar's contributions may be tripped up in thinking of an oxymoron, a reaction at equilibrium that also has a velocity forward!

Thus, in terms of partial current densities, let the reaction be proceeding in the forward direction under equilibrium conditions and let its velocity be represented by $\vec{j}$. The arrow represents direction. If you like, you can say, "From left to right." Now we choose to use electrochemical terminology, and then the full definition of $\vec{j}$ would be "the partial current density in one direction." (The "density" refers to the standardization with respect to the area one is considering, a square centimeter [cm^2] or square meter [m^2].) If we were talking about a chemical reaction, then we should use the symbol $\vec{v}$ and the definition then would be the rate of reaction in one direction. In a chemical reaction, $\vec{v}$ would be the number of gram moles per square centimeter per second (g mol cm^{-2} s^{-1}) (or per square meter per second [m^{-2} s^{-1}]).

But one aspect we have left out. At equilibrium there can be no *net* reaction rate in any direction. If, therefore, the $\vec{j}$ is the partial current density from left to right, there must be a partial current density drawn from right to left, so that what is measured is the difference between these two:

$$j = \vec{j} - \overleftarrow{j} = 0 \tag{1.5}$$

But, at equilibrium

$$j_0 = \vec{j}_{eq} = \overleftarrow{j}_{eq} \tag{1.6}$$

Hence the current density in one direction or the other is equal in magnitude but opposite in direction. There is indeed no net velocity at equilibrium. However, there is a reaction rate, but it consists of two reaction rates, and, thus, the two reaction rates will be equal in magnitude and opposite in direction.

1.1.7 Departing from Equilibrium

If one departs from equilibrium by a potential lower than RT/F, one can easily show* that the region of small net currents, called the "reversible region," follows:

$$j \propto \eta \tag{1.7}$$

As the electrode potential is increased in the negative (or "cathodic") direction, it follows

$$\overleftarrow{j}_{\text{cathodic}} \gg \overrightarrow{j}_{\text{anodic}} \tag{1.8}$$

and eventually, when $\eta_i = E_i - E_{\text{rev}}$ is sufficiently negative, that is, $\gg RT/F$, one can neglect numerically the anodic partial current density whereupon the relevant equation becomes

$$j \approx j_0 \exp\left(-\alpha F\eta/RT\right) \tag{1.9}$$

(η is negative here, and α is the "transfer coefficient of values between $1/2$ and 2").

Correspondingly, if we make E depart in the positive (anodic) direction from its value at equilibrium

$$\overleftarrow{j}_{\text{cathodic}} \ll \overrightarrow{j}_{\text{anodic}} \tag{1.8'}$$

and then

$$j \approx j_0 \exp\left[(1-\alpha)F\eta/RT\right] \tag{1.10}$$

(η is positive here).

Therefore, it is quite rational to say that j_0 is the (equal and opposite) current density in each direction at equilibrium. Thus, j_0 is a real finite number, for example, 10^{-3} A cm^{-2} for hydrogen evolution on platinum at 25°C. The dependence of the current on temperature would be helpful, too, but there is little of it as yet.

1.1.8 Measuring Electrocatalysis Numerically

Using a comparison of the exchange current densities for different metals is the usual way academics often measure the relation of the electrocatalytic power of varying catalysts. Although a series of j_0's corresponding to a given reaction is the usual way to rate a metal for its catalytic power, it has limitations. The proper thing to compare would be values of the rate constant, k, but to evaluate this from j_0 data is not always straightforward.

In Figure 1.7, the two metals compared have different β's; so, now, their relative electrocatalytic powers depends on the potential at which they are to function—one cannot say which is the better unless one states the potential. So, the method of measuring catalysis by comparing the exchange current densities may give different results than from an alternative method.

On the other hand, in Figure 1.8, I have plotted two imaginary Tafel lines (as η–$\log j$ plots are called). One can see that, if the two different metals being compared have the same value of symmetry factor, β, then the ratio of j_0's would give the same result as comparing the relative current densities at any potential, including the reversible equilibrium potential, which is the potential to which j_0 refers.

---

* See Bockris et al. [20], Vol. IIA, Section 7.2.2.

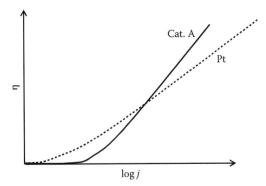

FIGURE 1.7 Tafel lines for a simple organic oxidation process comparing platinum with a better electrocatalyst A.

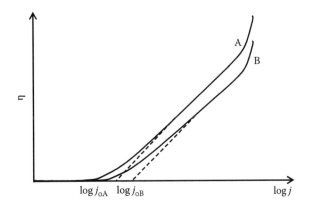

FIGURE 1.8 This diagram illustrates the practical engineer's method of determining which two electrodes, A and B, would be better for a certain electrocatalytic reaction with the rest of the experimental variables fixed. The electrochemical engineer will want to operate near the limiting current, after which, it is impractical to work since it will be a waste of energy and money. Electrocatalyst B will need less overpotential (less expensive) than electrocatalyst A near the limiting current density. Thus, the cost-attractive electrode will be B. The problem of taking the value of j_o only as a diagnostic criterion (Figure 1.7) will lead to erroneous results at high current densities as a consequence of their different Tafel slopes.

Those interested in electrocatalysis, from an economic standpoint, take a different attitude from that of the academic workers. Their outlook begins by the thought that the current densities employed in most industrial processes using electrochemistry are in the high milliamperes per square centimeter (mA cm^{-2}) range. Of course, in the economical aspect, the higher the current density the better the electrocatalyst, that is, more production per square meter of the plant. However, unless one uses some special means (stirring, etc.) for increasing the limiting current density, the highest practical value of this quantity is in the 100 mA cm^{-2} range. The industrialist's attitude is to choose the potential for judging the best electrocatalyst in the region in which he or she is to work, and then compare the relative current densities there. This gives a practical comparison of electrocatalytic power. The electrode material giving the higher current density at a potential near that of the limiting current density is clearly the best catalyst for the reaction concerned.

One has also to keep the same temperatures in comparing electrocatalysis and although this is obvious, it sometimes turns out that a desired reaction is only viable at a lower rather than at room temperature. An example would be the work of Hori on the reduction of carbon dioxide. He found that this only occurred to methanol near 0°C [11]. So, in experimental work, always working at near 25°C may lead to a loss of opportunity. Thus, the definition of electrocatalysis, although by no means obscure, has to be made with clear objectives in mind.

1.2　THE RATE-DETERMINING STEP

In electrode kinetics, at least for reactions with not more than three to four consecutive steps for one act of the overall reaction, there is usually one step that—when the reaction is running in the steady state—determines the rate of the whole reaction. Obviously, if one is to design a catalyst, this is the reaction, the rate of which has to be increased. Let me illustrate this—as Henry Eyring used to do—with a homely model.

Suppose we identify ourselves with a certain car running between two towns, A and D. In the first instance, one imagines an unlikely situation, all the bridges between A and D are open and there is no reason for any holdup along the route and nothing one could call an r.d.s. Then, in our second example, one imagines that A–B is okay (no bridge under repair) but B–C contains some repairs to a bridge and is down to one lane and men with red flags and walkie-talkies. A → B traffic can go as fast as the speed limit, and faster, but B → C goes at only 10 mph because of the road repairs.

As everything that gets to D has to pass through the blockage between B and C, it is at once clear that the number of cars reaching D per hour is the same as the number passing slowly through the road blockage.

The blockage is called the r.d.s.

If one wanted to make the overall reaction A → D go faster, one would have to speed up passage through the road building, that is, accelerate the r.d.s., B → C.

I need hardly say that similar things apply to molecules in reactions. When we say we are going to try to speed up the r.d.s., we are talking about catalysis.

There is still something to say about A → B when there is a holdup for B → C, the r.d.s. One can imagine an eager 80 mph driver approaching the man with the red flag who is going to slow him down to 10 mph while he or she trickles along the one narrow lane still open.

Naturally, there are many happy drivers who do not know about the holdup between B and C and when they arrive at B and are slowed down to 10 mph, some of them, the majority, start going back to A to see if it is possible to find another way to D that escapes the holdup at B.

Many cars do this—those which cannot get through at the pace they would like—but they find it is not much good going back to A (no alternative path), so they tear back to B, only to find it still blocked, whereupon the driver impatiently goes back to A again. In the end, the number waiting to get past the barrier becomes more or less constant. Most of them are tripping back to A and then returning to B. It is almost at an equilibrium as far as A → B (A ↔ B?) is concerned and in fact there is a term, pseudo-equilibrium, which applies to all the cars coming up to the barrier and going back to A again, with just a small chance that they will hit a time when the flag is withdrawn, the light green, and the drivers get through the r.d.s.

To summarize, an r.d.s. usually controls the rate of an overall reaction.

After the r.d.s., there is a steady flow of molecules, few in number per unit time, though now traveling at the speed limit toward D.

Finding the r.d.s. and trying to speed it up is one method of how one can find the best catalyst. It is a necessary preliminary to designing a good catalyst for the situation concerned. If one identifies it, one can *think* about designing a surface, which will speed up the r.d.s.

1.2.1　More on the Importance of a Rate-Determining Step

This chapter is about electrocatalysis, in practice, how to make an electrochemical reaction go faster (thus reducing costs, etc.).

Many electrode reactions are consecutive in nature and if one has a consecutive reaction—and it contains only a few steps—it is indeed a common experience that one of the successive reactions has a much greater influence on the rate than the others. In extreme cases, the one reaction in a sequence of, say, four, has a 99% control over the rates of the other reactions in the series.

Now, we are going to present several pathways to electrocatalysis, and one can see that, if this one reaction, the rate-determining one, is clearly the one to attempt to speed up and is the one that would make the entire reaction go faster. The other reaction steps will fit in. If we can make the r.d.s. go faster, the whole reaction will go faster, and that is the electrocatalysis we want.

But to be able to make the r.d.s. go faster, one needs to know much about the structure of the surface and how the atoms in the r.d.s. are bonded to the surface of the electrode catalyst. It may be a special feature of the surface, which is particularly catalytic, and perhaps this can be adjusted so that the r.d.s. is speeded up.

Of course, this is a crude sketch, a bare description of electrocatalysis in simple cases, but it does stress that to know which step is rate determining is more than half way to determining how to catalyze the overall reaction.*

1.2.1.1 High α's[†]

In the presentation, I have made of the relation of electrochemical rate (measured in electrochemistry as the current density), we have not explored the effect of changing the coefficient, α.

In Figure 1.9a and b, I show the effect of varying α[‡] from low values of one-half to a higher value of, say, -2 (Figure 1.9a and b). Coming back to Figure 1.9a, consider making α larger than

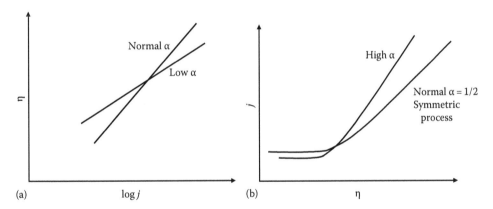

FIGURE 1.9 Most of the electrochemical reactions have symmetric profiles, that is, $\alpha \cong 1/2$, but few 2/3, 1 or even 2. From the electrocatalytic point of view and for industrial purposes, it is desired to use processes with large values of α, since they involve large current densities. (a) The figure shows the Tafel semilogarithm plot of the situation explained earlier. (b) This panel shows the exponential relationship between the symmetric process and that of large α values. Here, it clearly shows the problem of considering only the value of j_o at the reversible potential. The change in the kinetics is really determining at each overpotential, which is the best electrocatalyst.

---

* A naive misunderstanding by some beginners is that they think electrocatalysis gets its name because, by adjusting the settings on the potentiostat, one can vary the reaction rate. Now, all electrochemical reactions can have their speeds changed by changing the setting on the potentiostat, but electrocatalysis means more than that. It means making the reaction go faster on one surface more than one another, although one has not changed the setting on the potentiostat.

[†] CATHODIC means a reaction in which the electrons leave the electrode for an entity in the double layer next to the electrode. ANODIC means a reaction in which the reactants leave an entity in the double layer and enter the electrode. Examples: $H_3O^+ + e_{metal} \rightarrow MH$ would receive electrons from the electrode (cathodic); $H_2O + H \rightarrow H_3O^+ + e_{metal}$ would give them to the electrode (anodic).

[‡] Perhaps confusingly, I have started to use the symbol α in place of the β, I have been using Equation 1.2 with β earlier. No, it is not carelessness. Electrochemists tend to use β in simple reactions but in multi-step reactions, the corresponding symbol is α. This quantity contains β but also several other quantities (see Bockris et al. [20], Vol. II A, Section 7.6.11).

one-half, which one sometimes finds. The result is obvious: choose a certain overpotential and the graph shows one gets more current density (a higher rate for the larger α's).

However, nature sets the value of α for a given reaction on a given metal. Suppose we particularly want to use that metal (e.g., for economic reasons), is there anything we can do about making it give higher α's?

It is a fact of the situation that usually the α's are low and that is not desirable because a low α means a lower current density for a given overpotential, and poor economics.

The subject of how to change α for a given electrode reaction has not yet been touched on in the literature. A famous example is in the iron dissolution where with a value of α equal to 1.5, the anodic reaction goes faster at a given potential than if we had the usual value of 1/2. Of course, here we would prefer to slow the reaction (hence the rate of corrosion) down.

A more attractive example is with the dissolution of molecular oxygen on platinum and platinum alloys in concentrated phosphoric acid at temperatures above 100°C. Now here the value of α for the oxygen dissolution is 1.5, and that is helpful. It means the reaction goes faster; a fuel cell in which the cathodic reaction of oxygen is often an r.d.s. is faster for the same electrode potential, which is good economics.

I can say what one wants to do is get higher α's but I do not think we are as yet sufficiently deep into the understanding of electrocatalysis at the molecular level, to be able to take advantage of what we know, which is that, in a successive series of reactions, the α gets larger the further down the series of consecutive partial reaction stops one goes. It seems that much more investigation is needed for reaching these larger values of α. It all depends on government research funding and that depends often on the knowledge of the program managers. So, basically, our ability in the next decades to improve electrocatalysis, through fundamental work, depends on the education of the program managers in physical electrode kinetics in NSF.

1.2.2 What Happens When One Rate-Determining Step Is Not Dominant?

The way we have been writing our account so far, one reaction in a sequence controls the rate of the whole reaction. But this is an idealization and a much more general reaction, which at first does not speak of an r.d.s., is as follows [5]:

$$j = \frac{\prod K_j C_j}{1/k_1 + 1/k_2 + 1/k_3} \tag{1.11}$$

Here k_1, k_2, and k_3 are the rate constants of consecutive steps $1, 2, 3, \ldots$. One sees at once that if, say, k_2 of the second step in the sequence $k_2 \ll k_1, k_3, \ldots$, then as an approximation, the rate of the reaction is controlled by Step 2.

However, one has to admit that, sometimes, $k_i \cong k_j$ and both are much less than k_1. Then, from the electrocatalysis point of view, there will be a half way house, the structure of a good catalyst will have to be chosen with the properties of both i and j taken into account.

This is easy if k_2 and k_3 need the same surface to increase the rate of each step. But there are ways of meeting the situation even if (as one might say) there are two nearly equal r.d.s.'s. Could an electrode surface be made with tiny patches of metals favoring k_2 but also of metals favoring k_3?

1.2.3 Use of Selected XL Faces as an Aid to Electrocatalysis

During the last two decades, it has been discovered that, using single crystals as electrodes, one can find crystal planes on which reactions occur much more rapidly than on other planes. In some systems, a particular crystal face may give an advantage of an order of magnitude in rate constant compared with the results on a polycrystal. The effects are reaction specific, that is, the positive

TABLE 1.1
Electrochemical Rate Constants for the Oxygen Electroreduction on Gold Single Crystals in 1 M Sodium Hydroxide

$-E/V$	Au(100) $n = 4$	Au(110) $n = 2$	Au(111) $n = 2$
0.05	6.4		
0.07	8.5		
0.10	13.7		
0.15		9.9	5.0
0.18		13.7	5.73
0.19			6.5
0.20	32.7	22.3	7.07
0.21		28.3	8.5
0.22			9.43

Source: Adzić, R.R. and Marković, N.M., *J. Electroanal. Chem.*, 165, 121, 1984, Table 5.

effects for a certain metal catalyst of changing the crystal face and the effects of different planes depend on the reaction (Table 1.1, Figure 1.10).

The question is the economics of the practical development of single crystals. The advantages of using an optimal crystal face, which will increase a reaction rate by an order of magnitude, may cost considerably, and the economic advantage of the enhanced production rate must be compared with the extra cost of producing the fast crystal face upon a large scale. Clearly, in industrial practice, the size of crystals needed is relatively great compared with the small areas used in fundamental work. Right now, what we can tell of the economic balance suggests that the advantage of developing faces of crystals to be exposed to the solution, that is, to be the seat of electrochemical reactions,

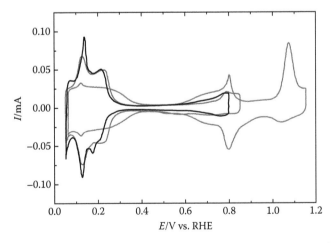

FIGURE 1.10 Current intensity versus electrode potential voltammetric profile for platinum single crystals, Pt(100) (cooled in air) with blue lines, Pt(110) with black lines and Pt(111) in red lines, in 0.5 M sulfuric acid solution run at 0.05 V s^{-1} at room temperature. A difference with respect to the early single crystal preparation method by Will [18] is that the ones reported here are those obtained by Clavilier early in 1980. After 50 years of the first paper by Will on the use of single crystals for the oxygen electroreduction reaction and his finding of the given preferred activities on (111) and (100) planes, the industry still uses polycrystalline surfaces.

may be too expensive to develop in an industrial scheme. However, the number of reactions where the economics of single crystals has been examined is small.

1.2.4 Langmuir and Tempkin

A Langmuir isotherm is the kind of relation of surface coverage to concentration in which it is justified to interpret observation by neglecting heterogeneity and intermolecular force effects. To determine whether your system corresponds better to Langmuir or, alternatively, to a Tempkin [12] isotherm (which takes heterogeneity into account), is important for mechanism analysis. It turns out that a complete determination of the isotherm is seldom necessary because at both low and high θ (the fraction of the surface covered with reacting atoms or molecules), the isotherms give rise to the same behavior. It is if $0.1 < \theta < 0.8$ that the reaction may need treatment in terms of Tempkin's thinking (heterogeneity and heats the potential dependence of adsorption). The reason why we should identify non-Langmuirian behavior is that if this is the case, it affects the values of the criteria for various mechanisms. One should, then, determine θ early on in the investigation of a given system because the meanings of the various mechanism indicating factors change with the isotherm.

1.3 CLEAN SYSTEMS IN INDUSTRY

Systems in which great care is taken to purify the solution do give better (i.e., catalysis at higher rates) results and hence have positive economic value than systems for which no special care is given to maintaining high purity. It is easy to make a system pure to begin with but more difficult to maintain it uncontaminated, particularly for electro-organic reactions where "gunk" from unintended intermediates or unexpected side reactions may accumulate on the electrode surface and block the catalytic sites on it.

1.3.1 Automated Mechanism Analysis

One may determine the Tafel constants, orders of reaction with respect to each reactant and product, the response to potential pulse jabs, and the results of attempts to introduce higher values of α's in favor of a specific mechanism. These criteria can then be compared with expected trends for the types of reactions concerned. From such data, an r.d.s. can sometimes be worked out or the number of possible r.d.s.'s reduced to two or three possible sequences can be determined. If one can reduce the possible r.d.s.'s to a low number, it is often possible to devise specific tests to distinguish between them. Such distinction methods may include the analysis of predicted and determined isotopic effects. Further, sometimes an r.d.s. can be eliminated on the basis of model calculations of the heat of reaction: impractical high values showing a pathway that is too difficult.

The idea of automating experiments that are programmed to show up a behavior typical of certain mechanisms and pathways, having experiments being carried out in the control of a computer program, and then comparing the results with those calculated theoretically for each supposed model is attractive and reduces time demand greatly.

1.3.2 Electrocatalysts Are Not Smooth Plates

There are books on physical electrochemistry, which present *electrocatalysis* with one of the main influential features left out. I refer to surface heterogeneity. Thus, when a surface is viewed under an electron microscope one can see that the apparently smooth electrode surface is in fact not smooth at all. Because the structure of electrode surfaces is, in fact, very much a part of the interpretation of electrocatalysis, I will give a brief sketch of some of the features of a metal surface that may affect surface reactions.

1. First, nearly all electrodes are in fact polycrystals. What this means is that although one cannot see them except by means of the use of sufficiently high magnification, a polycrystal is rather like a stone-covered beach, the pebbles on a beach are in fact analogous to the microcrystals that make up the untreated metal. Each microcrystal has several crystal "faces" each of which offers different bonding characteristics for intermediate radical atoms of the reaction that complete a sequence on the surface (Figure 1.11). The flatness appears so because the heterogeneity of the surface is less in magnitude than the wavelength of visible light. The light, therefore, does not "pick up" the roughness or "poly" nature of the metal and that is why the metal looks flat.

2. Next, there is a different kind of "roughness" that can be seen if one increases the degree of magnification until one is receiving information back from one crystal, one "pebble." Now, it is possible to detect several features of the crystal forms on the surface. One can see two levels, one a plane atomic dimension above another plane. Therefore, a "step" that rises gradually out of the surrounding micro flat surface. These irregularities such as the step I am describing are part of the structure of the surface. During electrodeposition these steps sometimes rotate around an axis and if the process goes on long enough it may form a spiral on the surface.

There are places where these irregular deformities meet and indeed the meeting place of two steps is a particularly active site on the surface where atoms of reactants may become more strongly adsorbed because the absorbed atoms can become bonded in several directions.

3. Another form of irregularity or "structure" on a surface is called a dislocation. Imagine three lines of atoms running exactly parallel together for thousands of atoms. Occasionally, one of the atoms gets left out, say, in the middle row. The two other rows continue but the one that is missing causes the two rows surrounding it to close in, and fill the missing row of atoms. Such a happening—a dislocation—has effects for a while; the rows of atoms are no longer straight. During this adjustment, a high local pressure develops near the dislocation. It can be shown that this is an area of much-enhanced chemical potential and therefore the solubility of gases, where absorbed atoms tend to get squashed together here [19]. Sometimes such sites tend to form the beginning of internal cracking (stress corrosion cracking). Such sites tend to be attended by H atoms and if these are concentrated enough they may be the origin of disasters, for example, when a bridge or a roof breaks down or when, in a sea with very big waves, two ends of a ship get lifted and a middle part subject to stress beyond its limits, literally cracks open, the ship rapidly sinks sometimes before

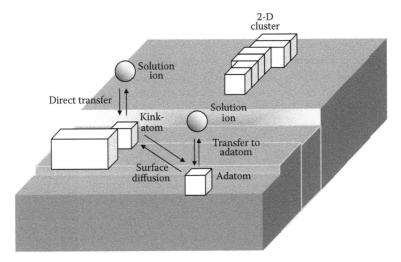

FIGURE 1.11 The structure of the face of a defect-containing single crystal surface with a face-centered cubic system. The spheres indicate the ions in solution, the cubes the metallic adatoms, and the rest the metal adsorbed ensembles.

contact with a possible rescue mechanism has been made. (Thus two or three ships are said to "disappear" each year, this may be the mechanism.)

We have seen, then, that even if we stick to a tiny area of a "single crystal," there are on each of these irregularities not only different planes but also dislocations and new planes. In a real surface, there is a jumble of differently oriented crystals that justify the term polycrystal.

Why do I go to some length to exemplify the different types of "structure" on the surface? You see, catalysis occurs by different degrees of bindings of different adsorbed atoms on the surface, weaker here, stronger there.

There is surface diffusion, too, of the adsorbed atoms deposited from solution and although the more strongly adsorbed ones are "fixed," that is, cannot diffuse about; the weakly absorbed entities do more. This surface diffusion involves atoms moving about, forming new microcrystals, temporally, usually shape into small subgroups on the surface. We find that, here and there, new bonding is of different strengths gives atomic groups of varying size and mobility. This is a strong reason for the hyperactivity on certain parts of the catalyst surface.

So far we have written about the surface of a pure metal. There are, however, still more reasons for heterogeneity; for now we may be dealing with surfaces containing two or three different kinds of atoms, which are again the origins of heterogeneity, of strain, planes where disturbances in the lattice cause the different properties to manifest themselves in regions, which again cause heterogeneity with its microstructure and differential stresses.

So what we so often see in our books are surfaces represented as flat, structureless planes, a great oversimplification that leaves out the most important part of a catalyst surface—irregularity.

1.4   HOW DOES ELECTROCATALYSIS WORK?

So far in this chapter we have looked at a few things about electrode surfaces. Except for Tafel's law, reaction rate as an exponential function of potential, all had some relation to electrocatalysis. I started with Tafel's law because it is the basis of much of the rest; for example, when one measures rates, current densities, as a function of potential, $dj/d\eta$ or $d\ln j/d\eta$ vary with the mechanism of the reaction, and although several mechanisms have the same $d\ln j/d\eta$ (i.e., the coefficient is not diagnostic), several do give values of that parameter, which indicates a specific r.d.s.

Now, yet one more general thing before some examples of real reactions and their catalysis. To know a mechanism, to find the r.d.s. therein, and the devising of the best path to good catalysis, one has to catch the reaction in what you might call its middle period.* Right at the beginning, just a little above zero overpotential, up to an overpotential of RT/F V, the observed current, and its type of behavior as a function of potential, is often hidden by the residual influence of the still significant back reaction. Then, for the middle (linear) part of the $\ln j$–η relation, there is a possibility that one can derive facts (e.g., dependent on concentration), which, if done for all the reactants and products, may make a certain mechanism more likely than others. The r.d.s. often peeps out here, though usually one has to work on for some more criteria† (pressure effects, coverage-potential, automated analysis of steps characteristic of a sequence). There is another region to keep away from, when the reaction ceases to be rate-controlled by charge transfer and successive reaction on the electrode surface because, finally, too fast for the supply of ions by diffusion to the electrode, transport becomes rate determining. So, to study electrocatalysis, keep

---

* I am not talking about the time the reaction has been running but to its current–potential relation.

† Automation in electrochemical research laboratories has not grown very fast, and the reasons are (1) automation needs a lot more instrumentation than using a graduate student (who thereby learns much) and (2) automation is economically most effective if one action is repetition. Although experiments in research programs certainly have to be repeated, doing it three to four times and getting the same result is counted as confirmatory in research, in production runs, and may need thousands, even hundreds of thousands of repetitions.

to that long central section, where Tafel's exponential law is in effect, and all that follows from it, and there one can get information that tends to lead to the determination of the r.d.s.

Mind you, like many of the authors of this book, I have the view that the reaction mechanism at the interface is important in electrode kinetics. This outlook originated from the fact that, in my education, I got to study, and become impressed by the breadth of electrode kinetics in the middle region from my earlier study of chemical kinetics.*

However, and particularly in the United States, there are many who call themselves electro-chemists and who take a different view of the optimal region to study. They concentrate on the transport dependent (limiting current) part and are interested more in electroanalysis and less in electrocatalysis. They want to use electrochemical kinetics as an analytical tool (rather than a way to investigate atomic movements in a reaction). Scientists who take the former view in electrode processes have generally originated in the analytical division of their chemistry departments and if you read accounts of the subject stressing that view, you gain much, but not about electrocatalysis.†

1.4.1 How to Approach a Case of Electrocatalysis

Suppose one considers an ion, situated in the double layer, and hence so near to the electrode surface (3–4 Å distant) that one might expect the first thing that will happen to the ion is that it will receive an electron from the electrode (or alternatively, give an electron to the metal). To find a good electrocatalytic material here, one needs the ion in the process to make a chemical bond, M–H, with the electrode, M. Nevertheless, as we want to see this initial discharge reaction as rate determining for the following sequence, we shall take it that the bond strength of the reactant to the metal catalyst M is intermediate in value. To make this situation easy to comprehend, look at Figure 1.3. Insofar as it does this, the heat of activation for the reaction will be pulled down and the reaction will be relatively fast (a good electrocatalyst).

Now, in an imaginary series of metal catalysts, let it be assumed that they are arranged in a series of increasing bond strengths. The atom radical, which has arisen from the first rate-determining discharge, will at first (weaker bonds) recombine quickly together and so the intermediate coverage in the steady state will remain small. However, as one progresses along the imaginary sequence with increasing bond strength, the rate constant of the discharge reaction will get faster (see Horiuti and Polanyi at Fig. 3) while that of the following reaction will get slower, atoms more difficult to desorb (θ, the intermediate coverage in the steady state will increase). At some point, the initial discharge reaction will get too fast to be rate determining and the slowing of the rate constant in the second reaction, that involving surface atoms will have slowed enough (θ raised enough) for it to become the r.d.s.

The circumstances of the last two examples should give rise to a maximum of rate at an intermediate strength of bonding of the surface radical when the r.d.s. is still discharged directly

---

* But some research labs I have visited show automation in action. I recall a Japanese University laboratory in the 1990s. The professor said he would like to show it to me. We entered a rather dark room but I could see about a dozen small circular tables, each supported by a central pole. On each there was a modest amount of equipment. Some seemed to contain beakers filled with solutions, electrodes, and other electrochemical impedimenta. One could hear clicks some times and even the sound of liquid being poured into a fresh beaker. From some of the mini tables there issued lengthy sheets of paper—some graphs—trailing out on the floor. But in one corner of the room, there was a dimly lit portion and a white-coated figure. "Solly," said my Japanese colleague, "still have one man work with hands…"

† This aspect of electrochemistry has been given a sound presentation by Bard and Faulkner in their well-known book, which follows a tradition spreading from two electrochemists of eminence that worked at Harvard University. One of my graduate students, Ray Richards, used humor to characterize the diffusion-oriented work of this kind (because it involves a specialized mathematics) as "erf differs," and the physical electrochemists, specializing in mechanisms on the catalyst surface as "Tafel slopes."

The former term refers to a mathematical technique and the latter term has an obvious origin, referring to what some think as the overuse of a $d(\ln j)/d\eta$ in the determination of the whole mechanism.

onto the catalyst, and before the surface fills and the determining step becomes desorption, discharge now having a very high rate constant.

1.4.2 STRUCTURES OF METAL SURFACES AND THEIR EFFECTS ON ELECTROCATALYSIS

In one sense, the reader might react to the section title by saying: "But we just found out about that in a previous section." It depends, we saw, on the bond strength to the metal of the intermediate radical, and there are two cases.

True, but there are other cases where the strength of the bonding is less important in determining the catalysis than the actual geometric properties of a surface. A heuristic model of a surface showing several features already discussed is shown in Figure 1.11. Suppose, for example, an r.d.s. that sometimes simply involves adsorbed atoms that have come from an earlier ionic discharge reaction. The rate of a chemical combination reaction on the surface depends on the distance between the two adsorbed atoms. Too far apart on the catalyst and the heat of activation for recombination A–A will be too high (see Figure 1.12). But if the internuclear distance is too small, the two atoms will not be able to settle down in nearest neighbor position with greatly reduced heat of activation: they will repel each other. The basis of such a situation is sketched in Figure 1.12 [20].

These play between the ions in the surface, attractive and repulsive forces, and each kind will speed up or slow down the rate of the reaction if the elements we are discussing happen to be in an r.d.s.

Of course, we are giving here an introduction to electrocatalysis. Adsorbed water, potential dependent with respect to its bonding to the surface, and orientation has not been touched upon, although, because everything is dependent upon the ubiquitous interatomic forces, the orientation of water on a surface affects the reaction rate. Active planes have not been brought into a specific mechanism, one, which experiment shows, is faster than other planes. And there is macroroughness that one can see, at least with a light microscope.

1.4.3 A SPECIFIC EXAMPLE OF ELECTROCATALYSIS: OXYGEN EVOLUTION ON PEROVSKITES

What has been written so far in this chapter has come primarily from the thoughts of the author, and has arisen from the ideas of many physical electrochemists and their students and collaborators, not forgetting my own. But now I want to tell you about some electrocatalysts called perovskites* and their electrocatalytic action on oxygen evolution from alkaline solution. One of the reasons I want

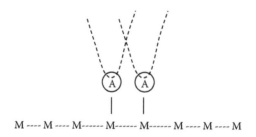

FIGURE 1.12 When the interatomic distance in the electrocatalyst is too large, the heat of activation (defined by the intersection between the potential energy curves) is too high to be reached. On the contrary, the atomic adsorbates will be repelled from the surface as a consequence of the contact to the metal.

---

* Perovskites are compounds described by the general formula RTO_3, where R is a rare earth element, say, lanthanum; and T is a transition metal, say, nickel. They conduct electrons poorly and have to be helped by a little doping.

to include this account is that it was first published in a symposium given in the honor of Henry Eyring, who dominated twentieth century physical chemistry in the United States.* The date of the work on perovskites is 1983, and it is admittedly not of the most recent vintage (2007–2008). But it has the advantage of being intermediate in complexity. It is, I hope, understandable to readers of this book, yet it may tax them a little, for the most likely mechanism is a good deal more than the simplest electrocatalytically active reactions involved, say, in hydrogen evolution.

On the other hand, the perovskites have not yet caught up in complexity with the far more complex bioreactions that often involve many consecutive steps compared with the three to four we shall consider here.

In contemplating how to tell you about the perovskites, I turned to an account I wrote in 2000 and after I had digested it, I really could not think of a clearer or even shorter way of explaining the conclusions. The rest of this chapter is new but the perovskite section is reprinted in [10]. I wrote the original paper with a very able graduate student called Taka Ottagawa from Japan. Part of the account runs as follows.

So far, the bonding and surface structure aspects of electrocatalysis have been presented in a somewhat abstract sort of way. In order to make electrocatalysis more real, it is helpful to go through an example that of the catalysis of the evolution of oxygen from alkaline solutions onto substances called perovskites. Such materials are given by the general formula RTO_3, where R is a rare earth element such as lanthanum, and T is a transition metal such as nickel. In the electrocatalysis studied, the lattice of the perovskite crystal was replicated with various transition metals, that is, Ni, Co, Fe, Mn, and Cr, the R remaining always La.

Figure 1.13 shows a view of the perovskite lattice. Of course, the diagram shows the surface of a perovskite, and before going any further, the student should spend sufficient time comprehending the surface on which the electrode reaction is to take place, with an understanding of the symbols showing where the La, the transition metal, and the Os are. This diagram clears up the puzzle. If one contemplates the general empirical formula for perovskites, RTO_3, it is not obvious how the OH^- ions in an alkaline solution would be able to discharge upon an oxide surface. It would seem that O might be on the surface the OH^- would face as it diffused in from the solution and clearly no bonding would occur. However, the slice through the crystal, exposing the surface

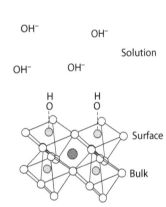

FIGURE 1.13 Schematic model for the active surface of the perovskite with an electrochemically active metal, M ●, anion lanthanide ●, and the lattice oxide ○ ion with hydroxide ions in solution and OH adsorbates at M. This figure is a simplified scheme of that in [23].

---

* Like all famous scientists, Henry Eyring attracted accounts about himself, which showed up aspects of his character. There is no doubt that the first word which comes to mind when I am asked to describe Eyring is "informality," in his teaching of classes and in private scientific discussions. Eyring was very relaxed and he tended to express the informality by using informal examples: "See, this fence there, it's being jumped over by these sheep but now and again, a smart fellow thinks he'll go right through, see, kind of tunneling." However, the thing I shall remember most about Henry occurred in a visit I made to Stockholm to proselytize the Nobel Prize capabilities of the Russian electrochemist, A.N. Frumkin. I chose to visit Stig Claeson, who at this time, it was in the late 1960s, was the Chairman of the Chemistry Committee which finally decided on the NOBEL PRIZE, who should get it. Cleason did not think much of Frumkin's candidacy; he evidently did not know his great position in the electrochemical world. The talk soon turned to Eyring. "Aah," said Claeson, "we much wanted to give it to Henry. Well, we had him over for a few lectures, you know. Of course, we Swedes are pretty formal fellows, so for the formal dinner we offered Henry at the beginning of his lecture tour, we dressed up as Swedish gentlemen (wives were present too, in their best long dresses). It was a banquet really. However, when Henry was fetched from the hotel in the Academy's Limousine, he had a surprise for us, he was very informally dressed. No tie, open neck sports wear, sandals, and would you believe it, short trousers as if for playing football." Well, that was the end of the Nobel Prize for Henry Eyring.

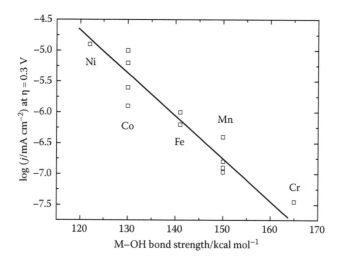

FIGURE 1.14 Current density related to real surface area for the oxygen evolution reaction on perovskites at 0.30 V of overpotential versus the M–OH bond strength. The transition metals, M, of the perovskites are indicated in the plot. This plot is a simplified representation of that in [21].

(Figure 1.13), shows that in fact the transition metal ion, with its multiple valences and strong bonding power, is indeed on the surface. So OH^- discharges onto the transition metal of the perovskite, M, not onto Os:

$$M^{Z+} + OH^-_{(aq)} \leftrightarrow [M - OH]^{Z+} + e^- \qquad (1.12)$$

Now, the next point to understand is the relation between the rate of the reaction (here measured uniformly for all perovskites studied at an overpotential of 0.3 V), and the strength of the OH bond to the transition metal. It is made clear from Figure 1.14 that the stronger the bond strength, the slower the reaction. This is a determinative piece of information that suggests that the reaction mechanism must involve OH in the r.d.s. The more difficult this becomes (the stronger the bonding), the more difficult it is for the reaction to occur.

Any mechanism suggested must involve the desorption of OH formed in the first step, the second and r.d.s. in the series going to molecular oxygen is therefore likely to be

$$M^{Z+} - OH_{(ads)} + OH^- \xrightarrow{\text{r.d.s.}} M^{Z+} \rightarrow H_2O_2 + e^- \qquad (1.13)$$

The first and second steps can be represented in some detail, as shown in Figure 1.15. Further steps finally to get oxygen are beyond the r.d.s. and hence not of primary importance in the electro-catalysis process. They could be

$$H_2O_2^- + OH^- \leftrightarrow (H_2O)_{(phys.ads.)} + H_2O + 2e^- \qquad (1.14)$$

$$H_2O_2 + (HO_2^-)_{(phys.ads.)} \leftrightarrow H_2O + OH^- + O_2 \qquad (1.15)$$

Such matters can be represented in a different way in terms of the d-electron configuration of transition metal ions in a molecular orbital scheme (Figure 1.16).

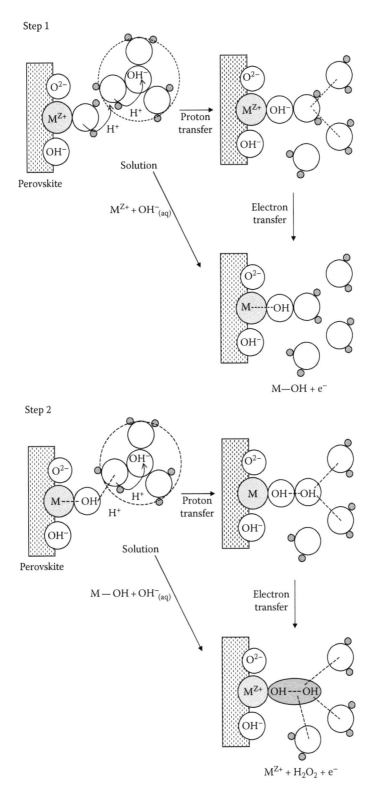

FIGURE 1.15 Schematic representation of the first and second (rate-determining) steps of the mechanism of oxygen evolution in perovskites.

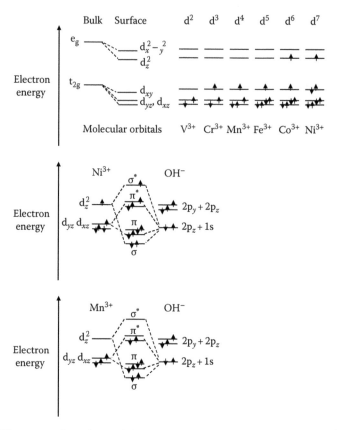

FIGURE 1.16 p-Electron configuration and d-electron configuration for the OH and M^{+n} (transition metal ions), respectively. The case of nickelates and magnates are explained in more detail. (From Bockris, J.O'M. and Ottagawa, T., *J. Phys. Chem.*, 87, 2964, 1983.)

The considerations of this reaction of molecular oxygen evolution on an oxide catalyst again show the importance of electronic factors and bonding. However, the discussion covers only the essentials; the reality of the catalysis of perovskites in oxygen evolution involves several other factors that can be referred to here only briefly.

One important matter is the availability of electrons in the oxide, which for a metal would not be a factor because there are always plenty. Pure perovskites are nonconductors and before they can be used as electrodes, it is necessary to add to them other substances that increase their conductances. BaO is one substance used. The process is similar to the addition of doping agents to semiconductors. Then, it is necessary to know the effect of the adsorbing OH upon the resulting electron density created by the addition of BaO. Could it be locally decreased by OH^- adsorption and desorption as though it were happening in isolation. What about the buildup of OH^- on the surface and the lateral repulsive interaction among adsorbed OH^- ions, which increase greatly with θ? What about the entropy of activation and how it will affect reaction rate on the various oxide catalysts?

Such elaborations are beyond the message it was desired to give here: conducting oxides can be electrocatalysts. The catalysis still depends on the M–OH bonding, but in inverse sense to that occurring, say, in OH^- discharge, when this is rate determining in the evolution of oxygen.

1.4.4 THE MECHANISM AND ELECTROCATALYSIS OF METHANOL TO CARBON DIOXIDE

One of the functions of methanol is as a fuel for fuel cells. The general concept is that the best fuel for that purpose is hydrogen, and there is no doubt that the high j_o ($\approx 10^{-3}$ A cm^{-2}) for hydrogen evolution on platinum cannot be approached by other fuels, which makes their use more expensive. On the other hand, there are increasing numbers of possibilities where fuel cells can be useful independent of high efficiency. Owing to the fact that the amount of electricity that fuel cells can produce is simply dependent upon the fuel (perhaps kept in a separate tank) available to it, a fuel cell can take the place of a battery but with a capacity for supplying electricity for much more prolonged durations than are available in batteries.* There is, particularly, the increasing use of fuel cells in situations such as driving computers.

Here, methanol could be a practical fuel while hydrogen, with its need for a high-pressure container, would not be. On the other hand, compared with hydrogen, methanol has certain defects, which might be overcome if we knew its mechanics of oxidation and could thus think about its catalysis, etc.

In contemplating such studies, an important characteristic has to be revealed. When an electrochemical reaction is carried out quickly (seconds) its mechanism may differ from that which operates when it is to function in hours or even months and years. There are two general methods of study of the kinetics of electrode reactions. In the first, the potential sweep approach, the current density is recorded while the potential is moved at a steady rate toward the more positive (or anodic) side. Thus, the results do not necessarily correspond to those which would be felt if the fuel cell was supposed to be the power source for long periods, driving cars, for example.

Another method (potentiostatic) involves keeping the potential constant and examining the current over seconds, minutes, etc. Now, one gets to know the behavior of the cell in a longer, more practical time zone in which it is to function in practice. The current–time relation is measured at a series of potentials corresponding to those of practice and from these results one can construct the $\log j$, $-\eta$ relation, which may help with mechanism determination.

This latter method can be called "potentiostatic," and its results in general correspond more to the steady-state condition of practical use than do the potential sweep treatments that involve potential changes all the time measurements are being made.

An early investigation of the mechanism of the methanol oxidation was made by two Russian workers, Bagotskii and Vasiliev [23], from the famous Moscow group directed by Frumkin. They suggested that the r.d.s. is not an electron-transfer reaction such as water discharge ($4OH^- \rightarrow 2H_2O + O_2 + 4e^-$), but that it occurs on the electrode surface between an OH (adsorbed) with radicals arising from the dissociative adsorption of methanol. Their mechanism would have an r.d.s.

$$-C - OH + OH_{(ads)} \rightarrow CO_{(ads)} + H_2O \qquad (1.16)$$

The adsorbed $CO_{(ads)}$ here was seen as a product of this r.d.s.

When this suggestion was first made, the *in situ* detection of adsorbed radicals by Fourier transform infrared spectroscopy was not available and was first conducted by Chandrasekaran and Wass in 1990 [24]. If one notes the frequency of the peaks that the adsorbed spectra gives one can identify the radicals present.

Thus, the conclusion from the spectroscopic work is that oxidation at the surface contains a single-bonded carbon monoxide at less positive potentials and a bridge-bonded carbon monoxide in the more positive region. A mechanism fitting that spectroscopic work is

$$CH_3OH_{(l)} \underset{k_{-1}}{\overset{k_1}{\rightleftarrows}} CO_{(ads)} + 4H^+{}_{(aq)} + 4e^-{}_{(metal)} \qquad (1.17)$$

---

* The reason is simple; batteries have to carry their fuel with them in the material on their plates. So when that is used up, a recharge is necessary. But fuel cells take their fuel from separate tanks ($+O_2$ from air) and that means that the range of a fuel cell driven car is several times more than one powered by a battery.

The surface concentration of carbon monoxide can be calculated for the reaction. Thus

$$-COOH_{(ads)} \underset{k_{-4}}{\overset{k_4}{\rightleftharpoons}} CO_{2(g)} + H^+_{(aq)} + e^-_{(metal)} \tag{1.18}$$

$$CO_{(ads)} + OH_{(ads)} \overset{k_3}{\rightarrow} -COOH_{(ads)} \tag{1.19}$$

$$H_2O_{(l)} \underset{k_{-2}}{\overset{k_2}{\rightleftharpoons}} OH_{(ads)} + H^+_{(aq)} + e^-_{(metal)} \tag{1.20}$$

$$j = FK_2(CH_3OH)\exp(4FV/RT) \tag{1.21}$$

followed by the independence of θ_{CO} at more positive potentials. The mechanism suggested therefore interprets the Tafel lines with two slopes verified in a number of studies of methanol.

This mechanism of methanol electro-oxidation avoids any mention of a "blocking" carbon monoxide, that is, a carbon monoxide radical that is tightly bonded to the surface, taking no more part in the oxidation of methanol to carbon dioxide. But possibly builds up and gradually decreases the amount of the surface available for reaction.

This situation brings up a point that often is relevant in mechanism analysis: surfaces are heterogeneous. Reaction may occur in patches of the surface favorable to the fastest mechanism. The reaction does not necessarily spread itself uniformly over the surface. In gas-phase catalysis, the concept of active sites has long been recognized.

The mechanism determination put forward here has been based upon only two types of measurements: the Tafel lines with the two slopes in different potential regions (not shown) and the spectroscopic analysis indicating the adsorbed radicals. It is an example of how spectroscopic measurements have improved mechanism analysis and preclude the determination of the optimal electrocatalyst.*

1.4.5 THE MECHANISM OF THE OXIDATION OF SATURATED HYDROCARBONS

There are two good reasons for investigating this mechanism, and hence the r.d.s. in the oxidation of saturated hydrocarbons. We still have more than four decades worth of saturated hydrocarbons left to give us energy, and it would be helpful were we to use them with the best catalyst possible, stretching the time during which we would have to develop inexhaustible (renewable) fuels. Further, the structure of the saturated hydrocarbons is simpler, the mechanism easier to fathom, than for less simple structures, such as methanol.

Much of the basic work here was carried out by a prominent electrochemist, Glenn Stoner [13], while he was a graduate student at the University of Pennsylvania. He found (for propane)

$$\frac{d\eta}{d\log j} = \frac{2.303RT}{F} \quad \therefore \alpha = 1 \tag{1.22a}$$

$$\left[\frac{d\log j}{d\log(H^+)}\right]_{\Delta\gamma} = 0 \tag{1.22b}$$

---

* One must guard against saying: "I understand you've told me about how you got information about the mechanism of the oxidation of methanol, applicable to a long time use of the electrochemical oxidation in a fuel cell. But what is the best catalysis? The answer is that, as the present author sees it, once you have the mechanism, and if the reaction is one which has an r.d.s., one can easily conclude what are the best and likely catalysis. That this final step is often not taken in the logical and scientific way is because, in industry, there is a very great pressure to turn out something practical and salvable so that "trying it out" becomes only too easily the program. Indeed, there is a good excuse. Material such as I am giving you, well understood, will lead you to the right address and where to knock at the door, but, finally, the competition between likely catalysts from the mechanism work but the final choice does involve testing.

$$\left[\frac{d\log j}{d\log(C_3H_8)}\right]_{\Delta\gamma} = 1 \tag{1.22c}$$

An observation that will help to find a probable mechanism is that if one plunges a platinum electrode into a solution saturated with propane and current passes at first and by knowing the coulombs involved, and assuming that the current comes from the following reaction:

$$C_3H_8 \rightarrow C_3H_7 + H^+ + e^- \tag{1.23}$$

From measuring the current time relation, one can make a calculation of $\theta_{C_3H_7}$ being present at 25% of full coverage.

There are several mechanisms that fit these facts. I give only the steps up to the one likely to be rate determining (all this material came originally from the PhD thesis of Glen Stoner; see Table 1.2) [13].

$$1.\; C_3H_8 \rightarrow C_3H_{7(ads)} + H^+ + e^- \tag{1.24a}$$

$$C_3H_{7(ads)} \rightarrow C_3H_6 + H \tag{1.24b}$$

$$2.\; C_3H_8 \rightarrow C_3H_{7(ads)} + H^+ + e^- \tag{1.24a}$$

$$C_3H_{7(ads)} \rightarrow C_2H_{4(ads)} + CH_{3(ads)} \tag{1.24c}$$

$$3.\; C_3H_8 \rightarrow C_3H_6 + 2H^+ + 2e^- \tag{1.24d}$$

TABLE 1.2
Possible Rate-Determining Steps (with Preceding Quasi-Equilibrium Steps Where Present) for the Oxidation of Propane and Platinum in Phosphoric Acid, 80°C–150°C and 0.30–5.0 V (Reversible Hydrogen Scale) (Work by G. Stoner)

1. $-C_3H_{8(sol)} \rightarrow C_3H_{7(ads)} + H^+ + e^-$

2. $-C_3H_{8(sol)} \leftrightarrow C_3H_{7(ads)} + H^+ + e^-$
    $C_3 + H_{7(ads)} \rightarrow C_3 + H_{6(ads)} + H^+ + e^-$

3. $-C_3H_{8(sol)} \leftrightarrow C_3H_{7(ads)} + H^+ + e^-$
    $C_3H_{7(ads)} \rightarrow C_3H_{6(ads)} + H_{(ads)}$

4. $-C_3H_{8(sol)} \leftrightarrow C_3H_{7(ads)} + H^+ + e^-$
    $C_3H_{7(ads)} \rightarrow C_2H_{4(ads)} + CH_{3(ads)}$

5. $-C_3H_{8(sol)} \leftrightarrow C_3H_{8(ads)}$
    $C_3H_{8(ads)} \rightarrow C_3H_{7(ads)} + H^+ + e^-$

6. $-C_3H_{8(sol)} \leftrightarrow C_3H_{7(ads)} + H^+ + e^-$
    $H_2O_{(l)} \leftrightarrow OH_{(ads)} + H^+ + e^-$
    $C_3H_{7(ads)} + OH_{(ads)} \rightarrow C_3H_7OH_{(ads)}$

7. $-C_3H_{8(sol)} \leftrightarrow C_3H_{7(ads)} + H^+ + e^-$
    $C_3H_{7(ads)} + H_2O_{(l)} \rightarrow C_3H_4OH_{(ads)} + CH_{3(ads)} + H^+ + e^-$

8. $-C_3H_{8(sol)} \leftrightarrow C_3H_{7(ads)} + H^+ + e^-$
    $C_3H_{7(ads)} \rightarrow C_3H_{6(ads)} + H^+ + e^-$
    $C_3H_{6(ads)} + H_2O_{(l)} \rightarrow C_3H_7OH_{(ads)}$

We can reject No. 3 because this mechanism involves two electrons transferred in one step, and there are quantum mechanical reasons against this. It comes down to whether one can distinguish between Nos. 1 and 2. There are no experimental data at present, which allow such a distinction, but an energetic argument in favor of No. 2 can be made. Thus, in No. 2 there is a C–C bond break and $(CH_3)_{ads}$ formation and this radical adsorption would bring the energy of the receiver state down so that the heat of activation would be smaller in reaction path 2, than in reaction path 1. Thus, the r.d.s. is the C–C rupture of the adsorbed C_3H_7 radical.

We do not have any information that would help us with the rest of the mechanism of the reaction to carbon dioxide, that is, after the r.d.s. However, from the catalytic point of view, finding the r.d.s. is the main thing with respect to forming a good catalyst.

It has been suggested that this mechanism (supposed to go in concentrated phosphoric acid at 80°C–150°C), could be made stronger if one examined the kinetics using normal H containing propane and then with the corresponding deuterium containing propane. One would expect an isotopic ratio of around five times faster with the hydrogen-containing molecules than with those containing deuterium due to the difference of the zero point energies of the molecular entities concerned. Were such a result to be found, $j_H/j_O \cong 5$, this would be consistent with the mechanism put forward earlier.

1.5 "PLATINUM IS ALWAYS THE BEST ELECTROCATALYST"

In this chapter, I am trying to speak to workers interested in physical electrochemistry, and I distinguish such workers from those who have interest in electrochemistry on the basis of its usefulness in analytical problems and to whom the electrode is primarily a source for electrons. This chapter is meant to give a basis for people who want to THINK about what is the best electrocatalyst for a given reaction.

Now, there is opposition to my views, and it is caricatured by those who say "Isn't platinum the best catalyst?" People who speak like this should not be regarded as poor in knowledge because it turns out that, in practice, platinum is often the best electrocatalyst available for anodic oxidation reactions.*

Let us look back at Figure 1.17. That is a so-called VOLCANO relation happens that if you plot on the ordinate axis the velocity of a reaction and on the ventricle abscissa axis bonding of the named substrates to an important radical bonded to the surface (as an intermediate). "Bonding" (including the heat of sublimation) of the "catalyst substrate," one often finds a volcano shape. Now, in Table 1.3, I have listed the melting points of a number of refractory (i.e., stable) metals. What do you see? Bonds to platinum will be in the middle of this group. It will bond well enough, but not too much so that reactants just adhere to the surface and do not desorb.

Apart from platinum's "intermediate" nature on bonding, another point in platinum's favor is availability; platinum can be purchased in various suitable forms at a "reasonable" price; some noble metals are difficult to find and purchase. The word noble means here stable and of course that is a first point one wants in an electrocatalyst. It must be a catalyst, not enter into the reaction. It is meant to accelerate the reaction. It must itself be stable, thermally and electrochemically. On the last point, platinum is only fairly good because oxide-free platinum does start itself to dissolve around 1.0 V on the normal hydrogen scale. By using it in anodic reactions in a potential range anodic to 1.0 V, Pt(II) is likely to get into the solution and may be deposited on the cathode.

---

* Concentrating, as I am doing, on the chemistry of electrocatalysis does leave out the chemical engineering, for example, how the electrode should be held, and its lifetime. I want the reader to know that this short chapter is intended to show electrocatalysis originated in fundamental electrode kinetics, and show examples of how one gets to the mechanisms, the knowledge of which helps one find design catalysts. But, be assured, there is much more. There is all the material that is dealt with by those useful people on the engineering side (e.g., Ralph White and John Weidner in the University of South Carolina). Such people are educated in chemical engineering, and we shall leave their contributions to further reading.

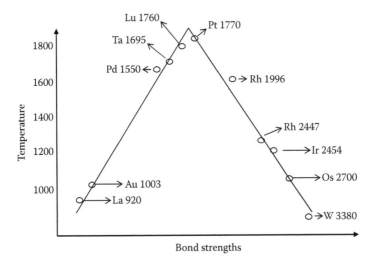

FIGURE 1.17 The increase in melting point with the bond strength. At first, it encourages the formation of radicals. When their bond strength to metals is greater than that with platinum, it is difficult to get the radical formation in the first step off the surface. In the second part of the curve, the bond strength is too strong that the radicals are not desorbed from the metal surface.

TABLE 1.3
Melting Point of Some Refractory Elements (°C)

Elements Lower or Equal to Platinum	MP (°C)	Elements Higher than Platinum	MP (°C)
La	920	Rh	1996
Au	1003	Ru	2427
Pd ~ Pt	1550	Ir	2454
Ta	1695	Os	2700
Lu	1760	W	3380

However, and particularly for anodic reactions, platinum has a few good points going for it with respect to its properties before one gets into talking about electrocatalysis, and virtually all experimental research in electrochemical laboratories starts with platinum. After that one can sometimes use the information in this chapter to do better. But let me be honest: doing better in the electrocatalysis of anodic reactions is a complex business, and one has to figure out the number of man years it will take to think the electrochemistry and then come up with rational suggestions (involving sometimes some informal guesses) as to what could be a better electrocatalyst than platinum. (There are certainly better catalysts, for example, for oxygen reductions, ruthenium and platinum ruthenium alloys.)

1.5.1 ELECTROCATALYSTS AND CATALYSIS

Sometimes in company laboratories, a boss realizes that he or she has a large group researching catalysis, and then there is a much smaller group working on electrocatalysis. Bosses are good at planning programs but do not always know too much about the subjects being worked on. Sometimes this leads to time-wasting misdirection. "On Wednesday I've arranged for three of you from the electrocatalysis group to sit with three of you from the catalysis group and just hammer things out together." It is the "together" that grinds a little because there is not much togetherness between the two groups. First of all, there are misunderstandings to be put right. The catalyst people start off with a feeling of superiority because they know their subject is a century or older, and the electrocatalyst people's subject were first named in 1959 by Grubb of G.E. All chemists know what catalysis means but few know what electrocatalysis is about. The catalysis people think at once that it means speeding chemical reactions by doing them under an electric field, and although there is a certain crude truth in that, in fact, nearly work on electrocatalysis means seeking a change of surface and the nature of the substrate on which a specific reaction goes faster than the same reaction does at the same electrode potential on some standard surface on, say, platinum. Some differences are huge. The most investigated electrode reaction in this respect is undoubtedly the hydrogen evolution reaction ($2H^+ + 2e \rightarrow H_2$), and it has an exchange current density as high as 10^{-3} A cm^{-2} on platinum and as low as 10^{-10} A cm^{-2} on mercury. A scientist deep into catalysis will say "Electrocatalysis refers to the phenomenon that, in carrying out an electrode reaction, one can cause large (e.g., 10^7 times) change in the reaction rate by changing the potential of the working electrode. That is quite right, but electro-chemists do not regard that quintessential fact, that is, that changes in rate can be brought about by the electric field across an interface as electrocatalysis. That is just fundamental electrochem-istry, Tafel's law, etc. They only bring in catalysis when the difference they examine is the effect of change of substrate. For hydrogen evolution, change from platinum to mercury causes a change of 10 million times in rate at the same potential: that is electrocatalysis. It is surprisingly difficult for electrochemists to get a chemist to grasp the point that one has not to count as electrocatalysis changes in reaction rate brought about directly by the change of electrode potential.

Another difficulty for chemists is the power of the change in potential in his hand. When electrochemists talk about changing an electrode potential by a volt, it seems trivial but when it is explained that this refers to a change of a volt delivered over 3–4 Å, that is, one is dealing with electric fields of $\sim 10^7$ V cm^{-1}, the light begins to shine.

There are sub-effects of these changes in electric field strengths too. The magnitude of such fields distort the shape of molecules in it and clearly this changes their reactivity or at least modifies the effect these squeezed and oriented molecules have on nearby reactants and products.

Yet another fact: the chemists would have to understand if they want to help the electrochemists to understand the presence of water. Of course, water has quadrupole properties and changes the local electric field within the overall electric field at the metal–solution interface.

So far, I have brought out points that seem to give electrically stimulated reactions a big advantage over chemical ones, why are there not 10 electrochemists to 1 chemist instead of the other way round? I think that the explanation of the discrepancy in numbers of workers is due largely to an absence of relevant knowledge. The low numbers of physical electrochemists in the United States does not apply in the same degree in, for example, Italy, Germany, Russia, or Japan (China?). Part of the gap here in America lies in the poor opportunities for training in physical electrochemistry in universities in the United States. I had expected that when fuel cells became prominent, there would be a substantial swing toward electrochemical research groups. However, many of the big things in which electrochemistry plays a leading role (space flight, fuel cells, biology, electrochemical condenser storage) some of which are all of recent vintage are not yet associated in people's minds with modern electrochemistry.

Let me finish this note by pointing out where electrochemical phenomena and hence electro-catalysis provides a number of hurdles to the student coming from chemical catalysis. I will name only three, but they are all considerable. First, the terminology. In chemical catalysis, the measure-ment of a reaction rate is in moles per square centimeter per second (mol cm^{-2} s^{-1}) or moles per square meter per second (mol m^{-2} s^{-1}). In electrocatalysis, the measurement of rate is in current density, and, without training, there is at once a curtain between the two fields, until one learns that j/nF means the number of times that the overall reaction has occurred; and that the unit of that magnitude means moles per square centimeter per second (mol cm^{-2} s^{-1}). Thus, electrocatalytic data can be converted to the catalytic and *vice versa*.*

Second, the control of the electrochemical reaction is much connected with electronic instrumentation and what one can do with electrochemical currents is to make them perform various tasks; one must have a good training in practical electronics, less needed by the catalytic worker. Advanced forms of spectroscopy specialized to observe surfaces in contact with water have come into electrochemistry, too, and demand much new knowledge and practice.

A final problem that separates the chemist from the electrochemist in interpretation and experimental technique lies in the presence of a solvent that is sometimes complexing. The major thing it does is to incapacitate the vacuum methods for looking at the surface, which are used in chemical catalysis. It was not until Reddy and I did apply ellipsometry to electrochemistry in 1964 that one began to be able to observe material adsorbed on the typical electrode surface while in contact with solution. Last of all, no chemist doing catalysis work should avoid the realization that the 10^7 V cm^{-1} means a radical reduction of the heat of activation of the reaction and the easily achievable changes of potential as the dominant variable in the place of the more complexly arrangeable temperature change. Such things should bring attention to an attractive field, in which the use of what seems a trivial amount of electricity makes it possible to achieve reactions at temperatures much lower than those used in pure chemistry.

1.5.2 ELECTROCATALYSIS WITH MORE COMPLEX MOLECULES

1.5.2.1 What Is the Field

Examples of mechanism determination and consequent catalytic design have so far been of simple molecules.

As one moves further away from the academic laboratory, the size of the molecules upon which one is asked to focus increases, and, finally gets to biomolecules, the molecular weight of which may be in the millions. For the moment, let us think about molecules the size of some amino acids as shown in Figure 1.18.

1.5.2.2 Problems in Dealing with Electrode Reactions of Larger Molecules

1. The larger the molecules, the easier it is to see it breaking up upon adsorption and go partly into confusing side reactions. These often form polymerized, unreactive junk ("gunk") blocking the electrode surface and sometimes blanking out sites that could have been reaction centers, for example, given catalysis. Such a difficulty can often be removed by momentary high current density anodic pulses tuned to a potential to oxidize suspected, illegitimate gunk and keep the catalyst surface clean.

2. More complex molecules may not have the strength for a safe landing when they adsorb but may simply fall to pieces on "landing."

3. Is the idea of an r.d.s. still useful with the bigger molecules? The more the number of steps making up a reaction, the less one step determines the rate of the overall reaction. However, there are sometimes reactions, in which the first step is an electron-transfer reaction, and that may be the slow one, because the adsorbed molecule is so big that the path length that the electron has to travel to the central cities is more than, say, 20 Å. Such

---

* But there may be complexities in the models engendered by such conversions. One recalls that an electrochemical surface is covered with adsorbed solvent molecules. Correspondingly, both in catalysis and electrocatalysis not the entire surface is active. The degree of occupancy and the frequency of "active sites" is likely to be different in each of the two subsciences.

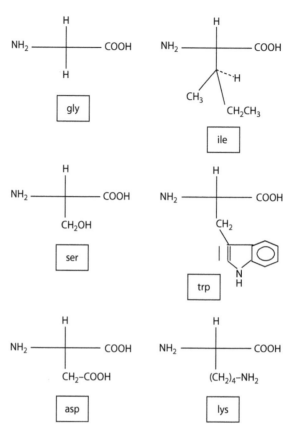

FIGURE 1.18   Formulae of some important amino acids. (From Bockris, J.O'M. and Kainthla, R.C., *J. Phys. Chem.*, 84, 2463, 1985.)

an electron transfer is likely to involve quantum mechanical tunneling, and if the distance needed exceeds about 20 Å, the reaction rate falls off sharply.

4. The larger the molecule, the easier it is to take up some alternative pathway. Thus, one starts the reaction with the aim of a certain product, only to find an efficiency of, say, 40%, while 60% goes to, an unwanted pathway.

   This is a bigger problem, one that is known in chemical catalysis too. Its solution can sometimes be reached if methods for suppressing the interfering reactions can be found.

5. Organic molecules are often involved in electrocatalysis studies and such entities dissolve to acceptably high concentrations only in organic solvents. This brings the difficulty that water may leak into the system from the atmosphere and "poison" the nonaqueous solution. The effects of 1%–2% water are far greater than one might think (water tends to congregate on the surface). A worse case comes if a metal reactive to water is involved (as in the lithium batteries). Lithium reacts with water producing molecular hydrogen, and if this contacts air the possibility of an explosion is born. Even when no explosion threatens, the small quantities of water are likely to migrate to the surface and displace the organic molecules adsorbed on the electrode. But such displacement and adsorption will change the electric field in the double layer at the electrode–solution interface, and the behavior of the electrode reaction changes correspondingly.

   Water leaking into a reaction containing an organic solvent can sometimes be defeated by keeping "coils" of sodium in the solution to "mop up" the water by reacting with it. However, this means hydrogen evolution, and resurrects the possibility of an explosion.

1.5.2.3 The Reach of the Bigger Molecules
The intention in this chapter is to focus the discussion on electrochemistry. It does not encompass biochemistry. The biochemicals of nature "work" with the aid of enzymes. These are the super catalysts of nature and the mechanism by which they function so effectively is still continuing to progress. How enzymes come into the functioning of a reaction we do not see as electrochemistry yet.

1.5.2.3.1 Spectroscopic Approaches
The role of techniques that are able to obtain spectroscopic data from adsorbed materials in solution was born in the 1990s and is still not generally a part of most electrochemistry laboratories. The equipment is expensive in a field where, otherwise, equipments are of low cost. There are now several techniques that are useful in electrochemistry and here we shall only mention two.

1. *Fourier transform infrared spectroscopy*: The Fourier transform refers to mathematical procedures, which are needed to convert the basic data, the actual quantities measured, back to the familiar data of absorption spectroscopy. Properly adjusted and understood, it is possible to find the spectrum of what is adsorbed on the electrode and how it varies with potential.

2. *Ellipsometric spectroscopy*: Mathematical programs involving Fourier transform can also help* ellipsometry. Ellipsometry was the first optical device to be used as an *in situ* method for observing at electrode surfaces in a group at the University of Pennsylvania in 1964. It is more usually employed with monochromatic light of fixed frequency for measuring the thickness of films on electrodes (down to submonolayers). However, the data also contains an absorption coefficient and if, then, appropriate wavelengths of light are used, spectroscopic data can be obtained and hence the nature of the adsorbing entities identified. Moreover, the use of Fourier transform ellipsometer allows this to be done quickly so that one can follow the buildup or decay of what is on the surface.

Used to identify structures, neither of these techniques is new; although only since the 1980s has it been possible to make them practical. To follow the reaction path and build up knowledge on an atomic scale that would contribute directly to the design of catalysts is not quite there: the time response needs to get down to the millisecond, and that, like so much else in fundamental science, depends on federal research programs. It is a reasonable speculation to suggest that methods such as the two referred to here will be in routine use in electrocatalytic studies well before 2050.

1.5.3 MECHANISM MAYBE, BUT ELECTROCATALYSIS?
Is finding an optimal electrocatalyst for a given electrochemical reaction the same as finding its mechanism, that is, its pathway and, if it is simple enough, it is r.d.s.? A short answer is no. But a longer answer would say that researching to find the best catalyst is greatly helped if one knows the mechanism, and if the reaction has an r.d.s. For, to find a catalyst, one has to know about bonding between the products of an electron transfer and the catalyst surface or the behavior of one or more radicals deposited within the functioning of the r.d.s. on the electrode surface.

But there is more one can know before one tries out indicated substrates, as empirical help in finding a catalyst. For example, it would not be wise to think that a reaction took place all over the surface. There are certainly active sites that have been written about in gas-phase catalysis for more than 50 years.

Then, we have said little about the reactivity of various crystal planes on the electrode surface (see Table 1.1 and Figure 1.10) should specific planes be used in experiments.

Correspondingly, anions will be adsorbed on the various planes to different extents and the degree will be potential dependent and their local electric fields will in turn attract or repel the

---

* This is an understatement. To operate an ellipsometer without computational help is too time-consuming.

presence of other ions and radicals on those planes. What a known r.d.s. does is to focus on a few sites in contact with reactants in adsorption or desorption during the functioning of an r.d.s.

Then there is surface diffusion.* Adsorbed ions and radicals arrive on adsorbing sites and their ability to get away from them to active catalytic sites is determined by surface diffusion on the various faces of critical surface. Obviously, all this occurs under high electric field strengths, which will change with the chosen potential at which the reaction is to be run.

Bonding to surfaces covered with adsorbed water can be an r.d.s. in a way that helps reaction rate. Something that has a parallel in gas-phase kinetics comes in here. Look back at Tafel's equation. The lower the b value $(RT/\alpha F)$ of Tafel's equation the less the overpotential developed for a certain rate of reaction—or the faster the reaction will go at a certain overpotential. Clearly, then, it is advantageous to try for an r.d.s. late in a consecutive series of reactants, where the α will be higher, and the b value at a given current density lower (Tafel's law). In fact, this is an insight in the computer-generated electrocatalyst. Can the researcher think out a scheme to bring about an r.d.s. that cuts down the influence of the $\log j$ term in Tafel's law[†] $(\eta = a + b \log j)$?

Another tool, which will help us produce optimal electrocatalysts, is the use of the creative computer program. The author used an elementary version of such a program in 1963 to determine the pathway in the early steps in metal deposition. But in 1960, computers available in university labs were little more than advanced electronic calculators. The concept of a program that calculates the energy of a representative point in the reaction as it seeks (by calculation) to find the minimal energy pathway is now possible, and again it depends on the complexity of the molecular paths involved. It would be in principle possible to model a catalyst surface with typical heterogeneity on several crystal faces. We know that nature will take the lowest energy pathway. If one has in his team a person who can convert the basic equations for the potential energy of a representative point into a program, it is possible (6 months work) to calculate the pathway of a reaction according to hypothetical pathways, and finally calculate the optimal surface for a catalyst (Figure 1.19).

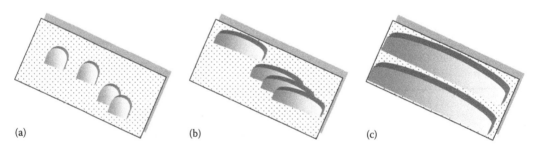

(a)   (b)   (c)

FIGURE 1.19 Schematic representation of glucose oxidase adsorption. (a) Situation where the potential is positive with respect to the zero potential value; (b) case where the potential is equal to that of zero charge; and (c) final state where the enzyme is unfolded.

---

* When I contemplate diffusion in electrochemistry, I am reminded of Carl Wagner, an Massachusetts Institute of Technology professor of the early 1950s. He was an extremely private man and disliked being interrupted in his thinking by the need, his colleagues had advised him, to attend faculty parties at which, yes, alcohol was being offered. Wagner always left it till halfway through the party to arrive (cutting down lost time). He would have an aim in his mind, to use the time in a discussion with a professor whose paper in Wagner's opinion contained a vital error. Walking near the wall of the room to progress toward the target, Wagner would finally arrive beside him, "Aha," Wagner would start, "Well, I have found an important error in your recent paper..." Carl Wagner was much respected for his contributions particularly to the basis of early discussions of the three-phase boundary in fuel cells. But the truth is that, socially, he was not very popular.

[†] Try to get used to thinking of overpotential as BAD. As it grows it will make the cost of driving a reaction greater, hence, the cost of the product greater. In a fuel cell, the efficiency of the conversion of a fuel decreases as the overpotential increases.

I am writing in 2008. I claim that experimental analyses of a surface and the reactants and products on it can lead to the determination of the optimal electrocatalyst. If one can have a program written to express the theory, one can calculate the optimal catalyst.* In a few years, intelligently chosen funding will be able to do just what is claimed here and truly design the optimal catalyst for a given reaction potential and temperature range.

1.5.4 The Fixing of Carbon Dioxide

One of the greatest dangers from our threatened world is the mounting carbon dioxide and methane concentration.

During the next half-century, the atmospheric concentration of these is likely to rise to 400 ppm and even 600 ppm if we continue to burn fossil fuels. Such concentrations will lead to methane from the tundra and the temperature rise from that may well make the surface of this earth uninhabitable.

Now, one way to deal with this undeniable world threat is to "fix" atmospheric carbon dioxide by converting it to the simplest (liquid) hydrogen carrier, methanol. It is not fanciful to consider "fixing" most of the carbon dioxide output in our present economy by turning it into easily handled methanol, providing our cars then with the methanol or derived hydrogen to drive their fuel cells.

Regarding suitable cathodes for the electrocatalytic reduction of carbon dioxide to methanol, indium, and tin have good prospects because each has a high overpotential for hydrogen evolution, the chief competitor in aqueous solution for carbon dioxide reduction to methanol:

$$CO_{2(g)} + 6H^+_{(aq)} + 6e^-_{(Metal)} \rightarrow CH_3OH_{(liq)} + H_2O_{(liq)} \tag{1.25}$$

with $E^o_{rev} = -0.38$ V.

However, Hori (1986) found the electrocatalytic reduction of carbon dioxide to methane on a molybdenum cathode to be electrocatalytic [11]. One hundred percent reduction of carbon dioxide to methanol at 0°C was reported. What remains open to research is the use of nonaqueous solvents, for example, acetonitrile, because there would be then no competition from hydrogen evolution. Carbon dioxide does undergo electrochemical reduction in acetonitrite to glycolic acid. Here is a research topic, which could have world consequences and cost very little (about 10^5 per year per researcher, even a dozen would make a great difference in five years).

Park, Anderson, and Eyring examined the reduction of carbon dioxide on mercury. The sequence that they determined was

$$CO_{2(g)} + e^-_{(Metal)} \rightarrow CO_2^-_{(ads)} \tag{1.26}$$

$$CO_2^-_{(ads)} + H_2O \rightarrow HCO_2^-_{(ads)} + OH^-_{(aq)} \tag{1.27a}$$

$$HCO_2^-_{(ads)} + e^-_{(Metal)} \rightarrow HCO_2^-_{(liq)} \tag{1.27b}$$

At the higher current density, Reaction 1.26 was rate determining. The adsorbed CO_2^- ion present in the mechanism was discovered in the electrode in an early application of FTIR spectroscopy applied to mechanism analysis.

---

* The advantage of computer situation is that once the program writer has taken electrochemistry molecular models and turned them into a program, which can be loaded on a computer, it is possible to plot the optimal path of the reaction in a few hours. And the real advantage comes when one uses the same program to find out controllable variables (potential catalyst surface, temperature) and what they do to the heat of activation for the reaction. Each of these calculations would take less than a day's work or less if one can program the computer to work at night.

In practice, the limitation is the money allowed for the grant to have a person sufficiently skilled, and hopefully knowing just a little physical chemistry, to write the program. In the present author's experience, one is sometimes forced to accept a program that has been previously written and which can be rented. But then, of course, such an available program may not contain the force fields you, the scientist, designated, and is unlikely to allow for the vital heterogeneity. As usual it all depends on the size of the grant one is working on.

Bockris and Wass [28] used a photocatalytic reduction. A 10^3 increase in rate was found in the photoreduction on a cadmium catalyst if 18 crown ether in dimethyl formamide 5% water was present. NR_4^+ ions and the appropriate crown ether were essential to the reaction, which fitted the requirements of the Tafel behavior.

1.5.4.1 Photoelectrochemical Fixing of Carbon Dioxide

The formation of methanol from carbon dioxide is usually thought to be the goal of carbon dioxide reduction. However, Uosaki and Nakebeyeski (1993) examined carbon dioxide reduction in a system containing acetonitrile and DMF to phenyl acetic acid. The electrode was GaP. The presence of magnesium is helpful during irradiation. The interpretation of the behavior of the photo electrode indicated a high degree of surface states.

1.5.4.1.1 An International Goal

The electroreduction of atmospheric carbon dioxide to methanol should be a community goal for it would stop the increase of greenhouse warming and provide inexhaustible fuel [29]. The chemical and electrochemical goals are probably attainable. Bockris and Wass discovered the use of crown ethers in the photoreduction of carbon dioxide to small amounts of methanol on illuminated GaP electrodes [30]. However, the key chemicals present were certain crown ethers in the presence of corresponding tetra alkyl ammonium nitrites. Making synthetic food from atmospheric carbon dioxide, nitrogen, and water using enzymes could be a goal of importance. The three compounds are all easily available. We need selected bacteria and semiconductor photo-electrodes. Research money with specific society goals is the main need.

1.6 THE ELECTROCHEMICAL REMOVAL OF WASTES

Factory wastewaters often contain metal ions of sufficient concentration, which makes their recovery worthwhile.

Fleischmann and Chu [29] were the first to develop a packed-bed type of electrode, one containing, for example, lead shot of a certain path length and holding an appropriate potential (Figure 15.23 in [30]). They proposed the following equation:

$$\frac{C_2}{C_0} = \exp(-L/\lambda) \qquad (1.28)$$

where C_2 is the concentration remaining in the original liquid after passage through a packed bed, for example, of lead shot of length, L. λ is a parameter that depends on the flow rate and other factors of a plant.

Suppose there are several metals in the original solution. These can be deposited on the packed bed, holding it at an appropriate potential. Then, the potential of the packed bed is reversed and such metals present are dissolved out and deposited onto an electrode potentiostat at a potential negative to the reversible value of the specific material. In this way, 99% of the metals to be recovered can be taken out separately. (It is necessary for the reversible potentials of the various metals to be separated to ca. 0.3 V for this process to be practical.)

1.6.1 The Destruction of Nitrates

One of the primary waste disposal problems of the world is that of radioactive materials. At present, the U.S. nuclear wastes are stored in solution at Hanford, Washington, and Savannah River, Georgia. These wastes are divided into two types of materials. The high-level radioactive waste are absorbed into porous solids and transported to permanent depositories within mountains.

There remain over the low-level radioactive wastes, largely ruthenium, mercury, and chromium. These metals are in the form of soluble nitrates: The problem is that any disposal scheme that allows the material to reach the ground water may result in contamination. Were those nitrates to be reduced eventually to ammonia and even molecular nitrogen, this would remove the hazard (Hobbs and White, 1992). If the economics justified it, some of the metals would be recovered. The rest would be stored as oxides.

The metals present can be recovered in a batch process using packed-bed deposition, dissolution, and potentiostatic deposition of individual ions described earlier (cf. [30]). The nitrates can be electrolyzed in a 24 h cycle in parallel plate cells with proton exchange membranes. Thus,

$$NO_3^- + H_2O + 2e^- \rightarrow NO_2^- + OH^- \tag{1.29a}$$

$$NO_2^- + 5H_2O + 6e^- \rightarrow NH_3 + 7OH^- \tag{1.29b}$$

$$2NO_2^- + 4H_2O + 6e^- \rightarrow N_2 + 8OH^- \tag{1.29c}$$

1.6.2 Catalytic Removal of the Electro Toxic Organic Wastes

Clarke and Kühn [31] have devised particularly complex electrochemical processes. The need to sketch out here some ideas as to how they work is given by the surprising, even amazing, range and effectiveness of waste destruction that Clarke and Kühn reported as early as 1975. They reported, for example, the destruction of rubber gloves, sewage sludge, phenols, and many other toxic wastes. Thus, for example, iron, nickel, or silver are made of anodes and produce ions of these materials in the right valence state. Solutions containing them are then raised to 200°C under pressure and under these conditions oxidize the wastes. The ions are now in their lower valence state whereupon they are reoxidized at an electrode and returned to the solution containing wastes to be oxidized.

Rubber gloves, gaskets, epoxy resin, kerosene, and oil waste have been oxidatively consumed in this simple electrochemical way [32]. There are significant environmental advantages of this electrochemical approach. Were one to burn the wastes, they would contaminate the atmosphere. Low current density chlorine evolution may be used in electrolysis involving oxides, particularly in dealing with sewage (Tennakoon, 1993) [33].

Other fields to conquer here include the destruction of Agent Orange, of explosives, and other agents intended for use in biological warfare. The work of Stoner on the destruction of bacteria has been known since 1975. The avoidance of the buildup of dead bacteria (thus reducing the heat conductivity in pipes associated with OTEC) is noteworthy. The electrochemical method produces hydrogen peroxide on the bacteria adsorbed on the pipe surfaces and eliminates their negative effects on the pipe's heat conductivity.

1.6.2.1 Electrocatalytic Development of the Oxidation of Hydrogen Sulfide

Hydrogen sulfide is generally regarded in the oil industry as a contaminant. However, one can regard hydrogen sulfide as a substance, the use of which may be profitable. Thus, if the potential for its decomposition is compared with that for water, it is found that the production of hydrogen from alkaline solution of hydrogen sulfide can be achieved at about three-fourths the electrical energy needed to obtain hydrogen from water. The overpotentials involved are also less than that in normal water decomposition where the high overpotential for the unnecessary evolution of oxygen is the origin of the legend that hydrogen is too expensive. Although Dandapani first investigated the electrochemistry of hydrogen sulfide decomposition in 1988 [34]. The process has been developed to an engineering level by Petrov and Srinivasan [35]. The sale of the sulfur contributes greatly to lessen the cost of hydrogen.

There are other ways of obtaining hydrogen from hydrogen sulfide, for example, a photo-electrochemical way [26] and the running cost of this process produces hydrogen and sulfur without the burden of the electricity costs needed in the electrochemical process. Hydrogen could be made at negative cost.

1.6.2.2 Electrocatalytic Sewage Disposal

In the present sewage disposal practice, the final result is a sludge, which defies further destruction (though cf. the work of Clarke and Kühn). This is an unsatisfactory state of affairs because the sludge contains mercury and lead, and these toxic materials are reaching the water table in response to normal rainfall.

Kaba and Hitchens made a discovery in 1989 [36]. They found that if fecal material was mixed with urine, the electrolysis of the mixture leads to the production of hydrogen at low cost. The production of a colorless, odorless liquid is due to the bleaching effect of the coevolution of chlorine at the anode. The complete elimination of the organic contents of the solid material is possible. (No sludge!)

This process was developed further with packed-bed techniques. A limitation is that the particle size had to be reduced and this was carried out by the ultrasonic irradiation of the mixture before electrolysis. The average particle size was 25 p.m. A detailed analysis of the material balance was made [37]. Ebonex reduced TiO_2 coated with SnO_2 doped with Sb_2O_3 was found to be the best anode materials. A 1996 model cell is shown in Figure 1.20 [38].

1.6.3 ELECTROCATALYSIS AND GLOBAL WARMING

There are opportunities for electrocatalytic contributions to delay the world-threatening process of global warming.

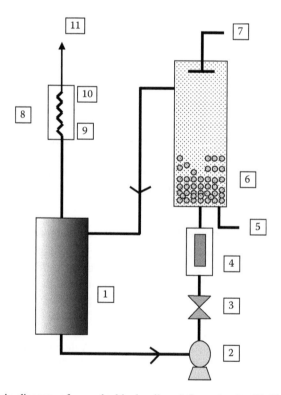

FIGURE 1.20 Schematic diagram of a packed-bed cell and flow circuit. (1) Tank-reservoir; (2) pump; (3) valve; (4) flow-meter; (5) anode current collector; (6) packed-bed anode; (7) cathode; (8) water condenser; (9) water inlet; (10) water outlet; (11) gas outlet. (From Tennakoon, C.K., *J. Appl. Electrochem.*, 26, 27, 1996.)

Two of the consequences of the temperature rise are not yet in sight: flooding to 200 ft above present sea level and, finally, the super greenhouse caused by the exuding of methane, from the tundra as the temperature rises. Positive feedback would lead rapidly to temperatures, which might make the earth's surface uninhabitable ($>50°C$).

There are only two escapes:

1. Total conversion to nuclear power (50 years)
2. Total conversion to inexhaustible clean energies (25 years)

Electrocatalysis comes into both: into nuclear power to take the electricity made in low-use periods and uses it to electrolyze water (or hydrogen sulfide) to hydrogen, thus decreasing the time it will take to build such an economy by approximately 15 years. As to the renewables and electrochemistry, the most well-known ones are wind and solar. Both give plenty of energy but it is sporadic. The energy has to be collected when available, converted to methanol, available for general use.

1.6.4 ELECTROCHEMISTRY AND A THREATENING FUTURE

I wish to finish this chapter then by writing about the future I see for our civilization and the part electrochemistry, electrocatalysis, may play in alleviating it.

1.6.4.1 Global Warming

Writing in electrochemical books about rising temperature and sea level dates back to 1970 and was generally rejected as being the work of "doomsters" [39,40]. The year 1971 saw the suggestion of eliminating carbon dioxide emitting fuels in favor of hydrogen ("The Hydrogen Economy"). And here one sees the beginning of a movement, which has spread, although more outside the United States than in it [40]. Replacing carbon dioxide emitting energy schemes by electricity generation and hydrogen was rejected, not because a flaw was found in the argument but because, in the 1970s, gasoline (leaving out the cost of environmental damage it is causing) could be bought at one-third the price of hydrogen. Electrochemistry was rejected too, it would have been the main way to make hydrogen without the coevolution of carbon dioxide.

1.6.5 ELECTROCATALYSIS AND THE FUTURE

There is little doubt that our greatest threat is the super greenhouse, for then the rise in temperature and danger to inhabitability would be in years and not in decades. Meadows' classical analysis of exhaustion (1972) [41] and the follow-up of 1992 [42] show the middle of the century to be the beginning of a general descent but now the reaction against global warming may have to be pushed in front of Meadows' mid-century date.

The *in situ* scanning tunneling microscopy and atomic force microscopy are relatively new techniques whereby surfaces in contact with solution can be observed. The usual technique reaches 10–100 Å. However, Szklarzcyck, Velev, and Bockris reported in 1989 [15] that they could resolve pictures at atomic scale. This and the corresponding technique of atomic force microscopy have made a very significant increase in our ability to watch surfaces during reactions. A goal of seeing atoms react does not seem (2007) too "far out."

The northern European states, Germany and the Scandinavian countries, are the only ones that have taken some beginning of measures that could conceivably get their countries ready in time to counter the threats that lie so near: global warming, flooding, super greenhouse. The United States, the giant polluter, is the only one that has not signed the Kyoto Protocol to limit greenhouse gases, and so far has suggested only the 1% efficiency alternative for using solar energy.

There is a tragic flabbiness about our resolve to save the inhabitants of our world from environmental disaster. In the United States, gigantic cars are still being sold to anyone with the money to buy them. What seems strange is the lack of hope, of dynamic action. The most popular date for oil exhaustion is now 2010. After that the roaring demand for more oil will be replaced by oil from tar sands while the carbon dioxide and methane grow exponentially.

The 1% efficiency of the collection of solar light to fuel plant growth should not be taken seriously when solar energy can be converted to electricity at 30%.

Logic dictates that we should quickly build wind and solar sources and perhaps if there is enough uranium, nuclear, as well. But to launch a Manhattan Project to build the distribution network in time may need a more quickly deciding form of government and not the ever arguing and often long-to-decide debates that we have now.

1.6.5.1 The Food Supply

Making cheap artificial food for everyone has been considered. To make it need water, hydrogen, carbon dioxide, nitrogen, we have a lot of all of those noted, and then certain enzymes we have to find. If a significant number of us survive global warming, energy and food exhaustion, and a possible asteroid, the forms of energy remaining will certainly involve electricity, hence electrocatalysis.

REFERENCES

1. J. Tafel, *Z. Phys. Chem.* **50** (1905) 641.
2. F.P. Bowden and E.K. Rideal, *Proc. R. Soc. Lond. Ser. A* **120**(784) (Aug. 1928), 59–79.
3. F.P. Bowden and E.K. Rideal, *Proc. R. Soc. Lond. Ser. A* **120**(784) (Aug. 1928), 80–89.
4. R.W. Gurney, *Proc. R. Soc. Lond.* **A134** (1931) 127.
5. J.O'M. Bockris and H. Mauser, *Can. J. Chem.* **37** (1959) 475.
6. J.A.V. Butler, *Proc. R. Soc. Lond.* **A157** (1936) 423.
7. J.N. Agar and F.S. Dainton, *Discuss. Faraday Soc.* 1947.
8. T. Grubb, U.S. Patent 2, 913, 511, 1959.
9. J. Horiuti and M. Polanyi, *Acta Physicochim. U.R.S.S.* **II**(4) (1935), 505.
10. J. O'M. Bockris, A.K.N. Reddy, and M. Gamboa-Aldeco, *Modern Electrochemistry*, 2nd edn., Plenum Press, New York, 2000.
11. Y. Hori, K. Kikuchi, A. Murah, and S. Suzuki, *Chem. Lett.* **34** (1986) 897.
12. N. Tempkin, *Z. Fizc. Khim.* **15** (1941) 296.
12a. K. Uosaki, *J. Appl. Phys.* **52** (1981) 808.
12b. E. Budevski, G. Staikov, and W.J. Lorenz, *Electrochemical Phase Formation and Growth*, John Wiley & Sons, New York, 1996, p. 17.
12c. H. Seiter, H. Fischer, and L. Albert, *Z. Elektrochem.* **63** (1959) 249.
13. G. Stoner, PhD thesis, University of Pennsylvania, Philadelphia, PA, 1970.
14. G. Stoner, U.S. Patent 3, 725, 326.
15. N. Szklarzcyck, O. Velev, and J.O'M. Bockris, *J. Electrochem. Soc.* **136** (1989) 2433.
16. W. Beck, J.O'M. Bockris, L. Nanis, and J. McBreen, *Proc. R. Soc.* **A290** (1966) 220.
17. R.R. Adzić and N.M. Marković, *J. Electroanal. Chem.* **165** (1984) 121.
18. F.G. Will, *J. Electrochem. Soc.* **112**(4) (1965) 451.
19. J.O'M. Bockris, Dislocations and high pressure, from P.K. Subramanian, *Comp. Treat Electrochem.* **7** (1986) 383.
20. J.O'M. Bockris, A.K.N. Reddy, and M. Gamboa-Aldeco, *Modern Electrochemistry*, 2nd edn., Plenum Press, New York, 2000, Figure 7.107, p. 1278.
21. J.O'M. Bockris and T. Ottagawa, *J. Phys. Chem.* **87** (1983) 2964.
22. J.O'M. Bockris and T. Ottagawa, *J. Electrochem. Soc.* **131** (1984) 2965.
23. V.S. Bagotskii and Y.B. Vasiliev, *Electrochim. Acta* **12** (1967) 1323.
24. K.C. Chandrasekaran, J.C. Wass, and J.O'M. Bockris, *J. Electrochem. Soc.* **137** (1990) 520.
25. *Acc. Chem. Res.*, **21**, copyright 1989, Fig 1, American Chemical Society.
26. J.O'M. Bockris and R.C. Kainthla, *J. Phys. Chem.* **84** (1985) 2463.
27. F.A. Armstrong, H.A. O'Hill, and N.J. Walton, *Counts Chem. Search* **21**, 408.
28. J.O'M. Bockris and J. Wass, *J. Electrochem. Soc.* **136** (1989) 2521.
29. M. Fleischmann and A.K. Chu, *J. Appl. Electrochem.* **4** (1974) 323.

30. S. Kim and J.O'M. Bockris, *J. Appl. Electrochem.* **27** (1997) 323.
31. R.L. Clarke, A.T. Kühn, and E. Okao, *Electrochem. Britain* (1975).
32. J.P. Steele, *Platinum Metals Rev.* **34** (1990) 1190.
33. J.O'M. Bockris, R.C. Bhardwaj, and C.K. Tennakoon, *J. Appl. Electrochem.* **28** (1990) 1341.
34. P. Dandapani, U.S. Patent 357891, 1985.
35. K. Petrov and S. Srinivasan, *Int. J. Hydrogen Energy* **21** (1996) 163.
36. L. Kaba and G.B. Hitchens, *J. Electrochem. Soc.* **137** (1990) 1341.
37. C.K. Tennakoon and R.C. Bhardwaj, *J. Appl. Electrochem.* **6** (1980) 15.
38. C.K. Tennakoon, *J. Appl. Electrochem.* **26** (1996) 27.
39. J.O'M. Bockris and A.K. Reddy, *Modern Electrochemistry*, Plenum Press, New York, 1970.
40. J.O'M. Bockris, *Environment* **13** (1971) 51.
41. D.H. Meadows, D.L. Meadows, J. Render, and W.W. Behrens. *Limits of Growth*, University Books, New York, 1972.
42. D.H. Meadows, D.L. Meadows, and J. Render, *Beyond the Limits*, Chelsea Green, Post Mills, VT, 1992.
43. J.O'M. Bockris, *The Solar-Hydrogen Alternative*, Australian, New Zealand Press, Sydney, Australia, 1975.

2 Fundamental Aspects of Electrocatalysis

C. Fernando Zinola

CONTENTS

2.1 INTRODUCTION

The term "catalysis" was coined by Berzelius in 1835 and is derived from the Greek *kata* (go down) and *lysis* or *lyein* (letting). The first authors to introduce the term "catalytic electrode reactions" were Bowden and Rideal in 1928 [1], who observed the different currents that appear for a certain reaction on distinct electrode surfaces but under the same electrode potentials. There is still some controversy over the first use of the term "electrocatalysis." It seems from the literature that the Soviets were the pioneers in the field of electrocatalysis since 1934 [2]. The first reported work in electrocatalysis was on fuel cell processes by Grubbs in the 1950s [3].

Electrochemical reactions are considered different from the heterogeneous processes, since the electrolyte conductor and the electronic conductor establish the so-called electrified interface. In the usual heterogeneous catalysis, the chemical changes that involve breakage of bonds with charge redistribution occur only over atomic distances. In the electrochemical reaction, besides the reactant concentration and temperature available in the heterogeneous catalysis, another degree of freedom of the electrode potential appears. In addition, the nature of the electrode surface, that is, the surface catalytic site, is also considered. It is important to mention that electrocatalysis happens on an active surface site and that this site involves an electron transfer pathway. The term electrocatalysis applies to the influence of the nature of the electrode material and the morphology of the electrode surfaces on the behavior of the chemical reactions. Interesting developments such as oxygen electroreduction and hydrocarbon electrooxidation in electrocatalysis appeared in the late 1960, especially in fuel cells [4–10].

2.1.1 Basic Concepts of Electrocatalytic Reactions

The difference between a true catalytic and a specific electrocatalytic reaction involves the addition of a certain type of additive inside or in another phase. According to the catalytic reaction theory, the increase in the rate of the reaction is the result of the fast formation and decomposition of the intermediates between two reactants and the additive known as the "catalyst." Considering the integrated kinetic law of Arrhenius, we can say that the rate of the catalytic reaction, v, is

$$v = \frac{kT}{h} \prod_{i}^{n} a_i^p \exp(-E/kT) \tag{2.1}$$

where

$\prod$ denotes the product of activities of the reactants to the reaction order, p, in the slowest step

E is the energy of activation (known to be the energy required to place the molecules in a particular position to react)

k is the Boltzmann constant and T is the absolute temperature [11]

This law together with the steady-state regime for the intermediates [12] gives rise to the conclusion that there is an invariant final concentration of the product that can be obtained after a certain time of reaction.

One of the most useful theories of activation is the collision theory of reactions. This theory assumes that the activated molecules are formed from collisions with other normal reactant species. Such thermally activated molecules decompose to or react with other molecules. The number and frequencies of collisions indicate that not all collisions are effective in producing an activated molecule. As it was shown before [13], the key point is the effective collision factor in the theory. In the case of a heterogeneous reaction, if the surface catalyst concentration is relatively small compared with the bulk concentration of the reactants, the number of active sites for the catalytic reaction will suffice for the reaction to occur. Therefore, the catalytic reaction has one more advantage; only a small quantity of the additive is enough for the reaction.

For electrocatalytic reactions, the charge transfer process to or from the charged particles has to be considered. The formation of a reactant-surface site complex is involved together with a fast charge transfer that produce a "product species" with the re-formation of the "active surface site." In most cases, the formation of the complex is rapid and the subsequent charge transfer process is slow. It is important to know whether the re-formation of the catalytic site is fast or slow. The pathway will depend on the height of the activation barriers for the individual steps and in most of the cases the latter is the rate-determining step. The rate expression for the case of an electrocatalytic reaction can, *a priori*, be

$$v = \kappa\rho \prod_{i}^{n} a_i^p \exp(-E(\Phi)/kT) \tag{2.2}$$

where

κ is the integral probability for the charge transfer process

ρ is the surface electron concentration at the electrode

$E(\Phi)$ is the activation energy barrier at the electrode potential Φ

Thus, it is important to underline the main differences between catalytic and electrocatalytic reactions. In fact, the only capable parameter that is able to affect either a catalytic or electro-catalytic reaction is temperature. According to the activation energy of each reaction, the temperature is able to change the rate of the reaction by many degrees. However, the problem of changing the rate of a chemical reaction in one order requires that the temperature varies by more than one decade of magnitude. Achieving such temperatures in liquid/solid interfaces is an important problem (the solvent evaporating from the solution is inevitable). Thus, the electrode potential parameter is easier to change—one order of magnitude in the reaction rate is possible by only changing the electric potential from 0 to 0.10 V for the case of a medium-activated barrier reaction. This leads to another important property, selectivity, required for the electrocatalytic reaction. This is especially important for electroorganic reactions as most of their pathways are very near to each other. By calculating the operating electrode potential at the working current density, one process can be distinguished from another. The reduction of nitrobenzene is an interesting case. This chemical catalytic reduction involves the complete reduction of nitrobenzene by zinc/hydrochloric acid yielding aniline, whereas using tin/acetic acid, the reduction to nitrosobenzene is observed. The calculation of the operating potential for each case makes a better selection of the desired process and also helps accurately select the electrode material. The side reaction effects are also important for electropolymerization reactions and the electrochemical conversion of energy. There are many examples for each topic.

A study of electrocatalysis is also important to understand the mechanism of the reaction that operates the process. Since the reaction mechanisms are very difficult to discern, the possibility of the variation of different parameters, such as the electrode potential, the nature of the electrode material, the electrolyte composition, etc., brings into focus a wide possibility of pathways. The fixing of the experimental conditions allows us to choose the appropriate route for substance preparation, complete energy conversion, inhibition of the corrosion process, and eliminating the side reaction pathways.

2.1.2 DISTINCTION BETWEEN ELECTROCHEMICAL AND ELECTROCATALYTIC REACTIONS: USE OF VARIABLE ELECTRODE POTENTIALS AND THE PRESENCE OF ADSORBATES

The most accepted modern activation theory for the outer electron transfer is that of Rudolph A. Marcus (Nobel Prize in Chemistry in 1992) [14], which is different from the transition state theory. His studies on unimolecular reactions and the transition and collision theories committed him to elaborate on the Rice–Ramsperger–Kassel–Marcus (RRKM) theory in 1952. This theory is an extension of the previous RRK theory proposed by Rice, Ramsperger, and Kassel between 1927 and 1928. Moreover, Hush and Marcus further extended the electron transfer theory of Marcus for inner electron transfers [15–17].

The resulting theory, named as the Marcus–Hush theory [17], has been the widest and most accepted theory for kinetics overviews since then. However, the theory is based basically on classical kinetics for electron transfer, and the quantum nature of the process is almost shielded by using other related concepts. This is rather strange since, between 1960 and 1970, electron quantum mechanics by Jortner and Kuznetsov [18–20] was well accepted in the specialized literature for non-radiant transitions.

Nevertheless, the theories have been referred to consider only the vibrational effects of the electron transfer and, particularly, the vibronic Piepho–Krausz–Schatz (PKS) Hamiltonian element in the theory. The main expression propounded by Marcus can be symbolized as

$$\vec{k} = \frac{2\pi}{\hbar} |H_{AB}|^2 \frac{1}{\sqrt{4\pi\lambda kT}} \exp\left(-[\lambda + \Delta G^{\pm}]^2 / 4\lambda kT\right) \qquad (2.3)$$

where

$\vec{k}$ is the electron transfer specific constant rate

$|H_{AB}|$ is the Hamiltonian matrix elements between the initial and final states A and B

λ is the reorganization free energy for either the inner and outer spheres' mechanisms

$\Delta G^{\pm}$ is the standard free energy of activation

The most interesting feature of Expression 2.3 is the quadratic expression for the free energy, which in a Morse potential energy curve for a given reaction leads to the Marcus inversion region. In this case, the quadratic potential energy variation of the initial and final states is responsible for the equation since the Morse curves really approach parabolas. Two quadratic dependencies with the reaction coordinates are then introduced into the Arrhenius expression between the activation energy and the rate constant.

In general, the rates of the electrocatalytic reactions on different electrode materials are given by the difference between the chemical potentials of the transition states at the same electrode potential. This means that the energy of activation in Equation 2.2 is referred to the value obtained in the rate-determining step as an energy difference between the reactants in their initial state and their catalyzed state prior to the determining path:

$$v = \kappa \rho a_i \exp\left(-\Delta G^{\pm}(\Phi)/RT\right) \tag{2.4a}$$

where

a_i is the activity of the adsorbed state

$\Delta G^{\pm}(\Phi)$ is the standard free energy of activation at the electrode potential Φ

This energy of activation is the electrochemical standard free energy of activation, $\Delta G^{\pm}(\Phi)$, which is purely chemical when $\Phi = 0$:

$$v = \kappa \rho a_i \exp\left(-\Delta \hat{G}^{\pm}(\Phi)/RT\right) \tag{2.4b}$$

The identification of the real zero value electrode potential is a problem that has been discussed in many papers [21,22]. Many authors have claimed that it is the hypothetic independent reference electrode potential at the zero charge potential of the metal [22,23]. At this potential, $\Delta\Phi_{\text{Metal-Solution}}$ is zero, since the potential of the zero charge is approximately proportional to the work function divided by the Faraday constant. The dipole potentials are practically independent of the electrode material. However, it has been demonstrated that the free energy of the electron in the metal is not a determining point for the electrocatalytic reaction rates [24], since most of the reaction rates are measured in the same electrolyte and at a fixed potential against a reference electrode (the latter being the reversible electrode potential). Within the reversible electrode potential, there is no net free energy for the overall process, and the forward and reverse reactions have the same probability of occurrence. On the other hand, the free energy in the transition state corresponds to the i-particle bonded to the substrate, which reconfigures to the adsorbed state attached to the electrode surface. Where the electron transfer occurs is a problem of quantum electrochemistry. Since it can be anywhere, we can arbitrarily put

$$\Delta \hat{G}^{\pm}(\Phi) = \Delta G^{\pm} \pm \beta F \Delta\Phi_{\text{M-S}} \tag{2.5}$$

For a symmetrical energy barrier, the value of β is 1/2, as has been commonly presented in the literature:

$$v = \kappa \rho a_i \exp\left(\frac{-2\Delta G^{\pm} - F\Delta\Phi_{\text{M-S}}}{2RT}\right) \tag{2.6}$$

These are the considerations that are always evaluated in an electrochemical process. But the problem in electrocatalysis goes a little further. If we consider that the species i in solution adsorbs on to a species j on the electrode, there is an extra stabilization energy, namely, ΔG_{ad}, the standard free energy of adsorption. Thus, for a Langmuir isotherm

$$[j] = [i] \exp\left(\frac{-\Delta G_{ad} - \delta F \Delta \Phi_{M-S}}{RT}\right) \qquad (2.7)$$

Considering that the rate-determining step is the charge transfer, then from adsorbate j, the kinetic law will be

$$v = \kappa \rho [i] \exp\left(\frac{-\Delta G_{ad} - \Delta G^{\pm} - (\delta + \beta) F \Delta \Phi_{M-S}}{RT}\right) \qquad (2.8)$$

This equation leads to extensive discussions. First of all, the chemical free energies of stabilization and activation will be calculated in a compressed form together with the "slope" value between the variations of the former electrochemical free energies with the electrode potential. If no mechanism of reaction is proposed, speculation arises. Perhaps in the real situation, both the reactant and the product are adsorbed on to the electrode surface. Hence, the free energy of stabilization is more likely to be a change between the adsorbed reactant and the adsorbed product, $\Delta\Delta G_{ad}$ [25–27]. Precisely, the overall kinetic equation contains a term that can be associated with the "charge transfer coefficient" of the entire process. With this "coefficient," many possibilities are likely, and a partial charge transfer of the electrons is possible. This possibility has been the subject of discussion by many authors since the early 1960s. However, it is certain that the Tafel slopes, E vs. log i, lower than 0.06 V decade^{-1} can only be easily explained through this complex adsorbed state charge transfer, or by a single charge transfer with the adsorbed state under a Temkin isotherm.

Figure 2.1 shows the changes in the activation energy due to the adsorption of the reaction product. It is evident as shown before [13,27] that when the product is adsorbed on to the electrode surface, the activation energy decreases in magnitude, and there is a displacement in the relative position due to the ground state modification of the configuration. This implication becomes contrary when the reactant is adsorbed on to the surface. The problem of both the reactant and the product attached to the catalyst defines the real change in the free energy of adsorption, $\Delta\Delta E_{ad}$, as the free energy or enthalpy. If we consider that the intersection curves are nearly straight, then the

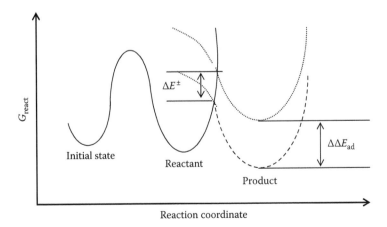

FIGURE 2.1 Effect of the energy of adsorption between reactants and products ($\Delta\Delta E_{ad}$) in the lowering (increasing) of the activation energy ($\Delta E^{\pm}$).

contribution to the activation energy will be $\pm\beta\Delta\Delta E_{ad}$ with β approximately constant over a wide range of $\Delta\Delta E_{ad}$. These considerations regarding the influence of adsorption energies on activation were introduced by Horiuti in the early 1930s [28].

The entropic factor is usually neglected when the activation energy is considered; however, it is considered in the case of electrocatalytic reactions that involve adsorbed compounds. There is an important contribution of the entropic factor, $T\Delta\Delta S_{ad}$, due to bond formations and bond breakages caused by adsorption. However, the thought that this magnitude can change similar to the enthalpic component is rather inconsistent. The change in the entropy of adsorption would probably involve a gradual variation of the rotations and vibrations together with the bond formations at the electrode surface. In the case of statistical thermodynamics, the change in the partition functions along the reaction coordinate is usually denoted as multiples of the frequency factor, along which there is some change in the entropy due to adsorption. On the other hand, we can also say that once the bond to the electrode is formed, it is probable that relaxation energy will be gained, and in some cases of strong adsorption the $T\Delta\Delta S_{ad}$ term will undergo maximum activation. Finally, the change in the activation energy, $\Delta E^{\pm}$, will be different from a single enthalpic influence along the reaction coordinate—both the height and intersection between the Morse curves of the adsorbed reactant and the products will be very different from the straight intersected lines. These facts are discussed in Ref. [29].

According to the above equations, if both the reactant and the product are adsorbed on the electrode, the pathway will probably not be the rate-determining step. However, if a proton or electron charge transfer occurs, it will be important to consider it as a low rate reaction step. The strong dependence of $\Delta\Delta E_{ad}$ on the θ values makes us think that it is very important for the special case of intermediate surface coverages. As usual, we can consider a linear variation of ΔG_{ad} with θ both for the reactants and the products, since there is no drastic change in the configuration of the adsorbed species (there will be some exception as always):

$$\Delta\hat{G}_{ad}(\Phi) = \Delta G_{ad} \pm r\theta \pm nF\Delta\Phi_{M-S} \qquad (2.9)$$

where
  r is the Frumkin interaction coefficient
  n is the number of electrons
  F is the Faraday constant

The expression of the isotherm will be

$$\ln\left(\frac{\theta}{1-\theta}\right) = \ln[i] - \frac{\Delta G_{ad} \pm r\theta \pm nF\Delta\Phi_{M-S}}{RT} \qquad (2.10)$$

Since it seems to be important, the change in the covered to the uncovered surface fraction in the isotherm with θ is

$$\frac{\partial \ln\left(\dfrac{\theta}{1-\theta}\right)}{\partial\theta} = \pm r$$

and for the intermediate surface coverages, r is approximately ± 4.

Then, the change in the electrochemical free energy of adsorption will be

$$\frac{\partial\Delta\hat{G}_{ad}}{\partial\Delta\Phi} = \frac{\partial\Delta\hat{G}_{ad}}{\partial\theta} \cdot \frac{\partial\theta}{\partial\Delta\Phi} \Rightarrow$$
$$\pm nF = \pm r\theta \cdot \frac{\partial\theta}{\partial\Delta\Phi} \Rightarrow \left(\frac{\partial\theta}{\partial\Delta\Phi}\right) \approx \pm nF/4\theta \qquad (2.11a,b)$$

Then, a rather complex behavior should be expected, so it would be interesting to calculate the change in the rate expression with the electrode potential, $\Delta\Phi_{\text{Metal-Solution}}$, to keep the influence of θ at the medium surface coverages. Thus, by reordering Equation 2.8 and defining δ as n, we have for the reduction reaction:

$$v = \kappa\rho[i]\exp\left(\frac{-\Delta G_{\text{ad}} - r\theta - \Delta G^{\pm} - (n+\beta)F\Delta\Phi_{\text{M-S}}}{RT}\right) \tag{2.12}$$

The change in $\ln v$ with respect to $\Delta\Phi_{\text{Metal-Solution}}$ has an implicit form in θ:

$$\ln v = cte - \frac{\Delta G_{\text{ad}} + (nF/\theta)\Delta\Phi_{\text{M-S}} + \Delta G^{\pm} + (n+\beta)F\Delta\Phi_{\text{M-S}}}{RT} \tag{2.13}$$

$$\frac{\partial\ln v}{\partial\Delta\Phi} = -\frac{(nF/\theta) + (n+\beta)F}{RT} \tag{2.14}$$

These differential effects show that the change in ΔG_{ad} with $\Delta\Phi_{\text{Metal-Solution}}$ involves two factors: the first is truly dependent on the electrostatic component and the other is indirectly dependent on the potential coming from the variation of θ. Since different factors can involve a certain Tafel slope, the simple table of 0.06 or 0.12 V decade^{-1} does not represent the true case of an electrocatalytic reaction.

2.2 ELECTROCATALYSIS OF INTERESTING PROCESSES

Understanding an electrocatalytic reaction is important as several factors are responsible for the changes in the kinetics of the process. The existence of several species with different bond energies on the surface of the electrode, reactants, intermediates, or products makes the reaction pathway difficult to interpret or predict. Some of the possible mechanisms can be simultaneous and occur with the same probability. Thus, some important questions arise: Which reaction intermediate defines the path of the process? Which type of active site is most likely to be considered? Which type is inhibiting? Which surface coverage has to be considered or neglected? Since it is impossible to provide a brief review of most of the interesting electrocatalytic reactions, we consider the most accepted mechanism in hydrogen evolution—the oxygen reduction and methanol oxidation reactions.

2.2.1 ELECTROCATALYSIS OF METHANOL ADSORPTION AND OXIDATION

Several cases involve the adsorption of reactants and products (shielding or blocking effects), but one interesting situation in electrocatalysis is that of the electrooxidation of an organic fuel such as methanol. It involves the formation of a carbon monoxide adsorbed residue that follows a progressive oxidation. We can first state that the reaction occurs without an electrocatalytic influence of the surface but only due to the carbon monoxide residue:

$$CH_3OH_{(ac)} \leftrightarrow CO_{\text{ads}} + 4H^+_{(ac)} + 4e^- \tag{I}$$

$$CO_{\text{ads}} + H_2O_{(l)} \leftrightarrow CO_{2(g)} + 2H^+_{(ac)} + 2e^- \tag{II}$$

$$CH_3OH_{(ac)} + H_2O_{(l)} \leftrightarrow CO_{2(g)} + 6H^+_{(ac)} + 6e^- \tag{III}$$

The reaction rates of the individual processes are

$$v_1 = (CH_3OH)(1-\theta)k_{\text{ox},1} - k_{\text{red},1}\theta(H^+)^4 \tag{2.15}$$

$$v_2 = k_{\text{ox},2}\theta - P_{CO_2}(H^+)^2(1-\theta)k_{\text{red},2} \tag{2.16}$$

$$v_3 = (CH_3OH)k_{\text{ox},3} - P_{CO_2}(H^+)^6 k_{\text{ox},3} \tag{2.17}$$

where

    θ is the surface coverage by the carbon monoxide adsorbed species

    v_i is the net rate of each reaction path evaluated as a difference between the forward and reverse reactions, electrooxidation and reduction, respectively

    $k_{red,i}$ and $k_{ox,i}$ are the electrochemical rate constants of the i-reaction and the rest of the symbols have the usual meanings

The net rate of the methanol electrooxidation can be designed by either $-d(CH_3OH)/dt$ or $d(PCO_2)/dt$, which has the form of the following expression:

$$-\frac{d(CH_3OH)}{dt} = (CH_3OH)(1-\theta)k_{ox,1} - k_{red,1}\theta(H^+)^4 + k_{ox,2}\theta - P_{CO_2}(H^+)^2(1-\theta)k_{red,2}$$
$$+ (CH_3OH)k_{ox,3} - P_{CO_2}(H^+)^6 k_{ox,3} \tag{2.18}$$

Working at the steady state, $-d(CH_3OH)/dt = 0$, we obtain

$$\theta = \frac{P_{CO_2}\left[(H^+)^6 k_{red,1} + (H^+)^6 k_{red,2}\right] + (CH_3OH)(k_{ox,1} - k_{ox,3})}{(CH_3OH)k_{ox,1} - P_{CO_2}(H^+)^2 k_{red,2} - k_{red,2} - (H^+)^4 k_{red,1}} \tag{2.19}$$

According to the literature [30,31], we can consider that at any electrode surface the reaction (II) is the rate-determining pathway, and then the reverse reduction reaction can be neglected:

$$CO + H_2O \rightarrow CO_2 + 2H^+ + 2e^- \tag{II'}$$

Thus, the value of the surface coverage by the carbon monoxide adsorbate is

$$\theta = \frac{P_{CO_2}(H^+)^6 k_{red,1} + (CH_3OH)(k_{ox,1} - k_{ox,3})}{(CH_3OH)k_{ox,1} + (H^+)^4 k_{red,1}} \tag{2.20}$$

We can further assume that at potentials lower than 0.9 V, step (III) is negligible compared to the others, that is, $k_3 \rightarrow 0$:

$$\theta = \left(1 + \frac{(H^+)^4 k_{red,1}}{(CH_3OH)k_{ox,1}}\right)^{-1} \tag{2.21}$$

Since step (II') is rate determining, $v_2 = k_{ox,2}\,\theta$.

    The bulk electrooxidation of methanol on platinum single crystals has been extensively studied since the late 1980s. The electrochemical profile of the reaction on Pt(111) in acid media is depicted in Figure 2.2:

$$v_2 = \frac{(CH_3OH)k_{ox,2}k_{ox,1}}{(CH_3OH)k_{ox,1} + (H^+)^4 k_{red,1}} \tag{2.22}$$

Moreover, considering that the overpotential exceeds the minimum for the Tafel lines to be considered, we can neglect the reduction reverse reactions: $k_{ox,1} \gg k_{red,1}$:

$$v_2 = k_{ox,2} \tag{2.23}$$

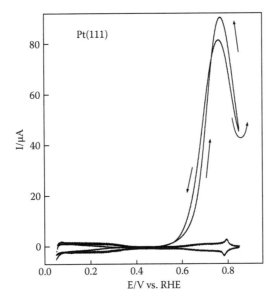

FIGURE 2.2 Forward and reverse bulk methanol electrooxidation on Pt(111) run at 0.01 V s^{-1} in 0.1 M methanol +0.1 M sulfuric acid at room temperature. The voltammetric profile of the single crystal in the supporting electrolyte is superimposed as dashed lines.

If the counter reactions have to be evaluated (large reaction times), then we will need to establish that

$$v_2 = \left(1 + \frac{(H^+)^4 k_{red,1}}{(CH_3OH)k_{ox,2}\,k_{ox,1}}\right)^{-1} \tag{2.22'}$$

However, the real significance of the electrocatalytic reaction has to be taken into account. One of the first authors who discussed the problem of the influence of the electrode surface was Parsons [32]. He pointed out that the main oxidizing agent was the adsorbed hydroxyl coming from the metal oxidation:

$$M + H_2O \leftrightarrow [M(OH)]_{ad} + H^+ + e^- \tag{IV}$$

The electrochemical oxidation of methanol has been extensively studied on *pc* platinum [33,34] and platinum single crystal surfaces [35,36] in acid media at room temperature. Methanol electrooxidation occurs either as a direct six-electron pathway to carbon dioxide or by several adsorption steps, some of them leading to poisoning species prior to the formation of carbon dioxide as the final product. The most convincing evidence of carbon monoxide as a catalytic poison arises from *in situ* IR fast Fourier spectroscopy. An understanding of methanol adsorption and oxidation processes on modified platinum electrodes can lead to a deeper insight into the relation between the surface structure and reactivity in electrocatalysis. It is well known that the main impediment in the operation of a methanol fuel cell is the fast depolarization of the anode in the presence of traces of adsorbed carbon monoxide.

The modification of platinum surfaces by foreign metal atoms promotes the oxidation of methanol either in UHV conditions or in the electrochemical environment. This promotion model has been mainly discussed in electrochemistry using the "third body model" [37], the "ligand effect" [38], or the "bifunctional effect" [9,39,40]. A theoretical review on the inclusion of metal reaction promoters was undertaken by Anderson et al. [41] and later discussed in [42].

There still stands valid for platinum alloys; the two metals that are able to promote methanol oxidation are ruthenium and tin. The case of ruthenium is interesting since it was also studied under UHV conditions [43,44]. The reaction of methanol on Pt/Ru alloys results in the production of carbon dioxide at lower potentials than on pure platinum. However, the presence of tin in Pt_3Sn alloys only enhances methanol oxidation at low potentials, increasing carbon dioxide production (and diminishing carbon monoxide production) [45]. The addition of tin (II) ions to previously adsorbed methanol produces a fast oxidation process, demonstrated by DEMS experiments [46].

In the presence of foreign metals, M, on platinum, the bifunctional mechanism proposes that M promotes water discharge at lower potentials than pure platinum:

$$M + H_2O \leftrightarrow [M(OH)]_{ad} + H^+ + e^- \qquad \text{(IV)}$$

These oxygenated species promote the oxidation of the adsorbate by either a direct recombination (reaction (V)) of the species through a Langmuir–Hinshelwood mechanism [47] (Scheme 2.1)

$$[Pt(CO)]_{ad} + [M(OH)]_{ad} \leftrightarrow CO_2 + H^+ + e^- + Pt + M \qquad \text{(V)}$$

or indirectly by the reverse Eley–Rideal mechanism when the intermediate is a methoxy species as follows (reaction (VI)), where no other metal is really needed [47]:

$$[Pt(OCH_3)]_{ad} + H_2O \leftrightarrow CO_2 + 3H^+ + 3e^- + Pt \qquad \text{(VI)}$$

According to the ligand effect theory, the energy level of the modified substrate is changed to weaken the bond energy of the carbon monoxide adsorbate to facilitate its oxidation. The best choice for the alloying or co-deposited elements depends on which step of the latter mechanism is the rate-determining step (normally reaction (V)). Also the molar fraction has to be 1:1 if this is true. As different results were obtained the discussion is still open. The effect can be observed in Figure 2.3 with the bulk electrooxidation of methanol on Pt(111) and Pt(110) with different surface coverages by ruthenium adatoms.

If we simply consider the two main reaction pathways,

$$CH_3OH_{(ac)} + Pt \leftrightarrow (k_{ox,1}, k_{red,1})[Pt(CO)]_{ad} + 4H^+_{(ac)} + 4e^- \qquad \text{(I)}$$

$$[Pt(CO)]_{ad} + [M(OH)]_{ad} \rightarrow CO_2 + H^+ + e^- + Pt + M \qquad \text{(V)}$$

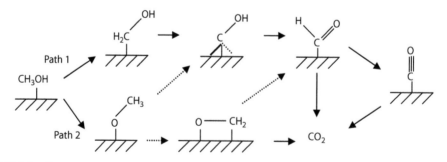

SCHEME 2.1 General scheme of methanol electrooxidation considering series and parallel pathways to form carbon dioxide as the product. Solid and dashed arrow lines indicate the demonstrated and possible reaction pathways, respectively. Path 1 denotes the "formyl intermediate" mechanism and Path 2 the "methoxyl intermediate" mechanism.

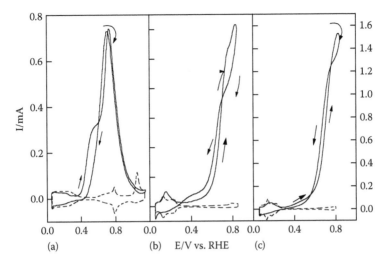

FIGURE 2.3 Forward and reverse bulk methanol electrooxidation on (a) Pt(111), (b) Pt(110)/Ru $\theta = 0.33$, and (c) Pt(110)/Ru $\theta = 0.50$ run at 0.01 V s^{-1} in 0.1 M methanol +0.1 M perchloric acid at room temperature. The voltammetric profile of each surface in the supporting electrolyte is superimposed as dashed lines.

we can check the solution through the following differential equation

$$I(t) = \sigma \frac{d\theta_{CO}}{dt} = B(CH_3OH)(1-\theta)k_{ox,1} - B(H^+)^4\theta k_{red,1} - A\theta k_{ox,2} + A\theta^2 k_{ox,2} \qquad (2.24)$$

which renders

$$\sigma \frac{d\theta}{dt} + \theta\left[B(CH_3OH)k_{ox,1} + B(H^+)^4 k_{red,1} + Ak_{ox,2}\right] - A\theta^2 k_{ox,2} - B(CH_3OH)k_{ox,1} = 0 \qquad (2.25)$$

If (CH_3OH) and (H^+) are constants, we will obtain the compressed expression of the equation

$$\sigma \frac{d\theta}{dt} + \theta K + \theta^2 L + M = 0 \qquad (2.26)$$

with

$$K \equiv B(CH_3OH)k_{ox,1} + B(H^+)^4 k_{red,1} + Ak_{ox,2}$$

and

$$L \equiv -Ak_{ox,2} \quad M \equiv -B(CH_3OH)k_{ox,1} \qquad (2.27)$$

This equation has the form of the Riccati differential equation with electrode potential dependent coefficients (see Appendix):

$$\frac{d\theta}{dt} + \theta\frac{K}{\sigma} + \theta^2\frac{L}{\sigma} + \frac{M}{\sigma} = 0 \qquad (2.28)$$

Since this type of equation cannot be solved by quadratic integrations, the following substitution can be proposed: $\theta = -\dfrac{h'\sigma}{hL}$, and the Riccati equation reduces to a linear equation of the second order:

$$\frac{L}{\sigma}h'' - \left[\left(\frac{L}{\sigma}\right)' + \frac{LK}{\sigma^2}\right]h' + \left(\frac{L}{\sigma}\right)^2 \frac{M}{\sigma}h = 0 \tag{2.29}$$

or

$$h'' - \left[1 + \frac{K}{\sigma}\right]h' + \frac{LM}{\sigma^2}h = 0 \tag{2.30}$$

Since the coefficients of the equation do not depend on time, if we work at a constant potential expression we can solve the characteristic equation:

$$r^2 - \left[1 + \frac{K}{\sigma}\right]r + \frac{LM}{\sigma^2} = 0 \tag{2.31}$$

The solution to the equation gives the roots

$$r_{1,2} = 1/2\left[1 + \frac{K}{\sigma}\right] \pm \sqrt{\left(\frac{1}{2} + \frac{K}{2\sigma}\right)^2 - \frac{LM}{\sigma^2}} = 0 \tag{2.32}$$

Thus, the general solution will be

$$h = C_1 \exp(r_1 t) + C_2 \exp(r_2 t) \tag{2.33}$$

Remembering that $\theta = -\dfrac{h'\sigma}{hL}$, we will have

$$h' = C_1 r_1 \exp(r_1 t) + C_2 r_2 \exp(r_2 t) \tag{2.34}$$

Then,

$$\theta = -\frac{h'\sigma}{hL} = -\frac{\sigma C_1 r_1 \exp(r_1 t) + \sigma C_2 r_2 \exp(r_2 t)}{C_1 L \exp(r_1 t) + C_2 L \exp(r_2 t)} \tag{2.35}$$

and substituting with $L \equiv -Ak_{ox,2}$

$$\theta = -\frac{h'\sigma}{hL} = \frac{\sigma C_1 r_1 \exp(r_1 t) + \sigma C_2 r_2 \exp(r_2 t)}{Ak_{ox,2}C_1 \exp(r_1 t) + C_2 Ak_{ox,2}\exp(r_2 t)} \tag{2.36}$$

We can look at the values of r_1 and r_2 by taking the approximate values of the specific reaction constants:

$$k_{ox,2}\,k_{ox,1} \approx 10^{-11}, \ (k_{ox,2})^2 \approx 10^{-12}, \ (k_{ox,1})^2 \approx 10^{-10}, \ (k_{red,1})^2 \approx 10^{-8}, \ k_{ox,1}\,k_{red,1} \approx 10^{-9}$$

We can obtain a simpler form:

$$r_{1,2} = 1/2\left[1 + \frac{B(H^+)^4 k_{red,1}}{\sigma}\right] \pm \frac{\sqrt{B^2(H^+)^8 k_{red,1}^2 + B^2(CH_3OH)(H^+)^4 k_{ox,1}k_{red,1}}}{2\sigma} \tag{2.37}$$

By taking $(k_{red,1})^2 > k_{ox,1}\, k_{red,1}$, the roots are

$$r_1 = 1/2 + \frac{B(H^+)^4 k_{red,1}}{\sigma} \quad \text{and} \quad r_2 = 1/2 \qquad (2.38)$$

and the values of θ will be

$$\theta = -\frac{h'\sigma}{hL} = \frac{\sigma C_1\left(1/2 + \dfrac{B(H^+)^4 k_{red,1}}{\sigma}\right)\exp\left(1/2 + \dfrac{B(H^+)^4 k_{red,1}}{\sigma}\right)t + \sigma C_2/2\,\exp(1/2t)}{Ak_{ox,2}C_1\exp\left(1/2 + \dfrac{B(H^+)^4 k_{red,1}}{\sigma}\right)_1 t + C_2 Ak_{ox,2}\exp(1/2t)} \qquad (2.39)$$

But r_1 is approximately equal to r_2 and to $1/2$, and then

$$\theta = -\frac{h'\sigma}{hL} = \frac{\sigma/2(C_1 + C_2)\exp(1/2t)}{(C_1 + C_2)Ak_{ox,2}\exp(1/2t)} = \frac{\sigma}{2Ak_{ox,2}} \qquad (2.40)$$

The steady-state surface coverage by the carbon monoxide residue can be studied by anodic stripping voltammetry. By this technique, it is possible to separate the adsorption residue contribution from the bulk electrooxidation process. The micro-flux cell is adapted with a big flask containing the supporting electrolyte, which is used to wash the cell until there is no trace of methanol in solution. The current vs. potential profile run from the adsorption potential upward is the tripping profile for the oxidation of the adsorbed residue. An example is presented in Figure 2.4.

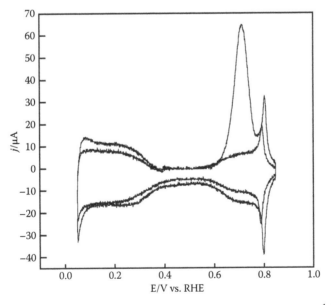

FIGURE 2.4 Adsorbed carbon monoxide electrooxidation on Pt(111) run at 0.05 V s^{-1} in 0.1 M perchloric acid at room temperature. The voltammetric profile of Pt(111) in the supporting electrolyte is superimposed. The adsorption of the residue was performed at 0.35 V for 5 min from a carbon monoxide saturated acid solution.

2.2.2 ELECTROCATALYSIS OF THE HYDROGEN EVOLUTION REACTION

Any combination of two or three elementary pathways will give the overall mechanism of the hydrogen evolution reaction. Thus, the electrochemical Volmer discharge of the proton, the electrodesorption Heyrovský step, and the chemical Tafel recombination of the H adatoms can serve as a combination for the hydrogen evolution process. The electrochemical rate constants can be estimated through different experimental conditions, such as their exchange current densities $j_{o,1} = 10^{-1}$, $j_{o,2} = 10^{-4}$ and $j_{o,3} = 10^{-2}$ A cm^{-2} at $V = 0$ V where $\Delta G_{ads} = 0$ with $\theta_H \approx 1/2$ [7,48].

The experimental determination of the kinetic parameters involved in the elementary steps of the reaction mechanism of the hydrogen electrode reaction is generally conducted through the calculation of the potential dependence of the current density at the high overpotential region. These current vs. potential (overpotential) relations have been taken from the hydrogen evolution process [49], at conditions without the mass transfer components. At the same time and with a similar scope, kinetic studies have been carried out for the hydrogen oxidation reaction [50], although the diffusion of the molecular hydrogen makes the calculations of the kinetic rate constants difficult. To overcome this problem, the rotating disc electrode technique is useful.

The combination of the three hydrogen containing steps leads to a mechanism with one rate-determining reaction and the others in quasi-equilibrium:

$$H^+_{(ac)} + e^- \leftrightarrow H_{ads} \tag{VII}$$

$$H_{ads} + H^+_{(ac)} + e^- \leftrightarrow H_{2(g)} \tag{VIII}$$

$$2H_{ads} \leftrightarrow H_{2(g)} \tag{IX}$$

A generalized Volmer–Heyrovský–Tafel mechanism explained the problem of the intermediate and product adsorption with a nearly zero coverage at potentials approximately equal to that of the net hydrogen evolution process [51]. The electrochemical reaction rates of the individual processes are

$$v_1 = (H^+)(1 - \theta)k_{red,1} - k_{ox,1}\theta \tag{2.24}$$

$$v_2 = (H^+)\theta k_{red,2} - P_{H_2}(1 - \theta)k_{ox,2} \tag{2.25}$$

$$v_3 = \theta^2 k_{red,3} - P_{H_2}(1 - \theta)^2 k_{ox,3} \tag{2.26}$$

where
  θ is the surface coverage by the atomic hydrogen adsorbate
  v_i is the net rate of the process as a difference between the forward and reverse reactions of each path, electrooxidation and reduction, respectively
  $k_{red,i}$ and $k_{ox,i}$ are the electrochemical rate constants of the i-reaction pathway

The net rate of the hydrogen evolution is $-d(H^+)/dt$ or $1/2\, d(P_{H2})/dt$, which has the form of

$$-\frac{d(H^+)}{dt} = (H^+)(1 - \theta)k_{red,1} - k_{ox,1}\theta + (H^+)k_{red,2}\theta - P_{H_2}(1 - \theta)k_{ox,2} + \theta^2 k_{red,3}P_{H_2}(1 - \theta)^2 k_{ox,3} \tag{2.27}$$

If we consider again a steady–state condition, we can obtain the expression of θ by solving a quadratic equation. However, we can consider another expression where the Tafel recombination chemical path (IX) can be neglected. Thus, ignoring k_3 we obtain

$$\theta = \left(1 + \frac{k_{ox,1} + (H^+)k_{red,2}}{(H^+)k_{red,1} + P_{H_2}k_{ox,2}}\right)^{-1} \tag{2.28}$$

It is generally assumed that the electrochemical desorption (the Heyrovský process) is the rate-determining step in the case of platinum-type noble metals [52], so we can neglect the forward path $k_{ox,2}$ and also $k_{red,1} \gg k_{red,2}$; then,

$$v = \frac{2(H^+)k_{red,2}}{1 + k_{ox,1}/(H^+)k_{red,1}} \tag{2.29}$$

and if $k_{red,1}(H^+) \gg k_{ox,1}$ we have the following rate of hydrogen evolution:

$$v = 2(H^+)k_{red,2} \tag{2.30}$$

Moreover, if we consider the value of the coverage by hydrogen we obtain

$$\theta \cong \left(1 + \frac{(H^+)k_{red,2}}{(H^+)k_{red,1}}\right)^{-1} \to 1 \tag{2.31}$$

One interesting problem is the participation of the underpotential deposited hydrogen as a reaction intermediate or an adsorbate in the hydrogen evolution process. The analysis of this type of species can be performed with the help of a cyclic voltammogram recorded on a platinum electrode in a strong acid solution [53]. On the one hand, the electrodesorption peaks far from the hydrogen evolution process (at potentials lower than 0.35 V vs. RHE) show that the hydrogen-containing adsorbates are most likely the adsorbates in the Volmer step without any further nucleation of the bubbles. One interesting treatment is Conway's [54,55] who proposed a nearly similar hydrogen intermediate to that of the hydrogen underpotential deposition, but with a negligible surface coverage. The values of the surface coverage by the hydrogen intermediate reaction are usually calculated from the limiting current density. Some studies on nickel and mercury electrodes [56,57] show kinetic limiting current densities of 10 A cm^{-2}, but strangely in platinum, limiting values not higher than 0.25 A cm^{-2} were found [58]. Under Langmuirian conditions of adsorption, not the best for strong electric potential changes in θ_H, and from the kinetic limiting current we can calculate the equilibrium value of θ_H. With these values, we can estimate a surface coverage by the supposed intermediate of less than 0.14. Under Frumkin conditions, it is probable that the equilibrium θ values for the intermediate will be less than 0.10.

One interesting and useful plot to study the electrocatalytic performance of the hydrogen evolution reaction is the η vs. ΔH_{ads} plot at galvanostatic conditions, according to data collected by Conway and Bockris [59]. According to these authors, an inverted volcano is observed with one line formed by those materials that show small values of ΔH_{ads} (those sp metals that exhibit a slow electrochemical hydrogen discharge) and those (with a change in the slope sign) that exhibit large values of ΔH_{ads} (those platinum-type and d metals with the electrodesorption process as rate determining). For mercury, ΔH_{ads} is approximately -32 kcal mol^{-1} and corresponds to $\eta \approx -1.1$ V, whereas in the same branch for gallium, ΔH_{ads} is approximately -42 kcal mol^{-1} and corresponds to $\eta \approx -0.8$ V. In the branch of the opposite slope, platinum and palladium reveal approximately -64 kcal mol^{-1} for approximately -0.05 to -0.08 V of η [29,60]. The interesting point is that there is a large gap between -40 and -60 kcal mol^{-1} with no sp or d metal or its alloys that can present hydrogen activity within this region. On both sides, we can also calculate the slope of the branches that are identical to $\pm 1/F$ since the electronic coefficients for η and ΔH_{ads} are the same.

The values coincide almost perfectly well with the experiment; however, some corrections for the entropy changes, proton tunneling, and potential dependent transfer coefficients were suggested. For example, Parsons made an approximate estimation of the pre-exponential term in the case of the entropic correction [61] based on statistical thermodynamics and mechanics. The surface mobility of the adsorbate or the low bonding strength hydrogen adatoms may be responsible for the low free energy of adsorption (less than kT). Besides, the effect of the surface coverage on the value of ΔH_{ads} was also considered by Conway and Bockris [59] for the d metals, because in these cases the electrodesorption process is the rate-determining step. They plotted $-\log j_o$ against $-\Delta G_{ads}$ and it can be seen that the value for platinum lies as expected, but the large Temkin region produces a shift of several kJ mol^{-1} that is more positive. This effect can explain the large difference between platinum and nickel upon hydrogen evolution reaction, since nickel exhibits low surface coverages by the hydrogen adatoms.

The hydrogen evolution reaction is an example where its electrocatalytic character shows that it is necessary for both the description of the adsorption process and the knowledge of the kinetic parameters. Most analysis of the electrocatalytic properties involve correlations from the estimated exchange current densities with characteristic electric potentials, free energies of adsorption, enthalpy of sublimations for the metal electrode, etc. [60].

A distinguishing representation for electrocatalytic mechanisms is the volcano plot: $\log j_o$ vs. ΔG_{ads} at the reversible potential. This plot is often used for purposes of comparison of the different possibilities that arise from each combination step [62–64]. The volcano plots usually depict an inverted and truncated negative concavity curve as shown qualitatively in Figure 2.5. It is clear that near $\Delta G_{ads} = 0$ there is more probability of the Tafel recombination step to occur as the rate-determining (with the Volmer discharge at steady state) step. According to the literature [64,65], adsorption data of hydrogen adatoms on platinum-type metals must have ΔG_{ads} close to zero at the reversible potential of the reaction. When the values of ΔG_{ads} are positive, we are at the ascending part of the plot, and it is likely that the rate-determining step will be the electrochemical desorption process, whereas for the negative values the situation is again the same with a slope of $\log j_o = \Delta G_{ads}/RT$. However, this representation should not be considered for the zero over-potential, whereas it can be for the cathodic polarization that yields negative overpotentials. A shift in the volcano plots toward the right direction is expected as shown in Figure 2.5. These comparisons were earlier recognized by Butler [66] and as a matter of example, we can say that

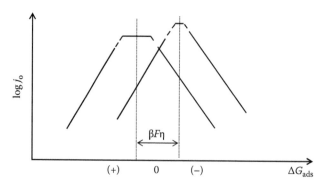

FIGURE 2.5   Qualitative volcano plots for $\log j_o$ vs. ΔG_{ads} on different metal electrodes at the reversible potential of the hydrogen evolution reaction. The cathodic polarization is plotted as a $\beta F\eta$ shift to the right and upward.

the extreme difference in the kinetics of the hydrogen evolution reaction on mercury and nickel can be evaluated through both the values of j_o and the extent in θ_H.

In order to evaluate the most likely reaction pathway, we have changed each volcano plot of the Volmer, Heyrovský, and Tafel steps with the electric potential toward negative values. The complex figure is not shown here because it is beyond the scope of this chapter. However, we can state that at relatively low overpotentials ($\eta \approx 0.12$ V), the chemical recombination reaction is not totally rate determining as at $\eta = 0$ V, but at higher overpotentials the volcano plot resembles qualitatively the original one with a slope approximately $\Delta G_{ads}/RT = 1/2$. In this case, the ascending branch denotes a Volmer discharge slow step up to $\Delta G_{ads} = F\beta\eta$, intersecting with the descending branch (with the same slope) for which the Heyrovský desorption becomes the rate-determining reaction. A complete treatment can be found elsewhere [13,67].

It seems that a better way to correlate ΔG_{ads} with the current density or the overpotential can be accomplished by introducing the problem of hydride formation instead of hydrogen atom chemisorption. It seems that this proposal better correlates those metals with which these compounds can be formed and also make the values of j_o more reliable [68]. From the analysis of these currents by considering the formation of the hydride, the calculations placed the ferrous metals in the ascending branch of the curve because of their low ΔG_{ads} value. Moreover, in the case of the other metals, the estimation of the kinetic data also agreed well with the calculations made from the covalent-type metal hydride. This type of correlation was also very interesting in gas phase measurements, since either reaction orders with respect to the proton or the Tafel lines are identical for the Volmer or the Heyrovský reactions as the rate-determining steps. According to the literature [69] on platinum and in gas-phase conditions, the bond frequencies for the stretching of the hydrogen adatoms show a bridge-type adsorption configuration on apolycrystalline substrate. In the electrochemical environment, the values seem to be too high to occur.

With the developments in UHV (ultra-high vacuum) technology and the development of extra-clean surfaces, it is necessary to check the experimental data, especially for single crystal surfaces of noble metals. Since some kind of relationship between the gas-phase and the electrochemical interphase conditions is expected, certain problems in relation to the existence of an electric field and the electrolyte solution will make this correlation rather complicated.

The location of the surface atoms and the strongly adsorbed atoms can be determined by low-energy electron diffraction (LEED) [70] and with the help of scanning tunneling microscopy (STM) [71] since the adsorbed atoms on the solid surfaces lead to ordered ad-layers [72,73]. The surface layer can be analyzed by Auger electron spectroscopy, and the combination of LEED and AES provides the surface chemist an opportunity to study chemisorption [72,74]. Hydrogen adsorption has been extensively studied [75–81], and many advances have been published since the work of Will [62]. Since most of the initial analysis came from electrochemical measurements on polycrystalline electrodes, the inferences toward the structure sensitivity of hydrogen adsorption have to be made with care. This is the essential reason to combine UHV techniques with electrochemical characterization. The cyclic voltammetry of the single crystal surface prepared by this technique can be used as a fingerprint of the electrochemical interface. The problem with this type of interface is that the counter-ions also affect the electrosorption of the hydrogen adatoms. In fact, immediately after the anion desorption the hydrogen species can be adsorbed or more precisely, simultaneously adsorbed, except for the thermally prepared Pt(111) [75], where both the processes can be separated (Figure 2.6).

Many papers have dealt with the problem of hydrogen chemisorption on platinum single crystals in acid or alkaline media, and detailed discussions can be found elsewhere [79–85]. Most of the interesting characteristics of hydrogen underpotential deposition were developed by using the flame-annealing technique either on platinum, as explained above, and on gold single crystal electrodes [86–88]. For the latter, since there is no hydrogen underpotential deposition in normal conditions, the problem of surface reconstruction can be easily studied [89,90]. The unusual results encountered during the preparation of single crystals, especially in the case of Pt(111), were

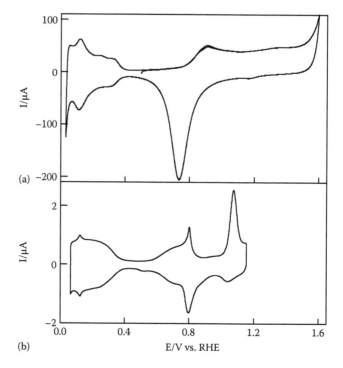

FIGURE 2.6 Cyclic voltammogram run at 0.05 V s^{-1} for (a) polycrystalline platinum in 0.1 M perchloric acid and (b) Pt(111) in 0.1 M sulfuric acid at room temperature. The preparation details are explained in the text.

obtained with the help of UHV techniques coupled with cyclic voltammetry in a pre-vacuum chamber or by STM [73,91].

2.2.3 Electrocatalysis of the Oxygen Electroreduction Reaction

The electrocatalysis of the oxygen electroreduction reaction has been studied since the early days of electrochemistry and surface science, and since the importance of corrosion technology and fuel cells was realized. In the early 1960s, it was proved that the reaction was not structure sensitive [92]; however, the problem of the preparation of clean and atomic ordered single crystal surfaces still remained.

Nowadays, it has been demonstrated that the reaction is indeed structure sensitive with a multi-electron transfer process that involves several steps and the possible existence of several adsorption intermediates [93–96]. The main advantage that we have with the new procedures with respect to cleanliness is that we have well-ordered surfaces to study a complex mechanism such as the oxygen electroreduction reaction [96–99]. In aqueous solutions, the four-electron oxygen reduction appears to occur by two overall pathways: a direct four-electron reduction and a "peroxide" pathway. The latter pathway involves hydrogen peroxide as an intermediate and can undergo either further reduction or decomposition in acid solutions to yield water as the final product. This type of generic model of a reaction has been extensively studied since the early 1960s by different authors [100–108].

The most appropriate experimental arrangement for the quantitative determination of the stationary curves for the reaction is the rotating disk electrode technique, in which the convective transport is controlled mechanically and thus a constant diffusion layer for each species is achieved [101–106]. Modifications of this technique, to primarily collect the intermediates of the reactions, such as ring-disk electrode techniques [93,94] and a hanging meniscus rotating disk electrode

[95,96], have also been used. An understanding of the reaction will help manage the problem of mass transfer under diffusion-convective conditions, but the enhancement of mass transport to the electrode implies higher requirements of cleanliness of the solution, especially for platinum-type single crystals.

The nature and the structure of the electrode surfaces are the fundamental aspects to be considered for an understanding of the catalytic activity for the reduction of oxygen. Although earlier studies have reported that oxygen reduction activity on the three low-index platinum surfaces at room temperature was identical in both perchloric and sulfuric acid solutions [109], further results show clear evidence that the reaction kinetics and mechanism are significantly dependent on the crystallographic orientation of the surface [95,98,99]. As reported for the polycrystalline platinum electrodes [110], oxide formation, even at low coverage, plays a significant role in the kinetics and activation energy of oxygen reduction, by varying the mechanism of the reaction because of the change in the adsorption sites for oxygen molecules.

It is also very important to monitor the effects of the upper potential limit since the potential at which oxygen reduction begins implies hydroxide or oxide co-formation on platinum. There are many studies of the reaction on the three low-index platinum surfaces [95,98]. The catalytic activity of these surfaces decreases in the order of Pt(110) > Pt(111) > Pt(100) in perchloric acid solution [96], while the order is Pt(110) > Pt(100) > Pt(111) in sulfuric acid solution [93]. In the case of Pt(111), the formation of a two-dimensional ordered ad-layer of specifically adsorbed (bi) sulfate anions is the main reason for the inhibition of oxygen reduction. Moreover, the direct four-electron mechanism was found for the three surfaces in acidic media, while the reaction mechanism varied to a two-electron reduction on the Pt(111) and Pt(100) due to the shielding of the hydrogen adatoms.

Electrochemical studies with a stepped surface in which the terrace size and the step symmetry can be controlled are analyzed in [111]. This can be achieved by introducing a series of stepped surfaces with different step densities to study the behavior of the oxygen electroreduction on these stepped-surface electrodes. In that paper, it was concluded that oxygen must compete for the adsorption sites with anions and oxide precursors, and the two effects together contribute to the structure sensitivity of this reaction.

2.2.3.1 Complexities of the Oxygen Electrocatalytic Reaction in Both Directions

Two main aspects make both the oxygen electroreduction and the evolution reactions difficult to study and understand, that is, the absence of a uniform (the ability of the substrate to have oxygenated species depends on the electrode potential) and a homogeneous (atomic roughness and ideal single crystal surfaces are required) substrate for examining the electron transfer reaction in both directions. Thus, the oxygen evolution reaction occurs on a partially or totally oxidized surface, and the oxygen electroreduction reaction occurs on two types of sites: surfaces covered and not covered by metallic hydroxides or oxides. The degree of surface oxidation depends on the nature of the metallic substrate, the electrode potential, and the pH of the electrolyte. It is interesting to note that it is impossible to apply the reversibility principle at the zero current potential, since the reaction occurs on different types of surfaces in two directions—one totally reduced and the other partially oxidized, that is, a rest potential is defined.

Another experimental aspect that can make the analysis rather difficult, in the case of working with a well-ordered single crystal, is the surface instability of the single crystal at potentials higher than 0.9 V, because oxygen electroreduction starts at potentials at approximately 1.2 V. Since single crystal surfaces reorient at these potential values, we have performed some experiments preferentially on the oriented surfaces [20,99,112,113]. These surfaces do not exhibit a pure crystalline orientation since they are prepared by repetitive dissolution and recrystallization of platinum crystallites on certain sites induced by potential cycling at high frequency [114–116]. The process produces a rather flat surface with exposed planes that result from the crystalline growth at characteristic positions defined by the upper and lower potential limits, the symmetry of the perturbation, and the frequency value [116]. These surfaces do not exhibit pure crystalline orientations, instead stepped planes are formed that are multiples of (110) and (100) low Miller indices. However, the studies performed on single crystal surfaces are only of academic interest since the oxygen electroreduction reaction in fuel cells probably occurs at a potential of approximately 0.9–1.0 V, which on different platinum morphologies will be of special significance.

Some reactions produce lateral products such as in the case of the oxygen reduction where peroxides and superoxides act as intermediates in the reaction mechanism by changing the values of the specific rate constants. The existence of the surface oxide or its precursors cannot be omitted in the analysis of the mechanism, so the existence of a preferentially oriented surface that is stable at high potentials makes the study of the oxygen electroreduction reaction possible with some limited inferences. It was possible to determine that almost 10% of the hydrogen peroxide (as a parallel product in the oxygen electroreduction reaction) on the platinum surface was preferentially oriented to the (100) plane and more properly on the (530) plane [112,113]. Negligible amounts of hydrogen peroxide were detected [112,113] on the stepped surface with (110) planes in acid solutions.

2.2.3.2 Electroadsorption of Molecular Oxygen and the Formation of Surface Oxides on Platinum

Oxygen electroreduction mostly occurs on partially oxidized surfaces as a consequence of the large electrode potentials developed in the reaction. Therefore, the formation of adsorbed molecular oxygen species would be shielded because of some platinum active sites that will be occupied by oxide precursors. In this case, the presence of the OH-containing species also acts as a catalyst for the dissociation of the molecule leading to the formation of atomic oxygen adsorbates (see Scheme 2.2). The distinction between one adsorbate and the other was ascertained with non-electrochemical *in situ* methodologies, such as ellipsometry [117,118]. It was concluded that there is a partially formed ad-layer of the OH species that originates from the electrochemically discharged water at underpotential conditions on platinum in 1 M H_2SO_4 at potentials between 0.70 and 0.95 V. At potentials larger than this value, large islands of flat two-dimensional formations have been observed, providing evidence for a new phase. This topic was extensively studied [119,120] on platinum, where a "replace and turn over" oxide conversion process is observed. The "turn over" step is proposed as the rate-determining step [120] that is run for the platinum oxide formation by only using voltammetric techniques. It was found that under 0.78 V, three current peaks can be found assigned to different OH arrangements on platinum, and for the (100) plane a calculated ratio of 1:4 (OH to platinum) was estimated. From this potential upward, there is no real order of growth, and at approximately 0.95 V a complete two-dimensional layer of OH is formed until the complete oxide is defined. The main point of this study is that the adsorption/desorption of the OH species is fast and reversible, and other processes occur by chemical or electrochemical ageing effects.

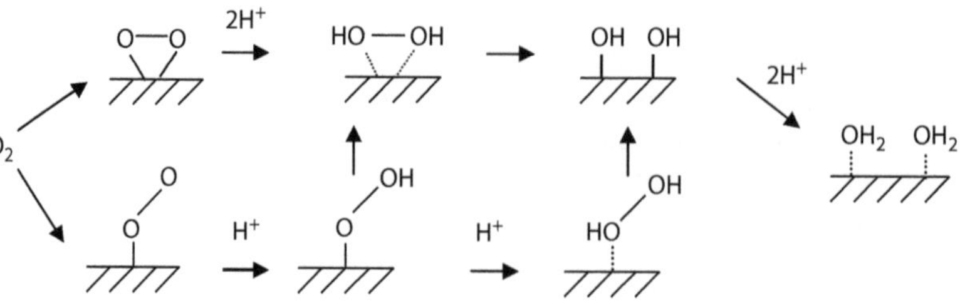

SCHEME 2.2   Main scheme for oxygen electroreduction on platinum considering series and parallel pathways to water and hydrogen peroxide as products.

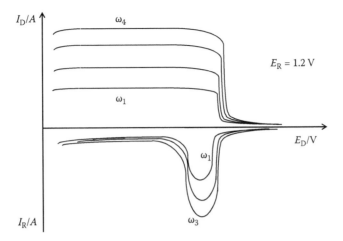

FIGURE 2.7 Characteristic rotating ring (I_R) and disk (I_D) currents vs. disk potential (E_D) curves at different rotation speeds (ω) and at a constant ring potential (E_R) for oxygen electroreduction.

It is clear that the existence of an oxide layer changes the kinetics of the oxygen reduction reaction. In this way, the steady-state polarization curves on platinum at any pH show two Tafel lines, one at potentials larger than 0.92 V of -0.06 V decade^{-1} (in the presence of the surface oxide) [101,102] and independent of the platinum surface morphology, and the other of -0.12 V decade^{-1} that truly depends on the platinum crystallography, the anion adsorption, and the solution temperature. The explanation for the high potentials of the Tafel slopes is given by different rate-determining step considerations, namely, the chemical decomposition of the adsorbate. However, a Frumkin adsorption isotherm based on the kinetic law was also proposed [101], but the surface coverage by the OH species does not exceed 1/4 of the monolayer at the rest potential of the interface. Thus, the only possibility is that a large interaction coefficient (named as $\partial \Delta G_{ad}/\partial \theta$) must be involved between the OH species that comes from the electrochemical water discharge and the oxygen-containing species that comes from the oxygen electroreduction.

The formation of hydrogen peroxide is strongly dependent on the presence of OH species. Thus, in the potential range below 0.9 V, with negligible possibilities of adsorption, it is possible to detect important amounts of hydrogen peroxide with the anions on platinum in acidic solutions [121]. However, Frumkin and Nekrasov [122] first detected hydrogen peroxide species on platinum because of oxygen electroreduction by using the rotating ring-disk electrode technique. Later, Damjanović et al. [103] made the first analysis of the disk and ring current ratios as a function of the rotation speed at constant ring potentials to distinguish between the intermediate and the reaction products (see Figure 2.7). This type of analysis is very useful for understanding oxygen electro-reduction on platinum in alkaline media because of the larger extent of hydrogen peroxide production.

2.2.3.3 Reaction Intermediates in Oxygen Electroreduction

The nature of the adsorption intermediate and the degree of the surface coverage by the oxygen-containing species in anodic or cathodic reactions at high electrode potentials is of main interest in electrocatalysis. In the case of the oxygen electroreduction reaction, understanding the process is of primary importance. Butler and Armstrong [123] were the first to measure molecular oxygen adsorption on polycrystalline platinum through the length of the current *plateau* from chronoam-perometric curves. Later, Will and Knorr [124] measured oxygen adsorption through the charge associated with the current peaks in potentiodynamic scans.

Any intermediate of a reaction in an electrocatalytic mechanism is commonly considered as an adsorbed species that takes part in a stoichiometrically in one or more reaction pathways. We can formally say that the reduction of molecular oxygen can take place as a result of

$$O_2(dis) + Pt(site) \leftrightarrow [Pt - O_2]_{ads} \tag{X}$$

$$[Pt - O_2]_{ads} + e^- \leftrightarrow [Pt - O_2]_{ads}^- \tag{XI}$$

$$[Pt - O_2]_{ads}^- + H_2O(l) + e^- \leftrightarrow [Pt - O_2H]_{ads}^- + OH_{(ac)}^- \tag{XII}$$

Three types of adsorbates can be considered as reaction intermediates. However, one of them (the peroxide-like adsorbate) can directly produce hydrogen peroxide, thus completing a whole sequence of reactions, and diffuse away from the electrode surface. The distinction between the intermediates produced during the direct four-electron pathway and that of the two electrons that produce hydrogen peroxide can be easily made using the rotating-ring disk electrode system. The type of reaction intermediate affects the kinetic laws, and the changes in the electrode potential, pH, temperature, and bulk concentration predict the possible mechanism of the reaction. The charge transfer process has to be evaluated by incorporating the surface coverage of the intermediates in the current density expression together with the influence of the presence of different adsorbed species.

2.2.3.4 Effect of Crystallographic Orientation in the Oxygen Electroreduction Reaction

The analysis of the polarization curves for the oxygen electroreduction reaction on platinum single crystals has been advanced by the rotating hanging-meniscus electrode technique since the past 15 years [93,95,98]. In this technique, first, the clean and well-prepared platinum single crystal is checked through the cyclic voltammetric profile of the interface in an oxygen-free perchloric acid solution. Second, the electrode is mounted in the rotating disk configuration and the steady-state polarization curve is recorded at different rotation rates by using a gold or platinum ring electrode to collect the peroxo-type species in the same electrolyte. The double plot of the disk and the ring currents against the disk electrode potential is represented. Third, at an intermediate, selected rotation speed value, the mass-transfer corrected Tafel lines are plotted to calculate the two slopes of the electroreduction reaction. Fourth, for the expected reaction order with respect to oxygen, the Koutecký–Levich plot at various disc potentials is represented against the reciprocal value of the square root of the angular rotation speed, to calculate the real kinetic current at each potential and the number of electrons involved in the process at the fixed electrode potential. Finally, the Tafel plots are again represented to check for the initial plot significance.

Actually, the values of the Tafel lines strongly depend on the platinum electrode's crystallographic orientation, the solution composition, and the temperature. According to the literature [125] on Pt(111) in 0.1 M perchloric acid or potassium hydroxide, a Tafel slope of approximately −0.06 V decade^{-1} was measured; however, in 0.05 M sulfuric acid the Tafel slope increases to −0.120 V decade^{-1}. As a distinguishing factor with respect to the other surfaces and as a consequence of the absence of oxygen adsorption, a single and unique Tafel slope is seen. On the other hand, on Pt(100) and in acid media at −0.12 V decade^{-1}, the Tafel slope is found at low electrode potentials, whereas large potential values of different slopes can be detected, for example, the classical −0.06 V decade^{-1} in sulfuric acid. The situation on Pt(100) in alkaline media is different since at low electrode potentials, Tafel slopes of −0.170 V decade^{-1} can be observed as a consequence of the larger extent of hydrogen peroxide formation. Besides, Tafel lines in the low electrode potential region for the oxygen electroreduction reaction on Pt(110) in alkaline media show larger values for the same reason. However, in sulfuric acid, the Tafel plot shows the −0.06 and −0.12 V decade^{-1} expected values, but not so in perchloric acid where the Tafel line in the high potential region is larger than −0.11 V decade^{-1}, probably because of minimal perchlorate anion adsorption [93,95,98].

The results of the oxygen electroreduction reaction on platinum single crystals in alkaline media are interesting as they assess the belongings of the OH adsorption on the reduction kinetics. The special case of the single Tafel slope on Pt(111) has to be considered along with the reversible character of OH adsorption/desorption. Further the order of activity of the oxygen reduction reaction shows that in a sulfuric acid solution Pt(111) shows the lowest of all low Miller index planes. The strong adsorption of the tetrahedral (bi)sulfate anion on the trigonal (111) planes reduces the extent of molecular oxygen adsorption and electroreduction [126,127,132].

The other interesting problem of adsorbate competition in the oxygen electroreduction reaction on platinum is hydrogen adsorption. Obviously, the problem of adsorption on single crystals is that the effect is structure sensitive. In the case of Pt(111), it can be observed that at potentials lower than 0.3 V vs. RHE in perchloric acid that there is a strong decrease in the disk current (due to hydrogen adatom formation) with a consequent increase in the value of the ring currents (due to hydrogen peroxide production) that reach a limiting current [125].

Considering these cases of adsorption and competition, we can say that the molecular oxygen adsorption is rather weak. In the early 1960s, it was considered that oxygen reduction on platinum in purified acid media does not exhibit peroxide radical formation [100]. However, the presence of impurities was not a problem, but rather an opportunity for the stabilization of the peroxide. This opportunity arises from the adsorption configuration of molecular oxygen on platinum single crystals or on the preferentially oriented surfaces. Hence, three main adsorption configurations as depicted in Scheme 2.3 can be distinguished.

Pauling's model was proposed for molecular oxygen adsorption in the presence of strongly adsorbable anions since the latter are important to understand surface competition at platinum. However, in the case of phosphate adsorption on polycrystalline platinum, the amounts of hydrogen peroxide detected by the conventional methodologies do not correspond to the proposed model of adsorption configuration [126]. The only possible explanation for this was the assumption of the Pauling's model along with a strong platinum–oxygen bond inducing the oxygen–oxygen dissociation.

For the bridge model, on the other hand, the oxygen molecule interacts with the platinum d orbitals and a back donation to the oxygen π orbitals occurs. In the case of the Griffiths'model, there is a direct interaction between the platinum d orbitals and the oxygen π orbitals with some back donation from the d orbitals to the oxygen π^* orbitals. When the production of hydrogen peroxide arising from the strong adsorption of anions, surface poisoning, etc. occurs, the Pauling's model is valid with a single platinum atom interaction (see Scheme 2.3). However, in most of the cases many experimental parameters define the nature of the desorption products, and so a generalization is not possible.

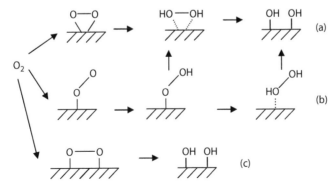

SCHEME 2.3 Three adsorption configurations for molecular oxygen on platinum. (a) Griffiths, (b) Pauling, (c) Bridge models.

2.2.3.5 Analysis of Experimental Data

There are two types of problems in the analysis of electrocatalytic reactions with mixed control kinetics: reactant adsorption and combined considerations of mass and charge transfer processes in the current vs. potential profiles. The dependence of the current density, j, with the overpotential, η, can be expressed under η values larger than 0.12 V (in absolute values) through the Tafel expression corrected by the mass transfer effects:

$$j = j_0 \left(\frac{C^*_{O_2}}{C^0_{O_2}} \right) \exp\left[\frac{-\bar{\alpha}F\eta}{RT} \right] \tag{2.32}$$

where

  j_0 is the exchange current density
  $\bar{\alpha}$ is the charge transfer coefficient for the cathodic reaction
  C_{O_2} is the soluble oxygen concentration at the interface (* super index) or in the bulk of the solution (° super index)
  the rest of the symbols have their usual meaning

When the reaction is arbitrarily considered with an experimental p order with respect to oxygen, the kinetic law will obey the formalism:

$$j = j_0 \left(\frac{C^*_{O_2}}{C^0_{O_2}} \right)^p \exp\left[\frac{-\bar{\alpha}F\eta}{RT} \right] \tag{2.33}$$

The main objectives of the electrocatalytic reaction kinetics are to find j_0, p, and β, which are characteristics of a given electrode system. In general, the procedure is as indicated above. We can define the kinetic current density, j_k, as that current that is not affected by mass transfer effects:

$$j_k = j_0 \exp\left[\frac{-\bar{\alpha}F\eta}{RT} \right] \tag{2.34}$$

The current density affected by the mass transfer arises from the limiting film layer model using the diffusion current density j_D:

$$j_D = nFD_{O_2} \frac{\left(C^0_{O_2} - C^*_{O_2} \right)}{\delta_D} \tag{2.35}$$

and the diffusion current density at limiting conditions $j_{D,lim}$ is

$$j_{D,lim} = nFD_{O_2} \frac{C^0_{O_2}}{\delta_D} \tag{2.36}$$

Combining both expressions, we have

$$\frac{C^*_{O_2}}{C^0_{O_2}} = 1 - \frac{j_D}{j_{D,lim}} \tag{2.37}$$

and

$$j = j_\mathrm{o}\left(1 - \frac{j_\mathrm{D}}{j_\mathrm{D,lim}}\right)^p \exp\left[\frac{-\bar{\alpha}F\eta}{RT}\right] \qquad (2.38)$$

and by substituting with the true kinetic current density, j_k

$$\frac{j}{j_k} = \left(1 - \frac{j_\mathrm{D}}{j_\mathrm{D,lim}}\right)^p \qquad (2.39)$$

or

$$\ln\left(\frac{j}{j_k}\right) = p\ln\left(1 - \frac{j_\mathrm{D}}{j_\mathrm{D,lim}}\right) \qquad (2.40)$$

Frumkin and Tedoradse [128], using a graphic method, could evaluate the kinetic parameters, p and j_k, for each working electrode potential from j vs. ω. At low rotation speed values, ω, the mass transfer processes predominate and then $j \approx j_\mathrm{D,lim}$, or in other words the current density is the diffusion limiting condition. On the other hand, for high ω values the electronic charge transfer is the rate determining $j_k \ll j_\mathrm{D,lim}$ and in this case the measured current is practically j_k. From the slope of this plot, we can calculate the reaction order, p, using the value of $j_\mathrm{D,lim}$ determined by the Levich equation [129,130]. In the case of $p = 1$, we have

$$\frac{1}{j} = \frac{1}{j_k} + \frac{1}{j_\mathrm{D,lim}} \qquad (2.41)$$

or

$$\frac{1}{j} = \frac{1}{j_k} + \frac{1}{B\omega^{1/2}} \qquad (2.42)$$

Thus, the true charge-transfer current can be calculated from the ordinate at the origin in the plot between the reciprocal of the measured current density, j^{-1}, as a function of ω^{-1}. The slope (B^{-1}) is the reciprocal value of the Levich constant, $0.620nFC_j^\circ D_j^{2/3}$, because it is the only portion that strictly depends on the ω value [107], where D_j is the coefficient of diffusion of the j-particle. With the currents corrected from the mass transport effects, we can depict the Tafel lines, from which the values of j_o and $\bar{\alpha}$ can be calculated.

2.2.3.6 Oxygen Electroreduction Mechanism

2.2.3.6.1 Damjanović's Scheme for Oxygen Electroreduction

Damjanović et al. [103] presented one of the first and most simple reaction schemes that was successfully applied in both acid and alkaline media on noble metals (Scheme 2.4).

Depending on the electrochemical rate constants, $\bar{k}_1$, $\bar{k}_2$, and $\bar{k}_3$, a certain amount of hydrogen peroxide can diffuse away from the electrode surface without a further reduction step, ($\bar{k}_3$). The simplified mechanism can be useful but the considerations implicit in it have to be taken into account [104,106], such as no chemical decomposition of adsorbed hydrogen peroxide, fast adsorption/desorption of hydrogen peroxide, slow electrochemical oxidation of hydrogen peroxide to oxygen again within the potential range of the oxygen electroreduction reaction.

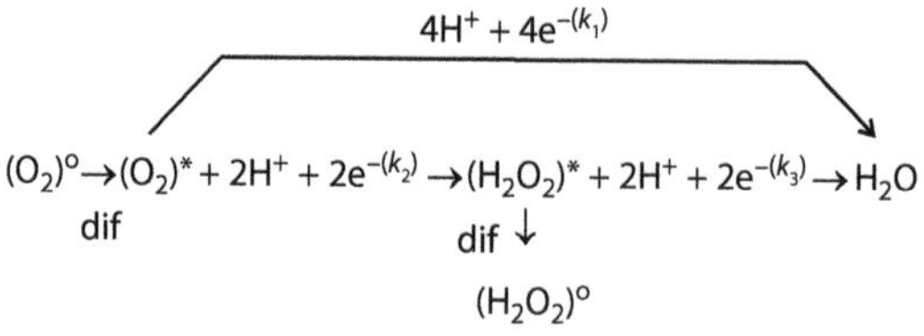

$$4H^+ + 4e^{-(k_1)}$$

$$(O_2)^0 \rightarrow (O_2)^* + 2H^+ + 2e^{-(k_2)} \rightarrow (H_2O_2)^* + 2H^+ + 2e^{-(k_3)} \rightarrow H_2O$$

dif   dif $\downarrow$

$$(H_2O_2)^0$$

SCHEME 2.4 Damjanović's oxygen electroreduction mechanism with three electrochemical rate constants: k_1 for the direct reduction to water, k_2 for the parallel formation of adsorbed hydrogen peroxide, and k_3 for the electrochemical decomposition of hydrogen peroxide to water. The species with the super index * and ° are located at the interface and in the bulk of the solution, respectively. The term dif symbolizes the diffusion to the bulk of the solution and vice versa.

From this reaction scheme, it is possible to derive the following expressions for the analysis of the electrochemical rate constants:

$$\frac{I_D}{I_R} = \frac{1}{N}\left[1 + \frac{2\bar{k}_1}{\bar{k}_2} + \frac{2\bar{k}_3(1 + \bar{k}_1/\bar{k}_2)}{Z_2\omega^{1/2}}\right] \tag{2.43}$$

$$\frac{I_D}{(I_{D,\text{lim}} - I_D)} = 1 + \frac{(1 + \bar{k}_1/\bar{k}_2)}{Z_1\omega^{1/2}} \tag{2.44}$$

where

N is the collection efficiency of the rotating electrode system

Z_1 and Z_2 are the Levich's slopes for the oxygen and hydrogen peroxide electroreductions, respectively

Using the above expressions that were obtained from the disk and ring currents at different rotation speeds as a function of the disk potential, we can obtain the rate constants as a function of the electrode potential:

$$\bar{k}_1 = Z_1 S_2 \frac{I_1 N - 1}{I_1 N + 1} \quad \bar{k}_2 = \frac{2Z_1 S_2}{I_1 N + 1} \quad \bar{k}_3 = \frac{NZ_2 S_1}{I_1 N + 1} \tag{2.45}$$

where I_1 and S_1 are the slope and the origin ordinate of the I_D/I_R vs. $\omega^{-1/2}$ plot at constant E_D, and S_2 is the slope of the $I_D/(I_{D,\text{lim}} - I_D)$ vs. $\omega^{-1/2}$ plot at constant E_D. It is important to state that the reaction order with respect to oxygen is that of the intercept at the origin for the $I_D/(I_{D,\text{lim}} - I_D)$ vs. $\omega^{-1/2}$ plot. Figure 2.8 shows different cases arising from the oxygen electroreduction reaction using the I_D/I_R vs. $\omega^{-1/2}$ plot.

The first case depicts the situation where there is no direct four-electron transfer process and no electrochemical decomposition of hydrogen peroxide. We can see that there is no slope because the I_D/I_R ratio is directly the reciprocal value of the collection efficiency. The case (b) is that of a two-step reaction process where no direct electroreduction to water occurs, with a main definition of the slope by $\bar{k}_3$. The third case shows a clear origin ordinate potential dependent situation; this is the consequence of $\bar{k}_3 = 0$, that is, the accumulation of hydrogen peroxide changes the collection efficiency of the rotating electrode system. Finally, for curve (d), all possible constant rates are considered, and the ordinate at the origin depends on the $\bar{k}_1/\bar{k}_2$ ratio, which also depends on the disk electrode potential. In summary, these curves serve as diagnostic criteria for deciding between a parallel vs. series reaction mechanism and also for the determination of the reaction constant rates.

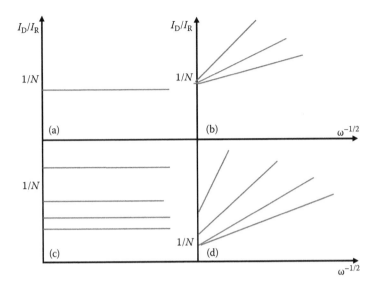

FIGURE 2.8 I_D/I_R vs. $\omega^{-1/2}$ plot at constant E_D for the oxygen electroreduction reaction (a) $k_1 = k_3 = 0$, (b) $k_1 = 0$, (c) $k_3 = 0$, (d) k_1, k_2, and $k_3 \neq 0$.

2.2.3.6.2 Wróblowa's Scheme for Oxygen Electroreduction

One of the main disadvantages of the Damjanović's scheme is that it does not consider possible weak adsorptions and the reversible adsorption/desorption of hydrogen peroxide at the interface. The mechanism proposed by Wroblowa et al. [104] considers the adsorption/desorption equilibrium either for the oxygen reactant or the hydrogen peroxide intermediate. They also proposed the chemical decomposition of the intermediate ($\bar{k}_4$), besides the electrochemical reduction to water through $\bar{k}_3$ (Scheme 2.5).

The I_D/I_R vs. $\omega^{-1/2}$ plot at constant E_D gives rise to the following expression:

$$\frac{I_D}{I_R} = \frac{1}{N}\left[1 + \frac{2\bar{k}_1}{\bar{k}_2} + A + \frac{\bar{k}_5 A}{Z_2 \omega^{1/2}}\right] \tag{2.46}$$

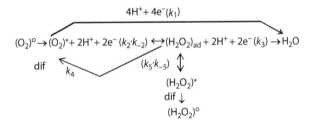

SCHEME 2.5 Wróblowa's oxygen electroreduction mechanism with electrochemical rate constants: k_1 for the direct reduction to water, k_2 and k_{-2} for the parallel electrochemical reduction and oxidation of adsorbed hydrogen peroxide, k_3 for the electrochemical decomposition of hydrogen peroxide to water, k_4 for the chemical catalytic decomposition of hydrogen peroxide to oxygen, and k_5 and k_{-5} for the parallel adsorption/desorption of the adsorbed hydrogen peroxide. The species with the super index * and ° are located at the interface and in the bulk of the solution, respectively. The sub index ads refers to the adsorbed state, and the term dif symbolizes the diffusion to the bulk of the solution and vice versa.

where

$$A = \left(\frac{2\bar{k}_1}{\bar{k}_2 \bar{k}_{-5}}\right)(\bar{k}_2 + \bar{k}_3 + k_4) + \left(\frac{2\bar{k}_3 + k_4}{\bar{k}_{-5}}\right) \tag{2.47}$$

As expected, the expressions are rather complicated and were simplified by Wroblowa, considering that k_{-2} is negligible within the potential region of oxygen electroreduction. Taking the expressions of the I_D/I_R vs. $\omega^{-1/2}$ plot at constant E_D and the $I_D/(I_{D,\text{lim}} - I_D)$ vs. $\omega^{-1/2}$ plot at constant E_D, we can obtain the equations for the calculation of the rate constants:

$$\bar{k}_1 = \frac{A_3(I_1 N - 1 - [Z_2 A_2 S_7 / A_1])}{I_5 N + 1 - [N Z_2 A_2 S_7 / A_1]} \quad \bar{k}_2 = \frac{2A_3}{I_5 N + 1 - [N Z_2 A_2 S_7 / A_1]} \tag{2.48}$$

$$\frac{\bar{k}_2}{\bar{k}_5} = \frac{[N Z_2 A_2 S_7 / A_1] - A_2(I_5 N - [Z_2 A_2 S_7 / A_1])}{2(I_5 N + 1 - [N Z_2 A_2 S_7 / A_1])} \quad \frac{k_4}{\bar{k}_5} = A_2 \frac{A_1}{A_2} = \bar{k}_{-5} \tag{2.49}$$

where

$$A_1 = \frac{k_4 \bar{k}_{-5}}{\bar{k}_5} \quad A_2 = \frac{k_4}{\bar{k}_5} \quad A_3 = \bar{k}_1 + \bar{k}_2 \tag{2.50}$$

Even if these relationships are analyzed in detail, in some cases it will be not possible to conclude on the possibility of a direct or parallel reaction pathway because the I_D/I_R vs. $\omega^{-1/2}$ plot will always show a potential dependent ordinate at the origin. Thus, Wroblowa et al. [104] decided to plot the ordinate at the origin, J, as a function of the slope, S, of the $N I_D/I_R$ vs. $\omega^{-1/2}$ curve at different E_D values. The generalization performed by Adzić and coauthors [131] based on Wróblowa's work led to $\bar{k}_1 = 0$ and $A \to 0$ (or in other words negligible desorption of hydrogen peroxide from the equilibrium constant), and thus we obtain

$$J = 1 + \frac{Z}{\bar{k}_5} S \tag{2.51}$$

It is clear that when we plot J vs. S, we have to obtain from the slope the value of $\bar{k}_5$, but this treatment is valid only when this rate constant is independent of the electrode potential. When there is a similar potential dependent situation for $\bar{k}_1$ and $\bar{k}_2$, the intercept of the J vs. S curve leads to $1 + \bar{k}_1/\bar{k}_2 > 1$. On the other hand, in the treatment for the different potential dependent shapes for $\bar{k}_1$ and $\bar{k}_2$, there is no linear behavior between J and S.

Though the model proposed by Wróblowa is rather limiting because of the problem of the number of equations and the required parameters, there is another model that was proposed even before that of Wróblowa's by a Russian group of electrochemists that needs to be considered.

2.2.3.6.3 Bagotskii's Scheme for Oxygen Electroreduction

In this scheme, Bagotskii et al. [106] developed a parallel reaction pathway taking into consideration fast adsorption/desorption of hydrogen peroxide and negligible oxidation of the intermediate to oxygen (Scheme 2.6).

The I_D/I_R vs. $\omega^{-1/2}$ plot at constant E_D gives rise to the following expression:

$$\frac{I_D}{I_R} = \frac{1}{N}\left[1 + \frac{2\bar{k}_1}{\bar{k}_2}\right] + \frac{(\bar{k}_3 + k_4)[1 + 2\bar{k}_1/\bar{k}_2] + \bar{k}_3 \omega^{-1/2}}{N Z_2} \tag{2.52}$$

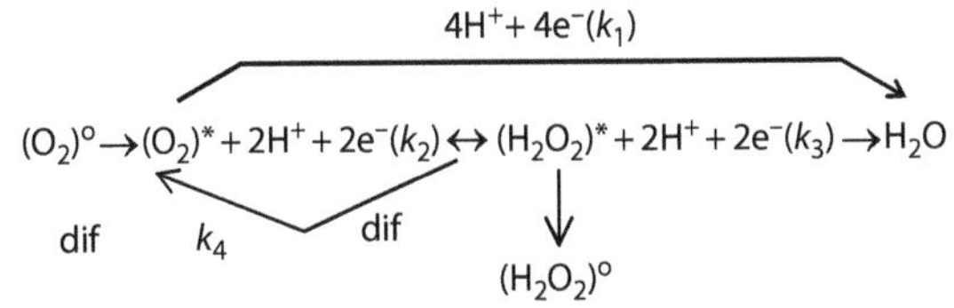

$$4H^+ + 4e^- (k_1)$$

$$(O_2)^\circ \rightarrow (O_2)^* + 2H^+ + 2e^-(k_2) \leftrightarrow (H_2O_2)^* + 2H^+ + 2e^-(k_3) \rightarrow H_2O$$

dif k_4 dif

$$(H_2O_2)^\circ$$

SCHEME 2.6 Bagotskii's oxygen electroreduction mechanism with electrochemical rate constants: k_1 for the direct reduction to water, k_2 for the parallel electrochemical reduction of hydrogen peroxide, k_3 for the electrochemical decomposition of hydrogen peroxide to water, and k_4 for the chemical catalytic decomposition of hydrogen peroxide to oxygen. The species with the super index * and $^\circ$ are located at the interface and in the bulk of the solution, respectively. The term dif symbolizes the diffusion to the bulk of the solution and vice versa.

and $I_D/(I_{D,lim} - I_D)$ vs. $\omega^{-1/2}$ gives rise to

$$\frac{I_D}{(I_{D,lim} - I_D)} = \frac{1 + \left[\dfrac{(\bar{k}_1 + \bar{k}_2)\omega^{-1/2}}{Z_1} \right]}{1 + \left[\dfrac{k_4 I_1}{2A_D F N Z_1 Z_2 C_{O2}^\circ} \right]} \tag{2.53}$$

where
  A_D is the disk electrode area
  F is the Faraday constant
  Z_i is the Levich slopes for oxygen (1) and hydrogen peroxide (2) species

With the same type of treatment, we can calculate the electrochemical and chemical rate constants from I_D/I_R vs. $\omega^{-1/2}$ and $I_D/(I_{D,lim} - I_D)$ vs. $\omega^{-1/2}$ plots at constant E_D.

APPENDIX: RICCATI'S EQUATION

The linear differential equation with quadratic terms

$$\frac{dy}{dx} = P(x) + Q(x)y + R(x)y^2 \tag{2.A.1}$$

is called the Riccati's equation. If y_1 is a known particular solution of (2.A.1), then the substitutions of

$$y = y_1 + u \quad \text{and} \quad \frac{dy}{dx} = \frac{dy_1}{dx} + \frac{du}{dx} \tag{2.A.2}$$

in (2.A.1) lead to the following differential equation for u:

$$\frac{du}{dx} - (Q + 2y_1 R)u = Ru^2 \tag{2.A.3}$$

Since (2.A.3) is a Bernoulli equation with $n = 2$,

$$\frac{dy}{dx} + P(x)y = f(x)y^n \tag{2.A.4}$$

it is reduced to the linear equation

$$\frac{dw}{dx} + (Q + 2y_1 R)w = -R \tag{2.A.5}$$

by substituting $w \equiv u^{-1}$. In many cases, the solution of a Riccati equation cannot be expressed in terms of elementary functions.

Example

Suppose that we have the following differential equation:

$$\frac{dy}{dx} = 2 - 2xy + y^2 \tag{2.A.6}$$

It can be easily verified that a particular solution of this equation is $y_1 = 2x$. From (2.A.1), we make the following identifications $P(x) = 2$, $Q(x) = -2x$, and $R(x) = 1$ and then solve the linear equation (2.A.6):

$$\frac{dw}{dx} + (-2x + 4x)w = -1 \quad \text{or} \quad \frac{dw}{dx} + 2xw = -1 \tag{2.A.7}$$

The integrating factor for the last equation is e^{x^2}, so

$$\frac{d}{dx}\left(e^{x^2}w\right) = -e^{x^2} \tag{2.A.8}$$

Now the integral

$$\int_{x_0}^{x} e^{t^2} dt \tag{2.A.9}$$

cannot be expressed in terms of the elementary functions. Thus, we write instead

$$e^{x^2}w = -\int_{x_0}^{x} e^{t^2} dt + c \quad \text{or} \quad e^{x^2}\left(\frac{1}{u}\right) = -\int_{x_0}^{x} e^{t^2} dt + c \tag{2.A.10}$$

so that

$$u = \frac{e^{x^2}}{c - \int_{x_0}^{x} e^{t^2} dt} \tag{2.A.11}$$

A solution of the equation is then $y = 2x + u$.

REFERENCES

1. F. P. Bowden and E. K. Rideal, *Proc. R. Soc. Lond. Ser.* **A120** (1928) 59.
2. J. O'M. Bockris, *Trans. Faraday Soc.* **43** (1947) 417.
3. P. Reutschi and P. Delahay, *J. Chem. Phys.* **23** (1955) 195.
4. N. Kobosev and W. Monblanowa, *Acta Physicochim URSS* **1** (1934) 611.
5. W. Grubb, *Nature* **198** (1963) 883.

6. P. Reutschi and P. Delahay, *J. Chem. Phys.* **23** (1955) 566.
7. R. Parsons, *Trans. Faraday Soc.* **54** (1958) 1053.
8. H. Gerisher, *Bull. Chim Soc. Belg.* **67** (1958) 506.
9. J. O'M. Bockris and H. Wroblowa, *J. Electroanal. Chem.* **7** (1964) 428.
10. A. N. Frumkin, B. B. Damaskin, and O. A. Petrii, *Elektrokhimiya* **12** (1976) 3.
11. S. Arrhenius, *Z. Phys. Chem.* **4** (1889) 226.
12. M. Bodenstein, *Z. Phys. Chem.* **85** (1913) 329.
13. A. J. Appleby, Electrocatalysis, in *Comprehensive Treatise of Electrochemistry* (B. E. Conway, J. O'M. Bockris, E. Yeager, S. U. M. Khan, and R. E. White, eds.), Vol. 7 (1985), Plenum Press, London, New York, pp. 173–239.
14. R. A. Marcus, *Faraday Discuss. Chem. Soc.* **74** (1982) 7.
15. S. H. Lin, K. H. Lau, W. Richardson, L. Volk, and H. Eyring, *Proc. Natl. Acad. Sci. USA* **69** (1972) 2778.
16. M. C. Flowerscan, *J. Chem.* **56** (1978) 29.
17. D. M. Wardlaw and R. A. Marcus, *J. Chem. Phys.* **83** (1985) 3462.
18. A. M. Kuznetsov and J. Ulstrup, *Faraday Discuss. Chem. Soc.* **74** (1982) 31.
19. A. M. Kuznetsov, *Faraday Discuss. Chem. Soc.* **74** (1982) 49.
20. M. Bixon and J. Jortner, *Faraday Discuss. Chem. Soc.* **74** (1982) 17.
21. D. C. Grahame, *Chem. Rev.* **41** (1947) 441.
22. S. Trasatti, *J. Electroanal. Chem.* **33** (1971) 351.
23. J. O'M. Bockris and S. D. Argade, *J. Chem. Phys.* **49** (1968) 5133.
24. P. Delahay and M. Kleinermann, *J. Am. Chem. Soc.* **82** (1960) 4509.
25. D. S. Gnanamuthu and J. V. Petrocelli, *J. Electrochem. Soc.* **114** (1967) 1036.
26. W. J. Plieth and K. Vetter, *Z. Phys. Chem.* **NF 61** (1960) 282.
27. C. F. Zinola, A. M. Castro Luna, and A. J. Arvia, *Electrochim. Acta* **39** (1994) 1951.
28. J. Horiuti and M. Polanyi, *Acta Physicochim. URSS* **2** (1935) 505.
29. A. J. Appleby, in *Modern Aspects of Electrochemistry* (B. E. Conway and J. O'M. Bockris, eds.), Vol. 9 (1974) Plenum Press, New York, pp. 369–389.
30. V. S. Bagotskii and Y. B. Vasilev, *Electrochim. Acta* **9** (1964) 69.
31. A. Capon and R. Parsons, *J. Electroanal. Chem.* **44** (1973) 239.
32. R. Parsons, *Discuss. Faraday Soc.* **45** (1968) 40.
33. W. Chrzanowski and A. Wieckowski, *Langmuir* **14** (1998) 1967.
34. M. Krausa and W. Vielstich, *J. Electroanal. Chem.* **379** (1994) 307.
35. M. Hogarth, J. Munk, A. Shukla, and A. Hamnett, *J. Appl. Electrochem.* **24** (1994) 85.
36. W. Chrzanowski, H. Kim, and A. Wieckowski, *Catal. Lett.* **50** (1998) 69.
37. B. Beden, F. Hahn, C. Lamy, J. M. Leger, N. Tacconi, R. Lezna, and A. J. Arvia, *J. Electroanal. Chem.* **261** (1987) 215.
38. C. Coutanceau, A. Ralotondrainibé, Al. Lima, E. Garnier, S. Pronier, and J. Léger, C. Lamy, *J. Appl. Electrochem.* **34** (2004) 61.
39. D. M. Kolb, in *Advances in Electrochemistry and Electrochemical Engineering* (H. Gerisher and C. W. Tobias, eds.), Vol. 11 (1978) Wiley-Interscience, New York.
40. M. Watanabe and S. Motoo, *J. Electroanal. Chem.* **60** (1975) 275.
41. A. B. Anderson, E. Grantscharova, and S. Seung, *J. Electrochem. Soc.* **143** (1996) 2075.
42. M. T. Koper, *Surf. Sci.* **548** (2004) 1.
43. H. A. Gasteiger, N. M. Marković, P. N. Ross, and E. J. Cairns, *J. Electrochem. Soc.* **141** (1994) 1795.
44. H. A. Gasteiger, N. M. Marković, P. N. Ross, and E. J. Cairns, *J. Phys. Chem.* **97** (1993) 12020.
45. A. Haner and P. N. Ross, *J. Phys. Chem.* **95** (1991) 3740.
46. B. Bittins-Cattaneo and T. Iwasita, *J. Electroanal. Chem.* **238** (1987) 151.
47. E. A. Batista, G. Malpass, A. J. Motheo, and T. Iwasita, *J. Electroanal. Chem.* **571** (2004) 273.
48. D. R. Flinn and S. Schuldiner, *Electrochim. Acta* **19** (1974) 421.
49. J. H. Barber and B. E. Conway, *J. Electroanal. Chem.* **461** (1999) 80.
50. J. X. Wang, S. R. Branković, Y. Zhu, J. C. Hanson, and R. R. Adzić, *J. Electrochem. Soc.* **150** (2003) A1108.
51. M. R. Gennero and A. C. Chialvo, *Phys. Chem. Chem. Phys.* **6** (2004) 4009.
52. J. O'M. Bockris and H. Mauser, *Can. J. Chem.* **37** (1959) 475.
53. H. Angerstein-Kozlowska, in *Comprehensive Treatise of Electrochemistry* (J. O'M. Bockris, E. Yeager, B. E. Conway, and S. Sarangapani, eds.), Vol. 9 (1984), Plenum Press, New York, p. 28.
54. L. Bai, D. A. Harrington, and B. E. Conway, *Electrochim. Acta* **32** (1987) 1713.
55. D. A. Harrington and B. E. Conway, *Electrochim. Acta* **32** (1987) 1703.
56. G. Kreysa, B. Hakansson, and P. Ekdunge, *Electrochim. Acta* **33** (1988) 1351.

57. O. Nagashima and H. Kita, *J. Res. Inst. Cat. Hokkaido Univ.* **15** (1967) 49.
58. P. M. Quaino, J. L. Fernández, M. R. Gennero, and A. C. Chialvo, *J. Mol. Catal. A* **252** (2006) 156.
59. B. E. Conway and J. O'M. Bockris, *J. Chem. Phys.* **26** (1956) 532.
60. D. J. Barclay, *J. Electroanal. Chem.* **44** (1973) 47.
61. R. Parsons, *Trans. Faraday Soc.* **59** (1960) 1340.
62. F. G. Will, *J. Electrochem. Soc.* **112** (1965) 451.
63. P. N. Ross, *J. Electroanal. Chem.* **76** (1977) 139.
64. K. Kinoshita and P. Stonehart, *Electrochim. Acta* **20** (1975) 101.
65. D. M. Novak, B. V. Tilak, and B. E. Coway, in *Modern Aspects of Electrochemistry* (J. O'M. Bockris and B. E. Conway, eds.), Vol. 14 (1981) Plenum Press, New York.
66. J. A. V. Butler, *Proc. R. Soc. Lond.* **A157** (1931) 423.
67. A. J. Appleby, G. Bronoel, M. Chemla, and H. Kita, Hydrogen, in *Electrochemistry of the Elements* (A. J. Bard, H. Lund, eds.), (1982), Marcel Dekker, New York.
68. L. I. Krishtalik, in *Advances in Electrochemistry and Electrochemical Engineering* (H. Gerisher and C. W. Tobias, eds.), Vol. 7 (1970), Wiley-Interscience, New York.
69. W. A. Pliskin and R. P. Eischens, *Z. Phys. Chem. N.F.* **24** (1960) 11.
70. L. H. Germer and C. Davidsson, *Nature* (London) **119** (1927) 558.
71. A. T. Hubbard, J. L. Stickney, S. D. Rosasco, M. P. Soriaga, and D. Song, *J. Electroanal. Chem.* **150** (1983) 165.
72. G. A. Somorjai, *Catal. Rev.* **7** (1972) 251.
73. A. T. Hubbard, *Chem. Rev.* **88** (1988) 633.
74. F. M. Propst and T. C. Piper, *J. Vac. Sci. Technol.* **4** (1967) 53.
75. J. Clavilier, R. Faure, G. Guinet, and R. Durand, *J. Electroanal. Chem.* **107** (1980) 205.
76. F. E. Woodward, C. L. Scortichini, and C. N. Reilley, *J. Electroanal. Chem.* **151** (1983) 109.
77. N. Marković, M. Hanson, G. Mc. Dougall, and E. Yeager, *J. Electroanal. Chem.* **214** (1986) 555.
78. B. Love, K. Seto, and J. Lipkowski, *Rev. Chem. Intermed.* **8** (1987) 87.
79. D. Armand and J. Clavilier, *J. Electroanal. Chem.* **263** (1989) 109.
80. J. Clavilier, K. El Achi, and A. Rodes, *Chem. Phys.* **141** (1990) 1.
81. R. R. Adzić, F. Feddrich, B. Z. Nikolik, and E. Yeager, *J. Electroanal. Chem.* **341** (1992) 287.
82. J. Clavilier, D. Armand, S. G. Sun, and M. Petit, *J. Electroanal. Chem.* **205** (1986) 267.
83. S. Motoo and N. Furuya, *Ber. Bunsen Ges. Phys. Chem.* **91** (1987) 457.
84. J. Clavilier, J. M. Feliú, A. Fernández-Vega, and A. Aldaz, *J. Electroanal. Chem.* **269** (1989) 175.
85. J. Clavilier, M. Wasberg, M. Petit, and L. H. Klein, *J. Electroanal. Chem.* **374** (1994) 123.
86. D. M. Kolb and J. Schneider, *Electrochim. Acta* **31** (1986) 929.
87. A. Hamelin, L. Doubova, D. Wagner, and H. Schirmer, *J. Electroanal. Chem.* **220** (1987) 155.
88. A. Hamelin, *Electrochim. Acta* **31** (1986) 937.
89. D. M. Kolb, W. Boeck, K. M. Ho, and S. H. Liu, *Phys. Rev. Lett.* **47** (1981) 1921.
90. K. P. Bohnen and D. M. Kolb, *Surf. Sci.* **407** (1998) 629.
91. K. Itaya, S. Sugawara, and K. Higaki, *J. Phys. Chem.* **92** (1988) 674.
92. P. N. Ross, *J. Electrochem. Soc.* **126** (1979) 78.
93. N. M. Marković, H. A. Gasteiger, and P. N. Ross Jr., *J. Phys. Chem.* **99** (1995) 3411.
94. K. Tammeveski, K. Kontturi, R. J. Nichols, R. J. Potter, and D. J. Schiffrin, *J. Electroanal. Chem.* **515** (2001) 101.
95. J. Perez, H. M. Villullas, and E. R. Gonzalez, *J. Electroanal. Chem.* **435** (1997) 179.
96. N. M. Marković, R. R. Adzić, B. D. Cahan, and E. Yeager, *J. Electroanal. Chem.* **377** (1994) 249.
97. H. Kita, H. W. Lei, and Y. Gao, *J. Electroanal. Chem.* **397** (1994) 407.
98. F. El Kadiri, R. Faure, and R. Durand, *J. Electroanal. Chem.* **301** (1991) 177.
99. C. F. Zinola, A. Castro Luna, W. E. Triaca, and A. J. Arvia, *J. Appl. Electrochem.* **24** (1994) 119.
100. J. O' M. Bockris and A. K. M. S. Huq, *Proc. Roy. Soc.* **237 A1** (1956) 1733.
101. A. Damjanović and V. Brusić, *Electrochim. Acta* **12** (1967) 1615.
102. E. B. Yeager, *Electrochim. Acta* **29** (1984) 1527.
103. A. Damjanović, M. A. Genshaw, and J. O' M. Bockris, *J. Chem. Phys.* **45** (1966) 4057.
104. H. S. Wroblowa, Y. C. Pan, and G. Razumney, *J. Electroanal. Chem.* **69** (1976) 195.
105. V. Vesović, N. Anastasijević, and R. R. Adzić, *J. Electroanal. Chem.* **218** (1987) 53.
106. V. S. Bagotskii, M. R. Tarasevich, and V. Yu Filinovskii, *Elektrokhimiya* **5** (1969) 1218.
107. R. W. Zurilla, R. K. Sen, and E. B. Yeager, *J. Electrochem. Soc.* **125** (1978) 1103.
108. M. R. Tarasevich, K. A. Radyushkina, V. Yu Filinovskii, and R. Kh. Burstein, *Elektrokhimiya* **6** (1970) 1522.

109. P. N. Ross Jr., *J. Electroanal. Chem.* **20** (1970) 78.
110. C. F. Zinola, G. L. Estiú, E. A. Castro, and A. J. Arvia, *J. Phys. Chem.* **98** (1994) 1766.
111. N. P. Lebedeva, M. T. M. Koper, E. Herrero, J. M. Feliu, and R. A. van Santen, *J. Electroanal. Chem.* **487** (2000) 37.
112. C. F. Zinola, A. Castro Luna, W. E. Triaca, and A. J. Arvia, *J. Appl. Electrochem.* **24** (1994) 531.
113. C. F. Zinola, A. Castro Luna, W. E. Triaca, and A. J. Arvia, *Electrochim. Acta* **39** (1994) 1627.
114. W. E. Triaca, T. Kessler, J. C. Canullo, and A. J. Arvia, *J. Electrochem. Soc.* **134** (1987) 1165.
115. R. M. Cerviño, A. J. Arvia, and W. E. Vielstich, *Surf. Sci.* **154** (1985) 623.
116. L. Vázquez, J. M. Gómez, J. Gómez, A. M. Baró, N. García, J. C. Canullo, and A. J. Arvia, *Surf. Sci.* **181** (1987) 98.
117. A. K. Reddy, M. A. Genshaw, and J. O'M. Bockris, *J. Electroanal. Chem.* **81** (1964) 406.
118. R. Greef, *J. Chem. Phys.* **51** (1969) 3148.
119. A. H. Lanyon and B. M. W. Trapnell, *Proc. Roy. Soc.* **227A** (1955) 387.
120. H. Angerstein-Kozlowska, B. E. Conway, and W. B. Sharp, *J. Electroanal. Chem.* **43** (1973) 9.
121. C. F. Zinola, W. E. Triaca, and A. J. Arvia, *J. Appl. Electrochem.* **25** (1995) 740.
122. A. N. Frumkin and L. N. Nekrasov, *Dokl. Akad. Nauk. SSSR* **126** (1959) 115.
123. J. A. V. Butler and G. Armstrong, *Proc. Roy. Soc.* **A137** (1932) 604.
124. F. G. Will and C. A. Knorr, *Z. Elektrochem.* **64** (1960) 258.
125. N. M. Marković and P. N. Ross, Electrocatalysis at well-defined surfaces, in *Interfacial Electrochemistry: Theory, Experiment and Applications* (A. Wieckowski, ed.), (1999) Chap. 46, Marcel Dekker, Inc., New York-Basel, pp. 821–841.
126. P. W. Faguy, N. Marković, and P. N. Ross, *J. Electrochem. Soc.* **140** (1993) 1638.
127. A. Wieckowski, P. Zelenay, and K. Varga, *J. Chim. Phys.* **88** (1991) 1247.
128. K. Vetter, *Elektrochemische Kinetik*, 2nd edn., Springer Verlag, Frankfurt, Germany (1961).
129. V. G. Levich, *Physicochemical Hydrodinamics*, Prentice Hall, Englewood, CO, London, U. K. (1962).
130. A. J. Arvía and S. L. Marchiano, Los Fenómenos de Transporte en Electroquímica, Universidad de La Plata Ed., Argentina (1971).
131. N. A. Anastasijević, V. Vesović, and R. R. Adzić, *J. Electroanal. Chem.* **229** (1987) 305.
132. M. R. Tarasevich, A. Sadkowskii, and E. B. Yeager, in *Comprehensive Treatise of Electrochemistry* (J. O'M. Bockris, E. B. Yeager, S. U. M. Khan, and R. E. White, eds.), Vol. 7 (1983), Chap. 6, Plenum Press, New York, pp. 301–398.

3 Quantum and Theoretical Electrocatalysis

C. Fernando Zinola

CONTENTS

3.1 DEVELOPMENT OF QUANTUM MECHANICS

Great advances to knowledge obtained from Newton's mechanics were made at the beginning of the twentieth century, particularly in the understanding of the quantum mechanical structure of matter. These developments made a new connection between mathematics and theoretical physics, producing a cooperative development that provided path-breaking scientific insights. Quantum mechanics, the new discipline in theoretical physics, provided a primarily new approach for restructuring mathematical concepts in the analysis of the main structure of matter with great success.

In the case of the discipline of electrocatalysis, it is important to state that the "reactants," the electrons, are microscopic particles that obey the new theoretical approaches that are presented below. The term "electrocatalysis" as stated in Chapter 2 involves a clear participation of the metal's properties, such as the electronic configuration, that can only be understood by considering quantum mechanics. The introduction of these novel mathematical methods into quantum physics and chemistry started with the use of group theory in spectroscopy and the study of chemical bonds [1–3]. Moreover, the study of all the concepts together as an interdisciplinary science is due to the contributions of Hawkins [4] and Rechenberg [5] to group theory and quantum mechanics.

Quantum mechanics involves the characterization of a physical system by a set of Hermitian operators, one for any observable quantity, in a state space S assumed to be a Hilbert space. In Schrödinger's perspective, S was viewed as a space of complex wave functions with differential operators as tools. In this sense, the operator characterizing the energy of the system, the *Hamilton operator* H, was one of the most important. However, linear momentum P, coordinated spatial positions Q, rotational (orbital) momentum L, the square of the total momentum L^2, and the *spin J* of

77

a particle (particle's proper rotation) were also considered. We are especially interested in characterizing the energy of an electron in an electrocatalyst given by H. The problem we have is that our system is multielectronic and the electrocatalyst usually acts as an interface system with a compromised region between the electrolyte and a metal.

For an atom, the eigenspaces of H could typify the stationary states of the electron systems, where the eigenvalues $E_1, E_2, \ldots$ of H symbolized the energy values under these eigenstates, which can be degenerated; that is, they fitted in to an eigenvalue of multiplicity >1. The differences between the two energy values, E_1 and E_2, were observable only by the frequency, ν, of the emitted radiation during the electron transition from one energy state to another:

$$h\nu = E_1 - E_2 \qquad\qquad (3.1)$$

The problem of characterizing a defined electron was not possible in science until the advent of quantum mechanics.

3.1.1   Ideas of Pauli: Pauli's Spinor

To acquire this new knowledge, no scientific methodology as accepted now was used then, for example, Pauli guessed that bound electron states in a molecule have an intrinsic two-valued configuration. This could not be demonstrated but only "assumed" to occur and that it was conveniently proposed that the electrons obey an *exclusion principle* with a forbidding rule: "different electrons were not able to occupy the same state of a system." Later, the hypothesis of electron *spin* was established, the spin assumed to be caused by a "proper rotation" of the electron so that the magnetic properties can be explained at last. Empirical evidence was used to clarify this intrinsic spin that was also quantized with respect to any specified spatial direction. Two possible states of spin "up" and spin "down" were proposed. Later, Pauli [6] rationalized this idea as a *spin state space*, extending the complex phase of the Schrödinger wave function, $\psi(x)$. He described a spinning particle by a two-valued wave function, $\psi = (\psi_1, \psi_2)$, called a *Pauli spinor*. The tensorial products of the complex two-dimensional space of the pure spin states' superposition and the spin state space have to be in a linear span of the spin up and spin down states. The total wave function of a collection of n electrons was expressed formally as a "product." Dirac realized that *Pauli's exclusion principle* implied that multielectron (fermion) states must be represented by *alternating products* [7].

3.1.2   Multielectron Systems

Just before Pauli's spin rationalization, Heisenberg started to consider the mechanical and electrostatic consequences in multielectron systems [8]. The division of energy terms in the spectrum of higher atoms into different subsets of fixed energies was proposed and the condition that no exchange of electrons really took place between them appeared for the first time. The effect was seen as that of comparing the "missing lines" with the observed spectral lines of all energy levels in the atom with higher mass. Heisenberg's proposal can be interpreted as that these lines are caused by the interaction of the orbital magnetic momentum of the electrons with the spin. An "interplay or resonances" phenomenon between the spin states of different electrons and their orbital momenta [9] could be deduced, that is, the *spin coupling effects*. Of course, after applying the H operator, without considering the spin, to n eigenfunctions, we are able to introduce the concept of degeneracy: equal energy states of single electrons without spin. The spin coupling as "resonance" in the interplay between electrons, however, had to be evaluated differently. First, we now know that for unperturbed electrons, the eigenfunction of the total system can be written as a product of n (all) solitary electronic functions. Thus, the unperturbed system is $n!$-fold degenerate, because the simple permutation of electrons always yields equal energies of the system [9]. All eigenvalues have to be identical for degeneracy.

For either the spin up or the spin down element, the space can be constructed by the tensorial product between each eigenfunction ($\psi_1 \psi_2 \dots \psi_n$) with an index $1 < i < n$ of electrons. Heisenberg was able to compute the electron permutation, S_u, as a subspace of S_n, symmetric group of n elements, by

$$S_u = \psi_{S_{(1)}} \psi_{S_{(2)}} \dots \psi_{S_{(n)}} \tag{3.2}$$

The state space of a sole electron $V(n)$ is the linear span of $\psi_1, \psi_2, \dots \psi_n$. The permutation between the components of any one product state u will be

$$V(n) = \langle S_u | S \in S_n \rangle \tag{3.3}$$

$V(n)$ was constructed to characterize the state space of an "unperturbed" system of n electrons distributed according to Pauli's principle on the n states of the linear span. In the case of n electrons of index $j/ 1 < j < n$, the tensorial product between the states and electrons is

$$V(n) = \psi_1^{(i1)} \otimes \psi_2^{(i2)} \otimes \psi_3^{(i3)} \otimes \dots \otimes \psi_n^{(in)} \tag{3.4}$$

However, mathematically speaking, the wave functions can be orthogonal, but in the physical space, this conception impeded an unfeasibility for electron transitions between the states. The decomposition of $V(n)$ into "non-combining" (orthogonal) spaces was possible only if spin resonance was taken as a kind of perturbation. Its basic structure could be analyzed even at the level of the unperturbed system without spin.

Here again guesses and suppositions based on intuitive concepts arose since Heisenberg believed that an eigenfunction of the total system has to be antisymmetric for an electron permutation, though any *perturbation* of transition probabilities, arising from spin coupling, should be symmetric under the transposition of two electrons. This means that the decomposition of the total space into orthogonal subspaces is not affected by spin resonance.

3.1.3 ROTATIONAL AND SPIN SYMMETRIES (HYPERFUNCTIONS): THE INTERNAL QUANTUM NUMBER

What happens with the interaction between the rotational and spin symmetries once the system is characterized as being defined by at least different spinors? Wigner and von Neumann [10] combined both types of symmetries with the permutation aspect [11]. They intuitively reached the idea using atomic spectroscopy that the H operator has to be constructed by two terms: H_1, resulting from the spatial motion of the single electron only (and the electromagnetic interaction with the field of the atomic core), and (H_2), which has to visualize the electron spin. For simplicity, we can consider the eigenvalue problem of the "spinless" wave function ψ without the second term as

$$H_1 \psi = \lambda \psi \tag{3.5}$$

and after that refine the effect by moving to the "hyperfunctions" including the Pauli spinor. The considerations of Pauli are now better understood since the interaction between different electrons is explicitly given in an energy operator.

They considered an increasing spin perturbation H_2 that may reduce the original symmetry to only the second operation, or in other words, the irreducible structure of subspaces for H_1 are decomposed into smaller non-decomposing components of $H_1 + H_2$. This theory, then, also explained the splitting of spectral terms by a perturbation that produces spin differences naturally. Empirically, such a phenomenon has been observed in the *anomalous Zeeman effect*, as spectral lines belonging to the same magnetic number m can split into different terms: m_l and m_m. Fine structure terms in a spectral line of a multielectron (n) system can be observed besides an azimuthal quantum number l. The simple (no repetition) combinatorial possibilities of the quantum numbers m_j of individual electrons are used to build the total magnetic quantum number $m = m_1 + \ldots + m_n$.

Then, the state was finally explained clearly. Dirac expected this but considered that the antisymmetric principle for the entire wave eigenfunctions in multielectron systems has to be the only real consideration occurring in physics [11]. The conclusion was that the momentum (including spin) of an n-electron system in such a state can be defined by a (integer or half-integer) value j, called the *internal quantum number* [10]. Many quantization ideas arose from the integer or half-integer values of this number and can really be considered as a simplistic mathematical reality of two totally different behaviors, later condensed by thermodynamic statistics.

3.1.4 RESUME OF QUANTA PARTICLES

An understanding of the mechanisms of the reactions in electrodics is provided by physical electrochemistry through the analysis of the electronic and ionic phases. For the first phase, the electronic character of the metals is important and hence solid state physics comes into focus. The quantal characteristic of the metal conductor defines the surface structure properties that are dealt by quantum electrochemistry. The concept of quantum particles is one of the main considerations of this chapter. The properties of the dual nature of this corpuscular wave produce equivocal understanding even in electrocatalysis. When a beam of electrons passes through a solid, the effective mass is the real quantity to be considered in the calculations, since the interactions of the electron with a nucleus are shielded by strong electrostatic interactions.

The quanta of lattice vibrations, *phonons*, are the result of the interactions of the ions in the crystal lattice or, in other words, the vibrations of an atom in the lattice as its individual property and not of the whole cluster. However, most of them are not treated as interacting particles but as noninteracting particles (*pseudo particles*). In the case of phonons, one of the movements is associated to an *acoustic* horizontal and symmetrical motion and the other to an *optical* asymmetrical (forward and backward motion) process. The latter vibrates in the infrared portion of the spectrum, and since they exhibit a weak interaction with the rest of the particles, they can be treated independently.

When the electrons are removed from the nucleus at a high temperature, a gas of ions and electrons forms as plasma. In this case, the kinetic energy exceeds the potential energy between each particle (electrons and metal ions). When they are very near, the electron and nucleus get attracted and the density of states increases, whereas the opposite repulsion energy occurs in electron interactions. In the first case, the number of charged particles, n, of the volumetric charge densities, ρ, produces the surface polarization, P. This property exhibits the polarizability tensor, α, in an electric field, $\vec{E}$:

$$P = \alpha \vec{E} \tag{3.6}$$

With

$$\vec{E} = \frac{n\rho\chi}{\varepsilon} \tag{3.7}$$

where
   χ is the interaction distance (collection of coordinates) between the electron and the position of zero polarization
   ε is the dielectric constant of the material

If we consider the crystal lattice as a harmonic vibrator without any amortization factor, we will have the equilibrium position when [12]

$$m\left(\frac{\partial^2 \chi}{\partial^2 t}\right) - \rho \vec{E} = m\left(\frac{\partial^2 \chi}{\partial^2 t}\right) - \frac{n\rho^2 \chi}{\varepsilon} \tag{3.8}$$

where m is the mass of the elemental charge. The solution for the radiation frequency at the ground state is known as *plasma frequency* and its quanta is known as *plasmons* (with 100 GHz):

$$\omega = \sqrt{\frac{n\rho^2}{m\varepsilon}} \tag{3.9}$$

Now, what happens with the interaction between the electronic/ionic conductors? The quanta resulting from the interaction between the surroundings (electron interchange of electrons in metals) and each ion are called *polarons*. In other words, it is the interaction of the electrostatic energy transfer from a local charge to a vibration in the crystal lattice. This interaction produces the phonon–electron coupling that deforms the crystal lattice. In this case, the *polaron* corresponds to the interaction energy between the electron and the electrically polarized medium [13]. The effective mass of interaction is higher than that of the electronic elemental charge, since the ion cores accompany the motion of electrons during the interaction. When long-range energy transfer occurs at the semiconductor electron–hole pair, the quantum associated with it is called the *exciton*. This behavior is typical of a pair of positive and negative charges surrounding a common centre of gravity, where the whole system is able to migrate or diffuse to the inner-part of the crystal lattice.

3.1.5 DEVELOPMENT OF FIRST CONCEPTS IN QUANTUM CHEMISTRY

Before progressing into electrocatalysis, we first need to develop some ideas from the discipline of quantum chemistry in order to characterize the formation of chemical bonds. Heitler and London published the quantum mechanical explanation of covalent bonds through valence electron pairs [14]. According to Pauling, their view can be considered as the greatest single contribution to the clarification of the chemist's conception of bonding since Lewis's suggestion in 1916—consisting of a pair of electrons held together by two atoms.

The research was accomplished using two hydrogen atoms and their electrons, at an infinite distance between the nuclei. They were characterized by identical Schrödinger functions with an energy eigenvalue E_0. Moreover, using the perturbation theory, the electron energies were calculated when the atomic distance d was reduced. They showed the existence of two solutions, ψ_1 and ψ_2, with a total energy E_i (E_1 and E_2) and interpreted the energy difference as a kind of *exchange energy* between the electrons:

$$\Delta E_i = E_i - 2E_0 \tag{3.10}$$

The negative exchange energy indicated that the compound system had a lower energy state than the two single systems—*energy splitting terms*. The conception of the system was such that the energy was strictly dependent on d up to some value d_1, where E_1 fell to a minimum. Moreover, it rises again from d_1 to ∞, while E_2 fell monotonically for $d > 0$ with increasing $d(d \to \infty)$. An analytical function with a smooth decay of energy, and after d_1, the fast increasing figures of E_1 have to be obtained as described in the final Section 3.1. Thus, ψ_1 represented a bound state for $d = d_1$, while ψ_2 characterized the repulsive forces for any value of the atomic distance [14].

The calculation for more than two electrons demonstrated that only the case of a repulsive interaction can be considered and that too only if the electron spin and the Pauli exclusion's principle were taken into account. Accordingly, the "exchange energy" appeared as an effect of the spin coupling which has to be positive. This is one of the first steps to quantum chemistry, but still the interpretation of long-distance covalent bonding and the cases of the metallic-bound species, which are of interest to our work remain. Furthermore, when an external electric field is applied (the discipline of electrochemistry), the metallic bonding has to be considered as a shifting electronic configuration for each particle and yet things are far from being explained. We are going to see that this is possible after calculating the energy of each state in the absence as well as the presence of an electric field. We will approach this by moving to the Fermi level of the metal, however, not arbitrarily but through empirical data.

Heitler explored the representation theory of the symmetric group for the determination of quantum bond states. His goal [15] was to extend the approach of his joint work with London to much larger molecules than helium. He defined the *valence electrons* as those with quantum numbers (l, m) in the outer "shell" that had no partner of equal quantum numbers with an opposite spin in the same atom. By this, we have for the first time the possibility of understanding more realistic systems, but we are still far from electrocatalysis since the simple equations for the interaction energy can only be considered as a very rough approximation. The "exchange molecules" represent only a part of the chemical molecules and in the absence of external forces, such as electric energy. Conversely, we have to say that Heitler was able to distinguish between the different kinds of chemical bonds, but even he could explain only some of them by spin coupling accessible to group-theoretic methods (*exchange molecules*).

The above result established a quantum mechanical explanation of certain nonionic bonds, which could not be explained in terms of electrostatic forces. An explanation of such "valences" by the pairing of electrons with opposite spin but with equal quantum numbers can be proposed. In this sense, Heitler's idea was to investigate the extent of the hypothesis that spin coupling of valence electron pairs are the basis for molecule formation or a metallic bond to occur [16]. He described the results as having established a "complete equivalence" of the quantum mechanical explanation of chemical bonds for molecules and the traditional explanation of chemical valences by Lewis' electron pairs. He introduced an integral expression derived by Heisenberg for the *exchange energy* between two systems [17] and concluded that it can be interpreted as a *valence bond*, symbolically denoted by a valence dash. Although his theory did not predict new or different effects in comparison to classical chemical knowledge, it claimed to explain the empirical information on valence bonds during those days (1920–1927).

3.1.6 Special Linear Group of Weyl: Invariant Subspaces of Tensor Powers

Taking into account the early findings of valence bonds into the discipline of quantum chemistry, many questions related to this challenging new field were carried into a second phase with an active involvement of mathematical physics. It is important to consider the question what happens to the connection between more than one or two chemical bonds. This second phase of work really does not involve chemistry but it was a follow-up to the first phase of activity in theoretical physics. When Weyl entered the field of quantum mechanics, he was particularly interested in the role of group representations. He had worked with classical groups during his first stint of work with mathematical physics. He found them to be crucial in answering two questions in this context: the use of tensors as universal tools in general relativity and in differential geometry and the *Pythagorean* nature of the metric in general relativity. It was important to use new mathematical tools to manage all the operators and to know the rules that govern the case of more than two bonds.

Weyl answered the first point with the insight that all irreducible representations of the special linear group can be made as invariant subspaces of tensor powers of the underlying standard representation. They were conceived as operations of the linear group transformations with a determinant on a geometrical "coordinate space." Any representation of the linear group can be characterized with a tensor product of the coordinate space by a symmetry property.

In general, symmetry conditions are part of the characterization of a definite type of quantity in a physical space. Tensors and tensor spaces were universal objects for the representation of the linear group transformations that are fundamental for the expansion of the chemical quantum theory of bonding. All the irreducible representations could then be characterized by some *symmetry condition* inside some tensor power of the state space, symbolized as $\otimes^k V$. Thus, a broad correspondence between the representations of the symmetric group and the irreducible representations inside the state space (representations of "order k") played an important role for the answer to the first question.

From the above result, it could be inferred exactly that such irreducible subspaces of the state space establish the proper mathematical domain of the classical physical field quantities. In fact, the demonstration was undertaken by using a relativistic electromagnetic field tensor, F_{ij}, and its antisymmetric property:

$$F_{ij} + F_{ji} = 0 \qquad (3.11)$$

Weyl thus answered the second question in his investigations of the mathematical analysis of the space problem. His researches with the representation theory of Lie groups started because of his diverse background, from the philosophy of mathematics to the natural sciences. Further, he clarified Heisenberg's non-commuting "physical quantities" in quantum mechanics, which were initially stated in a mathematical form rather than a physical form.

Later on, Born and Weyl worked together to make more progress in quantum mechanics, using the linear theory groups. Weyl proposed the relationship between unitary one-parameter groups and their anti-Hermitian generators, g and h:

$$gh - hg = \hbar l \qquad (3.12)$$

where
l is the identity
$\hbar$ is a number without the imaginary factor i

This could be related to a commutation relation among the integral operators. Typical relations among the infinitesimal operators can be derived from this approach. He had come close to a *derivation* of the canonical commutation relation from the definition of the derivative of an operator-valued function of a real variable. Before this canonical commutation, Born considered the assumption of a complex domain of numbers:

$$gh - hg = \hbar l / i \qquad (3.13)$$

Weyl's proposal was a first step to derive the Heisenberg relations from the basic properties of projective unitary representations. He indicated how an observable H given in terms of the conjugate observable g and h could be characterized [18] by $H = H(g,h)$, using the wave function expanded over the eigenvalues ε and κ:

$$\iint \exp[\varepsilon g + \kappa h]\Psi(\varepsilon, \kappa)d(\varepsilon, \kappa) \qquad (3.14)$$

which is less formal than the following nomenclature, but easier to understand:

$$\sum_{m,n} g^m h^n \qquad (3.15)$$

This was the first indication of what [19] became the proposal to use inverse Fourier transforms for quantization, the now so-called *Weyl quantization*. We do not rely on this suggestion as Born and Jordan had stated that Weyl's approach was too heavy for the introduction of quantum mechanics to physics. Having so stated, Born and Jordan made their own approach. Thus, it became more incomprehensible to the chemists.

3.1.6.1 Canonical or Heisenberg Commutations

We now introduce the ideas of Weyl to distinguish between *pure states* and *mixtures*. Pure states were mathematically represented by eigenvectors of observables, which described the properties of a particle or a dynamic state. On the other hand, mixtures were composed of "pure states" of a certain mixing relationship. These aspects are clearly important to chemists and obviously to the electrochemists too. The canonical variables, G and H [19], have to satisfy the *canonical* or *Heisenberg commutation* relation, derived from Equations 3.12 and 3.13:

$$[G, H] = \frac{\hbar}{i} l \tag{3.16}$$

Weyl considered a classical state space depicted by pairs of n conjugate observable quantities (p, q), for example, the spatial displacement q with respect to a frame and its conjugate momentum p, or (E, t), the energy and temporal observable magnitudes. After this, the state space can be considered as an abelian group G of two continuous parameters $(t, s) \in \Re^2 = G$ (in the case of $n = 1$ pairs). For the quantization, Weyl looked at a representation as a product of a complex number of unit norm. Subsequently, it was clear that in the quantum context, the commutation relation for the generation of one-parameter groups, $\exp(itP)$ and $\exp(isQ)$, have to be weakened. Commutation must be only up to a unitary factor, that is, $\exp(icst)$. Thus, we have to write

$$e^{(isP)} e^{(itQ)} = e^{(icst)} e^{(itQ)} e^{(isP)} \tag{3.17}$$

where c is a real constant normalized to $c = 1$ or $c = \hbar$ referred as the *Weyl commutation* for conjugate pairs of one-parameter groups in unitary projective representations. Hermitian infinitesimal generators, iP, iQ, in (3.17) deviate from strict commutability when

$$PQ - QP = -icl \tag{3.18}$$

that is, the Heisenberg commutation rule for a pair of conjugate observables.

There is a definite possibility of using classical physical quantities under operator companions, that is, an approach to *quantization* [19]. If a classical quantity was expressed by a function $f(p, q)$ of the canonical variables, p, q, the Fourier transform ξ of f can be used. Then, f is back-transformed from ξ by

$$f(p, q) = \int e^{i[ps+qt]} \xi(s, t) d(s, t) \tag{3.19}$$

The use of such Fourier transforms or other types of recurrent transforms is appropriate since the proposal of Weyl was to use the equivalent operator integral F

$$F = \int e^{i[Ps+Qt]} \xi(s, t) d(s, t) \tag{3.20}$$

as the quantum mechanical version of the physical quantity related to f. Instead of working with canonical variables, he proposed to use the conjugate observables as the Hermitian infinitesimal generators. For a real-valued function f, ξ satisfies an antisymmetric property:

$$\xi(-s,-t) = -\xi(s,t) \tag{3.21}$$

and leads to a Hermitian operator F.

The quantization of the classical observables, p for P and q for Q, and the non-commutative P and Q led to a fundamental difficulty for an observable given as a function $f(p,q)$ of the basic dynamical variables p and q. Weyl's unitary representation approach avoided this difficulty. The inverse operator of the Fourier transform (3.20) gave a unique well-determined assignment, f for F, of the Hermitian operators to the real-valued quantities. The same proposition can be advanced for electric variables such as "the amount of electric charge," σ as the classical observable, which after the quantization leads to $\sum$. Moreover, dynamical variables that are expressly related to the "current intensity" i as the classical observable yield after quantization to I. The same has to occur to the magnitude of the "electric potential" from v to V, after quantization of the classical observable.

Weyl considered the independent variables, such as coordinates x, by arbitrary conventions and the dependence of the physical quantities, such as current intensity i, on them could not be measured. According to him, the independent variables played the role of some kind of an *a priori* component in the construction of the theory. They were essential for the conceptual structure of building the entire representative, although they were not directly related to the observable quantities. This is supported by the fact that the distance to the electrode or the electrocatalyst, x, is really the base of its geometry, but the current intensity, i, is not the only observable that is defined from x, but the value of v and the concentration, c, are also defined.

In the time concept of the pre-relativistic mechanics, the observable quantities, time t and energy E, have to be considered as another canonically conjugate pair, as in classical mechanics. The dynamic law (time-dependent energy term) of the Schrödinger equation will then completely disappear [19]. A good occasion for Weyl to introduce the relativistic view would have been his contributions to Dirac's electron theory. His other colleagues developed the method of the so-called second quantization that seemed easier for the entire community of physicists and chemists to accept.

Weyl entered the field of quantum mechanics using group theory with a view centered on the fascinating interaction between the orthogonal groups and the permutation groups, which, about the same time, was advanced by Wigner and von Neumann. They established a common theoretical approach to groups in the quantum-mechanical rationalization of atomic spectra, and then, the interpretation of the spectra of higher atomic numbers such as metals became possible.

On the other hand, Weyl emphasized the structural role of group illustration that was established by the knowledge of quantum physics. Even he had appreciated the stochastic character of natural laws well before the birth of quantum mechanics. After Dirac's advance to the relativistic quantum theory with empirical successes, Weyl expected further changes to come. The theoretical methods appeared to him as part of a stable core of the knowledge in quantum mechanics. Weyl gave a complete introduction to the mathematics of the field and included the notions of unitary geometry, that is, the theory of Hilbert spaces and the diagonalization of the Hermitian forms, restricted to the finite dimensional case [20]. His approach also differed from that of quantum mechanics by integrating Schrödinger's view of dynamic rules in the non-relativistic case represented by Hermitian operators and their quantum stochastical interpretation [18].

In summary, on the one hand, classical mechanics was able to presume that the constructive properties were attributes of matter even if the experiments that were necessary for their determination were not accepted. On the other hand, in quantum physics, this was no longer possible due to the limitation of Heisenberg's indeterminacy relation, for any couple and conjugated variables. Weyl accepted it as a fundamental insight, different from Heisenberg's mathematical characterization of the commutation relation. In the case of electrochemistry and electrocatalysis, the fundamentals of

Weyl and Heisenberg's indeterminacy relation can be applied for the "amount of charge," σ, and the mass of the species in consideration, that is, m. After the corresponding quantizations, m to M and σ to $\sum$, we will have

$$M \sum - \sum M = -icl \tag{3.22}$$

3.2 DISTRIBUTION OF ELECTRONS AT ELECTRODE SURFACES

Many electron distributions are associated with the laws governing the density of states (atoms with the same energy), and the most well-known among them is the Fermi distribution. Boltzmann (non-interactive) and Bose–Einstein (attractive) or Fermi–Dirac (repulsive) statistics describe the Fermi distribution for gases. The latter does not give any important probability distinct from zero. In the case of Pauli's exclusion principle, the occupancy of any allowed state is restricted to one or zero electrons. Fermi's law gives the probability of the number of electrons, P, at a given energy state, E, in the system that will be occupied:

$$P = \left(\exp\left[\frac{E - E_f}{kT} \right] + 1 \right)^{-1} \tag{3.23}$$

where E_f is the Fermi energy of the electron in the metal, at which the probability of having the particle is half. The common interpretation is the maximum kinetic energy of the conduction of electrons in the metal at $T \rightarrow 0$.

It is important to calculate the number of electrons in order to evaluate the charge density and the current density through quantum electrochemistry. The number of electrons between the energy E and $E + dE$ can be calculated first considering the density of states, $\rho(E)$, at a given energy E:

$$\rho(E) = \frac{1}{2\pi^2} \left(\frac{2m}{\hbar^2} \right)^{3/2} E^{1/2} \tag{3.24}$$

Thus, the number of electrons will be

$$\rho(E)P(E)dE = \frac{1}{2\pi^2} \left(\frac{2m}{\hbar^2} \right)^{3/2} \frac{E^{1/2}dE}{\exp[(E - E_f)/kT] + 1} \tag{3.25}$$

By integrating the energy E between 0 and ∞,

$$n = \int_0^\infty \rho(E)P(E)dE = \frac{1}{2\pi^2} \left(\frac{2m}{\hbar^2} \right)^{3/2} \int_0^\infty \frac{E^{1/2}dE}{\exp[(E - E_f)/kT] + 1} \tag{3.26}$$

This equation cannot be integrated analytically but the plot of the occupied density of states as a function of energy is of the form presented in Figure 3.1. The value of E_f at $T = 0$ K decreases when the temperature increases.

The energy of the conduction electrons is given by $\hbar^2 k^2/2\mu$, where $\overline{k}$ is the wave vector number and μ is the effective mass of the electron–nucleus. In a real space of Cartesian coordinates $\overline{k} = [k_x, k_y, k_z]$, a *Fermi sphere* can be constructed with radius $k_f = (2\mu E_f)^{1/2}/\hbar$. The shape of this sphere is a clearly defined by the electrical properties of the metal. The current density obeys the change in the occupancy of states near the Fermi level, which separates the unfilled orbitals in the metal from the filled ones in the linear momentum space: $\overline{p} = \hbar \overline{k}$.

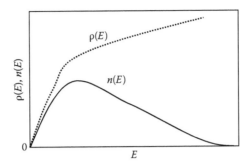

FIGURE 3.1 Density of states, $\rho(E)$, and the number of electrons, $n(E)$, as a function of energy.

The surface energy of a metal can be briefly considered as the contribution of the bulk of the solid and that of the surface. The energy separation between these two surfaces is the surface energy that is strongly affected by reconstruction (crystal rearrangement) and surface relaxation. However, the real interesting situation is that involving the kinetic energy of the valence electrons, E_k:

$$E_k = \sum_i \int_0^{r1} \varphi_i^*(r)\left(-1/2\nabla^2\right)\varphi_i(r)\mathrm{d}^3 r \tag{3.27}$$

where
$1/2\nabla^2$ is the kinetic energy operator
$r1$ is the radius of the useful bulk metal

We also have to consider the electrostatic energy, E_e, as the product of the charge density, $\rho(E)$, and the electric potential, $V_e(r)$, of the valence electrons:

$$E_e = 1/2 \int_0^{r1} \rho(r)V_e(r)\mathrm{d}^3 r - c \int_0^{r1} \left[1 + \frac{c'}{1 + c''[\rho(r)]^{1/3}}\right][\rho(r)]^{1/3}\mathrm{d}^3 r \tag{3.28}$$

and the core energy, E_c, is similarly found with the core electric potential, V_c:

$$E_c = \int_0^{r1} \rho(r)V_c(r)\mathrm{d}^3 r \tag{3.29}$$

The agreement was established for liquid metals, considering the jellium model; in the case of solid or high-density materials, the criteria do not give good results and more corrections for the interaction energies are needed [21].

The theoretical calculation of the surface states is considered by the linear combination atomic orbital (LCAO) method; however, in some cases the relation between the electrode potential and the distance from the electrode surface is used. These electric potentials obey the Schrödinger equation and the surface states are obtained as eigenvalues. Inside the crystal lattice, the equation is reduced to the Mathieu differential equation with known solutions. Using the LCAO method, the interaction between the surface states and the electron is avoided. In this case, the proposed wave function is of the form:

$$\Psi = \sum_r a(r)\varphi(r) \tag{3.30}$$

where

φ(r) are the atomic orbitals
a(r) are the linear, independent combined coefficients

The proper choice of φ(r) is the right clue for obtaining the wave function for a defined metal lattice. In this case, the Bloch functions are the right choice for the atomic orbitals and the energy orbital can be obtained as a mean value:

$$\langle E \rangle = \frac{\int \Psi^* H \Psi d\tau}{\int \Psi^* \Psi d\tau} = \frac{\int \Psi^* H \Psi d\tau}{\int |\Psi|^2 d\tau} \tag{3.31}$$

As commonly performed, the integrals are solved using the variation method by minimizing the energy with respect to each a(r) coefficient and finally replacing them in an entire determinant. The exact solution is not always obtained and a first approximation of the zero contribution for the overlapping integrals leads only to the coulombic and resonance integrals for the nearest neighbors. The difference between the surface and inner coulomb integrals produces the so-called deformation [21]. The solution of these integrals gives the surface states.

3.2.1 Electron Tunneling at an Electrochemical Interface

The Schrödinger equation of the electron can be solved by considering that the metal acts as a potential barrier (taking the electron kinetic energy as less than the potential energy of the barrier). We are going to consider only the one-dimensional case using the Eckart potential energy profile [22]. The expression of the latter can be given as a function of the reaction coordinate, x, but with the rationalization denoted by the barrier width 2L (Figure 3.2).

In spite of the advances in quantum mechanics between 1923 and 1927, the first paper on the quantum mechanical theory of electrode kinetics was presented by Gurney in 1931 [23]. In this

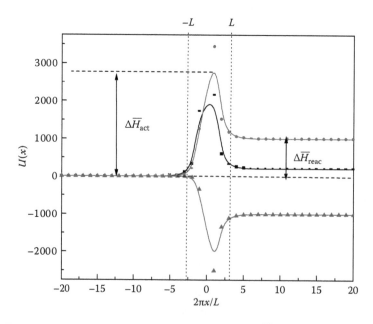

FIGURE 3.2 Eckart potential energy, $U(x)$, for different values of $\Delta \overline{H}_{reac}$ and $\Delta \overline{H}_{act}$ as a function of $\xi = 2\pi x/L$ (∘) $\Delta \overline{H}_{reac} = -1000$, $\Delta \overline{H}_{act} = +2000$; (■) $\Delta \overline{H}_{reac} = -200$, $\Delta \overline{H}_{act} = +2000$; (▲) $\Delta \overline{H}_{reac} = +1000$, $\Delta \overline{H}_{act} = -1000$ arbitrary units.

paper, the tunneling of the electron across an electrochemical interface was considered from the ground states of a metal to an ion in solution. However, the theorization reached scientific circles only during the middle of the twentieth century after Butler's correction of the activation energy barrier calculation by considering the attractive interaction between each particle [24].

We will consider the Eckart potential energy, $U(x)$, to be larger than the total energy of the electron, E:

$$U(x) = \frac{\Delta H_{\text{reac}} e^{2\pi x/L}}{1 + e^{2\pi x/L}} + \frac{\Delta H_{\text{act}} e^{2\pi x/L}}{\left(1 + e^{2\pi x/L}\right)^2} \tag{3.32}$$

Considering the electron of mass m moving in the potential barrier, the potential energy, $U(x)$, is zero for $x < -L$ and $x > L$. Then, outside the potential barrier

$$\frac{-\hbar^2}{2m} \frac{d^2\psi(x)}{dx^2} = E\psi(x) \tag{3.33}$$

The interesting case is that of Schrödinger's equation of an electron inside a potential barrier:

$$\frac{-\hbar^2}{2m} \frac{d^2\psi(x)}{dx^2} = (U(x) - E)\psi(x) \tag{3.34}$$

$$\frac{-\hbar^2}{2m} \frac{d^2\psi(x)}{dx^2} = \left(\frac{\Delta H_{\text{reac}} e^{2\pi x/L}}{1 + e^{2\pi x/L}} + \frac{\Delta H_{\text{act}} e^{2\pi x/L}}{\left(1 + e^{2\pi x/L}\right)^2} - E\right)\psi(x) \tag{3.35}$$

It is well known that the solution in the case of a constant $U(x) = U_{\text{o}}$ is [25,26]

$$\psi(x) = Ae^{kx} + Be^{-kx} \tag{3.36}$$

with

$$k = \sqrt{\frac{2m(U_{\text{o}} - E)}{\hbar^2}} \tag{3.37}$$

where $E < U_{\text{o}}$.

The probability of the electron transmission can be calculated from

$$P = \frac{v_{\text{outside}}(x)\psi_{\text{outside}}(x)}{v_{\text{incid}}(x)\psi_{\text{incid}}(x)} \tag{3.38}$$

where $v(x)$ and $\psi(x)$ are the rate and the wave function of the particle at the incidence side of the potential barrier and outside of it, respectively. This probability of transition is equivalent to, say,

$$P = \frac{|\psi_{\text{outside}}(x)|^2}{|\psi_{\text{incid}}(x)|^2} \tag{3.39}$$

In this way, we will need the wave function outside the barrier on both sides:

$$\psi_{\text{incid}}(x) = Ce^{ik'x} + De^{-ik'x}$$
$$\psi_{\text{transm}}(x) = Ee^{ik'x} + Fe^{-ik'x} \tag{3.40}$$

with

$$k' = \sqrt{\frac{2mE}{\hbar^2}} \tag{3.41}$$

Accordingly, the only component of wave functions (3.40) that represents the transmission refers to the positive exponential terms, that is, the first terms:

$$\psi_{incid}(x) = Ce^{ik'x}$$
$$\psi_{transm}(x) = Ee^{ik'x} \tag{3.42}$$

Therefore,

$$P = \frac{E^*E}{C^*C} = \frac{16E}{U_o}\left(1 - \frac{E}{U_o}\right)e^{-2kL} \tag{3.43}$$

In the case of not having a potential barrier independent of the distance, like in the Eckart potential, some approximations can be proposed. The Wentzel, Kramer, and Brillouin (WKB) approach is a clear example to overcome the problem. If the energy equation inside the potential barrier is

$$\frac{-\hbar^2}{2m}\frac{d^2\psi(x)}{dx^2} = (U(x) - E)\psi(x) \tag{3.44}$$

then the general solution of the differential equation can be written as

$$\psi(x) = C(x)e^{ikS(x)} \tag{3.45}$$

where $S(x)$ has the wave vector number, $\bar{k}(x)$. Substituting in Equation 3.44, we will have

$$C''(x) - C(x)[S'(x)]^2 - \frac{2m}{\hbar^2}(U(x) - E)C(x) + i[2C'(x)S'(x) + C(x)S''(x)] = 0 \tag{3.46}$$

Under both real and imaginary contributions, we have to state

$$C''(x) - C(x)[S'(x)]^2 - \frac{2m}{\hbar^2}(U(x) - E)C(x) = 0$$
$$2C'(x)S'(x) + C(x)S''(x) = 0 \tag{3.47}$$

or

$$C^2(x)S''(x) + 2C(x)C'(x)S'(x) = 0$$
$$\frac{d}{dx}\left[C^2(x)S'(x)\right] = 0 \tag{3.48}$$

Since the differential of a constant is zero, $C^2(x)S'(x)$ has to be constant (defined as C_o):

$$C(x) = \frac{C_o^{1/2}}{[S'(x)]^{1/2}} \tag{3.49}$$

According to this, it is not surprising that $C(x)$ will vary very little with x, since it includes the first derivate of $S(x)$. Thus, it is correct to state that $C''(x) \to 0$. This is the WKB approach.

By using this approximation,

$$-[S'(x)]^2 = \frac{2m}{\hbar^2}(U(x) - E) \tag{3.50}$$

or

$$[S'(x)] = \pm i \int \sqrt{\frac{2m}{\hbar^2}(U(x) - E)}\,dx \tag{3.51}$$

The wave function is under the WKB approximation:

$$\psi(x) = \frac{\sqrt{C_0/i}}{\left(\dfrac{2m}{\hbar^2}(U(x) - E)\right)^{1/4}} \exp\left[\pm \int \sqrt{\frac{2m}{\hbar^2}(U(x) - E)}\,dx\right] \tag{3.52}$$

with $E < U(x)$. It can be demonstrated that this approximation accounts for

$$k(x) \gg \frac{1}{k(x)}\left|\frac{dk(x)}{dx}\right| \tag{3.53}$$

which is equivalent to

$$p(x) \gg \frac{1}{k(x)}\left|\frac{dp(x)}{dx}\right| \quad \text{or} \quad \frac{p(x)}{\Delta p(x)} \gg \frac{\lambda}{2\pi}\left|\frac{1}{\Delta x}\right| \tag{3.54}$$

It is clear that $p(x) \gg \Delta p(x)$, and then $p(x)/\Delta p(x) \gg 1/2\pi$ if $\lambda \cong \Delta x$ (that is, when the particle wavelength is similar to the barrier width).

However, the real and exact solution of the expression is given in the Appendix by using the hypergeometric functions:

$$y(1 - y)\frac{d^2\psi(x, y)}{dy^2} + (1 - 2y)\frac{d\psi(x, y)}{dy} + \frac{2mL^2}{\hbar^2}\left(\frac{E}{y(1 - y)} - \frac{\Delta H_{\text{reac}}}{y} - \Delta H_{\text{act}}\right)\psi(x, y) = 0 \tag{3.55}$$

The solution for ψ is presented by considering that it obeys the product of an asymptotic function, $\Xi(y)$, and its convergential contribution:

$$\psi(x, y) = (z)^{-i\beta}\,[z - 1]^{i\alpha}\,\Xi(y)$$

The general solution is of the form

$$\Xi(y) = C_1 F(a, b, c, y) + C_2 y^{1-c} F(1 + a - c + 1 + b - c + 2 - c - y) \tag{3.56}$$

where
  C_1 and C_2 are arbitrary constants
  F is the hypergeometric function (see Appendix)

The polynomial expression of F obeys $F(a, b, c, 0) = 1$.

We are going to propose a linear combination of hypergeometric functions of the form

$$F(a, b, c, y) = MF(a, b, a + b - c + 1, 1 - y) + (1 - y)^{c-a-b}NF(c - a, c - b, c - a - b + 1, 1 - y) \tag{3.57}$$

where

$$M = \frac{\Gamma(c)\Gamma(c - a - b)}{\Gamma(c - a)\Gamma(c - b)} \quad \text{and} \quad N = \frac{\Gamma(c)\Gamma(a + b - c)}{\Gamma(a)\Gamma(b)} \tag{3.58}$$

where $\Gamma(x)$ is the gamma function, that is, $\Gamma(x) = \int_0^\infty e^{-t}t^{x-1}dt$. The hypergeometric function F is equal to 1 when $x \to \infty$ and $y \to 0$. The problem is the value of the function F when $x \to -\infty$ ($y \to 1$). In this case, the function acquires the same unity value for both sides:

$$F(a, b, c, y) = M + (1 - y)^{c-a-b}N = M + (1 - y)^{-2i\alpha}N \tag{3.59}$$

And in the entire non-asymptotic solution we will have

$$\Xi(y) = My^{i\beta}(1 - y)^{i\alpha} + Ny^{-i\beta}(1 - y)^{-i\alpha} \tag{3.60}$$

Thus, $a_1 = M$ and $a_2 = N$. We can define the transmission probability as the complement of the reflection coefficient, R, defined as

$$P = 1 - R = 1 - \frac{|a_2|^2}{|a_1|^2} = 1 - \frac{|N|^2}{|M|^2} \tag{3.61}$$

Taking (3.52) and substituting into (3.61), we will have

$$P = 1 - \left| \frac{\Gamma(a + b + c)\Gamma(c - b)\Gamma(c - a)}{\Gamma(a)\Gamma(b)\Gamma(c - a - b)} \right|^2 \tag{3.62}$$

But since $a + b - c = 2i\alpha$ and $c - a - b = -2i\alpha$, it is possible to find that

$$\left| \frac{\Gamma(a + b + c)}{\Gamma(c - a - b)} \right|^2 = \left| \frac{\Gamma(0 + 2i\alpha)}{\Gamma(0 - 2i\alpha)} \right|^2 = 1 \tag{3.63}$$

With α and β being real and $\gamma \equiv \frac{1}{2}(4B' - 1)^{1/2}$,

$$P = 1 - \frac{ch\pi(\alpha - \beta + \gamma)ch\pi(\alpha - \beta - \gamma)}{ch\pi(\alpha + \beta + \gamma)ch\pi(\alpha + \beta - \gamma)} \tag{3.64}$$

Otherwise, we can reduce (3.64) using Moivre's formulas:

$$ch(x + y)ch(x - y) = 1/2[ch(2x) + ch(2y)]$$

Thus,

$$P = 1 - \frac{ch[2\pi(\alpha - \beta)] + ch(2\pi\gamma)}{ch[2\pi(\alpha + \beta)] + ch(2\pi\gamma)} = \frac{ch[2\pi(\alpha + \beta)] - ch[2\pi(\alpha - \beta)]}{ch[2\pi(\alpha + \beta)] + ch(2\pi\gamma)} \tag{3.65}$$

The above expression is only applicable when γ is real ($B' > 1/4$), where the wavelength λ is small compared to $2L$. In the case of being imaginary, the method proposed in [30] has to be used.

APPENDIX: HYPERGEOMETRIC FUNCTIONS

3.A.1 General Considerations

At the beginning of the nineteenth century, Gauss published [27] "General discussions about infinite series" where a rigorous study on series is presented with the introduction of hypergeometric functions. The hypergeometric functions are of the form

$$F(a, b, c; x) = 1 + \frac{ab}{c}\frac{x}{1!} + \frac{a(a+1)b(b+1)}{c(c+1)}\frac{x^2}{2!} + \ldots + P(a+n)(b+n)x^n/(c+n)n! \quad (3.A.1)$$

where a, b, and c are parameters of the F function of variable x.

Later, Kummer [28] demonstrated that Gauss' hypergeometric functions are the solution for the hypergeometric differential equations:

$$x(x-1)\frac{d^2 F}{dx^2} + [c - (a+b+1)x]\frac{dF}{dx} - abF = 0 \quad (3.A.2)$$

We can say that the function $F(x)$ is hypergeometric when it can be written as a power series:

$$F(x) = \sum_{n\geq0} a_n x^n \quad (3.A.3)$$

where a_n are the hypergeometric coefficients, obeying the d'Alambert quotient a_{n+1}/a_n as a rational function of n.

Why are these functions important for electrochemistry and electrocatalysis? They are important because almost all the elemental functions can be reduced to hypergeometric relations or can be obtained from them [29].

Most of the hypergeometric functions are of two variables and can be decomposed as a power series of these two variables:

$$F(x, y) = \sum_{m,n\geq0} a_{n,m} x^n y^m \quad (3.A.4)$$

The conditions that have to be fulfilled refer to the ratios $a_{n+1,m}/a_{n,m}$ and $a_{n,m+1}/a_{n,m}$ as being rational functions of n and m, or in other words there exist bivariant polynomials P_1, P_2 and Q_1, Q_2 obeying

$$\frac{a_{n+1,m}}{a_{n,m}} = \frac{P_1(n, m)}{Q_1(n, m)}, \quad \frac{a_{n,m+1}}{a_{n,m}} = \frac{P_2(n, m)}{Q_2(n, m)} \quad (3.A.5)$$

This equation assumes certain compatibilities between each polynomial.

3.B.1 Solution for the Transmission Probability under the Eckart Potential

In the case of the Eckart barrier, when defining $y \equiv -\exp(2\pi x/L)$ for the Hamiltonian expression

$$\frac{d^2\psi(x, y)}{dx^2} + \frac{2m}{\hbar^2}\left(\frac{\Delta H_{\text{reac}}y}{1-y} + \frac{\Delta H_{\text{act}}y}{(1-y)^2} + E\right)\psi(x, y) = 0 \quad (3.B.1)$$

But

$$\frac{d^2\psi(x,y)}{dx^2} = \frac{d^2\psi(x,y)}{dy^2}\left(\frac{dy}{dx}\right)^2 + \frac{d\psi(x,y)}{dx}\left(\frac{d^2y}{dx^2}\right) \tag{3.B.2}$$

and then

$$y^2\frac{d^2\psi(x,y)}{dy^2} + y\frac{d\psi(x,y)}{dy} + \frac{2mL}{\hbar^2}\left(E - \frac{\Delta H_{reac}y}{1-y} + \frac{\Delta H_{act}y}{(1-y)^2}\right)\psi(x,y) = 0 \tag{3.B.3}$$

a. In the case of the regions near $x \rightarrow \pm\infty$, the expression reduces to

$$y^2\frac{d^2\psi(x,y)}{dy^2} + y\frac{d\psi(x,y)}{dy} + \frac{2mL}{\hbar^2}E\psi(x,y) = 0 \tag{3.B.4}$$

The solution of the (3.B.4) is

$$\psi(x) = a_1 e^{ikx} + a_2 e^{-ikx} \tag{3.B.5}$$

with

$$k \equiv \sqrt{\frac{2mE}{\hbar^2}} \tag{3.B.6}$$

The asymptotic form of (3.B.5) can be expressed as a function of y:

$$\psi(x) = a_1 e^{ikx} + a_2 e^{-ikx} = a_1(e^{2\pi x/L})^{ikL/2\pi} + a_2(e^{-2\pi x/L})^{-ikL/2\pi}$$
$$= a_1(-y)^{i\beta} + a_2(-y)^{-i\beta} \tag{3.B.7}$$

with

$$\beta \equiv \frac{kL}{2\pi} \tag{3.B.8}$$

In the case of $x \rightarrow -\infty$, a similar situation accounts:

$$\psi(x) = b_1 e^{ik'x} + b_2 e^{-ik'x} = b_1(-y)^{i\alpha} + b_2(-y)^{-i\alpha} \tag{3.B.9}$$

with

$$\alpha \equiv \frac{k'L}{2\pi} \tag{3.B.10}$$

b. In the case of $-L < x < L$ (inside the potential energy barrier), we propose the solution as a product of two independent functions:

$$\psi(x,y) = \Omega(y)\Xi(y) \tag{3.B.11}$$

where $\Omega(y)$ is the asymptotic behavior of the wave function $\Psi(x,y)$ at $x \to \pm\infty$. One proposed solution is

$$\Omega(y) = (1-y)^{i\beta}\left[\frac{y}{1-y}\right]^{i\alpha} \tag{3.B.12}$$

Now we need to determine $U(y)$, so we prefer to change the variables again:

$$z \equiv 1/(1-y)$$

$$\psi(x,y) = (z)^{-i\beta}[z-1]^{i\alpha}\,\Xi(y) \tag{3.B.13}$$

Considering Equation 3.B.3 by a factor of $(1-y)^{-1}$,

$$y(1-y)\frac{d^2\psi(x,y)}{dy^2} + (1-2y)\frac{d\psi(x,y)}{dy} + \frac{2mL^2}{\hbar^2}\left(\frac{E}{y(1-y)} - \frac{\Delta H_{\text{reac}}}{y} - \Delta H_{\text{act}}\right)\psi(x,y) = 0 \tag{3.B.14}$$

By considering the expression of (3.B.12) and (3.B.13) in (3.B.14),

$$y(1-y)\frac{d^2\Xi(y)}{dy^2} + [1 - 2i\beta - (2 + 2i\alpha - 2i\beta)y]\frac{d\Xi(y)}{dy}$$

$$+ \left[-2\alpha\beta - B' + (1-y)^{-1}(-\alpha^2 y + i\alpha y - i\alpha) + y^{-1}(-a' + i\beta y + \beta^2 y - \beta^2 + \frac{E'}{y(1-y)}\right]\Xi(y) = 0 \tag{3.B.15}$$

where $a' \equiv 2mL^2\Delta H_{\text{reac}}/\hbar^2$, $B' \equiv 2mL^2\Delta H_{\text{act}}/\hbar^2$, $E' \equiv 2mL^2E_t/\hbar^2$ and from the definition of α and β, we have $\alpha^2 = E'$, $\beta^2 = E' - \Delta H_{\text{reac}}$ and $\alpha^2 - \beta^2 = \Delta H_{\text{reac}}$.

Taking these considerations we will obtain

$$y(1-y)\frac{d^2\Xi(y)}{dy^2} + [1 - 2i\beta - (2 + 2i\alpha - 2i\beta)y]\frac{d\Xi(y)}{dy} + [(\alpha-\beta)^2 - B' + i(\beta-\alpha)]\Xi(y) = 0 \tag{3.B.16}$$

If we define $ab \equiv \Delta H_{\text{reac}} - (\alpha-\beta)^2 - i(\alpha-\beta)$, $c \equiv 1 - 2i\beta$ and $a + b + 1 = 2 + 2i\alpha - 2i\beta$, we will reduce (3.B.6) to

$$y(1-y)\frac{d^2\Xi(y)}{dy^2} + [c - (a+b+1)y]\frac{d\Xi(y)}{dy} - ab\Xi(y) = 0 \tag{3.B.17}$$

The general solution is of the form

$$\Xi(y) = C_1 F(a, b, c, y) + C_2 y^{1-c} F(1 + a - c + 1 + b - c + 2 - c - y) \tag{3.B.18}$$

where
  C_1 and C_2 are arbitrary constants
  F is the hypergeometric function

The polynomial expression of F obeys $F(a, b, c, 0) = 1$.

Then, for $y \to 0$ the function reaches the value $\Xi(0) = C_1 + C_2 y^{1-c}$ and then since $z = 1/(1-y)$

$$\Xi(y) = \left[C_1(1-z)^{i\beta} + C_2(1-z)^{-i\beta} \right] \left(\frac{z}{z-1} \right)^{i\alpha} \tag{3.B.19}$$

and

$$\Xi(y) = C_1(-z)^{i\beta} + C_2(-z)^{-i\beta} = \left[C_1 e^{ik'x} + C_2 e^{-ik'x} \right] \tag{3.B.20}$$

Since when $y \to 0$, $z \to -\infty$ and $z/(z-1) \to 1$, $\Xi(y) = F(a, b, c, y)$. We are going to propose a linear combination of the hypergeometric functions of the form:

$$F(a, b, c, y) = MF(a, b, a+b-c+1, 1-y) + (1-y)^{c-a-b} NF(c-a, c-b, c-a-b+1, 1-y) \tag{3.B.21}$$

where

$$M = \frac{\Gamma(c)\Gamma(c-a-b)}{\Gamma(c-a)\Gamma(c-b)} \quad \text{and} \quad N = \frac{\Gamma(c)\Gamma(a+b-c)}{\Gamma(a)\Gamma(b)} \tag{3.B.22}$$

where $\Gamma(x)$ is the gamma function, that is, $\Gamma(x) = \int_0^\infty e^{-t} t^{x-1} dt$ where $\Gamma(x+1) = x!$

The hypergeometric function F is equal to 1 when $x \to \infty$ and $y \to 0$. The problem is the value of the function F when $x \to -\infty$ ($y \to 1$). In this case, the function acquires the same unity value for both sides:

$$F(a, b, c, y) = M + (1-y)^{c-a-b} N = M + (1-y)^{-2i\alpha} N \tag{3.B.23}$$

And in the entire non-asymptotic solution, we will have

$$\Xi(y) = My^{i\beta}(1-y)^{i\alpha} + Ny^{-i\beta}(1-y)^{-i\alpha} \tag{3.B.24}$$

or

$$\Xi(z) = M(-z)^{i\alpha} + N(-z)^{-i\alpha} \tag{3.B.25}$$

Thus, $a_1 = M$ and $a_2 = N$.

We can define the transmission probability as the complement of the reflection coefficient, R, as

$$P = 1 - R = 1 - \frac{|a_2|^2}{|a_1|^2} = 1 - \frac{|N|^2}{|M|^2} \tag{3.B.26}$$

Taking Equation 3.B.22 and substituting into Equation 3.B.26, we will have

$$P = 1 - \left| \frac{\Gamma(a+b+c)\Gamma(c-b)\Gamma(c-a)}{\Gamma(a)\Gamma(b)\Gamma(c-a-b)} \right|^2 \tag{3.B.27}$$

But since $a+b-c = 2i\alpha$ and $c-a-b = -2i\alpha$, it is possible to find that

$$\left| \frac{\Gamma(a+b+c)}{\Gamma(c-a-b)} \right|^2 = \left| \frac{\Gamma(0+2i\alpha)}{\Gamma(0-2i\alpha)} \right|^2 = 1 \tag{3.B.28}$$

Therefore,

$$P = 1 - \left| \frac{\Gamma(c-b)\Gamma(c-a)}{\Gamma(a)\Gamma(b)} \right|^2 \tag{3.B.29}$$

Since $\Gamma(x+1) = x\Gamma(x)$ and $\Gamma(x)\,\Gamma(x-1) = \pi\,csc\,(\pi x)$, it also follows that $\Gamma(1/2 + ix) = 1/2(ch\,\pi x)^{-1/2}$

$$a = 1/2 + i(\alpha - \beta + \gamma)$$
$$b = 1/2 + i(\alpha - \beta - \gamma) \tag{3.B.30}$$

With α and β being real and $\gamma \equiv 1/2(4B' - 1)^{1/2}$,

$$P = 1 - \frac{ch\pi(\alpha - \beta + \gamma)ch\pi(\alpha - \beta - \gamma)}{ch\pi(\alpha + \beta + \gamma)ch\pi(\alpha + \beta - \gamma)} \tag{3.B.31}$$

Otherwise, we can reduce Equation 3.B.31 by using combinations of Moivre's formulas for hyperbolic trigonometry: $ch(x+y)ch(x-y) = 1/2\,[ch(2x) + ch(2y)]$
Thus,

$$P = 1 - \frac{ch[2\pi(\alpha - \beta)] + ch(2\pi\gamma)}{ch[2\pi(\alpha + \beta)] + ch(2\pi\gamma)} = \frac{ch[2\pi(\alpha + \beta)] - ch[2\pi(\alpha - \beta)]}{ch[2\pi(\alpha + \beta)] + ch(2\pi\gamma)} \tag{3.B.32}$$

The above expression is only applicable when γ is real ($B' > 1/4$), where the wavelength λ is small compared to $2L$. In the case of γ being imaginary, the method proposed in [30] must be used.

REFERENCES

1. H. Weyl, *The Theory of Groups and Quantum Mechanics*, Dutten, New York (1931).
2. E. Wigner, *Gruppentheorie und ihre Anwendung auf die Quantenmechanik der Atomspektren*, Vieweg, Wiesbaden, Germany (1931).
3. B.L. van der Waerden, *Die gruppentheoretische Methode in der Quantenmechanik*, Springer, Berlin, Germany (1932).
4. T. Hawkins, *Emergence of the Theory of Lie Groups. An Essay in the History of Mathematics 1869–1926*, Springer-Verlag, Berlin, Germany (2000).
5. J. Mehra and H. Rechenberg, *The Historical Development of Quantum Theory, Vol. VI: The Completion of Quantum Mechanics 1926–1941*, 2 parts, Chaps. III.4, III.5, Springer, Berlin, Germany (2000/2001).
6. W. Pauli, *Z. Phys.* **43** (1927) 601.
7. P.A.M. Dirac, *Proc. R. Soc. Lond. Ser. A* **112** (1926) 661.
8. W. Heisenberg, *Z. Phys.* **38** (1926) 411.
9. W. Heisenberg, *Z. Phys.* **41** (1927) 239.
10. E. Wigner and J. von Neumann, *Z. Phys.* **49** (1928) 73.
11. E. Wigner, *Z. Phys.* **40** (1926) 492.
12. R.T. Elliot and A.F. Gibson, *Solid State Physics*, Macmillan, New York (1974).
13. C.G. Kuper and G.D. Whitfield (Eds.), *Polarons and Excitons*, Plenum Press, New York (1963).
14. W. Heitler and F. London, *Z. Phys.* **44** (1927) 455.
15. W. Heitler, *Z. Phys.* **47** (1928) 835.
16. W. Heitler, *Z. Phys.* **51** (1928) 805.
17. W. Heisenberg, *Z. Phys.* **49** (1928) 619.
18. H. Weyl, *Mathem. Z.* **24** (1926) 789.
19. H. Weyl, *Z. Phys.* **46** (1927) 1.

20. H. Weyl, *Gruppentheorie und Quantenmechanik*, Hirzel, Leipzig, Germany (1928), Second German Edition (1931).
21. N.D. Jang and W. Kohn, *Phys. Rev.* **3** (1971) 1215.
22. C. Eckart, *Phys. Rev.* **35** (1930) 1303.
23. R. Gurney, *Proc. R. Soc.* **A134** (1931) 137.
24. J.A.V. Butler, *Proc. R. Soc.* **A125** (1936) 423.
25. R.M. Eisberg, *Fundamentals of Modern Physics*, John Wiley & Sons, New York (1967).
26. J.O'M. Bockris and S.U.M. Khan, *Quantum Electrochemistry*, Plenum Press, New York, London (1979).
27. J.C.F. Gauss, General discussions about infinite series, *Soc. Regiae Sci.* Gött (1812). Comment.
28. E.E. Kummer, Über die hypergeometrische Rehie F(a; b; c; x), *J. Reine Angew. Math.* **15** (1836) 39.
29. L.J. Slater, *Generalized Hypergeometric Functions*, Cambridge University Press, Cambridge, U.K. (1966).
30. N. Neilson, *Handbuch der Theorie der Gammafunction*, Teubner, Leipzig, Germany (1906).

4 Overview of Semiempirical Methods

C. Fernando Zinola

CONTENTS

4.1 GENERAL VIEW

Semiempirical molecular orbital (SEMO) methods have been used widely in computational studies [1,2]. Various reviews [3–6] describe the underlying theory, the different variations of SEMO methods, and their numerical results. Semiempirical approaches normally originate within the same conceptual framework as *ab initio* methods, but they overlook minor integrals to increase the speed of the calculations. The mistakes arising from them are compensated by empirical parameters that are introduced into the outstanding integrals and standardized against reliable experimental or theoretical reference data. This approach is successful if the semiempirical model keeps the essential physics and chemistry that describe the behavior of the process.

In this case, parameterization can account for all other effects in an average sense, and it is then a matter of validation to establish the numerical accuracy of a given approach.

Quantum derivations for SEMO methods are defined according to Thiel [7] as follows:

1. Most SEMO methods are based on molecular theory and occupy a reduced basis set for only valence electrons. Electron correlation is treated openly only if this is required for a suitable zero-order sketch.
2. Conventionally, there are three levels of integral approaches: CNDO (complete neglect of differential overlap), INDO (intermediate neglect of differential overlap), and NDDO (neglect of diatomic differential overlap). The last is the best of the approximations since it keeps the superior multipoles of charge distributions in the two-center interactions (unlike the others that shorten after the monopole).
3. At a known stage of the preliminary integral calculation, the integrals are either determined directly from experimental data by calculating from each analytical expression or computed from proper parametric terminology. The first choice is commonly viable for one-center integrals that result from spectroscopic data. The choice between the second option and the

third option is predisposed by the simplicity of the analytical formulas but mostly depends on an estimation of the essential interactions.

4. SEMO methods are parameterized to replicate the experimental literature data. The reference properties are best selected such that they are representative for the intended applications. The excellent semiempirical results obtained are strongly predisposed toward new attempts at parameterization.

4.1.1 CLASSIFICATION

A large number of methods with and singular acronyms have been reported, including CNDO/2 [8], INDO [9], MINDO/3 [10], SINDO1 [11], MNDO [12], AM1 [13], PM3 [14], MNDO/d [15], PM3/tm [16], and NDDO-G [17].

The most useful method for background studies is based on the MNDO model [18], which is a valence-electron self-consistent-field (SCF) MO treatment. It takes up a minimal basis of atomic orbitals (AOs) and the NDDO integral estimation. The molecular orbitals, ψ_i, and the corresponding orbital energies, ε_i, are obtained from the linear combination of the AO base functions, ϕ_u, and the solution of the secular equations with S_{uv}:

$$\psi_i = \sum_u c_{ui} \phi_u \tag{4.1}$$

with

$$\sum_v (F_{uv} - S_{uv}\varepsilon_i)c_{vi} \tag{4.2}$$

where $S_{uv} = \delta_{uv}$.

After assigning the subscripts u and v to an AO and the superscript A or B to an atom, the NDDO Fock matrix elements, F_{uv}, are given for the symmetric and antisymmetric elements:

$$F_{uv}^{AA} = H_{uv}^{AA} + \sum_{w,A}\sum_{x,A} P_{wx}^{AA}\left[\left((uv)^{AA}, (wx)^{AA}\right) - \frac{1}{2}\left((uw)^{AA}, (vx)^{AA}\right)\right]$$
$$+ \sum_B \sum_{w,B}\sum_{x,B} P_{wx}^{BB}\left((uv)^{AA}, (wx)^{BB}\right) \tag{4.3}$$

and

$$F_{uv}^{AB} = H_{uv}^{AB} - \frac{1}{2}\sum_{w,A}\sum_{x,B} P_{wx}^{AB}\left((uw)^{AA}, (vx)^{BB}\right) \tag{4.4}$$

where H_{uv} and P_{uv} are elements of the one-electron core Hamiltonian and the density matrices, respectively, and $((uv), (wx))$ denotes a two-electron integral.

The total energy, E_{tot}, of a molecule is the sum of its electronic energy, E_{el}, and the repulsions, E_{core}^{AB}, between the cores of the addition of A and B atoms:

$$E_{tot} = E_{el} + \sum_A \sum_B E_{core}^{AB} \tag{4.5}$$

with

$$E_{el} = \frac{1}{2}\sum_u \sum_v P_{uv}(H_{uv} + F_{uv}) \tag{4.6}$$

The MNDO model includes only single-center and two-center terms that account for much of its computational competency. New implementations of the MNDO model (e.g., in the MNDO, AM1, and PM3 methods) are rather similar: Conceptually, the single-center terms are taken from atomic spectroscopic data, with the modification that minor adjustments are allowed in the optimization to account for possible variations between free atoms and atoms inside a molecule. The one-center two-electron integrals resulting from atomic spectroscopic data are significantly less important than their analytically calculated values, which are attributed to a normal assimilation of electron correlation effects. For internal uniformity reasons, these integrals provide the one-center limit of the two-center two-electron integrals $((uv)^{AA}, (wx)^{BB})$, whereas its asymptotic limit is determined by classical electrostatics. The semiempirical calculation for $((uw)^{AA}, (vx)^{BB})$ conforms to these limits and evaluates these integrals from semiempirical multipole–multipole interactions. The relevant multipoles are represented by suitable point-charge configurations whose interaction is damped according to the Klopman–Ohno approximation formula. Therefore, at intermediate distances, the semiempirical two-electron integrals are smaller than their analytical counterparts, which again reflects some inclusion of electron correlation effects. Aiming for a reasonable balance between electrostatic attractions and repulsions within a molecule, the core–electron attractions and the core–core repulsions are treated in terms of the corresponding two-electron integrals, neglecting penetration effects. The additional effective atom-pair potential that is included in the core–core repulsions (with an essentially exponential repulsion in MNDO and a more flexible parameteric function in AM1 and PM3) attempts to compensate for errors introduced by the above assumptions but mainly represents the Pauli exchange repulsions. Covalent bonding arises from the two-center one-electron integrals, H_{uv}^{AB} (resonance integrals), that are often taken to be proportional to the corresponding overlap integrals.

4.2   EXTENDED HÜCKEL METHOD

This is a semiempirical all-valence electron quantum mechanical method, apart from the π-approximation and the neglect of overlap integrals, as those of Hückel molecular orbital (HMO) theory. The method reproduces, relatively well, the shapes and the order of the energy levels of molecular orbitals. To consider the overlapping, it is possible to describe the net destabilization caused by the interaction of the two doubly occupied orbitals, the effect of which is not reproduced by HMO theory.

The confluence of surface science and electrochemistry with the molecular orbital theory is bringing a better understanding of the molecular processes at electrode surfaces. One of the simplest methods used in theoretical electrochemistry is the extended Hückel approach [19]. The methodology is the advanced application of the atom superposition and electron delocalization molecular orbital method, where the molecule's energy is calculated by the addition of a pair-wise repulsive atom superposition and the electron delocalization bond formation terms [20]. However, the determination of accurate electrode surfaces has not been possible using this method because it fails to predict reasonably the geometries for large systems. According to Anderson [21], the extended Hückel method (EHMO) orbitals include too much information about the structure and energy levels than that about energy surfaces. By adding a two-body repulsive energy to the EHMO energy, good predictions have been obtained for bond lengths and force constants in diatomic molecules [22].

The surface is first considered as a composition of N rigid atoms. By Hellmann–Feynman theorem, the repulsive energy, E_R, is a pair-wise repulsive energy, $E_{\mu\eta}$:

$$E_R = \sum_{\mu<\eta}^{N} E_{\mu\nu} = \sum_{\mu<\eta} -Z_{\eta} \int \frac{\rho_{\mu}(\vec{r})}{|R_{\eta} - \vec{r}|} d\vec{r} \tag{4.7}$$

where

$\rho_\mu(\vec{r})$ is the electronic and nuclear charge density of the μ-atom

Z_η is the η-atom nucleus charge, being the electronegativity of the μ-atom larger than or equal to η

In this Equation 4.7, the charge density of the more electronegative atom is used.

The single-electron Hamiltonian can be expressed as

$$H_i = \left[-\frac{1}{2}\nabla_i^2 + \sum_\mu^N V_\mu(r_\mu) + E_R \right] \tag{4.8}$$

where

$$V_\mu(r_\mu) = \frac{-Z_\mu}{r_\mu} + \sum_j \left[2\left\langle \left| \Psi_j^\mu(2) \right| \frac{1}{r_{12}} \left| \right| \Psi_j^\mu(2) \right\rangle - \left\langle \left| \Psi_j^\mu(2) \right| \frac{1}{r_{12}} \left| \right| \Psi_i^\mu(2) \right\rangle \frac{\Psi_j^\mu}{\Psi_i^\mu} \right] \tag{4.9}$$

As usual Ψ_i^μ is the eigenfunction of the Fock operator $-\frac{1}{2}\nabla^2 + V_\mu(r_\mu)$ that is added all over the filled orbitals with eigenvalues, ε_i^μ. In the case of open-shell atoms, the electrostatic potential has a different form. To simplify the treatment, we have to consider that the off-diagonal matrix elements on the same atom are zero in the secular equation. A typical diagonal matrix element, H_{ii}, for the μ-atom is

$$H_{ii}^\mu = \varepsilon_i^\mu + \sum_{\mu \neq \eta}^N \left\langle \left| \Psi_j^\mu \right| V_\mu(r_\mu) \left| \right| \Psi_i^\mu \right\rangle + E_R \cong \varepsilon_i^\mu + E_R \tag{4.10}$$

Usually, the terms in the sum are neglected since they represent a small fraction of the electron affinity of each η-atom. The off-diagonal matrix element, $H_{ii}^{\mu\eta}$, has the form:

$$H_{ii}^{\mu\eta} = \varepsilon_i^\mu S_{ij}^{\mu\eta} + \left\langle \left| \Psi_j^\eta \right| V_\eta(r_\eta) \left| \right| \Psi_i^\eta \right\rangle + \sum_{\nu \neq \eta,\mu}^N \left\langle \left| \Psi_i^\mu \right| V_\nu(r_\nu) \left| \right| \Psi_j^\eta \right\rangle + E_R S_{ij}^{\mu\eta} \tag{4.11}$$

Or when written in a more condensed form

$$H_{ii}^{\mu\eta} \cong \frac{1}{2}K\left(\varepsilon_i^\mu + \varepsilon_j^\eta \right) S_{ij}^{\mu\eta} + E_R S_{ij}^{\mu\eta} \tag{4.12}$$

This expression is the contribution of both atoms with $K = 1 + 2/(2 - n)$, which is equal to 1.75–2.25. The real EHMO formalism [23] is a constant value of E_R. Thus, the integral: $\left\langle \left| \Psi_i^\mu \right| V_\nu(r_\nu) \left| \right| \Psi_j^\eta \right\rangle$ is approached by an $\left(\vec{r}_\mu^{-\eta} \right)$ expression of the atomic potential $V_\nu(r_\nu)$ by expanding Ψ_i^μ over the ν-atom by using the Virial theorem. Therefore, it is possible to demonstrate that $\left\langle \left| \Psi_i^\mu \right| V_\nu(r_\nu) \left| \right| \Psi_j^\eta \right\rangle = 2\left(\frac{\varepsilon_j^\eta}{2 - n} \right) S_{ij}^{\mu\eta}$.

Since the electron interactions in the overlap regions are omitted, the predicted binding energies are not precise, and we need to compare the surface or molecule geometries. By considering $\frac{K}{2}\left(\varepsilon_i^\mu + \varepsilon_j^\eta \right) S_{ij}^{\mu\eta} \exp(-\delta R)$, the dissociation energies are better approached. This effect is not observed for the ionic bonds as the attractive electrostatic energy is neglected. However, the binding energies are also improved when the inner electrons are incorporated with the inclusion of the attractive interaction energies. It is noteworthy that a charge transfer process has to be modeled as in the case of electrocatalysis. The energy levels have to be adjusted [24] by the iteration from the H_{ii} term, valence-state ionization potential energy until a fixed or experimentally proved situation can be mimicked. The single-electron procedure contains information for finding good estimates of the structural, vibration, and electronic properties of the molecule adsorbed at the electrode surface. The predictions of the equilibrium distances and planar and dihedral angles are similar to EHMO calculations without an E_R.

In some cases, a distance-dependent exponential factor, K_{AB}, different from that mentioned above, is also included to correct the off-diagonal EHMO matrix elements and is calculated by the empirical weighted Wolfsberg–Helmholz formula. It considers the Hamiltonian off-diagonal elements, H_{ij}, by the following expression:

$$H_{ij}^{\mu\eta} = \frac{1}{2} K_{AB} \left(H_{ii}^{\mu\mu} + H_{jj}^{\eta\eta} \right) S_{ij}^{\mu\eta} \qquad (4.13)$$

where
μ and η are the μ-th and the ν-th orbitals of the atoms A and B, respectively
$S_{ij}^{\mu\eta}$ are the elements of the overlap matrix

K_{AB} is expressed by two adjustable empirical parameters from the following formula:

$$K_{AB} = 1 + \kappa \exp[-\delta(R_{AB} - \vec{r}_o)] \qquad (4.14)$$

where κ varies as $0.4 < \kappa < 1.5$ and δ differs with $0.0 < \delta < 0.1$ nm^{-1} with r_o being the addition of the atomic radii of both atoms.

The other energy matrix elements, $H_{ii}^{\mu\mu}$ and $H_{jj}^{\eta\eta}$, are kept as in the conventional EHMO methodology, that is, the opposite value of the valence-state ionization potential (VSIP).

4.2.1 Some Examples of Electrocatalytic Reactions Based on EHMO Calculations

4.2.1.1 Interaction of a Single Benzoate Molecule with a Fe(111) or a Fe(100) Cluster Surface

Fe(111) and Fe(100) single crystals were simulated using geometric clusters, and the adsorption characteristics of the neutral benzoate species on these clusters are predicted by EHMO calculations. Since the EHMO methodology is rather limited, some improvements were made before [25,26]. This improved EHMO procedure was employed to analyze the adsorption of a single benzoate molecule on either a Fe(111) or a Fe(100) cluster surface.

Iron single crystals are simulated by constructing superimposed bilayer Fe$_N$ geometric clusters of $N = 25$ and 32 to model the Fe(111) or the Fe(100) surface, respectively (Figure 4.1). Clusters were geometrically built up, keeping the Fe—Fe bond length constant at 0.248 nm [27].

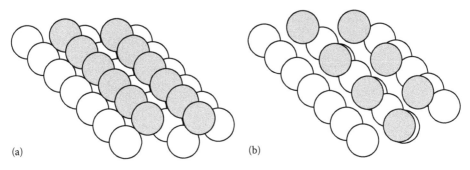

(a)   (b)

FIGURE 4.1   (a) Fe(100) ($N = 32$) and (b) Fe(111) ($N = 25$) clusters showing the first (white balls) and the second (gray balls) atom layers.

Different adsorption configurations resulting from the occupation of the various iron adsorption sites on the surface were considered, that is, the single interaction of the carboxylate group by changing its hybridization to a sp^3 carbon atom (Compound A), the aromatic and carboxylate interaction of the planar molecule parallel to the surface (Compound B), linear on-top (onefold), bridge (twofold), and hollow (higher coordination sites) configurations. In all of them, the complete aromatic ring is kept as an entire entity. Hollow sites are associated with a fivefold adsorbate coordination on Fe(100) (four iron atoms of the topmost layer and one iron atom from the underlying layer). In the case of Fe(111), either three or four iron atoms may define the hollow coordination, depending on the local symmetry of the adsorbate, that is, (3-1) or (3-3) hollow sites, respectively.

The equilibrium VSIP values were evaluated by the method proposed by Anderson and Hoffmann [22]. Benzoate adsorption on iron can involve Fe—C interactions through the carbon atoms of the >C=C< (from the aromatic ring) or >C=O moieties and the resulting geometries imply a different polarization of the surface. However, in each case, the original VSIP and Slater orbital exponents define the open-circuit potential of the adsorbed ensembles, and all of them are compared through a parameterization based on the Fe—C bond. The open-circuit potential can be correlated to the experimental open-circuit value of the interface, that is, the electrode potential that results from the interaction between benzoate and iron.

The open-shell configuration of the iron cluster surface has been considered in the definition of the spin magnetic moment of the $Fe_N \cdot C_7H_5O_2H$-adsorbed ensembles [20]. The neutral benzoate binding energies (BE) were calculated according to the following equation:

$$BE = E_{T,Fe_N C_7H_5O_2H} - E_{T,Fe_N} - E_{C_7H_5O_2H} \qquad (4.15)$$

where

$E_{T,Fe_N C_7H_5O_2H}$ and E_{T,Fe_N} are the total energies of the $[Fe_N \cdot C_7H_5O_2H]$ and $[Fe_N]$ clusters
$E_{C_7H_5O_2H}$ is obtained from the energy of the free benzoate molecule

The geometries of the adsorbed ensembles were optimized to the minimum energy (Figure 4.2). This implies the simultaneous change of C—C, C—O, C=O, O—H, and Fe—C bond lengths together with r_{CC}, r_{CO}, $r_{C=O}$, r_{OH}, and r_{FeC}, respectively; and Fe–C=C, O—C—C, and C—C (O)—C planar angles, $\alpha_{Fe-C=C}$, α_{O-C-C}, and $\alpha_{C-C(O)-C}$, respectively. No dihedral angles were optimized to avoid the distortion and oxidative disruption of the molecule.

The EHMO calculations conducted for the free benzoate molecule lead to the following equilibrium bond lengths: $r_{CC} = 0.142$ nm, $r_{C=C} = 0.126$ nm, $r_{C=O} = 0.117$ nm, $r_{CO} = 0.126$ nm,

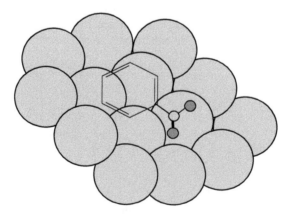

FIGURE 4.2 Benzoate adsorption configuration via the carboxylate group on a Fe(111) cluster. The white balls are the Fe atoms, the middle-dark gray ball is the C atom, and the dark gray balls are the O atoms. The H atoms are omitted for simplicity. The wide bars represent the double bonds and the narrow bars the single bonds.

$r_{OH} = 0.104$ nm, and $r_{CH} = 0.108$ nm. Besides, the planar angles are nearly those expected for a conjugated molecule, that is, $\alpha_{H-C=C} = 120°$ and $\alpha_{C-C(O)-O} = 117°$.

From the analysis of the charge populations on the different atoms of the metal and the adsorbate, we can conclude that the interaction of benzoate with an iron cluster involves the donation and back donation of electronic charge from the $>C=C<$ and $>C=O$ functional groups of the molecule to the metal and vice versa. The donation takes place via an electron transfer from the π orbitals of the adsorbate to the metal unoccupied d orbitals, whereas the back donation populates the π^* orbitals of the adsorbate with the electrons from the occupied metal orbitals. Both interactions and electron transfers are attractive processes and the contribution of the repulsion is only the result of the interaction between the occupied orbitals of the adsorbate and the metal.

The adsorption of the flat molecule parallel to the surface by the simultaneous interaction of $>C=C<$, $>C-C<$, and $>C=O$ groups involves at least five surface atoms (Figure 4.3). This type of adsorbate is interesting for the explanation of the high surface coverage seen for this molecule on our experiments.

The adsorption of the molecule through the carboxylate group only leads to the opening of the double carbonyl bond and to the formation of a sp^3 carbon atom. The binding of the new $>C<^{O^-}$ group to the surface involves two Fe atoms yielding a different structure, which is depicted in Figure 4.3.

Besides, three more coordinations have also been checked for the benzoate adsorption by a single interaction of the $>C=C<$ moiety with the Fe(111) or the Fe(100) cluster surface. The di-σ configuration involves the interaction of three iron atoms by σ bonds with the aromatic ring and one oxygen atom (Figure 4.4a). As a consequence of the sp^2 character of the molecule, a μ-bridging

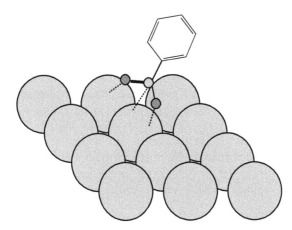

FIGURE 4.3 Flat adsorption configuration of the benzoate species on a Fe(111) cluster. The white balls are the Fe atoms, the middle-dark gray ball is the C atom, and the dark gray balls are the O atoms. The H atoms are omitted for simplicity. The wide bars symbolize the double bonds and the narrow bars the single bonds.

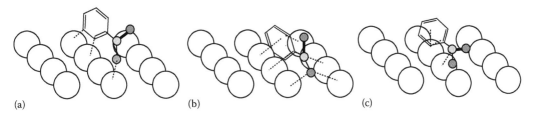

(a)   (b)   (c)

FIGURE 4.4 Adsorption modes of the benzoate species on the Fe clusters: Di-σ (a), μ-bridging (b), and π- (c) modes. White balls are the first layer of the Fe atoms, the middle-dark gray ball is the C atom, and the dark gray balls are the O atoms. The H atoms are not drawn.

configuration is likely where each carbon atom can bind to two iron atoms in a hollow position through π and σ bonds (Figure 4.4b). In the case of the π configuration (Figure 4.4c), the single interaction of the π bond of the aromatic moiety with one iron atom and one oxygen atom attached to another iron ensues. The donation and back-donation processes produce the weakening of the C—C bond strength (the one between the aromatic and carboxylate moieties), increasing the corresponding bond length and the simultaneous formation of Fe—C=O bonds [28].

The main difference among π, di-σ, and μ-bridging configurations arises from the existence of C=C and C—C bond lengths in the aromatic ring that can interact through σ and π orbitals with four iron atoms together with the carboxylate moiety with two more metal atoms. However, in all the three configurations, the carbonyl species has the lower interaction energy and screens only the adjacent iron atom.

For the first three configuration modes, larger BE values on Fe(100) cluster surfaces are observed; however, greater stability is achieved for the π configuration on both cluster surfaces. Thus, the >C=C< bond length in di-σ and μ-bridging configurations almost corresponds to the bond length of a saturated molecule, whereas in the π configuration, it is very near to an undistorted ethylene molecule. On the other hand, the >C=O bond length is much larger than expected on the two cluster surfaces.

The charges involved on the carbon, oxygen, and hydrogen atoms after the benzoate adsorption indicate an electron transfer during the adsorption on both iron clusters that originated from a back-donation process. All electrons from the back-donation process are located on the carbon atoms of the aromatic moiety. For example, the electrons donated from the π orbitals of the μ-bridging configuration to the iron atoms is half of the amount of the back-donated electrons to the π^* from the metal. In the case of the di-σ species, the contribution of the π^* back donation is only approximately 30% lower than the electron donation to the π orbitals, but both processes are stronger than those found on the μ-bridging configuration. Molecular orbital analysis shows that p_x and p_z orbitals (on the plane of the molecule) of the benzoate molecule are involved in the bonding interaction with d_{xz} and $d_{x^2-y^2}$ atomic-platinum orbitals. The larger Fe–C bond length in the μ-bridging configuration produces a lower net charge on the involved iron atoms than that produced on the other configurations.

The interaction of the carboxylate moiety lying normal to the surface with the iron atoms, that is, the interaction of >C=O with two iron atoms, yields another adsorbate as depicted in Figure 4.4. This adsorbate screens less iron surface atoms than that of the flat configuration, resulting in the lowest BE. However, the difference in the BE values between Fe(111) and Fe(100) is smaller than that with adsorbates involving the two combined interactions of the aromatic and carboxylate moieties. This is the consequence of the lower coordination number of this type of adsorbate.

4.2.1.2   Adsorption and Electroreduction of Oxygen at Platinum Electrodes

4.2.1.2.1   Generalities about the Electroreduction of Oxygen on Platinum

The molecular oxygen electroreduction reaction on platinum in acid solutions is one of the most relevant electrocatalytic reactions. This reaction has a complex mechanism that involves first the adsorption of molecular oxygen followed by a charge transfer step and the desorption of the electro-reduced adsorbates and finally leads to water as the main product and to hydrogen peroxide as the minor product. Electrode kinetic data have shown different current density vs. electrode potential profiles, depending on the platinum topography in both acid and alkaline solutions. The oxygen kinetics can be interpreted by considering that there is a competition between the four-electron electroreduction to water and the two-electron electroreduction to hydrogen peroxide, which can further decompose to water. The relative contribution of each reaction depends on the platinum-surface crystallography, the electrode potential, and the nature and concentration of the anions in the solution.

The formation of hydrogen peroxide in aqueous solutions is almost twice as much for Pt(100) than for Pt(111) and the polycrystalline platinum electrodes. The difference indicates either a more strongly adsorbed peroxide species or faster hydrogen peroxide decomposition on the two surfaces. In addition, the superior contribution of peroxo-intermediates to the oxygen reduction kinetics on Pt(100) can explain the greater surface coverages found in the oxygen-saturated acid solution. The difference in the electronic characteristics of the adsorption sites, resulting from the different crystallographic structures, leads to a topology dependence of the heterogeneous catalytic reaction kinetics.

Theoretical calculations are mainly related to lower-atomic-number transition metals. The interaction of an oxygen molecule with Pt(100) and Pt(111) cluster surfaces is analyzed at a molecular orbital level, as the first step to understand the influence of the electrode topology on the OERR mechanism. The selective generation of the peroxo-species on Pt(100) can be justified on the basis of the involved molecular orbital interactions and the different nature of the adsorbed intermediates on both surfaces.

4.2.1.2.2 On the Calculation Procedure

For the treatment of large polyatomic systems, computational methodologies deal with a compromise between an overall description of the entire system and a more detailed handling of a properly selected part of it. This situation particularly applies to the transition metal structures that have to be drastically minimized for an adequate *ab initio*, local density functional, or even semiempirical calculation at a good correlation level. In contrast to this simplification of the system, the improvements of the simpler methods, which are capable of handling the system as a whole, have regained acceptability. This is the case of the EHMO method developed by Hoffman [19], which was initially used for a reasonable description of the structural and electronic properties of the systems at a frozen geometry. Improvements of this method are mainly related to the addition of the (two-body electrostatic correction) term as explained above [20,21].

Values of the VSIP are experimentally based and theoretically considered as the opposite of the diagonal Hamiltonian matrix elements. Valence orbitals are of the Slater form for simplicity. The oxygen molecule–platinum site interactions are represented as an adsorbed ensemble that is characterized by its specific VSIP.

According to Pauling's ionicity relationship, the VSIP value that defines the equilibrium potential of the system—the zero applied potential condition—results when the charge transfer at the equilibrium distance of each internuclear bond is close to that predicted for the adsorbed ensemble from the electronegativity difference. The VSIP adjustment to the charge transfer conditions was made for different Wolfsberg–Helmholz parameters. Since the positive applied electric potential shifts the metal energy Fermi level downward and vice versa, changes in the electrode potential were simulated by either decreasing or increasing the absolute value of the metal VSIP from the reference equilibrium value, for negative or positive charging, respectively. The change in the value of the VSIP with the applied potential has to be considered using experimental evidences such as oxidative or reductive decomposition of the molecule with the electrode potential.

The open-shell configuration of the oxygen molecule $^3\Sigma_g^-$ has been considered in the definition of the spin magnetic moment of the $[Pt_N–O_2]$-adsorbed ensemble according to the d-rule. Different types of adsorption sites can be defined on platinum single crystals, namely, on-top (onefold), bridge (twofold), and hollow (higher coordinated sites). Hollow sites are associated with a fivefold coordination of an oxygen atom on Pt(100) (four platinum atoms of the topmost layer and one Pt atom from the underlying layer). Otherwise, either three or four Pt atoms may define the hollow coordination in Pt(111) depending on the *fcc* ((3-1) hollow site) and the *hcp* ((3-3) hollow site) local symmetries. The interaction of an oxygen molecule with these sites defines several adsorption configurations. The interaction of a single molecule with a platinum surface was initially considered, but the electroreduction reaction in aqueous environments also involves water and other coadsorbed intermediates on the platinum surface. In the 0.7–1.0 V potential range, oxygen adsorption takes place with an important contribution of the adsorbed OH species. Therefore, the stability of the adsorbed ensemble, constituted by the molecular oxygen, hydroxyl, and platinum sites was evaluated as a function of the applied potential.

4.2.1.2.3   On the Oxygen Adsorption and Electroreduction

Because of the important role of the oxygen frontier orbitals of the type π and π^* in the interaction with the platinum surface, adsorption geometries with the oxygen–oxygen interatomic bond parallel to the platinum surface plane are likely to occur. However, the "parallel" (side-on) and the "perpendicular" (end-on) configurations were compared in detail in [29]. Among the possible configurations, the most important are those that involve either the coordination of one oxygen atom to a single surface site (onefold, bridge and hollow coordination) or the simultaneous coordination of both oxygen atoms to a unique surface site (on-top, side-on, or bridge side-on configuration). Table 4.1 shows the main results with the optimized binding energies and geometric characteristics.

For the different adsorbate configurations, σ and π oxygen orbitals are involved, their relative weight depending on the specific geometry of the adsorption site. For the linear on-top end-on coordination, the interaction of the π molecular orbital with the (d_{xz}, d_{yz}) platinum atomic orbital is more important than that of the σ oxygen d_{z^2} platinum, with the z-axis being normal to the platinum surface. The contribution of the π^* oxygen molecular orbital to the stability of the d_{xy} platinum atomic orbitals is negligible. The net interaction can be described as a charge transfer to the platinum surface which, being similar for both Pt(111) and Pt(100), shows the local character of the linear bond. Otherwise, the bridge end-on coordination implies a greater stabilization of the σ oxygen molecular orbital, through a bonding interaction with the adjacent platinum atom $d_{x^2-y^2}$ orbitals, rather than a π-type interaction that involves a coordination of the π orbitals from both platinum and oxygen. However, the more important π interaction on Pt(100) than that on Pt(111) appears to be responsible for the greater stability of the bridge end-on coordination geometry in Pt(100), and the π orbitals are not really hybridized for a further contribution to Pt—Pt bonding. Both π and σ interactions imply a charge transfer to platinum, as can be inferred from the decrease in Mülliken populations in the oxygen atoms. These populations were defined according to the x-, y-, and z-axis contributions to the π and σ orbitals. The greater O—O bond distance (Table 4.1) for the oxygen molecule adsorbed on Pt(100) agrees with the larger π orbital Mülliken population, showing the relevance of the antibonding interactions.

TABLE 4.1

Molecular Oxygen Binding Energies, BE, and Optimized Pt—O and O—O Distances ($R_{Pt—O}$ and $R_{O—O}$), and Perpendicular Distance from the O Atom Closer to the Plane of the Platinum Cluster (D) for Different $[Pt(111)]_{18}O_2$ and $[Pt(100)]_{18}O_2$ Configurations at Equilibrium Potential

Configuration	BE (eV)	$R_{Pt—O}$ (nm)	$R_{O—O}$ (nm)	D (nm)
Pt(111)				
On-top end-on	−1.9213	0.189	0.120	0.189
Bridge end-on	−1.2655	0.217	0.116	0.168
Bridge side-on	−2.5920	0.185	0.138	0.171
On-top side-on	−1.3225	0.177	0.258	0.129
Hollow (3-1)	0.9211	0.167	0.220	−0.006
Hollow (3-3)	2.2900	0.167	0.220	−0.006
Pt(100)				
On-top end-on	−2.3653	0.185	0.120	0.185
Bridge end-on	−1.8283	0.207	0.130	0.155
Bridge side-on	−4.0749	0.180	0.119	0.161
On-top side-on	−3.3177	0.172	0.243	0.121
Hollow	1.5971	0.196	0.218	−0.019

The most stable configuration on both platinum surfaces implies a nearly dissociated state that can be simply described as an ionic pair with the positive end closer to the platinum surface. Together with a charge transfer from the oxygen atom closer to the surface, the π Mülliken population in the second oxygen atom increases. Besides, the σ contribution on Pt(100) also increases. The on-top side-on adsorbate coordination is characterized by a large stabilization of the π^* oxygen molecular orbital through the bonding interaction with d_{xy} Pt atom orbitals, as a consequence of the oxygen-adsorbed structure parallel to the surface. The larger O—O bond length in Pt(111) weakens the up-bond in the oxygen molecule and diminishes the importance of the O–Pt interaction as compared to Pt(100). Otherwise, the greater oxygen bond length results from a larger π^* orbital population due to a charge transfer from the most densely packed Pt(111) structure. This effect does not reflect in the energy level resulting from the molecular orbital interactions for the bridge side-on geometry, as these interactions involve optimized geometries where the interatomic distances are already stabilized, yielding a greater $R_{O—O}$ on Pt(111) resulting from the larger population of the π^* orbitals. For hollow coordination, the $R_{O—O}$ equilibrium value for the on-top side-on adsorbate coordination almost corresponds to a dissociated oxygen adsorbate, where the O atoms are bridge coordinated to the adjacent platinum sites. The stabilization of the bridge side-on geometry implies the interaction of π and σ oxygen atom orbitals with the d orbitals in the two closest symmetrically equivalent platinum atoms. The Mülliken population analysis indicates the oxygen-to-metal charge transfer without back bonding. The particular geometry of the platinum site facilitates the simultaneous interactions of both the oxygen atoms with the adjacent platinum atoms, leading to a greater stabilization of the $Pt_N O_2$ structure on Pt(100), where the charge transfer is favored because of the lower electronic density of Pt(100) as compared with Pt(111). Despite the similarity of the adsorption sites, the local Pt(100) and Pt(111), leads to an adsorbate geometry with a greater $R_{O—O}$ value on Pt(111). The electronic characteristics of Pt(100) and Pt(111) determine the specific charge-transfer contributions and the oxygen molecular orbital occupation. In particular, a greater occupation of oxygen π orbitals on Pt(111) favors larger $R_{O—O}$ values and the dissociation of O_2.

4.2.1.2.3.1   The Most Likely Intermediate Species Involved in the Oxygen Electroreduction Mechanism on Pt(111) and Pt(100) Clusters   The adsorption configuration study for the oxygen adsorbates on platinum, described above, provides the possibility of exploring further mechanistic aspects of the oxygen electroreduction, particularly in relation to the possible species involved in the reaction that account for the different behavior of Pt(111) and Pt(100) electrode surfaces.

As stated in Chapter 1, several mechanisms have been postulated to interpret the oxygen electroreduction kinetic data. Following oxygen dissociative adsorption on Pt(111), the reaction in aqueous acids implies a first coordination of the H^+-ions from the solution to the oxygen adatoms, which further desorb as water at negative overpotentials. The fact that only water molecules desorb from Pt(111) in the course of the oxygen reaction is consistent with the dissociative adsorption on this surface. By contrast, molecular oxygen adsorption would produce a peroxo-like adsorbate structure on Pt(100), which gives rise to hydrogen peroxide desorption at negative potentials. Peroxo-intermediates and hydrogen peroxide are experimentally detected mostly on Pt(100) electrodes with rotating ring-disc electrode techniques [30].

When the electrode potential is shifted toward negative values, peroxo-adsorbates (bridge side-on coordination) would dissociate, making water the final product on Pt(111) [29]. However, at potentials where oxygen adsorption on platinum occurs, other species resulting from the water electrochemical and chemical decomposition, such as OH and O adatoms, would also participate in the adsorbate structure, yielding, for instance, a $[Pt(100)]_N O_2$–OH ensemble at the equilibrium potential. The geometry of this ensemble was fully optimized for both oxygen and OH adsorbates. The adsorption of OH results in a linear configuration adjacent to the peroxo-group bridge side-on adsorbed species, rendering a structure that is 0.84 and 2.04 eV more stable than those associated with an OH bridge and an OH hollow coordination, respectively [29]. Stability calculations of these ensembles showed that an H atom transfer from the hydroxo to the peroxo group is favored.

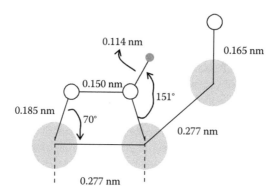

FIGURE 4.5   Geometric parameters of the fully optimized adsorbed ensemble from [Pt(100)]$_{25}$O$_2$OH, after the hydrogen atom transfer from the hydroxyl to the peroxo group. The oxygen atoms (the white balls), the hydrogen atom (the dark gray ball), and platinum atoms (the gray balls).

This H atom transfer implies an initial tilting of the Pt—O—H bond until it becomes parallel to the Pt—O bond in the peroxo group, with a simultaneous elongation of the Pt—O bond from 0.168 to 0.182 nm. Besides, the energy increases slightly while the OH bond is bent toward the peroxo group to be finally transferred with a net gain of 0.41 eV in stability. Then, in the 0.6 to 1.0 V potential range, the H atom transfer takes place and the adsorbate ensemble can be better described as a hydrogenated peroxo group and an oxygen atom coadsorbed on Pt(100). This adsorbed ensemble is a possible reaction intermediate in the electroreduction reaction on Pt(100). The H atom transfer implies an activation energy equal to 0.26 eV. The energy required for the transition from Pt(100) O$_2$OH to Pt(100)O$_2$HO might be provided by other simultaneous reactions, such as the underpotential deposition water discharge or the oxygen adsorption on platinum.

The stability of the O$_2$H–O coadsorbate on Pt(100) was also envisaged as a function of the electrode potential. Changes in the adsorbed structure due to the upward shifting of the d band, (potential negatively increased) have been studied after a full optimization of the geometry. Thus, the activation energy barrier for the hydrogen atom displacement from the hydroxyl oxygen to the peroxo-oxygen depends on the potential, and it decreases almost to zero for a −1 eV potential in the forward direction. Accordingly, it is reasonable to admit that the larger the cathodic overpotential, the lower the activation energy. Changes in the Mülliken atomic overlaps with the electrode potential show that although they both decrease for negative potentials, the destabilization of the Pt–O bond becomes more pronounced than that of the O—O bond for a given potential change. Hence, although a competition between the dissociation of the peroxo-adsorbate and the formation of a hydroperoxo-intermediate ensemble takes place, it is likely that the hydroperoxo-intermediate ensemble preferentially desorbs as hydrogen peroxide prior to dissociation.

The hydroperoxide formation on Pt(100) implies that at negative potentials, the remaining oxospecies (Figure 4.5) would be negatively charged and after the H$^+$-ion coordination, water desorption would occur. The mechanism for this reaction is similar to that proposed on Pt(111), and it justifies the simultaneous desorption of water from Pt(100) and the detection of the only oxygen reduction product from Pt(111).

4.3   ON THE CALCULATION OF ACTIVATION BARRIERS AND CURRENT DENSITIES

Bockris and Abdu [31] presented the adiabatic potential energy surface for the initial Pt$_N$O$_2$ + H$_3^+$O and the final Pt$_{N-1}$O and PtOH states for similar discussions and especially for calculating the activation energy of the oxygen electroreduction reaction. They juxtaposed both the complex ensembles together according to the geometry chosen to find out the classical transition-state geometry. The transition matrix element is calculated at the transition-state geometry using the relevant wave functions with appropriate perturbation energies. The electronic transmission factor, $\kappa_{el}(x)$, is calculated using the Landau–Zener formalism [32]. Finally, the reaction rate at an electrode potential was calculated with a general equation [33]:

$$j_{oxygen} = veQ^{-1} \iint C_{O_2}(x)C_{H^+}(x)\rho(E)f(E)P_{H^+,tunn}(E,x)\kappa_{el}(x)P_{act}(E)dEdx \qquad (4.16)$$

where
v is a frequency factor
e is the electronic charge
$C(x)$ is the concentration of the species at the double layer in mol cm^{-3}
$\kappa_{el}(x)$ is the electronic transmission factor
$P_{H^+,tunn}(E,x)$ is the probability of tunneling for the proton
$\rho(E)$ is the density of the states in the metal
Q^{-1} is the normalization factor for the electrons in various states

$$Q = \int P(E)f(E)dE \qquad (4.17)$$

where
$f(E)$ is the Fermi–Dirac distribution function
P is the probability of the activation of reactants

Bockris [31] instead of integrating the double-layer concentration with respect to distance made the following approach. He assumed that the contribution to the current comes only from the first layer of the reactants in the double layer. This allows us then to eliminate the integration with respect to the distance. The surface concentration of the reactants within a double-layer thickness of δ can be evaluated. P_{act} can be calculated by means of the time-dependent perturbation theory using Fermi's golden rule within a phonon–vibron-coupling formalism [34], which can be simply reduced to $\exp[-(\Delta G^{o,\#} + \beta FE)/RT]$, which is the chemical Gibbs energy of activation and the metal–solution inner potential difference calculated as usual. Under these conditions, the value of v can be replaced by the transition-state factor, kT/h:

$$j_{oxygen} = \frac{kT}{h}e\delta^2 Q^{-1}\kappa_{el}C_{O_2}C_{H^+} \int \rho(E)f(E)P_{H^+,tunn}(E)e^{[-(\Delta G^{o,\#}+\beta FE)/RT]}dE \qquad (4.18)$$

The calculation of the electronic-transmission factor currently involves three different methods, viz. the Landau–Zener formula, Fermi's golden rule [35], and electron tunneling formalism such as the Wentzel–Kramer–Brillouin method [36]. We used the Landau–Zener formula [37,38] to calculate it:

$$\kappa_{el} = 1 - \exp\left[-\frac{2\pi|H_{fi}|^2}{\hbar v|s_f - s_i|}\right] \qquad (4.19)$$

where
H_{fi} is the matrix element of transition, $\langle \Psi_f|U_{pert}||\Psi_i\rangle$, for the electron, under the influence of the perturbation potential, U_{pert}, from its initial state, Ψ_i, in the metal to a state, Ψ_f, at the transition state of the reactant
v is the relative velocity of the nuclei
s are the slopes of the potential energy near the transition state region

The most important and difficult quantity to calculate in the above equation is the electronic coupling matrix element. The initial state wave function, Ψ_i, of a conducting electron in the metal electrode is not taken as a Bloch wave function because the periodicity of the metal is no longer effective at the interface (normal to the surface). The wave function of a free particle is taken as that given in [35], but allowance is made for the interaction with the metal using an internal effective mass of the conducting electron [39]:

$$\Psi_i(\vec{r}) = L_o^{-3/2} e^{i(\vec{k} \cdot \vec{r})} \tag{4.20}$$

where
L_o is a normalization constant of the free-particle wave function
k is the wave vector given by

$$\left| \vec{k} \right| = \sqrt{\frac{2m^*E}{h^2}} \tag{4.21}$$

where
m^* is the effective mass of the electron
E is the kinetic energy of the electron at the Fermi level in the metal electrode

The $\Psi_i(\vec{r})$ can be expressed in terms of spherical waves:

$$\Psi_i(\vec{r}) = 4\pi l^{-3/2} \sum_{l=0}^{\infty} i^l \sum_{m=-l}^{l} j_l(kr) Y_{lm}(\vec{k}) Y_{lm}(\vec{r}) \tag{4.22}$$

where
l is the angular momentum quantum number
m is the magnetic quantum number
$j_l(kr)$ is the spherical Bessel function of order l and argument kr
$Y_{lm}(\vec{r})$ and $Y_{lm}(\vec{k})$ are spherical harmonics of order l and projection m with (n, l, m) the usual quantum numbers

We are only interested in the localized wave function of the electron, for example, the highest occupied molecular orbital (HOMO), of the transition state complex with the electron transferred. Hence, we may use the HOMO wave function of the transition state complex after the electron transfer, including only the nearby surface metal atoms that contribute significantly to this HOMO [40]. This wave function at the cluster is calculated using the EHMO method together with the parameters of VSIP and double-zeta orbitals given in [31]:

$$\Psi_f(\vec{r}) = \sum_{i=1}^{N} \sum_{j=l} c_{ij} \Phi_{ij}(\vec{r}) \tag{4.23}$$

where

$$\Phi_{ij}(n, l, m) = \frac{(2\zeta/a_o)^{n+1/2}}{\sqrt{2n!}} r^{n-1} e^{-r\zeta/a_o} Y_{lm}(\theta, \Phi) \tag{4.24}$$

where

$\Phi_{ij}(n, l, m)$ is the Slater-type jth atomic orbital of the ith atom multiplied by its associated spherical harmonics

ζ is the double-zeta orbital exponent

a_o is the Bohr radius

c_{ij} is the coefficient obtained after applying the EHMO modified method to $\Psi_f(\vec{r})$

Thus, the Hamiltonian elements, H_{fi}, are

$$\langle \Psi_f | U_{pert} | | \Psi_i \rangle = \left\langle \sum_{i=1}^{N} \sum_{j=l} c_{ij} \Phi_{ij}(\vec{r}) | U_{pert} | L_o^{-3/2} \sum_{l=0}^{\infty} i^l \sum_{m=-l}^{l} j_l(kr) Y_{lm}(\vec{k}) Y_{lm}(\vec{r}) \right\rangle \quad (4.25)$$

where the perturbing potential, U_{pert}, is

$$U_{pert} = e\vec{E} \cdot \vec{r} + \frac{1}{1/\Phi_M + 4\varepsilon_o r / e^2} \quad (4.26)$$

where

$\vec{r}$ is the distance between the electrode surface and the electron

$\vec{E}$ is the electric field across the double layer

ε_o is the optical dielectric constant of the medium at the interphase

Φ_M is the work function of the metal

The elements of H_{fi} are determined by numerical integration using $\vec{E} \approx 0$ at the potential of zero charge. However, since the electron is attracted to the proton at the potential, the perturbation potential has to be $\dfrac{1}{1/\Phi_M + 4\varepsilon_o r / e^2} - \dfrac{e^2}{\varepsilon_o r_{H^+}}$. Assuming a free electron model, the Fermi level is $E_F = \dfrac{h^2}{2m} \left(\dfrac{3n}{8\pi}\right)^{2/3}$, so the calculation of the H_{fi} [eV] $= 0.11 + 0.03\,i$ and $0.11 + 0.05\,I$, for Pt(111) and Pt(100), respectively.

Bockris [31] also determined the values of s_f and s_i, which are -5.25 and 4.0 eV Å^{-1}, respectively, for Pt(100); and -4.5 and 3.5 eV Å^{-1}, respectively, for Pt(111). According to Equation 4.19, the values of κ_{el} are 0.62 and 0.60 for Pt(100) and Pt(111), respectively.

The potential energy surface for the proton motion is not of a simple shape in our model calculation. Hence, the Eckart barrier formula is used for proton tunneling by adjusting the two variables in it [41]. Thus, the barrier for the proton transfer was fitted to the following Eckart formula:

$$E = A \frac{e^{2\pi \vec{r}/d}}{1 + e^{2\pi \vec{r}/d}} + B \frac{e^{2\pi \vec{r}/d}}{(1 + e^{2\pi \vec{r}/d})^2} \quad (4.27)$$

where

r is the distance from the initial (r negative) to the final (r positive) ground states

d is the half-width of the barrier

A is the energy of the reaction

B is a barrier constant

$$B = 2E_{act} - A + 2\sqrt{E_{act}(E_{act} - A)} \quad (4.28)$$

It should be noted here that the energy term in Eckart's equation is taken with respect to the initial ground-state energy.

Finally, to calculate the current, the entropy of the activation is added to the activation energy, and then the relevant part for the proton tunneling is evaluated numerically using the Eckart barrier formalism for $P_{H^+, tunn}(E)$:

$$P_{H^+, tunn}(E) = \frac{\cosh[2\pi(\alpha + \zeta)] - \cosh[2\pi(\alpha - \zeta)]}{\cosh[2\pi(\alpha + \zeta)] + \cosh[2\pi\gamma]} \tag{4.29}$$

Calculation shows that the contributions of the above or below Fermi-level electrons to the cathodic current are negligible compared to those of the electrons coming from the Fermi level. After this further simplification,

$$j_{oxygen} = \frac{kT}{h} e\delta^2 Q^{-1} \kappa_{el} C_{O_2} C_{H^+} P_{H^+, tunn}(E) e^{[-(\Delta G^{o, \#} + \beta FE)/RT]} \tag{4.30}$$

Since light particle tunneling can take place at any energy level between the ground level and col, one has to sum all the current contribution coming from the tunneling at each energy level in this range. In order to find the total current, one has to differentiate the above equation with respect to energy and integrate it from the ground level to the col. The total current expression found by Bockris is

$$j_{total} = \frac{kT}{h} e\delta^2 Q^{-1} \kappa_{el} C_{O_2} C_{H^+} \left[\frac{1}{kT} \int_{G_{init}}^{G_{col}} P_{H^+, tunn}(E) e^{[-(\Delta G^{o, \#} + \beta FE)/RT]} dG + e^{[-(\Delta G^{o, \#} + \beta FE)/RT]} \right] \tag{4.31}$$

The experimental activation energy extrapolated to zero current for the oxygen electrochemical reduction on platinum in acidic media is around 1 eV [42,43] and the experimental current density at 0.6V is approximately 1 mA cm^{-2} [44]. The extrapolation of Bockris results of activation energies to the open-circuit potential [31] is 1.38 eV for Pt(111) and 1.23 eV for Pt(100) for the proton. However, the currents are two orders of magnitude greater, but they can be reduced after considering the $(1 - \theta_{oxygen})$ term. This expression is appropriate for the electrocatalysis since it gives a coverage indication of the free-surface sites of O and OH species. They have found [31] also that the Pt(100) surface is more active than Pt(111). However, the experimental facts show the reverse in the absence of specific ion adsorption [45]. On Pt(111), there are more adsorbate sites in linear coordination to adsorb O and OH in their model calculation than on the Pt(100) surface. The model indicates the importance of proton tunneling.

REFERENCES

1. J.A. Pople and D.L. Beveridge, *Approximate Molecular Orbital Theory*, Academic Press, New York (1970).
2. J.N. Murrell and A.J. Harget, *Semiempirical Self-Consistent-Field Molecular Orbital Theory of Molecules*, Wiley, New York (1972).
3. M.J.S. Dewar, *Science* **187** (1975) 1037.
4. W. Thiel, *Tetrahedron* **44** (1988) 7393.
5. J.J.P. Stewart, in *Reviews in Computational Chemistry*, K.B. Lipkowitz and D.B. Boyd (Eds.), vol. 1, VCH Publishers, New York (1990), pp. 45–81.
6. M.C. Zerner, in *Reviews in Computational Chemistry*, K.B. Lipkowitz and D.B. Boyd (Eds.), vol. 2, VCH Publishers, New York (1991), pp. 313–365.

7. W. Thiel, Semiempirical methods, in *Proceedings of the Modern Methods and Algorithms of Quantum Chemistry*, 2nd edn., J. Grotendorst (Ed.), vol. 3, John von Neumann Institute for Computing, Jüllich, Germany, NIC Series (2000), pp. 261–283.

8. J.A. Pople and G.A. Segal, *J. Chem. Phys.* **44** (1966) 3289.

9. J.A. Pople, D.L. Beveridge, and P.A. Dobosh, *J. Chem. Phys.* **47** (1967) 2026.

10. R.C. Bingham, M.J.S. Dewar, and D.H. Lo, *J. Am. Chem. Soc.* **97** (1975) 1285.

11. K. Jug, R. Iffert, and J. Schulz, *Int. J. Quantum Chem.* **32** (1987) 265.

12. M.J.S. Dewar and W. Thiel, *J. Am. Chem. Soc.* **99** (1977) 4907.

13. M.J.S. Dewar, E. Zoebisch, E.F. Healy, and J.J.P. Stewart, *J. Am. Chem. Soc.* **107** (1985) 3902.

14. J.P. Stewart, *J. Comput. Chem.* **10** (1989) 209.

15. W. Thiel and A.A. Voityuk, *J. Phys. Chem.* **100** (1996) 616.

16. SPARTAN 4.0, Wavefunction Inc., Irvine, CA (1995).

17. A.A. Voityuk, M.C. Zerner, and N. Rösch, *J. Phys. Chem. A* **103** (1999) 4553.

18. M.J.S. Dewar and W. Thiel, *J. Am. Chem. Soc.* **99** (1977) 4899.

19. R. Hoffman, *J. Phys. Chem.* **39** (1963) 1797.

20. G. Calzaferri, L. Forss, and Y. Kamber, *J. Phys. Chem.* **93** (1989) 5366.

21. A.B. Anderson, *J. Chem. Phys.* **62** (1975) 1187.

22. A.B. Anderson and R. Hoffmann, *J. Chem. Phys.* **60** (1974) 4271.

23. G. Blyholder and C.A. Coulson, *Theor. Chim. Acta* **10** (1968) 316.

24. E. Clementi, *IBM J. Res. Dev.* **9** (1965) 2.

25. T.A. Gurner, *Methods in Molecular Orbital Theory*, Prentice-Hall, Englewood Cliffs, NJ (1974).

26. C.F. Zinola, G.L. Estiú, E.A. Castro, and A.J. Arvia, *J. Phys. Chem.* **98** (1994) 7566.

27. D.R. Lide (Ed.), *CRC Handbook of Chemistry and Physics*, CRC Press, Boca Raton, FL (1990–1991).

28. G. Blustein and C.F. Zinola, *J. Colloid Interface Sci.* **278** (2004) 393.

29. C.F. Zinola, G.L. Estiu, E.A. Castro, and A.J. Arvia, *J. Phys. Chem.* **98** (1994) 7566.

30. C.F. Zinola, A.M. Castro Luna, and A.J. Arvia, *Electrochim. Acta* **39** (1994)1951.

31. J.O'M. Bockris and R. Abdu, *J. Electroanal. Chem.* **448** (1998) 189.

32. S.U.M. Khan and Z. Zhou, *J. Chem. Phys.* **93** (1990) 8808.

33. R. Gurney, *Proc. R. Soc. Lond.* **50** (1931) 137.

34. S.U.M. Khan, *Appl. Phys. Commun.* **4**(2–3) (1984) 149.

35. S.U.M. Khan, P.Wright, and J.O'M. Bockris, *Elektrokhimiya* **13** (1977) 914.

36. J.O'M. Bockris and S.U.M. Khan, *Quantum Electrochemistry*, Plenum Press, New York (1979), p. 489.

37. L.D. Landau, *Z. Phys.* **2** (1932) 46.

38. C. Zener, *Proc. R. Soc. Lond. Ser. A* **137** (1932) 696.

39. E. Merzbacher, *Quantum Mechanics*, Wiley Interscience, New York (1970), p. 192.

40. S. Larsson, *J. Chem. Soc. Faraday Trans.* **2** (1983) 1375.

41. C. Eckart, *Phys. Rev.* **35** (1930) 1303.

42. A.J. Appleby, *J. Electrochem. Soc.* **117** (1970) 328.

43. A. Damjanović and D.B. Sepa, *Electrochim. Acta* **35** (1990) 11578.

44. H. Kita, H. Lei, and Y. Gao, *J. Electroanal. Chem.* **379** (1994) 407.

45. N.M. Marković, R.R. Adzić, B.D. Cahan, and B.E. Yeager, *J. Electroanal. Chem.* **377** (1994) 249.

5 Density Functional Theory

C. Fernando Zinola

CONTENTS

The development of molecular orbital (MO) techniques and their applications to structural and reactivity problems in surface chemistry were improved by density functional theory (DFT) methods. In the case of semiempirical MO methods, symmetry arguments in structural chemistry and reactivity were provided; however, DFT has allowed theoretical chemistry to predict accurately the structures of clusters in surfaces and organometallic compounds.

5.1 FUNDAMENTALS OF DFT

The fundamental basis of DFT has been investigated in many reviews and books [1–8]. We briefly summarize the changes to the fundamental Hamiltonian elements in the case of DFT, since the kinetic energy term of the nuclei is neglected. The operator can be written as

$$\hat{H} = -\sum_i \frac{\hbar^2}{2m_e} \nabla_i^2 + \sum_{i,A} \frac{Z_A e^2}{|r_i - R_A|} + \frac{1}{2}\sum_{i \neq j} \frac{e^2}{|r_i - r_j|} + \frac{1}{2}\sum_{A \neq B} \frac{Z_A Z_B e^2}{|R_A - R_B|} \tag{5.1}$$

This indicates that atoms A and B act as an external fixed electric potential acting on electrons. This potential is compensated by charge neutrality with the addition of the last term in Equation 5.1.

The total energy, E_{tot}, is calculated as usual from the wave function, Ψ:

$$E_{tot} = \frac{\langle \Psi | \langle \hat{H} \rangle | \Psi \rangle}{\langle \Psi | \Psi \rangle} = \langle \hat{H} \rangle = \langle \hat{T} \rangle + \langle \hat{V}_{int} \rangle + \int d^3 r V_{ext}(r) n(r) dr \qquad (5.2)$$

where $\Psi = \Psi(r_1, r_2, \ldots, r_N)$, considering the ground state, Ψ_o, as the lowest energy that obeys the symmetry of the particles and conservation laws:

$$E_o = \min \left(\frac{\langle \Psi | \langle \hat{H} \rangle | \Psi \rangle}{\langle \Psi | \Psi \rangle} \right) \qquad (5.3)$$

Among the different theorems, we summarize only the Hohenberg and Kohn theorem [9]. This theorem established that the ground state of an electronic system is just a function of its density so that, in principle, one only needs information on the density to work out all the properties of these structures. Splitting the energy functional into kinetic energy, potential energy, non-quantum Coulomb electrostatic repulsion energy, and exchange-correlation energy (everything else), E_{xc} is the first step for the description of the electronic system.

Unfortunately, the exchange-correlation energy functional that should be universal is not known, and practical (and approximate) solutions are obtained by the use of the so-called Kohn–Sham (KS) orbitals. The KS orbitals differ from other kinds of orbitals mostly by the fact that the sum of the squares of the occupied KS orbitals is the true density of the system, an assumption that is only approximated in other quantum chemical methods such as the Hartree–Fock method. The Kohn–Sham method, introduced in 1965 [10], allows us to solve the problem through the Schrödinger equation that differs from the well-known Hartree–Fock equations by the replacement of the exchange potential term by a more universal exchange-correlation potential term that is straightforward because it is merely a function of the density. According to Equations 5.1 and 5.2, the spin orbitals, $\Psi(r)$, are solutions for the KS equations:

$$\left[-\frac{1}{2} \nabla_i^2 + V_{eff}(r) \right] \Psi_i = \varepsilon_i \Psi_i \qquad (5.4)$$

where V_{eff} is the effective potential, which is the addition of the external potential $V(r)$, the Hartree potential for electrons $V_H(r)$, and the exchange-correlation potential, $V_{xc}(\vec{r})$, as in Equation 5.1. The latter can be formally presented as

$$V_{xc}(\vec{r}) = \left[\frac{\partial E_{xc}(\rho(\vec{r}))}{\partial \rho(\vec{r})} \right] \qquad (5.5)$$

with

$$\rho(\vec{r}) \equiv \sum_i |\Psi|^2 \qquad (5.6)$$

The main difference and the potential of this approach lies in the detail that $V_{xc}(\vec{r})$ includes not only the exchange in the Hartree–Fock (HF) equations, but also the correlation (referred to all that is missed by the Hartree–Fock approach) components. In addition, the difference between the exact kinetic energy of the system and the one calculated from the KS orbitals are included. This method states that $V_{xc}(\vec{r})$ is the best way to describe the fact that every electron aims to maximize the attraction from the nuclei and to minimize the repulsion from the rest of the electrons along its constant movement within an entity (atom or molecule). $V_{xc}(\vec{r})$ describes the exchange correlation hole, which is the region around each electron into which no other electron is allowed. Besides, the KS correlation should be larger (more negative) than the HF-defined correlation since the first is a functional of the exact density whereas the second is a functional of the orbitals.

However, in any case, the differences between the KS orbitals and the others are small, so that all the concepts that have been elaborated through more approximate semiempirical methods can be used [11]. In addition, relativistic corrections can be launched into the DFT equations through many approximate calculations according to the accuracy required [12,13].

Since $E_{xc}(\vec{r})$ is not exactly known, an accurate approximation has to be used. Thus, the elaboration of new functionals is still an area of intense work. Ziegler [14] has classified the exchange-correlation energy functionals into three generation groups.

The first generation is the local density approximation (LDA). This estimation involves the Dirac functional for exchange, which is nothing else than the functional proposed by Dirac [15] in 1927 for the so-called Thomas–Fermi–Dirac model of the atoms. For the correlation energy, some parameterizations have been proposed, and the formula can be considered as the limit of what can be obtained at this level of approximation [16–18]. The Xa approximation falls into this category, since a known proportion of the exchange energy approximates the correlation.

The second generation of functional uses both the density and its gradients. However, a simple gradient expansion, already tested by Sham [19] and Herman et al. [20], was not successful. Levy and Perdew [21,22] have shown that various relations, divided into sum rules, scaling properties, and asymptotic properties, are satisfied by the LDA, but not by a simple gradient expansion, so it is necessary to parameterize the gradient expansion. The first gradient-corrected energy functional was proposed by Becke [23], and Perdew and Wang [24] for the exchange, and Perdew [25] for the correlation. They are all often called generalized gradient approximations (GGAs).

The third generation of exchange-correlation energy functionals are "beyond the GGA." It comprises various options with more accuracy through sophistication and additional computational cost. The most important ones are the hybrid functionals and the functionals that are dependent on the density, its gradients, and its Laplacians [26]. Hybrid functionals are energy functionals that contain both a DFT exchange and a Hartree–Fock type exchange calculated from the orbitals. The justification of this approach lies in the so-called adiabatic connection scheme defined by Langreth and Perdew [27], and practically introduced by Becke [28]. In 1998, the most popular hybrid functional was definitely the so-called B3LYP functional, available for the first time in the Gaussian package [29] that introduced this family of acronyms:

$$E_{xc} = \varepsilon_0 E_{H-F} + \varepsilon_1 E_{X(LSD)} + \varepsilon_2 E_{X(GGA)} + \varepsilon_3 E_{Corr} \tag{5.7}$$

where

E_{H-F} is the pure exchange energy of Hartree–Fock's calculated with the KS orbitals
$E_{X(LSD)}$ is the local exchange energy, taken as the Dirac exchange
$E_{X(GGA)}$ is the gradient corrections to E_{xc}, namely the Be88 [30] form
E_{Corr} is the gradient corrected correlation, namely the Lee–Yang–Parr (LYP) functional [31]

To keep the Dirac exchange [15] of the electron gas part complementary to E_{xc} of Equation 5.7, the classic form of linear combinations is $\varepsilon_1 = 1 - \varepsilon_0$ for hybrid functionals. Thus, the number of fitted parameters adds up to three, as indicated in the acronym. B3LYP was proven to lead to reasonable results for coordination compounds, although the "pure DFT" functionals, that is, not containing the pure exchange (e.g., second generation GGAs), are comparable for some thermodynamic data.

On the other hand, the DFT is a dominant tool to optimize fixed geometrical structures and compute stationary points with their relative energies along a reaction path. It is also a powerful tool through *ab initio* molecular dynamics (MD) calculations, the methodology being first described in 1985 by Car and Parrinello (CP) [32]. These calculations are of molecular dynamics and involve the movement of nuclei and electrons, whereas standard molecular dynamics illustrates trajectories of atoms as rigid objects under a classical interaction potential. The DFT model is used normally at the LDA level and the kinetic energy of both electrons and nuclei is controlled with fictitious masses, in order to save computational time. Although more demanding in computational effort, *ab initio* molecular dynamics calculations are powerful tools for the localization of reaction pathways. For numerical efficiency, plane waves are usually used for the description of the valence orbitals, whereas frozen cores or pseudo-potential approximations [33,34] are used for taking into account the core electrons. CP dynamics allow efficient reaction path scans, free energy calculations, and simulations covering sub-picosecond fluxionality. A combined Car–Parrinello quantum mechanics/molecular mechanics (QM/MM) implementation for *ab initio* molecular dynamics (MD) simulation of extended systems has been successfully applied to transition metal catalysis [35]. In this approach, the core of the molecular system is treated as in DFT, while the substituted chains are treated with a molecular mechanics force field. QM/MM is an efficient way for considering bulky ligands for both "static" calculations and "dynamic" simulations. It is an active area of methodological development, particularly when the QM/MM boundary crosses the bonds.

5.2 HOHENBERG–KOHN THEOREMS

Theorem 5.1

For any electron system, the external potential is solely determined except for a constant, by the ground state density, $\rho(r)$. ■

Corollary 5.1

Since the Hamiltonians are fully determined except for a constant shift of energy, the full many-body wave function and the rest of the system properties are fully determined. ■

Theorem 5.2

There exists a *universal functional* for the energy of the fixed density, $\rho(r)$. For a given external potential, the minimum of energy is the exact ground state energy that occurs for the exact ground state density, $\rho(r)$. ■

Corollary 5.2

The functional of energy alone is sufficient to determine exactly the ground state energy and density. Electron excited states have to be determined by other methodologies. ■

5.2.1 KOHN–SHAM EQUATIONS

The theorem implies that $V_{ext}(r)$ has to satisfy that the eigenvalue of the minimum energy evaluated by the Kohn–Sham equations, E_{HK}, is

$$E_{HK} = \langle \hat{T} \rangle + \langle \hat{V}_{int} \rangle + \int d^3 r V_{ext}(r)\rho(r)dr = F_{HK} + \int d^3 r V_{ext}(r)\rho(r)dr \qquad (5.8)$$

where F_{HK} is the Hohenberg–Kohn functional. F_{HK} is defined for all densities that can be solutions to some Hamiltonians. Levy and Lieb have shown a way to define the functional more generally as a search over wave functions. In formula, it is

$$E_{LL} = \min_{\Psi \to \rho(r)} \left[\langle \Psi | \langle \hat{T} \rangle | \Psi \rangle + \langle \Psi | \langle \hat{V}_{int} \rangle | \Psi \rangle \right] + \int d^3 r V_{ext}(r) \rho(r) dr + E_{II} \tag{5.9}$$

where a Levy–Lieb functional, F_{LL}, is defined as

$$F_{LL} = \min_{\Psi \to \rho(r)} \left[\langle \Psi | \langle \hat{T} \rangle | \Psi \rangle + \langle \Psi | \langle \hat{V}_{int} \rangle | \Psi \rangle \right] \tag{5.10}$$

The only result is that the electron full density determines the potential, but there is still the original many-body problem. One possibility is to replace the original interacting-particle problem with one that can be more easily solved, that is, the Kohn–Sham auxiliary system. This alternative is a *non-interacting "electron"* system assumed to have the same density as the interacting system. The Kohn–Sham system presumes to have the same density as the true interacting system. Thus,

$$\hat{H}_{eff} = \frac{-\hbar^2}{2m_e} \nabla^2 + V_{eff}(\vec{r}) \tag{5.11}$$

with

$$\rho_{eff} = \sum_{i=1}^{N} |\Psi_i(\vec{r})|^2 \tag{5.12}$$

On the other hand, the effective kinetic energy is T_{eff}:

$$\hat{T}_{eff} = \frac{-\hbar^2}{2m} \sum_{i=1}^{N} \langle \Psi_i(\vec{r}) | \nabla^2 | \Psi_i(\vec{r}) \rangle \tag{5.13}$$

The K–S energy, E_{KS}, is the addition written below:

$$E_{KS} = \hat{T}_{eff} + \int V_{ext}(\vec{r}) \rho(\vec{r}) d\vec{r} + E_{HF} + E_{xc} \tag{5.14}$$

The K–S equations replace the interacting expression between electrons with an auxiliary non-interacting problem. Each term in Equation 5.14 is exclusively related to each other. Although the alternative has fulfilled several simple cases, it is not general. By minimizing this energy, we can check the "not general" cases:

$$\frac{\partial E_{KS}}{\partial \Psi_i(\vec{r})} = \frac{\partial \hat{T}_{eff}}{\partial \Psi_i(\vec{r})} + V_{ext}(\vec{r}) \frac{\partial \rho(\vec{r})}{\partial \Psi_i(\vec{r})} + \frac{\partial E_{HF}}{\partial \Psi_i(\vec{r})} + \frac{\partial E_{xc}}{\partial \Psi_i(\vec{r})} \tag{5.15}$$

The minimization implies

$$\frac{\partial E_{KS}}{\partial \Psi_i(\vec{r})} = 0 \tag{5.16}$$

Considering that the wave functions are orthonormal

$$\left\langle \Psi_i(\vec{r}) \middle| \Psi_j(\vec{r}) \right\rangle = \delta_{ij} \tag{5.17}$$

In this respect, after minimization the K–S Hamiltonian will be

$$\left(\hat{H}_{\text{eff}}(\vec{r}) - \varepsilon_i \right) \Psi_i(\vec{r}) = 0 \tag{5.18}$$

where ε_i are the eigenvalues of the Hamiltonian.

The expression of $\hat{H}_{\text{eff}}(\vec{r})$ can be written as

$$\hat{H}_{\text{eff}} = \frac{-\hbar^2}{2m_e} \nabla^2 + V_{\text{ext}}(\vec{r}) + \frac{\partial E_{\text{HF}}}{\partial \rho_{\text{eff}}(\vec{r})} + \frac{\partial E_{\text{xc}}}{\partial \rho_{\text{eff}}(\vec{r})} \tag{5.19}$$

This expression is condensed as

$$\hat{H}_{\text{eff}} = \hat{T}_{\text{eff}} + V_{\text{ext}}(\vec{r}) + V_{\text{HF}}(\vec{r}) + V_{\text{xc}}(\vec{r}) \tag{5.20}$$

Now the total energy is

$$E = \sum_{i=1}^{N} \varepsilon_i - \frac{1}{2} \int V_{\text{HF}}(\vec{r}) \rho_{\text{eff}}(\vec{r}) d\vec{r} + \left(E_{\text{xc}} - \int V_{\text{xc}}(\vec{r}) \rho_{\text{eff}}(\vec{r}) d\vec{r} \right) \tag{5.21}$$

5.2.2 EXCHANGE CORRELATION FUNCTIONAL ENERGY

The exchange correlation functional energy, E_{xc}, is defined as

$$F_{\text{HK}} = \hat{T}_{\text{eff}} + E_{\text{HF}} + E_{\text{xc}} \tag{5.22}$$

with

$$E_{\text{HF}} = \int d^3 r \rho(r) \frac{1}{2} \int d^3 r' \frac{\rho(r')}{|r - r'|} \tag{5.23}$$

and

$$E_{\text{xc}} = \int d^3 r \rho(r) \varepsilon_{\text{xc}}([\rho], r) \tag{5.24}$$

The latter is a functional of ρ calculated from the Hohenberg–Kohn theorems.

On the other hand, around each electron from a point $\vec{r}$, other electrons are excluded to delimit a hole at a point $\vec{r}'$ with the definition of a $\rho_{\text{xc}}(\vec{r}, \vec{r}')$. According to Pauli's principle, this hole has to "occlude" only one "missing" electron. From the correlation, it follows

$$\varepsilon_{\text{xc}}([\rho], r) = \frac{1}{2} \int d^3 r' \frac{\rho(r, r')}{|r - r'|} + T(r) - T_{\text{eff}}(r) \tag{5.25}$$

where the last two terms are positive kinetic expressions, whereas the first is a negative kinetic expression. Both kinetic terms can be neglected by using the Hellmann–Feynman theorem by a mean value of the coupling constants:

$$\varepsilon_{xc}([\rho], r) = \frac{1}{2} \int d^3 r' \frac{\bar{\rho}(r, r')}{|r - r'|} \tag{5.26}$$

5.2.3 Local Density Approximation, Optimized Effective Potential, and Exact Exchange Orbital Functionals

By the local density approximation, the exchange-correlation energy, E_{xc}, is the addition of the entire contributions around a point $\vec{r}$ independent of the rest of the points:

$$E_{xc}(\rho(r)) = \int d^3 r \rho(r) \varepsilon_{xc}(\rho(r)) \tag{5.27}$$

where $\varepsilon_{xc}(\rho(r))$ is the exchange correlation energy per electron. Since $\varepsilon_{xc}(\rho(r))$ has to be universal, it must be the same for all homogeneous electron gas of density ρ. The sole exchange energy term can be considered in general [36] as

$$\varepsilon_x(\rho(r)) = -\frac{0.458}{r_s} \quad \text{in Hartress} \tag{5.28}$$

where r_s is the average distance between the "spherical" electrons arrived at by using $\rho(r) = 4/3\pi r_s^3$.

The other important situation is the calculation of the exchange correlation potential $V_{xc}(\vec{r})$ that can be defined by

$$V_{xc}(\vec{r}) = \frac{\partial E_{xc}(\rho(\vec{r}))}{\partial \rho(\vec{r})} \tag{5.29}$$

This expression is transformed into

$$V_{xc}(\vec{r}) = \varepsilon_{xc}([\rho], \vec{r}) + \rho(\vec{r}) \frac{\partial \varepsilon_{xc}([\rho], \vec{r})}{\partial \rho(\vec{r})} \tag{5.30}$$

From the latter analytical expression, it is easily shown that the expression cannot be derived from the interaction of particles. On the other hand, the second term is discontinuous at densities corresponding to filled shells. As stated above, the potential of all the electrons in a crystalline solid changes discontinuously when another electron is added. This case is not predicted by the LDA method but it is predicted in the orbital-dependent forms. For an orbital functional, the kinetic energy is evaluated first:

$$T_{eff}(\rho(\vec{r})) \rightarrow T_{eff}[\Psi_i(V_{eff}(\vec{r}))] \tag{5.31}$$

Then, the "optimized effective potential," OEP, is defined as

$$E_{xc}(\rho(r)) \rightarrow E_{xc}[\Psi_i(V_{eff}(\vec{r}))] \tag{5.32}$$

Similarly, the exchange-correlation potential under the OEP will be

$$V_{xc}^{OEP}(\vec{r}) = \frac{\partial E_{xc}^{OEP}([\rho], \vec{r})}{\partial \rho(\vec{r})} \qquad (5.33)$$

This is simply solved as an integral equation.

The exact exchange, EXX, orbital functional is a K–S density functional theory that uses the Hartree–Fock-orbital-dependent exchange functional:

$$E_{xc}(\rho(r)) \rightarrow E_x[\Psi_i(V_{eff}(\vec{r}))] - \frac{1}{2} \sum_{i,j} \int dr dr' \Psi_i^*(\vec{r}) \Psi_j^*(\vec{r}') \frac{1}{|r - r'|} \Psi_j(\vec{r}) \Psi_i(\vec{r}') \qquad (5.34)$$

In the case of a one-electron system, the ground state is exactly that of the Hartree–Fock's. However, for the excited states, the eigenvalues are exactly those of the exact excitation energy for a single electron. The EXX eigenvalues correspond to the excitation energies with no change in the number of atoms except the addition or the removal of energies.

5.3 SEMIEMPIRICAL DFT: SCC-DFTB

In the case of electrode surfaces, DFT can be used to build up ~100 atoms in routine applications. In some cases, even more *ps* in MD simulations are possible. When we require larger systems, more approximations to the DFT are necessary.

5.3.1 NON-SELF-CONSISTENT SCHEMES

First, we will consider a case where we know the ground state density ρ_o with adequate accuracy. It is then possible to omit the self-consistent solution of the K–S equations and get the orbitals immediately through

$$\left[\frac{-\hbar^2}{2m_e} \nabla^2 + V_{eff}(\rho(\vec{r})) \right] \Psi_i = \varepsilon_i \Psi_i \qquad (5.35)$$

by using the prefixed value of ρ_o to define the starting point for further approximations.

Consider a minimal basis set consisting of atomic orbitals, that is, $\Phi\sigma = 2s, 2p_x, 2p_y, 2p_z$ for first row elements (we omit the core states for the following, since they are, in a good approach, chemically inactive) and $\Phi\sigma = 1s$ for H.

With the basis set expansion

$$\Psi_i = \sum_\sigma a_\sigma^i \Phi_\sigma \qquad (5.36)$$

and the Hamiltonian

$$\hat{H}_{eff}(\rho_0) = \hat{T}_{eff} + V_{eff}(\rho_0) \qquad (5.37)$$

leads to

$$\sum_\sigma a_\sigma^i \hat{H}_{eff}(\rho_0) |\Phi_\sigma\rangle = \varepsilon_i \sum_\sigma a_\sigma^i |\Phi_\sigma\rangle \qquad (5.38)$$

and after multiplication by $\langle\Phi_\eta|$ we have

$$\sum_\sigma a_\sigma^i \langle\Phi_\eta|\hat{H}_{\mathrm{eff}}(\rho_o)|\Phi_\sigma\rangle = \varepsilon_i \sum_\sigma a_\sigma^i \langle\Phi_\eta\|\Phi_\sigma\rangle \tag{5.39}$$

5.3.2 Empirical Tight-Binding: ETB or Hückel Theory

We have to diagonalize the Hamilton matrix $H_{\eta\sigma} = \langle\Phi_\eta|\hat{H}_{\mathrm{eff}}(\rho_o)|\Phi_\sigma\rangle$. However, our basis set is non-orthogonal, that is, the overlap matrix $S_{\eta\sigma} = \langle\Phi_\eta|\Phi_\sigma\rangle$ appears in the eigenvalue equations.

In empirical schemes, the basis functions are taken as orthogonal matrix elements, that is, $S_{\eta\sigma} = \delta_{\eta\sigma}$.

In the background, we have the so-called Löwdin orthogonalization. The introduction of the orthonormal orbitals produces an effective changing of the Hamiltonian.

The diagonalization leads to the one-particle energies ε_i, that is, to the so-called electronic energy:

$$E_{\mathrm{elec}} = \sum_i \varepsilon_i \tag{5.40}$$

If we compare this to the total energy in DFT,

$$E[\rho] = \sum_i^{\mathrm{occ}} \varepsilon_i - \frac{1}{2}\int \frac{\rho(\vec{r})\rho(\vec{r}')\mathrm{d}\vec{r}\mathrm{d}\vec{r}'}{|r-r'|} + E_{\mathrm{xc}}[\rho_o] - \int v_{\mathrm{xc}}(\vec{r})\rho_o(\vec{r})\mathrm{d}\vec{r} + \frac{1}{2}\sum_{\eta\sigma}\frac{Z_\eta Z_\sigma}{R_{\eta\sigma}} \tag{5.41}$$

From Equation 5.41, we can see that a big part of the energy is missing, that is, the so-called double counting and core-core repulsion terms in DFT. Mainly, the double counting terms depend on the reference density ρ_o only. The XC parts are hard to evaluate, however, we can say that they decay exponentially due to the exponential decay of the density overlap.

If we assume an atomic density decomposition, $\rho = \sum_\eta \rho_\eta$, the coulomb contributions shown below demonstrate an exponential contribution:

$$\frac{1}{2}\sum_{\eta\sigma}\left[\frac{Z_\eta Z_\sigma}{R_{\eta\sigma}} - \int \frac{\rho_\eta(\vec{r})\rho_\sigma(\vec{r}')\mathrm{d}\vec{r}\mathrm{d}\vec{r}'}{|r-r'|}\right] \tag{5.42}$$

Therefore, the first ETB model has to be

$$E_{\mathrm{tot}} = \sum_i^{\mathrm{occ}} \varepsilon_i + \frac{1}{2}\sum_{\eta\sigma} U_{\eta\sigma} \tag{5.43}$$

with the two-body terms $U_{\eta\sigma}$ being the exponentials fitted to reproduce the geometries, vibrational frequencies, and reaction energies of suitable systems.

5.3.3 Density Functional–Based Tight Binding: DFTB

The derivation of parameters by fitting is a complicated process. If it were possible to derive the parameters from the DFT calculations, more flexibility and simplified parameterizations would be gained. At first, a basis set is required. In tight binding theory, the fundamental functions are atomic orbitals, which are calculated from the atomic KS equations:

$$\left[\frac{-1}{2}\nabla^2 + V_{\mathrm{eff}}(\rho_{\mathrm{atom}}(\vec{r}))\right]\Psi_i = \varepsilon_i\Psi_i \tag{5.44}$$

Atomic orbitals have a disadvantage in that they are diffuse. In solids, large molecules or clusters that are the size of the orbitals are "compressed" due to the interaction with the neighbors. A measure for the distance between neighbors is given by the so-called covalent radius, r_0, and is empirically determined for all atoms. Therefore, it is wise to use orbitals that somehow incorporate this information. To enhance this effect, an additional harmonic potential is added to the atomic Kohn–Sham equations that leads to compressed atomic orbitals, or optimized atomic orbitals (O-LCAO):

$$\left[\frac{-1}{2}\nabla^2 + V_{\text{eff}}(\rho_{\text{atom}}(\vec{r})) + \left(\frac{\vec{r}}{r_o}\right)^2 \right]\Psi_i = \varepsilon_i\Psi_i \tag{5.45}$$

As a result of the atomic calculations, we get the orbital Ψ_i, the electron density at atom $\eta\rho_\eta = \sum_i |\Psi_i|^2$, and the overlap matrix $S_{ij} = \langle\Psi_i|\Psi_j\rangle$. To solve the eigenvalue problem (Equation 5.38), we only need the Hamiltonian matrix. This leads to further approximations; although we have the complete input density, $\rho_0 = \sum_i \rho_i$, the Hamiltonian evaluation would be very complicated:

$$H_{ij} = \left\langle\Psi_i|\hat{H}[\rho_0]|\Psi_j\right\rangle = \left\langle\Psi_i\left|\hat{H}\left[\sum_\mu \rho_\mu\right]\right|\Psi_j\right\rangle \tag{5.46}$$

We, therefore, usually make the so-called two-center approximation for $i \neq j$:

$$H_{ij} = \left\langle\Psi_i|\hat{H}[\rho_\eta + \rho_\mu]|\Psi_j\right\rangle \tag{5.47}$$

where orbital η is located on atom i and orbital μ is located on atom j. The diagonal Hamiltonian elements, $H_{ii} = \varepsilon_i$, are taken from Equation 5.45.

H_{ij} and S_{ij} are tabulated for various distances between atom pairs up to 10 Å, where they vanish. For any molecular geometry, these matrix elements are based on the distance between the atoms and then oriented in space by using the Slater-Koster sin/cos combination rules. Then, the generalized eigenvalue problem Equation 5.38 is solved and the first part of the energy can be calculated. It should be emphasized that this is a non-orthogonal TB scheme, which is more transferable due to the appearance of the overlap matrix.

The second part, namely,

$$E_{\text{rep}}[\rho_0] = \frac{1}{2}\sum_{\eta\sigma} U_{\eta\sigma} \tag{5.48}$$

is calculated piont-wise as follows: To get the repulsive potential, for example, for the first atom one could take the second atom as a dimmer, $\eta\mu$, stretch its bond, and for each distance calculate the total energy with DFT and the electronic TB part $\sum_i \varepsilon_i \cdot U_{\eta\mu}(R_{\eta-\mu})$ is the given point-wise for every $R_{\eta-\mu}$ by

$$U_{\eta\sigma}[R_{\eta-\sigma}] = E_{\text{tot}}^{\text{DFT}}(R_{\eta-\sigma}) - \sum_i \varepsilon_i \tag{5.49}$$

The resulting density functional–based tight binding (DFTB) method works well for homonuclear systems, where the charge transfer between the atoms in the system does not occur or is very small. As soon as the charge starts flowing between atoms because of an electronegativity difference, the resulting density is no better approximated by the superposition of the atomic densities, $\rho_0 = \sum_i \rho_i$. However, the formalism works very well, when the charge flow is small, therefore an extension will try to start from the non-self-consistent scheme. However, the effective Kohn–Sham potentials contain only the neutral reference density, ρ_0, that does not account for the charge transfer between the atoms. With a Taylor series expansion of the potential with a ground state density, ρ, around the reference density, ρ_0

$$V_{\text{eff}}[\rho] = V_{\text{eff}}[\rho_0] + \int \frac{\mathrm{d}V_{\text{eff}}[\rho]}{\mathrm{d}\rho} \mathrm{d}\rho \mathrm{d}\vec{r} \qquad (5.50)$$

In this expression, the potential inserted into the K–S equations will lead to the same matrix elements of $H_{ij}[\rho_0]$, depending on the reference density as above, which has to deal with the functional derivative.

5.3.4 Self-Consistent Solution of the KS Equations

The total energy above the K–S equations requires the addition of other terms. Thus, we can start with the functional expansion of the DFT total energy. The self-consistent charge (SCC)-DFTB method is derived from density functional theory by a second-order expansion of the DFT total energy functional with respect to the charge density fluctuations, $\mathrm{d}\rho$, around a given reference density, ρ_0. From these values, we define an equivalent charge density and integral at $\rho'_0 = \rho_0(r')$ and $\int' = \int \mathrm{d}r'$:

$$E = \sum_i^{\text{occ}} \langle \Psi_i | \hat{H}^\circ | \Psi_i \rangle + \frac{1}{2} \int' \int \left(\frac{1}{|r-r'|} + \frac{\partial^2 E_{\text{xc}}}{\partial \rho \partial \rho'}\bigg|_{n_o} \right) \mathrm{d}\rho(\vec{r}) \mathrm{d}\rho(\vec{r}')$$

$$- \frac{1}{2} \int' \int \left(\frac{\rho_0 \rho'_0}{|r-r'|} \right) + E_{\text{xc}}[\rho_0] - \int V_{\text{xc}}[\rho_0] n_o + E_{\text{cc}} \qquad (5.51)$$

After introducing the LCAO basis from Equation 5.36, the first term becomes

$$\langle \Psi_i | \hat{H}^\circ | \Psi_i \rangle = \sum_{\sigma\nu} a_\sigma^i a_\nu^i H_{\sigma\nu} \qquad (5.52)$$

The last four terms depend only on the reference density, ρ_0, and represent the repulsive energy contribution, E_{rep}, discussed above. Thus, we just have to deal with the second-order terms. The second-order term in the charge density fluctuations $\mathrm{d}\rho(\vec{r})$, that is, the second term in Equation 5.51, is approximated by writing $\Delta\rho$ as a superposition of atomic contributions, $\Delta\rho_0 = \sum_\nu \Delta\rho_\nu$. This approach decays quickly with the increasing distance from the corresponding center. To simplify the second term further, Elstner applied a monopole approximation:

$$\Delta\rho_\nu \approx \Delta q_\nu F_{00}^\nu Y_{00} \qquad (5.53)$$

$\Delta\rho_\nu$ is assumed to look like a 1s orbital. F_{00} denotes the normalized radial dependence of the density fluctuation on the atom ν that is constrained to be spherical (Y_{00}). In this case, the angular deformation of the charge density change in second order is neglected:

$$E^{\text{2ndterm}} \approx \frac{1}{2} \sum_{\nu\eta} \Delta q_\nu \Delta q_\eta \int' \int \left(\frac{1}{|r-r'|} + \frac{\partial^2 E_{\text{xc}}}{\partial \rho \partial \rho}\bigg|_{n_o} \right) F_{00}^\eta F_{00}^\nu Y_{00}^2 \mathrm{d}(\vec{r}) \mathrm{d}(\vec{r}') \qquad (5.54)$$

For large distances, $R_{\nu\eta} = |r - r'| \to \infty$, the XC terms vanish, and the integral describes the coulomb interaction of two spherical, normalized charge densities, which reduce basically to $1/R_{\nu\eta}$, that is, we get

$$E^{\text{2ndterm}} \approx \frac{1}{2} \sum_{\nu\eta} \frac{\Delta q_\nu \Delta q_\eta}{R_{\nu\eta}} \tag{5.55}$$

For the vanishing interatomic distance, $R_{\nu\eta} = |r - r'| \to \infty$, the integral describes the e–e interaction on atom ν. We can approximate the integral as

$$E^{\text{2ndterm}} \approx \frac{1}{2} \frac{\partial^2 E_\nu}{\partial q_\nu^2} = U_\nu \tag{5.56}$$

where U_ν is known as the Hubbard parameter ("chemical hardness"). It describes how much the energy of a system changes upon adding or removing electrons.

Now we need a formula to interpolate between these two cases. A very similar situation appears in semiempirical quantum chemical methods, where $\gamma_{\nu\eta}$ has a simple form, given by the Klopman–Ohno approximation:

$$\gamma_{\nu\eta} = \frac{\gamma_{\nu\eta}}{\sqrt{R_{\nu\eta}^2 + \frac{1}{4}\left(\frac{1}{U_\nu} + \frac{1}{U_\eta}\right)^2}} \tag{5.57}$$

By approaching the charge density fluctuations with spherical charge densities, the Slater distributions are

$$F_{00}^\nu = \frac{\tau_\nu}{8\pi} \exp(-\tau_\nu |\vec{r} - R_\nu|) \tag{5.58}$$

The function $\gamma_{\nu\eta}$ depends on the parameters τ_ν and τ_η that determine the extension of the charge densities of both atoms. This function has $1/R_{\nu\eta}$ dependence for large $R_{\nu\eta}$ and approaches a finite value for $R_{\nu\eta} \to 0$. For zero interatomic distances, $\nu = \eta$, one finds that

$$\tau_\nu = \frac{16}{5}\gamma_{\nu\nu} \tag{5.59}$$

The expansion of the charge distribution is inversely proportional to the "chemical hardness" of the respective atom, that is, the size of an atom is inversely related to its chemical hardness.

After integration, E^{2nd} becomes a simple two-body expression depending on atomic-like charges:

$$E^{\text{2ndterm}} \approx \frac{1}{2} \sum_{\nu\eta} \Delta q_\nu \Delta q_\eta \gamma_{\nu\eta} \tag{5.60}$$

The diagonal terms $\gamma_{\nu\nu}$ model the dependence of the total energy on charge density fluctuations of the second order. The monopole approximation restricts the change of the electron density considered, and no spatial deformations are included. Only the change of energy with respect to the change of charge on the atom ν is considered. By neglecting the effect of the chemical environment on atom ν, the diagonal part of γ can be approximated by

$$\gamma_{\nu\nu} = U_\nu = \frac{\partial^2 E_\nu}{\partial q_\nu^2} \tag{5.61}$$

The Hubbard parameter can be approximated by the difference of the ionization potential and the electron affinity of the atom.

5.4 PROBLEM OF ELECTROCATALYTIC ADSORPTION OF HYDROGEN ON PLATINUM

5.4.1 INTRODUCTION TO EXPERIMENTAL APPROACHES FOR THE ADSORPTION OF HYDROGEN ON PLATINUM

Hydrogen adsorption and hydrogen ion electroreduction at platinum electrodes is a system of great importance in electrochemistry. Apart from being involved in most electrode processes, this has significance in many applications, such as fuel cells, water electrolysis, organic electroreductions, embrittlement, and acidic corrosion processes. It is not surprising to see that various kinds of experimental methods ranging from conventional electrochemical techniques to modern surface and *in situ* technologies have been intensively employed to study this system as explained in chapter 2.

At least five types of hydrogen adsorption states on platinum electrodes have been postulated. These are the so-called strongly bound hydrogen, the weakly bound hydrogen, the on-top hydrogen, the dihydride state, and the sub-surface state. Most of these can be detected by cyclic voltammetry and characterized at single crystal electrodes [37–42].

Terminal hydrogens at approximately 2090 cm^{-1} were claimed to be overpotential deposition (OPD) hydrogens that could only be observed at the onset potential of hydrogen evolution [39]. This fact led to the conclusion that the on-top adsorbates are the active intermediates directly involved in the hydrogen evolution reaction. Ogasawara and Ito, however, found that while the terminal hydrogens exhibit a remarkable dependence on surface crystallographic orientation, the rate of hydrogen evolution is not sensitive to the surface orientation [43]. This indicated that the terminal hydrogen might not be an intermediate that is directly involved in the hydrogen evolution process. By using the sum-frequency generation (SFG) technique, it was possible to conclude that the terminal hydrogens in the frequency region of 1800–2020 cm^{-1} were associated with the under-potential deposited (UPD) hydrogens rather than the OPD hydrogens. Further, the dihydride adsorption state observed at approximately 1770 cm^{-1} by SFG was attributed to the intermediate in the hydrogen evolution reaction [41]. The possible existence of H sub-surface states was studied using platinum electrodes of different topographies that were characterized by *ex situ* scanning tunneling microscopy (STM). Based on the latest results of voltammetry and *ex situ* STM, Martins et al. [42] suggested that the H-adatom subsurface state is produced in the potential range where an anodic current hump appears between the strong and weak hydrogen peaks. This process occurs under either a potential holding or a potentiodynamic fast cycling within the 0.01–0.07 V domains. Using a highly sensitive confocal microprobe Raman method and a sole surface pretreatment procedure, high-quality surface Raman spectra of hydrogen adsorbed on top of the platinum electrodes were obtained within the potential region of hydrogen evolution [44,45]. The results showed that, with the negative shift of the potential, there is a red shift of Pt–H vibrational frequency, accompanying an increase of the band intensity. No Pt–H band could be detected in the strongly bound hydrogen adsorption region. A very weak and broad s band for the terminal Pt–H vibration in the weakly bound hydrogen adsorption region was discernible. The full width at half-maximum intensity of the Pt–H vibrational band broadens when the potential was swept into the H region, while the intensity decreased sharply.

5.4.2 INTRODUCTION TO THEORETICAL APPROACHES FOR HYDROGEN ELECTROADSORPTION ON PLATINUM

The Pt–H atom interaction plays a key role in electrochemistry, particularly at the Pt/aqueous solution interface in the range of the potentials related to the H-adatom electrosorption equilibrium and hydrogen evolution reaction. The situation outlined above suggested the convenience of attempting a quantum chemistry approach to surface species that are likely formed at a simulated platinum/aqueous electrochemical interface in order to discriminate the structure and energy of possible H-adsorbates. This is a relevant issue in dealing with, for instance, the interpretation of the complex electrosorption spectra of H-atoms on platinum in an aqueous solution, as well as to provide a more realistic approach to the nature of H-atom intermediates involved in the hydrogen evolution reaction.

Some semiempirical quantum chemical studies of the H–Pt system at the electrochemical interface [46–48] have been carried out. The effect of the external electric potentials was simulated by shifting the VSIP values from the Fermi energy level of platinum upward or downward (as explained above in EHMO methods). Thus, Leban and Hubbard performed an EHMO iterative in which a $Pt_5(4,1)$ cluster, consisting of four platinum atoms in the first layer and one platinum in the second, was used to imitate the Pt(111) surface. Predictions of the most stable orientations on the Pt(111) surface of species such as H_2O, OH^-, and H inside the plane of the three adjacent platinum surface atoms were envisaged [46]. Zinola and Arvia presented an EHMO calculation with large clusters of $Pt_{22}(14,8)$ and $Pt_{25}(16,9)$. The adsorption of atomic H on hollow sites co-adsorbed with an on-top OH species for Pt(111) and H atom adsorption on bridge sites for Pt(100) were found as the main adsorbates. Hydrogen adsorption on subsurface platinum atoms could also occur when the potentials are lower than the hydrogen-electrode equilibrium potential [47]. On the other hand, density functional calculations have been performed to elucidate the nature of on-top Pt–H bonding and to examine the spectroscopic properties of the terminal adsorbed hydrogens [48]. For the latter, calculations were carried out with Gaussian 94 and the theoretical method employed was B3LYP (hybrid method with a mixture of Hartree–Fock exchange with DFT exchange correlation) [49,50].

The single on-top Pt–H interaction is described by an adcluster of $[Pt_5–H]$ with the hydrogen atom sitting on the central platinum atom. To simulate the effect of high adsorption coverage, four additional on-top H atoms were added, which led to a $[Pt_5–H_5]$ adcluster. In all cluster models, the shortest Pt–Pt distance was 0.274 nm, which was taken from the experimental value of the bulk crystal and was kept frozen during the model calculations. The adsorption geometry, that is, the equilibrium bond distances $r_{Pt–H}$, and the vibrational frequencies, $\nu_{Pt–H}$, in an e harmonic approximation were obtained by a fourth-degree polynomial fit to eight points around the minimum of the corresponding potential curves.

For $[Pt–H]$, our B3LYP calculation results can be compared favorably with those from the all-electron calculations [51,52], from which $^3\Delta$ is the ground state. The computed Pt–H bond distance, bond energy, and Pt–H vibrational frequency are 1.537 Å, 3.21 eV, and 2321 cm^{-1}, respectively, which are close to the corresponding values obtained by Dyall [51]. The calculations in Ref. [48] indicated that the Pt–H bond length is shortened and the Pt–H vibrational frequency is increased as the Pt–H is negatively charged. The computed Pt–H bond energy in $[Pt–H]^-$ is around 3.15 eV.

Both the charged and field models have been employed in Ref. [48] to simulate the effects of the electrode potentials. Since H atoms have no permanent dipole and the dipole moment of Pt–H is small, it is not surprising that the field model predicts a very small change in the Pt–H bonding properties with a change of the electrode potential. Since the strong electric fields across the double layer are due to the surface charge, the presumed uncharged surface cluster in the field model may lead to unreal results.

The main effect of charging a cluster is to alter its HOMO and LUMO levels that correctly reflect the fact that the electrode potentials change the Fermi level of the electrode to control the electrochemistry of the system. The observed large tuning rate of the Pt–H frequency should be attributed to the work-function shift with the change of electrode potential. Vibrational studies of hydrogens adsorbed on platinum single crystal surfaces have been made under UHV. Vibrations from threefold hydrogen adsorbed on the Pt(111) surface were observed at wavenumbers 1200 cm^{-1} [53]. The band at 2000 cm^{-1} due to the terminal hydrogens has not been reported under UHV. On the other hand, no Pt–H band due to the multi-coordinated hydrogens could be detected spectroscopically in the electrochemical environment. Only a band for the terminal Pt–H vibration is discernable in the potential region of hydrogen evolution.

5.4.3 ADSORPTION MODES OF NITROGEN OXIDES ON PT(111) SINGLE CRYSTAL

The effective remediation of NO_x ($x = 1, 2$), generated as side products of combustion, is a major challenge in environmental catalysis and electrocatalysis [54,55]. The catalytic oxidation of NO(g) to NO_2(g) is a common element of most NOx removal strategies, including NO_x storage and the reduction catalyst and NO_x-selective catalytic reduction [56]. Nitrous Oxide (NO) oxidation on platinum indicates that the rate of NO oxidation per active surface-platinum site increases with the increasing particle size, suggesting that the low-index platinum plane faces to be the most active for the catalysis [57]. The electrooxidation activity is associated with high coverages of oxygen-containing species, although very high coverages appear to inhibit activity.

NO form a number of higher oxides, such as NO_2 and NO_3 that compete with adsorbed NO and atomic oxygen for platinum surface sites. Several atomistically detailed models of the NO oxidation reaction based on DFT-derived parameters for the reaction kinetics have been reported [58,59]. These models are successful in describing the sensitivity of the reaction kinetics to surface coverage. They are somewhat limited in terms of the surface species and the reaction steps considered.

Schneider et al. [54] use large super cell DFT simulations to characterize the stable and metastable states of adsorbed atomic nitrogen, atomic oxygen, and the oxides mentioned above on a Pt(111) surface. Charge density analysis is used to characterize the nature and extent of interactions between the adsorbates and the surface. Calculated harmonic vibrational frequencies are used to correlate the adsorbates with experimental observation, and the adsorption energies of all the species are compared. In the low coverage limit, all adsorbates are thermodynamically unstable with respect to adsorbed NO, thereby reinforcing the critical role of local surface coverage in promoting the conversion of NO to NO_2 on the Pt(111) surface.

The DFT calculations were performed using the periodic super cell plane-wave basis approach. Electron cores were described with the projector-argumented wave (PAW) method [60], and plane waves were included at an energy cutoff of 400 eV. Electronic energies were computed with the PW91 implementation of the generalized gradient approximation (GGA) [61].

The close-packed Pt(111) surface offers a number of potential adsorption sites for a NO molecule, including onefold coordination (on-top) at a single platinum, twofold coordination at a bridge site, and threefold coordination for hollow (3–3) and (3–1) coordinations. There are clear evidences of NO binding at bridge sites at low coverage from different but more recent techniques [62,63]. It seems that NO adsorbs in the threefold hollow sites at low coverage on Pt(111). All the theoretical predictions by DFT [64–66] coincide with this coordination to be the most stable at low coverage.

Some results in the literature show that nitrogen monoxide (NO) is bound N-down and is normal to the platinum surface at hollow sites with an increase in the N–O bond length by 0.046 Å over the gas phase. The GGA-calculated adsorption energies in [54] at the 1/16 and 1/4 monolayer are 2.00 and 1.93 eV, respectively. Previously reported values at the 1/4 monolayer are from 1.75 to 2.10 eV [64,67]. These differences arise from the difference in the number of atoms simulating the cluster surface that affect the absolute value of adsorption energies. The adsorption of NO causes the platinum atoms local to the adsorbate to relax laterally or parallel to the surface and away from the adsorbate, as well as vertically, or normal to, and upward from the surface. These relaxations can result in substantial changes in the platinum–platinum separations near the adsorbates. In the case of

NO adsorption in a hollow site, the three platinum atoms nearest to N separate to a Pt–Pt distance of 2.95 Å, or 0.13 Å greater than on the clean Pt(111) surface, and vertically by 0.153 Å with respect to the platinum surface. The vertical distance between the three local platinum and nitrogen atoms is 1.119 Å. In general, the lateral platinum relaxations are smaller at higher surface coverage and the platinum-adsorbate distances are greater. As surface coverage increases, more surface platinum becomes involved in the platinum-adsorbate bond, causing less pronounced vertical platinum relaxation.

At large NO coverages, other adsorption configurations are reported for platinum [68–70]. We have characterized two other isomers of surface-bound NO, both of which are metastable at 1/16 ML coverage. The isomer bonds N-down in a threefold hollow site. The bond length variations and charge distributions are quite similar, but the adsorption energy decreases by 0.14 to 1.86 eV, comparable to the previous study [64].

On the other hand, NO can bind N-down and be bent on-top a single platinum at high NO coverages, as has been observed experimentally [69,70] and by the DFT calculations [71–73]. At 1/16 ML, the on-top configuration has a N–O bond length of 1.179 Å, an increase of 0.010 Å over the gas phase but significantly shorter than the threefold isomer bond lengths and is placed at an angle of 53.8° with respect to the surface normal. The adsorption causes the on-top platinum to rise a rather large 0.29 Å above the surface. The on-top adsorption energy is 1.60 eV at 1/16 monolayer, 0.40 eV less than the other isomer. Consistent with this difference, the net charges transferred to the on-top NO adsorption is only 0.15 electrons and the charge distribution between N and O is quite similar to the gas-phase molecule.

The larger GGA predicted favorite for hollow over the on-top adsorption for NO (0.40 eV) than for CO (0.15 eV) [74] as well as the experimentally observed NO preference for hollow sites over the on-top, can both be traced to the π-acceptor strength of NO. On the basis of the comparisons of the integrated DOS, π back-bonding from platinum to NO constrained to adsorb on-top and perpendicular to the platinum surface is 0.8 electrons less effective than in a hollow site, due to poorer overlap between the $2\pi^*$ and Pt d states in the on-top position. In order to compensate for the lack of donated d-density, on-top NO bends with respect to the platinum surface, enhancing σ bonding at the expense of π bonding and increasing the binding energy by 0.5 eV. The NO bending allows the 5σ and the partially filled in-plane $2\pi_y$ orbitals to hybridize into an N-centered lone pair and a half-filled orbital directed toward the on-top platinum. The latter hybrid σ bonds effectively with the Pt d_{z^2} orbitals oriented normal to the platinum surface. The π back-donation is weak and localized to the out-of-plane $2\pi_x$ [75]. The strong π acceptor NO thus prefers hollow over on-top sites at low coverage because of better π interactions with the surface in the former. When forced to compete with other adsorbates for Pt d density, as occurring at higher coverage, the bent NO σ bonding at the on-top sites becomes more favorable. The weaker π-acceptor/better σ-donor CO discriminates less strongly between the two sites, and in fact, the GGA appears to reverse the actual on-top over the hollow site preference because it exaggerates the contributions of π back-bonding [76].

5.4.4 Adsorption Modes of Carbon Monoxide on Pt(111) Single Crystal

In electrocatalysis, the presence of surface defects is important to activate electron transfer in complex reactions. This is the consequence of the adsorbed species involved during the electrocatalytic process. Thus, Pt(110), the low Miller index plane, deserves special attention in theoretical studies.

The clean Pt($1\bar{1}0$)-(1×2) surface consists of close-packed rows running in the [$1\bar{1}0$] direction with every second top row missing, as shown in Figure 5.1. At room temperature and low surface coverage, the metal species fills the missing rows of the 1×2-reconstructed Pt(110) surface both as mobile "adatoms" and in some cases as alloyed chains; see for example [77,78]. The alloy chains can be formed by a reaction that creates vacancies and pits in the surface. The vacancy defects on the Pt(110)-(1×2) surface is thus able to catalyze a multitude of potential adsorption sites for CO, a model molecule for reactions in the electrochemical environment. The structure and bonding of CO

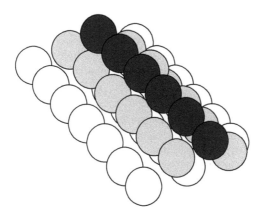

FIGURE 5.1 Pt(110) (1 × 2) (N = 37) cluster showing the first (white balls), the second (gray balls), and the third (black balls) atomic layers.

on the clean Pt(111) surface is well-known, with CO binding up-right with the carbon down in an on-top position at low coverage, which at a higher coverage is accompanied by adsorption in a bridge site position [79,80]. At CO saturation both the on-top and the bridge are filled with the ($\sqrt{3} \times \sqrt{3}$) R30° and the c(4 × 2) adlayers [80,81].

Some of the calculations to check for CO adsorption on the single crystal by DFT require periodic boundary conditions using the DACAPO code [82]. The gradient corrected PW91 exchange-correlation functional [83] can be used to calculate the geometries; however, energies can be evaluated using the RPBE functional [84]. Wave functions were expanded in terms of the plane waves up to a kinetic energy of 340.145 eV. The Brillouin zone was sampled by 5 × 4 × 2 k-points [85]. The adsorbed CO molecules were relaxed at the on-top positions at 1/4 ML coverage with respect to the unreconstructed Pt(110) surface. The surface model layers, resulting from the periodic nature of the calculations, were separated by 20 Å.

It is relevant to compare the calculated energies and the adsorption trends with the experimental data. For CO/Pt(110) systems, the reported adsorption energies for the CO adsorption range is from 35 to 43 kcal mol^{-1} for on-top sites [86,87]. Von Schenck [77] found adsorption energies of 42.9 and 42.7 kcal mol^{-1} for the on-top adsorbates depending on the layer of platinum atoms to which they were attached. The authors by comparing calculated energies to experimental values found an overestimate of the adsorption energy. Investigating the adsorption of small molecules such as NO and CO on transition metal surfaces (Ni(100), Ni(111), Rh(100), Pd(100), and Pd(111)), Nørskov et al. found an average over-binding of ~5.8 kcal mol^{-1} [84].

In the presence of a metal adspecies on the surface, it was found that CO also adsorbs on top of any platinum atom. However, in the case of tin metal species, the CO coverage decreases as the tin coverage increases. Although only the on-top adsorption geometry is observed, several platinum atoms show very similar characteristics with respect to CO adsorption.

5.4.5 ADSORPTION MODES OF METHANOL, FORMALDEHYDE, AND FORMIC ACID ON PT(111) SINGLE CRYSTAL

The oxidation of small organic molecules has been extensively studied due to their relevance to electrocatalysis [88]. Among these molecules, methanol, formaldehyde, and formic acid have attracted the attention of the scientific community because of their simple structure and the relative simplicity of the investigation and interpretation of the results. There are several results which show that methanol on group VIII transition metal surfaces (Pt(111) [89], Pd(111) [90], Pd(100) [91], Rh(111) [92], and Ru(001) [93]) decomposes directly to adsorbed CO (or related compounds) and hydrogen atoms.

The methoxy intermediate was only observed on Pt(111) [89] and Rh(111) [92]. On Pd(111), formaldehyde species were created on the surface during the decomposition. The adsorbed methanol altered into the formate in the presence of molecularly adsorbed oxygen [89]. The formaldehyde can organize on the metal surface through the oxygen lone pair in η^1 and in an $\eta^2(C,O)$ dihapto-configuration. Formaldehyde easily decomposes and polymerizes on clean Pt(111) [94], Rh(111) [95], and Pd(111) [96] surfaces. It is interesting to note that η^1 formaldehyde is formed during the decomposition of methanol on Pt(111) in the presence of oxygen. On the other hand, formic acid decomposed at low coverage via dehydrogenation and formed catamers upon the adsorption at high coverage. The formation of the hydrogen-bonded catamers is suppressed in the presence of atomic oxygen on Pd [97], Pt [98], and Rh(111) [99] surfaces.

Bakó and Pálinkás [100] have conducted a complete study of these three low-carbon oxygenated species, which are of significance to electrocatalysis. The adsorption energies as well as the electronic and vibrational properties were calculated by using the Vienna *Ab Initio* Simulation Package (VASP) [101], a density functional theory (DFT) code with a plane-wave basis set. The electron–ion interactions were described using the PAW [101] method, which was expanded within a plane-wave basis set up to a cutoff energy of 400 eV. Electron exchange and correlation effects were described by the Perdew–Burke–Ernzerhof (PBE) [102] GGA type exchange-correlation functional. The total energy of platinum had a minimum at lattice constant of 3.989 Å, which is close to the experimental value of 3.912 [103].

Geometry optimizations were performed on a super-cell structure using periodic boundary conditions. The (111) surfaces were generally modeled using a $3 \times 2\sqrt{3}$ super cell. The metal slab was chosen to be three atomic layers thick, and a 15 Å vacuum layer was used to ensure that there were no interactions between the surface adsorbates on one layer and the next slab. The first metal layer was allowed to relax, while the bottom two layers of the platinum atoms were held fixed in their bulk position. All atomic coordinates of the adsorbed species and the metal atoms in the relaxed metal layers were optimized to a force of less than 0.025 eV Å^{-1} on each atom.

Methanol binds through its oxygen atom in an on-top position to the transition metal surfaces [104,105]. At the 1/12 monolayer, the calculated binding energy is 0.32 eV, which agrees well with the results of Greely and Mavrikakis [106] (0.33 eV). Figure 5.2 represents the binding mode of methanol on Pt(111).

In this most favorable configuration, two methyl hydrogens are pointing toward the Pt(111) surface, whereas in the metastable state (for only one methyl hydrogen pointing to the surface) the binding energy is about 0.05 eV smaller than in the global minimum. The electron density of the adsorbate minus the sum of the electron densities of the isolated surface and the methanol

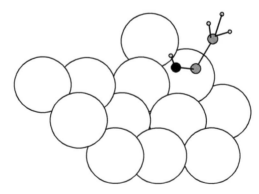

FIGURE 5.2 Pt(111) ($N = 12$) cluster showing only the first atomic layers. The calculated binding mode of methanol is shown on Pt(111).

molecule in the same configuration can be considered. Charge depletion can be observed on the lone pair of oxygen and on the d_{z^2} orbital of platinum, and charge accumulation on the d_{xz} and d_{yz} orbitals of platinum and oxygen. The change in the surface work function with respect to the clean surface is about $\Delta\Phi = -1.02$ eV, which refers to a charge transfer to the surface. The most significant difference in the normal mode of the adsorbed methanol can be seen in the OH stretching frequency around 3564 cm^{-1}. This frequency has a red shift of almost 200 cm^{-1} on the surface compared to the free molecule, where the ν_{OH} is at about 3750 cm^{-1}. The most intensive peak in the simulated spectra can be seen at about 976 cm^{-1}, which corresponds to the CO stretching mode of methanol.

On the other hand, formaldehyde can bind to the platinum surfaces in different configurations; however, Bakó and Pálinkás [100] considered only the most stable structure. It has been already shown that formaldehyde prefers the top-bridge-top (di-sigma) configuration [107,108]. In this configuration, the molecular axis of the formaldehyde molecule is parallel to the surface, and the Pt–C and Pt–O bonds have nearly the same lengths, while the hydrogen atoms are pointed away from the surface. The calculated binding energy is 0.58 eV, which is slightly different from the previous results. This discrepancy could be due to the fact that smaller coverage (1/12 monolayer) was used in their calculations, while in other studies 1/4 to 1/9 coverages were applied [107,108]. The adsorption energy of formaldehyde on Pt(111) was experimentally estimated to be 0.54 eV by Abbas and Madix [109]. Platinum atoms underneath the oxygen and carbon are raised above the platinum top layer by 0.12 and 0.14 Å, respectively. The carbonyl bond of formaldehyde is elongated by about 0.15 Å, and the carbon atom that is sp^2 hybridized in the gas phase partially re-hybridizes to sp^3 upon interaction with the metal. The charge depletion around the CO bond of formaldehyde and an increase in the electron density along the Pt–O and Pt–C bonds are demonstrated. The work function is decreased by a value of $\Delta\Phi = -0.11$ eV, with respect to the clean surface. The small negative value of $\Delta\Phi$ indicates that the magnitude of donation and back-donation effects is nearly the same. The vibrational properties of the adsorbate show a loss of the double bonds by the decrease of the corresponding mode, that is, 1767 cm^{-1} in the gas phase in contrast to 1144 cm^{-1} in the adsorbed state.

Formic acid exhibits rotational isomerism, representing two structures defined by the rotation of the OH group around the CO bond. In the gas phase, the trans conformer is dominant, and the energy difference between the two conformers is about 0.16–0.17 eV [110]. The energy difference according to the calculation is about 0.16 eV, which is in good agreement with the experimental value. The calculated structural and energetic properties of trans formic acid on Pt(111) surface indicates a preference of the adsorption in the di-sigma configuration with its O–H bond pointing to the surface. In this case, the OH bond is elongated by about 0.05 Å, compared to the gas phase value. This data revealed that this OH bond is weak, and we can assume that our calculated structure is a metastable structure on the surface during the OH bond-breaking process. The adsorption energy of formic acid over the surface was found to be -0.42 eV. The work function is decreased by a value of $\Delta\Phi = -0.73$ eV, with respect to the clean surface. The OH stretching frequency has about 1000 cm^{-1} red shift that corresponds to the strong interaction of the OH bond with the surface. The interaction of the carbonyl oxygen lone pair with the Pt d_{z^2} orbital is also clear for the di-σ configuration.

REFERENCES

1. H. Chermette, *Coord. Chem. Rev.* **178–180** (1998) 699.
2. R.G. Parr and W. Yang, *Density-Functional Theory of Atoms and Molecules*, Oxford University Press, New York, 1989.
3. R. Dreizler and E.K.U. Gross, *Density Functional Theory*, Springer, Berlin, Germany, 1990.
4. E.K.U. Gross and R. Dreizler, *Density Functional Theory*, Plenum Press, New York, 1995.

5. R.F. Nalewajski, *Topics in Current Chemistry: Density Functional Theory*, Springer, Berlin, Germany, 1996.
6. J. Labanowski and J. Andzelm, *Theory and Applications of Density Functional Approaches to Chemistry*, Springer, Berlin, Germany, 1995.
7. D.E. Ellis, *Density Functional Theory of Molecules, Clusters and Solids*, Kluwer, Dordrecht, the Netherlands, 1995.
8. M. Springborg, *DFT Methods in Chemistry and Material Science*, Wiley, New York, 1997.
9. P. Hohenberg and W. Kohn, *Phys. Rev. A* **136** (1964) 864.
10. W. Kohn and L.J. Sham, *Phys. Rev. A* **140** (1965) 1133.
11. M. Morin, A.E. Foti, and D.R. Salahub, *Can. J. Chem.* **63** (1985) 1982.
12. J.G. Snijders and E.J. Baerends, *Mol. Phys.* **33** (1977) 1651.
13. M. Mayer, O.D. Haberlen, and N. Rösch, *Phys. Rev. A* **54** (1996) 4775.
14. T. Ziegler, *Can. J. Chem.* **73** (1995) 743.
15. P.A. Dirac, *Proc. Camb. Philos. Soc.* **26** (1930) 376.
16. S.H. Vosko, L. Wilk, and M. Nusair, *Can. J. Phys.* **58** (1980) 1200.
17. J.P. Perdew and A. Zunger, *Phys. Rev. B* **23** (1981) 5048.
18. J.P. Perdew and Y. Wang, *Phys. Rev. B* **45** (1992) 13244.
19. L. Sham, in: P.M. Marcus, J.F. Janak, and A.R. Williams (Eds.), *Computational Methods in Band Theory*, Plenum Press, New York, 1971.
20. F. Herman, J.P. van Dyke, and I.B. Ortenburger, *Phys. Rev. Lett.* **22** (1969) 807.
21. M. Levy, in: E.K.U. Gross and R.M. Dreizler (Eds.), *Density Functional Theory*, Plenum Press, New York, 1995.
22. M. Levy and J.P. Perdew, *Int. J. Quantum Chem.* **49** (1994) 539.
23. A.D. Becke, *J. Chem. Phys.* **84** (1986) 8524.
24. J.P. Perdew and Y. Wang, *Phys. Rev. B* **33** (1986) 8800.
25. J.P. Perdew, *Phys. Rev. B* **33** (1986) 8822; **38** (1986) 7406. (Erratum.)
26. E.I. Proynov, A. Vela, and D.R. Salahub, *Chem. Phys. Lett.* **230** (1994) 419.
27. D.C. Langreth and J.P. Perdew, *Phys. Rev. B* **15** (1977) 2884.
28. A.D. Becke, *J. Chem. Phys.* **98** (1993) 5648.
29. M.J. Frisch, G.W. Trucks, H.B. Schlegel, P.M.W. Gill, M.A. Robb, J.R. Cheeseman, T.A. Keith, G.A. Petersson, J.A.Montgomery, K. Raghavachari, M.A. Al-Laham, V.G. Zakrzewski, J.V. Ortiz, J.B. Foresman, J. Cioslowski, B.B. Stefanov, A. Nanayakkara, M. Challacombe, C.Y. Peng, P.Y. Ayala, W. Chen, M.W. Wong, J.L. Andres, G. Johnson, E.S. Replogle, R. Gomperts, R.L. Martin, D.J. Fox, J.S. Binkley, D.J. Defrees, J. Baker, J.P. Stewart, M. Head-Gordon, C. Gonzalez, and J.A. Pople, *GAUSSIAN 94*, Gaussian, Inc., Pittsburgh, PA, 1995.
30. A.D. Becke, *Phys. Rev. A* **38** (1988) 3098.
31. C. Lee, W. Yang, and R.G. Parr, *Phys. Rev. B* **37** (1988) 785.
32. R. Car and M. Parrinello, *Phys. Rev. Lett.* **55** (1985) 2471.
33. P.E. Blöchl, *Phys. Rev. B* **50** (1994) 17953.
34. D.K. Remler and P.A. Madden, *Mol. Phys.* **70** (1990) 921.
35. T.K. Woo, P.M. Margl, P.E. Blöchl, and T. Ziegler, *J. Phys. Chem. B* **101** (1997) 7877.
36. N.W. Ashcroft and N.D. Mermin, *Solid State Physics*, Saunders, New York, 1976, p. 411.
37. K. Christmann, in: J. Lipkowski and P.N. Ross (Eds.), *Electrocatalysis*, Chap. 1, Wiley–VCH, New York, 1998.
38. A. Bewick and J.W. Russel, *J. Electroanal. Chem.* **142** (1982) 337.
39. R.J. Nichols and A. Bewick, *J. Electroanal. Chem.* **243** (1988) 445.
40. Clavilier, D. Armand, S.G. Sun, and M. Petit, *J. Electroanal. Chem.* **205** (1986) 267.
41. A. Peremans and A. Tadjeddine, *J. Chem. Phys.* **103** (1995) 7197.
42. M.E. Martins, C.F. Zinola, G. Andreasen, R.C. Salvarezza, and A.J. Arvia, *J. Electronanal. Chem.* **445** (1998) 135.
43. H. Ogasawara and M. Ito, *Chem. Phys. Lett.* **221** (1994) 213.
44. Z.Q. Tian, B. Ren, and B.W. Mao, *J. Phys. Chem. B* **101** (1997) 1338.
45. B. Ren, X. Xu, X.Q. Li, W.B. Cai, and Z.Q. Tian, *Surf. Sci.* **427–428** (1999) 157.
46. M.A. Leban and A.T. Hubbard, *J. Electroanal. Chem.* **76** (1976) 253.
47. C.F. Zinola and A.J. Arvia, *Electrochim. Acta* **41** (1996) 2267.
48. X. Xu, D.Y. Wu, B. Ren, H. Xian, and Z.-Q. Tian, *Chem. Phys. Lett.* **311** (1999) 193.
49. A.D. Becke, *J. Chem. Phys.* **98** (1993) 5648.
50. C. Lee, W. Yang, and R.G. Parr, *Phys. Rev. B* **37** (1988) 785.

51. K.G. Dyall, *J. Chem. Phys.* **98** (1993) 9678.
52. S.W. Wang and K.S. Pitzer, *J. Chem. Phys.* **79** (1983) 3851.
53. A.M. Baro, H. Ibach, and H.D. Bruchmann, *Surf. Sci.* **88** (1979) 384.
54. R.B. Getman and W.F. Schneider, *J. Phys. Chem. C* **111** (2007) 389.
55. M. Shelef and R.W. McCabe, *Catal. Today* **62** (2000) 35.
56. W.S. Epling, L.E. Campbell, A. Yezerets, N.W. Currier, and J.E. Parks II, *Catal. Rev.* **46** (2004) 163.
57. S.S. Mulla, N. Chen, L. Cumaranatunge, W.N. Delgass, W.S. Epling, and F.H. Ribeiro, *Catal. Today* **114** (2006) 57.
58. L.D. Kieken, M. Neurock, and D. Mei, *J. Phys. Chem. B* **109** (2005) 2234.
59. D. Mei, Q. Ge, M. Neurock, L. Kieken, and J. Lerou, *Mol. Phys.* **102** (2004) 361.
60. G. Kresse and D. Joubert, *Phys. Rev. B* **59** (1999) 1758.
61. J.P. Perdew, J. Chevary, S.H. Vosko, K.A. Jackson, M.R. Pederson, D.J. Singh, and C. Fiolhais, *Phys. Rev. B* **46** (1992) 6671.
62. J.L. Gland and B.A. Sexton, *Surf. Sci.* **94** (1980) 355.
63. M.E. Bartram, B.E. Koel, and E.A. Carter, *Surf. Sci.* **219** (1989) 467.
64. Z.-P. Liu, S.J. Jenkins, and D.A. King, *J. Am. Chem. Soc.* **125** (2003) 14660.
65. H. Tang and B.L. Trout, *J. Phys. Chem. B* **109** (2005) 17630.
66. Q. Ge and D.A. King, *Chem. Phys. Lett.* **285** (1998) 15.
67. D.C. Ford, Y. Xu, and M. Mavrikakis, *Surf. Sci.* 587 (2005) 159.
68. F. Esch, T. Greber, S. Kennou, A. Siokou, S. Ladas, and R. Imbihl, *Catal. Lett.* **38** (1996) 165.
69. M. Matsumoto, K. Fukutani, T. Okano, K. Miyake, H. Shigekawa, H. Kato, H. Okuyama, and M. Kawai, *Surf. Sci.* **454** (2000) 101.
70. M. Matsumoto, N. Tatsumi, K. Fukutani, and T. Okano, *Surf. Sci.* **513** (2002) 485.
71. F. Esch, T. Greber, S. Kennou, A. Siokou, S. Ladas, and R. Imbihl, *Catal. Lett.* **38** (1996) 165.
72. H. Aizawa, Y. Morikawa, S. Tsuneyuki, K. Fukutani, and T. Ohno, *Surf. Sci.* **514** (2002) 394.
73. H. Tang and B.L. Trout, *J. Phys. Chem. B* **109** (2005) 17630.
74. P.J. Feibelman, B. Hammer, J.K. Nørskov, F. Wagner, M. Scheffler, R. Stumpf, R. Watwe, and J. Dumesic, *J. Phys. Chem. B* **105** (2001) 4018.
75. B. Hammer and J.K. Nørskov, *Adv. Catal.* **45** (2000) 71.
76. S.E. Mason, I. Grinberg, and A.M. Rappe, *Phys. Rev. B* **69** (2004) 161401.
77. H. von Schenck, E. Janin, O. Tjernberg, M. Svensson, and M. Göthelid, *Surf. Sci.* **526** (2003) 184.
78. E. Janin, H. von Schenck, S. Hellden, O. Tjernberg, U.O. Karlsson, and M. Göthelid, *Surf. Sci.* **515** (2002) 462.
79. H. Hopster and H. Ibach, *Surf. Sci.* **77** (1978) 109.
80. H. Steininger, S. Lehwald, and H. Ibach, *Surf. Sci.* **123** (1982) 264.
81. I. Zasada and M.A. van Hove, *Surf. Rev. Lett.* **7** (2000) 15.
82. DaCapo, http://www.fysik.dtu.dk/CAMPOS/ by CAMP (Center for Atomic-scale Materials Physics).
83. J.P. Perdew, J.A. Chevary, S.H. Vosko, K.A. Jackson, M.R. Pederason, D.J. Singh, and C. Fiolhaism, *Phys. Rev. B* **46** (1992) 6671.
84. B. Hammer, L.B. Hansen, and J.K. Nørskov, *Phys. Rev. B* **59** (1999) 7413.
85. H.J. Monkhorst and J.D. Pack, *Phys. Rev. B* **13** (1976) 5188.
86. J.R. Engstrom and W.H. Weinberg, *Surf. Sci.* **201** (1988) 145.
87. C.E. Wartnaby, A. Stuck, Y. Yeo, and D.A. King, *J. Phys. Chem.* **100** (1996) 12483.
88. G. Somorjai, *Introduction to Surface Chemistry and Catalysis*, John Wiley & Sons, New York, 1994.
89. M. Endo, T. Matsumoto, J. Kubotam, K. Domen, and C. Hirose, *Surf. Sci.* **441** (1999) L931.
90. J.A. Gates and J.L. Kesmodel, *J. Catal.* **83** (1983) 437.
91. I. Kovacs, J. Kiss, and F. Solymosi, *Surf. Sci.* **566** (2004) 1001.
92. F. Solymosi, A. Berko, and T.I. Tarnoczi, *Surf. Sci.* **141** (1984) 533.
93. J. Hrbek, R.A. dePaola, and F.M. Hoffmann, *J. Chem. Phys.* **81** (1984) 2818.
94. M.A. Henderson, G.E. Mitchell, and J.A.M. White, *Surf. Sci.* **188** (1987) 206.
95. C. Houtman and M. Barteau, *Surf. Sci.* **248** (1991) 57.
96. J.L. Davis and M.A. Barteau, *J. Am. Chem. Soc.* **111** (1999) 1782.
97. J. Davis and M.A. Barteau, *Surf. Sci.* **256** (1990) 50.
98. M.R. Columbia and P.A. Thiel, *Surf. Sci.* **235** (1990) 53.
99. F. Solymosi, J. Kiss, and I. Kovacs, *Surf. Sci.* **192** (1987) 47.
100. I. Bakó and G. Pálinkás, *Surf. Sci.* **600** (2006) 3809.
101. P. Blöchl, *Phys. Rev. B* **50** (1994) 17953.
102. J.P. Perdew, K. Burke, and M. Ernzerhof, *Phys. Rev. Lett.* **77** (1996) 3865.

103. *CRC Handbook of Chemistry*, 76th edn., CRC Press, New York, 1996.
104. S.K. Desai, M. Neurock, and K. Kourtakis, *J. Phys. Chem. B* **106** (2002) 2559.
105. J. Greely and M. Mavrikakis, *J. Am. Chem. Soc.* **126** (2004) 3910.
106. J. Greely and M. Mavrikakis, *J. Am. Chem. Soc.* **124** (2002) 7193.
107. F. Delbecq and P. Sautet, *J. Catal.* **211** (2002) 398.
108. R. Hirschl, A. Eichler, and J. Hafner, *J. Catal.* **226** (2004) 273.
109. N.M. Abbas and R.J. Madix, *Appl. Surf. Sci.* **7** (1981) 241.
110. W.H. Hocking, *Z. Naturforsch.* **31** (1976) 1113.

6 General Considerations of Periodic Crystals

C. Fernando Zinola

CONTENTS

Bloch electrons in a perfect periodic potential can sustain an electric current even in the absence of an external electric field. This infinite conductivity is limited by the imperfections of the crystals, which lead to deviations from a perfect periodicity. The most important deviation is the atomic thermal vibration from the equilibrium position in the lattice; however, electric perturbations can also promote this type of vibration. A quantitative treatment of the external electric perturbation of a crystal, therefore, starts with the observation of the change in the lattice vibrations [1]:

$$U_{\text{per}} = \sum_R \phi(r - R) \tag{6.1}$$

where
   $\phi(r)$ is the electrostatic energy resulting from the attraction ($\propto r^{-1}$) balanced against the kinetic potential ($\propto r^{-2}$)
   r is the atomic position

In the case of a square-wave potential perturbation,

$$\phi(r, t)_{\text{per}} = \begin{cases} \phi_1(r) \\ \phi_2(r) \end{cases} \tag{6.2}$$

where $\phi_1(r)$ and $\phi_2(r)$ are the electric potentials for odd and even N half-cycle values, respectively. Considering the generalized term for a periodic potential energy,

$$U(r) = \sum_R \phi(r - R - u(R)) = U_{per} - \sum_R u(R) \cdot \nabla\phi(r - R) + \cdots \quad (6.3)$$

where $u(r)$ is the atomic displacement in the lattice.

The difference between the two forms is the perturbation that causes the transitions between the Bloch levels that initiates the lower conductivity. This process can be considered as a transition of the emitting electrons to absorbing phonons due to a change in their energy by the phonon wave vector. The result is the electron scattering by the lattice vibrations. In most cases, the crystal conduction is dominated by electronic transitions of a single phonon absorption or emission. The energy and crystal momentum conservation laws require that the phonon energy satisfy

$$E_k = E_{k+q} \pm \hbar\omega[k - (k + q)] \quad (6.4)$$

where the positive sign obeys the phonon emission of the q wave vector between the transitions of the electron with the k wave vector. It has to be remembered that the surface of the allowed q for a given k vector will be very close to the set of vectors connecting k to all other points of the constant energy: $E_k \approx E_{k+q}$. Considering that the static model works for the platinum electrode, the total potential energy of the crystal will be

$$U(r) = \frac{1}{2}\sum_{RR'} \phi[r(R) - r(R')] = \frac{1}{2}\sum_{RR'} \phi[R - R' + u(R) - u(R')] \quad (6.5)$$

The following Hamiltonian governs the resolution of the dynamic system:

$$H = \sum_R \frac{P^2(R)}{2M} + U \quad (6.6)$$

where
$P(R)$ is the linear momentum
M is the mass
R is the position of the crystal atom

For a pair potential, ϕ, of the form $(1/r^2 - 1/r)$, it is not difficult to obtain clear information from the Hamiltonian, but with a periodic electric potential the problem gets more complicated. We will consider that the platinum atoms will not deviate much from the equilibrium positions, that is, small $u(r)$ values. Thus, the Taylor expansion of the term $U(r)$ is of the following expression:

$$U(r) = \frac{1}{2}\sum_R \phi(R) + \frac{1}{4}\sum_{RR'} [u(R) - u(R')] \cdot \nabla\phi(R - R') + \frac{1}{8}\sum_{RR'} ([u(R) - u(R')] \cdot \nabla)^2\phi(R - R') + \cdots \quad (6.7)$$

6.1 PHONON DISPERSION IN THE ABSENCE OF A PERIODIC ELECTRIC POTENTIAL

When no net force is applied to the crystal atom, the linear term vanishes because the coefficient $u(R)$ is $\nabla\phi(R - R')$. The latter is minus the force exerted by other atoms in the equilibrium position. The non-vanishing term is the quadratic contribution, which is called the *harmonic approximation*.

Moreover, $\frac{1}{2}\sum_R \phi(R)$ is a constant contribution and is independent of the linear momenta and atomic displacement (see Appendix 6.B):

$$U_{\text{harm}}(R) = \frac{1}{4} \sum_{RR',ij} [u_i(R) - u_i(R')] \cdot \phi_{ij}''[u_j(R) - u_j(R')] \qquad (6.8)$$

with

$$\phi_{ij}''(r) = \frac{\partial^2 \phi(r)}{\partial r_i \partial r_j} \qquad (6.9)$$

If we consider a set of platinum atoms (for example) along a line, so that the 1-dimensional Bravais lattice vector is $R = na$, where a is the platinum interatomic distance and n an integer, the harmonic potential energy has the form [2]

$$U_{\text{harm}}(R) = \frac{1}{2} \sum_n [u(na) - u((n+1)a)]^2 \phi''(a) \qquad (6.10)$$

where $\phi''(a)$ is the second derivative of the interaction potential between two platinum atoms separated by an equilibrium distance equal to a.

The same situation is applied to the rest of the coordinates.

The boundary conditions are a problem because there is a finite number of atoms, N, in any real crystal. The most adequate choice is the Born–von Kárman boundary condition, that is [2,3],

$$\begin{cases} u(Na) = u(0) \\ u[(N+1)a] = u(a) & \text{for a 1-dimensional Bravais lattice} \\ u(N_i a_i + R) = u(R) \end{cases} \qquad (6.11)$$

The last condition obeys the situation where each primitive a_i vector is in a 3-dimensional R-lattice. This means that the end of the right extreme of the chain is connected with the left part of it. This restricts the allowed wave vectors k to those related to

$$\vec{k} = \frac{n_1}{N_1}\hat{b}_1, \frac{n_2}{N_2}\hat{b}_2, \frac{n_3}{N_3}\hat{b}_3 \qquad (6.12)$$

where n_i is the integral and b_i are the reciprocal lattice vectors, satisfying $\hat{b}_i \cdot \hat{b}_j = 2\pi\delta_{ij}$.

Only k, within a single primitive cell of the reciprocal lattice, yields different solutions; then, there will be N non-equivalent values of k. This is why they can be chosen to be any primitive cell of the reciprocal lattice adequately selected in the first Brillouin zone. In the case of a 3-dimensional harmonic interaction potential, we can use the following expression [2]:

$$U_{\text{harm}}(R) = \frac{1}{2} \sum_{RR'} \mathbf{u}(R)\mathbf{D}(R - R')\mathbf{u}(R') \qquad (6.13)$$

where $\mathbf{D}(R - R')$ is the matrix of the interionic forces that obey different symmetric conditions:

$$\begin{cases} D_{ij}(R - R') = D_{ji}(R' - R) \\ D_{ij}(R - R') = D_{ij}(R' - R) \\ \sum_R D_{ij}(R) = 0 \end{cases} \qquad (6.14)$$

which are the order independence condition for the second differentiation, the inversion symmetry condition, and zero distortion in spite of displacement from equilibrium, for the first, second, and third conditions, respectively, in Equations 6.14.

Therefore, the dynamic matrix, $\mathbf{D(k)}$, will satisfy

$$M\omega^2 \boldsymbol{\varepsilon} = \mathbf{D(k)}\boldsymbol{\varepsilon} \tag{6.15}$$

with

$$\mathbf{D(k)} = \sum_R \mathbf{D(R)} \exp\left[i(\mathbf{k} \cdot \mathbf{R})\right] \tag{6.16}$$

where $\boldsymbol{\varepsilon}$ is the polarization vector of the normal mode that determines the directions in which the atoms move. The values of $\boldsymbol{\varepsilon}$ are real and satisfy

$$\mathbf{D(k)}\varepsilon_i(\mathbf{k}) = d_i(\mathbf{k})\varepsilon_i(\mathbf{k}) \tag{6.17}$$

where $d_i(\mathbf{k})$ are the eigenvalues of the symmetric $\mathbf{D(k)}$ matrix that obeys

$$\varepsilon_i(\mathbf{k}) \cdot \varepsilon_j(\mathbf{k}) = \delta_{ij} \tag{6.18a}$$

and

$$\varepsilon_i(\mathbf{k}) = \varepsilon_i(-\mathbf{k}) \tag{6.18b}$$

From the symmetric conditions, it follows

$$\mathbf{D(k)} = -2\sum_R \mathbf{D(R)} \operatorname{sen}^2\left[\frac{1}{2}(\mathbf{k} \cdot \mathbf{R})\right] \tag{6.19}$$

The values of the normal modes with the wave vector $\mathbf{k}$ will then have the following expression:

$$\omega_i(\mathbf{k}) = \sqrt{\frac{d_i(\mathbf{k})}{M}} \tag{6.20}$$

Consequently, in the case of a 1-dimensional crystal, it follows

$$\omega(\mathbf{k}) = \sqrt{\frac{4K\operatorname{sen}^2\left(\frac{1}{2}kna\right)}{M}} \quad \text{and} \quad \lim_{k\to\frac{\pi}{2}}\omega(\mathbf{k}) = \sqrt{\frac{4K}{M}} \tag{6.21}$$

in the Brillouin zone. The solutions of the dynamic matrix are of a 3-dimensional planar wave as it is predicted by Bloch's theorem (Appendix 6.D):

$$\mathbf{u}(\mathbf{r}, t) = \frac{1}{\sqrt{NM}} \sum_{j\mathbf{R}} \varepsilon_j \exp\left[i(\mathbf{k} \cdot \mathbf{r} - \omega t)\right] \tag{6.22}$$

where ε_{ij} are the polarization components of the $\boldsymbol{\varepsilon}$ vector with $\mathbf{k}$ of the first Brillouin zone. It has to be mentioned that there are $3N$ ($N_1 N_2 N_3$) motion equations corresponding to the three components of the N metallic atoms.

6.2 PHONON DISPERSION IN THE PRESENCE OF A PERIODIC ELECTRIC POTENTIAL

Since the electric perturbation vibrates along a unique axis, that is, perpendicular to the surface of the electrode, a simplified contribution of the new wave function with the interlayer distance is expected. This fact leads to longitudinal waves having an $u(r)$ atomic displacement in the lattice. However, in spite of the absence of the other transversal components, three modes of propagation are expected: one longitudinal and the other two degenerating in a single transversal mode.

The first Brillouin zone of a face-centered cubic system as in the case of platinum has a characteristic point where three Bragg planes coincide: (200), (111), and ($1\bar{1}1$). The wave vector **k** is $2\pi/a(1, 1/2, 0)$ where a is the platinum interatomic distance. This means that the application of the perturbation discussed earlier to a *fcc* crystal can produce (210), (420), etc., faceted planes. The important fact is that the application of a periodic potential program produces the faceting of the electrode surface, as was demonstrated some years ago [4–8]. The application of a square-wave potential at a metal surface at potentials larger than 1 V produces a charge transfer involving solvent molecules with the formation of oxygen-containing species. These species are related directly to the change in the crystallographic plane distribution on the metal. Previous results have shown that it is possible to change the crystalline distribution of platinum by applying a constant potential below 0.08 V or above 0.85 V [4]. However, the time involved during the process, at least that observable by *ex situ* XRD patterns, was very large. We decided to test periodic perturbations instead, because they are faster and can be modulated by appropriate upper (E_u) and lower (E_l) potentials at relatively large frequencies (1–10 kHz). At these large frequencies, a mean potential value can be defined by the E_u and E_l values [4].

Suppose that we introduce a square-wave perturbation of the form

$$\phi(r, t)_{\text{per}} = \begin{cases} \phi_1(r) \\ \phi_2(r) \end{cases} \tag{6.2}$$

where $\phi_1(r)$ and $\phi_2(r)$ are the electric potentials for odd and even N half cycles' values, respectively. The nonhomogeneous solution depends on the hemiperiod as follows:

$$u_{\text{NH}}(r, t)_{\text{per}} = \begin{cases} e\phi_1(r)/Kr \\ e\phi_2(r)/Kr' \end{cases} \quad \begin{cases} 0 < t < T \\ T < t < 2T \end{cases} \tag{6.23}$$

The atomic displacement $u(r)$ produces an excess of energy (eV) that surpasses the mean thermal energy (meV) leading to a change in the equilibrium position from r to r'. Thus, in the dynamic matrix we have to add the periodic perturbation:

$$M\omega^2 \varepsilon = \mathbf{D}(\mathbf{k})\varepsilon + \frac{e\phi_i}{r}\varepsilon \tag{6.24}$$

The interactions of the phonons with the electrons produce a change in the expression for the motion equation:

$$M\omega^2 \varepsilon = \mathbf{D}(\mathbf{k})\varepsilon + eE(t)\varepsilon \tag{6.25}$$

The homogeneous solution will be (see Appendix 6.C)

$$\mathbf{u}_{\text{H}}(\mathbf{r}, t) = \frac{1}{\sqrt{NM}} \sum_{j\mathbf{R}} \varepsilon_j \exp\left[i(\mathbf{k} \cdot \mathbf{r} - \omega_o t)\right] \tag{6.26}$$

with the vibration mode:

$$\omega_o(\mathbf{k}) = \sqrt{\frac{4K\operatorname{sen}^2\left(\frac{1}{2}kr\right)}{M}} \tag{6.27}$$

The nonhomogeneous solution can be evaluated within the two ranges: $0 < t < T$ and $T < t < 2T$. In both cases, the proposed solution will be of the form

$$u_{NH}(r, t) = \text{constant} \tag{6.28a}$$

for which it obeys

$$\frac{\partial u_{NH}(r, t)}{\partial t} = \frac{\partial^2 u_{NH}(r, t)}{\partial t^2} = 0 \tag{6.28b}$$

Thus,

$$u_{NH}(r, t) = \frac{e\phi_i}{Kr_i} = \frac{eE_i}{K} \quad \text{with } \phi_i = \phi_1 \text{ or } \phi_2 \tag{6.29}$$

Then, the global solution is the contribution of both $u_H(t)$ and $u_{NH}(t)$:

$$\mathbf{u}_1(\mathbf{r}, t) = \frac{1}{\sqrt{NM}} \sum_{j\mathbf{R}} \varepsilon_j \exp\left[i(\mathbf{k} \cdot \mathbf{r} - \omega_o t)\right] + \frac{e\phi_1}{Kr} \tag{6.30a}$$

$$\mathbf{u}_2(\mathbf{r}, t) = \frac{1}{\sqrt{NM}} \sum_{j\mathbf{R}} \lambda_j \exp\left[i(\mathbf{k}' \cdot \mathbf{r}' - \omega_o' t)\right] + \frac{e\phi_2}{Kr'} \tag{6.30b}$$

The differential equation solution must obey the following continuity conditions:

$$u_1(0) = u_2(T) \tag{6.31a}$$

and

$$\frac{\partial u_1(0)}{\partial t} = \frac{\partial u_2(T)}{\partial t} \tag{6.31b}$$

Considering the first continuity condition,

$$\sum_{j\mathbf{R}} \varepsilon_j \exp\left[i(\mathbf{k} \cdot \mathbf{r})\right] + \frac{e\phi_1}{Kr} = \sum_{j\mathbf{R}} \lambda_j \exp\left[i(\mathbf{k}' \cdot \mathbf{r}' - \omega_o' T)\right] + \frac{e\phi_2}{Kr'} \tag{6.32}$$

For the second continuity condition,

$$\sum_{j\mathbf{R}} \varepsilon_j \exp\left[i(\mathbf{k} \cdot \mathbf{r})\right] = \frac{\omega_o}{\omega_o'} \sum_{j\mathbf{R}} \lambda_j \exp\left[i(\mathbf{k}' \cdot \mathbf{r}' - \omega_o' T)\right] \tag{6.33}$$

Upon operation, we can obtain both polarization coefficients:

$$\varepsilon_j = \frac{\dfrac{e}{K}\left(\dfrac{\phi_2}{r'} - \dfrac{\phi_1}{r}\right)}{\dfrac{\omega_0}{\omega_0'} - 1}\exp[-i(\mathbf{k}\cdot\mathbf{r})] \tag{6.34}$$

and

$$\lambda_j = \frac{\dfrac{e}{K}\left(\dfrac{\phi_2}{r'} - \dfrac{\phi_1}{r}\right)}{\dfrac{\omega_0}{\omega_0'} - 1}\exp[-i(\mathbf{k}\cdot\mathbf{r} - \omega_0' T)] \tag{6.35}$$

Therefore, the global solutions of the differential equations are

$$\mathbf{u}_1(\mathbf{r}, t) = \frac{1}{\sqrt{NM}}\sum_{j\mathbf{R}}\frac{\dfrac{e}{K}\left(\dfrac{\phi_2}{r'} - \dfrac{\phi_1}{r}\right)}{\dfrac{\omega_0}{\omega_0'} - 1}\exp[-i\omega_0 t] + \frac{e\phi_1}{Kr} \quad \text{for } 0 < t < T \tag{6.36a}$$

$$\mathbf{u}_2(\mathbf{r}, t) = \frac{1}{\sqrt{NM}}\sum_{j\mathbf{R}}\frac{\dfrac{e}{K}\left(\dfrac{\phi_2}{r'} - \dfrac{\phi_1}{r}\right)}{\dfrac{\omega_0}{\omega_0'} - 1}\exp[-i\omega_0'(t - T)] + \frac{e\phi_2}{Kr'} \quad \text{for } T < t < 2T \tag{6.36b}$$

These solutions are rather similar but completely different from that for a typical planar wave, since they are only dependent on the vibration mode and the external frequency perturbation. The new equilibrium position, r', can be obtained from the calculation of the potential energy and its derivation with respect to r'. After performing this operation, we can obtain [9]

$$r' = r\sqrt{\frac{\phi_2}{\phi_1}} \quad \text{with } \phi_2 > \phi_1 \tag{6.37}$$

The new equilibrium position depends on the square root of the ratio between the upper and the lower potential limits of the potential perturbation.

6.3 QUANTUM MECHANICS IN HARMONIC CRYSTALS

The analysis of the electronic levels in a periodic potential can be first treated in 1-dimension. The advantage is obtaining an exact solution to the problem and not an approximated one as in the generalization. Suppose that we have a periodic crystal lattice of a periodic potential energy of the form

$$U(x)_{\text{harm}} = \sum_{-\infty}^{\infty} v(x - na) \tag{6.38}$$

We can fix the position of the metal atoms at the minimum energy, that is, arbitrarily fixed as zero-potential energy. We can also consider the periodic potential as a superposition of the barriers with a width, $v(x)$, centered at points $x = \pm na$ (Bravais lattice of constant a), that is, the electron tunneling between each metal atom separated by na. For symmetry, we consider a symmetric line of atoms, $v(x) = v(-x)$. The energy of the electron in the lattice is

$$\varepsilon_{harm} = \frac{\hbar^2 K^2}{2M} \tag{6.39}$$

Since outside the barrier $v(x) = 0$, we have for $x \leq -a/2$:

$$\Psi_{ref}(x) = e^{iKx} + re^{-iKx} \tag{6.40}$$

and for $x \geq a/2$

$$\Psi_{trans}(x) = te^{iKx} \tag{6.41}$$

where r and t are the reflection and transmission coefficients, respectively, that indicate the amplitude of the electron tunneling probability.

If the function $v(x)$ is even, we will have $\Psi_{harm}(x) = \Psi_{harm}(-x)$ that is also a solution of the Schrödinger equation of energy equal to that of Equation 6.39.

For a particle getting in from the right, we can say that in the region of $x \leq -a/2$

$$\Psi_{ref}(x) = te^{-iKx} \tag{6.42}$$

and for $x \geq a/2$

$$\Psi_{trans}(x) = e^{iKx} + re^{-iKx} \tag{6.43}$$

Since $\Psi_{ref}(x)$ and $\Psi_{trans}(x)$ are independent solutions of the Schrödinger equation, the general solution has to be the linear combination of them:

$$\Psi(x) = A\Psi_{ref}(x) + B\Psi_{trans}(x) \quad a/2 \geq x \geq -a/2 \tag{6.44}$$

According to Bloch's theorem, we can choose a value of K such as

$$\Psi(x+a) = e^{iKa}\Psi(x) \tag{6.45}$$

Then,

$$\frac{d\Psi(x+a)}{dx} = e^{iKa}\frac{d\Psi(x)}{dx} \tag{6.46}$$

From these expressions, we can arrive at Bloch's electron energy related to the wave vector K:

$$\cos(Ka) = \frac{t^2 - r^2}{2t}e^{iKa} + \frac{1}{2t}e^{-iKa} \quad \text{and} \quad \varepsilon = \frac{\hbar^2 K^2}{2M} \tag{6.47}$$

On the other hand, the conservative electron energy requires

$$|t|^2 + |r|^2 = 1 \tag{6.48}$$

Therefore, it follows that for a δ phase change from the transmitted to the reflected wave

$$t = |t|e^{i\delta} \tag{6.49}$$

Suppose that we have two different solutions to the Schrödinger equation, $\Psi_1(x)$ and $\Psi_2(x)$, the Hamiltonian in 1-dimensional will be

$$\frac{-\hbar^2}{2M}\Psi_i''(x) + v(x)\Psi_i(x) = \frac{\hbar^2 K^2 \Psi_i(x)}{2M} \tag{6.50}$$

We can use the Wronskian, $w(\Psi_1(x), \Psi_2(x))$, between both the solutions:

$$w(\Psi_1(x), \Psi_2(x)) = \Psi_1'(x)\Psi_2(x) - \Psi_1(x)\Psi_2'(x) \tag{6.51}$$

and to the complex conjugated wave function

$$w(\Psi_{\text{trans}}(x), \Psi_{\text{ref}}^*(x)) = \Psi_{\text{trans}}'(x)\Psi_{\text{ref}}^*(x) - \Psi_{\text{ref}}^{*\,'}(x)\Psi_{\text{trans}}(x) \tag{6.52}$$

Then,

$$r = \pm i|r|e^{i\delta} \tag{6.53}$$

Finally, the energy and the wave vector of the Bloch electrons can be related to

$$\frac{\cos(Ka + \delta)}{|t|} = \cos(Ka) \quad \text{and} \quad \varepsilon = \frac{\hbar^2 K^2}{2M} \tag{6.54}$$

Since $|t| < 1$ and the value of K is nearly 1, the barrier gets less effective, and the incident wave is enhanced.

The specific heat for the metallic crystals $(c = AT + BT^3)$ is lower at very low temperatures than that calculated by considering quantum mechanics. The calculation of the real specific heat can be performed using energy eigenvalues from the stationary states:

$$H_{\text{harm}} = \sum_R \frac{P(R)^2}{2M} + \frac{1}{2}\sum_{RR'}[u_i(R)D_{ij}(R - R')u_j(R')] \tag{6.55}$$

However, we can make a correlation of this crystal lattice with a harmonic oscillator. The quantum theory of a single 1-dimensional oscillator predicts a Hamiltonian function of the form:

$$H = \frac{P^2}{2M} + \frac{M\omega^2 q^2}{2} \tag{6.56}$$

where
q is the position of the oscillator
$P = mv = m\omega q$

To consider the double effects of the position and the linear momentum in the Hamiltonian, we are going to define the ascending and descending operators, a^+ and a^-, respectively:

$$a^+ \equiv \sqrt{\frac{M\omega}{2\hbar}}q + i\sqrt{\frac{1}{2\hbar M\omega}}P \quad \text{and} \quad a^- \equiv \sqrt{\frac{M\omega}{2\hbar}}q - i\sqrt{\frac{1}{2\hbar M\omega}}P \tag{6.57}$$

Since these quantum operators can be considered as canonical relations for commutation $[q, P] = i\hbar$, we will have $[a^+, a^-] = 1$. Thus, expressing the Hamiltonian of Equation 6.56 with the a^+ and a^-:

$$a^+ a^- = \frac{1}{\hbar\omega}\left[\frac{P^2}{2M} + \frac{M\omega^2 q^2}{2}\right]$$

(6.58)

Then, the Hamiltonian arises as

$$H = \hbar\omega\left(a^+ a^- + \frac{1}{2}\right)$$

(6.59)

Since $[a^+, a^-] = 1$, the eigenvalues of H will be $(n + 1/2)\hbar\omega/n = 0, 1, 2, 3$.

Using Dirac's nomenclature, we can define the fundamental state's operator as $|0\rangle$ and that of the n-excited state as $|n\rangle$, so it accounts that

$$|n\rangle = \frac{1}{\sqrt{n!}}(a^+)^n|0\rangle$$

(6.60)

This operator satisfies

$$a^+ a^- |n\rangle = n|n\rangle$$

(6.61a)

$$\hbar|n\rangle = (n + 1/2)\hbar\omega|n\rangle$$

(6.61b)

The matrix elements of a^+ and a^- are given for these states by the following expressions:

$$\langle n'|a^-|n\rangle = 0 \quad \text{with } n' \neq n - 1$$

(6.62a)

$$\langle n - 1|a^-|n\rangle = \sqrt{n}$$

(6.62b)

$$\langle n'|a^+|n\rangle = \langle n|a^+|n'\rangle$$

(6.62c)

Considering Equation 6.38 again, we need to transform the Hamiltonian expression. Thus, if $\omega_s(k)$ and $\varepsilon_s(k)$ are the frequency and the polarization vector for the classic modes with polarization s and wave vector $\mathbf{k}$, respectively, we can define the phonon creation (a_{ks}^+) and annihilation (a_{ks}^-) operators as

$$a_{ks}^- = \frac{1}{\sqrt{N}}\sum_R \exp[-i(\mathbf{k}\cdot\mathbf{r})]\boldsymbol{\varepsilon}_s(\mathbf{k})\left[\sqrt{\frac{M\omega_s(\mathbf{k})}{2\hbar}}u(\mathbf{R}) - i\sqrt{\frac{1}{2\hbar M\omega_s(\mathbf{k})}}P(\mathbf{R})\right]$$

(6.63)

$$a_{ks}^+ = \frac{1}{\sqrt{N}}\sum_R \exp[-i(\mathbf{k}\cdot\mathbf{r})]\boldsymbol{\varepsilon}_s(\mathbf{k})\left[\sqrt{\frac{M\omega_s(\mathbf{k})}{2\hbar}}u(\mathbf{R}) + \sqrt{\frac{1}{2\hbar M\omega_s(\mathbf{k})}}P(\mathbf{R})\right]$$

(6.64)

If $\omega_s(\mathbf{k}) = 0$, the definition of the quantum operators will not account for this treatment. Since the crystal lattice develops in three dimensions, we have only three of the N modes ($\mathbf{k} = 0$ acoustic modes). These modes are associated with three freedom degrees of linear translation. Therefore, the atomic displacement vector is related to the polarization vector by the expression:

$$\mathbf{u}(\mathbf{R}, t) = \boldsymbol{\varepsilon}\, \exp[-i(\mathbf{k}\cdot\mathbf{r} - \omega t)]$$

(6.65)

For this equation, the following commuting relations are given:

$$[\mathbf{u}_i(\mathbf{R}, t), \mathbf{P}_j(\mathbf{R}', t)] = i\hbar\delta_{ij}\delta_{RR'} \tag{6.66}$$

$$[\mathbf{u}_i(\mathbf{R}, t), \mathbf{u}_j(\mathbf{R}', t)] = [\mathbf{P}_i(\mathbf{R}, t), \mathbf{P}_j(\mathbf{R}', t)] = 0 \tag{6.67}$$

and also the identity condition

$$\sum_R \exp[-i(\mathbf{k} \cdot \mathbf{r} - \omega t)] = \begin{cases} 0 & \mathbf{k} \in \text{reciprocal crystal lattice} \\ N & \mathbf{k} \notin \text{reciprocal crystal lattice} \end{cases} \tag{6.68}$$

On the other hand, the orthonormal polarization vectors are

$$\boldsymbol{\varepsilon}_s(\mathbf{k})\boldsymbol{\varepsilon}_{s'}(\mathbf{k}) = \delta_{ss'} \quad s, s' = 1, 2, 3, \dots \tag{6.69}$$

Then, it follows the subsequent properties:

$$[a_{ks}{}^+, a_{k's'}{}^-] = \delta_{ss'}\delta_{kk'} \tag{6.70}$$

$$[a_{ks}{}^+, a_{k's'}{}^+] = [a_{ks}{}^-, a_{k's'}{}^-] = 0 \tag{6.71}$$

In order to define the operators and their reciprocal magnitudes, we need to transform the position's coordinate and the linear momentum:

$$\mathbf{u}(\mathbf{R}) = \frac{1}{\sqrt{N}} \sum_{ks} \sqrt{\frac{\hbar}{2M\omega_s(\mathbf{k})}} (a_{ks}{}^+ + a_{-ks}{}^-) \exp[i(\mathbf{k} \cdot \mathbf{r})]\boldsymbol{\varepsilon}_s(\mathbf{k}) \tag{6.72}$$

$$\mathbf{P}(\mathbf{R}) = \frac{1}{\sqrt{N}} \sum_{ks} \sqrt{\frac{\hbar M\omega_s(\mathbf{k})}{2}} (a_{ks}{}^+ - a_{-ks}{}^+) \exp[i(\mathbf{k} \cdot \mathbf{r})]\boldsymbol{\varepsilon}_s(\mathbf{k}) \tag{6.73}$$

Since the polarization vectors are orthonormals, for their three components

$$\sum_{s=1}^{3} \boldsymbol{\varepsilon}_{si}(\mathbf{k})\boldsymbol{\varepsilon}_{sj}(\mathbf{k}) = \delta_{ij} \quad i, j = 1, 2, 3 \dots \tag{6.74}$$

By substituting Equations 6.72 and 6.73 into Equation 6.38 and considering the identity and orthonormal conditions, we obtain the following kinetic and potential energy expressions:

$$\frac{1}{2M} \sum_R \mathbf{P}^2(\mathbf{R}) = \frac{1}{4} \sum_{ks} \hbar\omega_s(\mathbf{k})(a_{ks}{}^- - a_{-ks}{}^+)(a_{ks}{}^+ - a_{-ks}{}^-) \tag{6.75}$$

$$U = \frac{1}{4} \sum_{ks} \hbar\omega_s(\mathbf{k})(a_{ks}{}^- + a_{-ks}{}^+)(a_{ks}{}^+ + a_{-ks}{}^-) \tag{6.76}$$

By considering both the energies, we obtain the Hamiltonian:

$$H = \frac{1}{2M} \sum_R \mathbf{P}^2(\mathbf{R}) + U = \frac{1}{2} \sum_{ks} \hbar\omega_s(\mathbf{k})(a_{ks}{}^- a_{ks}{}^+ + a_{ks}{}^+ a_{ks}{}^-) \tag{6.77}$$

$$H = \sum_{ks} \hbar\omega_s(\mathbf{k})\left(a_{ks}^{-}a_{ks}^{+} + \frac{1}{2}\right) \tag{6.78}$$

since Equation 6.70 can be applied.

The Hamiltonian above is the addition of the $3N$ independent Hamiltonian of each of them with their polarization and wave vector. When this occurs, the eigenfunctions are the product of these $3N$ independent Hamiltonians and their eigenvalues are the sum of the products. The eigenfunction is a set of $3N$ quantum numbers n_{ks} with independent $3N$ oscillators of the Hamiltonian: $\sum_{ks} \hbar\omega_s(\mathbf{k})(a_{ks}^{-}a_{ks}^{+} + 1/2)$. The energy of each state is

$$E = \sum_{ks} \hbar\omega_s(\mathbf{k})\left(n_{ks} + \frac{1}{2}\right) \tag{6.79}$$

This simple expression is enough for determining the elementary properties for metallic crystals, but when there is some interaction of the vibration modes in the crystal leading to external radiations, it is necessary to go to Equations 6.72 and 6.73 with some inharmonic terms. Suppose that we have a polyatomic crystal with p number of ions

$$\sum_{s=1}^{3p} [\boldsymbol{\varepsilon}_{si}(\mathbf{k})]_\mu [\boldsymbol{\varepsilon}_{sj}(\mathbf{k})]_\nu = \frac{1}{M}\delta_{ij}\delta_{\mu\nu} \tag{6.80}$$

and

$$[\boldsymbol{\varepsilon}_{si}(-\mathbf{k})] = [\boldsymbol{\varepsilon}_{sj}(\mathbf{k})]^* \tag{6.81}$$

For specifying the energy levels of an N-ionic harmonic crystal, we have to consider them as $3N$ independent oscillators, whose frequencies are those of the $3N$ normal classic modes. The contribution to the total energy of a particular normal mode of angular frequency $\omega_s(\mathbf{k})$ can only have discrete numbers, $(n_{ks} + 1/2)\,\hbar\omega_s(\mathbf{k})$, with n_{ks} being the normal excitation mode (from $0, 1, 2, \ldots$). The total energy is simply the addition of the individual energies of the normal modes as given by Equation 6.79.

The normal excitation mode interchanges energy in its transitions similar to the electromagnetic radiation between molecule bands, leading to the emission or absorption of photons that are named *phonons*. This species is the quanta for the ionic displacement in the metal lattice that normally characterizes the classic sonic wave. The value n_{ks} is the normal mode of the wave vector $\mathbf{k}$ in its branch s for the n-dim excited state. Taking the particle point of view, we can say that we have n_{ks} phonons of the s-type with a $\mathbf{k}$ wave vector.

6.4 ELECTRON TRANSITIONS IN THE VIBRONIC STATES OF THE H$^+$–H$_2$O SYSTEM UNDER A PERIODIC POTENTIAL PERTURBATION OF WEAK ENERGIES

The H$^+$–H$_2$O system is characterized by different electronic states $E_o, E_1, \ldots, E_n$, which, at $t = 0$, are given by the Hamiltonian, $\mathbf{H}^o$:

$$\mathbf{H}^o\Psi_n = E_n\Psi_n \tag{6.82}$$

where Ψ_n is the eigenfunction for the stationary state.

Suppose that we are now perturbing the system with a periodic external potential as explained above, the eigenstates and eigenvalues are solved by the time-dependent Hamiltonian:

$$\mathbf{H}\Psi_n = i\hbar \frac{\partial \Psi_n}{\partial t} \tag{6.83}$$

where the new Hamiltonian is $\mathbf{H} = \mathbf{H^o} + \mathbf{H'}$, where $\mathbf{H'}$ is the time-dependent perturbation contribution. The following solution applies for $\mathbf{H^o}$:

$$\Psi_n = \Psi_n^o \exp\left[\frac{iE_n t}{\hbar}\right] \tag{6.84}$$

where Ψ_n^o is the eigenfunction of the non-time-dependent Hamiltonian, $\mathbf{H^o}$. To find the solutions $\Psi_n(t)$ of $\mathbf{H}$, we have to expand it in Ψ_n^o with a time-dependent series of coefficients $a_k(t)$:

$$\Psi_{n,k} = \sum_k a_k(t)\Psi_{n,k}^o \tag{6.85}$$

Since the Ψ_n^o eigenfunctions are orthonormals, this operation can be performed:

$$\frac{da_k(t)}{dt} = \frac{-i}{\hbar} \sum_k a_k(t)H'_{nk} \exp\left[\frac{iw_{nk}t}{\hbar}\right] \tag{6.86}$$

where

$$H_{nk} = \left\langle \Psi_{nk}^o \middle| H' \middle| \Psi_{nk}^o \right\rangle \tag{6.87a}$$

and

$$w_{nk} = \frac{E_n - E_k}{\hbar} \tag{6.87b}$$

H'_{nk} is the transition matrix involving two states responsible for the construction of $a_k(t)$, that is, the n state formed by an initial state among the set of k values.

From the above equation and the integration of the $a_k(t)$ values, we can calculate the expressions of the eigenfunctions. We will use the example of the square-wave potential program, since the effects of its crystalline arrangement was clearly demonstrated. First, the vibronic and electronic states of the H^+–H_2O system perturbed by an ac electric potential are subjected to a pulsating electric field:

$$\mathbf{E} = -\nabla\phi = \begin{cases} -\partial\phi_1(y)/\partial y \\ -\partial\phi_2(y)/\partial y \end{cases} \quad \begin{cases} 0 < t < T \\ T < t < 2T \end{cases} \tag{6.88}$$

Suppose now that the proton in the H^+–H_2O system is separated from the negative charge density by a distance y, both magnitudes define the dipolar moment, $\boldsymbol{\mu}$. Therefore, the electrostatic energy, U, will be

$$U = \sum_i q_i E y_i = \mathbf{E} \sum_i q_i y_i = \mathbf{E} \cdot \boldsymbol{\mu} \tag{6.89}$$

This assumption is possible since we are considering that at an instant t, there is a uniform electric field that arises from the difference between the wavelength of the atomic crystal electronic spectra (approximately 10^3 Å) and the molecular (vibronic) dimensions (a few Å). Finally, the time-dependent Hamiltonian, $\mathbf{H}'$, is

$$\mathbf{H}' = \mathbf{E} \cdot \boldsymbol{\mu} \tag{6.90}$$

or expressing it by the three components (Cartesian coordinates for the case of a planar geometric electrode symmetry):

$$\mathbf{H}' = \sum_{i(x,y,z)} E_i \mu_i = \frac{1}{2}\left(E_{ox}\mu_x + E_{oy}\mu_y + E_{oz}\mu_z\right) \tag{6.91}$$

Only the normal component, E_{oy}, of the electric field of the electrode surface has an important contribution since the electrocatalytic reaction occurs along this axis:

$$\mathbf{H}' = \begin{cases} E_1(\mu_y)_{nk}/2 & \{0 < t < T \\ E_2(\mu_y)_{nk}/2 & \{T < t < 2T \end{cases} \tag{6.92}$$

where

$$(\mu_y)_{nk} \equiv \langle \Psi_n | \mu_y | | \Psi_k \rangle \tag{6.93}$$

Substituting $(H')_{nk}$ in the time-dependent expression of $a_k(t)$ (Equation 6.42):

$$\frac{da_k(t)}{dt} = \frac{-iE_i}{2\hbar} \sum_k a_k(t)(\mu_y)_{nk}[\exp(iw_{nk}t)] \tag{6.94}$$

If the time variation of this equation is taken at the first stages, the coefficients $a_k(t)$ can be nearly equal to the initials. For any k value different to n, we can consider that at $t = 0$, the $a_k(0) = 0$ and $a_n(0) = 1$, whereas at $t \neq 0$, $a_k(t) \approx 0$ and $a_n(t) \approx 1$. In this time-dependent perturbation theory, the time at which the perturbation occurs is small. Since the energy $eE_{o,y} \leq 0.1$ eV, because of Heisenberg's principle, $t \leq (h/eE_{o,y}) \approx 10^{13}$ s. Then, it is admissible to consider $a_n(t) \approx 1$. On the other hand, we take into account that only one state mostly contributes to the sum and with this arbitrary approximation, we restrict the series to only one component:

$$\int_0^{a_k} a_k(t) = \frac{-iE_i}{2\hbar}(\mu_y)_{nk} \int_0^t \exp(iw_{nk}t)dt \tag{6.95}$$

$$a_k(t) = \frac{-iE_i}{\hbar}(\mu_y)_{nk}\frac{1}{i}\left[\frac{\exp(iw_{nk}t) - 1}{w_{nk}}\right] \tag{6.96}$$

The angular frequency of the transition from E_n to E_k is w_{nk}. When $w_{nk} > 0$, $E_n < E_k$ and the H^+-H_2O system absorbs energy and vice versa

$$\Psi_{n,k} = \sum_k \frac{-iE_i}{\hbar}(\mu_y)_{nk}\frac{\Psi_{n,k}^0}{i}\left[\frac{\exp(iw_{nk}t) - 1}{w_{nk}}\right] \tag{6.97}$$

From the wave function, we can calculate one important parameter, that is, the transition probability. Since $\Psi_{n,k}$ are eigenfunctions, it is physically related to only $|a_k|^2$. This value implies the probability of having an electronic transition from the stationary Ψ_n state to another Ψ_k at an instant t in the presence of a radiation frequency ω (emission or absorption):

$$|a_k(t)|^2 = \frac{1}{2}\left(\frac{(\mu_y)_{nk}E_i}{\hbar}\right)^2 \left[\frac{\exp(iw_{nk}t)-1}{w_{nk}}\right]^2 \tag{6.98}$$

But considering $|e^{it}-1|^2 = 4\,\mathrm{sen}^2(t/2)$, we will reduce the above equation to

$$|a_k(t)|^2 = \left(\frac{4(\mu_y)_{nk}E_i}{\hbar}\right)^2 \frac{\mathrm{sen}^2\left(\frac{w_{nk}t}{2}\right)}{w_{nk}^2} = \frac{(\mu_y)_{nk}^2 E_i^2 t^2}{2\hbar}\frac{\mathrm{sen}^2\left(\frac{w_{nk}t}{2}\right)}{\left[\frac{w_{nk}}{2}\right]^2} \tag{6.99}$$

However, the probability is frequency dependent, and the total transition probability has to be considered as the integral expression of $|a_k|^2$ over the entire space (quasi-continuum final states):

$$P_{nk} = \int_{-\infty}^{\infty} |a_k(t)|^2 \mathrm{d}N_n = \frac{\pi(\mu_y)_{nk}^2\rho(E_n)E_i^2 t^2}{2\hbar^2} \tag{6.100}$$

since

$$\int_{-\infty}^{\infty} \frac{\mathrm{sen}^2\left(\frac{w_{nk}t}{2}\right)}{\left[\frac{w_{nk}}{2}\right]^2}\,\mathrm{d}t = \pi$$

and E_o is constant. The number of the final states, N_n, and the density of these states, $\rho(E_n)$, of the final energy, E_n, are defined as

$$\rho(E_n) = \frac{\mathrm{d}N_n}{\mathrm{d}E_n} \tag{6.101}$$

The transition probability per time will be

$$\frac{\mathrm{d}P_{nk}}{\mathrm{d}t} = \frac{\pi(\mu_y)_{nk}^2\rho(E_n)E_i^2 t}{\hbar^2} \tag{6.102}$$

If we have N vibrons of the H^+–H_2O form, which are initially in the Ψ_n state and in the presence of a frequency υ go to another Ψ_k state, the velocity for the process will be

$$\frac{\mathrm{d}(\Psi_n \rightarrow \Psi_k)}{\mathrm{d}t} = \frac{\pi N(\mu_y)_{nk}^2\rho(E_n)E_i^2}{\hbar^2} \tag{6.103}$$

APPENDIX 6.A: GENERAL CONCEPTS IN PLANAR WAVES

A longitudinal wave is called a *planar wave* if the scalar potential $\phi = -\int \vec{E}\,d\vec{q}$ or each parameter that is able to define the oscillatory displacement depends only on time and a single Cartesian coordinate. The d'Alembert operator in this case will be

$$\frac{\partial^2 \phi}{\partial x^2} = \frac{1}{c^2}\frac{\partial^2 \phi}{\partial t^2} \tag{6.A.1}$$

The longitudinal wave is called *stationary* when it results from the interference of waves that propagate in opposite directions but with the same frequency and amplitude as those of the coordinates. The wave is *transversal* when the polarization is the same as the wave.

When the propagation of the planar waves defines parallel planes that are normal to the propagation line of the wave, these planes are called *rays*. In any case, the general solution to the d'Alembert equation is

$$\phi(x, t) = f_1(ct - x) + f_2(ct + x) \tag{6.A.2}$$

The equation that corresponds to a planar wave is sinusoidal

$$\phi(x, t) = A \operatorname{sen}(kx - \omega t + \phi) \tag{6.A.3}$$

where
A is a constant independent of the propagation direction (when the continuum medium is ideal)
ϕ is the initial propagation phase

However, it is common to write the function as

$$\phi(x, t) = A' \exp i(kr - \omega t) \tag{6.A.4}$$

where
the complex amplitude $A' = A \exp i(\pi/2 - \phi)$
r is the vector of the radius measured at the midpoint
the wave vector $k = 2\pi\hat{n}/\lambda$, $\hat{n}$ being the normal vector that indicates the direction sign of the propagation and λ is the wavelength

It is important to say that the physical meaning of the function is the real part of Equation 6.A.4:

$$\phi(x, t) = \operatorname{Re}\{A' \exp i(kr - \omega t)\} \tag{6.A.5}$$

Some cases refer to a *spherical wave*, since the spatial coordinate is the radius vector, and in this case, the d'Alembert operator is

$$\frac{1}{r^2}\left(r^2 \frac{\partial^2 \phi}{\partial r^2}\right) = \frac{1}{c^2}\frac{\partial^2 \phi}{\partial t^2} \tag{6.A.6}$$

In this case, we will have

$$\phi(x, t) = \frac{A'}{r} \exp i(kr - \omega t) \tag{6.A.7}$$

and in the case of a *cylindrical wave*

$$\phi(x, t) = \frac{A'}{\sqrt{r}} \exp i(kr - \omega t) \qquad (6.A.8)$$

APPENDIX 6.B: NORMAL MODES IN A 3-DIMENSIONAL MONOATOMIC BRAVAIS LATTICE

Let us consider a set of metal atoms of 1-dimensional configuration so that the Bravais lattice vector is a set of $\mathbf{R} = na$, a is the metal interatomic distance, and n an integer; as stated above, the harmonic potential energy, U_{harm}, is

$$U_{\mathrm{harm}}(R) = \frac{1}{2} \sum_n [u(na) - u((n+1)a)]^2 \phi''(a) \qquad (6.B.1)$$

where $\phi''(a)$ is the second derivate of the interaction potential between two atoms separated by the equilibrium distance (a).

The same situation is applied to the rest of the coordinates, so we can formulate the matrix element formulation:

$$U_{\mathrm{harm}}(R, R') = \frac{1}{2} \sum_{RR'} u_i(R) D_{ij}(R - R') u_j(R') \qquad (6.B.2)$$

with

$$D_{ij}(R - R') \equiv \delta_{RR'} \sum_{R''} \phi_{ij}(R - R'') - \phi_{ij}(R - R') \qquad (6.B.3)$$

Equation 6.B.2 is the vectorial product of $u_i(R)$ with the vector that is obtained from the operation between $u_j(R')$ and the matrix $D_{ij}(R - R')$. Then, it follows

$$U_{\mathrm{harm}} = \frac{1}{2} \sum_{RR'} \mathbf{u}(\mathbf{R}) \mathbf{D}(\mathbf{R} - \mathbf{R}') \mathbf{u}(\mathbf{R}) \qquad (6.B.4)$$

Three symmetry cases follow from Equation 6.B.4

$$1. \quad D_{ij}(R - R') = D_{ji}(R' - R) \qquad (6.B.5)$$

there is a free selection of coefficients to obey that condition where

$$D_{ij}(R - R') = \frac{\partial^2 \phi}{\partial u_i(R) \partial u_j(R')}\bigg|_{u=0} \qquad (6.B.6)$$

with successive derivations of the interaction potential

$$2. \quad D_{ij}(R - R') = D_{ij}(R' - R) \qquad (6.B.7a)$$

or

$$D_{ij}(R) = D_{ij}(-R) \qquad (6.B.7b)$$

that can be reunited in

$$D_{ij}(R - R') = D_{ji}(R - R') \tag{6.B.8}$$

This symmetry condition is an inversion matrix condition. An atom at an $\mathbf{R}$ site displaced by $\mathbf{u}(\mathbf{R})$ has the same energy as a displaced $-\mathbf{u}(-\mathbf{R})$ at the same position $\mathbf{R}$.

$$3. \quad \sum_{R} D_{ij}(R) = 0 \vee \sum_{R} \mathbf{D}(\mathbf{R}) = 0 \tag{6.B.9}$$

In this case, if $\mathbf{u}(\mathbf{R}) = d$ all the crystals will be displaced without any variation in the internal energy of distortion. Thus, the value of U_{harm} will be the same as that of $\mathbf{u}(\mathbf{R}) = 0$.

$$\sum_{\substack{RR' \\ ij}} d_i D_{ij}(R - R') d_j = \sum_{ij} N d_i d_j \sum_{R} D_{ij}(R) \tag{6.B.10}$$

With these three symmetries, we can obtain $3N$ movement equations, one for each of the three components of N ions in the metal lattice:

$$M \ddot{u}_i(\mathbf{R}) = -\frac{\partial U_{harm}}{\partial u_i(\mathbf{R})} = -\sum_{Rj} D_{ij}(\mathbf{R} - \mathbf{R}') u_j(\mathbf{R}) \tag{6.B.11}$$

or in matrix nomenclature

$$M \ddot{\mathbf{u}}(\mathbf{R}) = -\sum_{R'} \mathbf{D}(\mathbf{R} - \mathbf{R}') \mathbf{u}(\mathbf{R}') \tag{6.B.12}$$

The solutions in the case of 3-dimensional spaces are also of planar waves:

$$\mathbf{u}(\mathbf{R}, t) = \boldsymbol{\varepsilon} \exp[i(\mathbf{k} \cdot \mathbf{R} - \omega t)] \tag{6.B.13}$$

where the parameters and factors have the same meaning as in the text. Thus, the Born–von Kárman periodicity contour condition has to be obeyed:

$$\mathbf{u}(\mathbf{R}) = \mathbf{u}(\mathbf{R} + N_i \mathbf{a_i}) \tag{6.B.14}$$

where
  $\mathbf{a_i}$ are the three primitive vectors
  N_i are integer numbers of $N = N_1 N_2 N_3$

This condition reduces $\mathbf{k}$ wave vectors to

$$\mathbf{k} = \frac{n_1 \mathbf{b}_1}{N_1} + \frac{n_2 \mathbf{b}_2}{N_2} + \frac{n_3 \mathbf{b}_3}{N_3} \tag{6.B.15}$$

where
  n_i are the integer numbers
  $\mathbf{b_i}$ are the reciprocal lattice vectors

$$\mathbf{b}_i \cdot \mathbf{a}_i = 2\pi \delta_{ij} \tag{6.B.16}$$

Similar to the 1-dimension approach, only $\mathbf{k}$ inside the primitive cell initiates different solutions, but when a vector of the reciprocal lattice, $\mathbf{k}^{-1}$, is added to Equation 6.B.13, the displacement of atoms $\mathbf{u}(\mathbf{R})$ will not be altered since $\exp\left[i(\mathbf{k}^{-1} \cdot \mathbf{R})\right] = 1$. Therefore, we will have N values of $\mathbf{k}$ that are not equivalent in form, $n_i\mathbf{b}_i/N_i$ arbitrary selected to be inside the primitive cell of the reciprocal lattice. Most of the times the first Brillouin zone is considered:

$$M\omega^2\varepsilon\mathbf{u}(\mathbf{R}) = -\mathbf{D}(\mathbf{k})\varepsilon\mathbf{u}(\mathbf{R}') \tag{6.B.17}$$

where

$$\mathbf{D}(\mathbf{k}) = \sum_R \mathbf{D}(\mathbf{R})\exp\left[-i(\mathbf{k} \cdot \mathbf{R})\right] \tag{6.B.18}$$

The three solutions for each N value admitted for $\mathbf{k}$ originates $3N$ normal modes. However, for symmetry we know that: $\mathbf{D}(\mathbf{R}) = \mathbf{D}(-\mathbf{R})$ (symmetric matrix) and $\sum_R \mathbf{D}(\mathbf{R}) = 0$

$$\mathbf{D}(\mathbf{k}) = \frac{1}{2}\sum_R \mathbf{D}(\mathbf{R})(e^{-i(\mathbf{k}\cdot\mathbf{R})} + e^{i(\mathbf{k}\cdot\mathbf{R})} - 2) \tag{6.B.19}$$

$$\mathbf{D}(\mathbf{k}) = \sum_R \mathbf{D}(\mathbf{R})[\cos(\mathbf{k}\cdot\mathbf{R}) - 1] = -2\sum_R \mathbf{D}(\mathbf{R})\left[\sin^2\left(\frac{\mathbf{k}\cdot\mathbf{R}}{2}\right)\right] \tag{6.B.20}$$

Since both $\mathbf{D}(\mathbf{R})$ and $\mathbf{D}(\mathbf{k})$ are symmetric, we will have three eigenvalues of polarization, $\varepsilon_i(\mathbf{k})$, obeying

$$\varepsilon_i(\mathbf{k})\mathbf{D}(\mathbf{k}) = d_i(\mathbf{k})\varepsilon_i(\mathbf{k}) \tag{6.B.21}$$

where $d_i(\mathbf{k})$ are the eigenvalues of $\mathbf{D}(\mathbf{k})$. The eigenvalues of the polarization vector $\varepsilon_i(\mathbf{k})$ are orthonormals:

$$\varepsilon_i(\mathbf{k}) \cdot \varepsilon_j(\mathbf{k}) = \delta_{ij} \quad i, j = 1, 2, 3 \tag{6.B.22}$$

The three normal modes of the $\mathbf{k}$ wave vector and the ε polarization vector are of real numbers of frequency, $\omega_i(\mathbf{k})$:

$$\omega_i(\mathbf{k}) = \sqrt{\frac{d_i(\mathbf{k})}{M}} \tag{6.B.23}$$

When $d_i(\mathbf{k})$ are negative, the values of $U_{\mathbf{harm}}$ are also negative, contrary to the minimum energy in the equilibrium potential energy. For real numbers of $\omega_i(\mathbf{k})$, we only take the positive values of the potential energy.

The values of $\omega_i(\mathbf{k})$ tend to disappear when $\mathbf{k} \to 0$ or $\mathbf{k} \cdot \mathbf{R} \to 0$. In this case, $\sin^2\left(\frac{\mathbf{k}\cdot\mathbf{R}}{2}\right) \approx \left(\frac{\mathbf{k}\cdot\mathbf{R}}{2}\right)^2$, and then

$$\mathbf{D}(\mathbf{k}) = -\frac{\mathbf{k}^2}{2}\sum_R \mathbf{D}(\mathbf{R})(\hat{\mathbf{k}} \cdot \mathbf{R})^2 \tag{6.B.24}$$

with $\hat{\mathbf{k}} = \mathbf{k}/k$. In the situation of k being small, $\omega_i(\mathbf{k}) = C_i(\hat{\mathbf{k}})k$, where

$$C_i(\hat{\mathbf{k}}) = \sqrt{\frac{-1}{2M}\sum_R \mathbf{D}(\mathbf{R})(\hat{\mathbf{k}}\cdot\mathbf{R})^2} \tag{6.B.25}$$

These values depend on the direction, $\hat{\mathbf{k}}$, of the wave propagation at the i-branch. In addition, the propagation of the wave depends on the relation between the polarization and wave vectors. In an isotropic medium, the three solutions for a single value of $\mathbf{k}$ can be selected to be a longitudinal i-branch along the propagating direction, $(\varepsilon /\!/\mathbf{k})$, and the remaining transversal branches can be polarized perpendicularly to the propagation, $(\varepsilon \perp \mathbf{k})$.

APPENDIX 6.C: HOMOGENEOUS AND PARTICULAR SOLUTION OF THE DYNAMIC MATRIX IN A METAL LATTICE: THE APPLICATION OF AN EXTERNAL VOLTAGE

In the case of an external electric voltage the dynamic equation will be

$$M\ddot{u} + ku = V_o e \cos \omega t \tag{6.C.1}$$

The homogeneous solution is

$$M\ddot{u}(na) + k[2u(na) - u(n-1)a - u(n+1)a] = 0 \tag{6.C.2}$$

If the solution is

$$u(na, t) = A\exp[i(kna - \omega t)] \tag{6.C.3}$$

the second differentiation will be

$$\ddot{u}(na, t) = -A\omega^2\exp[i(kna - \omega t)] \tag{6.C.4}$$

Therefore,

$$-AM\omega^2 e^{[i(kna-\omega t)]} + k\left[2Ae^{[i(kna-\omega t)]}(na) - Ae^{[i(kna-\omega t)]}(n-1)a - Ae^{[i(kna-\omega t)]}(n+1)a\right] = 0 \tag{6.C.5a}$$

Or alternatively,

$$M\omega^2 = 4k\mathrm{sen}^2\left[\frac{ka}{2}\right] \tag{6.C.5b}$$

then

$$\omega_{o1,2}(k) = \sqrt{\frac{4k}{M}\mathrm{sen}^2\left[\frac{ka}{2}\right]} \tag{6.C.6}$$

is the normal mode where only the real part has to be considered and the homogeneous solution is

$$[u(na,t)]_H = A\exp[i(kna - \omega t)] \tag{6.C.7}$$

In the case of applying an external voltage, we will have a nonhomogeneous solution. We propose

$$u(na, t) = A \cos[(kna - \omega t)] \qquad (6.C.8)$$

Equation 6.C.1 uses the second derivative of the atom displacement:

$$u(na, t) = -A\omega^2 \cos[(kna - \omega t)] \qquad (6.C.9)$$

$$(\omega_0^2 - \omega^2) A \cos[(kna - \omega t)] = \frac{V_o e}{M} \cos \omega t \qquad (6.C.10)$$

The values of A can be determined from the initial and counter conditions. Thus, after operating the proposed solution

$$\cos[(kna - \omega t)] = \cos[kna] \cos[\omega t] + \sin[kna] \sin[\omega t] \qquad (6.C.11)$$

We will obtain the value of A:

$$A = \frac{V_o e}{M(\omega_0^2 - \omega^2) \cos[kna]} \qquad (6.C.12)$$

We have two different situations:

1. $\omega_0^2 \gg \omega^2$ $A = \dfrac{V_o e}{M\omega_0^2 \cos[kna]}$ with no oscillations, (A is not a function of ω).

2. $\omega_0^2 \ll \omega^2$ $A = \dfrac{V_o e}{M\omega^2 \cos[kna]}$ with oscillations because of being a function of ω.

Then, the particular solution is

$$[u(na, t)]_P = \frac{V_o e \cos[(kna - \omega t)]}{M(\omega_0^2 - \omega^2) \cos[kna]} \qquad (6.C.13)$$

The global solution is the addition of Equations 6.C.13 and 6.C.7:

$$[u(na, t)]_T = A \exp[i(kna - \omega t)] + \frac{V_o e \cos[(kna - \omega t)]}{M(\omega_0^2 - \omega^2) \cos[kna]} \qquad (6.C.14)$$

According to Euler's formula,

$$\exp[i(kna - \omega t)] = \cos(kna - \omega t) + i \sin(kna - \omega t) \qquad (6.C.15)$$

we can obtain

$$[u(na, t)]_T = \left\{ A + \frac{V_o e}{M(\omega_0^2 - \omega^2) \cos[kna]} \right\} \cos(kna - \omega t) + iA \sin(kna - \omega t) \qquad (6.C.16)$$

We are interested in the real part of the atomic displacement:

$$\mathrm{Re}\{[u(na, t)]_T\} = \left\{ A + \frac{V_o e}{M(\omega_0^2 - \omega^2) \cos[kna]} \right\} \cos(kna - \omega t) \qquad (6.C.17)$$

One interesting property is the group velocity of the planar waves, $v = \partial\omega/\partial k$, propagating along one axis in the crystal. Thus,

$$\omega = \sqrt{\omega_o^2 - \frac{V_o e}{MA \cos[kna]}} \qquad (6.C.18)$$

And then,

$$v = \frac{\partial\omega}{\partial k} = \frac{2}{\sqrt{\omega_o^2 - \frac{V_o e}{MA}}} \quad \text{for } n = 0 \quad \text{and} \quad kna = 2\pi \qquad (6.C.19)$$

APPENDIX 6.D: ELECTRONIC LEVELS IN PERIODIC POTENTIALS AND BLOCH'S THEOREM

The Hamiltonian describing the electron interaction in a metallic lattice is a complex function since each electron is subjected to a potential $U(r)$ with a periodicity given by the Bravais lattice $U(\mathbf{r} + \mathbf{R}) = U(r)$ where $\mathbf{R}$ are the vectors in the lattice. The metal lattice is considered as a pure electronic conductor. However, the presence of impurities decreases the infinite electrical conductivity. Therefore, we are going to consider independent electrons, that is, those simply described by $U(r)$.

$$H\Psi = \left[\frac{-\hbar^2\nabla^2}{2M} + U(r)\right]\Psi = \varepsilon\Psi \qquad (6.D.1)$$

In the case of an electric potential applied to the metal, as in electrocatalysis, $U(r) = \mathbf{U}(\mathbf{r}) + eV$. However, the modified energy is the total energy, so the kinetic operator has to be also modified in an unknown form. We define the *Bloch electrons* as those obeying the periodic Schrödinger equation and the *free electrons* as those obeying a zero periodic potential.

Bloch's theorem is a corollary of the lattice periodicity applied to the potential energy. The eigenfunctions of the Hamiltonian is given in Equation 6.D.1 with $\mathbf{U}(\mathbf{r} + \mathbf{R}) = \mathbf{U}(\mathbf{r})$ with $\mathbf{R}$ of the Bravais lattice, which can be selected as planar waves that are recurrent to the Bravais lattice.

$$\Psi_{nk}(\mathbf{r}) = \mathbf{u}_{nk}(\mathbf{r}) \exp(i\mathbf{k} \cdot \mathbf{r}) \qquad (6.D.2)$$

where
  n is the band index
  $\mathbf{k} = 2\pi/\lambda$ is the wave vector
  $\mathbf{u} = U/V$

Bloch's theorem means

$$\mathbf{u}_{nk}(\mathbf{r}) = \mathbf{u}_{nk}(\mathbf{r} + \mathbf{R}) \qquad (6.D.3)$$

or

$$\Psi_{nk}(\mathbf{r}) = \exp(-i\mathbf{k} \cdot \mathbf{r})\Psi_{nk}(\mathbf{r} + \mathbf{R}) \qquad (6.D.4)$$

A set of $\mathbf{R}$-values repeated over the entire lattice produce the Bravais lattice with a planar propagation wave of the $e^{(i\mathbf{k}\cdot\mathbf{r})}$ form. However, only some values of $\mathbf{k}$ can produce the periodicity of the wave. The *reciprocal lattice* is defined as all of the wave vectors **k** that are able to originate with the periodicity of the Bravais lattice:

$$\exp(i\mathbf{k} \cdot [\mathbf{r} + \mathbf{R}]) = \exp(i\mathbf{k} \cdot \mathbf{r}) \Rightarrow \exp(i\mathbf{k} \cdot \mathbf{R}) = 1 \qquad (6.D.5)$$

The Born–von Kárman contour condition demonstrates that the Bloch wave vector of free electrons in a cubic lattice is, according to Sommerfeld, constituted only by real components. The number of **k** values ($\mathbf{k} = \mathbf{p}/\hbar$) admitted in a primitive cell of a reciprocal lattice is equal to the number of sites in the crystal. The linear momenta operator, **p**, is

$$\mathbf{p} = \frac{\hbar\nabla}{i} \qquad (6.D.6)$$

When this operator is applied to the wave function, Ψ,

$$\mathbf{p}\Psi_{nk}(\mathbf{r}) = \frac{\hbar\nabla\Psi_{nk}(\mathbf{r})}{i} = \frac{\hbar\nabla \exp(-i\mathbf{k} \cdot \mathbf{r})\mathbf{u}_{nk}(\mathbf{r})}{i} \qquad (6.D.7)$$

$$\mathbf{p}\Psi_{nk}(\mathbf{r}) = \hbar\mathbf{k}\Psi_{nk}(\mathbf{r}) + \exp(i\mathbf{k} \cdot \mathbf{r})\frac{\hbar}{i}\nabla\mathbf{u}_{nk}(\mathbf{r}) \qquad (6.D.8)$$

For each **k** value, there are n solutions of the Schrödinger equation. Thus, using Bloch's theorem

$$\Psi_{nk}(\mathbf{r}) = \exp(-i\mathbf{k} \cdot \mathbf{r})\mathbf{u}_{nk}(\mathbf{r}) \qquad (6.D.9)$$

The values of $\mathbf{u}_{nk}(\mathbf{r})$ can be determined from its eigenvalues:

$$H_k\mathbf{u}_k(\mathbf{r}) = \frac{\hbar}{2M}[(1/i\nabla + \mathbf{k})^2 + U(\mathbf{r})]\mathbf{u}_k(\mathbf{r}) = \varepsilon_k\mathbf{u}_k(\mathbf{r}) \qquad (6.D.10)$$

where ε_k is the anti-Hermitian eigenvalue restricted to the primitive cell. This means that **k** emerges as a parameter of the Hamiltonian and the energy levels vary with **k**. However, there will be infinite solutions of n band indices with discrete eigenvalues. In the case of a periodic potential, it follows

$$\begin{cases} \varepsilon_{n,k+K} = \varepsilon_{nk} \\ \Psi_{nk+K}(\mathbf{r}) = \Psi_{nk}(\mathbf{r}) \end{cases} \qquad (6.D.11)$$

These values are not restricted to the primitive cell but are expanded through the entire **k**-space. The reciprocal lattice only produces identical values of the **k** vector, since the displacement does not lead to any changes in the properties of the cell.

Each function $\Psi(\mathbf{r})$ has to obey Bloch's theorem, so it can be expanded into a set of planar waves:

$$\Psi(\mathbf{r}) = \sum_q c_q \exp(i\mathbf{q} \cdot \mathbf{r}) \qquad (6.D.12)$$

The expansion of the planar wave only results in planar waves with the same periodicity and wave vectors, **q**, of the reciprocal lattice:

$$U(\mathbf{r}) = \sum_K U_K \exp(i\mathbf{K} \cdot \mathbf{r}) \qquad (6.D.13)$$

Because of Bloch's theorem

$$\Psi_{nk}(\mathbf{r}) = e^{(i\mathbf{k}\cdot\mathbf{r})}\mathbf{u}_{nk}(\mathbf{r}) = e^{(i\mathbf{k}\cdot\mathbf{r})}\sum_q c_{nq-K}e^{(-i\mathbf{K}\cdot\mathbf{r})} \qquad (6.D.14)$$

We can expand the eigenvalue and Hamiltonian terms:

$$\varepsilon_n(k+q) = \varepsilon_n(k) + \sum_i \frac{\partial \varepsilon_n(k_i)}{\partial k_i} q_i + \frac{1}{2}\sum_{ij} \frac{\partial^2 \varepsilon_n(k_{ij})}{\partial k_i \partial k_j} q_i q_j + R(q^3) \qquad (6.D.15)$$

$$H_{k+q} = H_k \varepsilon_n(k) + \frac{\hbar^2}{M}\mathbf{q}\cdot[(1/i\nabla + \mathbf{k})^2] + \frac{\hbar^2}{2M}\mathbf{q}^2 \qquad (6.D.16)$$

This treatment is similar to that proposed by perturbation theory. If $H = H_o + V$, the normalized eigenvectors will be

$$\begin{cases} E_n = E_n^o + \int \Psi_n{}^*(\mathbf{r})V\Psi_n(\mathbf{r})d\mathbf{r} + \sum_{n'\neq n} \frac{\left|\int \Psi_n{}^*(\mathbf{r})V\Psi_n(\mathbf{r})d\mathbf{r}\right|^2}{(E_n^o - E_{n'}^o)} \\ +\cdots H_o\Psi_n(\mathbf{r}) = E_n^o \Psi_n(\mathbf{r}) \end{cases} \qquad (6.D.17)$$

To a first approximation, we can take only the linear component of Equation 6.D.16, and inserting it into Equation 6.D.17 we have

$$\frac{\hbar^2}{M}\mathbf{q}\cdot[1/i\nabla + \mathbf{k}] = \int \Psi_n{}^*(\mathbf{r})V\Psi_n(\mathbf{r})d\mathbf{r} \qquad (6.D.18)$$

and

$$\sum_i \frac{\partial \varepsilon_n(k_i)}{\partial k_i}q_i = \sum_i \int u_{nk}{}^*(\mathbf{r})\frac{\hbar^2}{M}q_i[1/i\nabla + k]_i u_{nk}(\mathbf{r})d\mathbf{r} \qquad (6.D.19)$$

But the orthonormal condition for $u(\mathbf{r})$ is $\int |u_{nk}(\mathbf{r})|^2 d\mathbf{r} = 1$, and considering Equation 6.D.19 in full vectorial nomenclature

$$\frac{\partial \varepsilon_n}{\partial \mathbf{k}} = \frac{\hbar^2}{M}\int u_{nk}{}^*(\mathbf{r})[1/i\nabla + k]_i u_{nk}(\mathbf{r})d\mathbf{r} \qquad (6.D.20)$$

So applying Bloch's theorem $\Psi_{nk}(\mathbf{r}) = e^{(i\mathbf{k}\cdot\mathbf{r})}\mathbf{u}_{nk}(\mathbf{r})$ we will obtain

$$\frac{\partial \varepsilon_n}{\partial \mathbf{k}} = \frac{\hbar^2}{M}\int \Psi_{nk}{}^*(\mathbf{r})[\nabla/i]\Psi_{nk}(\mathbf{r})d\mathbf{r} \qquad (6.D.21)$$

On the other hand, the rate operator is defined as

$$\mathbf{v} = \frac{\partial \mathbf{r}}{\partial t} = \frac{\mathbf{p}}{M} = \hbar\nabla/iM \qquad (6.D.22)$$

Using Equation 6.D.21, we can obtain the rate of propagation of the planar waves of the Bloch electrons given by the band index n and the wave vector $\mathbf{k}$.

$$\frac{\partial \varepsilon_n}{\partial \mathbf{k}} = \hbar \int \Psi_{nk}*(\mathbf{r})\mathbf{v}\Psi_{nk}(\mathbf{r})d\mathbf{r} = \hbar \mathbf{v} \qquad (6.D.23a)$$

Then, the rate operator will be

$$\mathbf{v} = \frac{\partial \varepsilon_n}{\hbar \partial \mathbf{k}} \qquad (6.D.23b)$$

REFERENCES

1. N.W. Ashcroft and N.D. Mermin, *Solid State Physics*, Saunders, New York (1976) Chaps. 2, 12, 17, and 22.
2. C. Kittel, *Introduction to Solid State Physics*, 4th edn., Reverté, Ed., John Wiley, New York (1975).
3. P.V. Pávlov and A.F. Jojlov, *Solid State Physics*, Mir, Moscow (1987) Chaps. 5 and 7.
4. W.E. Triaca and A.J. Arvia, *J. Appl. Electrochem.* **20** (1989) 347.
5. A.J. Arvia, R.C. Salvarezza, and W.E. Triaca, *Electrochim. Acta* **34** (1989) 1057.
6. A. Visintín, J.C. Canullo, W.E. Triaca, and A.J. Arvia, *J. Electroanal. Chem.* **239** (1988) 67.
7. W.E. Triaca, T. Kessler, J.C. Canullo, and A.J. Arvia, *J. Electrochem. Soc.* **134** (1987) 1165.
8. A.C. Chialvo, W.E. Triaca, and A.J. Arvia, *J. Electroanal. Chem.* **146** (1983) 93.
9. C.F. Zinola and C. Bello, *J. Colloid Interface Sci.* **258** (2003) 259.

7 Electrode Catalysis versus Enzyme Catalysis: Theoretical Modeling of a Biological Enzyme

Guillermina L. Estiu

CONTENTS

7.1 INTRODUCTION

Biological enzymes naturally catalyze many important charge transfer and redox reactions that occur in the electrode surface. This is the case of [Ni–Fe] and [Fe–Fe] hydrogenases, which catalyze one of the simplest redox reactions in nature, the reversible two-electron oxidation of H_2.[1–4] These enzymes provide, thence, a way to use H_2 as a "clean fuel." For the reductive process, the ideal source of hydrogen is water that can, for example, be split electrolytically using electricity from renewable, nonfossil energy sources. In nature, green plants can do it, and some bacteria (photosynthetic microorganisms) also contain the enzyme hydrogenase to further convert the released protons into dihydrogen.

The biological and industrial catalysts differ structurally and mechanistically. [NiFe] hydrogenases, for example, are large molecules (98 kDa) that contain a buried bimetallic active site composed of only the base metals, nickel and iron, in a sulfur- and carbon-rich coordination environment. The active site is connected to the surface via a series of Fe–S clusters that transfer electrons through the protein matrix. By contrast, fuel cells typically employ platinum, a precious metal.[4] The comparatively greater prevalence of high-spin metal sites at the active centers of enzymes is connected to the dominance of first-row transition metal ions, and to the type of ligands bound. Ligands can be amino acid side chains (N, O, or S bonded to the metal), main chain peptide groups, cofactors, and/or simple groups like sulfides, oxides, hydroxides, or substrate molecules. Second-shell, third-shell, and more extended group interactions can also be very important, particularly when the active site is charged, as in the case of FeS proteins. From a theoretical/computational standpoint, these long-range charged or polar interactions are difficult to treat with proper accuracy, but major progress has been made on this problem by combined quantum mechanical and electrostatic methods. There are related efforts with combined quantum mechanics (QM)/molecular mechanics (MM) methods, and some advances have been seen utilizing molecular dynamics (MD).[5] Solvation often plays an essential role, involving either discrete bonded waters, waters filling small cavities, or bulk water. Solvation effects are one major source of nonadditive in cofactor–protein and protein internal interactions. Other major sources of complexity include proton-coupled electron transfer and cooperative charge-coupled proton transfer. Many metalloenzymes are redox active with one or multiple electron transfers intrinsically associated with the catalytic transformation. In some cases, first- or second-shell ligands (or more distant groups) can influence the catalytic or activation cycle. The chemistry of a low-spin type is much less prevalent, but it does occur at the catalytic center of hydrogenase, in nitrile hydratase, and some other enzymes. In other enzymes, both high-spin states and low-spin states can be involved at different stages in the catalytic cycle, as in cytochrome P450s and related heme enzymes, or in carrier proteins like hemoglobin and myoglobin. Where high-spin metal sites are involved, there may be one, two, or more metal sites, which allow for possible spin coupling between sites aided by the bridging ligands.

As for any catalytic process, the study of enzyme catalysis focuses on the determination of reaction energies and kinetic parameters, which requires understanding ground state and transition state (TS) geometries. Computational methods are valuable in this matter, as the TS is hard to be trapped experimentally due to the short-lived nature of the species. However, the intrinsic complexity of the bio-enzymatic reaction, previously described, as well as the way it influences the kinetics have to be modeled in the calculations.

Many proposals have been put forward to rationalize the catalytic power of enzymes:[6]

1. Electrostatic pre-organization. The ability of enzymes to provide a pre-organized electrostatic environment has been found to account for the major part of the catalytic effect in many enzymatic reactions.[7,8] Other studies also have supported the view that electrostatic stabilization of the TS plays a major role in catalysis.[9]

2. Steric strain. The early idea that enzyme catalysis results from the destabilization of the ground state[10] was examined and concluded that the actual amount of energy associated with steric strain is small, due to the inherent flexibility of proteins.[11] Nevertheless, the strain proposal has been invoked in several recent studies.

3. Near-attack conformations (NAC). This model considers that enzymes catalyze reactions by favoring configurations in which the reactants are pushed to a close interaction distance that optimizes reactivity.[12] In several cases, the energy associated with moving the reacting fragments from their average configuration in water to the average configuration in the enzymes was small, indicating that the corresponding catalytic effect was relatively minor.

4. Entropic effects. The idea that a loss of entropy upon substrate binding decreases the activation entropy for the rate-limiting catalytic step was proposed in the seventies, and has gained some support in recent computational studies. However, it has been argued that the activation entropy in solution is usually relatively small, as the formation of the TS does not require losing many degrees of freedom.[6]

5. Desolvation. This model considers that enzymes reduce the activation barrier by desolvating and destabilizing the ground state of the reacting fragments.[13] However, systematic analyses have shown that the TS is solvated much more strongly in many enzymes than in solution.[6] The desolvation proposal can be tested computationally through the calculation of the actual binding energies of the reactants in the ground and TSs.

6. Low-barrier hydrogen bonds (LBHB). Some enzymes have been proposed to catalyze their reactions by forming a so-called low-barrier, partially covalent (delocalized) hydrogen

bonds (LBHBs) with the reactants.[14] Electron valence bond (VB) studies and molecular orbital QM/MM studies have failed to support the LBHB idea,[15] and even found an anticatalytic effect of the LBHB in some cases, as enzymes appear to stabilize the TS more effectively with localized charges than with delocalized charges.

7. Dynamical effect. The dynamical effect, as defined by Karplus andMcCammon,[16] considers that the enzyme has evolved to optimize a particular vibrational mode for moving the system to the TS, or for converting a system at the TS to the product state. The essence of the proposal is that the motions of the reacting groups are different in enzymatic and nonenzymatic reactions and, specifically, that the motions in the enzyme are more directional than the random thermal fluctuations that establish a Boltzmann equilibrium between the reactants and the TS for a reaction in solution. Thus, dynamical effects must be at work if a system at the TS has a higher probability of decaying to products in the enzyme than it does for the same reaction in solution. Warshel, Olsson, and Parson have thoroughly analyzed this effect and found that the rate constant is determined by the probability of reaching the TS rather than by the time dependence of fluctuations along the reaction path.[17]

There are no single explanations for enzyme activity. Different enzymes catalyze their reactions in different ways, and one approach can describe more closely the mechanism for one than for another. The main idea underneath is the necessity to build a model as close to reality as possible, including an environmental and chemical effect that modulates both reactants and TSs. The catalytic power has to be measured relative to a reference state, and a careful inspection of the reaction coordinate, including solvent and entropy effects, is necessary for its accurate determination.[18,19] The reference state allows the definition of the proficiency of the enzyme as a quantitative measure of its catalytic power. For a generic catalytic reaction

$$E + S \underset{k_{-1}}{\overset{k_1}{\rightleftarrows}} ES^+ \xrightarrow{k_{cat}} EP \longrightarrow E + P \tag{7.1}$$

where

E, S, and P are the enzyme, the substrate, and the product, respectively
ES, EP, and ES^+ are the enzyme–substrate complex, the enzyme–product complex, and the TS
The proficiency P is defined as k_{cat}/K_M, where $K_M = (k_{-1} + k_{cat})/k_1$

In this way, the main computational challenge in the determination of the enzyme catalytic power is reduced to the calculation of the rate constant, which can be expressed, according to Eyring's transition state theory (TST) as

$$k = \kappa \frac{k_B T}{h} \exp\left[\frac{-\Delta G^{\#,0}(T)}{RT}\right] \tag{7.2}$$

where
$\Delta G^{\#,0}$ is the standard state free energy of activation
k_B is the Boltzmann constant
h is Plank's constant
κ is the transmission constant

The κ factor takes into account the probability of trajectory recrossing over the TS ridge as well as nonequilibrium and quantum effects. However, it is often taken as 1. The challenge is, thence, redirected to the calculation of the activation free energy.

In conventional TST, $\Delta G^{\#,0}$ is evaluated at the saddle point of the potential energy surface (PES) using standard expressions from statistical mechanics based on the relative energies, vibrational frequencies, total masses and moments of inertia of reactants, TSs, and products.[20] The vibrational frequencies are readily available from electronic structure calculations within the harmonic approximation. In the enzyme, the environment affects all these magnitudes, and a high level of theory, as the one applicable to the study of inorganic reactions is sometimes computationally unreachable. Nevertheless, coordination chemistry offers this possibility through the use of biomimetic analogues, which will be discussed in a later section.

7.2   THEORETICAL METHODOLOGIES APPLIED TO THE STUDY OF BIOCATALYSTS

The study of catalytic effects in enzymes requires quantitative methods capable of calculating the rate constant of a reaction given the structure of the enzyme. Any such method requires evaluating the PES that connects the reactant and product states and finding the activation free energy for reaching the TS. Combined QM/MM methods provide a generic way of obtaining potential surfaces and, in principle, activation free energies of chemical processes in enzymes. However, the implementation of rigorous *ab initio* QM/MM approaches in quantitative calculations of activation free energies is still extremely challenging. The somewhat less rigorous empirical valence bond (EVB) method is also capable of quantifying catalytic effects in general and dynamical contributions in particular.[6,17,18,21,22] MD simulations can be also used to calculate free energies, using potential of mean force (PMF) or free energy perturbation (FEP) approaches.

The accurate prediction of enzyme kinetics from first principles is one of the central goals of theoretical biochemistry. Currently, there is considerable debate about the applicability of TST to compute rate constants of enzyme-catalyzed reactions. Classical TST is known to be insufficient in some cases, but corrections for dynamical recrossing and quantum mechanical tunneling can be included. Many effects go beyond the framework of TST, as those previously discussed, and the overall importance of these effects for the effective reaction rate is difficult (if not impossible) to determine experimentally. Efforts are presently oriented to compute the quasi-thermodynamic free energy of activation with chemical accuracy (i.e., 1 kcal mol^{-1}), as a way to discern the importance of other effects from the comparison with the effective measured free energy of activation.

7.2.1   MOLECULAR DYNAMICS SIMULATIONS

MD simulations are all atom models that use ordinary differential equation of Newtonian dynamics.[23] Modern MD programs (such as CHARMM[24] and AMBER[25]) depend on force fields and can simulate the atomic level motions of an enzyme solvated in an accurately described water environment (TIP3P[26]) under well-defined conditions of temperature and pressure. MD simulations of enzyme structures and motions are often calculated using the method of periodic boundary condition with the particle mesh Ewald (PME) algorithm.[27] PME treatment is particularly important for an accurate description of long-range electrostatic interactions. In this procedure, the enzyme–substrate complex is immersed in a pre-equilibrated box of TIP3P water. Since this box infinitely repeats itself in all three dimensions, one can simulate an infinite solvent environment in which the enzymes complex is periodically placed. Some concerned has been expressed on the accuracy of this approach for non-periodic systems, and the inclusion of polarization terms in the force fields is strongly recommended.

The trajectories obtained from the simulations can be used to evaluate electrostatic free energies. Several methods have been derived, the FEP method[28] being the most extensively used. This approach involves gradually changing the solute charge from zero to its actual value $Q = Q_0$ using a series of mapping potentials of the following form:

$$V_m = (1 - \lambda_m)V(Q = 0) + \lambda_m V(Q_0) \tag{7.3}$$

where λ_m is changed gradually from 0 to 1. Then the corresponding free energy change is evaluated by the following expression:

$$\exp\{-\Delta G(\lambda_m \to \lambda_{m+1})\beta\} = \langle \exp\{-(V_{m+1} - V_m)\beta\}\rangle_{V_m} \tag{7.4}$$

where $\langle\ \rangle_{V_m}$ designates an MD average over V_m and $\beta = 1/k_B T$. The overall charging free energy is obtained by collecting the $\Delta G(\lambda_m \to \lambda_{m+1})$. A very useful alternative is to use the linear response approximation (LRA), in which the free energy is given by[29]

$$\Delta G(Q = 0 \to Q = Q_0) = \frac{1}{2}\left[\langle V(Q_0) - V(Q = 0)\rangle_{V(Q_0)} + \langle V(Q_0) - V(Q = 0)\rangle_{V(Q=0)}\right] \tag{7.5}$$

Both FEP and LRA can involve major convergence problems when applied to charges in proteins, and they have been explored systematically in only a relatively small number of studies.[30,31] Another approach for studies of electrostatic energies in proteins and, in particular, for studying the free energy profile of ions in ion channels, is to evaluate the so-called PMF. The PMF reflects the free energy of moving a given charged group from the bulk solvent to a specific protein site, and is typically evaluated by using umbrella sampling (US) or related approaches.[32] In the PMF approach, one commonly uses a mapping potential of the following form:

$$\varepsilon_m = (1 - \lambda_m)\varepsilon_1 + \lambda_m\varepsilon_2 \tag{7.6}$$

with

$$\varepsilon_1(z) = E_g(z) + K\left(z - z_0^{(1)}\right)^2 = E_g(z) + E_{cons}^{(1)}$$

and

$$\varepsilon_2(z) = E_g(z) + K\left(z - z_0^{(2)}\right)^2 = E_g(z) + E_{cons}^{(1)}$$

Here
$z_0^{(1)}$ and $z_0^{(2)}$ are neighboring points on a reaction coordinate (z) such as the position of solvated Na^+ ion in a transmembrane channel
K is a quadratic constraint that holds the system at a specified point
$E_g(z)$ is the total potential of the system without the constraints

The free energy can then be obtained by the FEP/US formula[32]

$$\Delta g(z) = -\left(\frac{1}{\beta}\right)\ln\left[\exp\{-\Delta G(\lambda_m)\beta\}\langle\delta(z' - z)\exp\{E_{m,cons}(z)\beta\}\rangle_{\varepsilon_m}\right] \tag{7.7}$$

where
$\Delta G(\lambda_m)$ is the FEP energy associated with the change of λ from zero to λ_m
$E_{m,cons}(z)$ is $E_g - E_m$

The $\Delta G(\lambda_m)$ term in Equation 7.7 is sometimes replaced by a term obtained by the "weighted histograms analysis method" (WHAM), which is intended to provide the best overlap of results from simulations at different z'.[33] PMF approaches have been blamed for suffering from major hysteresis problems, and for having the enormous challenge of calculating the reversible work of moving the ion from the bulk solvent to the protein interior. Warshel and coworkers[32] compared FEP-based PMF to the adiabatic charging (AC) approach, a nonstandard PMF method that involves pulling the ion using Equations 7.6 and 7.7. The comparison was done for a well-defined benchmark composed of small part from the gramicidin A channel with various degrees of solvation and indicated that the PMF approach converges to AC results very slowly.

MD simulations have traditionally used a large number of explicitly treated solvent molecules. These represent the most detailed methods for the study of the effect of solvent on complex biomolecules. Unfortunately, in explicit solvent simulations, the major portion of the computational time is spent on integrating the trajectories of thousands of solvent molecules, when one is interested only in the motions of the protein atoms. Moreover, statistical convergence is an important issue because the net influence of solvation results from an averaging over a large number of configurations. An alternative approach consists of incorporating the influence of the solvent implicitly. Implicit solvent models are computationally efficient compared to the explicit solvent models, and they are straightforward to interpret, as the water degrees of freedom are absent. The implicit solvent model describes the instantaneous solvent dielectric response and, therefore, eliminates the need for lengthy solvent equilibration steps. Also, the possible artifacts of the replica interactions, observed in periodic boundary explicit solvent simulations, can be avoided, since the solvent is considered as one continuous medium with infinite volume. Its applicability is sometimes limited when the properties of the water molecules are substantially different from those of the bulk solvent, for example, when there is an explicit water channel to the active site and when water is one of the reactants (e.g., hydrolysis type reactions), or when there are solvent-mediated interactions during protein folding. On the other hand, the accuracy of an implicit solvent model is not always comparable to that of explicit solvent calculations. As an improvement of this model, a new hybrid explicit/implicit solvent method has been developed that partitions the simulation system into two regions, separated by an elastics boundary.[34] This method models explicitly the hydration of the solute by either a layer or a sphere of water molecules, and then generalized Born (GB) theory is used to treat the bulk continuum solvent outside the explicit simulation volume.

In order to set up an MD simulation based on the implicit solvation model, a set of atomic radii is required, which is an additional set of input parameters compared to the explicit solvent case. A number of implicit solvent models have been developed, and the effect of the solvent is treated as an average potential acting on the solute. The implicit solvent model based on a finite difference solution of the Poisson–Boltzmann (PB) theory provides a rigorous theoretical framework and captures the polar component of the free energy of solvation for a given biomolecule quite well.[35] However, it is a computationally expensive method and, therefore, has limited application in MD simulations.

PB models[36,37] solve Equation 7.8 or its nonlinear extension for stronger fields by using a finite difference grid and treating the shape of the protein in detail, while continuing to use macroscopic dielectric constants for both the protein and the solvent.

$$\nabla[\varepsilon(r)\nabla U(r)] = -4\pi\rho(r) + \kappa^2 U(r) \qquad (7.8)$$

Here
  κ is a function of the ionic strength of the system
  ρ is the charge distribution
  ε is the dielectric constant

The corresponding potential $U(r)$ can be determined by finite difference methods on a grid. This approach has been widely accepted, in part because of its relative simplicity.

Another simpler and faster implicit continuum model based on the PB equation is the so-called GB model.[38] In fact, a very good agreement between the GB and PB models can be achieved if the effective Born radii match those computed exactly using the PB approach. Therefore, one can improve the accuracy of the GB model by improving the way the effective radii are computed. Due to its relative simplicity and computational efficiency, this methodology has become especially popular in MD simulations, compared to the more standard numerical solution of the PB equation.

The GB approach considers that the electrostatic free energy is a set of charges in an infinite homogeneous medium with dielectric constant ε_{eff}, which can be written as the sum of the gas-phase energy (ΔU_{QQ}) and the solvation free energy (G_{sol}):

$$G_{\text{elec}} = \Delta U_{QQ} + G_{\text{sol}} \tag{7.9}$$

$$\Delta U_{QQ} = 332 \frac{Q_i Q_j}{r_{ij}} \tag{7.10}$$

Alternatively, the electrostatic free energy can be expressed as the sum of the solvation free energies of the individual charges at infinite separation (G_{sol}^{∞}) plus the free energy of bringing the charges in from infinity to their actual positions ($\Delta U_{QQ}/\varepsilon_{\text{eff}}$):

$$G_{\text{elec}} = G_{\text{sol}}^{\infty} + \frac{\Delta U_{QQ}}{\varepsilon_{\text{eff}}} \tag{7.11}$$

The solvation free energy of an individual charge (Q_i) at infinite separation can be related to the solvation free energy of a charged sphere of radius, a_i (the "Born radius"), embedded in an infinite medium of dielectric constant ε_{eff}, through the following relationship:

$$G_{\text{sol}(i)}^{\infty} = -332 \left(\frac{1-1}{\varepsilon_{\text{eff}}} \right) \left(\frac{Q_i^2}{2a_i} \right) \tag{7.12}$$

Combining Equations 7.10 through 7.12 and summing the individual Born energies, we have

$$G_{\text{sol}} = G_{\text{sol}}^{\infty} - \Delta U_{QQ} \left(1 - \frac{1}{\varepsilon_{\text{eff}}} \right)$$
$$= -332 \left\{ \sum_{i=1}^{n} \left(\frac{Q_i^2}{2a_i} \right) + \sum_{i=1}^{n-1} \sum_{j=i+1}^{n} \left(\frac{Q_i Q_j}{r_{ij}} \right) \right\} \times \left(1 - \frac{1}{\varepsilon_{\text{eff}}} \right) \tag{7.13}$$

In GB treatments, the sums in Equation 7.13 are combined further by writing

$$G_{\text{sol}}^{\text{GB}} = -166 \left(1 - \frac{1}{\varepsilon_{\text{eff}}} \right) \sum_{i=1}^{n} \sum_{f=1}^{n} \left(\frac{Q_i Q_j}{f_{ij}} \right) \tag{7.14}$$

where f_{ij} is an empirical function of distance. A commonly used function is

$$f_{ij} = \left[r_{ij}^2 + a_i a_j \exp \left(\frac{-r_{ij}^2}{4 a_i a_j} \right) \right]^{1/2} \tag{7.15}$$

where a_i and a_j are "effective" Born radii for atoms i and j. Replacing the interatomic distance (r_{ij}) by f_{ij} has the effect of decreasing the contribution of the term $(Q_i Q_j/r_{ij})(1 - 1/\varepsilon_{\text{eff}})$ to the solvation energy as r_{ij} becomes small. The effective dielectric screening thus increases with the interatomic distance. It should be noted in this respect that the considerations that led to Equation 7.13 involved the assumption that ε_{eff} is the same for charge–charge interaction and Born's energy. This assumption should be kept in mind while assessing the meaning of the GB results.

GB models have given encouraging results in MD simulations of proteins and for solvation energies of simple ions in solution, but have not yet been tested extensively for other problems in protein electrostatics. It is important to note that Equation 7.14 is not a mathematical solution to the problem of charges in a multicavity continuum, or of a system with multiple dielectric regions such as a protein surrounded by water.

7.2.2 QM/MM Methodologies

For the modeling of enzymatic reactions and other biomolecular processes that involve changes in the electronic structure, such as charge transfer or electronic excitation, QM/MM approaches are making rapid progress, both methodologically and with respect to their application range. The basic idea is to use a QM method for the chemically active region (e.g., substrates and cofactors) and combine it with an MM treatment for the surroundings (e.g., the full protein and solvent). Because the two regions generally (strongly) interact, it is not possible to write the total energy of the entire system simply as the sum of the energies of the subsystems. Coupling terms have to be considered, and it is necessary to take precautions at the boundary between the subsystems, especially if it cuts through covalent bonds. The exact form of the coupling terms and the details of the boundary treatment define a specific QM/MM scheme.

Following the previous idea, large chemical systems are partitioned into an electronically important region, which requires a quantum chemical treatment, and a remainder that only acts in a perturbative fashion and thus admits a classical description (Figure 7.1). The mathematical foundations can be expressed according to Equation 7.18.

The Hamiltonian for the molecular system in the Born–Oppenheimer approximation

$$\hat{H} = -\frac{1}{2}\sum_i^{Elec.}\nabla^2 - \sum_i^{Elec.}\sum_j^{Nuc.}\frac{Z_i}{R_{ij}} + \sum_i^{Elec.}\sum_{j<i}^{Elec.}\frac{1}{r_{ij}} + \sum_i^{Nuc.}\sum_{j<i}^{Nuc.}\frac{Z_iZ_j}{R_{ij}} \qquad (7.16)$$

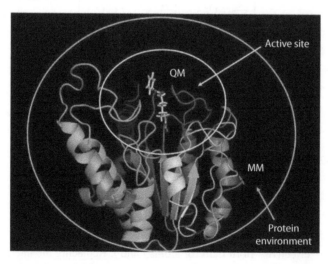

FIGURE 7.1 Illustration for the QM/MM method in the enzyme system. The active center is treated at the QM level and the surroundings is treated at the MM level.

In the presence of the external charges, we have two additional terms in the Hamiltonian:

$$\hat{H} = -\frac{1}{2}\sum_{i}^{\text{Elec.}}\nabla^2 - \sum_{i}^{\text{Elec.}}\sum_{j}^{\text{Nuc.}}\frac{Z_i}{R_{ij}} + \sum_{i}^{\text{Elec.}}\sum_{j<i}^{\text{Elec.}}\frac{1}{r_{ij}} + \sum_{i}^{\text{Nuc.}}\sum_{j<i}^{\text{Nuc.}}\frac{Z_iZ_j}{R_{ij}}$$

$$-\sum_{i}^{\text{Elec.}}\sum_{k}^{\text{Char.}}\frac{Q_k}{R_{ik}} + \sum_{i}^{\text{Nuc.}}\sum_{k}^{\text{Char.}}\frac{Z_iQ_k}{R_{ij}} \qquad (7.17)$$

These additional terms model the influence of the atoms in the MM region on the QM subsystem, that is, the polarization of the QM wave function by its MM environment. However, even in the most approximate expression, electrostatic and van der Waals (WdV) interactions between the QM and MM regions must be included in a description of the influence of the enzyme and the surrounding solvent on the QM region. The WdV terms on the MM atoms often provide the only difference in the interactions of one atom type versus another, that is, both chloride and bromide ions have unit negative charge and only differ in their WdV terms. In this way, it is quite reasonable to attribute the WdV parameters (as it is in the MM method) to every QM atom and the Hamiltonian describing the interaction between the QM and MM atoms can have the following form:

$$\hat{H}_{\text{QM/MM}} = -\sum_{i}^{\text{Elec.}}\sum_{j}^{\text{MM atoms}}\frac{Q_i}{r_{ij}} - \sum_{i}^{\text{Nuc.}}\sum_{j}^{\text{MM atoms}}\frac{Z_iQ_j}{R_{ij}} + \sum_{i}^{\text{Nuc.}}\sum_{j}^{\text{MM atoms}}\left\{\left(\frac{A_{ij}}{R_{ij}}\right)^{12} - \left(\frac{B_{ij}}{R_{ij}}\right)^{6}\right\} \qquad (7.18)$$

The WdV term also models electronic repulsion and dispersion interactions, which do not exist between QM and MM atoms because MM atoms possess no explicit electrons. Warshel and Levitt suggested such form of the Hamiltonian for the first time.[11] The hybrid QM/MM Hamiltonian is expressed as

$$\hat{H} = \hat{H}_{\text{QM}} + \hat{H}_{\text{QM/MM}} + \hat{H}_{\text{MM}} \qquad (7.19)$$

A "standard" MM force field can be used to determine the MM energy. For example, AMBER-like force field has the following form[25]:

$$E_{\text{total}} = \sum_{\text{Electrons}}\left[\left(\frac{A_{ij}}{R_{ij}}\right)^{12} - \left(\frac{B_{ij}}{R_{ij}}\right)^{6} + \frac{q_iq_j}{R_{ij}}\right] + \sum_{\text{H-bonds}}\left[\left(\frac{C_{ij}}{R_{ij}}\right)^{12} - \left(\frac{D_{ij}}{R_{ij}}\right)^{10} + \frac{q_iq_j}{R_{ij}}\right]$$

$$+ \sum_{\text{bonds}}K_b(R - R_0)^2 + \sum_{\text{angles}}K_\theta(\theta - \theta_0) \sum_{\text{dihedrals}}\frac{V_\varphi}{2}(1 + \cos(n\varphi)) \qquad (7.20)$$

The partition used in QM/MM methods lowers the computational expense, making it possible to treat large enzyme systems and the surrounding solvent, and to sample the phase space. Nevertheless, the number of QM atoms is still relatively large, and until now only low levels of QM theory, such as semiempirical methods or density functional theory (DFT), have been feasible. Semiempirical methods, though applicable to large systems, are generally not accurate enough because computed free energies of activation may have an error of 10 or more kilocalories per mole. DFT, especially with the B3LYP functional, offers improved accuracy but still lacks key physical interactions (e.g., dispersion). Often, DFT underestimates barrier heights by several kilocalories per mole, which cannot be systematically improved. Thus, when theoretical barriers do not agree with those from experiment, it is not clear whether the discrepancy arises from deficiencies in the electronic structure theory and the sampling, in the experimental observations, or from deficiencies in the underlying theoretical framework of QM/MM and TST. Consequently, there is a need for high-level electronic structure calculations for reliable predictions of enzyme reactivity. The *ab initio* electron correlation methods MP2 (Møller–Plesset second-order perturbation theory), CCSD (coupled-cluster theory with single and double excitations), and CCSD(T) (CCSD with a perturbative treatment of triple excitations) provide a well-established hierarchy that converge reliably to give high accuracy, and rate constants of gas phase reactions involving only a few atoms can be predicted with error bars comparable to those found experimentally.[39–41] However, the computational expense of these methods increases very rapidly with the number of atoms. Doubling the system size increases the cost of CCSD(T) by two orders of magnitude, and this has restricted the applications of such methods to small molecules. The steep increase in the computational cost is mainly a consequence of the delocalized character of the molecular orbitals from the Hartree–Fock (HF) theory. However, in covalent molecules, dynamic electron correlation is a short-ranged phenomenon, and by localizing the molecular orbitals it is possible to introduce a hierarchy of approximations that lead to linear scaling of all computational resources with size.

Perhaps the most accurate calculations performed to date are the MP2, LMP2, and LCCSD(T0) calculations on chorismate mutase (CM) and para hydroxy-benzoate-hydroxylase (PHBH) (the L in the acronyms indicates that local approximations were used, and T0 is an approximate triples correction).[41,42] These are coupled-cluster calculations that account for the effects of conformational fluctuations through an averaging over multiple pathways (16 for CM and 10 for PHBH). Initial structures were sampled from semiempirical QM/MM dynamics, using B3LYP/MM optimized reaction pathways.

Another method that looks promising is the SCC-DFTB (self-consistent charge density-functional tight-binding) method, a semiempirical, DFT-inspired approach.[43] It promises, within the validity domain of the parameterization, accuracy comparable to DFT at the cost of semiempirical methods. Although not new, it is being used in several attractive applications of biomolecular QM/MM simulations.[39,44] It is particularly useful to calculate free energy differences along reaction pathways. Indicating the two "solutes" as A and B and the environment as C, the free energy of conversion between A and B in the presence of C is conveniently calculated according to the standard thermodynamic integration approach[39,44]:

$$\Delta F_{A \to B} = \int_0^1 \frac{\delta F}{\delta \lambda} \, d\lambda = \int_0^1 \left\langle \frac{\delta U^\lambda}{\delta \lambda} \right\rangle_\lambda d\lambda \qquad (7.21)$$

where
λ is the coupling parameter
$\langle \ \rangle \lambda$ indicates ensemble average at a specific λ value

The potential function used in thermodynamic integration is usually written as

$$U^\lambda \lambda = (1 - \lambda)U_A(X_A X_C) + \lambda U_B(X_B X_C) + U_{CC}(X_C) \qquad (7.22)$$

where $U_A(U_B)$ includes both intramolecular A–A (B–B) interactions and intermolecular A–C (B–C) interactions.

A practical difficulty that arises in using the potential function in Equation 7.21 is that as λ approaches end points (0, 1), the contribution from either A or B to $U^\lambda(\lambda)$ vanishes, which causes sampling and convergence issues in the free energy derivative ($\delta F/\delta i$) due to large structural distortion and ideal gas–like nature of the corresponding "solute." To circumvent such "end-point" problems, the traditional solution is to use a revised version of the coupling potential:

$$U^\lambda \lambda = (1 - \lambda)U_{AC}(X_A X_C) + \lambda U_{BC}(X_B X_C) + U_{AA}(X_A) + U_{BB}(X_B) + U_{CC}(X_C) \qquad (7.23)$$

Although such decoupling of intra- and intermolecular interactions is straightforward with an MM force field, it is less so with a QM/MM potential function due to the non-separability of QM energies.[45] For example, it is possible to scale (by λ, $1 - \lambda$) only the QM/MM interactions so that as A is switched from the gas phase to the environment C, species B is switched from interacting with C to the gas phase. Thermodynamic integration following this protocol gives the free energy difference between A and B in the presence of C relative to the gas phase (i.e., the relative "solvation" or transfer free energy of A and B from the gas phase to C). Although this quantity is of interest in many cases, the complication is that the calculation of the free energy derivative requires extra computations due to the non-separability of the QM/MM energy.

Another approach for finding the minimum-energy path (MEP) is the nudged elastic band (NEB),[46] which includes, in the path definition, only selected degrees of freedom and makes sure that the environment follows the reaction smoothly. Given the QM/MM MEP, a reaction path potential is constructed using the energies, vibrational frequencies and electronic response properties of the QM region along the path. One crucial approximation in these schemes is the replacement of the QM density by ESP charges (atomic charges fitted to the electrostatic potential) in the evaluation of the electrostatic QM–MM coupling. This approximation also forms the basis for the QM/MM FEP method to calculate free-energy differences along a reaction path.[47] The MD sampling along the path, which is predetermined by QM/MM optimizations, can be performed effectively at the MM level; the QM part is kept fixed.

The previously discussed QM/MM methodologies consider the electrostatic influence of the secondary region on the primary region, an approach that is known as electrostatic embedding. A different approach only includes a mechanical embedding (ME), treating the interactions between both regions at an MM level. These interactions usually include both bonded (stretching, bending, and torsional) interactions and nonbonded (electrostatic and WdV) interactions. The original integrated molecular-orbital molecular-mechanics (IMOMM) scheme by Morokuma and coworkers,[48,49] which is also known as the two-layer ONIOM(MO:MM) method, is an ME scheme. It is based on the assumption that the subsystems can be treated additively, and has the advantage that several layers can be modeled:

$$E(\text{ONIOM}, \text{real}) = E(\text{high}, \text{model}) + E(\text{low}, \text{real}) - E(\text{low}, \text{model}) \qquad (7.24)$$

However, such a treatment has drawbacks. On one side, it requires an accurate set of MM parameters such as atom-centered point charges for both the regions, not easily available for the QM one, where the reaction takes place. Moreover, as the charge distribution in the active site usually changes as reaction progresses, the error in using a single set of MM parameters could be very serious. The second drawback of an ME scheme is that it ignores the potential perturbation of the electronic structure of the QM region due to the electrostatic interaction between the QM and MM regions. This is especially a problem if charge transfer accompanies the enzymatic reaction. This problem led to consider the mechanically embedded three-layer ONIOM (MO:MO:MM) method.[49,50] This method attempts to overcome the drawbacks of a mechanically embedded two-layer ONIOM(MO:MM) by introducing a buffer (middle) layer, which is treated by an appropriate lower-level QM theory (e.g., semiempirical molecular orbital theory), which is computationally less expensive than the method used for the innermost primary subsystem. One can label such a treatment as QM1:QM2:MM or QM1/QM2/MM. The second QM layer is designed to allow a consistent treatment of the polarization of the active center by the environment. The new treatment does improve the description, but, with ME, it does not solve the problem completely, since the QM calculation for the first layer is still performed in the absence of the rest of the atoms.

The quantitative accuracy of a QM/MM simulation depends on many factors including the reliability of the QM method, the size of the QM region, the way that the QM and MM atoms are partitioned and the scheme for computing their interactions, the amount of configurational sampling, as well as the quality of the MM force field. There is certain agreement that the key issue in achieving accurate results with a QM/MM approach is the specification of the interface between the QM and MM regions. If the QM and MM regions are not covalently linked (e.g., a QM solute and an MM solvent), the problem is reduced to appropriately parameterize nonbonded interactions between QM and MM atoms. However, if a covalent linkage between the regions is present, the specification of a total energy function that properly treats the interface region encounters fundamental problems with regard to functional form. To date, three approaches—simple fragment methods, link-atom methods, and frozen orbital methods—have been extensively pursued. There are also more complex approaches, which introduce a significant number of new terms (beyond what is normally present in a standard fixed charge MM force field); these methods are quite interesting, but will not be considered here because of space limitations. In a simple fragment method,[49,50] the QM regions are capped with hydrogen and their energies are evaluated independently at both the QM and MM levels. In a link-atom approach, a link atom is used to saturate the dangling bond at the "frontier atom" of the solute. This link atom is usually taken to be a hydrogen atom,[49] or a parameterized atom, for example, a one-free valence atom in the "connection atom,"[51,52] "pseudobond,"[53] and "quantum capping potential"[54] schemes, which involve a parameterized semiempirical Hamiltonian[51] or a parameterized effective core potential (ECP)[53,54] adjusted to mimic the properties of the original bond being cut. The second class of QM/MM methods consists of methods that use localized orbitals at the boundary between the QM and MM regions. An example is the so-called local self-consistent field (LSCF) algorithm,[55] where the bonds connecting both regions are represented by a set of strictly localized bond orbitals (SLBOs) that are determined by calculations on small model compounds and assumed to be transferable. The SLBOs are excluded from the self-consistent field (SCF) optimization of the large molecule to prevent their admixture with other QM basis functions. Another approach in the spirit of the LSCF method is the generalized hybrid orbital (GHO) method.[56] In this approach, a set of four sp^3 hybrid orbitals is assigned to each MM boundary atom. The hybridization scheme is determined by the local geometry of the three MM atoms to which the boundary atom is bonded, and the parameterization is assumed to be transferable. The hybrid orbital that is directed toward the frontier QM atom is called the active orbital, and the other three hybrid orbitals are called auxiliary orbitals. All four hybrid orbitals are included in the QM calculations, but the active hybrid orbital participates in the SCF optimizations, while the auxiliary orbitals do not (Figure 7.2).

Each kind of boundary treatment has its strength and weakness. The link-atom method is straightforward and is widely used. However, it introduces the artificial link atoms that are not present in the original molecular system, and this makes the definition of the QM/MM energy more complicated. It also presents complications in optimizations of geometries. In the original versions of the link-atom method, the polarization of the bond between the QM frontier atom and the link atom is unphysical due to the nearby point charge on the MM boundary atom. Special treatments are applied to the MM charges near the boundary so as to avoid this unphysical polarization.[53,57] The methods using local orbitals are theoretically more fundamental than the methods using link atoms, since they provide a quantum mechanical description of the charge distribution around the QM/MM boundary. The delocalized representation of charges in these orbitals helps to prevent or reduce the over-polarization that, as mentioned above, is sometimes found in the link-atom methods. However, the local orbital methods are much more complicated than the link-atom methods. The local orbital method can be regarded as a mixture of molecular-orbital and VB calculations; a major issue in these studies is the implementation of orthogonality constraints of MOs.[56] Moreover, additional work is required to obtain an accurate representation of the local orbitals before the actual start of a QM/MM calculation. For example, in the LSCF method, the SLBOs are predetermined by calculations on small model compounds, and specific force field parameters are needed to develop in order to work with the SLBOs. In the GHO method, extensive parameterization for integral scaling factors in the QM calculations is needed. Such parameters usually require reconsideration if one switches MM scheme (e.g., from CHARMM to OPLSAA), QM scheme

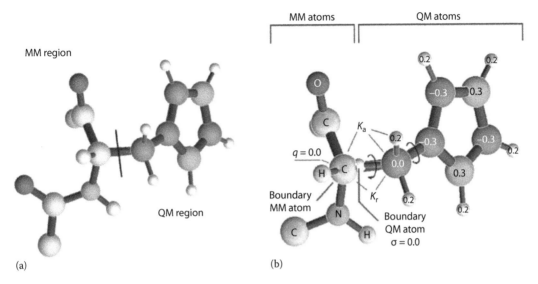

FIGURE 7.2   (a) Dividing covalent bonds across the QM and MM regions. (b) Using "link" atoms (hydrogens, halogens). For QM region, this is a hydrogen atom that interacts with the MM region only electrostatically. The Charge of "link" MM atom is set to zero to avoid double counting the electrostatic interactions. The van der Waals interaction between QM and MM atoms is not calculated. Bond stretching, angle bending, and torsion interactions between QM and MM regions are calculated as those in MM if 1–2, 1-2-3, or 1-2-3-4 terms contain at least one MM atom and one QM atom.

(e.g., from semiempirical molecular orbital methods to DFT or post-HF *ab initio* methods), or QM basis set. The low transferability limits the wide application of the local orbital methods.

The performance of both the link-atom and local-orbital approaches has been examined by extensive test calculations. The conclusion is that reasonably good accuracy can be achieved by both approaches if they are used with special care.

7.2.3   EMPIRICAL VALENCE BOND METHOD

EVB[6,11,21,22] is a method that describes reactions by mixing resonance states (or more precisely diabatic states) that correspond to VB structures, which describe the reactant, intermediate (or intermediates), and product states. The potential energy of these diabatic states is represented by a classical MM force field of the following form:

$$\varepsilon_i = \alpha_{gas}^i + U_{intra}^i(R, Q) + U_{Ss}^i(R, Q, r, q) + U_{ss}(r, q) \qquad (7.25)$$

Here R and Q represent the atomic coordinates and charges of the diabatic states, and r and q are those of the surrounding protein and solvent. α_i gas is the gas phase energy of the diabatic state (where all the fragments are taken to be at infinite separation), $U_{intra}(R, Q)$ is the intramolecular potential of the solute system (relative to its minimum); $U_{Ss}(R, Q, r, q)$ represents the interaction between the solute (S) atoms and the surrounding (s) solvent and protein atoms. $U_{ss}(r, q)$ represents the potential energy of the protein/solvent system ("ss" designates surrounding–surrounding). The ε_i of Equation 7.25 forms the diagonal elements of the EVB Hamiltonian (H_{ii}). The off-diagonal elements of the Hamiltonian, H_{ij}, are represented typically by simple exponential functions of the distances between the reacting atoms. The H_{ij} elements are assumed to be the same in the gas phase, in solutions and proteins. The ground state energy, E_g, is obtained by solving

$$H_{EVB}C_g = E_g C_g \qquad (7.26)$$

Here

C_g is the ground state eigenvector
E_g provides the EVB potential surface

To express the adiabatic energy surface of the solute–solvent system, it is useful to define a generalized reaction coordinate as the energy gap between the diabatic reactant and product EVB states:

$$x = \Delta\varepsilon_{1,2} = \varepsilon_2 - \varepsilon_1 \qquad (7.27)$$

This coordinate can be divided into a solute coordinate, R, for internal bonds of the reacting EVB structures and a solvent coordinate, S, for interactions of the solute with the solvent.

The analytical derivatives of the potential surface allow a sampling of the EVB energy surface by MD simulations. In principle, running MD trajectories on the EVB surface of the reactant state can provide the free energy function that is needed to calculate the activation free energy. In order to reach the TS, it becomes necessary to run trajectories on a series of potential surfaces ("mapping" potentials) that drive the system adiabatically from the reactant to the product state. In the simple case of two diabatic states the mapping potential (ε_m) can be written as a linear combination of the reactant and product potentials, ε_1 and ε_2:

$$\varepsilon_m = (1 - \lambda_m)\varepsilon_1 + \lambda_m\varepsilon_2 \qquad (7.28)$$

where λ_m changes from 0 to 1 in $n + 1$ fixed steps (λ_m) ($0/n$, $1/n$, $2/n$, ..., n/n) (and apply FEP).

The EVB method satisfies some of the main requirements for reliable studies of enzymatic reactions. Among the obvious advantages of the EVB approach is the efficient way to obtain proper configurational sampling and converging free energy calculations. Furthermore, the EVB benefits from treating the solute–solvent coupling consistently and conveniently. This feature is essential not only for physically consistent modeling of charge-separation reactions but also for the effective use of experimental information and *ab initio* data in calibrating PESs.

7.2.4   CLUSTER MODELS

The use of relatively small quantum chemical clusters to study enzyme-catalyzed processes has been justified on the consideration that the energies involved in chemical reactions, bond breaking and formation, are usually much higher than long-range electrostatics. The effect of the catalyst, in this case the active site of the enzyme, is thus to a large extent local, and environmental effects are usually of lower order. This approach has been applied to many different metalloenzymes systems by the groups of Himo and Sieghban.[5,58,59] In most cases, the hybrid B3LYP functional was used with a relatively small basis set (typically of double zeta quality) for geometry optimizations, as it was found to give quite reliable geometries. More accurate energies are derived as single point calculations with larger basis sets (typically of triple zeta quality and including polarization and diffuse functions).

The effect of the environment is considered from a twofold standpoint: by long-range polarization and by imposing steric restraints on the active site. To account for the polarization effects caused by the part of the surrounding enzyme that is not explicitly included in the quantum model, cavity techniques are used, assuming that the surrounding is a homogenous polarizable medium with some dielectric constant, usually chosen to be $\varepsilon = 4$ for protein environment. In order to account for steric effects, a freezing scheme is used: groups not bound to a metal center or linked by bonds or hydrogen bonds to some other groups at the active site have to be restrained in their movement, as they move typically in shallow potential wells. In order to keep the various groups in place to resemble the crystal structure as much as possible, certain atoms in the model, typically where the truncation is done, are kept frozen to their x-ray positions. This approach ensures structural integrity of the model, yet allows some flexibility of the various groups.

The cluster model often includes only the first coordination shell of the metal on top of which a polarizable continuous model (PCM) calculation is performed to evaluate the solvation. In situations where some second-shell residues are known from experiments, or can be suspected to influence the reactions, they are included explicitly. In instances where the metal site is particularly large, including several metal centers, some of the ligands that do not directly participate in the chemistry can be replaced by simpler ligands, such as water or ammonia, as a first approximation.

This simplification of the systems has been extensively applied in the 1990s, but has been surpassed by the use of QM/MM models due to two mean reasons: the higher accuracy of the latter and the enhanced computational resources available nowadays. The use of cluster models can be justified in some cases, when the x-ray structure is not available, as an initial calculation that provides initial structures for a further QM/MM.

7.2.5 Biomimetics

An insight into the structures and mechanisms of action of enzymes is often obtained by studying synthetic analogues, that is, small molecules that resemble the structural and functional sites of the enzymes.[5,60–62] Such studies are important because synthetic analogues are more amenable to structural, spectroscopic, and mechanistic studies than are the enzymes themselves. Furthermore, synthetic analogues are also more amenable to fine-tuning by systematic substituent effects than are the active sites of their enzyme counterparts so that it is possible to examine a variety of factors that influence reactivity.

An appropriate analogue simulates or achieves the coordination sphere composition and stereo-chemistry of the native site. A *structural* analogue allows the deduction of site characteristics common to the site and itself by possessing sufficiently similar features. A *functional* analogue sustains a catalytic reaction that transforms substrate to product as does the enzyme, albeit at a different rate and not necessarily with the natural stereochemical outcome. A functional model is not ineluctably a structural model, but, ideally, a good structural model is a functional model. The construction of an analogue complex can be an iterative process, until the desired level of similarity with the site is achieved. Nevertheless, a mechanism of enzyme action can only be won from the enzyme itself.

Site analogues have contributed significantly to physicochemical methodology by facilitating the correlation of composition and structure to spectroscopic and magnetic observable. Because of their relative simplicity, synthetic analogues provide suitable molecular models for theoretical calculations of geometry and electronic structure. Moreover, the detailed correlation of electronic structure with geometric structure can provide fundamental insight into reactivity.

The relative simplicity of the molecules allows the use of high-level calculations. Levels of theory similar to those described in the cluster approach are used, albeit they are now free from structural approximations.

7.3 CASE STUDY: HYDROGENASES

Metalloproteins are simply metal complexes, with remarkably intricate and complex ligands, which perform a variety of relevant functions in biological systems. Approximately one-half of all known protein crystal structures in the protein data bank contain metal cofactors, which play vital roles in charge neutralization, structure, and function.[61] The catalysis of several key chemical transformations occurs at metal centers embedded in the active sites of metalloenzymes.[62] The active sites usually involve one or more metal ions that, coordinated with several ligands of a macromolecule, define a biological metallocenter assembly.[60] Metal ions vary in their charge, radius, ligand exchange lability, and ligand preference. Similarly, the fold of the protein, its rigidity, and the properties of its side chains dictate its interactions with the substrate, forming a binding pocket near the metal active site, providing specific charge and H-bond residues to assist the catalysis, and imposing a unique geometry on the metal site, which can activate it for catalysis (entatic state).[61]

Hydrogenases catalyze one of the simplest redox reactions in nature, the reversible two-electron oxidation of H_2.[63–65] A mechanistic knowledge of the hydrogen conversion and consumption process in microbes is of utmost biotechnological importance. In addition to its relevance in basic and applied research, this knowledge would provide the necessary fundamental tools for designing biomimetic or bioinspired artificial "hydrogenase catalysts" for large-scale hydrogen production in the future. Both Fe–Ni and Fe–Fe hydrogenases are promising electrocatalysts. For this reason, they have been extensively studied by electrochemical methodologies.[4]

The structure of the active sites of these enzymes is complex, and the intimate mechanism of catalysis is currently a very active area of investigation.[63,66–71] Moreover, the study of these reactions is further complicated by the existence of three different types of hydrogenases, and therefore three potentially different mechanisms.[1,3,4] The three known structural types can be distinguished by their metal content. The most common type is the Ni–Fe hydrogenase,[65,72–75] followed by the Fe-only enzyme[76] and the metal-free enzyme,[77,78] which has been exclusively found in the *Methanobacterium* species. For the latter, recent studies have found Fe in its composition,[78] an issue that will not be solved until the x-ray structure becomes available. An overriding problem in establishing mechanisms arises from the nature of the substrates: H-species are too light to be located directly by x-ray diffraction methods, and the enzymes are too large to be studied with NMR. The detection of hydrogen species by magnetic resonance methods is, therefore, restricted to those states in which the metal center has an odd number of electrons, so that EPR-based methods, such as ENDOR (electron nuclear double resonance) and ESSEM (electron spin echo envelope modulation) can be used.[2,79] Partly as a consequence of these experimental limitations, there has been great interest in applying theoretical methods, particularly DFT, to determine the structures that are likely for each state of the enzyme.[2,66,67,70,80–85]

Despite the wealth of information available that has mainly resulted from structural, spectroscopic, and theoretical studies, the detailed reaction mechanism remains a matter of debate. This is in part originated in the ambiguity of the redox states involved, and in the apparent necessity of coupling a one-electron redox chemistry of the active site with a two-electron redox process. This situation, plus the unknown role of protons in determining the charge of the active site, has resulted in theoretical studies that do not agree on the redox states involved in catalysis or whether or not protonated cysteine residues are present.[66,67,70,80–85] There is some consensus in an initial heterolytic cleavage of the H_2 molecule, which diffuses through the protein to the active site. Theory has provided a number of possibilities for the structures of intermediates that can be tested by spectroscopic and other physical probes.[81] The DFT studies focus on the electronic structure of the active site, and one concern is how much the calculations need to be extended in order to include electrostatic and more specific details of the surrounding protein environment.[64,66] According to theoretical and experimental findings, nearby side chains, such as carboxylate and imidazole, are good candidates to play a mechanistic role as catalytic bases.[64,66,86]

Based on the present knowledge of the enzyme function and mechanism, present and future theoretical research effort has to be oriented to examine larger parts of the protein environment and to include all the residues that may be probably involved in the proton transfer mechanism. Protons are transported inside proteins via motion of water molecules and amino acid residues with acid–base properties, such as histidine, glutamic acid and aspartic acid.[87] Several routes have been proposed for hydrogenases, and most likely, proton transfer does not stick to a single one. In Ni–Fe hydrogenases, recent theoretical and experimental studies have pointed to a glutamate residue, bonded to a terminal cysteine ligand of the active site, as involved in proton transfer, being, thus, relevant for catalysis.[64,66,86] MD simulations have underscored the importance of the tunnels for substrate access to the active site and hydrogen storage inside the [Fe–Ni] hydrogenases.[1,65,88]

7.3.1 The Relevance of Hydrogenases

Hydrogenases are found in a large number of bacteria and archaea, as well as in some simple eukaryotes, where they catalyze the oxidation or the formation of molecular hydrogen according to the reaction $H_2 \rightleftarrows 2H^+ + 2e^-$.[63,64] This feature allows many microorganisms to use H_2 as an energy source. It provides low-potential electrons for a variety of metabolic pathways and final electron acceptors.[89] In addition, hydrogenases may be involved in the formation of a proton transfer gradient used for the generation of metabolic energy.[90] Alternatively, the formation of H_2 is used in many anaerobes as a sink for low-potential electrons produced by fermentative processes.[91]

Both environmental and strategic concerns have generated an increased interest in hydrogen as a clean, potentially renewable energy source. Hydrogenase studies contribute to the biotechnology of both hydrogen production and oxidation: there are ongoing efforts to couple photosynthesis, which uses solar energy, to hydrogen bioproduction. Many photosynthetic algae use hydrogenase to get rid of low-potential electrons through the reduction of protons.[89] One major problem with associating photosynthesis to hydrogenase activity is the high sensitivity of the enzyme to oxygen, a by-product of the water-splitting reaction. As far as hydrogen oxidation concerns, recent experiments using a hydrogenase-coated electrode have shown that the enzyme can be as effective as Pt in catalyzing that reaction.[92] The problem is the relative fragility of protein molecules when compared to inorganic catalysis. One way to circumvent this situation would be to synthesize small biomimetic molecules based on the active site of hydrogenases. The design of biomimetics opens a relevant research field that points to the $NiFeCO(CN^-)_2$ unit as promising for the design of fuel cell catalysis or to improve hydrogen production from biomass or from solar energy.

As a way of exemplifying the application of theoretical methodologies, we will restrain the present discussion to Fe–Ni hydrogenases.

7.3.2 Overall Structure and Active Site Metallocenter of Ni–Fe Hydrogenases

Crystal structures of Ni–Fe hydrogenases of five different bacteria have been deposited in the protein data bank. For the enzymes of *Desulfovibrio* (*D.*) *gigas, D. fructovorans, D. vulgaris Miyazaki F, D. desulfuricans,* and *D. baculatum,* the highest reported resolutions are 2.5,[74] 1.8,[72] 1.4,[75] 1.8,[93] and 2.7 Å,[73] respectively. Both *D. gigas* and *D. fructovorans* display the oxidized form whereas the active site of *D. vulgaris Miyazaki F* has been elucidated in the reduced state.[75] *D. baculatum,* on the other hand, contains a seleno-cysteine ligand.[73]

The folding patterns are essentially identical in *D. gigas* and *D. vulgaris,* reflecting a high sequence homology.[64] Two subunits are differentiated, the largest one containing the active site metallocenter. Main characteristics are shown in Figure 7.3.

The structures around the Ni–Fe active sites are also very similar. In the oxidized (inactive) form, shown in Figure 7.4a for *D. gigas,* two thiolate side chains of cysteine residues are coordinated to the Ni atom; two other thiolate side chains of cysteine residues and one monoatomic ligand coordinate to both the Ni and Fe atoms as bridges; and three diatomic ligands are coordinated to the Fe atom.[74] Differences have been reported for the chemical species of the monoatomic bridge and diatomic ligands: the *D. gigas* enzyme was reported to have an oxygen species as a monoatomic bridge and one CO and two CN molecules as diatomic ligands,[74] whereas the *D. vulgaris Miyazaki F* enzyme was reported to have a sulfur species as a monoatomic bridge and SO, CO, and CN molecules as diatomic ligands.[75] The monoatomic S bridge disappears in the latter upon reduction with H_2 (Figure 7.4b).[73] The reduced form has also been identified for *D. baculatum.*[73] The Fe atom has only been detected in the 2^+-oxidation state and is diamagnetic. The Ni atom exhibits variable oxidation states III, II, and I. In the EPR detectable Ni_r states, a high-spin paramagnetic (d_8) configuration for Ni(II) is largely accepted.[63,79] Three FeS clusters are well defined in the electron density maps (see Figure 7.3), one of which (Fe_4S_4) is located nearby, on a second subunit, helping to channel electrons in and out.[64]

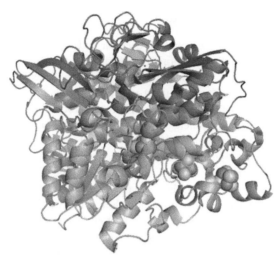

FIGURE 7.3 Polypeptide fold of *D. gigas* NiFe hydrogenase. The small and the large subunits are depicted in pale and dark gray, respectively, using ribbons for α helices and arrows for β strands. Metals and inorganic sulfur atoms are depicted as spheres, showing the presence of three iron–sulfur clusters in the small subunit.

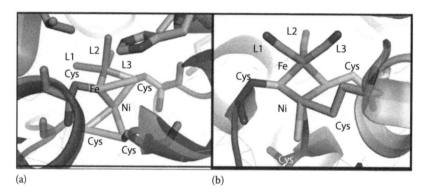

(a)   (b)

FIGURE 7.4 Active site structure the NiFe hydrogenases of *D. gigas* (a) (oxidized form) and *D. baculatum* (b) (reduced form), refined at 2.54 and 2.15 Å resolution, respectively. The second structure lacks the bridge O atom.

7.3.3 PROPOSED REACTION MECHANISMS: RELEVANT THEORETICAL WORK

In the process of enzyme "activation" and during the catalytic cycle, the [NiFe] hydrogenase passes through a number of intermediate states. They have first been observed and characterized by EPR spectroscopy, which showed that the enzyme cycles between EPR-silent and EPR-detectable (paramagnetic) nickel-centered states. They were named Ni-A, Ni-B, Ni-C (paramagnetic, EPR active), Ni-L (light-induced, EPR active), Ni-SI (EPR silent), and Ni-R (reduced, EPR silent). For historical reasons, these "names" have been retained until today. The catalytic cycle is represented in Figure 7.5, including also the inactivation of CO, not discussed here.

Several intermediate states have been identified by means of EPR and IR methodologies.[69] Despite numerous experiments, a consensus on the details of the oxidation states of Fe and Ni, as well as the degree of protonation of the residues involved in the intermediates are still unclear. Furthermore, the structural details of most of the [FeNi] forms are even more uncertain. Based on the experimental results, several theoretical studies have been carried out to determine the molecular

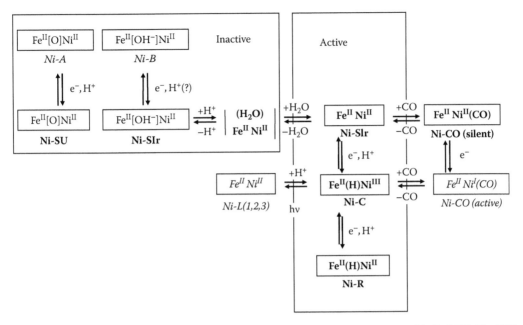

FIGURE 7.5   Schematic view of the [Ni–Fe] hydrogenase catalytic cycle, as suggested in Ref. [2]. The EPR detectable states are shown in italics; the EPR silent states are shown with bold fonts. The exact mechanism and the transient species involved are not yet known. More details in the text.

structures of these intermediate forms.[66,67,70,80–85] The structures derived from different theoretical studies have been summarized in Ref. [83] (see Figure 7.6).

On the basis of the consideration of the different intermediates, several mechanistic schemes have been proposed for the catalytic cycle of Ni–Fe hydrogenase. Nevertheless, there is little agreement between them. Hypothesis addressing plausible mechanisms of hydrogen metabolism by FeNi hydrogenases must explain the role of the different active components and the function of the protein environment and integrate the known structural data.

The activation of H_2 is believed to involve binding to the active site followed by heterolytic cleavage to form a hydride intermediate.[63,70,79] Molecular hydrogen is thought to diffuse from the medium to the active site through a hydrophobic channel.[65] The binding site has been associated with both the Ni and Fe atoms of one of the Ni-SI silent active forms with a vacant NiFe bridging site.[64] The initial formation of an Ni–H_2 complex is consistent with inhibition by exogenous CO (a competitive inhibitor of molecular hydrogen),[68] and also with MD simulations.[65] Nevertheless, theoretical calculations, based on the application of DFT, favor H_2 bonding to the Fe center.[66,79,81,84,85,94] What happens to the substrate after this step, which should generate the Ni-R state, has not been clearly established.

After binding to the Fe center, hydrogen heterolytic cleavage may lead to a metal bound hydride with the protonation of the bridging cysteine, which eventually ends up as a terminal Fe ligand (Figure 7.7).[95] The resulting hydride binds to the NiFe center, bridging both metal ions in the Ni-C form. The cycle is closed by the transfer of one proton and one electron from the active site and the regeneration of the Ni-SI species.

The protonation of an Ni-coordinated cysteine after heterolytic cleavage has also been proposed.[85] From the resulting intermediate, the reduction mechanisms evolve through a structurally different Ni-C intermediate (compare Figures 7.7 and 7.8), but similarly regenerates the Ni-SI intermediate. No corresponding low-energy pathway has been found for the initial coordination of H_2 to the Ni center.

As a way of taking into account the protein environment, QM/MM calculations have been performed for a QM region defined by the NiFe center, the cysteine side chains, and the

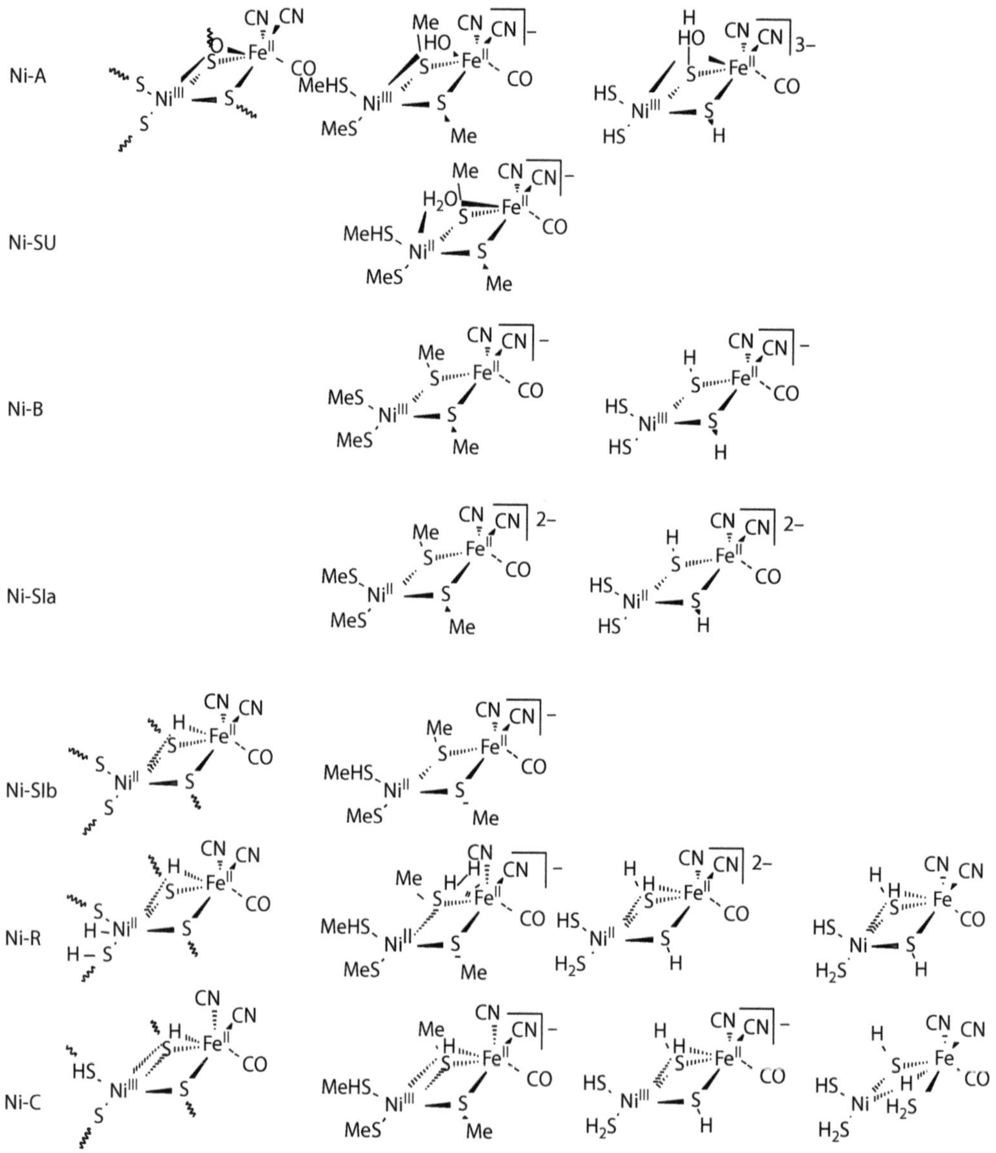

FIGURE 7.6 Summary of the proposed structures for the various observed form of Ni–Fe hydrogenase according to different research groups. The structures shown in the first column have been taken from Ref. [84], those shown in second column belong to the calculations done in Ref. [102], whereas the third and fourth column correspond to Refs. [94,95], respectively.

three non-proteic ligands.[84] The QM/MM hybrid method was composed of a Kohn–Sham DFT functional for the QM region, and the CHARMM force field for the MM one. Some structures of the intermediates have been analyzed but no TS structures were proposed (see Figure 7.9).

The QM/MM studies found a negligible effect of the protein environment. Nevertheless, theoretical DFT calculations that have included protein residues proximal to the active site in the cluster model have demonstrated the influence of near-neighbor residues in the calculation of the g-tensors for the Ni_r and Ni_u states.[66] Those calculations have even associated the different behavior of Ni_r and Ni_u with different interactions with a nearby glutamic acid residue, which are originated in different orientations of a cysteine residue that bridges it to the active site. It has been

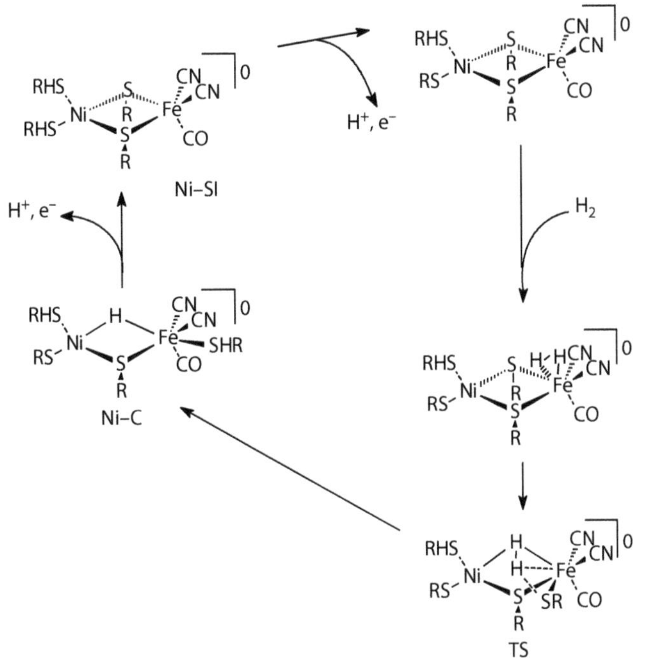

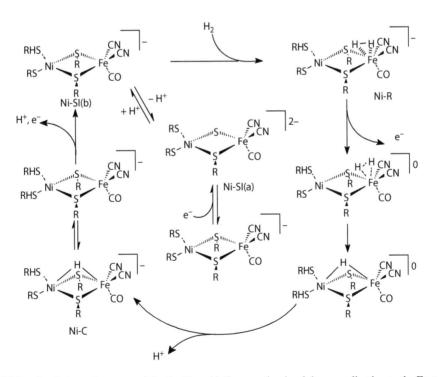

FIGURE 7.7   Catalytic cycle proposed for the H_2 oxidation reaction implying coordination to the Fe atom and protonation of a bridging cysteine residue.

FIGURE 7.8   Catalytic cycle proposed for the H_2 oxidation reaction implying coordination to the Fe atom and protonation of a Ni-coordinated cysteine residue.

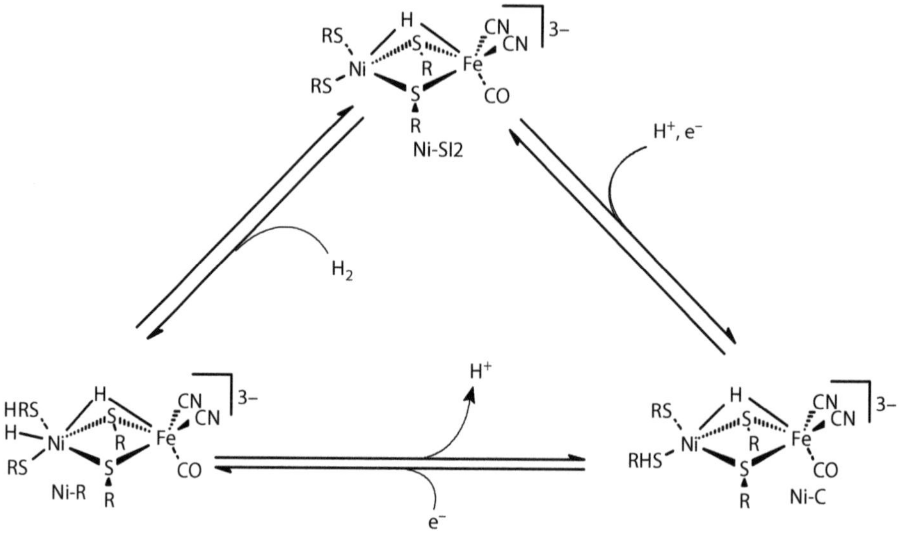

FIGURE 7.9 Structures proposed for the intermediates involved in the Ni–Fe hydrogenase catalyzed H$_2$ oxidation reaction, derived from QM/MM calculations. More details in the text.

recently proposed that a proton can be transferred to this glutamic acid residue after heterolytic cleavage between the Ni ion and the Cys residue.[86] This possibility is supported by the higher-than-average temperature factors of these residues, indicative of a degree of disorder that could arise from various degrees of protonation/deprotonation of the cysteine and glutamate side chains.[64]

Whereas the mechanistic details of the reversible hydrogen reduction reaction are still under debate; increasing theoretical and experimental evidences point to the necessity of properly modeling the protein environment in order to attain a trustable picture of the catalytic process.

The spectroscopic investigation of the above-discussed redox states gives information about the electronic structures of all intermediates, and has been used for contrasting the theoretical results. In early studies, the CO and CN vibrational frequencies were calculated for the proposed reaction intermediates as a way for experimental validation, and the site of hydrogen splitting was proposed to be iron.[82,83] In a different approach, the EPR g values and hyperfine coupling constants were calculated using the zeroth-order regular approach (ZORA), with good agreement considering the accuracy of the DFT method for the computation of g values.[80,81] At that time, a μ-oxo bridge was proposed for the active site structure, though recent experiments have shown that the bridge is probably protonated thus suggesting a hydroxo or hydroperoxo bridge.[96] A similar study, using DFT calculations, included more amino acids in the model to better investigate the influence of the environment,[66] suggesting an OH− as bridging ligand. The effect of the environment on the g values indicated that the overall structure is primarily determined by the Ni-coordinated ligands, and is further fine-tuned by surrounding amino acids. An improved model of the environment was attained by QM/MM calculations,[84] that found a special role of the terminal cysteine near the H$_2$ channel (Cys-530 for *D. gigas*) from the comparison of the data with FTIR spectroscopy. The spin density distribution for the Ni-A, Ni-B, and Ni-C states confirm the 3dz^2 character of the wave function of the unpaired electron.[80] In total, for the Ni-C state, 51% spin density was found on Ni and 29% on the cysteine of the axial sulfur (Cys-549 for *D. vulgaris Miyazaki F*), which gives rise to hyperfine splitting in the EPR spectrum of 33S enriched [NiFe] hydrogenase. The terminal cysteine near the gas channel (Cys-546) carries about 10% spin density and the remaining 5% is distributed over the other atoms. The spin density at the iron was found to be small (about 1%), indicating the presence of a low-spin Fe(II). Essentially, the same spin density distribution was found for the [NiFeSe] hydrogenase of *D. baculatum* in the Ni-C state.[80] A similar picture is obtained for the Ni-B state. Concerning the EPR-silent states, for which obviously no EPR or ENDOR data are available, a DFT study has been performed by Bruschi et al.[97] It was found that the low-spin (S)0 and high-spin (S)1 states for both Ni-SI and Ni-R are very close in energy, and are dependent, thence, on the level of theory.

The spectroscopic characteristics support the catalytic cycle shown in Figure 7.5.[2]

7.3.4 METALLOENZYME MIMETIC COMPLEXES

The study of model complexes as metalloenzyme mimetics has a long tradition in bioinorganic chemistry (see *Chem. Rev.* 2004, 104, issue 2 for a complete revision). Several Ni–Fe-based complexes have been prepared, with structural features similar to those found in the FeNi active site.[98] Nevertheless, none of them has been tested in its ability to coordinate or decompose H_2. Present research is mainly centered in mimicking the thiolate ligands in the Fe–Ni coordination. It has to be mentioned that Fe-only hydrogenase mimics able to electrocatalyze proton reduction have been recently reported.[99]

The dinuclear complex shown in Figure 7.10a was one of the first well-behaved structural analogues that has been synthesized.[100,101] The core of the complex is dinuclear with Ni bound to Fe by a bis-thiolate bridge and the Fe atom binding to carbon monoxide groups. The Ni–Fe distance of 3.31 Å is in good agreement with the one theoretically predicted for the active states of the enzyme.[102] However, this complex failed to reproduce the pseudo-tetrahedral geometry about the Ni atom that has been found in various forms of the enzyme. The pseudo-tetrahedral geometry has been reproduced in the trinuclear complex [Ni{Fe(N CH$_2$CH$_2$S$_3$)(CO)-S-S′}$_2$] (Figure 7.10b),[100] and in [{Ni(SCH$_2$CH$_2$CH$_2$S)(dppe)-S,S′}Fe(CO)$_3$].[103] In the latter, the Ni centers are linked through thiolate units and the metal atoms are 2.47 Å apart.

In the above-mentioned complexes, the coordination about the Ni atom has been completed by phosphine ligands. In contrast, a compound has been prepared in which the Ni center is coordinated only by thiolates (Figure 7.10c).[104] Nevertheless, the phosphine ligands are, in this case, on the iron atom.

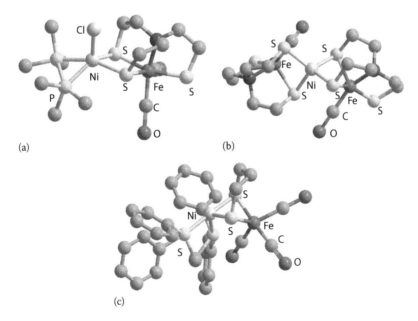

(a)   (b)

(c)

FIGURE 7.10   Schematic representation of biomimetic Ni–Fe complexes. (a) A molecule of [{Fe(NS$_3$)(CO)$_2$-S,S′}NiCl(dpp)], phenyl groups omitted for clarity. (b) [Ni{Fe(N CH$_2$CH$_2$S$_3$)(CO)-S-S′}$_2$]. (c) [{Ni (SCH$_2$CH$_2$CH$_2$S)(dppe)-S,S′}Fe(CO)$_3$].

Unfortunately, in spite of the significant efforts being made by several research groups, no biomimetic small molecule with catalytic properties that could be used in fuel cells has been yet synthesized. The complexity of the structure of the NiFe hydrogenases active site, in relation to both the chelating ligands and the local geometry around the metal centers, orientates future research to solid state–based chemistry, focusing on the consideration of polymer-supported Ni/Fe/S materials for the design of biocatalysts.

REFERENCES

1. J. Fontecilla-Camps, A. Volbeda, C. Cavazza, Y. Nicolet, *Chem. Rev.* **107** (2007) 4273.
2. W. Lubitz, E. Reijerse, M. van Gastel, *Chem. Rev.* **107** (2007) 4331.
3. P. Vignais, B. Billoud, *Chem. Rev.* **107** (2007) 4206.
4. K.A. Vincent, A. Parkin, F.A. Armstrong, *Chem. Rev.* **107** (2007) 4366.
5. L. Noodleman, T. Lovell, W.-G. Han, J. Li, F. Himo, *Chem. Rev.* **104** (2004) 459.
6. A. Warshel, P. Sharma, M. Kato, Y. Xiang, H. Liu, M. Olsson, *Chem. Rev.* **106** (2006) 3210.
7. I. Feierberg, J. Aqvist, *Biochemistry* **41** (2002) 15728.
8. S.N. Rao, U.C. Singh, P.A. Bash, P.A. Kollman, *Nature* **328** (1987) 551.
9. K. Soman, A.S. Yang, B. Honig, R. Fletterick, *Biochemistry* **28** (1989) 9918.
10. C. Blake, L. Johnson, G. Mair, A. North, D. Phillips, V. Sarma, *Proc. R. Soc. Ser. B* (1967) 368.
11. A. Warshel, M. Levitt, *J. Mol. Biol.* **103** (1976) 227.
12. S. Hur, T. Bruice, *J. Am. Chem. Soc.* **125** (2003) 1452.
13. J. Lee, K.S. Houk, *Science* **276** (1997) 942.
14. P. Frey, S. Whitt, J. Tobin, *Science* **264** (1994) 1927.
15. M. Garcia-Viloca, A. Gonzalez-Lafont, J.J.A.C. Lluch, *J. Am. Chem. Soc.* **123** (2001) 709.
16. S. Northrup, M. Pear, C. Lee, J. McCammon, M. Karplus, *Proc. Natl. Acad. Sci. USA* **79** (1982) 4035.
17. M. Olsson, W. Parson, A. Warshel, *Chem. Rev.* **106** (2006) 1737.
18. A. Warshel, J. Florian, *Proc. Natl. Acad. Sci. USA* **95** (1998) 5950.
19. G. Estiu, K.M.J. Merz, *J. Phys. Chem. B* **111** (2007) 6507.
20. B. Garrett, D. Truhlar, *J. Chem. Phys.* **70** (1979) 1593.
21. M. Olsson, J. Mavri, A. Warshel, *Philos. Trans. R. Soc. B* **361** (2006) 1417.
22. A. Warshel, P. Sharma, M. Kato, W. Parson, *Biochim. Biophys. Acta* **1764** (2006) 1647.
23. A. McCammon, B. Montgomery Petitt, L. Ridgway Scott, *Math. Appl.* **28** (1994) 319.
24. B.R. Brooks, R. Bruccoleri, B. Olafson, D. States, S. Swaminathan, M. Karplus, *J. Comp. Chem.* **4** (1983) 187.
25. D.A. Case, T.A. Darden, T.E. Cheatham, III, C.L. Simmerling, J. Wang, R.E. Duke, R. Luo, M. Crowley, R.C. Walker, W. Zhang, K.M. Merz, B. Wang, S. Hayik, A. Roitberg, G. Seabra, I. Kolossváry, K.F. Wong, F. Paesani, J. Vanicek, X. Wu, S.R. Brozell, T. Steinbrecher, H. Gohlke, L. Yang, C. Tan, J. Mongan, V. Hornak, G. Cui, D.H. Mathews, M.G. Seetin, C. Sagui, V. Babin, P.A. Kollman (2008), *AMBER 10*, University of California, San Francisco, CA.
26. W.L.C.J. Jorgensen, J. Madura, R.W. Impey, M.L. Klein, *J. Chem. Phys.* **79** (1983) 926.
27. V. Essman, L. Perera, M.L. Berkowitz, T. Darden, H. Lee, L.G. Pedersen, *J. Chem. Phys.* **103** (1995) 8577.
28. P.A. Kollman, *Chem. Rev.* **93** (1993) 2395.
29. F. Lee, Z. Chu, M. Bolger, A. Warshel, *Protein Eng.* **5** (1991) 215.
30. G. Del Buono, F. Figueirido, R. Levy, *Protein: Struct. Funct. Genet.* **20** (1994) 85.
31. Y. Sham, Z. Chu, A. Warshel, *J. Phys. Chem. B* **101** (1997) 4458.
32. M. Kato, A. Warshel, *J. Phys. Chem. B* **109** (2005) 19516.
33. A. Ferrenberg, R. Swendsen, *Phys. Rev. Lett.* **63** (1989) 1195.
34. V. Tsui, D.A. Case, *J. Am. Chem. Soc.* **122** (2000) 2489.
35. J.E. Shea, J.N. Onuchic, C.L. Brooks, *Proc. Natl. Acad. Sci. USA* **99** (2002) 16064.
36. W. Orttung, *J. Phys. Chem.* **89** (1985) 3011.
37. K. Sharp, B. Honig, *Annu. Rev. Biophys. Biophys. Chem* **19** (1990) 301.
38. T. Simonson, J. Carlsson, D.A. Case, *J. Am. Chem. Soc.* **126** (2004) 4167.
39. H. Senn, W. Thiel, *Curr. Opin. Chem. Biol.* **11** (2007) 182.
40. A. Mulholland, *Drug Discov. Today* **10** (2005) 1393.
41. A. Mullholland, *Chem. Cent. J.* **1** (2007) 19.

42. F. Claeyssens, J. Harvey, F. Manby, R. Mata, A. Mulholland, K. Ranaghan, M. Schütz, S. Thiel, W. Thiel, H.-J. Werner, *Angew. Chem. Int. Ed.* **45** (2006) 6856.
43. M. Elstner, *Theor. Chem. Acc.* **116** (2006) 318.
44. D. Riccardi, P. Schaefer, Y. Yang, H. Yu, N. Ghosh, X. Prat-Resina, P. Konig, G. Li, D. Xu, H. Guo, M. Elstner, Q. Cui, *J. Phys. Chem. B* **110** (2006) 6458.
45. G. Li, X. Zhang, Q. Cui, *J. Phys. Chem. B* **107** (2003) 8643.
46. L. Xie, W. Yang, H. Liu, *J. Chem. Phys.* **120** (2004) 8038.
47. Y. Zhang, H. Liu, W. Yang, *J. Chem. Phys.* **112** (2000) 3483.
48. T. Kerdcharoen, K. Morokuma, *Chem. Phys. Lett.* **355** (2002) 257.
49. K. Morokuma, *Philos. Trans. R. Soc. Lond. A* **360** (2002) 1149.
50. M. Svensson, S. Humbel, R.D.J. Froese, T. Matsubara, S. Sieber, K. Morokuma, *J. Phys. Chem.* **100** (1996) 19357.
51. I. Antes, W. Thiel, *J. Phys. Chem. A* **103** (1999) 9290.
52. S. Dapprich, I. Komiromi, K.S. Byun, K. Morokuma, M. Frisch, *J. Mol. Struct. TEOCHEM* **1** (1999) 461.
53. Y. Zhang, T.-S. Lee, W. Yang, *J. Chem. Phys.* **110** (1999) 46.
54. G.A. DiLabio, M.M. Hurley, P.A. Christiansen, *J. Chem. Phys.* **116** (2002) 9578.
55. N. Ferré, X. Assfeld, J.-L. Rivail, *J. Comp. Chem.* **23** (2002) 610.
56. J. Pu, J. Gao, D.G. Truhlar, *J. Phys. Chem. A* **108** (2004) 5454.
57. P. Amara, M.J. Field, *Theor. Chem. Acc.* **109** (2003) 43.
58. F. Himo, *Theor. Chem. Acta* **116** (2006) 232.
59. F. Himo, P. Siegbahn, *Chem. Rev.* **103** (2003) 2421.
60. J. Kuchar, R.P. Hausinger, *Chem. Rev.* **104** (2004) 509.
61. E.I. Solomon, R.K. Szilagyi, S. DeBeer George, L. Basumallick, *Chem. Rev.* **104** (2004) 419.
62. E. Tshuva, S.J. Lippard, *Chem. Rev.* **104** (2004) 987.
63. F.A. Armstrong, *Curr. Opin. Chem. Biol.* **8** (2004) 133.
64. A. Volbeda, J.C. Fontecilla-Camps, *Dalton Trans.* **2003** (2003) 4030.
65. Y. Montet, P. Amara, A. Volbeda, X. Vernede, E.C. Hatchikian, M.J. Field, M. Frey, J.C. Fontecilla-Camps, *Nat. Struct. Biol.* **4** (1997) 523.
66. C. Stadler, A.L. de Lacey, Y. Montet, A. Volbeda, J.C. Fontecilla-Camps, J.C. Conesa, V.M. Fernandez, *Inorg. Chem.* **41** (2002) 4424.
67. C. Stadler, A.L. de Lacey, B. Hernandez, V.M. Fernandez, J.C. Conesa, *Inorg. Chem.* **41** (2002) 4417.
68. A.L. DeLacey, C. Stadler, V.M. Fernandez, E.C. Hatchikian, H.J. Fan, S. Li, M.B. Hall, *J. Biol. Inorg. Chem.* **7** (2002) 318.
69. A. DeLacey, V.M. Fernandez, M. Rousset, C. Cavazza, E.C. Hatchikian, *J. Biol. Inorg. Chem.* **8** (2003) 129.
70. P.E.M. Sieghban, M. Blomberg, M.W. Pavlov, R.H. Crabtree, *J. Biol. Inorg. Chem.* **6** (2001) 460.
71. H. Ogata, Y. Mizoguchi, N. Mizuno, K. Miki, S. Adachi, N. Yasuoka, T. Yagi, O. Yamauchi, S. Hirota, Y. Higuchi, *J. Am. Chem. Soc.* **124** (2002) 11628.
72. A. Volbeda, Y. Montet, X. Vernede, E.C. Hatchikian, J.C. Fontecilla-Camps, *Int. J. Hydrogen Energy* **27** (2002) 1449.
73. E. Garcin, X. Vernede, E.C. Hatchikian, A. Volbeda, M. Frey, J.C. Fontecilla-Camps, *Structure* **7** (1999) 557.
74. A. Volbeda, E. Garcin, C. Piras, A.L. de Lacey, V.M. Fernandez, E.C. Hatchikian, M. Frey, J.C. Fontecilla-Camps, *J. Am. Chem. Soc.* **118** (1996) 12989.
75. Y. Higuchi, H. Ogata, K. Miki, N. Yasuoka, T. Yagi, *Structure* **7** (1999) 549.
76. M. Darensbourg, E. Lyon, I.P. Georgakaki, *Proc. Natl. Acad. Sci.* **100** (2003) 3683.
77. A. Berkessel, *Curr. Opin. Chem. Biol.* **5** (2001) 486.
78. E.J. Lyon, S. Shima, G. Buurman, S.A.B. Chowdhuri, K. Steinbach, R.K. Thauer, *Euro. J. Biochem.* **271** (2004) 195.
79. M.J. Maroney, P.A. Bryngelson, *J. Biol. Inorg. Chem.* **6** (2001) 453.
80. M. Stein, E. van Lenthe, E.J. Baerends, W. Lubitz, *J. Am. Chem. Soc.* **123** (2001) 5839.
81. M. Stein, W. Lubitz, *Curr. Opin. Chem. Biol.* **6** (2002) 243.
82. H.J. Fan, M.B. Hall, *J. Am. Chem. Soc.* **124** (2001) 394.
83. H.J. Fan, M.B. Hall, *J. Biol. Inorg. Chem.* **6** (2001) 467.
84. P. Amara, A. Volbeda, J.C. Fontecilla-Camps, M.J. Field, *J. Am. Chem. Soc.* **121** (1999) 4468.
85. S. Niu, L.M. Thomson, M.B. Hall, *J. Am. Chem. Soc.* **121** (1999) 4000.
86. S. Dementin, B. Burlat, A.L. de Lacey, A. Pardo, G. Adryanczyk-Perrier, B. Guigliarelli, V.M. Fernandez, M. Rousset, *J. Biol. Chem.* **279** (2004) 10508.

87. R.J. Williams, *Nature* **376** (1995) 643.
88. V.H. Teixeira, A.M. Baptista, C. Soares, *Biophys. J.* **91** (2006) 2035.
89. A. Melis, T. Happe, *Plant. Physiol.* **127** (2001) 740.
90. M.W. Adams, L.E. Morteson, J.S. Chen, *Biochim. Biophys. Acta* **594** (1981) 105.
91. J.M. Odom, J.H.D. Peck, *Annu. Rev. Microbiol.* **38** (1984) 551.
92. C. Leger, A.K. Jones, W. Roseboom, S.P. Albratch, *Biochemistry* **41** (2002) 15736.
93. P.M. Matias, C.M. Soares, L.M. Saraiva, R. Coelho, J. Morais, J. Le Gall, M.A. Carrondo, *J. Biol. Inorg. Chem.* **6** (2001) 63.
94. L. DeGioia, P. Fantucci, B. Guigliarelli, P. Bertrand, *Inorg. Chem.* **38** (1999) 2658.
95. M.W. Pavlov, M. Blomberg, P.E.M. Sieghban, *Int. J. Quantum Chem.* **73** (1999) 197.
96. M. van Gastel, C. Fichtner, F. Neese, W. Lubitz, *Biochem. Soc. Trans.* **33** (2005) 7.
97. M. Bruschi, G. Zampella, P. Fantucci, L. De Gioia, *Coord. Chem. Rev.* **249** (2005) 1620.
98. D.J. Evans, C.J. Pickett, *Chem. Soc. Rev.* **32** (2003) 268.
99. C. Greco, G. Zampella, L. Bertini, M. Bruschi, P. Fantucci, L. De Gioia, *Inorg. Chem.* **46** (2007) 108.
100. M.C. Smith, J.E. Barclay, S.P. Cramer, S.C. Davies, W. Gu, D.L. Hughes, S. Longhurst, D.J. Evans, *J. Chem. Soc. Dalton Trans.* **2002** (2002) 2641.
101. S.C. Davies, D.J. Evans, D.L. Hughes, S. Longhurst, J.R. Sanders, *Chem. Commun.* **1999** (1999) 1935.
102. S. Li, M.B. Hall, *Inorg. Chem.* **40** (2001) 18.
103. A.C. Marr, D.J.E. Spencer, M. Schroeder, *Coord. Chem. Rev.* **1055** (2001) 219.
104. D. Sellmann, F. Geipel, F. Lauderbach, F.W. Heinemann, *Angew. Chem. Int. Ed.* **41** (2002) 632.

8 Electrochemically Formed Film Layers of Importance in Electrocatalysis

B. F. Giannetti, C. M. V. B. Almeida, and S. H. Bonilla

CONTENTS

8.1 INTRODUCTION

One of the most intensively studied interfaces is the electronic conductor/ionic conductor where the interest is motivated by attempts to prevent corrosion and to improve the catalytic properties of metallic deposition. Both corrosion prevention and catalysis development can be described using an electrochemical and engineering approach, including film formation and growth and its optimization in the cell reactor.

In the case of interdisciplinary areas of study, layers and films with different properties are expected to grow in different ways to fulfill diverse requirements. Thus, the desired characteristics of a film with protective properties against corrosion will differ from those needed to perform as a good electrocatalyst. In this context, it is impossible to think that a unique model would be complete enough to embrace such different films.

As an effort to understand film formation and with the purpose of optimization (predicting, modifying, and/or improving this formation), several models supported by theoretical considerations were developed. In this way, the development of a physical model leads to the derivation of mathematical equations, which enables the development of a theory and comparison with experimental results.

In a general way, the reference point will restrict the way of focusing on the phenomenon and as a consequence it will determine the choice of the variables to be used in it. For the study of film growth and in order to establish a theoretical basis, two different reference points can be equally considered in reality. Figure 8.1 shows two reference points: Observer **B**, denoted with light grey arrows, at a normal position to the metal surface, will be able to describe the process of film growth by choosing variables that will differ, certainly, from those from Observer **A**, placed parallel to the surface, denoted with dark grey arrows.

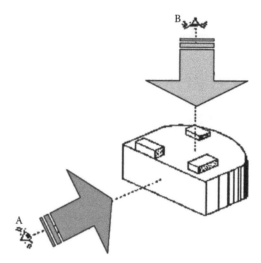

FIGURE 8.1 The aspect of film growth with the point of view of different imaginary observers: A (dark grey) along a parallel direction of movement and B (light grey) through a perpendicular position to the surface.

The rate of metal or metallic oxide nuclei formation during charge transfer, or the temporal variation of the degree of surface coverage will be useful to follow, describe and quantify the film formation for Observer B. On the other hand, Observer A, with a different perception of the same process, will be able to describe film growth in terms, for example, of ion transport through (or to) the substrate and/or the electrolyte. As a consequence, different growth models can be projected, making the selection of the most plausible model very difficult. The singular use of mathematical considerations to support the selection of the most satisfactory model should be avoided, and special emphasis should be put to verify the physical connotation of the parameters.

We are going to present a review of the main models of electrochemically formed films with the purpose of giving an initial reference to the subject and then illustrate the selection procedure to choose the most suitable model with an emphasis on the physical meaning of the determining variables.

8.2 CLASSIFICATION OF FILM MODELS

In general, films can be grouped as continuous and noncontinuous according to the morphology of the obtained surface. The former assumes that the metal surface is completely covered by the film. Continuous films form in systems where the mass transport of ion species toward the solution is hindered. On the other hand, systems in which the transport of ions to the solution is allowed give a noncontinuous film, that is, first a vacancy substrate site is defined, which in the progress of the growth, can lead to an irregular or porous film.

8.2.1 Models Developed for Continuous Films

According to the basic statement of the models we are going to summarize, the metal is conceived as a network of cations immersed in a "cloud" of free electrons in a crystalline structure. The transport of the ions controls the growth of the new phase (Figure 8.2). The ionic transport will depend on the nature of the system and on experimental conditions, such as temperature, local electric field, local concentration excess, etc. To better understand the continuous-film models, the main ionic transport mechanisms in crystalline solids are presented [1] as follows.

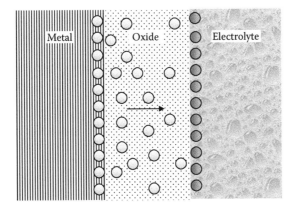

FIGURE 8.2 The growth of a new oxide phase on the metal with the process controlled by the mass transport of cations, symbolized as ○ toward the solution, where the oxygen anions are symbolized as ◐.

1. *Interstitial mechanism* (Figure 8.3A). The displacement of metal atoms takes place from one interstitial site to the other without permanently displacing the matrix atoms in the crystal cell. The distortion caused by the movement of the matrix atoms to allow "the jump" constitutes the energy barrier that must be surmounted. This mechanism is frequently used to describe film formation on metal alloys when metal atoms normally occupy interstitial positions.
2. *Vacancy mechanism* (Figure 8.3B). The movement of adjacent atoms occurs to an unoccupied site (vacancy) because of the surface energy difference between them. The lattice distortion caused by the movement of a neighboring atom through the vacancy is lower than that predicted by the interstitial mechanism. This mechanism is well known and is often used to describe film formation on metals, metal alloys, ionic compounds, and oxides.

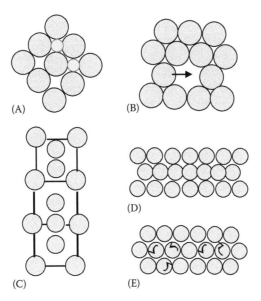

(A)   (B)

(C)   (D)

(E)

FIGURE 8.3 Simplified schemes for the main mechanisms of transport of atoms in solids. (A) Interstitial; (B) vacancy; (C) interstitially; (D) crowed-ion; (E) ring mechanisms.

3. *Interstitially mechanism* (Figure 8.3C). The atoms placed at the interstitial sites can move, pushing the nearest neighboring atom and displacing it. This kind of mechanism occurs in "metallic solutions" when solute atoms are the same size as matrix ones. The displacement of several atoms placed in a close-packed direction from their equilibrium position by means of an interstitially mechanism is called the *crowd-ion mechanism* (Figure 8.3D). This kind of distortion spreads out only along one dimension.

4. *Ring mechanism* (Figure 8.3E). The atoms can exchange places with their neighbors, and the deformation caused by the rotation depends on the number of atoms that are able to rotate.

The electrochemical growth of a continuous and homogeneous film can be well thought-out as a result of the migration of metallic cations by means of one of the above-mentioned transport processes. It is worthy to note that the dislocation of ions breeds defects (vacancies, interstitial ions, etc.). Hence, either the dislocation of defects or the migration of atoms can be used to describe the transport process [1]. Mathematical details are found in the cited references and also in the reviews [2,3].

Lanyon and Trapnell [4] and, later, Green [5] and Law [6] described the initial stages of metal oxidation by using the ring mechanism (known as "place exchange"). According to these authors, an oxygen atom adsorbed on the metal surface (M–M–O) can rotate, changing places with an adjacent metal atom (M–O–M). This process, successively repeated, would increase the thickness of the oxide film. Sato and Cohen [7] developed mathematical equations for the model when stationary electrochemical techniques were applied to generate the film. The authors assumed that all the rotations were simultaneous and that the activation energy for the place exchange increased linearly with film thickness. For increasing thickness, the activation energy would reach values excessively high, and another growth mechanism would become dominant, for example, the interstitial mechanism. Asakura and Nobe [8] solved the mathematical equations for potentiodynamic conditions, resulting in a linear relationship between the current intensity and the potential scan rate.

The model developed by Verwey [9] assumed that film growth occurred through an interstitial mechanism where the distortion of the lattice was the same for all the position changes; thus, the energy barrier is symmetric and periodic. The presence of a uniform electric field, however, could diminish the barrier promoting the migration of ions in a given direction. The rate-determining step of the process was considered as the migration of interstitial atoms inside the lattice. Two limiting cases were discussed according to the electric field value: the *low-field model*, where the total current presents an ohmic relationship with the field, and the *high-field model*, where the current is an exponential function of the electric field. Some years later, Cabrera and Mott [10], using similar considerations, considered that the diffusion of defects causes an accumulation of ions at the metal/film interface determining the rate of the process. The Verwey model [9] and the model of Cabrera and Mott [10] were developed for steady-state stationary electrochemical conditions. Mathematical expressions developed for potentiodynamic conditions are grouped elsewhere [11].

Lohrengel [12] proposed an extension of the high-field model where the concentration of defects varies with the electric field strength. When the concentration reaches a constant value, the high-field equations may be valid. D'Alkaine et al. [13] developed a model where migration is the main type of transport, applied for both stationary [13,14] and potentiodynamic [14,15] conditions of film growth. In this case, it was assumed that the movement of ions inside the film presents characteristics similar to their movement in solution and that the potential gradient across the film obeys a linear relationship.

In all the previous models, the effect of the space charge was neglected. Dewald [16] was the first to take into account the effect of the space charge. Moreover, Young [17] completed an extension of the theory by assuming a uniform space charge within the film; however, the equations resulting from Dewald's theory [16] do not relate current density with the electric potential. The hypothesis of Verwey, Cabrera, and Mott corresponds to limiting situations of the general case.

Bean et al. [18] considered that mobile interstitial ions are created within the film when the metal ions move out of their normal sites, leaving vacancies. The creation of mobile ions is a field-dependent process, so a fixed concentration of interstitial ions exists at characteristic field strength. This kind of transport occurs frequently in amorphous oxides.

Chao et al. [19] proposed a model that explains the growth of a film under steady-state conditions. It was considered that the passive film contains a high concentration of no recombining point defects. Metal/film and film/solution interfaces were assumed to be at electrochemical equilibrium. This theory successfully accounts for the linear dependencies of both the steady-state film thickness and the logarithm of the passive current on the applied voltage.

Bojinov et al. [20] considered the retarding, as well as the enhancing, effect of the defect migration within the film due to the excess of charge in the metal/film or the film/electrolyte interfaces. A high electric field that assisted the migration of defects was assumed for film growth. The model was developed considering electrochemical impedance measurements. The decrease of the electric field within the film as a consequence of the accumulation of interfacial charges is considered in the equations.

Several models were developed to describe the growth of continuous films. It is worthy of attention that those theories are based on the point of view of Observer A, placed parallel to electrode surface (Figure 8.1).

8.2.2 Models for Noncontinuous Films

Models that describe the growth of noncontinuous films use the point of view of Observer B (Figure 8.1) and can be classified into three main groups:

1. *Nucleation theories.* Nuclei are generated onto the electrode surface and grow with a given orientation and shape.
2. *Theories of variable coverage.* The surface coverage is the main parameter for the description of the film-formation kinetics.
3. *Combined theories.* Models that combine aspects of both of the previous theories are able to describe the formation of the film.

8.2.2.1 Nucleation Theories

The first scientific reports regarding nucleation appeared at the beginning of the twentieth century [21–24]. According to these theories, the formation of micro-phases, nuclei, takes place at active sites that are energetically favored. The study of nuclei formation and growth requires the use of techniques that enable very fast recording of the electrochemical response in order to register the first instants of film formation. For this reason, successful experimental results were only accomplished after the 1950s due to progress in electronics.

Fleishmann and Thirsk [25–27] derived analytical expressions for a nucleation model under potentiostatic conditions relating current with time (chronoamperometric plots). They assumed that there are N_0 active sites available on the substrate and that N nuclei are generated. Both parameters considered there were related through an exponential expression, which involved time and a factor depending on the overpotential. Two limiting cases of nucleation rates were described. The assumption of a large dependence on overpotential led to an instantaneous coverage of all the active sites (*instantaneous nucleation*). On the other hand, a slight dependence on the overpotential led to a lower nucleation rate (*progressive nucleation*). The nucleation rate also depends on the surface area occupied by the growing nuclei; thus, nuclei can grow without area changes, and this process is called one-dimensional growth. Bidimensional growth occurs, on the other hand, when lateral area increases and, in the case of three-dimensional growth, the whole superficial area of the nuclei is affected during growth. In addition, the geometry of the nucleus determines the electrical current expression.

When contact between the nuclei is neglected during growth, an independent type of growth is assumed. This assumption may be true in the early stages of growth; however, as radial growth proceeds, the occupied area of a growing nucleus will be limited by the presence of neighboring nuclei. As a consequence, overlapping of the nuclei can occur. Bewick et al. [28] applied Avrami's theorem [29] to estimate the overlapped area.

Although the latter models assumed that the slowest step involves the incorporation of species at the periphery of expanding nucleus, other nucleation models proposed that mass transfer in solution controlled the new phase growth [30–32].

Scharifker [33] considered that two processes interact in the film formation kinetics: the activation and the deactivation of sites on the substrate. Thus, the deactivation, caused by the development of a hemispherical diffusion field centralized at each nucleus, inhibits film growth in the neighborhood of the diffusion centers.

Barradas and Bosco [34] presented three novel models of two-dimensional nucleation. They involve the coupling of nucleation with diffusion in the electrolyte, with and without additional metal dissolution from the electrode surface.

The effect of potentiodynamic perturbations was only considered in the early seventies [35–39]. The delay was possibly due to the difficulty in dealing with the mathematical expressions, in which the nucleation rate depends on the applied overpotential, which varies with time. For this reason, limiting conditions of low potential scan rate ($v \to 0$) and high potential scan rates ($v \to \infty$) were considered. The limiting situations were called, rather arbitrarily, "reversible" and "irreversible" [39], and the model describes a nucleation-growth-overlap process in two steps: instantaneous and progressive. A scheme of the nucleation-growth-overlap model is shown in Figure 8.4. This model can be applied either for the study of the formation of a new phase (Figure 8.4A) or substrate dissolution (Figure 8.4B) [36,40].

8.2.2.2 Theories Based on Variable Coverage

Among the models developed for steady-state stationary conditions, the time-dependent surface coverage can be the real description of the growth of the new phase. Ogura [41] assumed that the film was generated from the active dissolution of the metal. The author derived expressions in order to determine Tafel coefficients corresponding to the following rate-determining steps: metal dissolution, film formation, and film development for the cases of high and low overpotentials. On the other hand, Ebersbach et al. [42] proposed a mechanism where the film growth would inhibit the metal dissolution (passivation). The kinetics of passivation is then explained as a competition between the formation of the oxide film and metal dissolution.

In 1931, Müller [43] explained the formation of a porous, insoluble, and poor conducting film by means of a simplified model based on potentiostatic experiments. In this way, it was considered that film formation occurs through the following simple, fast reaction:

$$M(s) + X^- \leftrightarrow MX(s) + ne^- \qquad (I)$$

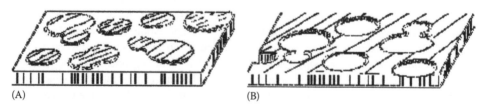

(A)   (B)

FIGURE 8.4 Schematic illustration of film formation (A) and dissolution (B) on the substrate according to the nucleation-growth-overlap model.

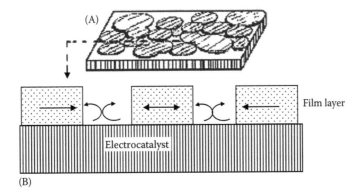

FIGURE 8.5   Passive film growth according to Müller's model. (A) Observer's B view and (B) Observer's A view.

where MX are solid species able to form nuclei, subsequently spreading laterally and finally generating a porous film. As the degree of coverage increases, the metallic area exposed to dissolution decreases, and the reaction rate would be limited by the resistance of the solution in the pores of the film (Figure 8.5). When the film covers approximately 99% of the surface, the resistance reaches its maximum value, inhibiting the film formation.

Calandra et al. [44] adapted Müller's model to potentiodynamic conditions. Mac Donald [45] corrected a typographical error found in the mathematical expressions in the article. Devilliers et al. [46] developed a general model for the formation of low-conductivity films, considering a process controlled by the solution resistance in the pores of the film. The authors simulated the potentiodynamic curves for the following particular cases: constant film thickness (bidimensional growth), three-dimensional growth, and a decomposition/dissolution process coupled to the electrochemical reaction. The potentiodynamic curves simulated for constant thickness are identical to those obtained by Calandra et al. [44].

Srinivasan and Gileadi [47] described the formation of a new phase under potentiodynamic conditions in terms of an adsorption process. The validity of the Langmuir isotherm was assumed for the following adsorption process:

$$A^- \leftrightarrow A_{ads} + e^- \tag{II}$$

Two situations were considered in the derivation of the equations: the *quasi-equilibrium* and the *irreversibility* of the reaction of the formation of adsorbed species.

Several authors [39,47–53] considered a more realistic hypothesis taking into account the long-range interactions between the adsorbed species, that is, the verification of the Frumkin isotherm. The mathematical relations, however, were obtained only for reversible processes [39,51,52].

8.2.2.3   Combined Theories

The description of film formation in terms of the combination of both nucleation and adsorption processes under potentiostatic [54,55] and potentiodynamic [39] conditions was later adopted. Mathematical simulations of the potentiodynamic curves [39] indicated that only one peak is recorded when the sites on the substrate are the same for both nucleation and adsorption (competitive process); however, when the surface sites for both processes are different (noncompetitive process), the potentiodynamic curves exhibit two current peaks.

8.3   GENERAL DISCUSSION

As it was mentioned in Section 8.1, an experiment is included in order to illustrate the selection procedure. Each model was developed for specific experimental conditions. Sometimes, a description can be modified, extended, and corrected in order to cover other experimental conditions. Thus, a model initially developed with the purpose of describing a film formed under potentiostatic conditions can be adapted, via mathematical derivations, to potentiodynamic conditions. In the present experiment, the film considered was generated under potentiodynamic conditions by the use of voltammetric techniques. As a consequence, only the models developed for potentiodynamic conditions were considered [56–58].

8.4   DIAGNOSTIC PARAMETERS' CRITERIA FOR THE DIFFERENT GROWTH MODELS

A mechanically polished tin disk immersed in 0.5 M citric acid was prepared following the procedure described in the literature [59]. Because high-scan-rate voltammetric curves were recorded, a calomel reference electrode is not recommended because of its slow response [60]. The reversible hydrogen electrode (RHE) was used as the reference electrode at 25°C. The required materials and reagents are a three-compartment electrochemical cell with a platinized platinum auxiliary electrode and a 0.5 M (pH = 1.8) citric acid solution prepared with triple-distilled water.

In order to obtain reproducible results, the working electrode must be pretreated as explained in the following text: the electrode used herein was polished with 600 emery paper and alumina suspension (0.3–0.2 μm granularity), washed in distilled water, immersed in a previously deaerated solution, and cathodically polarized at −0.9 V for 5 min. The latter step is important to reduce the oxide layer formed during polishing; however, polarization times longer than this period have to be avoided, otherwise hydride formation and metal lattice reconstruction turn up as a consequence of molecular hydrogen inclusion.

Figure 8.6 shows the voltammetric profile run at 0.1 V s^{-1} for polycrystalline tin in citric acid as reported before [61–65] between −0.9 and 3.0 V with the following main features:

1. Hydrogen evolution occurs at potentials lower than −0.5 V.
2. An anodic current, A_I, appears near 0 V.
3. A low stationary current (passive state) is observed after the anodic peak, until approximately 3.0 V.

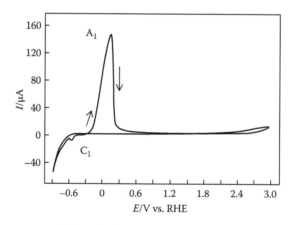

FIGURE 8.6   Current versus potential profile for polycrystalline tin in 0.5 M citric acid (pH = 1.8) run at 0.1 V s^{-1} between −0.9 and 3.0 V. A = 0.8 mm^2.

4. Oxygen evolution occurs above 3.0 V, indicative of the conductivity of the formed passive film.
5. A cathodic current peak, C_I, appears during the potential scan toward negative potentials, around the potential at which hydrogen evolution begins, similarly to other non-noble metals in acid solution.

The significantly higher anodic charge and the sudden current fall immediately after the peak are evidence for a net oxidation process with little redeposition on the electrode surface.

Accordingly, the oxidation current can be written as the sum of two contributions:

$$I = I_D + I_F \tag{8.1}$$

with I_D being the current related to the active dissolution of the metal and I_F being the current related to the film formation.

The first step considered in the selection procedure corresponds to the verification of the mathematical relationships resulting from the models. For this purpose, the experimental data is analyzed according to the mathematical relationships of the most suitable theoretical models. Table 8.1 will be very useful as it collects the mathematical relationships satisfied by models developed for potentiodynamic conditions.

Obviously, experimental data should verify all the mathematical expressions of the proposed model as a necessary condition to satisfy. Besides, it is not rare that several model expressions present linear relations involving the same variables. In these cases, it is risky to choose a model based solely on mathematical criteria. In these cases, the calculation of the model's parameters from the experimental data is a necessary condition to assume or to reject a model. If the values calculated from experimental data do not present physical meaning, the model is not applicable.

Graphs corresponding to peak current (I_p) versus v as well as peak potential (E_p) versus log v for peak A_I (Figures 8.7 and 8.8) show three regions, labeled A, B, and C. Region A is comprised of the 0–0.2 V s^{-1} domain, followed by region B that extends up to 7 V s^{-1}. The last, region C, falls between 7 and 22 V s^{-1}. The linear dependence of I_p versus v in region A (Figure 8.7), together with the constant behavior observed in the E_p versus log v plot (Figure 8.8), could support the Srinivasan–Gileadi model [47] (Table 8.1).

According to this model, a film formation controlled by a quasi-reversible adsorption seems to be suitable to describe the process occurring in the scan rate domain of region A. On the other hand,

---

TABLE 8.1
Diagnostic Parameters' Criteria for the Different Growth Models: Nucleation-Growth-Overlap (I) [39]; Müller–Calandra (II) [43,44]; and Srinivasan–Gileadi (III) [47] under Potentiodynamic Conditions

Parameter	(I) ($v \rightarrow 0$)	(I) ($v \rightarrow \infty$)	(II)	(III) Quasi-Reversible	(III) Irreversible
I_p/v	—	Constant	—	Constant	Constant
$I_p/v^{1/2}$	—	Constant	—	Constant	Constant
E_p/v	—	—	—	No	—
$E_p/v^{1/2}$	Constant	—	Constant	Varies on v	—
$E_p/\log v$	—	Constant $= 59/\alpha n$ mV	—	$E_p(v) = $ constant	Constant $= 0.086T/n$ mV K^{-1}
$E_{1/2}/v$	Constant	Constant $= 31/\alpha n$ mV	—	—	—
θ_p	—	0.63	0.999	0.5	0.63

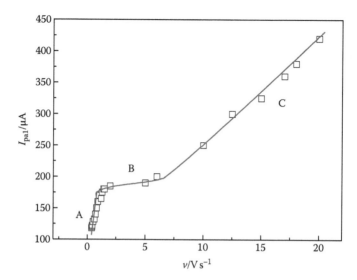

FIGURE 8.7   Current peak (I_p) versus log scan rate (log v) for peak a_I in the case of Figure 8.6.

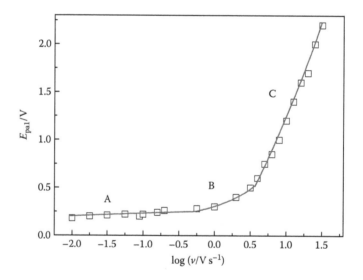

FIGURE 8.8   Potential peak (E_{pa1}) for peak A_I versus scan rate (v) in the case of Figure 8.6.

region C presents a linear behavior for both peak potential and current. Also, the Srinivasan–Gileadi model [47] (but in its limited application for an irreversible adsorption) seems reasonable (at least from a mathematical standpoint) to explain the film-formation process under region C conditions.

Two of the relationships compiled in Table 8.1 (corresponding to two applications related to the Srinivasan–Gileadi model [47]) reflect the mathematical dependence shown in the graphics; thus, the film formation process could be controlled by quasi-reversible adsorption in region A and by irreversible adsorption in region C. It is, therefore, advisable to examine the physicochemical processes involved, that is, to verify if the theoretical model is consistent with the observed phenomenon. In this way, complementary results are obtained.

The variation of the total anodic and cathodic charges with scan rate (Figure 8.9) indicates that for $v < 0.2$ V s^{-1} (region A), the current contribution corresponding to dissolution (I_D) is significantly higher than that corresponding to film formation (I_F). From the experiments for stirred and

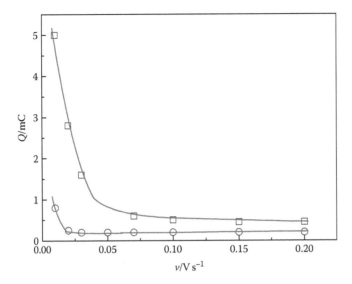

FIGURE 8.9 Absolute charge (Q) versus v for the ($\bigcirc$) anodic and ($\square$) cathodic components in the cyclic voltammetric profile of Figure 8.6.

unstirred solutions, it can be stated that at relatively low scan rates the hydrodynamic perturbation affects the peak A_I profile. On the other hand, it can be concluded that under high–scan rate conditions, transport is not the controlling step of anodic process and no changes are proven in the voltammetric profile.

In view of these last results, the 0.2 V s$^{-1} < v < 7$ V s^{-1} region (region B in Figure 8.7) could represent a transition between regions A and C. A review of these last results shows that the suitable model tested by the mathematical correspondence with experimental results for region A does not reflect the real performance of the system because current contributions from dissolution are neglected. As a consequence, the model was discarded.

For region C, the lack of a physical meaning for the parameters calculated using the proposed model was evident, resulting in a need to disregard of the proposed model. To discuss the discrepancies between the theoretical and the calculated parameters, the following equations (and corresponding to the employed models) will be useful.

The slope, $dE_p/dlog\ v$, of the straight line in Figure 8.8 (region C) equals 1.55. Assuming the validity of the Srinivasan–Gileadi model and considering a symmetry factor of 0.5, the number of electrons results in $n = 0.03$ from (Equation 8.6), a number without physical meaning. Thus, it can be concluded that, despite the linearity observed in region C, the Srinivasan–Gileadi model is not applicable to the studied system.

In order to continue with the selection procedure, another mathematical relationship presented in Table 8.1 was tested. Figures 8.10 and 8.11 show that both I_p and E_p vary linearly with the square root of v, only for $v > 7$ V s^{-1}. The diagnostic criteria in Table 8.1 indicate that the film formation can be described by the dissolution–precipitation model proposed by Müller and Callandra [43,44]. According to this model, the resistance remains low until the film covers 99% of the surface; therefore, it is reasonable to assume that the anodic current reaches its maximum when the entire surface is practically covered. As the total anodic current is the combined result of two contributions, namely, dissolution and formation of the film, the degree of coverage cannot be graphically evaluated from the voltammograms. Assuming that the film consists basically of SnO_2 [66], and that $\rho = 6$ g cm^{-3}, $\chi = 6 \times 10^{-3}$ S m^{-1}, and $S = 8 \times 10^{-3}$ cm^2, and that the value of $dI_p/dv^{1/2}$ calculated from Figure 8.10 is 84.6×10^{-6} A (V s$^{-1})^{-1/2}$, the coverage is $\theta = 0.9989$. The expressions for the peak current and potential are defined in the original article [44].

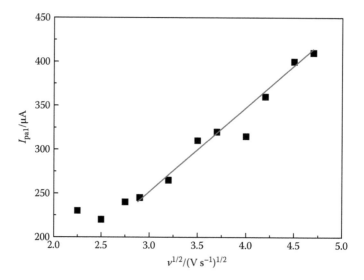

FIGURE 8.10 I_{pa1} versus $v^{1/2}$ plot for peak A_I at $v > 8$ V s^{-1}. Data from Figure 8.6.

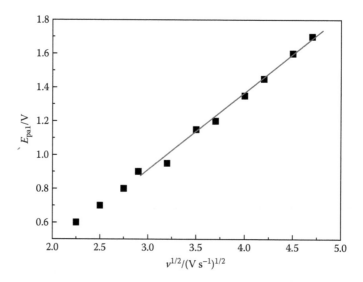

FIGURE 8.11 E_{pa1} versus $v^{1/2}$ plot for peak A_I at $v > 8$ V s^{-1}. Data from Figure 8.6.

In the Müller–Calandra model for noncontinuous growth on a porous surface, the limiting condition in a potentiodynamic experiment is the resistance of the solution within the pores. The following parameters rise up for a 0.999 of surface coverage (θ) for the growth of a *PM* (molecular weight) formation:

$$I_p = \sqrt{\frac{nF\rho\chi}{PM}}S(1 - \theta)v^{1/2} \tag{8.2}$$

$$E_p = \sqrt{\frac{nF\rho\chi}{PM}}\left[\frac{\delta}{\chi} + R_o S(1 - \theta)\right]v^{1/2} \tag{8.3}$$

where
   δ is the thickness of the film
   ρ is the solution density
   χ is the conductivity of the solution in the pores
   S is the real surface area of the electrode
   R_o is the total resistance

Besides, two cases can be found for the Srinivasan–Gileadi model using variable conditions of the surface coverage under potentiodynamic experiments. In the case of a reversible adsorption and for the standard condition $\theta = 0.5$

$$I_p = \frac{\sigma F}{4RT} v \tag{8.4}$$

$$E_p = -\frac{RT}{F} \ln K \tag{8.5}$$

where K is the equilibrium constant of the film formation process. On the other hand, in the case of the irreversible adsorption, the standard condition of $\theta = 0.63$

$$I_p = \frac{(1 - \theta)\beta n \sigma F}{4RT} v \tag{8.6}$$

$$E_p = \frac{RT}{\beta n F} \ln \left(\frac{\beta n F \sigma}{k^\circ RT} \right) + \left(\frac{RT}{\beta n F} \right) \ln v \tag{8.7}$$

where
   σ is the charge density required to form a monolayer of the film
   β is the symmetry factor
   k° is the specific rate constant for the forward process (film formation)

The nucleation theories for potentiodynamic conditions [44] lead mainly an instantaneous or progressive nucleation also for reversible and irreversible limiting conditions. The following dependences of I_p, E_p, and $\Delta E_{1/2}$ with v can be obtained for four limiting cases.

1. Instantaneous nucleation for a reversible process

$$I_p = 8.94S \left(\frac{N_T}{N_o} \right)^{1/4} \sigma \sqrt{nk_g v} \tag{8.8}$$

$$E_p = E^\circ + \left(\frac{0.1584}{\sqrt{nk_g}} \right) \left(\frac{N_T}{N_o} \right)^{1/4} v^{1/2} \tag{8.9}$$

$$\Delta E_{1/2} = \left(\frac{0.2060}{k_g \sqrt{n}} \right) \left(\frac{N_T}{N_o} \right)^{1/4} v \tag{8.10}$$

where
   N_T is the total number of sites on the substrate (number cm^{-2})
   N_o is the number of active sites on the substrate (number cm^{-2})

2. Instantaneous nucleation for an irreversible process

$$I_p = 28.63 S \sigma \alpha n v \qquad (8.11)$$

$$E_p = E^{o'} - \frac{0.7936}{\alpha n} + \frac{0.02959}{\alpha n} \log \left[\left(\frac{N_T}{N_o} \right) \left(\frac{nv}{k_g} \right)^2 \right] \qquad (8.12)$$

$$\Delta E_{1/2} = \left(\frac{0.0314}{\alpha n} \right) \qquad (8.13)$$

3. Progressive nucleation for a reversible process

$$I_p = 0.3679 K^2 S \sigma v / a \qquad (8.14)$$

$$E_p = E^{o'} + \frac{a}{K} \qquad (8.15)$$

$$\Delta E_{1/2} \rightarrow 0 \qquad (8.16)$$

where a is the potential dependent nucleation parameter in volts and K is

$$K = -9.04 + 2.303 \log \left[\left(\frac{N_o}{N_T} \right) \frac{k_N k_g^2}{n^2 a^3 v^3} \right] \qquad (8.17)$$

4. Progressive nucleation for an irreversible process

$$I_p = 28.63 S \sigma \alpha n v \qquad (8.18)$$

$$E_p = E^{o'} + \frac{0.0147}{\alpha n} + \frac{0.02959}{\alpha n} \log \left[\left(\frac{\alpha n v}{k_N} \right) \left(\frac{\alpha n v}{k_g} \right)^2 \right] \qquad (8.19)$$

$$\Delta E_{1/2} = \left(\frac{0.0314}{n} \right) \qquad (8.20)$$

where
k_N is the nucleation velocity (s^{-1})
k_g is the growth rate of the nuclei (s^{-1})

The graph of I_p versus $v^{1/2}$ (Figure 8.10) gives a straight line that does not pass through the origin as expected from the mathematical expression derived for the model [43,44]. The nonzero intercept results from the presence of dissolution current, not considered in the model, where only the actions of I_F versus v were taken into account.

Scanning electron microscopy results showed that the surface aspect, after a potential sweep from −0.9 to 2.0 V at 0.1 V s^{-1} and a holding at the last potential for 10 min, supported the idea that the film is formed by precipitation (noncontiguous aspect) and presented a porous aspect (see micrograph in Ref. [62]).

The complete diagnostic criteria related to the model was satisfied and considering that other experimental evidence also support it, it is reasonable to conclude that at high scan rates (region C) the film growth is controlled by pore resistance, as stated by the Müller and Calandra model [43,44].

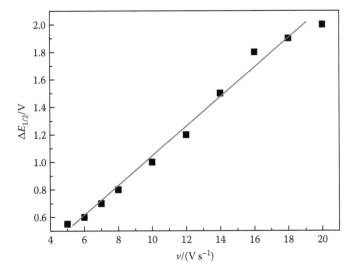

FIGURE 8.12 $\Delta E_{1/2}$ versus v plot for peak A_{I} at $v > 8$ V s^{-1}. Data from Figure 8.6.

On the other hand, it was verified that the half-peak width $\Delta E_{1/2}$ varies linearly with scan rate (Figure 8.12). This linear relationship together with those observed in Figures 8.10 and 8.11 are predicted by the nucleation-growth model for potentiodynamic conditions [39] when the nucleation process is fast and irreversible.

The instantaneous nucleation-growth-precipitation model [39] assumes that the film is formed directly on the substrate, without previous dissolution; however, it was observed that active dissolution of the metal occurs. Therefore, Equations 8.11 through 8.13 were examined and rewritten considering metal dissolution, that is, terms corresponding to dissolution were added to the mathematical expressions:

$$I_{\mathrm{p}} = I_{\mathrm{p}}' + 28.63(\alpha n)\,\sigma v \tag{8.21}$$

where
α is the charge transfer coefficient (linear combination of β)
σ is the total surface charge density

$$E_{\mathrm{p}} = E_{\mathrm{p}}' + \left\{ E^{o'} - 0.7936/(\alpha n) + [0.02959/(\alpha n)] \log\left[(N_{\mathrm{T}}/N_{\mathrm{o}})(nv/k_{\mathrm{g}})^2\right] \right\} \tag{8.22}$$

$$\Delta E_{1/2} = \Delta E_{1/2}' + 0.0314/(\alpha n) \tag{8.23}$$

And now rearranging the equations and dividing (8.21) by (8.22) and a part (8.22) with (8.23) we can obtain the following equation:

$$\frac{\left(I_{\mathrm{p}} - I_{\mathrm{p}}'\right)}{\left(E_{\mathrm{p}} - E_{\mathrm{p}}'\right)} = 56.44\left(\frac{N_{\mathrm{o}}}{N_{\mathrm{T}}}\right)^{1/2} nS\sigma k_{\mathrm{g}} \tag{8.24}$$

$$\left(\Delta E_{1/2} - \Delta E_{1/2}'\right)\left(E_{\mathrm{p}} - E_{\mathrm{p}}'\right) = \frac{1.3v^{1/2}}{k_{\mathrm{g}}^{1/2}} \tag{8.25}$$

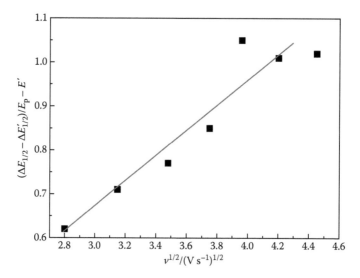

FIGURE 8.13 $\Delta E_{1/2} - \Delta E'_{1/2}$ versus $v^{1/2}$ plot for peak A_I at $v > 8$ V s^{-1}. Data from Figure 8.6.

The value of k_g calculated using (8.25) from the slope of the line in Figure 8.13 (0.24 s$^{1/2}$) is 29 s^{-1}. Because the growth rate depends on the nature of the system and on the characteristics of the applied perturbation, it is not possible to check the value of k_g with values given in the literature. The value of k_g is useful, however, for the evaluation of the number of active sites on the substrate, N_0. With this value, it is possible to verify the suitability of the model, as it is easily comparable with other system values.

The mean value of $\left(I_p - I'_p\right)/\left(E_p - E'_p\right)$, 1.8×10^{-4} s, obtained from Figure 8.13, and (8.24), gives $k_g = 29$ s^{-1}, with $S = 8 \times 10^{-3}$ cm^{-2}. Assuming that the film is formed, basically, by SnO$_2$ (then, $n = 4$ and $\sigma = 0.6$ mC cm^{-2} [66]), the ratio between the number of active sites and the total number of sites, N_0/N_T, equals 3.2×10^{-5}.

The total number of sites, $N_T = 9.4 \times 10^{14}$ sites cm^{-2}, was calculated by [39]

$$N_T = \frac{\sigma N_A}{nF} \tag{8.26}$$

The number of sites participating in the formation of SnO$_2$ is $N_0 = 3.0 \times 10^{10}$ cm^{-2}. This value is similar to those usually found for active sites in the case of metals [67]. As the diagnostic criteria for the instantaneous nucleation-growth-overlap model was fulfilled and physicochemical data obtained for N_0 is physically plausible, we may conclude that instant nucleation with lateral growth of the film occurs. Taking into account that the atomic density of the substrate is of the order of 10^{15} cm^2 [67], we conclude that the density of the active sites on the surface can severely limit the nucleation process.

The fact that the experimental data fit both models, the dissolution–precipitation and the instantaneous nucleation-growth-overlap models, confirms the assertion made at the beginning of this chapter, that is, some models are complementary to each other.

REFERENCES

1. P. G. Shewmon, *Diffusion in Solids*, McGraw-Hill: New York (1963), Chapter 2.
2. M. J. Dignam, in *Comprehensive Treatise of Electrochemistry* (J. M. Bockris, B. E. Conway, E. Yeager, R. E. White, Eds.), Plenum Press: New York (1981), Chapter 5, p. 4.
3. D. A. Vermilyea, in *Advances in Electrochemistry and Electrochemical Engineering* (P. Delahay, C. W. Tobias, Eds.), Interscience: New York (1963), Vol. 3, Chapter 4.

4. A. H. Lanyon and B. M. W. Trapnell, *Proc. Roy. Soc.* **227A** (1955) 387.
5. M. Green, *Prog. Semicond.* **4** (1960) 37.
6. J. T. Law, *J. Phys. Chem. Solids* **4** (1958) 91.
7. N. Sato and M. Cohem, *J. Electrochem. Soc.* **111** (1964) 512.
8. S. Asakura and K. Nobe, *J. Electrochem. Soc.* **118** (1971) 536.
9. E. J. W. Verwey, *Physica* **2** (1935) 1059.
10. N. Cabrera and N. F. Mott, *Rep. Prog. Phys.* **12** (1948) 63.
11. A. Wieckowski and E. Ghali, *Electrochim. Acta* **30** (1985) 1423.
12. M. M. Lohrengel, *Mat. Sci. Eng.* **R11** (1993) 243.
13. C. V. D'Alkaine and M. N. Boucherit, *J. Electrochem. Soc.* **144** (1997) 1331.
14. C. V. D'Alkaine, L. M. M. De Souza, and F. C. Nart, *Corros. Sci.* **34** (1993) 129.
15. C. V. D'Alkaine and M. A. Santanna, *J. Electroanal. Chem.* **457** (1998) 13.
16. J. F. Dewald, *J. Electrochem. Soc.* **102** (1955) 1.
17. L. Young, *Can. J. Chem.* **37** (1959) 276.
18. C. P. Bean, J. C. Fisher, and D. A. Vermilyea, *Phys. Rev.* **101** (1956) 551.
19. C-Y. Chao, L. F. Lin, and D. D. Mac Donald, *J. Electrochem. Soc.* **128** (1981) 1187.
20. M. Bojinov, I. Kanazirsky, and A. Girginov, *Electrochim. Acta* **41** (1996) 2695.
21. M. Volmer, *Phys. Z.* **22** (1921) 646.
22. W. Kossel, *Nachr. Ges. Wiss.* (1927) 135.
23. I. Stranski, *Z. Phys. Chem.* **136** (1928) 259.
24. T. Erdey-Grúz and M. Volmer, *Z. Phys. Chem.* **A159** (1931) 182.
25. M. Fleischmann and H. R. Thirsk, in *Advances in Electrochemistry and Electrochemical Engineering* (P. Delahay, C. W. Tobias, Eds.), Interscience: New York (1963), Vol. 3, Chapter 3.
26. M. Fleischmann and H. R. Thirsk, *Electrochim. Acta* **1** (1959) 146.
27. M. Fleischmann and H. R. Thirsk, *Electrochim. Acta* **2** (1960) 22.
28. A. Bewick, M. Fleischmann, and H. R. Thirsk, *Trans. Faraday Soc.* **58** (1962) 2200.
29. M. Avrami, *J. Chem. Phys.* **9** (1941) 177.
30. D. J. Astley, J. A. Harrisson, and H. R. Thirsk, *Trans. Faraday Soc.* **64** (1968) 192.
31. G. J. Hills, D. J. Schiffrin, and J. Thompson, *Electrochim. Acta* **19** (1974) 657.
32. J. W. M. Jacobs, *J. Electroanal. Chem.* **247** (1988) 135.
33. B. R. Scharifker, *J. Electroanal. Chem.* **240** (1988) 61.
34. R. G. Barradas and E. Bosco, *Electrochim. Acta* **31** (1986) 949.
35. S. K. Rangarajan, *Faraday Symp. Chem. Soc.* **12** (1977) 101.
36. H. Angerstein Kozlowska, B. E. Conway, and J. Klinger, *J. Electroanal. Chem.* **87** (1978) 301.
37. E. Bosco and S. K. Rangarajan, *J. Chem. Soc. Faraday Trans.* **77** (1981) 483.
38. E. Bosco and S. K. Rangarajan, *J. Electroanal. Chem.* **129** (1981) 25.
39. M. Noel, S. Chandrasekaran, and C. Ahmed Basha, *J. Electroanal. Chem.* **225** (1987) 93.
40. J. Wojtowicz, N. Marincić, and B. E. Conway, *J. Chem. Phys.* **48** (1968) 4333.
41. K. Ogura, *Electrochim. Acta* **25** (1980) 335.
42. U. Ebersbach, K. Schwabe, and K. Ritter, *Electrochim. Acta* **12** (1967) 927.
43. W. J. Müller, *Trans. Faraday Soc.* **27** (1931) 737.
44. A. J. Calandra, N. R. Tacconi, R. Pereiro, and A. J. Arvia, *Electrochim. Acta* **19** (1974) 901.
45. D. D. Mac Donald, *Transient Techniques in Electrochemistry*, Plenum Press: New York (1977), Chapter 8.
46. D. Devilliers, F. Lantelme, and M. Chemla, *Electrochim. Acta* **31** (1986) 1235.
47. S. Srinivasan and E. Gileadi, *Electrochim. Acta* **11** (1966) 321.
48. B. E. Conway, E. Gileadi, and M. Dzieciuch, *Electrochim. Acta* **8** (1963) 143.
49. H. Angerstein-Kozlowska, J. Klinger, and B. E. Conway, *J. Electroanal. Chem.* **75** (1977) 45 and 61.
50. A. Bewick and B. Thomas, *J. Electroanal. Chem.* **85** (1977) 329.
51. H. Angerstein-Kozlowska, J. Klinger, and B. E. Conway, *J. Electroanal. Chem.* **95** (1979) 1.
52. A. Sadkowski, *J. Electroanal. Chem.* **97** (1979) 283.
53. H. Angerstein-Kozlowska and B. E. Conway, *J. Electroanal. Chem.* **113** (1980) 63.
54. E. Bosco and S. K. Rangarajan, *J. Chem. Soc. Faraday Trans.* **77** (1981) 1673.
55. S. A. Arulraj and M. Noel, *Electrochim. Acta* **33** (1988) 979.
56. G. A. Mabbott, *J. Chem. Educ.* **60** (1983) 697.
57. P. T. Kissinger and W. R. Heineman, *J. Chem. Educ.* **60** (1983) 702.
58. D. H. Evans, K. M. O'Connell, R. A. Petersen, and M. J. Kelly, *J. Chem. Educ.* **60** (1983) 290.
59. T. F. Sharpe and S. G. Meibhur, *J. Chem. Educ.* **46** (1969) 103.
60. F. G. Will, *J. Electrochem. Soc.* **133** (1986) 454.

61. B. F. Giannetti, P. T. A. Sumodjo, and T. Rabockay, *J. Appl. Electrochem.* **20** (1990) 672.
62. B. F. Giannetti, P. T. A. Sumodjo, T. Rabockay, A. M. Souza, and J. Barbosa, *Electrochim. Acta* **37** (1992) 143.
63. B. F. Giannetti and T. Rabockay, *J. Braz. Chem. Soc.* **4** (1993) 61.
64. C. M. V. B. Almeida and B. F. Giannetti, *Mat. Phys. Chem.* **69** (2001) 261.
65. C. M. V. B. Almeida, B. F. Giannetti, and T. Rabóczkay, *J. Electroanal. Chem.* **29** (1999) 123.
66. Y. M. Chen, T. J. O'Keefe, and J. James, *Thin Solid Films* **129** (1985) 205.
67. V. Tsakova and A. Milchev, *J. Electroanal. Chem.* **253** (1987) 237.

9 Surface Physical Properties and the Topology of Single Crystals: Atomic Rearrangements

C. Fernando Zinola

CONTENTS

The adsorption of species on ordered metal surfaces in the electronic conductor–ionic conductor interface produces a dramatic effect on the structure of the (1×1) surface. The presence of an adsorbed ad-layer changes the interlayer spacing, inducing a reconstruction at the topmost and, sometimes, also in the underlying layers. Even though the most interesting and applicable interactions emanate from the electrochemical interfaces, it is the UHV studies on platinum single crystals that provide an understanding of the adsorption-induced restructuring.

The adsorption of molecular hydrogen on platinum is produced without barriers and dissociatively (432 kJ mol^{-1} of dissociation energy) in different configurations that are able to reconstruct the surface. This makes the hydrogen adsorption state attractive for reduction reactions because of its low stabilization energy.

In the case of hydrogen adsorption on platinum, the reported heats of adsorption on Pt(111) are -60 kJ mol^{-1} and even as high as -90 kJ mol^{-1} [1,2] due to the presence of defects at the surface such as kinks and steps. The adsorption configuration of hydrogen on the Pt(111) unreconstructed (1×1) surface is on the hollow (3–1) or hollow (3–3) site as the higher coordination sites are achieved with this type of adsorbate (see Chapter 2 for more details). These configurations lead to a large degree of surface coverage, reaching a hydrogen to platinum stoichiometric ratio of 2. This large value implies a lateral repulsion between each other, which decreases the heat of adsorption.

However, this adsorption configuration is the only one that minimizes the repulsion energy at this extremely high hydrogen coverage. Some other authors have also found the adsorption of hydrogen at on-top sites but with little difference in the adsorption energy with respect to the theoretically predicted hollow sites [3,4]. Besides, no surface reconstruction after hydrogen adsorption on (111) planes has been found (as expected for its low adsorption energy) [2].

A different situation occurs for Pt(100) planes where hydrogen adsorption deconstructs the *hex*-reconstructed Pt(100) surface. This fact that above a critical temperature of 35 K, the surface is partially deconstructed by molecular hydrogen was first found by some authors [5,6] and later by Wandelt et al. [7]. By the analysis of low-energy electron diffraction (LEED) data, a surface expansion of 3% can be obtained for the deconstruction of Pt(100) from the (1×1) to the *hex* structure. The initial heat of adsorption (at a negligible coverage) for the unreconstructed surface is -90 kJ mol^{-1}, whereas it is -98 kJ mol^{-1} for the *hex* structure [8]; the difference attributed to the presence of steps during reconstruction.

On the other hand, Pt(110) does not deconstruct the (1×2) structure because a large adsorption energy would be required to move platinum atoms from the rows that cannot be gained from the low value of the hydrogen stabilization energy. However, it has been suggested that the initial adsorption of hydrogen on the missing row of the (110) surface is in a hollow position at the second layer [9]. This produces a 20% expansion from the topmost layer leading to a mean distance between hydrogen at the subsurface state and platinum atoms of approximately 2.1 Å, very similar to what was experimentally found [9]. The coverage dependence of hydrogen on Pt(110) is rather complex, especially at large surface coverage values [10]. The initial heat of adsorption on the (1×2) reconstructed surface is -74 kJ mol^{-1}, whereas it is -115 kJ mol^{-1} on the (1×1) unreconstructed surface. Besides, due to an attractive lateral interaction at coverages lower than 15%, the bond energy increases to 115 kJ mol^{-1}, becoming repulsive until it reaches saturation with a 55 kJ mol^{-1} bond energy (more proper for the weakly adsorbed states) [9].

The case of oxygen adsorption on platinum is even more complex because of the possibility of oxide formation in UHV conditions. Three different adsorption configurations can be characterized in the case of oxygen adsorption on Pt(111): molecular oxygen below 120 K (with 37 kJ mol^{-1} of adsorption heat), atomic oxygen down to 500 K, and oxide formation between 1000 and 1200 K [11]. The possibility of having atomic oxygen without any previous molecular adsorption has also been presented [12]. Theoretical calculations using *ab initio* presented two energetically distinct adsorption configurations: molecular oxygen parallel to the surface (paramagnetic superoxo compound) and a peroxo-species normal to the surface adsorbed at hollow sites but with some tilt toward the adjacent platinum atom [13].

As expected, the heat of adsorption decreases with the coverage value because of the repulsive lateral interaction. In a three-hollow configuration, atomic oxygen adsorbs in an ordered p(2×2) structure at 25% coverage in the 300–500 K temperature range [14]. At higher temperatures, subsurface oxygen forms up to 1200 K, achieving 25% coverage on the Pt(111)surface [15]. On Pt(100), two energetically different stages can be found: one on (1×1) and the other on the *hex* structure. It seems that the unreconstructed surface is more active, placing atomic oxygen at low temperatures (240 K) in the (2×1) arrangements [16]. In the case of the *hex* structure, molecular oxygen adsorption with low activation energies for the process occurs. Despite having low coverages (25%), the increase in temperature and partial pressure of oxygen produces large coverages for the monolayer, with no evidence for subsurface or normal oxide formation [15].

The adsorption of molecular oxygen on the Pt(110) (2×1) surface is more interesting. It is dependent on the partial pressure of oxygen and the operating temperature range. For low dosages, it adsorbs on the (111) terraces and at larger coverages on the (111) steps (superoxo and peroxide forms) [17]. The increase in the temperature produces the partial dissociation and desorption, the superoxo desorbing at lower temperatures, and at 300 K, all species are dissociatively adsorbed as atomic oxygen. It seems that the atomic oxygen is attracted to the upper side of the step edges [17]. However, it is interesting to note that the heat of adsorption does not change significantly on the three basal planes of platinum, that is, 250 kJ mol^{-1} [18], indicating the possible presence of the same type of atomic oxygen.

9.1 SURFACE STRUCTURES OF PLATINUM SINGLE CRYSTALS

The preparation of well-defined platinum single crystals for the electrochemical environment is discussed in Chapter 10. However, the stability of these arrangements is the main focus of this section. The LEED analysis of the emerged surfaces in UHV conditions is difficult [19,20] because of the water impurities and solids left on the metal surface after their evaporation in the main chamber. Thus, many potential control experiments (as scanning tunnel microscopy (STM) and surface enhanced x-ray spectroscopy (SEXRS)) have been conducted to gain reliable information on the electrode interface characteristics [21,22].

The stabilization and adsorption energies for hydrogen and oxygen species on platinum single crystals can be rapidly measured from their equilibrium coverage as a function of potential and temperature. The problem is the co-adsorption of anions that makes the calculation of the adsorption parameters and even the degree of surface coverage difficult [23]. Similar to the UHV conditions of environment, the adsorption of both species can cause the surface relaxation of platinum single crystals as observed by SXS [22,24,25]. Since the scattering from hydrogen or hydroxyl is negligible as compared with the diffraction of the platinum atoms at the top layer, the evaluation rises from the change in the interplanar distance or atom displacements. However, the largest expansion can be seen near the hydrogen evolution potential and mostly in Pt(110) (which is approximately 25%) either in acid or basic solutions.

The degree of surface coverage for the hydrogen adsorbates on Pt(111) can be obtained as a function of the potential from the temperature dependence of the cyclic voltammogram [23] similar to that on polycrystalline surfaces [26]. In most of the cases, the free energy of adsorption on Pt(111) varies linearly with the surface coverage [27–29]. The temperature dependence of the surface coverage determines the free energy of adsorption at zero coverage. The values of the heat of adsorption are evaluated from the variation of the ratio between the free energy of adsorption and temperature with the reciprocal of the temperature. On the other hand, the entropy of adsorption is the opposite of the differentiation between the free energy of adsorption and temperature. From these parameters, the interaction between hydrogen and platinum is obtained (240 kJ mol^{-1}), and this value is independent of the type of anion but dependent on the pH [23]. These considerations are unfailing of considering a weak interaction between platinum and water as it has been stated before [30]. If the same temperature and potential treatment are considered for hydroxyl adsorption on platinum, the situation becomes difficult because of anion competition and oxide formation. For half a monolayer of surface coverage, the heat of adsorption is 200 kJ mol^{-1} on Pt(111) at all pH values, and then the estimated bond energy between both species is 136 kJ mol^{-1}. This value is much lower than the platinum–atomic oxygen interaction energy at the gas–solid interface (350 kJ mol^{-1}) [31]. Theoretical calculations using DFT show that up to 1/3 of a monolayer of hydroxyl bound to the bridge and on-top sites leads to a chemisorption energy of 225 kJ mol^{-1}. Starting with 50% to almost 100% of coverage, hydroxyl tends to adsorb as on-top configurations with a little enhancement of the adsorption energy (approximately 15%), approaching what was experimentally determined on Pt(111) [32]. On the other hand, the anodically formed platinum oxide in UHV conditions produces the formation of water and oxygen in two different pathways from hydroxyl adsorbates of different critical temperatures. These results clearly show the totally different characteristics of both oxygen-containing species [33]. Finally, in spite of not having reliable data on the hydroxyl adsorption on the other two low Miller indices, electrochemical data are able to consider that Pt(110) and Pt(100) would have larger bond energies in alkaline media.

9.1.1 SURFACE RELAXATION OF PLATINUM SINGLE CRYSTALS INDUCED BY WATER ADSORBATES

The problem of surface relaxation induced by the adsorption of hydrogen or hydroxyl adsorbates was also studied theoretically using DFT calculations. In the case of hydrogen, on-top sites on

Pt(111) have been found with very little adsorption energies as compared with the other configurations [4]. On the Pt(100) surface, the hollow-square position is the favored one, whereas, because of the relatively large surface relaxation of Pt(110) (1×2), hydrogen adsorbates can be positioned at the hollow configuration below the platinum row as proposed by UHV measurements [34]. These results show that the surface relaxation heavily depends on the crystal orientation of the electrode and the adsorbate configuration type. Many attempts have been made to experimentally determine the configuration of hydrogen species on platinum single crystals using infrared vibration spectroscopy. However, it seems that completely different situations occur for the Pt(111) and the Pt(100) surfaces [35]. Interestingly, Tadjeddine et al. [36,37], using a more sensitive technique such as infrared SFG (sum frequency generation), found that surface on-top configurations are independent of the single crystals in the hydrogen sorption voltammetric region.

It seems that the situation of reversible adsorption of hydroxyl species on platinum is also quite complex, and the system is far from proven. The surface relaxation of Pt(110) (1×2) caused by hydrogen and hydroxyl species is nearly the same, and hence the proposed geometry of adsorption will be the same, that is, threefold hollow sites. Since the transition from the reversible to the irreversible adsorption produces surface roughening, the place-exchange mechanism of oxide formation is accepted. The condition that all the platinum sites—those of the top and on the missing row—have to be occupied by hydroxyl species must be fulfilled. Interestingly, the surface relaxation in the oxygen potential region is very little on the Pt(100) surface and almost none on the Pt(111) surface suggesting that it is likely that the reversible hydroxyl adsorbates will be at the topmost layer with no presence of it in the underlying layer. There is a contradiction between theory and experiment since the former proposes high-coordination sites such as threefold hollow positions [38], but the latter indicates only on-top positions [39].

9.1.2 Surface Relaxation of Platinum Single Crystals Induced by Anion Adsorbates

The adsorption of anions on metal electrodes has been studied in electrochemistry since the mid-fifties by many authors. The two main problems with these studies were the cleanliness of the electrolyte and the crystal orientation and topology of the noble-metal surfaces. At the beginning of the twenty-first century, the incorporation of *in situ* surface sensitive techniques became a reality among the electrochemists. The case of the specific adsorbable anions deserves special attention because their interaction with the surface can induce surface relaxation of the metal. One of the most interesting and important interfaces is the (bi)sulfate/platinum system.

9.1.2.1 (Bi)sulfate Anion Adsorption on Platinum Single Crystal Surfaces

The firsts attempts to quantify the adsorption of the (bi)sulfate anion were conducted with the help of *in situ* infrared vibration spectroscopy, where a potential-dependent strong signal at 1200 cm^{-1} was integrated to obtain the amount of anion adsorbed on Pt(111). The signal was interpreted as the asymmetric stretching mode of sulfuric anhydride in the anion [40] or directly as the symmetric stretching mode of bisulfate [35]. However, Iwasita et al. [41] denote that the interaction would correspond to a C_{3V} symmetry of adsorption through the three oxygen atoms of sulfate. The best methods for anion quantification are the radiotracer and the chronocoulometry methods, in strictly clean electrolyte solutions [42,43]. They show that bisulfate adsorption is a two-stage process, since two plateaus can be observed in the surface excess vs. electrode potential curve. The first plateau at 0.33 ML corresponds to one anion per three platinum atoms and the second at 0.4 ML (a value different from that found by the radiotracer methods) [42,43]. The first structure was later defined by LEED experiments [44,45] and assigned to $(\sqrt{3} \times \sqrt{3})$ R30°, that is, a structure responsible for the sharp spike due to the order–disorder transition in the cyclic voltammogram of Pt(111) in sulfuric acid. However, between 0.5 and 0.7 V, STM images demonstrate a new structure of $(\sqrt{3} \times \sqrt{7})$ at 0.22 ML, assigned to the second plateau from the integration of the chronocoulometry, due to co-adsorption between water and anions [46]. It was later corroborated by other radio-labeling experiments [47]. It is important to note that the $(\sqrt{3} \times \sqrt{7})$ structure can turn into $(\sqrt{3} \times \sqrt{3})$ R30°, after losing water in UHV conditions [45,48].

Because of the lower interaction energy of the (bi)sulfate on the Pt(110) and the Pt(100) surfaces, anion adsorption is shifted to lower potentials, denoting a lower value of the potential of zero charge of the interface [46]. The low interaction energy is due to a partial mismatch between the tetrahedral geometry of the sulfate anion and the square or rectangular geometry of the single crystals. A single interaction between one oxygen atom and the surface is expected. However, no ordered structure for these anions on Pt(110) and Pt(100) has been reported yet.

9.1.2.2 Iodide Anion Adsorption on Platinum and Palladium Single Crystal Surfaces

The pioneering works of Hubbard on halide anion adsorption on noble metals [49–52] has created an interest in the role of the system involving adsorbed iodine on platinum single crystals in the study of surface rearrangements in electrocatalysis. When the noble-metal single crystal is immersed in a much diluted potassium iodide solution at open circuit, a spontaneous oxidative discharge occurs with the formation of a full monolayer of atomic iodine, thereby creating an interesting situation.

Experiments in UHV showed that the adsorbed structures of iodine on Pt(111) or Pd(111) depends on the surface coverage and the preparation of the surface. When the surface is covered by 1/3 ML of iodine, $(\sqrt{3} \times \sqrt{3})$ R30° is obtained to minimize its interaction with the noble-metal atoms; however, it reaches the $(\sqrt{7} \times \sqrt{7})$ R19.1° for 3/7 ML [49]. At higher coverages (4/9 of a ML), the (3×3) structure is defined and demonstrated by STM measurements. On the other hand, the transition between the latter structure and the $(\sqrt{7} \times \sqrt{7})$ is evidenced by a sharp reversible spike, denoting the change from a hydrophobic to a hydrophilic behavior of the interface. The results shown by x-ray photoelectron spectroscopy (XPS) also demonstrate the occlusion of alkaline ions and water [50,51]. The (3×3) structure is characterized by two different packing arrangements: one with three iodine atoms at the bridge, one at on-top sites, and the other with four atoms close to on-top sites [53]. On the other hand, the adsorption of iodine vapor on stepped platinum surfaces reveals no particular affinity for these sites, but for the presence of $(\sqrt{3} \times \sqrt{3})$ R30° and (3×3) structures on terraces [51].

The strong interaction of this halide with the surface inhibits the hydroxyl adsorption and later the oxide formation. However, a hysteric effect from the 3/7 of the ML structure associated to a transition from the ordered to disordered structure can be seen [54]. This does not occur with the (3×3) structure as demonstrated by the diffraction patterns.

The effects of chloride and bromide are slightly different because of the lower adsorbability of these anions on the noble-metal surfaces. For example, bromide produces (3×3) and (4×4) structures, voltammetrically characterized on Pt(111) by a sharp reversible spike denoting the rearrangement from 4/9 of ML to 7/16 of a ML [55]. It is interesting to note that in the case of a weaker interaction such as that of chloride on platinum, a diffuse structure can be advanced on Pt(111), but an ordered structure can be obtained on Pt(100). Only when the anion concentration exceeds 10 mM, the Pt(111) (3×3) ad-layer is found [56].

In the case of Pd(111), the LEED pattern in UHV conditions shows a clear order of the surface after 10 min of sputtering with 5.0 purity argon current (pressure of 10^{-5} Torr) and 5 min of annealing as shown in Figure 9.1. The Pd(111) (1×1) unreconstructed ordered surface is observed.

The XPS spectrum of the clean and ordered Pd(111) single crystal before any electrochemical experiment is observed in Figure 9.2. The palladium peaks are clearly seen: 3s orbitals at 673 eV; $3p_{1/2}$ and $3p_{3/2}$ at 559.6 and 530.7 eV, respectively; $3d_{3/2}$ and $3d_{5/2}$ at 340.1 and 334.9 eV, respectively; and, finally, 4s and 4p orbitals at 88.4 and at 53.8 eV, respectively.

After this categorization, the single crystal is transferred to the pre-chamber to run electrochemical experiments. The first and second potential scans run at 0.5 V s^{-1} of the clean Pd(111) (1×1) in 0.5 M sulfuric acid solution is shown in Figure 9.3. The starting potential value is 0.60 V.

FIGURE 9.1 LEED pattern of Pd(111) after Ar sputtering and annealing.

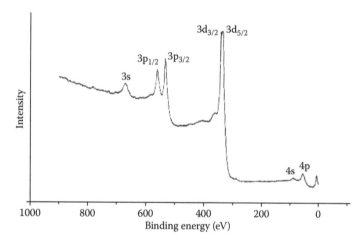

FIGURE 9.2 XPS spectrum for a clean and ordered Pd(111) (1 × 1) in UHV conditions (5.2 × 10^{-8} Torr). Photometer: 10K, 1s range 0.7% × 1 error, rate multiplier 0.1 eV s^{-1}, electron multiplier device of 10 kV and 24 mA, x-ray power 4.0 A Mg, analyzer control 6 eV pp, time response 5 ms, and pass energy 100 eV.

The first positive potential scan shows the early stages of the palladium oxygen adsorption by a hump at approximately 0.90 V, which develops a large peak at 1.1 V, denoting the formation of a hydroxyl monolayer on the palladium single crystal. After the second cycle, two adjacent anodic peaks of the first and second stages of the hydroxyl and atomic oxygen adsorption can be seen, with a monotonously increasing current up to 1.2 V of palladium dissolution.

When the potential is fixed again at 0.60 V, the electrolyte is replaced by the diluted solution of potassium iodide (1 mM) at pH = 13 for 5 min. The first anodic potential scan clearly shows the

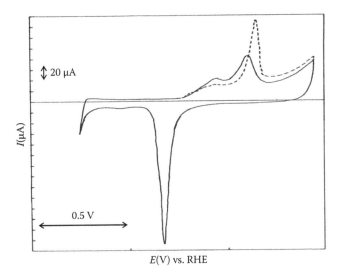

FIGURE 9.3   First (dashed lines) and second (continuous line) currents vs. potential scans for a clean and ordered Pd(111) (1 × 1) in 0.5 M sulfuric acid run at 0.05 V s^{-1}.

transition from one ordered structure of iodine to another, the features can be demonstrated again using LEED patterns. They show a superstructure of iodine responding to Pd(111)/I ($\sqrt{3} \times \sqrt{3}$) R30° that reconstructs after 1 min to a triplet structure. These triplet superlayers are immediately observed when the adsorption potential reaches 0.90 V (Figure 9.4). It is also interesting to note that when the palladium single crystal is transferred again to the main chamber to perform the LEED methodology, the XPS spectrum is still showing the signals attributed to the iodine adsorption (Figure 9.5). At potentials larger than 1.0 V, the triplet Pd(111)/I ($\sqrt{3} \times \sqrt{3}$) R30° structure is observed, which rapidly reconverts to the Pd(111)/I (4 × 4) superstructure.

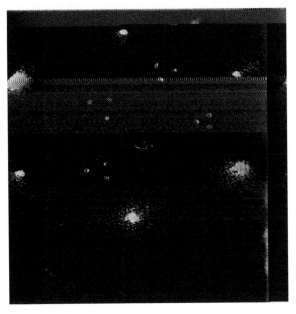

FIGURE 9.4   LEED pattern of Pd(111)/I($\sqrt{3} \times \sqrt{3}$) R30° triplet overlayer after iodine adsorption at 0.90 V vs. RHE.

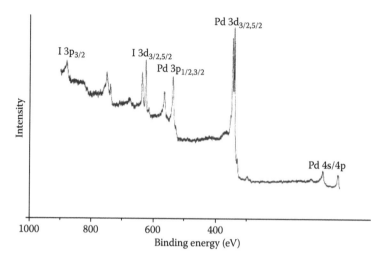

FIGURE 9.5 XPS spectrum after an adsorption experiment with potassium iodide at 0.60 V on Pd(111) in UHV conditions (8.6×10^{-8} Torr). Photometer: 10K, 1s range 0.7% × 1 error, rate multiplier 0.1 eV s^{-1}, electron multiplier device of 10 kV and 24 mA, x-ray power 4.0 A Mg, analyzer control 6 eV pp, time response 5 ms, and pass energy 100 eV.

The XPS spectrum of the Pd(111) single crystal after iodine adsorption at 0.60 V of Figure 9.5 shows palladium double peaks of $3p_{1/2}$, $3p_{3/2}$ and $3d_{3/2}$, $3d_{5/2}$ at those of the 4s, 4p orbitals. However, those of iodine exhibit features for 3p and 3d orbitals that are characteristic of the double $3d_{3/2}$, $3d_{5/2}$ signals (at 628.0 and 636.6 eV). The integration of the signals gives the "intensity," which when divided by the atomic sensitive factor can be used to obtain the degree of surface coverage by each elemental species (iodine, potassium, carbon, water, etc.) as a function of the adsorption potential [57]. Similarly, it was observed [58] from the integration of the ($3d_{3/2}$; $3d_{5/2}$) iodine signal that an approximately 0.32 coverage is obtained up to 0.80 V at pH 10 corresponding to the single Pd(111) /I ($\sqrt{3} \times \sqrt{3}$) R30° structure. The triplets and (4×4) structures can be observed from 0.80 to 1.10 V, whereas a diffused overlayer structure can be presented from 1.20 V upward. Potassium occlusion obeys a similar surface coverage, but between 0.60 and 0.80 V, the value goes under 0.10; however, it raises again to 0.40 from 1.0 V upward.

9.2 CHANGES IN THE SURFACE STRUCTURE OF PLATINUM-TYPE METAL INDUCED BY METAL UNDERPOTENTIAL DEPOSITION AND ANION ADSORBATES

According to the above explanations, it seems that the best methodology to prevent poisoning as a consequence of poor cleanliness is the preparation of the noble-metal surface in UHV conditions. Moreover, the characterization of the single crystal surfaces by *ex situ* surface sensitive techniques provides reliable information about the composition and surface configuration of adsorbed species. However, the use of *in situ* techniques gives the real overview to the potential-dependent overlayer structure of the interface. The use of Fourier transform infrared spectroscopy (FTIRS) and STM/atomic force microscopy (AFM) under potential control in the electrochemical environment has made the deposition methods turn out to be similarly vital in the preparation of binary and ternary alloys. Bulk electrodeposition and underpotential deposition methods together with spontaneous deposition and irreversible metal adsorption seem to be equally important in conventional metallurgy.

The electrodeposition at overpotentials (bulk deposition) is the fastest and the most applicable method of all, but only with a partial control on the growth of the metal layer, since the process occurs at an extremely high current density.

The underpotential deposition, on the other hand, is a reversible deposition on a different substrate, at potentials fairly positive to the reversible potential value of the electrode reaction [59]. A lot of metals can be deposited this way, but not for technical applications [60–74]. However, the case of irreversible metal adsorption seems to be also interesting because of the long-term surface stability even in the absence of adjacent solution cations [75–78]. In all of them, the application of potentiostatic or galvanostatic pulses or signals are required; however, good results (especially at low coverages) have been obtained using the spontaneous deposition process [69,79–82]. It is a very simple process that does not require any sophisticated electrochemical equipment and does not obey ohmic drops at low concentration depositions. Large values of coverage can be reached by multiple spontaneous depositions as demonstrated for ruthenium on Au(111) and for tin on polycrystalline platinum [83,84]. Besides ruthenium and osmium, there is lack of information about the physicochemical properties of spontaneous deposition [85,86].

9.2.1 SILVER DEPOSITION AND THE INFLUENCE OF THE PRESENCE OF ADSORBABLE ANIONS

Silver deposition is important for technical applications and is achieved by electroless deposition (from a silver-cyanide bath) or bulk electrodeposition from the same bath. It has been studied by a large number of techniques [59,62,63,87], and most of them on gold and platinum in sulfate or perchlorate electrolytes [88–95]. However, there is lack of information on the problem of co-adsorption of metal and anions and the surface restructuring caused by the presence of strong adsorbable anions such as halides.

Ageing potentiostatic effects were also studied [94,95] for silver electrochemical deposition on Pt(111) and pc platinum. Two different stages for silver deposition on platinum were observed: one at 1.1 V responding to a silver–platinum alloy electrodissolution (overlapped with the oxygen electroadsorption at free platinum sites) and the other at 0.65 V due to the silver oxidation (from the onset of the bulk deposition process) deposited on the former surface alloy [93,96]. The former process splits into two peaks when potentiostatic ageing is conducted. Spectroscopic techniques such as XPS and ARXPS (angle resolved x-ray photoelectron spectroscopy) were used to determine the chemical composition of silver films on platinum in an acid solution [97]. The technique was not able to discern between the presence of silver oxides and the presence of sulfates; only an energy shift of the clean silver $3d_{5/2}$ band at *upd* (under potential deposition) level of -0.5 eV was detected. A silver monolayer on Pt(100) was detected using Auger electron spectroscopy and cyclic voltammetry, which tends to form a three-dimensional structure [98]. Besides, using *online* cyclic voltammetry with *STM* images [99], a double voltammetric contribution was found on an unreconstructed Pt(100) just before the bulk deposition process, that is, the two silver layers form by two distinct and independent processes with characteristic charge densities. Similar results on Pt(111) were found by Herrero et al. [100], denoting that the first deposition consists of a 1.25 ML in the 0.91–1.07 V range, the second deposition (0.2 ML) occurs between 0.67 V and 0.91 V, and the third peak of 0.75 ML between 0.58 and 0.67 V. On the Pt(110) surface, the adsorption and desorption of silver and oxygen are not distinguishable in the more positive region of the voltammogram.

9.2.2 COPPER DEPOSITION AND THE INFLUENCE OF THE PRESENCE OF ADSORBABLE ANIONS

One of the most important processes in metallurgy is copper deposition. Galvanoplastic conditions imply a copper-acid bath over a pre-multilayer of copper from a copper–alkaline cyanide bath. The study of copper underpotential deposition on platinum is of special interest in electrochemistry in spite of its catalytic properties being rather poor with respect to fuel cell applications [101–104]. The mechanism of copper deposition and the presence of copper(I) ions as an intermediate have been studied with the help of the rotating ring-disk technique [105]. Before the onset of bulk deposition of copper, a 0.88 ML of copper underpotential deposition in a three-dimensional growth mechanism is observed [106]. Moreover, with the help of *ex situ* and *in situ* techniques, it has been demonstrated that copper underpotential deposition is a multistep process, where the crystal orientation of the platinum surface and the nature of the anion in solution play fundamental roles. In the presence of the perchlorate anion, the interaction of copper with platinum defines the island formation by maintaining almost the same platinum interatomic distance during the process, showing the absence of surface reconstruction (pseudomorphic ad-layer) [104]. In the case of sulfate anions in the supporting electrolyte, there are a wide range of ordered layers (containing the anion) in a multistep deposition.

The interaction model between copper, anions, and the platinum substrate can be explained with simplicity as in [107], for chloride as anion is explained in detail in [106] and for sulfate in [108].

For low surface coverages (lower than ½) by copper adatoms and at potentials where copper is underpotentially deposited, the anion-covered platinum surface is firstly perturbed by copper(II) ions deforming the anion overlayer structure. The resulting structure consists of an adsorbed ensemble (characterized by STM images) where the copper adatom is directly interacting with platinum active sites. For spherical anions, two main possible arrangements can be observed on single crystals, depending on the crystal orientation of the surface. Thus, an ordered and homogeneously distributed copper overlayer with ordered anions at a second layer (mostly on Pt(111) planes) or an ordered alternate structure of copper and anions in the topmost and second layers (mostly on Pt(100) planes) is observed. Both structures can be assisted or inhibited by water-discharged products whose complex ensemble depends on the applied potential. On Pt(111), a bridge position for copper is observed with anions positioned in a threefold hollow overlayer structure. On Pt(100), a fourfold hollow position for copper is demonstrated with the presence of anions in a second layer in alternate bridge positions.

At higher surface coverages (more than ½ but less than 1), anions can be entirely displaced by copper adatoms from the surface or both form a two-layer structure in which anions are adsorbed on both the platinum and the copper sites. The final step is the total filling of the copper monolayer to form a bilayer phase with a disordered anion ad-layer on the topmost of Cu–Pt(111) [106] or an ordered (2×2) bilayer of copper-halide structure on Pt(100) [104]. The same physical models can be used in the case of bromide and chloride with little differences between the anion distances with a surface structure like that of a honeycomb ad-layer. The situation accounting for iodine adsorption is very different because of its large atomic radius and specific adsorption on noble metals.

On the other hand, in a sulfuric acid solution, a ($\sqrt{3} \times \sqrt{3}$) R30° copper bisulfate structure can be observed for Pt(111) surfaces. This structure consists of 0.22 ML of anions and 0.66 ML of copper in the unit cell [108]. The same features were demonstrated on Au(111) [109]. In all of these models, a potential window exists, where it is possible to co-adsorb products of water and proton discharge, causing an electrocatalytic activity toward oxidation or reduction, respectively.

Chierche et al. [110] were first to study copper underpotential deposition on polycrystalline palladium. They assigned different copper voltammetric peaks to the contributions of the different low Miller index crystallographic planes of palladium. They also proposed that copper underpotential deposition obeys a Temkin-type isotherm with a saturation charge density of 420 μC cm^{-2}. The electrochemical response on polyfaceted palladium is observed in Figure 9.6.

The interaction of iodine with palladium and platinum electrodes was studied in different electrolyte solutions and on single crystals with electrochemical techniques and UHV spectroscopic *ex situ* measurements [111]. The chemisorption of atomic iodine on palladium researched extensively because of its ability to protect the surface from air and water interactions. Moreover, it is able to induce a surface reconstruction from a stepped surface to a (1×1) unreconstructed one.

On the Pd(111) single crystal, iodide adsorption leads the ($\sqrt{3} \times \sqrt{3}$) R30° I superlattice at potentials lower than 0.80 V at pH = 10, whereas on Pd(100), a c(2×2) I superlattice is found [111]. Copper underpotential deposition on noble metals is a relatively slow process, and the electrochemical response is strongly dependent on the potentiodynamic perturbation, because of

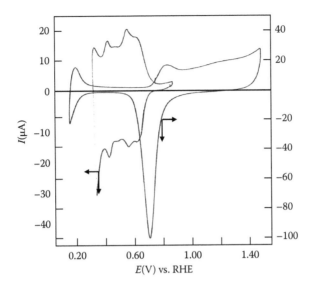

FIGURE 9.6 Cyclic voltammetry in the 0.35–0.80 V range for copper underpotential deposition on poly-crystalline palladium electrodes in 0.5 M NaClO$_4$, 10^{-3} M HClO$_4$, 0.01 M Cu(ClO$_4$)$_2$ run at 0.01 V s^{-1}. The electrochemical response of palladium electrodes in 0.5 M HClO$_4$ is superimposed from 0.10 to 1.45 V but is run at 0.10 V s^{-1}.

the less noble characteristics of these metals [112]. On the other hand, the deposition of small amounts of copper on platinum surfaces enhances the adsorption of halides from the solution, as confirmed by *in situ* surface extended x-ray absorption fine structure (EXAFS) and x-ray standing wave spectroscopy [107,113]. The voltammetric profiles, peak potential, and current values of the entire process depend on the nature of the halide, since the specific adsorption defines the electrochemical properties of the copper deposition and the superlattice structure of the halide-adsorbed layer.

Five different current peak contributions can be seen for copper electrodeposition and dissolution processes that were formerly characterized as the contribution of (111) and (100) crystallographic planes [111] at potentials higher than the massive copper electrodeposition potential.

The evaluation of the charge densities as a function of potential for copper electrodeposition was obtained from the stable low-scan-rate voltammetric profiles. A constant charge density value of 440 μC cm^{-2} was observed. The onset potential for massive copper electrodeposition in 0.5 M NaClO$_4$, 10^{-3} M HClO$_4$, 0.01 M CuSO$_4$ was located at 0.3 V.

The iodide ion is adsorbed at the open circuit on palladium from 1 mM KI, pH 10 for 180 s. In this process, iodide is oxidized, forming a chemisorbed monolayer of zero-valence iodine atoms, while H$^+$ or water molecules are reduced to produce molecular hydrogen. It can be observed from the potentiodynamic profile of palladium that the H-atom adsorption region is totally inhibited and the onset potential for the oxide formation is shifted toward more positive values. Besides, a large anodic peak current is developed as a result of the iodine electrooxidation to iodate. Complete iodine desorption can be accomplished by the application of negative potentials at pH 10. Thus, after 5 min of potential holding at -1.0 V in pH 10 and several cycles within the potential limits of water stability, the repetitive current potential profile of iodine-free palladium can be obtained.

Copper underpotential deposition on palladium in 0.5 M NaClO$_4$, 10^{-3} M HClO$_4$, and 0.01 M CuSO$_4$ in the presence of little amounts of iodide in solution (such as 10^{-6} M) produces drastic changes in the voltammetric profiles. In Figure 9.7, the repetitive voltammetric profile at 0.01 V s^{-1} of the process in 3×10^{-4} M KI can be seen.

The most remarkable feature is that the peaks representing copper underpotential adsorption disappear in the presence of iodide in solution when its concentration reaches values higher than 0.5 mM, but the anodic peak is still present due to the bulk deposition of copper. However, the

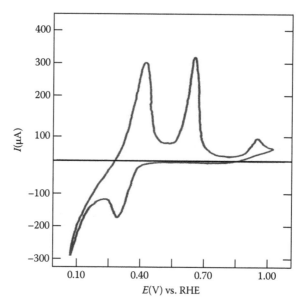

FIGURE 9.7 Repetitive cyclic voltammetric profiles for polycrystalline palladium electrodes in 0.5 M NaClO$_4$, 10^{-3} M HClO$_4$, 0.01 M Cu(ClO$_4$)$_2$, and 3 × 10^{-4} M KI run at 0.01 V s^{-1} in the 0.08–1.00 V potential range.

cathodic charge density for copper in the presence of different iodide concentrations remains constant (440 μC cm^{-2}). The massive electrodeposition onset potential of copper in the presence of iodide is shifted negatively with respect to copper deposition in the absence of the anion. Moreover, Figure 9.8 shows this situation for 3 × 10^{-4} M KI, where two large anodic peaks can be described; the most negative peak is assigned to the bulk deposited oxidation of metallic copper separated by 0.25 V from the unique anodic copper underpotential deposited peak. The difference between the bulk metal dissolution current and the most positive underpotential peak is proportional to the difference between

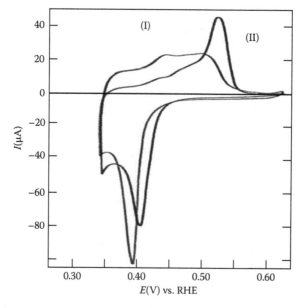

FIGURE 9.8 Competition between copper and iodine for palladium sites observed through cyclic voltammetry in the 0.35–0.62 V potential range. Polycrystalline palladium electrodes in 0.5 M NaClO$_4$, 10^{-3} M HClO$_4$, 0.01 M Cu(ClO$_4$)$_2$, 3 × 10^{-4} M KI in curve (I), and 4 × 10^{-5} M KI in curve (II) run at 0.01 V s^{-1}.

the electronic work functions of the substrate and those of the foreign metal [114]. A close inspection of the cathodic peak shows that it is really a double peak contribution. The voltammetric peaks for copper underpotential deposition, avoiding the potential region where bulk copper deposition takes place, are given in Figure 9.8.

When the potential is scanned toward more positive values (Figure 9.8), it is possible to see the oxidation of atomic iodine to soluble iodine peaking at 0.95 V. Moreover, when the potential scan is extended toward more positive values, it is possible to see a large and broad anodic peak (not shown) corresponding to the formation of soluble iodate species. In the case of having lower amounts of specifically adsorbed anions, the reversible couple due to copper electroadsorption/desorption is slightly distorted. In fact, a new system has to be proposed as a consequence of the interaction between the anion and the metallic species.

According to the voltammetric results, copper deposition takes place through a sharp current peak denoting a two-dimensional phase formation [115]. The width of the half maximum of the underpotential desorbed peak for copper is 58 mV, denoting a Langmuir desorption process if the electronic exchange is 2. On the other hand, the changes of the peak currents and potentials with the scan rate for the potentiodynamic response of copper on palladium in the presence of iodide help determine the nature of the controlling step in the whole process. Accordingly, the intensities of the deposition and dissolution peaks fit a linear relationship with the square root of the scan rate, indicating a diffusion-control process.

The study of the voltammetric changes of the iodide–copper interaction on palladium allows an examination of the control process during underpotential deposition. The electrochemical desorption of copper produces an increase in the current value with almost no variations in the peak potential. The fitting of the current yields again a diffusion control for Cu^{2+} desorption from the surface. On the other hand, the cathodic peak not only changes the current intensities but also produces a negative shift in the potential values. Also, a diffusion-control process for copper deposition on palladium is found after fitting the experimental data.

Moreover, the potential shift toward negative values corresponding to the onset potential of massive copper deposition due to the presence of iodine on the surface leads us to conclude that there is a surface formation of the CuI species, which becomes the new active electrode. If we consider that the standard potential of the Cu/Cu^{2+} electrode is 0.342 V vs. NHE, and the Cu^{2+} concentration is 0.01 M, the reversible potential for copper deposition will be 0.282 V. In the presence of iodide, a new electrode is formed, which is a $Cu/CuI/I^-$ interface. Considering its standard potential, -0.185 V vs. NHE in the 3×10^{-4} M iodide concentrations, the reversible potential of the $Cu/CuI/I^-$ electrode will be -0.032 V. The difference between the reversible potential of $Cu/CuI/I^-$ and the reversible potential of Cu/Cu^{2+} is 0.250 V, that is, the potential shift found experimentally. Therefore, the presence of iodide in solution, after adsorption, produces a new interface with completely distinct electrochemical properties.

The difference between the anodic and the cathodic contributions in the course of copper *upd* under the influence of iodide specific adsorption can be calculated from a theoretical point of view. In a mixed control system—adsorption/desorption and quasi-reversible charge transfer—the current density profile, j_i, can be assayed considering the metallic adsorbate under Langmuirian conditions at a stationary surface excess that is related to the stationary surface coverage, θ_i, by

$$\vec{j}_i(x, t) = F \vec{k}_f \theta_i(x, t) \tag{9.1}$$

with both j_i and θ_i being a function of the local position and time (x, t). The value of $\vec{k}_f$ is the cathodic rate constant (forward) for the metal ion adsorption and subsequent discharge (assuming order one in s^{-1}):

$$\vec{k}_f = k_f^o \exp\left(\frac{\vec{\alpha} F v t}{RT}\right) \tag{9.2}$$

where
$\vec{\alpha}$ is the electrochemical charge transfer for the cathodic process
v is the potential scan rate
vt is the local electrode potential
k_f^o is defined as usual as

$$k_f^o = k^o \exp\left(-\frac{\vec{\alpha}F(E_i - E^o)}{RT}\right) \tag{9.3}$$

where
E_i is the initial potential used in the cyclic voltammogram
E^o is the normal electrode potential

Thus, under Langmuirian conditions, we have

$$\frac{d\theta_i(x, t)}{dt} = -k_f^o \exp\left(\frac{\vec{\alpha}Fvt}{RT}\right)\theta_i(x, t) \tag{9.4}$$

For the resolution of this equation, we use the initial condition that at $t = 0$, $\theta_i = \theta_i(t)$ and at t where the complete surface coverage is reached, $\theta_i = \theta_i^o = 1$:

$$\int_{\theta_i(t)}^{1} \frac{d\theta_i}{\theta_i(t)} = -k^o \exp\left(-\frac{\vec{\alpha}F(E_i - E^o)}{RT}\right)\int_0^t \exp\left(\frac{\vec{\alpha}Fvt}{RT}\right)dt \tag{9.5}$$

$$\ln\theta_i(t) = k^o \exp\left(-\frac{\vec{\alpha}F(E_i - E^o)}{RT}\right)\left(\frac{RT}{\vec{\alpha}Fv}\exp\left(\frac{\vec{\alpha}Fvt}{RT}\right) - 1\right) \tag{9.6}$$

We can use E_i values far from E^o where it is possible to take $k^o e^{\left(-\frac{\vec{\alpha}F(E_i - E^o)}{RT}\right)} \rightarrow 0$:

$$\ln\theta_i(t) = k^o \exp\left(-\frac{\vec{\alpha}F(E_i - E^o)}{RT}\right)\frac{RT}{\vec{\alpha}Fv}\exp\left(\frac{\vec{\alpha}Fvt}{RT}\right) \tag{9.7}$$

but considering Equations 9.2 and 9.3,

$$\ln\theta_i(t) = k_f^o \frac{RT}{\vec{\alpha}Fv}\exp\left(\frac{\vec{\alpha}Fvt}{RT}\right) \tag{9.8}$$

and then

$$\ln\theta_i(t) = \vec{k}_f(t)\frac{RT}{\vec{\alpha}Fv} \tag{9.9}$$

and finally

$$\theta_i(t) = \exp\left(\frac{\vec{k}_f(t)RT}{\vec{\alpha}Fv}\right) \tag{9.10}$$

where the exponential term indicates an electrochemical adsorption process since $\vec{k}_f$ is a function of the electrode potential, E.

The current density profile is given as an implicit function of E:

$$\vec{j}_i(t) = F\vec{k}_f(t) \exp\left(\frac{\vec{k}_f(t)RT}{\vec{\alpha}Fv}\right) \tag{9.11}$$

with $\theta_i^o = 1$. Here it is possible to calculate the value of j_i at the minimum after the substitution of $\vec{k}_f(t)$ given by (9.2) and (9.3) at an electrode potential, $E(t)$:

$$\vec{j}_i(t) = Fk^o \exp\left(-\frac{\vec{\alpha}F(E(t) - E^o)}{RT} - \frac{k^oRT}{\vec{\alpha}Fv} \exp\left[-\frac{\vec{\alpha}F(E(t) - E^o)}{RT}\right]\right) \tag{9.12}$$

From the derivation of the above equation and after setting the result as equal to zero, we can obtain the peak potential for the cathodic metallic deposition and its current density, E_{pc} and j_p, respectively:

$$\frac{d\vec{j}_i(t)}{dE} = Fk^o \frac{d\left\{\exp\left(-\frac{\vec{\alpha}F(E(t) - E^o)}{RT} - \frac{k^oRT}{\vec{\alpha}Fv}\exp\left[-\frac{\vec{\alpha}F(E(t) - E^o)}{RT}\right]\right)\right\}}{dE} \tag{9.13}$$

$$\frac{d\vec{j}_i(t)}{dE} = Fk^o\left[\frac{k^o}{v}e^{-\frac{\vec{\alpha}F(E(t)-E^o)}{RT}} - \frac{\vec{\alpha}F}{RT}\right]\exp\left(-\frac{\vec{\alpha}F(E(t) - E^o)}{RT} - \frac{k^oRT}{\vec{\alpha}Fv}\exp\left[-\frac{\vec{\alpha}F(E(t) - E^o)}{RT}\right]\right) \tag{9.14}$$

Then, we consider

$$\frac{d\vec{j}_i(t)}{dE(t)} = 0$$

$$\frac{d\vec{j}_i(t)}{dE} = Fk^o\left[\frac{k^o}{v}e^{-\frac{\vec{\alpha}F(E(t)-E^o)}{RT}} - \frac{\vec{\alpha}F}{RT}\right]\exp\left(-\frac{\vec{\alpha}F(E(t) - E^o)}{RT} - \frac{k^oRT}{\vec{\alpha}Fv}\exp\left[-\frac{\vec{\alpha}F(E(t) - E^o)}{RT}\right]\right) \tag{9.15}$$

And after substitutions and taking $E(t) = E^o$,

$$\vec{j}_p = \frac{\vec{\alpha}F^2v}{RT}\frac{1}{e} \tag{9.16a}$$

or

$$j_p = \frac{\vec{\alpha}F^2v}{2.7RT} \tag{9.16b}$$

It is clear that j_p depends on v and it increases with its magnitude.

It is easy to calculate the peak potential, E_p, for a quasi-Nernstian behavior when we are proposing a similar law with adsorbed species:

$$E_{(M^+/M)}(t) = E^o_{(M^+/M)} + \frac{RT}{nF}\ln\frac{\theta_{M^+}(t)}{\theta_M(t)} \tag{9.17}$$

with

$$\theta_{M^+}^o = 1 = \theta_{M^+}(t) + \theta_M(t) \tag{9.18}$$

After combining these two equations,

$$\theta_{M^+}(t) = 1 - \frac{1}{1 + \exp\left[-\dfrac{nF(E(t) - E^o)}{RT}\right]} \tag{9.19}$$

The condition of the current density derivation and the electrode potential very near E^o gives rise to the calculation of the magnitude of $\theta_{M^+}(t)$ at the peak potential:

$$\theta_{M^+}(t) = \frac{1}{2} \tag{9.20}$$

since $\theta_{M^+}^0 = 1$. Considering the above equation, we can derive the peak potential value for the cathodic process, E_{pc}:

$$E_{pc} = E^o + \frac{RT}{\vec{\alpha}F} \ln\left(\frac{RT}{\vec{\alpha}F} \frac{k_{cat}^o}{v}\right) \tag{9.21}$$

We can also derive the difference expected between the anodic and the cathodic potentials, ΔE_p, in the cyclic voltammetry of an adsorbed species under a quasi-reversible performance:

$$E_{pa} = E^o + \frac{RT}{\vec{\alpha}F} \ln\left(\frac{RT}{\vec{\alpha}F} \frac{k_{an}^o}{v}\right) \tag{9.22}$$

after subtracting both terms and considering the ratio between k_{an}^o/v and k_{cat}^o/v,

$$\Delta E_p = \frac{2.44RT}{\vec{\alpha}F} \tag{9.23}$$

which is approximately $63/\alpha$ mV at 298 K. For a single electron transfer, α can be considered as the symmetry factor that is generally ½. According to Figure 9.8, the difference between the two voltammetric peaks is approximately this value.

REFERENCES

1. K. Christmann and G. Ertl, *Surf. Sci.* **41** (1976) 365.
2. K. Christmann, in *Electrocatalysis* (J. Lipkowski and P. N. Ross, Eds.) Wiley-VCH, New York (1998) p. 1.
3. A. Baro, H. Ibach, and H. Bruchmann, *Surf. Sci.* **88** (1979) 384.
4. R. A. Olsen, G. J. Kroes, and E. J. Baerends, *J. Chem. Phys.* **111** (1999) 11155.
5. X. Hu and Z. Lin, *Phys. Rev. B* **52** (1995) 11467.
6. P. R. Norton, J. A. Davies, D. K. Creber, C. W. Sitter, and T. E. Jackman, *Surf. Sci.* **108** (1981) 205.
7. B. Penneman, K. Oster, and K. Wandelt, *Surf. Sci.* **249** (1991) 35.
8. W. T. Lee, L. Ford, P. Blowers, H. L. Nigg, and R. I. Masel, *Surf. Sci.* **416** (1998) 141.
9. E. Kirsten, G. Parschau, W. Stocker, and K. H. Rieder, *Surf. Sci. Lett.* **231** (1990) L183.
10. J. R. Engstrom, W. Tsia, and W. H. Weimberg, *J. Chem. Phys.* **87** (1987) 3104.
11. J. L. Gland, B. A. Sexton, and G. B. Fisher, *Surf. Sci.* **95** (1980) 587.
12. C. T. Campbell, G. Ertl, H. Kuipers, and J. Segner, *Surf. Sci.* **107** (1980) 220.
13. A. Eichler and J. Hafner, *Phys. Rev. Lett.* **79** (1997) 4481.

14. K. Mortensen, C. Klink, F. Jensen, F. Besenbacher, and I. Stensgaard, *Surf. Sci.* **220** (1989) L701.
15. G. N. Derry and P. N. Ross, *Surf. Sci.* **140** (1984) 165.
16. P. R. Norton, P. E. Bindner, and K. Griffiths, *J. Vac. Sci. Technol.* **A2** (1984) 1028.
17. P. J. Feibelman, S. Esch, and T. Michely, *Phys. Rev. Lett.* **77** (1996) 2257.
18. T. Engel and G. Ertl, *Adv. Catal. Lett.* **28** (1979) 1.
19. A. T. Hubbard, R. M. Ishikawa, and J. Katekaru, *J. Electroanal. Chem.* **86** (1978) 271.
20. F. R. Wagner and P. N. Ross, *J. Electroanal. Chem.* **150** (1983) 181.
21. K. Itaya, *Prog. Surf. Sci.* **58** (1998) 121.
22. M. G. Samant, M. F. Toney, G. L. Borges, K. F. Blurton, and O. R. Melroy, *J. Phys. Chem.* **92** (1988) 220.
23. G. Jerkiewicz, *Prog. Surf. Sci.* **57** (1998) 137.
24. I. M. Tidswell, N. M. Markovič, and P. N. Ross, *Phys. Rev. Lett.* **71** (1993) 1601.
25. I. M. Tidswell, N. M. Markovič, C. Lucas, and P. N. Ross, *Phys. Rev. B* **47** (1993) 16542.
26. M. Breiter, *Electrochim. Acta* **7** (1962) 25.
27. B. E. Conway, H. Angerstein-Kozlowska, and W. B. Sharp, *J. Chem. Soc. Faraday Trans.* **74** (1978) 1373.
28. E. Gileadi, in *Electrode Kinetics for Chemical Engineers and Material Scientists*, Wiley-VCH, London, U.K. (1993).
29. N. M. Marković, B. N. Grgur, and P. N. Ross, *J. Phys. Chem. B* **101** (1997) 5405.
30. N. M. Marković and P. N. Ross, *Surf. Sci. Rep.* **45** (2002) 117.
31. D. D. Wagman, W. H. Evans, V. B. Parker, R. H. Schumm, I. Halow, S. M. Bailey, K. L. Churney, and R. L. Nuttall, *J. Phys. Chem. Ref. Data* **11** (suppl 2) (1982) 9.
32. A. Michaelides and P. Hu, *J. Chem. Phys.* **114** (2001) 513.
33. F. T. Wagner and P. N. Ross, *J. Electroanal. Chem.* **250** (1988) 301.
34. N. M. Marković, B. N. Grgur, C. A. Lucas, and P. N. Ross, *Surf. Sci.* **384** (1997) L805.
35. R. J. Nichols, in *Frontiers of Electrochemistry* (J. Lipkowski and P. N. Ross, Eds.) Wiley-VCH, New York (1999) p. 99.
36. A. Peremans and A. Tadjeddine, *J. Chem. Phys.* **103** (1995) 7197.
37. A. Tadjeddine and A. Peremans, *J. Electroanal. Chem.* **409** (1996) 115.
38. M. T. M. Koper and R. A. van Santen, *J. Electroanal. Chem.* **476** (1999) 64.
39. K. Bedurftig, S. Volkening, Y. Wang, J. Winterline, K. Jacoby, and G. Ertl, *J. Chem. Phys.* **111** (1999) 11147.
40. P. W. Faguy, N. M. Marković, R. R. Adzić, C. A. Fierro, and E. B. Yeager, *J. Electroanal. Chem.* **289** (1990) 245.
41. F. C. Nart, T. Iwasita, and M. Weber, *Electrochim. Acta* **39** (1994) 961.
42. P. Zelenay and A. Wieckowski, *J. Electrochem. Soc.* **139** (1992) 2552.
43. W. Savich, S. G. Sun, J. Lipkowski, and A. Wieckowski, *J. Electroanal. Chem.* **388** (1995) 233.
44. S. Thomas, Y. E. Sung, H. S. Kim, and A. Wieckowski, *J. Phys. Chem.* **100** (1996) 11726.
45. Y. Shingaya, K. Hirota, H. Ogasawara, and M. Ito, *J. Electroanal. Chem.* **409** (1996) 103.
46. A. M. Funtikov, U. Stimming, and R. Vogel, *J. Electroanal. Chem.* **482** (1997) 147.
47. A. Kolics and A. Wieckowski, *J. Phys. Chem. B* **105** (2001) 2588.
48. P. W. Faguy, N. M. Marković, and P. N. Ross, *J. Electrochem. Soc.* **140** (1993) 1638.
49. A. T. Hubbard, *Chem. Rev.* **88** (1988) 633.
50. H. Baltruschat, M. Martínez, S. Lewis, F. Lu, D. Song, D. Stern, A. Datta, and A. T. Hubbard, *J. Electroanal. Chem.* **217** (1987) 111.
51. A. Wieckowski, B. C. Schardt, S. D. Rosasco, J. L. Stickney, and A. T. Hubbard, *Surf. Sci.* **146** (1984) 115.
52. G. A. Garwood and A. T. Hubbard, *Surf. Sci.* **92** (1980) 617.
53. S. C. Chang, S. L. Yua, B. C. Schardt, and M. J. Weaver, *J. Phys. Chem.* **95** (1991) 4787.
54. C. A. Lucas, N. M. Marković, and P. N. Ross, *Phys. Rev. B* **55** (1997) 7964.
55. G. A. Garwood and A. T. Hubbard, *Surf. Sci.* **112** (1982) 281.
56. T. Solomun, A. Wieckowski, S. D. Rosasco, J. L. Stickney, and A. T. Hubbard, *Surf. Sci.* **147** (1984) 241.
57. M. P. Soriaga, *Prog. Surf. Sci.* **39** (1983) 325.
58. G. Samjeské, Untersuchungen zur elektrochimischen Iod Adsorption auf Palladium Elektroden mit UHV-methoden und zyklisher Voltammetrie, Diploma thesis, Universität Bonn, Bonn, Germany (1996).
59. D. M. Kolb, in *Advances in Electrochemistry and Electrochemical Engineering* (H. Gerischer and C. W. Tobias, Eds.) Wiley Int., New York (1978) p. 125.
60. W. Chrzanowski and A. Wieckowski, *Langmuir* **14** (1998) 1967.
61. M. Krausa and W. Vielstich, *J. Electroanal. Chem.* **379** (1994) 307.
62. J. W. Ndieyira, A. Ramadan, and T. Rayment, *J. Electroanal. Chem.* **503** (2001) 28.
63. D. Oyamatsu, H. Kanemoto, S. Kuwabata, and H. Yoneyama, *J. Electroanal. Chem.* **497** (2001) 97.
64. M. C. Santos, L. H. Mascaro, and S. A. S. Machado, *Electrochim. Acta* **43** (1998) 2263.

65. E. Herrero and H. D. Abruña, *Langmuir* **13** (1997) 4446.
66. J. Inukai, S. Sugita, and K. Itaya, *J. Electroanal. Chem.* **403** (1996) 159.
67. S. Abaci, L. Zhang, and C. Shannon, *J. Electroanal. Chem.* **571** (2004) 169.
68. C. Coutanceau, A. Ralotondrainibé, A. Lima, E. Garnier, S. Pronier, J. Léger, and C. Lamy, *J. Appl. Electrochem.* **34** (2004) 61.
69. M. Janssen and J. Moolhuysen, *Electrochim. Acta* **21** (1976) 861.
70. W. S. Li, L. P. Tian, Q. M. Huang, H. Li, H. Y. Chen, and X. P. Lian, *J. Power Sources* **104** (2002) 281.
71. M. Nakayama, H. Komatsu, S. Ozuka, and K. Ogura, *Electrochim. Acta* **51** (2005) 274.
72. F. W. Nyasulu and J. M. Mayer, *J. Electroanal. Chem.* **392** (1995) 35.
73. H. Zhang, Y. Wang, E. Rosim, and C. Cabrera, *Electrochem. Solid State Lett.* **2** (1999) 437.
74. H. Massong, H. Wang, G. Samjeské, and H. Baltruschat, *Electrochim. Acta* **46** (2000) 701.
75. J. Clavilier, J. M. Feliú, and A. Aldaz, *J. Electroanal. Chem.* **243** (1998) 419.
76. A. Fernández-Vega, J. M. Feliú, A. Aldaz, and J. Clavilier, *J. Electroanal. Chem.* **305** (1991) 229.
77. E. Herrero, A. Fernández-Vega, J. M. Feliú, and A. Aldaz, *J. Electroanal. Chem.* **350** (1993) 73.
78. E. Herrero, A. Rodes, J. M. Pérez, J. M. Feliú, and A. Aldaz, *J. Electroanal. Chem.* **412** (1996) 165.
79. E. Spinacé, A. Neto, and M. Linardi, *J. Power Sources* **129** (2004) 121.
80. A. Crown, C. Johnston, and A. Wieckowski, *Surf. Sci.* **506** (2002) L268.
81. S. Strbac, C. Johnston, G. Lu, A. Crown, and A. Wieckowski, *Surf. Sci.* **573** (2004) 80.
82. V. Colle, M. Giz, and G. Tremiliosi Filhio, *J. Braz. Chem. Soc.* **14** (2003) 601.
83. S. Strbac, R. J. Behm, A. Crown, and A. Wieckowski, *Surf. Sci.* **517** (2002) 2078.
84. P. A. Thiel and T. E. Madey, *Surf. Sci. Rep.* **7** (1987) 211.
85. N. M. Marković and P. N. Ross Jr., *Surf. Sci. Rep.* **45** (2002) 117.
86. D. M. Kolb, M. Przasnyski, and H. Gerisher, *J. Electroanal. Chem.* **54** (1974) 25.
87. K. Ogaki and K. Itaya, *Electrochim. Acta* **40** (1995) 1249.
88. S. Sugita, T. Abe, and K. Itaya, *J. Phys. Chem.* **97** (1993) 8780.
89. M. J. Esplandiú, M. A. Schneeweiss, and D. M. Kolb, *Phys. Chem. Chem. Phys.* **1** (1999) 4847.
90. S. Garcia, D. Salinas, C. Mayer, E. Schmidt, G. Staikov, and W. J. Lorenz, *Electrochim. Acta* **43** (1998) 3007.
91. M. A. van Hove, R. J. Koestner, P. C. Stair, J. P. Biberian, L. L. Kesmodel, I. Bartos, and G. A. Somorjai, *Surf. Sci.* **103** (1981) 189.
92. R. C. Salvarezza, D. V. Vásquez Moll, M. C. Giordano, and A. J. Arvia, *J. Electroanal. Chem.* **213** (1986) 301.
93. M. E. Martins, R. C. Salvarezza, and A. J. Arvia, *Electrochim. Acta* **41** (1996) 2441.
94. A. Vaskevich, M. Rosenblum, and E. Gileadi, *J. Electroanal. Chem.* **383** (1995) 167.
95. A. Vaskevich and E. Gileadi, *J. Electroanal. Chem.* **442** (1998) 147.
96. M. E. Martins, R. C. Salvarezza, and A. J. Arvia, *Electrochim. Acta* **36** (1991) 1617.
97. C. Palacio, P. Ocón, P. Herrasti, D. Díaz, and A. Arranz, *J. Electroanal. Chem.* **545** (2003) 53.
98. F. El Omar, R. Durand, and R. Faure, *J. Electroanal. Chem.* **160** (1984) 385.
99. A. M. Bittner, *J. Electroanal. Chem.* **431** (1997) 51.
100. E. Herrero, L. J. Buller, and H. D. Abruña, *Chem. Rev.* **101** (2001) 1897.
101. J. H. White and H. D. Abruña, *J. Phys. Chem.* **94** (1990) 894.
102. R. Michaelis, M. S. Zei, R. S. Zhai, and D. M. Kolb, *J. Electroanal. Chem.* **339** (1992) 299.
103. R. Durand, R. Faure, D. Aberdam, C. Salem, G. Tourillon, D. Gauy, and M. Ladouceur, *Electrochim. Acta* **37** (1992) 1977.
104. N. M. Marković and P. N. Ross, *Langmuir* **9** (1993) 580.
105. N. M. Marković, B. N. Grgur, C. A. Lucas, and P. N. Ross, *Electrochim. Acta* **44** (1998) 1009.
106. I. M. Tidswell, C. Lucas, N. M. Marković, and P. N. Ross, *Phys. Rev. B* **51** (1995) 10205.
107. R. Gómez, H. S. Yee, G. M. Bommarito, J. M. Feliú, and H. Abruña, *Surf. Sci.* **335** (1995) 101.
108. C. A. Lucas, N. M. Marković, and P. N. Ross, *Phys. Rev. B* **56** (1997) 3651.
109. Z. Shi and J. Lipkowski, *J. Electroanal. Chem.* **364** (1994) 289.
110. P. M. Rigano, T. Chierche, and C. Mayer, *Electrochim. Acta* **35** (1990) 1189.
111. T. Solomun, *J. Electroanal. Chem.* **302** (1991) 31.
112. C. F. Zinola and A. Castro Luna, *J. Colloid Interface Sci.* **209** (1999) 392.
113. R. Gómez, A. Rodes, J. M. Pérez, J. M. Feliú, and A. Aldaz, *Surf. Sci.* **344** (1995) 85.
114. D. Kolb, R. Koetz, and K. Yamamoto, *Surf. Sci.* **87** (1979) 20.
115. J. H. White and H. D. Abruña, *J. Electroanal. Chem.* **300** (1991) 521.

10 Problem of Surface Relaxation and Atom Rearrangement on Electrode Surfaces

C. Fernando Zinola

CONTENTS

10.1 INTRODUCTION

In the case of electronic conductor–gas phase interfaces, comparative studies of the behavior of single crystal surfaces of pure metals and bulk alloys with electrocatalysts have been carried out for several years. The use of true single crystals for the development of electrocatalytic reactions has provided insights into the sensitivity of the reaction.

The use of preferentially oriented surfaces or, more appropriately, single crystals invokes the idea of retaining the atoms' bulk position at the surface. The problem is that most of the metals do not fulfill this consideration, especially in the electrochemical environment. In this case, the excess of surface energy due to missing rows or the presence of strong adsorbable ions causes a rearrangement at the top of the layer, yielding a different surface structure with respect to that of the bulk. Two different general situations arise:

1. Missing of a local position at the topmost layers (lower coordination number) due to the inward relaxation of the topmost layers and the outward expansion of the lower layers.
2. Variation of bond length and relative angles between the first and the second rows caused by the local-position-missing row of the first.

Thus, a change in the crystallographic distribution of the surface atoms with respect to those of the bulk occurs. Surface restructuring can yield macroscopic variations of the surface morphology, manifesting as steps formation, surface roughening, and crystal faceting (*surface reconstruction*).

In the case of alloy electrocatalysts, the identification of the alloy constituent (at the topmost layers) during the electrocatalytic reaction is rather difficult. Therefore, the assumption of stability after the reaction makes the study rather simpler. In this case, the UHV conditions can be applied only in the *ex situ* variation, and then an idea of the process mechanism is also required. Not many techniques can be used for the identification of the alloy constituents. However, techniques under a high vacuum condition are applied: x-ray photoelectron spectroscopy (XPS), Auger spectroscopy, low-energy ion scattering, and low-energy electron diffraction.

The definition of the surface arrangement and atomic order of the single crystals is of tremendous importance in electrocatalysis and surface science. The findings of Clavilier et al. [1,2] on the "hydrogen–anion" complex wave for the Pt(111) face by a cyclic voltammogram open up a chapter on surface electrochemistry. The existence of a two-dimensional long-range order supported the discussion on the voltammetric response of Pt(111) in sulfuric acid solutions. The unusual adsorption states, appearing as sharp reversible spikes in the voltammograms, depend on the degree of the bidimensional order achieved on the (111) domains to which the adsorption/desorption of the supporting electrolyte also contributes [1,2]. In this respect, on stepped platinum surfaces with (111) terraces, the sharp spike decreases as the number of rows on the terrace decreases and finally vanishes when the steps are shorter than 10 atoms wide, irrespective of the orientation [3–5]. These characteristic features were formerly used as fingerprints of the crystalline orientation of the platinum surface. Similarly, the irreversible adsorption of different molecules and adatoms [6,7] on the well-ordered Pt(111) electrodes causes a disruption of the surface long-range order, as previously found for gold surfaces, even in the absence of the anions that are susceptible to adsorption.

As stated before, well-defined single crystal surfaces are essential for research on electrochemical characterization [1,2,8–11] and for the study of electrocatalysis [12–18] either in industrial electrochemistry or for academic purposes. Electrocatalysis by the oxidation and reduction of organic compounds has been studied on platinum single crystals, since they are affected by the topology of the electrode [13–15]. Potential cycles—between hydrogen and oxygen evolution—are usually applied to polycrystalline platinum etched with aqua regia or sulfuric acid–nitric acid mixture for the study of electrocatalytic activity. The potential cycling enhances the electrocatalytic activity at the electrode and the voltammograms at the chemically etched platinum electrodes become reproducible because of the electrochemical activation by the repetitive platinum oxide formation and reduction. The reduced surface exhibits a stepped crystalline orientation with (100) and (110) facets. Structural changes at the Pt(111) surface during electrochemical activation have been investigated by *in situ* STM [19]. Randomly oriented islands with a height of a few atomic steps were observed on the Pt(111) terrace after the electrochemical activation [19]. Similarly, some studies on Pt(100) during electrochemical activation were conducted by *in situ* STM [20].

The characterization of the surface can be usually managed with underpotential-deposited metals of lower nobility than the substrate. Recently, many papers have gained the scientific community about the deposition of noble metals not at underpotentials but at overpotentials of low value. It has been shown that palladium can be deposited on Pt(111) electrode surfaces [21,22] and is exemplified by an epitaxial growth with *ex situ* [23] and *in situ* methods [24]. Voltammetric characterization, carried out in sulfuric acid, agrees with the island growth of palladium on the substrate because the sharp spikes of the palladium-free Pt(111) substrate domains at 0.45 V are observed at relatively high-palladium coverage [21]. The behavior of adsorbed palladium on Pt(111) is the opposite of that observed with adsorbed rhodium on Pt(111) in perchloric acid, where the peaks at 0.80 V are present [25]. In fact, the high-potential adsorption state is quite featureless once palladium is adsorbed on the Pt(111) substrate. It should be noted that this high-potential adsorbate is only observed when working in perchloric acid solutions but disappears in the presence of specific anion adsorption.

For the surface characterization of the electrocatalyst, STM images allow the direct determination of the composition and the short-range order behavior of the clean surfaces of noble metals. The use of alloys instead of surface-modified noble metal catalysts produces a large mechanical and electrochemical stability that makes the substrate adequate for industrial purposes. For example, in the case of $Pt_{25}Rh_{75}(111)$, (110), and (100) alloy surfaces, it was also possible to evaluate the surface rearrangement [26]. In the top layers of both $Pt_{25}Rh_{75}(111)$ and $Pt_{25}Rh_{75}(110)$-(1×2), they found a preferential order of the even nearest neighbors. While $Pt_{25}Rh_{75}(100)$ exhibits a preference for clustering, the ordering tendencies appear at $Pt_{25}Rh_{75}(111)$ and (110). The number of hollow sites surrounded by the Rh atoms only can be significantly affected by the short-range order. $Pt_{25}Rh_{75}(110)$ exhibits a (1×2) missing-row reconstruction after annealing above 700°C. After the first annealing of the sputtered surface, it is accompanied by mesoscopic long-range "waves" with a height of approximately 2 nm and a wavelength of up to 200 nm, depending on the preparation temperature.

The problem in the electrode surface arises due to a clean and order surface. An interesting possibility of the use of UHV techniques in electrochemistry, that is, the *ex situ* approach, appears with the use of clean and unreconstructed surfaces. In the beginning, structure-sensitive techniques were not available for *in situ* studies. Therefore, electrochemists adopted UHV techniques, particularly low-energy electron diffraction (LEED) and Auger spectroscopy or XPS, to check the quality of the UHV-prepared single crystal before and after electrochemical cycling. The use of the UHV atmosphere to prepare single crystal surfaces was confined to strongly bound adsorbates, such as iodide, metal, or metal oxides [27]. UHV atmosphere is also employed to compare the results of former electrochemical experiments with those carried out the UHV environment, since these electrochemical interfaces do not undergo any change when they are transferred to UHV.

After the pioneering observation by Hansen [28] that electrodes could be removed from the solution by maintaining their electric double layer it became possible to investigate the water adsorption configuration on metal surfaces in the UHV environment [29]. Although his studies were accomplished under very different circumstances from those in electrochemistry, the information on the energies of metal–water interactions was extremely useful to check the accuracy of the classical double-layer Bockris–Devanathan–Müller model. The interaction of water with the commonly used electrode metals was much less than previously assumed.

The use of experimental physics and the implementation of new theoretical concepts and methods from solid-state physics or statistical mechanics to electrochemistry contributed to the development of surface electrochemistry. This was particularly important for a better understanding of the electric double layer or, more generally speaking, of the solvent structure near a charged metal (by shifting the Fermi level upward or downward). Important results came from computer simulations of the electric double layer that yielded new information about the spatial distribution of ions and water molecules toward the electrode surface [30].

The above-mentioned concepts lead to the evaluation of the possibility of positioning single atoms or molecules on single crystal surfaces. The positioning of single atoms or molecules on single crystal surfaces was accomplished with the help of STM techniques considering the tip–substrate interaction at a very close distance to manipulate individual atoms in order to conduct them to preset positions. This kind of "nanostructuring" of surfaces under UHV conditions and at low temperatures has been reported [31–33]. In an electrochemical environment, the tip-generated entities involved more than single atoms because of the room temperature at metal–electrolyte interfaces and the presence of the solvent and ions. The most common approach to produce an electrochemical nanostructure was by exploiting the tip to create surface defects, which plays a nucleation role for metal deposition on these prefixed locations [34]. A more recently developed strategy involves a two-step process in which the metal was first deposited from solution onto the tip, followed by a "source" of dissolution and new deposition on the sample [35]. Such a procedure left the surface undamaged, but the final cluster size was rather large, for example, in the tens of nm^2 range. Another mechanism [36] involves metal digital deposition. In this process, the first metal is deposited on the tip by applying a tip potential negative to the reversible deposition value. Next, the metal-loaded tip approaches the surface of the sample. On contact with the surface, it deposits the metal. The tip is then drawn back and while doing so the connective collar breaks, leaving a small metal cluster on the surface. The direction of the material transfer depends on the cohesive energy of both sides. Because of the cathodic overpotential on the tip, it is continuously deposited with new metal from the solution and is ready with another cluster formation for deposition on the surface. According to the literature [35,36], the contact requires an approach of approximately 0.3 nm tunnel gap.

10.2 SURFACE RECONSTRUCTION AND STRUCTURE SENSITIVITY

10.2.1 GENERAL FEATURES

Substantial advances in the understanding of the electronic/ionic conductors' interfaces have been made in recent years due to the incorporation of UHV techniques together with STM in *in situ* and *ex situ* modes. The structure of the single crystal metals in those environments can undergo a minimization of their surface energy as explained above. The processes involved in surface reconstruction are surface relaxation and rearrangement of atoms. The following crystallographic arrangements for the three low Miller indices are expected for the unreconstructed noble metals of true single crystals. Figure 10.1 shows a scheme of the crystalline arrangement for the unreconstructed surfaces that are electrochemically characterized by the cyclic voltammograms in Figure 10.2.

It seems that Pt(111) (1×1) with an atomic pack density of 1.53×10^{15} atoms cm^{-2} has no tendency to reconstruct. However, the topmost layer distance tends to contract at 2.5% with respect to the bulk [37,38]. On the other hand, Pt(100) is less dense than Pt(111) and is expected to have a larger tendency to reconstruct (1.28×10^{15} atoms cm^{-2}). Thus, there is an irreversible transformation from the (1×1) structure above 390 K by the contraction of the topmost layer to a quasi-hexagonal lattice (*hex*) of 1.55×10^{15} atoms cm^{-2} [39]. Above 1000 K, the surface is rotated 0.7° [39]. In the case of Pt(110), there is a larger reconstruction all along the [001] direction resulting in (1×n) reconstructions. However, at room temperature, the stable (1×2) structure can be detected, which is known as the "missing row" structure. It is interesting to note that there is a loss of the packing

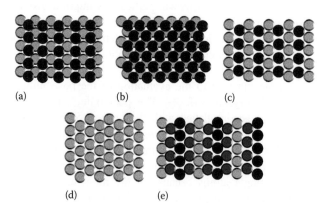

(a)   (b)   (c)

(d)   (e)

FIGURE 10.1 Unreconstructed (1×1) noble metal surfaces for (a) (100), (b) (111), and (c) (110) planes. Reconstructed (d) quasi-hexagonal structure (only the topmost layer is shown) and (e) (1×2) (110) plane. Black, dark-gray, and light-gray balls denote the upper (top), second, and third lower rows of the crystal lattice, respectively.

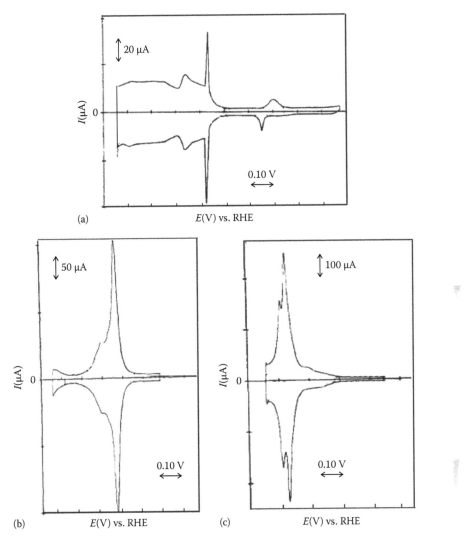

FIGURE 10.2 Voltammetric profile of (a) Pt(111), (b) Pt(100), and (c) Pt(110) single crystals run at 0.05 V s^{-1} in 0.5 M sulfuric acid prepared by flame annealing and cooled in a nitrogen:hydrogen atmosphere.

density from 9.40 to 5.10×10^{14} atoms cm^{-2}; however, the surface exhibits a back transition to the (1×1) structure at temperatures above 855 K [40]. According to the literature, the alternative expansion/compression repetitive process can reach the fourth or fifth layer either for (1×1) or for (1×2) structures [41].

In the case of gold, different situations upon reconstruction can be classified. Under UHV conditions and when the gold single crystals are subjected to thermal treatments, the three low indices show different tendencies to create more densely packed surfaces. For Au(100) (1×1), the reconstructions to a quasi-hexagonal close-packed form is evident [42] like that explained for the Pt(100) surface. On the other hand, Au(111) can reconstruct from (1×1) to $(\sqrt{3} \times 22)$, there being no direct underlying to the bulk [43]. This slight lateral compression of the surface in the [011] direction reduces the mechanical stress by approximately 4.4%. Similar to Pt(110), gold reconstructs the "missing row" (1×2) structure, and the second row yields a faceting to the (111) planes in the [011] direction [44]. These micro-faceting processes can continue to higher $(1 \times n)$ in the same [011] direction (Figure 10.3).

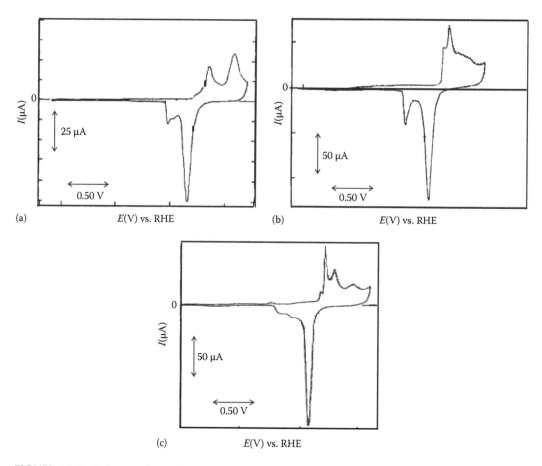

FIGURE 10.3 Voltammetric profile of (a) Au(111), (b) Au(110), and (c) Au(100) single crystals run at 0.05 V s^{-1} in 0.01 M perchloric acid prepared as explained in the text.

When physically well-constructed and clean gold faces are obtained, the analysis of the cyclic voltammograms leads to a large background of information. The cyclic current vs. potential profile can be used as a fingerprint of the gold face exposed to the electrolyte. Some discrepancies can be found in the literature between the reported current and potential profiles for several faces having the same crystallographic orientation under identical experimental conditions. As reported by Hamelin et al. [45,46], the reproducibility and stability of the cyclic voltammetric profile is taken as the main criteria of quality. Though spectroscopic [47,48], diffraction [49,50], and imaging [51] techniques are used for the *in situ* examination of electrodes, cyclic votammetry is a simpler and faster way to check the quality of the crystal.

The problem of reconstruction has also been studied in the case of the platinum surfaces belonging to the [0$\bar{1}$1] zone, which can be expressed as composed of (111) terraces and a regular array of (100) monatomic steps [52,53]. Preparing clean and ordered surfaces of this type at atmospheric pressure, when the surface reacts with atmospheric oxygen, can induce structural changes, such as surface faceting. The electrochemical behavior of platinum surfaces vicinal to Pt(111) in the [0$\bar{1}$1] zone has been investigated before [9,10]. The ordered structure for these surfaces, composed of a regular array of (111) terrace sites and (110) step sites, resists faceting when oxygen is adsorbed under UHV conditions [53]. Under these conditions, the surfaces prepared by the flaming technique showed a distribution of hydrogen adsorption sites obeying the conditions that were deduced from the hard sphere model of the Pt(S)[$(n-1)(111) \times (110)$] stepped surface [4,9].

In some cases, Pt(200) surfaces, after flame treatment and cooling in air, exhibit an anomalous distribution of the hydrogen adsorption states obtained after a long annealing time [10]. The strange hydrogen adsorption states corresponding to (110) surface sites indicated that the surface has a structure not predicted by the step-terrace model that does not include sites of (110) symmetry [54]. This behavior has been explained by a disruption of the order along the (100) step rows under the effect of oxygen adsorption [10] already detected in the gas phase [55].

10.2.2   Influence of Adsorbed Species on Reconstruction and Structure Sensitivity

It has been found that oxygen adsorption on Pt(100) perturbed the surface at long-range orders [55]. However, the interaction of the surface with hydrogen favors the development of a long-range bid [56]. Independently, it was proved that the interaction of the Pt(111) orientation with electroadsorbed oxygen creates mainly (110) step sites [57]. Besides, platinum planes with (111) terraces and (110) steps have a step distribution unaffected by thermal treatment, including a cooling step in the presence of oxygen [58]. These two evidences explain that (110) sites are those which participate in the equilibrium configuration of the surface in the presence of adsorbed oxygen after cooling [59].

In electrocatalysis, the structure sensitivity of an adsorption process at the platinum–electrolyte interface is evidenced by the effect of surface orientation on the distribution of the configuration states in current vs. potential profiles [1–4]. The electrolyte composition at constant pH affects these adsorption states because of the anion-specific adsorption [2].

In the case of Pt(111), voltammograms for flame-cleaned electrodes were obtained under UHV conditions [53] providing evidence that such voltammograms correspond to well-ordered (1×1) structures. Since the same voltammetric features were obtained with various surface preparation procedures, there is a large stability of the (111) (1×1) structure [60]. Conversely, UHV studies have shown the existence of various surface structures for Pt(100) [42] or Pt(110) [61]. Such structural changes may influence the electrochemical response of these orientations. Similar to Pt(111), voltammograms obtained with Pt(100) electrodes have been found to depart on the electrolyte composition; that is, a decrease of the anion-specific adsorption shifts the hydrogen adsorption states to more positive potentials.

On the other hand, for a given electrolyte, the distribution of the adsorption states up to 0.8 V vs. RHE depends on the degree of the surface order of the platinum single crystal. Adsorption states at higher potentials are related to the presence of wide two-dimensional ordered (100) domains [11,62]. The maximum development of these states was obtained after cooling a flame-treated sample in an oxygen-free atmosphere [3,11,62,63].

In the case of strongly adsorbed anions such as chloride [8] or bioxalate [64], the population of the adsorption states is higher than that observed in sulfuric acid. This was earlier explained as a result of the formation of a more densely packed structure of the topmost layer [62] and was found before [65] in the voltammogram of Pt(100) in sulfuric acid with a sample previously cycled in a chloride containing solution, which was also explained as the presence of a defect [65].

A method to elude those defects, induced reconstructions, or anion adsorption is to transfer the electrodes under well-controlled conditions including atmosphere. Thus, undesirable effects from oxygen adsorption or impurities as a source of voltammogram modifications can be avoided. These requirements are fulfilled in the "iodine–carbon monoxide" substitution method which was proposed for the preparation of clean and well-ordered Pt(111) [66] and applied to Pt(100) clean surface preparation [67]. An interesting alternative to this method would be to find experimental conditions that maintain a carbon monoxide adlayer for surface protection during the transfer, assuming that this adsorption is innocuous for the surface structure itself. If this efficient protection makes no detectable surface-order modifications for Pt(100) electrodes as deduced from the cyclic voltammetric contour, we can conclude that this protection method is convenient for studying the influence of anion adsorption on the surface structure in transfer experiments.

10.2.2.1 Influence of Carbon Monoxide–Adsorbed Species on Surface Reconstruction

The electrochemical performance of carbon monoxide adsorbed on platinum basal planes has been extensively studied with the help of various physical techniques such as LEED, AES, and *in situ* infrared spectroscopy of the surface species [68,69]. The common feature of these studies is the use of a non-adsorbing anion electrolyte such as perchloric acid. The problem arises from the coincident potential region for adsorbed carbon monoxide oxidation and desorption that only on carbon monoxide surface saturation, we can neglect the influence of the hydrogen-adsorbed states.

Electrochemical results for adsorbed carbon monoxide oxidation were firstly analyzed from the linearly or bridge-bonded modes of bonding of the molecule to the surface sites by considering the fractional values obtained for the number of electrons involved in the process. From the data available in the literature, it was unclear how the voltammetry of carbon monoxide showed two oxidation peaks and electron numbers were determined ascribing each peak to one mode of bonding of the carbon monoxide molecule [70,71]. However, this interpretation has been demonstrated as false. In some papers [72], they have shown that the appropriate use of sulfuric acid instead of perchloric acid demonstrates two ranges of potentials wherein the carbon monoxide may be oxidized and desorbed from the surface. The lower range of potential is obtained with adsorption conditions, leading directly to low coverage, that is, lower than 0.5. Surprisingly, a single oxidation peak is found where nearly equal amounts of carbon monoxide are linearly bonded and bridge bonded. The hydrogen adsorption–desorption voltammetric contour indicates that (111) planes of a few atomic diameters are carbon monoxide free.

The intermediate case of coverage is that wherein the voltammetric profile for carbon monoxide oxidation and desorption shows two peaks in both potential ranges. The hydrogen adsorption state distribution indicates the presence of small ordered (111) planes on the surface. Once the carbon monoxide that is oxidizable at the lower potential is desorbed, the distribution of the hydrogen adsorption states is observed with very wide (111) surface ordered domains free of carbon monoxide. This suggests that the remaining carbon monoxide is dispersed over the broad dense domains of the same type as that existing for the complete blocking of hydrogen adsorption, since oxidation occurs in the same potential range.

Two distributions of carbon monoxide molecules exist on Pt(111): one being a surface dispersion of small islands whose oxidation gives rise to a single oxidation peak at 0.75 V, corresponding to narrow carbon monoxide-free surface domains, and another being a surface dispersion of wide carbon monoxide islands and large carbon monoxide-free surface domains, which gives rise to an oxidation peak at 0.83 V, both with similar surface coverage and approximately 1.5 electrons per carbon monoxide species. The multiplicity of the carbon monoxide oxidation peak is kinetic in nature and cannot be associated with the adsorption energies of the two modes of bonding.

It has to be said that carbon monoxide species can be formed from a dissociative adsorption of formic acid, formaldehyde, methanol, ethylene glycol, etc. and are species that are formed as those of the first type of distribution. This suggests that the surface structure is an open structure, since dissociative adsorption of the organic molecule requires adjacent free platinum sites and that at the electrochemical–environment interface, once carbon monoxide is formed, there is almost no mobility at all.

Many groups have shown that the amount of adsorbed oxygen that is initially present on a (100) platinum surface can be dosed by changing the thermal or the electrochemical pretreatments of the sample [56]. It has been observed that an increase in adsorbed oxygen brings on an enhancement of hydrogen adsorbed without an appreciable change in the total electric charge in the hydrogen adsorption region during the first voltammetric scan. This permits a pre-evaluation of the initial hydrogen distribution on the different adsorption states. Besides, the latter can also be modified by an electrochemical procedure without any change in the number of total hydrogen adsorption sites on the surface [73]. The voltammogram profiles have been correlated successfully with the existence of more or less extended flat domains on the electrode surface.

For instance, it has been concluded that the weakly bound hydrogen adatom is more populated, as the surface is more perturbed. On the contrary, the strongly bound hydrogen adatom has been attributed to the presence of atomically flat domains with a long-range two-dimensional order [74,75]. Thus, the hydrogen adatoms are construed either as a short-range or as a long-range, order-sensitive check for platinum surfaces.

The specific adsorption of anions can play an important role in the potential distribution of the multiple hydrogen adsorption states, which can be modified when the nature or the concentration of these anions is changed [65,76,77]. It has also been observed that Pt(110) undergoes a (1×2) surface reconstruction in an acid medium and proposed that anion-specific adsorption can be the origin of the surface reconstruction at the initial stages of the Pt(110) unreconstructed surface [78]. In the case of Pt(100), the competition between the electrosorption in acid solution by the anion and the proton induces a narrowing of and a shift of the hydrogen adsorption states toward lower potentials and a shift of the oxygen electroadsorption process toward higher potentials. In the absence of specific adsorption, a new hydrogen adsorption state is noticed at a higher potential from 0.5 to 0.7 V vs. RHE.

It has been established for Pt(100) and Pt(111) that when the specific adsorption of hydrochloric acid is strong enough, anions are partially displaced from the surface-adsorbed oxygen with negligible transient currents. For Pt(100), this strong specific adsorption also induces a surface atomic rearrangement toward a more stable surface with a more densely packed structure similar to the (5×20) reconstruction observed in the gas phase.

Contrary to Pt(100) and Pt(110) surfaces, no reconstruction seems to occur with Pt(111), which might be a more stable surface structure than the other two crystallographic orientations.

10.2.3   Problem of the Preparation of Platinum Surfaces in Electrocatalysis

Electrocatalysis on single crystal surfaces has attracted the attention of the scientific and technological community because it is considered as the main method to optimize surface reactions for applications in electrochemical or chemical technology. Thus, the structure sensitivity of the reactions is outlined on well-defined single crystal surfaces. The use of platinum as catalyst is justified by its large activity for many reactions, previously demonstrated on polycrystalline samples. However, the preparation and the conservation of clean and well-defined single crystal surfaces run into many technical hitches. A surface of a single crystal exposed by cutting in a direction parallel to the net plane of a specified crystallographic index is not always the net plane with that index. The surface could be roughened by treatment of various kinds. The roughness factor of a true single crystal surface should ideally be unity, except in the case of palladium.

10.2.4   Platinum Pretreatments: The Preparation of the Samples and the Comparison between Annealing and Cooling Procedures

The distinction in the shape of the cyclic voltammograms seems to start off from the differences in the annealing and cooling procedures of platinum single crystals. One of the challenges is in the removal of the frozen-prepared layer of nearly polycrystalline nature (resulting from the crystal cutting and mechanical polishing to a mirror finish), to expose the surface of the desired crystallographic index. This layer is usually removed by sputtering and annealing in an UHV chamber to expose an atomically smooth single crystal surface. During this annealing procedure, the small crystallites in the first worked layer reconstitute on the substrate single crystal. Some of them may be chemically transported into the vapor phase in an atmosphere containing oxygen. Thus, a clean single crystal surface of the desired index can be obtained. Then, this surface should be brought to room temperature and at the same time has to be kept clean.

The following description for obtaining a clean and well-defined surface has been reported elsewhere [79]. Four types of Pt(111) surfaces can be obtained by the procedures given below:

1. The surface is annealed in vacuum at 680°C for 24 h and is then subjected to more than two hundred potentiodynamic sweeps between the solvent electrochemical stability potentials [80]. It produces a stepped platinum surface that resembles the polycrystalline one.
2. The samples can be annealed at approximately 950°C in a helium atmosphere for 3 h, and afterward the temperature is varied to lower values linearly with time until room temperature after approximately 6 h [81]. Because of little contamination, no hydrogen adsorption peaks can be seen in the cyclic voltammetric runs.
3. When the sample is annealed at 1050°C for 8 h in pure oxygen and is reduced at 1050°C for 1 h in a mixture of 10% hydrogen:argon and is then cooled down for 12 h in the same atmosphere, it results in the voltammogram of Ref. [82], which is similar to that of [2].
4. It is also possible to anneal platinum in a hydrogen–oxygen flame of 1100°C for 10 min, and then protect it by a droplet of ultrapure water resulting in a potentiodynamic curve of that in [82], similar to that found by Clavilier in 1980.

Many other procedures of annealing and sputtering in different atmospheres and cooling temperatures produce, depending on the vacuum chamber and potential cycling, other types of voltammograms. It has to be remembered that the residual gas in the case of prolonged cooling or prolonged exposure to a vacuum could contaminate the surface.

In all cases, a bead of platinum single crystal is prepared by the method of Kaischev modified by Clavilier et al. [7]. The crystallographic axes of the crystal were determined optically, using as a guide, the laser beam spots projected on a wall, as reflected by the (*hkl*) facets on the bead surface. Then, the crystal is cut with a diamond-cutting wheel to give it the (*hkl*) face. The surface is mechanically polished to optical flatness, using a 0.1 μm diamond paste for the finishing. A platinum–lead wire was welded onto the sample at the opposite side of the surface. The (*hkl*) face is annealed at 1100°C in a hydrogen–oxygen flame under an argon flow and is then cooled by a droplet of ultrapure water.

The heat furnace is shown in Figure 10.4 with an electrode holder (right part) put in contact with a small furnace (left part)—the narrow gap at the contact between Teflon. Atmospheric gas was introduced from the gas inlet of the furnace (I) on the left, and inert gas was introduced from the gas inlet of the electrode holder (II) on the right.

After being cooled, the sample was put into the electrode holder and was then transferred under the protection of an argon flow into the electrochemical cell. The single crystal surface was brought into contact with the solution as done by Clavilier et al. [7].

10.2.5 PLATINUM PRETREATMENTS BY POTENTIAL CYCLING

An interesting case of the Pt(111) surface is considered here as an example. When the surface is subjected to a potential cycling between 0.8 and 0.05 V vs. RHE at 0.05 V s^{-1}, independently of the

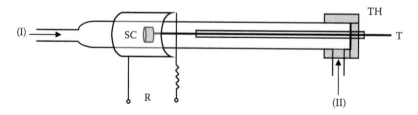

FIGURE 10.4 Electrode holder connected to the heat treatment furnace at the left: (I), inlet for atmospheric gas; (II), inlet for inert gas; R, the furnace holder; SC, the single crystal; T, the thermocouple; and TH, the Teflon holder.

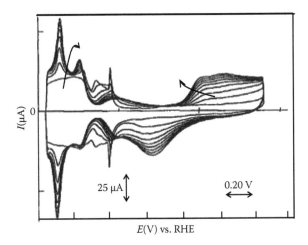

$I(\mu A)$ 0

25 μA   0.20 V ↔

E(V) vs. RHE

FIGURE 10.5 Voltammetric profile of unreconstructed Pt(111) after subjecting to a repetitive cycling between 0.05 and 1.2 V vs. RHE at 0.05 V s^{-1} in 0.5 M sulfuric acid prepared as explained in the text.

times that we cycle the electrode, the repetitive current vs. potential profile will be that of Figure 10.1. On the other hand, when the upper potential limit exceeds 1.3 V, each cycle tends to reconstruct the (111) plane of the polycrystalline platinum as shown in Figure 10.5. The peak height ratio changes with the number of cycles, and it decreases faster after cycling up to 1.5 V.

The voltammograms observed on Pt(111) reconstructed by potential cycling can be restored when the sample is heated in a hydrogen atmosphere. The sharp spike, typical of the Pt(111) surface, becomes larger with increasing temperature. These current vs. potential profiles can also be recovered by heating in the other inactive atmospheres such as argon, nitrogen, and helium. A recovered Pt(111) surface was also observed when the sample was heated in an oxygen atmosphere, but this requires higher temperatures than when the sample is heated in hydrogen. In the latter atmospheres, the temperature of the complete recovery is 270°C, but in the former, it is 540°C.

However, it seems that the use of the combined electrochemical pre-chamber/UHV main chamber device seems to be the best for these treatments. One of the methods consists of applying the same treatment to platinum single crystals for separate experiments. For characterization, low-energy electron diffraction, Auger electron spectroscopy, or XPS was used. Conventional electrochemical experiments were followed with the analysis of the results by taking into account the effect of the unavoidable differences in the required conditions.

After the flame annealing of the noble metal surface at the main chamber, the sample must be transferred to the pre-chamber covered with a droplet of pure water. This reduced the surface contamination by residual atmospheric contaminants. Then, by using the dipping technique [83] that avoids new risks of contamination by the elimination of nitrogen, sulfur, carbon, etc., the surface is put in contact with the solution by careful dipping to avoid water from the lateral side of the hemispherical crystal. The contact of the electrode with the solution is performed potentiostatically at a select potential, where a negligible transient current is observed. This overall experimental process allows the control of surface purity in the system for the electrochemical experiment. This special technique, adopted for the studies on platinum surfaces, provided the novel observations on the behavior of platinum electrodes reported previously [1].

The device (Figure 10.6) necessary for the sample preparation consisted of a blowpipe incorporated into an appendix chamber of the spectroscope under a flow of a pure argon atmosphere during the treatment in a hydrogen/oxygen flame. Water vapor was trapped on a platinum wall situated in the direct vicinity of the sample and cooled with liquid nitrogen. Immediately after the flame treatment, the chamber was pumped for 2 min, the sample was introduced into the spectroscope, and the LEED–Auger characterizations were carried out. The LEED pattern of the defined

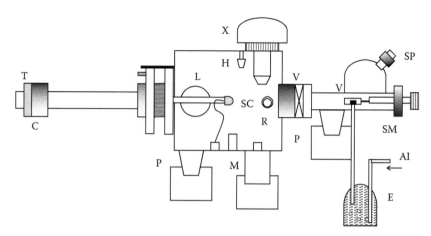

FIGURE 10.6 UHV, electrochemical device; C, electrochemical cell, with sample SC, auxiliary and reference electrodes; E, electrolyte; AI, argon inlet; V, valve for the separation between the electrochemical pre-chamber and the main UHV chamber; SM, sample manipulator; SP, sorption pump; P, the turbomolecular pumps; M, mass spectrometer; S, sputter gun; SC, sample of single crystals; R, x-ray emission tube; L, low-energy electron diffraction system; H, heat lamp; X, x-ray photoelectronic spectrometer; and T, transfer rod with sample holder.

electron energy is able to indicate the order of the surface structure and the Auger spectra showed all characteristic lines of platinum and perhaps the impurity lines of carbon, sulfur, oxygen, etc., of not more than 0.1 ML.

When a Pt(100) surface is subjected to ultrahigh vacuum conditions, the clean surface spontaneously reconstructs [84] from the (1×1) structure to the reconstructed state at room temperature by the adsorption of impurities, such as water molecules to the (5×20) structure [84]. The present (1×1) structure means that the state of the sample treated this way, which gave an LEED pattern of the (100) plane without further treatment under vacuum, has a satisfactory initial crystallographic surface structure from an LEED point of view for the electrochemical experiment.

The Auger spectra for the carbon line can be used for checking the surface contamination that can be corroborated with the voltammograms [85]. Nevertheless, with the help of only cyclic voltammetry, we can observe and analyze easily a platinum surface that is free of the various atmospheric contaminants. Contamination is avoided because of the very short time required for the transfer of the sample from the flame treatment to the cell. This rather unusual requirement of quick transfer of the samples shows that voltammetry should be a good tool for the study of metallic surfaces in the presence of a defined and complex environment.

A typical experimental setup that allows both UHV and electrochemical measurements is shown in Figure 10.6.

Stainless steel UHV chambers are usually equipped with the standard facilities for the cleaning and characterizing of the metal samples. Moreover, the integration of conventional electrochemical measurements to study the *in situ* current vs. potential profiles of the interfaces required the employment of a pre-chamber under high vacuum conditions. The samples with a small area (approximately 1 cm^2) are single crystal disks or cylinders mounted in a rod manipulator to allow a precise positioning to perform all the surface-analytical measurements. Before those measurements, it is required that a clean and ordered sample is prepared by sputtering with argon and annealing to high temperatures. The characterization of the metal–solution interface is achieved in the pre-chamber where high vacuum conditions are maintained by constantly evacuating atmosphere. The constant evacuation is achieved by using turbo-molecular and titanium sublimation pumps as detailed elsewhere [57,68]. However, in order to avoid corrosion problems, the study of electroadsorption and potentiodynamic measurements of only a limited number of cycles can be performed.

Structural and chemical cleanliness must be maintained with special care since imperfections, such as kinks, steps, and vacant sites, can act as trapping agents or sticking centers for adsorption. The presence of carbon and sulfur as contaminants on the surface produces drastic changes in the adsorption configuration of molecules and they can even change the degree of surface coverage. A lowering in the adsorption energy can occur, especially in the case of low adsorbable species. The presence of admetals, even at low coverages, can change the adsorption configuration of a molecule because of electronic variations in the work function.

10.2.5.1 Platinum Faceted Surfaces

Many studies have been carried out on the electrocatalysis of platinum single crystal electrodes [1–4,54,55]. However, the performance of an electrocatalyst is strongly affected by the structure of the surface. Therefore, it has been found that instantaneous heat treatment at 1400°C allows the formation of a thin polycrystalline electroplated film; that is, a single crystal film with the (100) surface from a galvanostatic deposition can be obtained [86]. We are going to call these types of electrodeposited layers "faceted surfaces" for the reasons given below. The conditions that prevail in electroplating, the concentration and the temperature of the electrolytic bath, and the current density of electroplating affect the preferred orientation of the electroplated film. However, the new peak at approximately 0.35 V vs. RHE, besides the typical of 0.25 V, shows the existence of a reconstructed surface. A galvanostatic control of deposition could not be as precise as that performed by using potentiodynamic conditions. Arvia et al. [87–96] have extensively studied the effects of potential controlled techniques on the growth of ordered faceted surfaces. The potentio-dynamic growth (commonly of square-wave programs) of electrodeposited platinum layers was reported before [88]. In platinum ion-containing solutions, the following redox couples should be considered in acid solution:

$$Pt + 4Cl^-_{(aq)} \rightarrow [PtCl_4]^{2-}_{(aq)} + 2e^- \tag{10.1}$$

$$[PtCl_4]^{2-}_{(aq)} + 2Cl^-_{(aq)} \rightarrow [PtCl_6]^{2-}_{(aq)} + 2e^- \tag{10.2}$$

with the equilibrium potentials being 0.76 and 0.68 V, respectively [89].

The electrochemical faceting of platinum electrodes, either polycrystalline or polyfaceted single crystal electrodes, can be conducted by applying a periodic potential treatment in a complexing acid solution [88–95]. The processes leading to faceting are associated with either the cyclic electro-dissolution/electrodeposition of a substrate or the cyclic formation/reduction of a platinum oxide layer [94,95]. The first method produces the formation of metal faceting to the stepped surfaces proportional to (110) and (100) planes. On the other hand, the second method favors the develop-ment of rough platinum surfaces, particularly when the supporting electrolyte is in a strong basic medium.

Then, when the limits of the periodic potential routine are larger than 0.76 V and lower than 0.68 V, the main anodic and cathodic processes should be related to the electrodissolution and the electrodeposition of platinum, respectively, as those of reactions (10.1) and (10.2).

For the discussion of the different situations arising from the irreversible character of reactions (10.1) and (10.2), it would be convenient to define the average potential applied to the metal–solution interface under the repetitive square-wave potential signal as $\langle E \rangle = (E_u + E_l)/2$, where E_u and E_l are the upper and lower potential limits of the potential routine, respectively. It has been found that the maximum faceting performance is produced when $\langle E \rangle = 0.55$ V, where the faradaic charge involved in the anodic and the cathodic half-cycles are equal. The value of maximum faceting is shifted negatively with respect to the reversible potential of the $Pt/[PtCl_6]^{2-}$ redox couple.

It has been demonstrated in other papers that the characteristics of the electrochemical processes depend on the frequency of the signal, which is related to the average thickness of the pulsating diffusion layer associated with the transport of the soluble reacting species through the solution side [96].

The average thickness of the diffusion layer decreases with the reciprocal of the square root of the frequency [96,97]. Thus, low values of frequencies such as 50 Hz (which indicates diffusion layers of 10^{-3} cm) should imply that the overall process would be diffusion controlled. Thus, the platinum electrodeposits would show large roughness and no preferred crystallographic orientation. Conversely, for a high frequency value (more than 1 kHz), the thickness decreases to 10^{-4} to 10^{-5} cm, and the overall process will be charge transfer controlled. In this case, the treated platinum surface will show a clear crystallographic orientation and no development of roughness.

There is so much to say about this simple electrochemical technique, but one of the main aspects is the fast and stable formation of crystallites of a preferred crystalline orientation. When the overall process becomes activation controlled, E_u values negative with respect to 0.55 V, the prevalence of the electrodeposition reaction in the overall potential routine yields platinum crystallites toward (110) faceting and high roughness. When E_u is set highly positive, the platinum electrodissolution reaction constitutes the main process leading to the development of crystallographic pitting around the [111] pole region and to a cubic-like faceting, but with (100) plane faceting, together with a decrease in the electrode roughness. A particular situation arises as $\langle E \rangle$ approaches the value of 0.55 V; a considerable influence of the properties of each crystallographic face is revealed in the surface morphology of treated platinum. Then, the SEM micrographs show a highly facetted surface in the region around the [111] pole, constituted by a network of triangular pyramids around the square holes with trihedral faces with predominantly stepped (110) plane orientations.

Arvia et al. [87,90–95] also studied the morphology changes of platinum surfaces by electrochemical perturbations in chloroplatinate-ion-free solutions. Cyclic voltammetry and SEM micrographs were used to characterize the treated platinum electrodes, which were prepared by making an adequate selection of the parameters of the periodic potential routine in electrolytes of strong ionic forces. According to the nomenclature on the matter [98,99], the obtained voltammograms resemble those shown in Figure 10.2b and, more properly, can be indicated as the (210) preferred crystallographic orientation. Such a structure can be described as $(210) = (100) \times (110)$. On the other hand, the voltammograms such as those illustrated in Figure 10.2c suggest the appearance of (320), (430), or (540) structure, depending on the experimental conditions [99]. In this case, different terrace and border structures can be described as follows: $(n20) = (n-1)(110) \times (100)$.

10.2.6 Single Crystal Platinum Prepared in Different Cooling Atmospheres

There was some controversy about the effect of cooling in air the surface structure at Pt(111) electrodes. According to Clavilier et al., no difference should be observed for cooling Pt(111) in air or in hydrogen [56,57]. On the other hand, Motoo and Furuya [63] noted that cooling in air results in a rough and dirty surface because of the adsorption of oxygen and other contaminants. In this sense, Kolb et al. [100] made an *in situ* STM investigation in different experimental conditions. On the one hand, when Pt(111) is cooled in air and immersed in sulfuric acid, the STM image reveals a wrinkled surface with fractal properties. The cyclic voltammogram exhibits two peaks associated with hydrogen adsorption at the defect sites and a very small "butterfly." The STM results demonstrate a roughening of the surface by oxygen, as already reported before [63], and they show the necessity of using an oxygen-free cooling atmosphere in order to prepare a smooth well-ordered Pt(111) surface.

On the other hand, the *in-situ* STM image of a Pt(111) electrode cooled in a carbon monoxide–nitrogen mixture and the corresponding cyclic voltammogram after oxidative stripping of the monocarbonaceous adlayer reveal that the layer remains on the surface even when the electrode is brought in contact with air. In this case, STM images reveal a surface consisting of atomically flat terraces. The corresponding cyclic voltammogram shows the classical "butterfly," the sharpness of which indicates a well-ordered Pt(111) surface [62]. It is important to say that they also found almost the same results when cooling the Pt(111) electrode in pure nitrogen or in a mixture of hydrogen and nitrogen.

On the other hand, Pt(100) in the (5×20) quasi-hexagonal is the most stable configuration at room temperature, contrary to the unreconstructed Pt(100) (1×1) [101,102]. It is known that Pt(100) at the quasi-hexagonal structure transforms above 1100 K into the so-called Pt(100)–hex–R 0.7°, where the densely packed hexagonal structure has been slightly rotated, as explained above [39]. It has been tabled that Pt(100)–hex surfaces prepared in UHV are stable in sulfuric solutions at potentials negative to the onset of surface oxidation [103]. It was also declared [104] that Pt(100)–hex surfaces can be prepared by flame annealing and cooling down in an inert atmosphere, where the reconstruction was enhanced by hydrogen adsorption at potentials more negative than 0.4 V vs. Pd/H$_2$. Again Kolb and coauthors [100] made an *in situ* STM investigation after cooling a Pt(100) electrode in air. The most striking feature in the STM image was the stripes running perpendicular to the step edges and separated by approximately 3.5 nm from each other. From a closer inspection, these stripes appear to be formed by many small islands that are seen more clearly at the step edges. These stripes were obtained only when the sample was cooled in air. Such long and narrow islands on Pt(100) that was prepared by flame annealing and cooling in air were already seen in the *ex situ* STM studies [105]. The cyclic voltammogram is dominated by the presence of a large peak at approximately 0 V and a smaller one at 0.2 V. According to Clavilier, the peak at 0 V corresponds to hydrogen adsorption at the defect sites caused by oxygen adsorption during the cooling step [56]. The small islands and the corresponding stripes most likely result from the enhancement of the thermally induced hex-reconstruction and further roughening of the surface by oxygen adsorption. When the sample of Pt(100) was cooled in a hydrogen atmosphere after flame annealing, the STM image reveals a composition of flat terraces separated by monoatomic high steps and covered by monoatomic high, square-shaped islands of 10–20 nm in size. The presence of such islands on the surface is a clear indication that the surface was reconstructed in the beginning and that the adsorption of hydrogen has transformed the initially reconstructed surface into Pt(100) (1×1). *Ex situ* STM images similar to that explained above were reported for Pt(100) prepared by flame annealing and cooling in hydrogen [105].

As reported before, the cooling in carbon monoxide also results in an unreconstructed Pt(100) (1×1) surface, as suggested by the well-known fact that CO adsorption enhances the hex reconstruction under UHV conditions [106]. However, no islands were seen in this case on the surface [100]. Sometimes, the islands cannot be observed by STM images due to their small size, but the peak at 0.2 V in the cyclic voltammogram points toward this direction.

Pt(110) is known to be reconstructed under UHV conditions [107,108]. This (1×2) missing-row reconstruction was reported to be quite stable within the double-layer charging region, that is, when oxide formation was avoided [109,110]. It has been also reported that the missing-row structure was enhanced after the immersion of the flame-annealed electrode into the electrolyte. However, in that study, the Pt(110) crystal was cooled in air, which is known to induce a high density of surface defects [111], and above all, the electrode was quenched in the electrolyte at elevated temperatures. Such quenching is known to destroy the crystal structure due to the temperature shock [112].

Marković et al. [24] have studied the significance of the annealing temperature to ascertain whether the (1×2) or the (1×1) structure is obtained. The phase transition of Pt(110) from (1×2) to (1×1) occurs in UHV at approximately 810°C [37,113]. However, the cooling conditions after annealing appear to be much more decisive for the surface structure than the annealing temperature. In contrast to carbon monoxide adsorption on Pt(111) and on Pt(100), a similar layer on Pt(110) is oxidized by air, as was shown by a partial blocking of the voltammetric peaks in the hydrogen adsorption region. After this treatment, Kolb and coauthors [100] found a pronounced peak at 0.04 V and smaller ones at 0.06 and 0.07 V. Cooling in nitrogen yielded only one main voltammetric peak in the hydrogen adsorption region at 0.03 V. The peaks in the hydrogen adsorption region are clearly different, demonstrating a significant influence of the cooling procedure on the surface structure of Pt(110).

Numerous studies on carbon monoxide adsorption on Pt(100) *hex* structure have shown that this results in a reconstructed (1×1) overlayer of CO adsorbed on Pt(100). The deconstruction process follows a mechanism of migration, cluster formation, and surface conversion from *hex* ($106 \, \text{kJ} \, \text{mol}^{-1}$) to (1×1) layers ($156 \, \text{kJ} \, \text{mol}^{-1}$) [43].

REFERENCES

1. J. Clavilier, R. Faure, G. Guinet, and R. Durand, *J. Electroanal. Chem.* **107** (1980) 205.
2. J. Clavilier, *J. Electroanal. Chem.* **107** (1980) 211.
3. S. Motoo and N. Furuya, *Ber. Bunsenges. Phys. Chem.* **91** (1987) 457.
4. J. Clavilier, K. El Achi, and A. Rodes, *Chem. Phys.* **141** (1990) 1.
5. B. Alvarez, J. M. Feliu, and J. Clavilier, *Electrochem. Commun.* **4** (2002) 379.
6. S. G. Sun and J. Clavilier, *J. Electroanal. Chem.* **236** (1987) 95.
7. J. Clavilier, J. M. Feliu, and A. Aldaz, *J. Electroanal. Chem.* **243** (1988) 419.
8. D. Armand and J. Clavilier, *J. Electroanal. Chem.* **270** (1989) 331.
9. J. Clavilier, K. El Achi, and A. Rodes, *J. Electroanal. Chem.* **272** (1989) 253.
10. A. Rodes, K. El Achi, M. A. Zamakhchari, and J. Clavilier, *J. Electroanal. Chem.* **284** (1990) 245.
11. A. Rodes, M. A. Zamakhchari, K. El Achi, and J. Clavilier, *J. Electroanal. Chem.* **305** (1991) 115.
12. J. Clavilier, C. Lamy, and J. M. Leger, *J. Electroanal. Chem.* **125** (1981) 249.
13. J. Clavilier, R. Parsons, R. Durand, C. Lamy, and J. M. Leger, *J. Electroanal. Chem.* **124** (1981) 321.
14. C. Lamy, J. M. Leger, and J. Clavilier, *J. Electroanal. Chem.* **135** (1982) 321.
15. S. G. Sung, J. Clavilier, and A. Bewick, *J. Electroanal. Chem.* **210** (1988) 147.
16. J. Clavilier, A. Fernandez-Vega, J. M. Feliu, and A. Aldaz, *J. Electroanal. Chem.* **258** (1989) 89.
17. S. Sun and J. Clavilier, *J. Electroanal. Chem.* **263** (1989) 109.
18. J. Clavilier, J. M. Orts, R. Gomez, J. M. Feliu, and A. Aldaz, *J. Electroanal. Chem.* **404** (1996) 281.
19. N. Kimizuka, T. Abe, and K. Itaya, *Denki Kagaku* **61** (1993) 796.
20. N. Furuya, M. Ichinose, and M. Shibata, *J. Electroanal. Chem.* **460** (1999) 251.
21. G. A. Attard and A. Bannister, *J. Electroanal. Chem.* **300** (1991) 467.
22. J. Clavilier, M. J. Llorca, J. M. Feliu, and A. Aldaz, *J. Electroanal. Chem.* **310** (1991) 429.
23. G. A. Attard, R. Price, and A. Al Akl, *Electrochim. Acta* **39** (1994) 1525.
24. N. M. Marković, C. A. Lucas, V. Climent, V. Stamenković, and P. N. Ross, *Surf. Sci.* **465** (2000) 103.
25. R. Gomez and J. M. Feliu, *Electrochim. Acta* **44** (1998) 1191.
26. E. L. D. Hebenstreit, W. Hebenstreit, M. Schmid, and P. Varga, *Surf. Sci.* **441** (1999) 441.
27. A. S. Dakkouri and D. M. Kolb, Reconstruction of gold surfaces. In: A. Wieckowski (Ed.), *Interfacial Electrochemistry*, Marcel Dekker, New York (1999) pp. 151–173.
28. W. N. Hansen, *J. Electroanal. Chem.* **150** (1983) 133.
29. P. A. Thiel and T. E. Madey, *Surf. Sci. Rep.* **7** (1987) 211.
30. E. Spohr, Computer simulations of electrochemical interfaces. In: R. C. Alkire and D. M. Kolb (Eds.), *Advances in Electrochemical Science and Engineering*, vol. 6, Wiley-VCH, Weinheim, Germany (1999) pp. 1–75.
31. G. Meyer, S. Zophel, and K. H. Rieder, *Appl. Phys.* **A 63** (1996) 557.
32. I.-W. Lyo and Ph. Avouris, *Science* **253** (1991) 173.
33. M. T. Cuberes, R. R. Schlittler, and J. K. Gimzewski, *Surf. Sci.* **371** (1997) L231.
34. W. Li, J. A. Virtanen, and R. M. Penner, *Appl. Phys. Lett.* **60** (1992) 1181.
35. W. Schindler, D. Hofmann, and J. Kirschner, *J. Appl. Phys.* **87** (2000) 7007.
36. D. M. Kolb, G. E. Engelmann, and J. C. Ziegler, *Solid State Ionics* **131** (2000) 69.
37. N. M. Marković and R. N. Ross, *Surf. Sci. Rep.* **45** (2002) 117.
38. P. R. Norton, J. A. Davies, D. K. Creber, C. W. Sitter, and T. E. Jackman, *Surf. Sci.* **108** (1981) 205.
39. P. Heilmann, K. Heinz, and K. Muller, *Surf. Sci.* **83** (1979) 487.
40. U. Korte and G. Meyer-Ehmsen, *Surf. Sci.* **271** (1992) 616.
41. E. C. Sowa, M. A. van Hove, and D. L. Adams, *Surf. Sci.* **199** (1998) 174.
42. M. A. van Hove, R. J. Koestner, P. C. Stair, J. P. Biberian, L. L. Kesmodel, I. Bartoš, and G. A. Somorjai, *Surf. Sci.* **103** (1981) 218.
43. G. Binnig, H. Rohrer, Ch. Gerber, and E. Stoll, *Surf. Sci.* **144** (1984) 321.
44. J. Möller, H. Niehus, and W. Heiland, *Surf. Sci.* **166** (1986) L111.
45. A. Hamelin. In: B. E. Conway, R. E. White, and J.O'M. Bockris (Eds.), *Modern Aspects of Electro-chemistry*, vol. 16, Plenum Press, New York (1985) Chapter 1.

46. A. Hamelin, S. Morin, J. Richer, and J. Lipkowski, *J. Electroanal. Chem.* **285** (1990) 249.
47. A. Tadjeddine, A. Peremans, and P. Guyot-Sionnest, *Surf. Sci.* **335** (1995) 210.
48. S. Mirwald, B. Pettinger, and J. Lipkowski, *Surf. Sci.* **335** (1995) 264.
49. C. A. Melendres and A. Tadjeddine (Eds.), *Synchrotron Techniques in Interfacial Electrochemistry*, NATO ASI, Ser. C, vol. 432, Kluwer Academic, Dordrecht, the Netherlands (1994).
50. H. D. Abruña. In: J.O'M. Bockris, R. E. White, and B. E. Conway (Eds.), *Modem Aspects of Electrochemistry*, vol. 20, Plenum Press, London, U. K. (1989) Chapter 4.
51. R. Sonnenfeld, J. Schneir, and P. K. Hansma. In: R. E. White, J.O'M. Bockris, and B. E. Conway (Eds.), *Modern Aspects of Electrochemistry*, vol. 21, Plenum Press, London, U. K. (1990) Chapter 1.
52. B. Lang, R. W. Joyner, and G. A. Somorjai, *Surf. Sci.* **30** (1972) 440.
53. D. W. Blakely and G. A. Somojai, *Surf. Sci.* **65** (1977) 419.
54. N. M. Marković, N. R. Marinković, and R. R. Adzić, *J. Electroanal. Chem.* **241** (1988) 309.
55. F. T. Wagner and P. N. Ross, *Surf. Sci.* **160** (1985) 305.
56. J. Clavilier, D. Armand, and B. L. Wu, *J. Electroanal. Chem.* **135** (1982) 159.
50. J. Clavilier. In: M. P. Soriaga (Ed.), *Electrochemical Surface Science: Molecular Phenomena at Electrode Surfaces*, ACS Symposium Series 378, American Chemical Society, Washington, DC (1988) Chapter 14.
58. J. Clavilier, K. El Achi, M. Petit, A. Rodes, and J. Clavilier, *J. Electroanal. Chem.* **295** (1990) 333.
59. A. Rodes and J. Clavilier, *J. Electroanal. Chem.* **344** (1993) 269.
60. D. Aberdam, R. Durand, R. Fame, and F. El Omar, *Surf. Sci.* **171** (1986) 303.
61. T. Gritsch, D. Coulman, R. J. Behm, and G. Ertl, *Phys. Rev. Lett.* **63** (1989) 1086.
62. J. Clavilier, A. Rodes, K. El Achi, and M. A. Zamakhchari, *J. Chim. Phys.* **88** (1991) 1291.
63. S. Motoo and N. Furuya, *J. Electroanal. Chem.* **167** (1984) 309.
64. J. M. Orts, J. M. Feliu, A. Aldaz, J. Clavilier, and A. Rodes, *J. Electroanal. Chem.* **281** (1990) 199.
65. C. L. Scortichini and C. N. Reilley, *J. Electroanal. Chem.* **139** (1982) 233.
66. D. Zurawski, L. Rice, M. Hourani, and A. Wieckowski, *J. Electroanal. Chem.* **230** (1987) 221.
67. L. Palaikis, D. Zurawski, M. Hourani, and A. Wieckowski, *Surf. Sci.* **199** (1988) 347.
68. N. Furuya, S. Motoo, and K. Kummatsu, *J. Electroanal. Chem.* **239** (1988) 347.
69. L. W. H. Leung, A. Wieckowski, and M. J. Weaver, *J. Phys. Chem.* **92** (1988) 6985.
70. J. M. Leger, B. Beden, C. Lamy, and S. Bilmes, *J. Electroanal. Chem.* **170** (1984) 305.
71. B. Beden, C. Lamy, N. R. de Tacconi, and A. J. Arvia, *Electrochim. Acta* **35** (1990) 691.
72. J. M. Feliu, J. M. Orts, A. Femandez-Vega, A. Aldaz, and J. Clavilier, *J. Electroanal. Chem.* **296** (1990) 191.
73. J. Clavilier and D. Armand, *J. Electroanal. Chem.* **199** (1986) 187.
74. D. Armand and J. Clavilier, *J. Electroanal. Chem.* **225** (1987) 205.
75. D. Armand and J. Clavilier, *J. Electroanal. Chem.* **233** (1987) 251.
76. F. T. Wagner and P. N. Ross, *J. Electroanal. Chem.* **150** (1983) 141.
77. N. Markovic, M. Hanson, G. McDougall, and E. Yeager, *J. Electroanal. Chem.* **214** (1986) 555.
78. D. Armand and J. Clavilier, *J. Electroanal. Chem.* **263** (1989) 109.
79. S. Motoo and N. Furuya, *J. Electroanal. Chem.* **172** (1984) 339.
80. F. G. Will, *J. Electrochem. Soc.* **112** (1965) 451.
81. K. Yamamoto, D. M. Kolb, R. Kötz, and G. Lehmpfuhl, *J. Electronanal. Chem.* **96** (1979) 133.
82. R. R. Adzić, W. E. O'Grady, and S. Srinivasan, *Surf. Sci.* **94** (1980) L191.
83. D. Dickertmann, F. D. Koppitz, and J. W. Schultze, *Electrochim. Acta* **21** (1976) 967.
84. R. M. Lshikawa and A. T. Hubbard, *J. Electroanal. Chem.* **69** (1976) 317.
85. J. Clavilier and J. P. Chauvineau, *J. Electroanal. Chem.* **100** (1979) 461.
86. M. P. Sumino and S. Shibata, *J. Electroanal. Chem.* **322** (1992) 391.
87. R. M. Cerviño. W. E. Triaca, and A. J. Arvia, *J. Electroanal. Chem.* **182** (1985) 51.
88. W. A. Egli, A. Visintín, W. E. Triaca, and A. J. Arvia, *Appl. Surf. Sci.* **68** (1993) 583.
89. G. Milazzo, *Electrochemistry. Theoretical Principles and Practical Applications*, Elsevier, Amsterdam, the Netherlands (1963).
90. J. C. Canullo, W. E. Triaca, and A. J. Arvia, *J. Electroanal. Chem.* **175** (1984) 337.
91. R. M. Cerviño, W. E. Triaca, and A. J. Arvia, *J. Electrochem. Soc.* **132** (1985) 266.
92. J. C. Canullo, W. E. Triaca, and A. J. Arvia, *J. Electroanal. Chem.* **200** (1986) 397.
93. W. E. Triaca, T. Kessler, J. C. Canullo, and A. J. Arvia, *J. Electrochem. Soc.* **134** (1987) 1165.
94. A. Visintín, J. C. Canullo, W. E. Triaca, and A. J. Arvia, *J. Electroanal. Chem.* **239** (1988) 67.
95. A. Visintin, J. C. Canullo, W. E. Triaca, and A. J. Arvia, *J. Electroanal. Chem.* **267** (1989) 191.
96. S. L. Marchiano, L. Rebollo Neira, and A. J. Arvia, *Electrochim. Acta* **35** (1990) 483.
97. A. R. Despic and K. I. Popov, *J. Appl. Electrochem.* **1** (1971) 275.
98. G. Somorjai, *Surf. Sci.* **92** (1980) 489.

99. N. Furuya and S. Koide, *Surf. Sci.* **220** (1989) 18.
100. L. A. Kibler, A. Cuesta, M. Kleinert, and D. M. Kolb, *J. Electroanal. Chem.* **484** (2000) 73.
101. J. A. Davies, T. E. Jackman, D. P. Jackson, and P. R. Norton, *Surf. Sci.* **109** (1981) 20.
102. E. Lang, W. Grimm, and K. Heim, *Surf. Sci.* **177** (1982) 169.
103. K. Wu and M. S. Zei, *Surf. Sci.* **415** (1998) 212.
104. A. Al-Akl, G. A. Attard, R. Price, and B. Timothy, *J. Electroanal. Chem.* **467** (1999) 60.
105. J. Clavilier, J. M. Orts, and J. M. Feliu, *J. Phys. IV* **4** (1994) 303.
106. A. Borg, A.-M. Hilmen, and E. Bergene, *Surf. Sci.* **306** (1994) 10.
107. L. D. Marks, *Phys. Rev. Lett.* **51** (1983) 1000.
108. Y. Kuk, L. C. Feldman, and I. K. Robinson, *Surf. Sci.* **138** (1984) L168.
109. E. Yeager, A. Homa, B. Cahan, and D. Scherson, *J. Vac. Sci. Technol.* **20** (1982) 628.
110. R. Michaelis and D. M. Kolb, *J. Electroanal. Chem.* **328** (1992) 341.
111. G. Beitel, O. M. Magnussen, and R. J. Behm, *Surf. Sci.* **336** (1995) 19.
112. Y. Uchida and G. Lehmpfuhl, *Surf. Sci.* **243** (1991) 193.
113. E. Vlieg, I. K. Robinson, and K. Kern, *Surf. Sci.* **233** (1990) 248.

11 Surface Chemistry of Bimetallic Catalysts

C. Fernando Zinola

CONTENTS

While attempting to use platinum in fuel cells, it has been demonstrated that its surface exhibits important electrocatalytic activities toward the oxidation of organic compounds. However, this effect can sometimes be enhanced by the use of bimetallic surfaces [1–10]. The physical mixture and the electronic interaction of the alloy components lead to a modification in the interaction between the adsorbate and the substrate in an electrocatalytic reaction. As a consequence of the structural changes at the single crystal surfaces during the electrochemical activation (examined with *in situ* STM) [11], it has been demonstrated that most of the catalysts are constituted by randomly oriented islands [12–14].

The modification of platinum catalysts by the presence of ad-layers of a less noble metal such as ruthenium has been studied before [15–28]. A cooperative mechanism of the platinum:ruthenium bimetallic system that causes the surface catalytic process between the two types of active species has been demonstrated [18]. This system has attracted interest because it is regarded as a model for the platinum:ruthenium alloy catalysts in fuel cell technology. Numerous studies on the methanol oxidation of ruthenium-decorated single crystals have reported that the Pt(111)/Ru surface shows the highest activity among all platinum:ruthenium surfaces [21–26]. The development of carbon-supported electrocatalysts for direct methanol fuel cells (DMFC) indicates that the reactivity for methanol oxidation depends on the amount of the noble metal in the carbon-supported catalyst.

11.1 INTRODUCTION

The preparation of bimetallic surfaces can be performed by physical, chemical, and electrochemical methods depending on the proportions of both the metals that are used. Electrodeposition methods are advantageous because of their simplicity and selectivity, as it is possible to screen rapidly the different

245

"metal 1" to "metal 2" ratios based upon their reactivity. The metallurgical method is the most direct for preparing bimetallic electrodes by melting the elements in a furnace until homogeneity is achieved. Though this method is the simplest, it is the most expensive as well. However, the samples resulting from this method are useful for either technical or fundamental studies, and the platinum alloys were used for this purpose three decades ago [29]. The references for the structure of the alloys can be found elsewhere [30].

A significant problem in alloys is the difference in the composition of their surfaces with respect to their bulk. The enrichment of one element with respect to the other on the surface compared to the bulk is called "surface segregation" and is a widespread phenomenon in bimetallic compounds. The most important property that has to be determined in an alloy is its exact composition at the surface and in the bulk, in relative number of atoms per unit area.

11.1.1   UHV *Ex Situ* Spectroscopic Techniques

There are many methods for studying the composition of the outermost layer of alloys that involve low- and medium-energy electrons emitted or scattered from the surface. The most important methods that are surface sensitive are low energy ion scattering (LEIS) and x-ray photoelectron spectroscopy (XPS). The first method is extremely sensitive owing to its large cross section since the ions scattering from the first layer is almost one million times more sensitive than those of any other layer below. In this method, a small fraction of ions are scattered elastically from the atoms at the surface with an energy loss that is described by a binary single scattered model [31]. The energy loss of the scattering ion is related to the masses of both the surface atom and the positive incident ion, and the angle of the emergent ion. The resolution increases with the mass of the incident cation (neon being the best), and the backscattering can be minimized by using an angle higher than 90° for the scattered ion. The use of argon gives rise to problems because of the reduced cross section and the multiple scattering caused by the multiple sputtering.

In the case of XPS, the surface sensitivity is characterized by the mean free-path length of the photoelectrons of a few nanometers that applies for up to 10 atomic layers. On the other hand, LIES occurs when a beam of low electron energy (from 100 eV up to 1 keV) is applied to the surface, and the energy of the elastically scattered ions is analyzed later. The surface sensitivity is largely enhanced because of the large ion neutralization that penetrates into the bulk of the sample. Only the top layer is able to respond to this signal.

Apart from ascertaining the true composition of the binary alloy at its surface, it is important to identify the distribution of the atoms, which depends on the preparation method and the condition of the surface. There are three main situations: (a) ordered and systematic arrangements, (b) statistical distributions, and (c) island formations. The formation of one or the other type of situation depends on the nature of both the metals and the presence of the adsorbable species in the environment. As described in other chapters, the strong adsorption of a species can perform a surface reconstruction so that the transformation from one case to another occurs. In addition, the degree of surface coverage by the new species on the surface can change the structure from a standard to that of an island formation, when going from low to large surface coverage values. Further, the configuration of the adsorbate (ontop, bridge, or hollow positions) can change because of the different values of the surface coverage. Moreover, the electronic structure and the chemical properties of these adsorbates will be different when the substrate has an ordered arrangement or an island formation.

Among all cases of interest in electrocatalysis, platinum alloys deserve special attention. An interesting report on approximately 70% of platinum:ruthenium alloys is found in [32]. The annealed surface in this alloy, as shown by LIES, gives a spectrum of approximately 92% of platinum on the surface that is sputtered with helium. The same surface to bulk concentration can be obtained by using argon ions. Even in this case, we can have a catalytic effect (despite having almost entirely platinum in the first layer) because the second layer confers different electronic properties to the uppermost stratum.

11.1.2 SURFACE SEGREGATION MODELS

For academic purposes, it is worth mentioning the use of single crystal alloys, where low-energy electron diffraction (LEED) determines the composition and crystallographic structure at the topmost layer and some lower layers as well. Gautier et al. [33] provide an example of the use of LEED on nearly exact surface-energy alloys of $Pt_{50}Ni_{50}$ and $Pt_{78}Ni_{22}$ on Pt(111). They found that the alloy exhibits a highly structured composition oscillation on the first three atomic layers of its single crystal. This is a consequence of the effect of size, where the atom near the surface tends to be the larger of the other two [34]. Therefore, surface segregation is important in electrocatalysis, and the composition profiles below the surface of an alloy explain the reactivity of the catalysts.

There are many theories on the mechanism of the segregation effect that suggest either a chemical or an electronic mechanism or both types of mechanisms. However, it seems that the most reliable mechanism is electronic as proposed by Mukherjee and Morán [35]. This electronic model calculates the chemical properties of the pure constituents from their physical parameters and then estimates those of the alloys. It employs the tight-binding electronic theory, the band filling of the density of states, and the bandwidth of the pure components for the calculations. However, it seems that the 2D Monte Carlo simulations produce better results by using the embedded atom and superposition methods. The latter allows for the calculation of the compositions from the relative atom positions, and the strain and the vibrational energies has been reviewed for 25 different metal combinations in [36]. It was also possible to predict composition oscillations as a consequence of the size mismatch.

On the other hand, the chemical models are of two types: the classical macroscopic thermodynamics that considers the excess surface energy, the enthalpy of combination, and mixing entropy; and the bond-breaking model that considers a minimization of the bond energies and the nearest coordination number to the resulting energy. From these, the segregated metal at the surface is estimated along with the lowest sublimation energy enthalpy. In order to apply these chemical models, some chemical data for the pure components are provided by the electronic theories. Some theoretical predictions on the surface segregations in platinum alloys were made in various papers [32,35–39]. Not many of these predictions were corroborated experimentally, but it seems that enrichment in platinum is a general outcome. One such prediction is for Pt_3Sn that exhibits one of the largest activities toward carbon monoxide oxidation. The alloy is an extremely exothermic compound with a formation enthalpy of -50.2 kJ mol^{-1} [40,41]. These studies validated the predictions of the bond-breaking model for the segregation of 50% of tin enrichment on the surface, first for polycrystalline samples [41] and then for (111) and (200) single crystal surfaces [38]. For the first single crystals, the absence of segregation was well predicted, but tin enrichment from the second layer atoms to the Pt(200) surface was proved wrong. However, it seems that the crystallization is ended by the formation of double-height steps from a (100) plane. Similar results were found for Pt(220) surfaces.

The bond formation between an electrode surface and an adsorbate is accompanied by a substantial change of the electronic properties of this surface, and by the change of the local density of the states including the density of the vacant state structure. The latter indirectly participates in the adsorption processes and the electrocatalytic reactions. In particular, one could expect that the peculiarities of the density of the vacant states of an electrocatalyst would correlate with its catalytic properties. The calculation of the local density of the states for platinum single crystals can be done with the modified tight-binding equation method [42]. The same can be performed with the second metal, and then a combined semi-empirical methodology can be used to obtain the predictions. However, the semi-empirical weight contribution will depend on the nature of each metal.

The threshold excitation phenomenon implies the presence of vacant electronic states at the sample surface that are available for the allocation of both the primary and the excited electrons and directly relates to the local density of the states. The most prominent features of these local states can arise from the methodology used in their determination such as *disappearance potential spectroscopy* (DAPS) or *ultraviolet photoelectron spectroscopy* (UPS). In the first case, the prominent features are attributed to the 5d-electrons with almost all of them settling below the Fermi level. The local density of the state at the Fermi level is relatively low, and it further decreases [43]. At these circumstances, the p-states play a significant role. In the second case, photoemission occurs from the core levels and from the valence band of the substrate (due to the overlapping of the outer-shell wave functions). When this is compared with the XPS technique, the fine structure of the valence band is accessible due to high resolution. The spectrum is characterized by the spin orbit and the crystal-field-split 5d states. The high density of the states is again at the Fermi level and involves a large valence band with a fine structure due to contributions of the different metals. In the presence of a second atom, significant changes appear on the electronic population of both the components. The decrease in the amounts of the platinum produces a splitting of the 5d electron states, and the density of the states as the Fermi level decreases. The opposite occurs in the other metal because of its binding energies as an outcome of their charge redistribution. When the adsorbed species are present, there is a photoemission effect in the same valence band. The change in the electronic properties of the metal is called the *ligand effect*, which is responsible for the modification in the catalytic properties of the surface [44].

Besides the composition of the sample, the electronic population density in each state will also depend on the geometric distribution of each of the components. The changes in the electronic distribution at the local density of the states due to a second metal was certainly expected because the presence of the hydrogen on the clean platinum (involved in the formation of the hydrogen: platinum bond) affects the vacant states of the platinum atoms that are arranged in the second layer [45]. In particular, the hydrogen adsorption state is assumed to occupy the hollow sites of the single crystal surface.

11.2 PROBLEM OF POTENTIAL CONTROL AND DOUBLE LAYER IN PLATINUM–METAL ALLOYS

A lot of research on platinum–rhodium alloys has been done with different experimental methods and by new studies on the segregation behavior of $Pt_{25}Rh_{75}$(111), (110), and (100) with LEED and LEIS [46,47]. However, the problem is whether the different components in the bulk reveal an ordering tendency [48] or show a preference for the phase separation [49] is particular of each system. The calculations using the local density approximation method indicated that the density of the states at the Fermi level is very similar in the platinum and rhodium atoms, causing no chemical contrast. Platzgummer et al. [47] studied $Pt_{25}Rh_{75}$(110) to analyze the composition and structure of the surface at room temperature as well as at elevated temperatures. They found a strong platinum enrichment up to 80% of the surface, which exhibits a (1×2) missing-row reconstruction. Annealing above 750°C leads to a reversible phase transition on the hot surface from (1×2) to (1×1).

The spectroscopic verification of a stable alloy surface in UHV (ultrahigh vacuum conditions) has to be done at an open circuit, where, in the case of a single crystal surface, it means a stable double layer. However, when the single crystal has to be transferred to the electrochemical cell to perform adsorption or electrode reactions, the persistence of this situation is not obvious. A constant correlation between the open circuit potential and the electron binding energy of the double layer components (anion, solvent, etc.) has to be maintained [50]. When the rest potential lies at the onset of the oxide formation, there is a further stabilization, but this can produce spontaneous dissolution of the less noble metal of the alloy. Thus, air contact has to be avoided.

The persistence of the double layer at the platinum alloy interface can be followed by the UPS or the XPS signals of oxygen (1s) (at 530 and 532 eV) arising from the water and the supporting electrolyte anion. The shift of the signal toward lower energies (i.e., 530 eV) is an indication of the oxygen in the metal oxide structure. The decrease in the binding energies as a function of the immersed rest potential is expected [50]. The problem of the oxygen-superposed signals is important, as the functionality with the electrode potential is not possible to determine until the rest potential reaches the onset of the oxide formation.

The electrochemical characterization of platinum alloys is performed by cyclic voltammetry, with a careful cycling procedure of no more than three profiles. Moreover, the upper potential limit of the scan has to be recorded with care, and is defined by that of the less noble component of the alloy. The scan potential value has to be always lower than 0.05 V s^{-1} to avoid large pseudocapacitive currents that promote surface enrichment on the platinum. The electrochemical description of the surface can be also performed on single crystal alloys. Besides this type of characterization, it is also possible to use the same experimental setup for surface alloy preparation by a potential-controlled deposition. Better cleanliness can be achieved with the help of a UHV chamber as studied for the Sn/Pt(111) surface alloy that shows the same structure as that of the bulk after annealing [51].

It is well known that when a metal film of a few atomic layers is deposited on the surface of a second metal, the diffusion phenomena lead to the formation of the surface alloys. This can happen when the thickness of an atomic layer or the multilayer epitaxial alloys is very thin [52]. The case of platinum:tin is special because it produces exothermic intermetallic compounds. Some LEED studies have been performed, where the formation of a single atomic layer in the alloy phases has been demonstrated [53]. It has been also confirmed later [51] that the moderate annealing of the deposited tin larger than a few monolayers causes the formation of an ordered multilayer surface alloy. This phase at the termination of the (111) oriented surface of the platinum:tin intermetallic compound has the same structure and composition as the bulk. After evaporating the tin in amounts of up to 5 monolayers at the room temperature, a progressive attenuation of the Pt(111) LEED pattern until its complete disappearance was observed. The x-ray photoelectron diffraction for the tin 3d peak of the deposited films showed no detectable modulation in the intensity at any coverage. Thus, under these conditions, the tin develops a layer with no well-defined epitaxial relation with the substrate. The annealing of the deposited films did not lead to the ordering of the tin layers but to the formation of the various alloy phases. Alloying was detectable by XPS from comparing the tin 3d peaks of the Sn/Pt(111) system with those of the bulk of the Pt$_3$Sn(111) sample. For the bulk alloy, the tin 3d peaks were found to be shifted by 0.3 eV toward the higher binding energy with respect to the same peaks in the bulk tin. After the deposition of 0.2 ML of tin and the subsequent annealing at 1000 K, a (2×2) LEED pattern was observed. Under these conditions, the tin 3d peaks were shifted by 0.3 eV with respect to the deposited tin, which is typical of alloying. After annealing 0.3 to 1 ML of tin at temperatures within the 800–1000 K range, a ($\sqrt{3} \times \sqrt{3}$) R30° LEED pattern and the same shift of the tin 3d peaks were observed. A (2×2) LEED pattern was observed after depositing the tin in the 3–5 monolayer range and at annealing temperatures from 400 to 600 K. Under these conditions, evidence for alloying due to the shift of the tin 3d peaks was observed [51].

The case of Sn–Pt(111) helps to understand the phenomena that occur in the interplay of the single-layer and the multilayer alloy formation. Other known cases are Au–Cu(100) [52] and Al–Ni(100) [54]. In other cases, such as Co–Pt(111) [55], only the multilayer surface alloys have taken known to form.

11.3 SURFACE DECORATION BY UNDERPOTENTIAL AND SPONTANEOUS DEPOSITION

Many efforts have been undertaken to enhance the electrocatalytic performance of catalytic reactions that have technological importance. The case of organic fuel electrooxidation is a major point for study, especially the possibility of achieving long-term, less-polluting fuel cells. In the case of methanol electrooxidation, the reaction occurs by a self-poisoning mechanism, so it is clear that the catalysts' performances must be improved to impede the formation of carbon

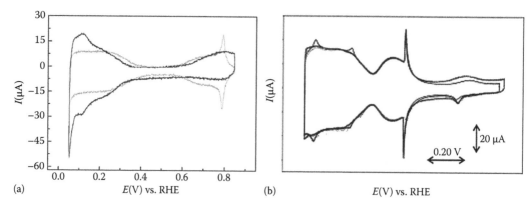

FIGURE 11.1 Cyclic voltammograms (dashed lines) run from 0.05 to 0.80 V at 0.050 V s^{-1} of (a) spontaneously deposited ruthenium submonolayer on clean Pt(111) (solid lines) from 120 s. 0.5 mM Ru(III) Cl$_3$ solution in 0.1 M perchloric acid; (b) spontaneously deposited osmium submonolayer on clean Pt(111) (solid lines) from 60 s. 0.5 mM Os(III) solution in 0.2 M sulfuric acid.

monoxide-type species. The oxidation of these adsorbed residues is accomplished by the co-adsorption of the oxygen-containing species arising from the discharge of water [56,57].

Platinum metal is the most efficient pure catalyst for the sequential dehydrogenation reaction in an acid medium. However, for the complete oxidation process, the water-discharge products are present at potentials larger than 0.6 V [58]. Therefore, the surface modification of the polycrystalline and the single-crystal platinum surfaces by using metal deposits at the sublayer or the monolayer levels, instead of the metal alloys, has also been developed. These new electrocatalysts are able to oxidize organic fuels, such as methanol, at lower potentials than bare platinum. The better performance of binary catalysts, such as platinum:tin, platinum:rhenium, platinum:molibdenum, platinum:ruthenium, and platinum:osmium, can be explained by two different effects: the low onset potentials for water adsorption on the foreign metal and the new electronic surface configuration of platinum atoms that weaken the chemical bonds between platinum and the adsorbed residue (Figure 11.1) [59,60].

11.3.1 Metal Modifiers at an Electrocatalyst

As explained above, the presence of foreign metals on noble surfaces can be stabilized for technical uses by the formation of alloys [61–64]. However, the preparation of the surface metal modifiers is also attractive for academic purposes and technical uses. The surface metal modifiers can be accomplished by different methods: underpotential deposition (*upd*) [24,65], overpotential deposition (*opd*), co-deposition [66,67], and metal adsorption [24]. In all of these, the application of a potentiostatic or galvanostatic pulse is required; however, good results (especially for achieving low coverages) have been obtained using the spontaneous deposition process [68–74]. This is a very simple process that does not require any sophisticated electrochemical equipment and does not obey the ohmic drops at low concentration depositions. Large values of coverage can be reached by consecutive spontaneous depositions, as has been demonstrated for ruthenium on Au(111) and for tin on *pc* platinum [74,75]. Much research has been conducted on ruthenium and osmium, and some on other ad-metals since the pioneering paper of Janssen and Moolhuysen [68].

11.3.1.1 Silver and Mercury Deposition on Noble Metals

Silver deposition has been studied by a large number of techniques [76–82], and most of them are on gold and platinum in sulfate or perchlorate electrolytes [76,82–86,87–91]. The ageing effects were also studied by using potentiostatic tests [89,90] for silver electrochemical deposition on Pt(111) and *pc* platinum, where the splitting of the *upd* peaks was observed.

Two different stages for silver deposition on platinum can be described: one at 1.1 V vs. RHE responding to a silver–platinum alloy electrodissolution (overlapped with the oxygen electroadsorption at free platinum sites) and the other at 0.65 V due to the silver oxidation (from the onset of the bulk deposition process) deposited on the former surface alloy [88,89]. The former process splits into two peaks when the potentiostatic ageing is performed. The spectroscopic techniques such as XPS and ARXPS (angle resolved x-ray photoelectron spectroscopy) were used to determine the chemical composition of the silver films on the platinum in an acid solution [92]. The technique was not able to discern between the presence of silver oxides and sulfates, only an energy shift of the clean silver $3d_{5/2}$ band at a *upd* level of -0.5 eV was detected.

A silver monolayer on Pt(100) that was detected using Auger electron spectroscopy and cyclic voltammetry tends to outline a three-dimensional structure [93]. Besides, by using online cyclic voltammetry with STM images [94], a double voltammetric contribution was found on an unreconstructed Pt(100) just before the bulk deposition process. Thus, two silver layers with characteristic and distinct charge densities form by two distinct and independent processes. Similar results on Pt(111) were found by Herrero et al. [95] where the first deposition of a 1.25 monolayer occurs between 0.85 and 0.69 V vs. Ag/AgCl, the second deposition of a 0.2 monolayer occurs between 0.69 and 0.45 V, and the third peak of a 0.75 monolayer occurs between 0.45 and 0.36 V. On the Pt(110) surface, the adsorption/desorption of silver and oxygen is not distinguishable in the more positive region of the voltammogram. The response on the polycrystalline surfaces is shown in Figure 11.2, for the electrochemically deposited ($\theta_{Ag} = 0.80$) and for the spontaneously deposited ($\theta_{Ag} = 0.30$) silver. The difference arises because of the large surface coverage achieved by the electrochemical deposition.

Silver electrodissolution produces a sharp peak from the silver–silver domains at 0.65 V along with three distinct voltammetric peaks corresponding to the *upd* processes: one at 0.89 V, and two near each other at 1.04 and 1.15 V. The latter peaks are associated with the splitting of the 1.1 V peak due to the first stages of silver–platinum alloy formation. The positive shift peak corresponds to the oxidation of silver from the inner platinum lattice (from the alloy), and the peak at 1.04 V is probably due to the silver oxidation process from the outer silver–platinum interaction. The peak

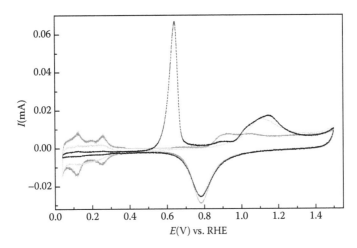

FIGURE 11.2 Complete cyclic voltammetric profile of stabilized spontaneous silver deposition (light gray line) and electrochemical deposition (dashed gray line) on polycrystalline platinum run from 0.05 to 1.50 V at 0.010 V s^{-1} in 1 M sulfuric acid solution. The continuous black lines represent the cyclic voltammetry of smooth platinum. The 60 s spontaneous and electrochemical depositions (at 0.10 V for 5 min) of silver were performed using 0.1 mM silver sulfate $+1$ M sulfuric acid.

at 0.89 V probably corresponds to the early stages of the platinum oxide formation in the presence of silver adatoms. On the other hand, the spontaneous deposition of silver at low scan rates shows similar voltammetric features when compared with those obtained at 0.10 V s^{-1} (only a single peak at 1.1 V). The inhibition of platinum oxide formation is clear in the cyclic voltammetric response, that is, within the 0.85–1.05 V range, and, hence, an important suppression of the early stages of platinum oxide caused by the presence of silver adatoms is observed. The voltammetric profile of the hydrogen *upd* region is slowly restored, while the peak associated with the oxidation of silver *upd* adatoms decreases. Spontaneously deposited silver exhibits larger stability upon potential cycling toward the high positive potentials.

Mercury deposition on highly oriented carbon at low overpotentials takes place as a progressive nucleation and a diffusion-controlled three-dimensional growth [96]. It has to be noted that the mercury *upd* on *pc* platinum exhibits similar features of silver even at a different surface roughness [88]. In the case of smooth platinum, mercury deposition from a 0.1 mM mercurous salt in perchloric acid shows two anodic peaks at 0.10 V s^{-1}, one at approximately 0.66 V (due to bulk oxidation) and the other at 1.07 V (assigned to the mercury *upd*). However, only a single cathodic peak at 0.75 V, including a shoulder assigned to the simultaneous oxygen electrodesorption and the initiation of the mercury *upd*, is observed. On the other hand, on the columnar surfaces, four anodic peaks are seen at low scan rates (0.005 V s^{-1}). The first is due to bulk anodic stripping, the second is due to oxygen adsorption, the third is due to mercury *upd*, and the last, presumably, is due to the mercury dealloying. For increasing scan rates, the columnar electrodes show only one anodic peak due to the overlapping of the oxygen electroadsorption and the stripping of the mercury *upd* [88,91]. It was also found that the surface and the subsurface diffusion processes into the bulk of the metal occur in the case of mercury.

Mercury deposition on Au(111) from the mercuric-ion-containing solution covers a wider potential range than on platinum [97–100]. The multiple voltammetric peaks reported in the literature are associated with crystallographic orientation, which, in the case of the single crystals, are strongly modified by the presence of adsorbable anions, such as (bi) sulfate or halides [101,102]. An ordered mercurous sulfate bilayer structure was found to be formed by the partial charge transfer at more negative potentials. When mercury is totally discharged, two additional ordered hexagonal mercury ad-layers develop.

11.3.1.2 Molybdenum and Tungsten Deposition on Noble Metals

Molybdenum and tungsten depositions on platinum have not been deeply studied; however, the electrochemical features at the beginning of the hydrogen sorption region have been well documented [101,102]. This behavior has been formerly attributed to an electrochemical redox couple of oxygen-containing either Mo(VI)/Mo(V) or W(VI)/W(V) species for high solution concentrations in acid media [103–105]. These redox couples are able to promote the oxidation of small organic molecules such as methanol. Some authors have explained [106] the catalytic effect by oxidation induction, caused by the presence of the Mo(VI)/Mo(IV) redox couple in the substoichiometric lower valence oxides such as MoO_x ($2 < x < 3$) along with the proton spillover effect. Accordingly, optimum activity can be reached in a 3.7 M sulfuric acid solution, where methanol oxidation takes place at 0.2 V, more negative than that on bare platinum.

Step decoration of molybdenum on platinum single crystals has also been studied [27]. Carbon monoxide is able to oxidize to carbon dioxide on molybdenum/Pt(111) at 0.2 V (demonstrated by the DEMS experiments). Vapor deposition of molybdenum on Pt(111) at 400 K produced more ordered molybdenum over-layers, with an improved layer-by-layer growth resulting from the higher molybdenum adatom mobility. The first layer can be imagined as growth through a fast migration of adatoms over platinum to the edges, yielding molybdenum islands that are nucleated at the steps [107]. The molybdenum/platinum alloys exhibit catalytic performance toward the oxidation of carbon monoxide and hydrogen fuel in acid media [108]. It was demonstrated that $Pt_{77}Mo_{23}$ has a similar reactivity toward both the fuels when compared with platinum/ruthenium. Moreover, a large pseudocapacitance contribution is observed to be assigned to the $MoO(OH)_2/MoO_3$ redox surface couple as proved by XPS measurements [108]. However, it is stated that when molybdenum is used in the form of adatoms, the catalytic activity is not improved [109].

In the case of tungsten species, adsorption from the tungstate ions on solid electrodes is strongly dependent on the nature and the atomic structure of the substrate. Conflicting views on the reduction mechanism of tungstic oxide (WO_3) exist, as the reduction potential lies in the vicinity of the hydrogen-evolution reaction [110]. Poltorak [111], Tsirlina et al. [112], and Kobozev et al. [113] indicate that the reduction mechanism depends upon the primary formation of free hydrogen atoms at the electrode surface, while Lukovtsev [114] and Bagotskii and Iofo [115] show that the electrode mechanism involves a direct electron acceptance. In addition, differing degrees of reversibility have been reported for the oxidation states of tungsten [116]. The tungstate ions may be reversibly reduced at platinum, rhodium, and mercury electrodes in phosphoric acid according to the reaction $WO_4^{2-} + e^- \leftrightarrow WO_4^{3-}$. Similarly, for molybdenum, the catalytic effect is assigned to the redox systems as explained above.

The addition of tungsten by electrochemical deposition has been studied elsewhere [74]. In this study, the lowest amount of carbon monoxide adsorbates were detected together with the largest methanol oxidation currents with a surface coverage of 15%. Since the total anodic charge density for methanol oxidation was less than that obtained on bare platinum, this indicates that the surface does not have a tendency to create surface poisons.

11.3.1.3 Ruthenium and Osmium Deposition on Noble Metals

Ruthenium and osmium depositions on platinum are of special interest with reference to methanol anodic oxidation on fuel cells and were first reported in the early 1960s [116,117]. Among the different methods of deposition of ruthenium or osmium on platinum [116–121], the spontaneous deposition method is attractive because of its simplicity and the fast achievement of a surface concentration plateau (reached in seconds) [26,122,123]. Contrary to the reversible behavior of the electrochemically deposited ruthenium on noble metal substrates, the films formed by spontaneous deposition normally do not dissolve easily from the metal surface. They are very stable and normally change to stable hydroxides and oxides when the electrode potential is increased.

The study of a sublayer or a monolayer of ruthenium or osmium on platinum allows for an understanding of the role of the surface composition in the electrocatalysis of organic fuels, such as methanol oxidation. Interesting papers about ruthenium and osmium deposition on platinum single crystals and *pc* surfaces have been published [26,124–126].

Ex situ examination of the electrode surfaces by STM images, especially in the case of Pt(111) surfaces at the submonolayer levels, shows the formation of islands [127] with maximum of 0.20 ML of coverage for ruthenium (at 120 s of exposure) and 0.15 ML of coverage for osmium (at 60 s of exposure). Depending on the osmium coverage, the island diameter varies from 2 to 5 nm [128], but in the case of ruthenium [129] the islands appear to be uniformly distributed along the surface with an average value of 3 nm. The island density increases with the ruthenium coverage values, but in the case of osmium, there is no optimum island size [124,128]. On the other hand, knowledge of foreign metal adatom concentration and the nature of the deposited species at the potentials of interest is important for the understanding of the mechanism of methanol oxidation. Thus, interfacial and surface analysis techniques were used for this purpose [125,129] and, particularly, for the electrodeposited and melted phases of these metals with platinum.

Figure 11.3 shows the cyclic voltammetric profiles of the spontaneously deposited ad-layers at 0.20 ML of coverage for ruthenium (at 120 s of exposure) and 0.15 ML of coverage for osmium (at 60 s of exposure) from freshly prepared diluted acidic solutions. The coverage was calculated from the voltammetric contour up to only 0.60 V vs. RHE in the supporting electrolyte. Figure 11.3a clearly shows the co-adsorption of the chloride as a current breaking slope at 1.4 V. Moreover, it can also be seen in Figure 11.3b that the oxidation peak of the osmium species is only seen in the first positive incursion, probably due to the dissolution of the Os^{8+} adatoms at 1.50 V.

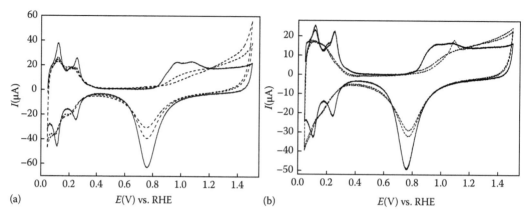

FIGURE 11.3 Cyclic voltammetric profiles of (a) ruthenium and (b) osmium ad-layers (dashed lines) on clean polycrystalline platinum (continuous lines) in 1 M sulfuric acid between 0.05 and 1.50 V at 0.10 V s^{-1}. The foreign metal ad-layers were prepared as explained in the text.

This peak was observed by Wieckowski and coauthors [124] on Pt(111) and was assigned to an intermediate state between the Os^{4+} and Os^{8+} species according to the binding energies of the *XPS* measurements.

The spontaneously deposited ruthenium layer has to be electrochemically stabilized by potential cycling in the hydrogen adsorption/desorption region at low scan rates. Thus, the following situation can occur [124]:

$$H^+_{(ac)} + Pt + e^- \leftrightarrow [HPt]_{ad} \tag{I}$$

$$RuCl_2(H_2O)_4^+ \leftrightarrow RuO_2 + 2H_2O + 2Cl^- + 4H^+ + e^- \tag{II}$$

In another paper [26], it is proposed that the first specific adsorption of the chloride (as it was found voltammetrically in Figure 11.3) after the interaction of the Ru(III) ions with three platinum atoms was through

$$Cl^-_{(ac)} + Pt \leftrightarrow [PtCl]^-_{ad} \tag{III}$$

$$Ru(H_2O)_6^{+3} + 3[PtCl]^-_{ad} + 3e^- \leftrightarrow [3Pt - Ru]_{ad} + 3Cl^- \tag{IV}$$

Any of these mechanisms are likely, but further research is required to achieve clarity.

11.4 BIMETALLIC ELECTROCATALYSIS: THE PLATINUM–RUTHENIUM AND PLATINUM–TIN ALLOYS

11.4.1 Carbon Monoxide Oxidation Electrocatalysis

One of the most attractive applications of bimetallic surfaces is for the analysis of their performance on the adsorption and oxidation reactions of interest in fuel cells. It is well known that one of the main limitations during the operation of a methanol fuel cell is the fast depolarization of the anode in the presence of traces of the adsorbed carbon monoxide.

There is a strong difference between the voltammetric profiles of the carbon monoxide oxidative desorption on a pure platinum single crystal in the presence of an ad-metal species and the same in the absence of an ad-metal species. The carbon monoxide adsorption and oxidation are also different depending on the type of the deposited metal as shown in Figure 11.4.

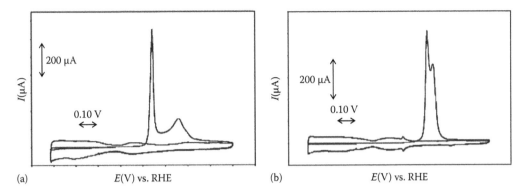

FIGURE 11.4 (a) Adsorbed carbon monoxide anodic stripping from Pt(111)/Ru run in 0.1 M sulfuric acid run at 0.05 V s^{-1}. The ruthenium was spontaneously deposited on Pt(111) from 0.5 mM Ru(III) in 0.1 M perchloric acid for 120 s. (b) Anodic stripping from Pt(111)/Os at similar conditions. The osmium was deposited during 60 s. In both the cases, the carbon monoxide was adsorbed at 0.05 V for 5 min from a saturated sulfuric solution followed by nitrogen purging.

On a clean Pt(111) surface, a single oxidation peak is expected at approximately 0.8 V with a pre-oxidation component of the disordered carbon monoxide adsorbates extending from 0.5 to 0.65 V. In the presence of the ad-layers of ruthenium or osmium, different carbon monoxide oxidation kinetics can be evidenced in the cyclic voltammograms as shown above. The promotional effect caused by the metals is associated with the activation of water to produce the oxidizing agents. The oxidation can occur at the edge of the hydroxyl-containing species or in wider domains far from the carbon monoxide adsorbates, where a surface mobility is later required. This means a slower kinetics of oxidation, and that a second oxidation voltammetric peak has to be observed at higher potentials (Figure 11.4). However, it has been also observed that the same experiment in perchloric acid shifted the onset potentials of both the regions toward lower values (approximately below 0.10 V). The presence of sulfate adsorbates inhibits the carbon monoxide adsorbates' diffusion to the appropriate regions, thus following the Langmuir–Hinshelwood mechanism. It is not as simple as it was thought to be, since the presence of strong adsorbable anions changes the current vs. potential anodic stripping profile drastically.

The atomic distribution of the various metal atoms also plays an important role in the electro-catalysis. Two different distributions containing 8% ruthenium were achieved by two distinct preparation methods: sputtered Pt$_{90}$Ru$_{10}$, and sputtered and annealed Pt$_{70}$Ru$_{30}$ [65,130]. The latter exhibits a completely different carbon monoxide anodic stripping profile, with sharp anodic peak at higher potentials. This was attributed to the ruthenium clustering through the coupling of the carbon monoxide vibration frequencies [131] after annealing as it has also been predicted by the Monte Carlo 2D-simulations [132]. The case of the dissolved carbon monoxide oxidation on Pt$_3$Sn alloys is also interesting [133] as the onset for the carbon monoxide oxidation lies approximately 0.45 V more toward the cathode than on the bare platinum. The problem is that tin dissolution cannot be avoided in the adsorption experiments. The reaction is also structure sensitive since the oxidation on the (111) plane is negatively shifted by approximately 0.1 V with respect to the (110) plane for the same amount of tin. The difference with respect to the platinum:ruthenium alloys is that the carbon monoxide is not adsorbed on the tin, even under UHV conditions [137]. The large activity of this surface has attracted the attention of many authors, and it has been proposed that this activity is the result of the absence of a true competition for carbon monoxide for the same site as in the other cases. Thus, a weakly adsorbed state is expected for the species; that is, there is a reversible behavior of adsorption at low potentials. This effect is more pronounced on the (111) plane than on the other low Miller indices.

11.4.2 METHANOL OXIDATION ELECTROCATALYSIS

A study of methanol adsorption on platinum under UHV conditions or at a gas/solid interface is also of interest. Not many papers dealing with methanol adsorption in a UHV chamber [135,136] are available. The adsorption takes place without a reaction on Pt(111) at low temperatures (100 K), and based on thermal desorption experiments it was concluded that a monolayer of methanol adsorbate desorbs at 180 K. The heat of adsorption of molecular methanol was estimated to be 46 kJ mol^{-1} on unreconstructed Pt(111) [137]. Infrared spectroscopy has been applied for the study of methanol adsorption on Pt(111) [138], and it was shown that a 0.36 monolayer of methanol corresponds to the saturation of the desorption peak found at 180 K. The methanol multilayer coverages were also found, but had different infrared frequencies that were associated with the methyl and C–O stretching modes (Scheme 11.1).

Under UHV conditions, methanol can be desorbed dissociatively as carbon monoxide and molecular hydrogen at temperatures between 200 and 300 K. A stable surface methoxy intermediate ($CH_3O–$) is formed by the scission of the O–H bond [115]. The Pt(110) (1×2) reconstructed surface is the only plane to show the stable methoxy species co-adsorbed with the oxygen, whereas on the (111) and (100) platinum planes the surface combination leads to the formation of carbon dioxide [138].

The first paper on methanol electrocatalysis under UHV conditions was published by Attard et al. [139] on the most active surface, Pt(110). Similar results to those on Pt(111) were found, that is, carbon monoxide and molecular hydrogen, but with a slightly larger methanol surface coverage of $\theta = 0.10$. It was the first time that methoxy species were proposed as intermediates and were different from the carbon monoxide or formyl species proposed earlier by Bagotskii et al. [140]. However, traces of the formyl species were also detected on reconstructed Pt(110) using vibrational spectroscopy, which was able to co-adsorb this species with atomic oxygen [117].

The electrochemical oxidation of methanol on *pc* platinum [25,66] and platinum single crystal surfaces [24,67] in acid media at room temperature has been extensively studied. The methanol electrooxidation occurs either as a direct six-electron pathway to the carbon dioxide or by several adsorption steps, some of them leading to poisoning species, prior to the occurrence of carbon dioxide as the final product. The most convincing evidence of carbon monoxide as a catalytic poison arises from *in situ* IR fast Fourier spectroscopy. An understanding of methanol adsorption and the oxidation processes on modified platinum electrodes can lead to a deeper insight into the relation between the surface structure and reactivity in electrocatalysis. It is well known that the main impediment in the operation of a methanol fuel cell is the fast depolarization of the anode in the presence of traces of the adsorbed carbon monoxide.

The identification of the reaction intermediate has gained the attention of many scientists. The use of infrared vibration spectroscopy has been extensively used for this purpose. The presence of the carbon monoxide–adsorbed intermediate as a poison plays a fundamental role in the understanding of monocarbonaceous oxidation reactions.

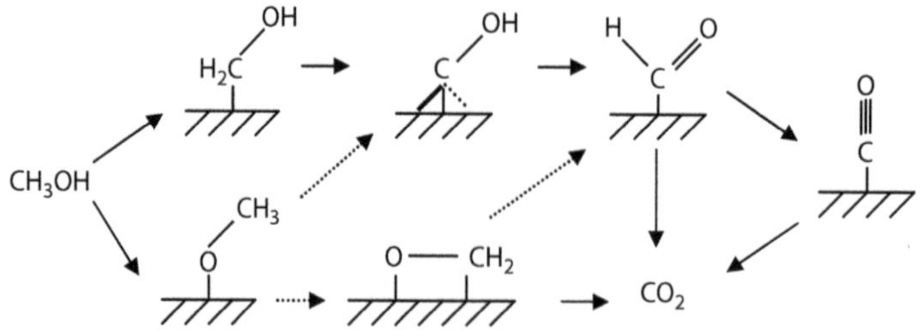

SCHEME 11.1 General scheme of methanol electrooxidation considering series and parallel pathways to form carbon dioxide as the product. The solid and dashed arrow lines are the demonstrated and possible reaction pathways, respectively.

The modification of platinum surfaces by foreign metal atoms promotes the oxidation of methanol either under UHV conditions or in the electrochemical environment. This promotion model has been mainly discussed in electrochemistry by using the "third body model" [140], the "ligand effect" [102], or the "bifunctional effect" [117,118]. A theoretical review on the inclusion of the metal reaction promoters was performed by Anderson et al. [141] and is also discussed in [60]. Similar to platinum alloys, ruthenium and tin are other metals that promote methanol oxidation. The case of ruthenium is interesting since it was also studied under UHV conditions [62,63]. The reaction of methanol on platinum:ruthenium alloys results in the production of carbon dioxide at lower potentials than that on pure platinum. However, the presence of tin in Pt_3Sn alloys only enhances methanol oxidation at low potentials, thereby increasing the carbon dioxide production (and diminishing carbon monoxide production) [141,142]. The addition of tin(II) ions to previously adsorbed methanol produces a fast oxidation process as demonstrated by DEMS experiments [143]. Very little information is available about methanol oxidation on platinum surfaces that are promoted by tungsten [25].

The adsorption of methanol takes place mostly on platinum free sites at potentials lower than 0.6 V and higher than 0.10 V. The mechanism for the spontaneous deposition on noble metals has not been clarified yet. Nevertheless, according to the literature, the simultaneous deposition of platinum and ruthenium from $PtCl_6^{2-}$ and $Ru(H_2O)^{3+}$ occurs first by a hydrogen adatom reduction of the ruthenium complex to metallic ruthenium, which subsequently reduces the chloroplatinate anion to the metallic platinum by the surface oxidation of the ruthenium to RuO_xH_y species [67]. Moreover, Hubbard [150] proposed a sequence for the case of the spontaneous deposition of tin: from tin(II) species, auto disproportion leads to metallic tin and tin(IV) interface species followed by the subsequent surface oxidation of metallic tin to $Sn(OH)_2$.

In general, we can propose that the early stage is the adsorption of methanol, which will be mostly on platinum free sites:

$$Pt + CH_3OH \leftrightarrow Pt(CH_3OH)_{ad} \tag{V}$$

After increasing the electrode potential, the methanol adsorbates oxidize as per the following reactions, depending on the nature of the catalyst and the experimental conditions [125]:

$$Pt(CH_3OH)_{ad} \leftrightarrow Pt(HCO)_{ad} + 3H^+ + 3e^- \tag{VI}$$

$$Pt(HCO)_{ad} \leftrightarrow Pt(CO)_{ad} + H^+ + e^- \tag{VII}$$

The bifunctional mechanism proposes that a foreign metal, M, promotes water decomposition at lower potentials than that of platinum, by the following reactions:

$$M + H_2O \leftrightarrow M(OH)_{ad} + H^+ + e^- \tag{VIII}$$

$$M + H_2O \leftrightarrow M(H_2O)_{ad} \tag{IX}$$

These species also promote the complete oxidation of the organic adsorbate by a recombination of these species (from reactions (IV–V) and reactions (VI–XVII)) through a Langmuir–Hinshelwood mechanism [131]. For example,

$$Pt(CO)_{ad} + M(OH)_{ad} \leftrightarrow CO_2 + H^+ + e^- + Pt + M \tag{X}$$

According to the ligand effect, the energy level of the modified substrate is changed to weaken the bond energy of the carbon monoxide adsorbate to facilitate its oxidation.

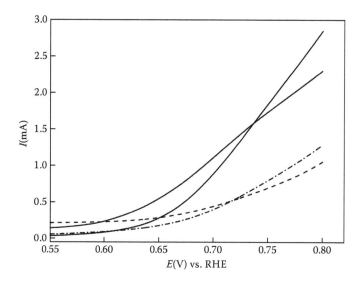

FIGURE 11.5 Positive potential scans of methanol oxidation on different platinum surfaces in 0.5 M methanol +1 M sulfuric acid at 0.01 V s^{-1}; (continuous line) bare polycrystalline platinum; (dotted line) platinum:ruthenium; (dash-dotted line) platinum:osmium; (dashed line) and ternary compound platinum:ruthenium:osmium (obtained by simultaneous spontaneous deposition after 60 s of immersion in a mixed solution).

The best choice of the alloying or co-deposited elements depends on which step of the latter mechanism is the rate-determining step (normally reaction (VIII)). Different results have been obtained with ternary and quaternary combinations of these metals with rhodium, molybdenum, and iridium [104,144–156], so there is a lot left to be explored with regard to this interesting process. In the next figure, the current vs. potential profiles of a positive going scan for the different platinum modified surfaces in acidic methanol solution are shown.

Some interesting features of the spontaneous deposition of metals on platinum are detailed here. In the case of platinum/ruthenium, the onset of methanol solution oxidation lies at the potentials applicable for methanol fuel cells (approximately 0.5 V) and is the largest of all. On the other hand, the oxidation currents observed at platinum/osmium are larger at 0.65 V, after which there is a crossing point, probably due to the start of surface osmium oxidation. A cross point at 0.72 V is observed, from which methanol oxidation on bare platinum exhibits larger currents Figure 11.5.

In spite of the fact that conventional electrochemistry gives valuable information, hybrid *in situ* techniques such as FTIRS (Fourier transform infrared spectroscopy) help in the understanding of the mechanism of electrocatalytic reactions. Figure 11.6 shows for the p-polarized light, the sequential infrared spectra for methanol oxidation at increasing electrode potentials taking 0.05 V as a reference on clean Pt(111) in 0.1 M methanol:perchloric acid. The following signals were found: 2344 cm^{-1} for the production of carbon dioxide starting at 0.5 V, the 2060, 1830, and 1795 cm^{-1} vibrations of on-top, bridge, and triply coordinated carbon monoxide on Pt(111), respectively, but the major compound seems to be the on-top carbon monoxide. Finally, the 1111 cm^{-1} vibration of the perchlorate anion can be clearly seen. From 0.75 V upward, the three types of carbon monoxide adsorbates convert to the bridge bound species. These experiments have been performed using D$_2$O to avoid water species interferences.

In the case of a Pt(111) single crystal covered by spontaneously deposited ruthenium of 50%, the infrared spectra change to those of that in Figure 11.7. As a difference to the infrared spectra on clean Pt(111), the linearly bonded carbon monoxide adsorbate is shifted to 2040 cm^{-1} due to the surface mobility of the adsorbate on platinum:ruthenium islands. On the other hand, there is a

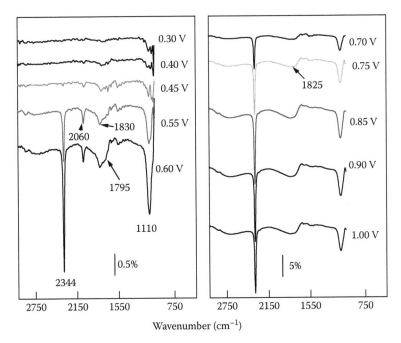

FIGURE 11.6 FTIRS signals for the p-polarized light on clean Pt(111) in 0.1 M methanol:perchloric acid in heavy water. Reference potential 0.05 V.

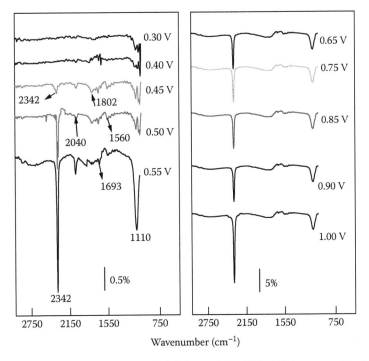

FIGURE 11.7 FTIRS signals for the p-polarized light on Ru:Pt(111) (50% coverage) in 0.1 M methanol: perchloric acid in heavy water. Reference potential 0.05 V.

lower concentration of the bridge carbon monoxide (which appears at 1802 cm^{-1}), and negligible signals were found due to the triply bonded carbon monoxide. However, there is a wide band between 1693 and 1560 cm^{-1}, attributed to the hydroxyl species that is adsorbed on the ruthenium islands. There is no signal that could be attributed to the carbon monoxide adsorbate on ruthenium (no signal appears at 2002 cm^{-1}). One of the most important features is the appearance of carbon dioxide at 0.45 V, that is, at 0.05 V more than the negative potential values, with a decrease in the carbon monoxide adsorbate coverage, which indicates that it acts as an intermediate of the reaction. In any case, there was no detectable amount of the formyl species and no partial oxidation products at all the potentials. However, it has to be taken into account that the experimental conditions, especially the measured time and traces of the oxygen, are definite factors for the detection of this species.

FTIR experiments were carried out with a Bruker Vector 22 spectrometer equipped with a MCT detector. The electrochemical IR cell, fitted with a 60° CaF$_2$ prismatic window, was provided with an inlet and an outlet for the solutions to allow the electrolyte exchange under potential control. For each spectrum, 128 interpherograms were collected at a resolution of 8 cm^{-1}. Parallel (p) and perpendicular (s) polarized IR light were obtained from a BaF$_2$-supported Al grid polarizer. The spectra were represented as a ratio, R/R_o, where R and R_o are the reflectances corresponding to the sample and reference spectra, respectively.

Different electrochemical and spectroelectrochemical spectra were obtained depending on the degree of the surface coverage by the ruthenium species. It seems that three distinct coverages can render a good catalytic performance. The coverages were lower than 0.1, approximately 0.3, and had a mean value of 0.50. Some authors found the highest rate of carbon monoxide oxidation to occur at 50% ruthenium coverage with only negligible amounts of carbon monoxide coverage. Therefore, this surface seems to be permanently unpoisoned, but this does not mean that it is the most active of all. Many papers have found that when the ruthenium coverage is less than 10% [62,63,134] methanol electro-oxidation reaches maximum current densities. The increase in the presence of ruthenium (which at low temperatures does not adsorb carbon monoxide) lowers the surface coverage by the surface intermediate (which can be the same carbon monoxide). Thus, the specific rate constant for the intermediate electrooxidation is decreased. These arguments make us think that the first adsorption and dehydrogenation of methanol to be the main rate-determining step. However, many authors have found that 50% or 33% of ruthenium as the most active surfaces [56,62,63,67,146], since the presence of the second metal can explain both types of the determining step or the first three hydrogen ion discharges, or the oxidation of the carbon monoxide adsorbate can explain the rate-determining step. However, the distribution of ruthenium atoms is probably different from the ordered and alternate arrangement. A real statistical distribution is expected where the maximum amount of the real active centers (three platinum atom sites) account for the presence of only 10% of the second metal. If there is a large population of three contiguous platinum atoms, they serve as sites for methanol dehydrogenation (even if the three protons are not free at once). Then, the new rate-determining step is now the oxidation of the adsorbed carbon monoxide by the hydroxyl groups on ruthenium sites. In summary, two different situations can explain the results and are equally well suited for the experimental results. The objective is in achieving a platinum:ruthenium alloy of 50% coverage with a proper distribution of atoms, thereby rendering the presence of the three platinum active centers.

The other problem is the removal of the fourth hydrogen atom. Some authors have found formyl species [147–153] (see more references above) as the true intermediate, but the real time of measuring the vibrational signals makes the detection quite impossible. Some evidence of this is the detection of partially oxidized products such as the formic acid or methyl formate when the inert atoms modify the platinum surface (iron germanium, cobalt, etc.). The presence of a second active atom is also required in order to avoid total coverage by hydrogen adatoms that inhibit methanol dehydrogenation (methanol adsorbates are not able to compete with hydrogen adsorbates).

REFERENCES

1. J. Clavilier, C. Lamy, and J.M. Leger, *J. Electroanal. Chem.* **125** (1981) 249.
2. J. Clavilier, R. Parsons, R. Durand, C. Lamy, and J.M. Leger, *J. Electroanal. Chem.* **124** (1981) 321.
3. C. Lamy, J.M. Leger, and J. Clavilier, *J. Electroanal. Chem.* **135** (1982) 321.
4. C. Lamy, J.M. Leger, J. Clavilier, and R. Parsons, *J. Electroanal. Chem.* **150** (1983) 71.
5. S.G. Sung, J. Clavilier, and A. Bewick, *J. Electroanal. Chem.* **210** (1988) 147.
6. J. Clavilier, A. Fernandez-Vega, J.M. Feliu, and A. Aldaz, *J. Electroanal. Chem.* **258** (1989) 89.
7. A. Fernandez-Vega, J.M. Feliu, A. Aldaz, and J. Clavilier, *J. Electroanal. Chem.* **258** (1989) 101.
8. J. Clavilier, A. Fernandez-Vega, J.M. Feliu, and A. Aldaz, *J. Electroanal. Chem.* **261** (1989) 113.
9. S. Sun and J. Clavilier, *J. Electroanal. Chem.* **263** (1989) 109.
10. J. Clavilier, J.M. Orts, R. Gomez, J.M. Feliu, and A. Aldaz, *J. Electroanal. Chem.* **404** (1996) 281.
11. N. Kimizuka, T. Abe, and K. Itaya, *Denki Kagaku* **61** (1993) 796.
12. M.S. Zei, N. Batina, and D.M. Kolb, *Surf. Sci.* **306** (1994) L519.
13. J. Clavilier, J.M. Orts, and J.M. Feliu, *J. Phys.* **IV** (1994) C1–303.
14. N. Furuya, M. Ichinose, and M. Shibata, *J. Electroanal. Chem.* **460** (1999) 251.
15. K.A. Friedrich, K.P. Geyzers, U. Linke, U. Stimming, and J. Stumper, *J. Electroanal. Chem.* **402** (1996) 23.
16. S. Cramm, K.A. Friedrich, K.P. Geyzers, U. Stimming, and R. Vogel Fresenius, *Anal. Chem.* **358** (1997) 189.
17. K.A. Friedrich, K.-P. Geyzers, F. Henglein, A. Marmann, U. Stimming, W. Unkauf, and R. Vogel, in: A. Wieckowski and K. Itaya (Eds.), *ECS Proceedings*, Vol. 96–8, The Electrochemical Society, Pennington, NJ, 1996, p. 119.
18. K.A. Friedrich, K.-P. Geyzers, F. Henglein, A. Marmann, U. Stimming, and R. Vogel, *Z. Phys. Chem.* **208** (1999) 137.
19. B.E. Hayden, A.J. Murray, R. Parsons, and D.J. Pegg, *J. Electroanal. Chem.* **409** (1996) 51.
20. B.E. Hayden, J.C. Davies, and D.J. Pegg, *Electrochim. Acta* **44** (1998) 1181.
21. E. Herrero, J.M. Feliu, and A. Wieckowski, *Langmuir* **15** (1999) 4944.
22. A. Crown, I.R. Moraes, and A. Wieckowski, *J. Electroanal. Chem.* **500** (2001) 333.
23. G. Tremiliosi-Filho, H. Kim, W. Chrzanowski, A. Wieckowski, B. Grzybowska, and P. Kulesza, *J. Electroanal. Chem.* **467** (1999) 143.
24. W. Chrzanowski, H. Kim, and A. Wieckowski, *Catal. Lett.* **50** (1998) 69.
25. W. Chrzanowski and A. Wieckowski, *Langmuir* **14** (1998) 1967.
26. T. Iwasita, H. Hoster, A. John-Anacker, W.F. Lin, and W. Vielstich, *Langmuir* **16** (2000) 522.
27. H. Massong, H. Wang, G. Samjeské, and H. Baltruschat, *Electrochim. Acta* **46** (2000) 701.
28. J.C. Davies, B.E. Hayden, and D.J. Pegg, *Surf. Sci.* **467** (2000) 118.
29. H. Binder, A. Koehling, and G. Sanstede, in: G. Sanstede (Ed.) *From Electrocatalysis to Fuel Cells*, University of Washington Press, Seattle, WA, 1972.
30. P.R. Watson, M.A. van Hove, and K. Herman, *Atlas of Surface Structure*, Vol. IA, Monograph No. 5, American Chemical Society Publications, Washington, DC, 1995.
31. W. Heiland and E. Taglauer, Ion scattering and secondary ion mass spectroscopy, in: R.L. Park and M. Lagally (Eds.), *Methods of Experimental Physics*, Vol. 22, *Solid State Physics: Surfaces*, Academic Press, Orlando, FL, 1985.
32. H.A. Gasteiger, P.N. Ross, and E.J. Cairns, *Surf. Sci.* **293** (1993) 67.
33. Y. Gautier, Y. Joly, R. Baudoing, and J. Rundgren, *Phys. Rev. B* **31** (1985) 6216.
34. F. Abraham, *Phys. Rev. Lett.* **46** (1981) 546.
35. S. Mukerjee and J.L. Morán López, *Surf. Sci.* **189–190** (1987) 1135.
36. S. Folies, M.I. Baskes, and M.S. Daw, *Phys. Rev. B* **33** (1986) 7983.
37. A.N. Hanner, P.N. Ross, and U. Bardi, *Catal. Lett.* **8** (1991) 1.
38. A. Atrei, U. Bardi, M. Torrini, E. Zanazzi, G. Rovida, and P.N. Ross, *Phys. Rev. B* **68** (1992) 1649.
39. J. Paul, S.D. Cameron, D.J. Dwyer, and F.M. Hoffman, *Surf. Sci.* **177** (1986) 121.
40. P.N. Ross, The science of electrocatalysis on bimetallic surfaces, in: J. Lipkowsky and P.N. Ross (Eds.), *Electrocatalysis; Frontiers in Electrochemistry*, Chap. 2, Wiley-VCH, New York, 1998.
41. W.M.H. Sachtler and R.A. van Santen, *Adv. Catal.* **26** (1977) 69.
42. V.M. Tapilin, *Phys. Rev. B* **52** (1995) 14198.
43. A.R. Cholach and V.M. Tapilin, *J. Mol. Cat. A* **158** (2000) 181.
44. U. Schneider, G.R. Castro, and K. Wandelt, *Surf. Sci.* **287–288** (1993) 146.
45. B. Pennemann, K. Oster, and K. Wandelt, *Surf. Sci.* **249** (1991) 35.
46. E. Platzgummer, M. Sporn, R. Koller, S. Forsthuber, M. Schmid, W. Hofer, and P. Varga, *Surf. Sci.* **419** (1999) 236.

47. E. Platzgummer, M. Sporn, R. Koller, M. Schmid, W. Hofer, and P. Varga, *Surf. Sci.* **423** (1999) 134.
48. R.E. Lakis, C.E. Lyman, and H.G. Stenger Jr., *J. Catal.* **154** (1995) 261.
49. M. Cyrot and F. Cyrot-Lackmann, *J. Phys. F* **6** (1976) 2257.
50. R. Kötz, H. Neff, and K. Müller, *J. Electroanal. Chem.* **215** (1986) 331.
51. M. Galeotti, A. Atrei, U. Bardi, G. Rovida, and M. Torrini, *Surf. Sci.* **313** (1994) 349.
52. D. Naumovic, A. Stuck, T. Greber, J. Ostenvalder, and L. Schlapbach, *Surf. Sci.* **269–270** (1992) 719.
53. M.T. Paffett and R.G. Windham, *Surf. Sci.* **208** (1989) 34.
54. S.H. Lu, D. Tian, Z.Q. Wang, Y.S. Li, F. Jona, and P.M. Marcus, *Solid State Commun.* **67** (1988) 325.
55. M. Galeotti, A. Atrei, U. Bardi, G. Rovida, and M. Torrini, *Surf. Sci.* **297** (1993) 202.
56. T. Iwasita, *Electrochim. Acta* **47** (2002) 3663.
57. P.V. Samant, C.M. Rangel, M.H. Romero, J.B. Fernandes, and J.L. Figuereido, *J. Power Sources* **151** (2005) 79.
58. V.B.G. Ramershkrishnan, L. Todd, J. Renxuan, L. Kevin, and E. Smotkin, *J. Phys. Chem. B* **102** (1998) 9997.
59. B. Beden, J.M. Leger, and C. Lamy, in: J. O'M. Bockris, B.E. Conway, R.E. White (Eds.), *Modern Aspects of Electrochemistry*, Vol. 22, Plenum Press, New York, 1992, pp. 97–247.
60. M.T. Koper, *Surf. Sci.* **548** (2004) 1.
61. T. Iwasita, F.C. Nart, and W. Vielstich, *Ber. Bunsenges. Phys. Chem.* **94** (1990) 1030.
62. H. Gasteiger, N. Marković, P. Ross, and E. Cairns, *J. Phys. Chem.* **97** (1993) 12020.
63. H. Gasteiger, N. Marković, P. Ross, and E. Cairns, *J. Electrochem. Soc.* **141** (1994) 1795.
64. N. Marković, H. Gasteiger, P.N. Ross, I. Villegas, and M. Weaver, *J. Electrochim. Acta* **40** (1995) 91.
65. K. Friedrich, K. Geyzers, U. Linke, U. Stimming, and J. Sumper, *J. Electroanal. Chem.* **402** (1996) 123.
66. M. Krausa and W. Vielstich, *J. Electroanal. Chem.* **379** (1994) 307.
67. M. Hogarth, J. Munk, A. Shukla, and A. Hamnett, *J. Appl. Electrochem.* **24** (1994) 85.
68. M. Janssen and J. Moolhuysen, *Electrochim. Acta* **21** (1976) 861.
69. E. Spinacé, A. Neto, and M. Linardi, *J. Power Sources* **129** (2004) 121.
70. A. Crown, C. Johnston, and A. Wieckowski, *Surf. Sci.* **506** (2002) L268.
71. S. Strbac, C. Johnston, G. Lu, A. Crown, and A. Wieckowski, *Surf. Sci.* **573** (2004) 80.
72. V. Colle, M. Giz, and G. Tremiliosi Filhio, *J. Braz. Chem. Soc.* **14** (2003) 601.
73. S.H. Bonilla, C.F. Zinola, J. Rodríguez, V. Díaz, M. Ohanian, S. Martínez, and B. Gianetti, *J. Colloid Interface Sci.* **288** (2005) 377.
74. S. Martínez and C.F. Zinola, *J. Solid State Electrochem.* **11** (2006) 947.
75. S. Strbac, R.J. Behm, A. Crown, and A. Wieckowski, *Surf. Sci.* **517** (2002) 2078.
76. J.W. Ndieyira, A.R. Ramadan, and T. Rayment, *J. Electroanal. Chem.* **503** (2001) 28.
77. D. Oyamatsu, H. Kanemoto, S. Kuwabata, and H. Yoneyama, *J. Electroanal. Chem.* **497** (2001) 97.
78. M.C. Santos, L.H. Mascaro, and S.A.S. Machado, *Electrochim. Acta* **43** (1998) 2263.
79. P. Mrozek, Y.-E. Sung, and A. Wieckowski, *Surf. Sci.* **335** (1995) 44.
80. M. Seo, M. Aomi, and K. Yoshida, *Electrochim. Acta* **39** (1994) 1039.
81. D.M. Kolb, in: H. Gerisher and C.W. Tobias (Eds.), *Advances in Electrochemistry and Electrochemical Engineering*, Vol. 11, Wiley-Interscience, New York, 1978.
82. K. Ogaki and K. Itaya, *Electrochim. Acta* **40** (1995) 1249.
83. S. Sugita, T. Abe, and K. Itaya, *J. Phys. Chem.* **97** (1993) 8780.
84. M.J. Esplandiú, M.A. Schneeweiss, and D.M. Kolb, *Phys. Chem. Chem. Phys.* **1** (1999) 4847.
85. S. Garcia, D. Salinas, C. Mayer, E. Schmidt, G. Staikov, and W.J. Lorenz, *Electrochim. Acta* **43** (1998) 3007.
86. M.A. van Hove, R.J. Koestner, P.C. Stair, J.P. Biberian, L.L. Kesmodel, I. Bartos, and G.A. Somorjai, *Surf. Sci.* **103** (1981) 189.
87. R.C. Salvarezza, D. Vásquez Moll, M.C. Giordano, and A.J. Arvia, *J. Electroanal. Chem.* **213** (1986) 301.
88. M.E. Martins, R.C. Salvarezza, and A.J. Arvia, *Electrochim. Acta* **41** (1996) 2441.
89. A. Vaskevich, M. Rosenblum, and E. Gileadi, *J. Electroanal. Chem.* **383** (1995) 167.
90. A. Vaskevich and E. Gileadi, *J. Electroanal. Chem.* **442** (1998) 147.
91. M.E. Martins, R.C. Salvarezza, and A.J. Arvia, *Electrochim. Acta* **36** (1991) 1617.
92. C. Palacio, P. Ocón, P. Herrasti, D. Díaz, and A. Arranz, *J. Electroanal. Chem.* **545** (2003) 53.
93. F. El Omar, R. Durand, and R. Faure, *J. Electroanal. Chem.* **160** (1984) 385.
94. A.M. Bittner, *J. Electroanal. Chem.* **431** (1997) 51.
95. E. Herrero, L.J. Buller, and H.D. Abruña, *Chem. Rev.* **101** (2001) 1897.
96. D.R. Salinas, E.O. Cobo, S.G. García, and J.B. Bessone, *J. Electroanal. Chem.* **470** (1999) 120.
97. J. Li, E. Herrero, and H.D. Abruña, *Colloids Surf. A* **134** (1998) 113.
98. E. Herrero and H.D. Abruña, *J. Phys. Chem. B* **102** (1998) 444.
99. E. Herrero and H.D. Abruña, *Langmuir* **13** (1997) 4446.

100. J. Inukai, S. Sugita, and K. Itaya, *J. Electroanal. Chem.* **403** (1996) 159.
101. S. Abaci, L. Zhang, and C. Shannon, *J. Electroanal. Chem.* **571** (2004) 169.
102. H.D. Abruña, J.M. Feliú, J.D. Brock, L.J. Buller, E. Herrero, J. Li, R. Gómez, and A. Finnefrock, *Electrochim. Acta* **43** (1998) 2899.
103. C. Coutanceau, A. Ralotondrainibé, A. Lima, E. Garnier, S. Pronier, J. Léger, and C. Lamy, *J. Appl. Electrochem.* **34** (2004) 61.
104. M. Nakayama, H. Komatsu, S. Ozuka, and K. Ogura, *Electrochim. Acta* **51** (2005) 274.
105. F.W. Nyasulu and J.M. Mayer, *J. Electroanal. Chem.* **392** (1995) 35.
106. H. Zhang, Y. Wang, E. Rosim, and C. Cabrera, *Electrochem. Solid State Lett.* **2** (1999) 437.
107. D.W. Bassett, *Surf. Sci.* **325** (1995) 121.
108. B.N. Grgur, N.M. Marković, and P.N. Ross Jr., *J. Phys. Chem. B* **102** (1998) 2494.
109. M. Barroso de Oliveira, L.P. Roberto Profeti, and P. Olivi, *Electrochem. Commun.* **7** (2005) 703.
110. P. Stonehart, *Anal. Chim. Acta* **37** (1967) 127.
111. O.M. Poltorak, *Zh. Fiz. Khim.* **27** (1953) 599.
112. G.A. Tsirlina, M.I. Borzenko, S. Yu Vassiliev, and E.V. Timofeeva, First Spring Meeting of the International Society of Electrochemistry, Poster No. 11, March 2–6, 2003, Alicante, Spain.
113. N.I. Kobozev, V.V. Monblanova, and S.V. Krillova, *Zh. Fiz. Khim.* **20** (1946) 653.
114. P.D. Lukovtsev, *Dokl. Akad. Nauk SSSR* **88** (1953) 875.
115. V.S. Bagotskii and Z.A. Iofa, *Dokl. Akad. Nauk SSSR* **53** (1946) 439.
116. V.S. Bagotskii, Z.A. Iofa, and A.N. Frumkin, *Zh. Fiz. Khim.* **21** (1947) 241.
117. A.N. Frumkin and B.I. Podlovchenko, *Ber. Akad. Wiss. USSR* **150** (1963) 34.
118. J.O.M. Bockris and H. Wróblowa, *J. Electroanal. Chem.* **7** (1964) 428.
119. M. Watanabe and S. Motoo, *J. Electroanal. Chem.* **60** (1975) 259.
120. O.A. Petrii and V.D. Kalinin, *Rus. J. Electrochem.* **35** (1999) 627.
121. M.M.P. Janssen and J. Moolhuysen, *Electrochim. Acta* **21** (1976) 869.
122. J.C. Davies, B.E. Hayden, D.J. Pegg, and M.E. Rendall, *Surf. Sci.* **496** (2002) 110.
123. D.M. Kolb, *Surf. Sci.* **500** (2002) 722.
124. P. Waszczuck, T.M. Barnard, C. Rice, R.I. Masel, and A. Wieckowski, *Electrochem. Commun.* **4** (2002) 599.
125. K.A. Friedich, K.P. Geyzers, A.J. Dickinson, and U. Stimming, *Surf. Sci.* **402–404** (1998) 571.
126. C.K. Rhee, M. Wakisaka, Y.V. Tolmachev, C.M. Johnston, R. Haasch, K. Attenkofer, G.Q. Lu, H. You, and A. Wieckowski, *J. Electroanal. Chem.* **554–555** (2003) 367.
127. T. Frelink, W. Visscher, and J.A.R. van Veen, *Langmuir* **12** (1996) 3702.
128. F. Colom and M. González-Tejera, *J. Appl. Electrochem.* **24** (1994) 426.
129. A. Crown, I. de Moraes, and A. Wieckowski, *J. Electroanal. Chem.* **500** (2001) 333.
130. A. Crown and A. Wieckowski, *Phys. Chem. Chem. Phys.* **3** (2001) 3290.
131. R. Liu, H. Iddir, Q. Fan, G. Hou, A. Bo, K.L. Ley, E.S. Smotkin, Y.E. Sung, H. Kim, S. Thomas, and A. Wieckowski, *J. Phys. Chem. B* **104** (2000) 3518.
132. H. Gasteiger, N. Marković, P. Ross, and E. Cairns, *J. Phys. Chem.* **98** (1994) 617.
133. R. Ianniello, V. Schmidt, U. Stimming, J. Stumper, and A. Wallau, *Electrochim. Acta* **39** (1994) 1863.
134. G. Vurens, F. Van Delft, and B. Nieuwenhuys, *Surf. Sci.* **192** (1987) 438.
135. H. Gasteiger, N. Markovic, P. Ross, and E. Cairns, *J. Phys. Chem.* **99** (1995) 16757.
136. A. Haner, P.N. Ross, U. Bardi, and A. Atrei, *J. Vac. Sci. Technol. A.* **10** (1992) 2718.
137. N. Kizhakevariam and E.M. Stuve, *Surf. Sci.* **286** (1993) 246.
138. I. Villegas and M.J. Weaver, *J. Chem. Phys.* **103** (1995) 2295.
139. B.A. Sexton, K.D. Rendulić, and A.E. Hughes, *Surf. Sci.* **121** (1982) 181.
140. J. Wang and R.I. Masel, *Surf. Sci.* **235** (1991) 199.
141. G.A. Attard, K. Chibane, H.D. Ebert, and R. Parsons, *Surf. Sci.* **224** (1989) 311.
142. V.S. Bagotskii, Y.B. Vassiliev, and O.A. Khazova, *J. Electroanal. Chem.* **81** (1977) 229.
143. K. Franaszczuk, E. Herrero, P. Zelenay, A. Wieckowski, J. Wang, and R.I. Masel, *J. Phys. Chem.* **96** (1992) 8509.
144. B. Beden, F. Hahn, C. Lamy, J.M. Leger, N. Tacconi, R. Lezna, and A.J. Arvia, *J. Electroanal. Chem.* **261** (1987) 215.
145. A.B. Anderson, E. Grantscharova, and S. Seung, *J. Electrochem. Soc.* **143** (1996) 2075.
146. A. Haner and P.N. Ross, *J. Phys. Chem.* **95** (1991) 3740.
147. B. Bittins-Cattaneo and T. Iwasita, *J. Electroanal. Chem.* **238** (1987) 151.
148. P.K. Shen and A.C.C. Tseung, *J. Electrochem. Soc.* **141** (1994) 3082.
149. S.R. Branković, J. McBreen, and R.R. Adzić, *J. Electroanal. Chem.* **503** (2001) 99.
150. A.T. Hubbard, *Chem. Rev.* **88** (1988) 633.

151. E. Reddington, A. Sapienza, B. Gurau, R. Viswanathan, S. Sarangapani, E.S. Smotkin, and T. Mallouk, *Science* **280** (1998) 1735.
152. K.L. Ley, R. Liu, C. Pu, Q. Fan, N. Layarovska, C. Segre, and E.S. Smotkin, *J. Electrochem. Soc.* **144** (1997) 1543.
153. B. Gurau, R. Viswanathan, R. Liu, T.J. Lafrenz, K.L. Ley, E.S. Smotkin, E. Reddington, A. Sapienza, B.C. Chan, T.E. Mallouk, and S. Sarangapani, *J. Phys. Chem. B* **102** (1998) 9997.
154. H. Herrero, K. Franaszczuck, and A. Wieckowski, *J. Phys. Chem.* **98** (1994) 5074.
155. S. Wilhelm, T. Iwasita, and W. Vielstich, *J. Electroanal. Chem.* **238** (1987) 383.
156. T. Iwasita, E. Santos, and W. Vielstich, *J. Electroanal. Chem.* **229** (1987) 367.
157. J. Willsau, O. Wolter, and J. Heitbaum, *J. Electroanal. Chem.* **185** (1985) 163.
158. T. Iwasita and F.C. Nart, *J. Electroanal. Chem.* **317** (1991) 291.
159. B. Beden, C. Lamy, A. Bewick, and K. Kunimatsu, *J. Electroanal. Chem.* **121** (1981) 343.
160. N. Sheppard, T.T. Nguyen, in: R.H.H. Clark and R.E. Hester (Eds.), *Advances in Infrared and Raman Spectroscopy*, Vol. 5, Heyden Editorial, London, U.K., 1978, p. 67.
161. R.J. Nichols and A. Bewick, *Electrochim. Acta* **33** (1988) 1691.

12 Surface Modifications of Ferrous Metals and Alloys Induced by Potential and Thermal Perturbations

C. Fernando Zinola

CONTENTS

12.1 SURFACE MODIFICATIONS INDUCED BY POTENTIAL AND THERMAL PERTURBATIONS

Different ferrous metal oxides or defined characteristic structures, like spinel and bronzes, have been investigated as substrates for electrocatalytic processes [1–3], since they offer large activities and a rather good selectivity. Spinels are usually prepared by thermal procedures starting from either a homogenized mixture of solid metal oxides, or by the evaporation of a solution containing the nitrates of the corresponding metals followed by the thermal decomposition of the salt residue, or by the co-precipitation of the metal hydroxides and the subsequent thermal decomposition of the solid [1,4]. The preparation procedures can be carried out on different substrates such as titanium, iron, and carbon, with the objective of developing highly dispersed materials [5,6]. The oxidation of the substrate during the course of the thermal treatment modifies the composition of the spinel-type metal oxide [5]. Depending on the preparative method, the electrocatalytic properties of the spinel oxides exhibit different features upon electrode reactions, such as hydrogen and oxygen evolution, chloride evolution, etc. The results obtained on the iron-group metal hydroxide electrodes show that their electrochemical behavior changes substantially on potential cycling as should be expected when the changes in the oxide layer composition and structure occur [7–10]. Furthermore, those changes are considerably dependent upon the entire history of the electrode, but under certain circumstances, the metal hydroxide electrodes attain a very reproducible and stabilized voltammetric response, which could be attributed to the formation of a definite oxide layer structure. Relatively thick hydrous cobalt oxide overlayers have been grown on cobalt electrodes that are subjected to a triangular potential cycling at low potential sweep rates in an alkaline solution [11].

The experimental preparation procedure consists of the application of periodic potential routines to pure metal wires that were previously chemically etched to achieve a polished surface before immersion in the electrolyte solution. The most common electrolyte solutions that can be employed are diluted or high concentrated sodium or potassium hydroxide.

The initial repetitive voltammogram of the metal in the alkaline solution is recorded between the upper and the lower potential limits, defined by the electrochemical stabilization of the solvent. In the positive sweep, a very thin oxide layer is formed on the electrode, which is almost totally electroreduced in the negative sweep. The potential cycling is then continued for almost 15 min to bring about the stabilized voltammogram, characterized by the same anodic and cathodic voltammetric charge values. Then, the electrode is subjected to a repetitive square-wave perturbing program from E_l (mainly negative values vs. RHE) to E_u (usually larger than 2.5 V) at a frequency ranging from 0.025 to 3.5 kHz during a given time to accumulate an anodic oxide layer on the electrode's surface. Most of the results correspond to a symmetric potential program, with the same residence time at $E_l(T_l)$ and at $E_u(T_u)$, that is, $T_u = T_l$, unless otherwise stated. Finally, the voltammogram of the resulting oxide is run under the same conditions as described above. This voltammogram provided the first insights into the possible structure of the resulting oxide overlayer as well as its charge-storage capacity.

In the case of cobalt, the cobalt oxide overlayer exhibits a large charge-storage capacity and a high catalytic activity for the oxygen electroreduction reaction in an alkaline solution [12]. The chemical properties and the electrocatalytic behavior of the overlayer confirm the assignment of this layer to a Co_3O_4 spinel. The potential limits, the frequency of the perturbation, and the electrolyte composition play a critical function. According to the standard potentials, of the cobalt species [13] in the potential ranges covered by the perturbing potentials, various reactions are possible. It has been demonstrated that the reactions involving Co, $Co(OH)_2$, and $CoOOH$ occur as distinguishable voltammetric conjugated peaks [9]. On the other hand, the formation of Co_3O_4 is difficult to be reduced later even at extremely low potentials (-2.5 V). This means that the formation of Co_3O_4 must be conceived as a side reaction that takes place along the main $Co(0)/Co(II)$ and $Co(II)/Co(III)$ redox processes during the potential treatment.

Interesting questions are, When is the spinel oxide formed? and Is it during the electrooxidation or during the electroreduction half cycle? The roughness factor shows an increase that does not depend on how positive the upper potential is, but depends on the residence time within a certain potential range. Then, the optimum value of E_u appears to be related to the accumulation of the oxide layer rather than to the development of the Co_3O_4 structure. To some extent, there is a sort of compensation effect between the E_u and the frequency for the accumulation of the anodic layer. On the other hand, the roughness developed at different E_l values shows that the greater the time spent at the electroreduction potential range the greater is the efficiency for the increasing surface roughness. Hence, the greater the frequency, the more negative is the value of E_l. Further, there is a compensation effect between the E_l and the frequency that results in the development of the spinel Co_3O_4 overlayer and in the increase in the roughness factor.

This analysis can be modeled in terms of the formation of $Co(OH)_2$, $CoOOH$, and CoO_2 during the anodic half cycle and its gradual reduction to metal cobalt and a side reaction that renders Co_3O_4 during the cathodic half cycle:

$$3CoOOH_{(s)} + H^+_{(aq)} + e^- \leftrightarrow Co_3O_{4(s)} + 2H_2O_{(l)} \qquad (I)$$

When these processes occur repetitively, an accumulation of the Co_3O_4 is produced. A reaction of the second type has also been described for the formation of Fe_3O_4 by the electroreduction of iron oxides in alkaline solutions [14,15]. It is also known that the electroreduction of this type of oxides is not complete in the case of iron, even at potentials within the hydrogen evolution reaction.

The accumulation of stable Co_3O_4 as a surface layer should produce a decrease in the metal surface area that is available for the reaction, and therefore the maximum efficiency should be related to an optimum treatment time. Furthermore, a symmetric perturbing potential becomes the most efficient because the overall reaction takes place within the oxide layer thickness that is the same for the anodic and the cathodic reactions, and at a constant frequency, the yields of both the reactions are the same. In the case of cobalt, the largest efficiency for the increase in the surface roughness was obtained at 0.1 kHz.

Increase in the alkaline concentration up to 1 M produces a diminished efficiency, and the maximum value of the roughness is observed at a larger frequency. At higher pH, a new side reaction has to be included, that is, the cobalt oxide-hydroxide dissolution of the anodic layer according to

$$CoOOH_{(s)} + OH^-_{(aq)} \leftrightarrow COO_2^-_{(aq)} + H_2O_{(l)} \tag{II}$$

It is known that either the electroreduction of the hydrous metal oxides involves phase changes that take place under two limiting situations, at "constant volume," with the roughness development, that is, producing a void-containing overlayer, or under "maximum change in volume," that corresponds to the formation of a compact layer. Whether one or the other type of mechanism prevails depends on the mass transport processes and the particle dynamics operating at the oxide layer itself during the electroreduction process. Thus, when an excess of the inert electrolyte is present, the electric field effects are smoothened down and the migration is replaced by the diffusion of the reacting particles. The overall effect is to cancel the increase in the surface roughness.

Once the spinel-type oxide layer is formed, the stabilized current vs. potential profile at 1.3 V obeys

$$Co_3O_{4(s)} + OH^-_{(aq)} + H_2O_{(l)} \leftrightarrow 3CoOOH_{(s)} + e^- \tag{III}$$

As reaction (III) is carried back and forth during the voltammetric cycles, there will be a gradual decrease in the surface roughness. This stabilization effect is also reflected in the SEM micrographs (see Ref. [12]).

The porous and amorphous structure of the resulting oxide overlayer is also interesting to discuss. The differential thermal analysis showed that at least six water molecules per Co_3O_4 are involved in the overlayer structure. This is not surprising when one deals with a hydrous metal hydroxide layer, and the fact that such a structure behaves as amorphous in x-ray diffractometry does not preclude the existence of the crystalline domains of dimensions lower than 5×5 nm. The catalytic activity of this system is probably explained better in terms of the local interactions of the oxygen molecules with the cations of the oxide by considering a microscopic approach based on the quantum-chemical theory of the chemical bond in the small-sized solid clusters.

Other methods can be employed to obtain metal oxides with interesting catalytic properties as amorphous alloys. These compounds promise low cost electrocatalysts for water electrolysis in alkaline solutions [16–18]. Electroless deposition appears as a relatively easy way for tailoring the desired amorphous alloy compositions on either conducting or insulating materials [19–21]. Hydrous oxide coatings with a controlled thickness can be grown by the cyclic potential routines applied to the metals or the amorphous alloys immersed in aqueous solutions [11,22–25]. For cobalt:nickel amorphous alloys, the hydrous oxide films have been discovered as the precursors of the spinel-structures [25] that exhibit a high electrocatalytic activity toward the oxygen evolution reaction [26].

It is usual to compare the infrared spectra of the electrochemically treated amorphous alloys to those of the chemically or thermally (at 380°C) prepared spinels to ascertain the nature and composition of the compound. With this methodology, it can be demonstrated that the hydrous oxide coating acts as a precursor of a spinel structure either from the electrolessly deposited amorphous alloys or after the square-wave potential treatment.

On the other hand, the electrochemical performance of the spinels can be ascertained by the nature of the compounds through the cyclic voltammetry of the polycrystalline pure metals such as nickel, cobalt, or iron. The accumulation of the charge under potential cycling conditions can explain the use of these compounds as batteries. Surface compounds such as M(OH) or MO(OH), the formation of relatively large amount of soluble M^{+z} species, and the precipitation and accumulation of the former hydroxides are favored under potential cycling.

One of the main applications of the accumulation of charge arises from the electrocatalytic behavior of the oxide coatings in the course of the oxide formation process. The Tafel slope (b) for the spinet-type electrodes that are based on cobalt, nickel, and/or iron can change considerably, such as in the case of Co–Ni–P from 0.04 to 0.06 V decade^{-1}, to a typical Tafel slope for the Co_3O_4 spinel in the same solution [27].

This change in b can be either due to a change in the surface oxide-layer composition or due to a change in the active site characteristics caused by the surface interactions. Otherwise, the change in b reported for crystalline Ni–Co alloys in alkaline solutions has also been explained by the different nature of the surface oxides [27]. It is also important to say that the nickel and cobalt-spinel voltammetric behavior is attained after potential cycling at a low scan rate, where $b = 0.04$ V decade^{-1}. However, a pre-anodization at 1.8 V vs. RHE led to a Co_3O_4 spinel electrode for which $b = 0.06$ V decade^{-1}.

Burke and O'Sullivan [28] have reported an improvement in the oxygen evolution reaction. Accordingly, the oxide catalytic layers are produced by the potential cycling because of the hydrous structure of the oxide coatings produced during the potential routine with a relatively large active surface area for the reaction. The value of the surface roughness might be an important factor to consider in the catalytic process since it is related to the total amount of the oxide coating instead of the true active surface area for the oxygen evolution reaction. Accordingly, Trasatti [29] made a comparison between the meaning of the roughness factor values as determined from the voltammetry and the observed capacitive current densities.

12.2 PASSIVITY AND CORROSION-RESISTANT SURFACES OF NON-NOBLE METALS

12.2.1 A New Approach to Passive Layers of Non-Noble Metals

Scanning tunneling microscopy (STM) is increasingly applied to the studies of surface topography at an atomic level resolution in aqueous solution [30]. The *in situ* STM technique is capable of giving information on the surface structure. The mechanistic interpretation of the electrochemical processes of interest in technology such as corrosion often depends on the structural details of the respective surfaces.

Among the *ex situ* methods that can be employed in surface analysis, low-energy electron diffraction (LEED) and x-ray photoelectron spectroscopy (XPS) can give the crystal structure and the nature of the surface ad-layers after the electrochemical and adsorption experiments as explained in this chapter [31,32]. Among the *in situ* non-electrochemical techniques, the radiotracer method [33] gives information about the adsorbed quantities; however, infrared spectroscopy in FTIR mode [34] allows the identity of the bonding of the adsorbed molecules, and finally ellipsometry [35] makes possible the study of extremely thin films. Recently, some optical methods such as reflectance, x-ray diffraction, and second harmonic generation (SHG) [36] have been added to this list.

It has been demonstrated that STM can operate not only in vacuum but also in air [37] and in electrolytes [38]. For example, Sonnenfeld et al. [37] monitored the *in situ* deposition of silver on graphite and Arvia et al. [39] monitored the effect of repetitive potential cycling on gold. On the other hand, Fan and Bard [40] studied the *in situ* STM measurements for the corrosion of stainless steel and the dissolution of nickel under potentiostatic conditions. Szklarczyk and Bockris [41] have published an account of the potential dependence of the crystal reconstruction on platinum in the sodium perchlorate solution at the angstrom scale. The atomic lattice of Al(111) has been reported under the UHV conditions [42] and the STM imaging of aluminum and its oxide under an *in situ* electrochemical environment has also been carried out [43].

The electrochemical process of aluminum in the alkaline solution involves mechanistic activation and passivation processes [44]. It has been observed with the help of the *in situ* STM images [45] that the reduction at -1.135 V of the oxide formed at the potentials is more negative than the -0.975 V reproduced at the original surface. When the anode potential is increased in the positive direction to -1.05 V, an oxide begins to grow as small humps that later fuse together in a longer time and at more positive potentials. Thus, -1.05 V marks the formation of the active region.

For example, it is possible to find with the help of the STM images that the highly crystalline nature of the oxide film on the nickel surfaces is Ni(100) in the alkaline solutions [46]. At low potentials, a well-ordered rhombic structure is formed, which is resistant to reduction and is assigned to the irreversible $Ni(OH)_2$ formation. At higher potentials, it is possible to see a quasi-hexagonal structure consistent with NiO(111). In other words, a crystalline oxide is formed of the NiO(111) order independently of the crystal orientation of nickel. Moreover, it is very useful and practical to obtain a reduced surface by the cathodic treatments, where the hydrogen evolution produces the chemical and electrochemical reductions of nickel. However, this has to be done with careful attention and in light alkaline solutions to avoid nickel dissolution. When this is performed, monoatomic steps mostly oriented in the (100) and (111) directions are observed [47].

As observed before, the electrochemical pretreatment of the electrode can affect the surface morphology. Thus, the passivized nickel film was also studied at different electrode potentials and a grain-like or a crystalline ordered structure was found. The first grain structure is assigned to the outermost hydroxide nickel, whereas the crystalline structure to the innermost nickel oxide. Many aspects are related to the effect of ion tunneling that produces higher or lower currents. It seems that when the current is low the tip is directly approached toward the oxide and is probably inside the outermost layer and thus, the tip only images the inner oxide. Other techniques such as the XPS [48], corroborate the conclusions drawn from these images. STM has been used to study the Ni(111) electrodes passivated in sulfuric acid at 0.65 V vs. NHE in air. Before passivation, the Ni(111) develops steps along the (110) direction with terraces of a few hundreds of nanometers. After passivation, a mosaic structure is formed with crystallites of 2–3 nm in size along the $[1\bar{2}1]$ direction whose width ranges from 2 to 3 nm. The triangular crystallites have smooth step-edges along the $[\bar{1}01]$ and $[01\bar{1}]$. The roughness of the surface is increased with respect to that of the non-passivated surface. The authors have also found that the preferential ordering along the $[1\bar{2}1]$ direction corresponds to a preferential growth that is perpendicular to a close packed direction [48].

In the case of chromium, the effect of the electrochemical pretreatments is more noticeable. The XPS measurements were also conducted since they give reliable data on the surface composition of the oxide layer and the same can be correlated with the STM images [49] as well. A comparative study, between the electrochemical polishing, mechanical polishing, and annealing at 1175 K for several hours in a hydrogen atmosphere, was carried out. After the potential was applied in the sulfuric acid within the passive region, the time required for the process was also measured to get additional information. In this case also a duplex layer model was considered with the outer layer thickness increasing with the passivation time. The inner parts of the passive films were composed of Cr_2O_3, whereas the outer layer was of $Cr(OH)_3$ as expected. The chemical effect of aging under potentiostatic polarization was the decrease of the outer layer of hydroxides and the increase of in the dehydrated compact inner oxide layer. No changes in the surface composition were found by this methodology.

There is substantial interest in the electrochemistry of copper surfaces because of their importance in electrometallurgy, catalysis, and microelectronic applications. Considerable efforts have been put into the inhibition of corrosion and the performance of copper toward the underpotential deposition. *In situ* atomic force microscopy (AFM) images of Cu(100) single crystals in dilute acid solutions reveal that a $(\sqrt{2} \times \sqrt{2})R45°$ oxygen ad-layer can be formed at the negative potentials, which is removed by applying 0.5 V more negative than the open circuit value. The ad-layer does not change when the electrolyte is modified or by using a different surface preparation. A hexagonal copper lattice can be demonstrated with an overlayer of the adsorbed oxygen or hydroxyls, in which the oxygen is chemisorbed in an alternate fourfold hollow site [50]. The authors found that the $(\sqrt{2} \times \sqrt{2})R45°$-oxygen structure is different from the $(2\sqrt{2} \times \sqrt{2})R45°$-oxygen lattice found to occur in the UHV in several studies in UHV conditions [51]. STM, LEED, x-ray diffraction and a combination of photoelectron diffraction and NEXAFS [52] also suggest the $(2\sqrt{2} \times \sqrt{2})R45°$-oxygen lattice with a missing row reconstruction, in which every fourth row of the copper atoms is missing. However, the $(2\sqrt{2} \times \sqrt{2})R45°$-oxygen structure can be formed only after high temperature cycling while the low temperature phase is the $(\sqrt{2} \times \sqrt{2})R45°$-oxygen [53].

12.2.2 MORPHOLOGICAL STABILITY OF METAL LAYERS DURING DEPOSITION AND DISSOLUTION

The development of a well-ordered metal surface during electrodeposition is of considerable theoretical and practical importance since it affects the electrocatalytic processes that occur later on it. In this sense the porosity, the surface roughness and the compactness of the deposit define the formation of the oxide layers, especially, in the case of the non-noble metals.

From the electrochemical engineering point of view, the problem of the current and potential distributions is also involved since a very large roughening (more than 10,000) can lead to short-circuiting between the cathode and the anode in the reactor. Moreover, in the case of anode corrosion, a large ion flux current produces a high local concentration of the metal ions at the kinks and edges, producing a chemical local deposition with completely different properties.

The initiation and growth of the surface roughening can be associated with the existence of a critical overpotential with mass and charge transfer terms [54] that are extended later by migration transport [55]. It is possible to conduct a stability analysis by considering a global mass transport control (Nernst–Planck equation) to the case of an electrocatalytic reaction under potentiostatic conditions. The derivation of the mass and charge transfer overpotentials leading to the real exchange current density is presented below.

12.2.2.1 Determination of Real Surface Area during Deposition and Dissolution

The determination of the real surface area of the electrocatalysts is an important factor for the calculation of the important parameters in the electrochemical reactors. It has been noticed that the real surface area determined by the electrochemical methods depends on the method used and on the experimental conditions. The STM and similar techniques are quite expensive for this single purpose. It is possible to determine the real surface area by means of different electrochemical methods in the aqueous and non-aqueous solutions in the presence of a non-adsorbing electrolyte. The values of the roughness factor using the methods based on the Gouy–Chapman theory are dependent on the diffuse layer thickness via the electrolyte concentration or the solvent dielectric constant. In general, the methods for the determination of the real area are based on either the mass transfer processes under diffusion control, or the adsorption processes at the surface or the measurements of the differential capacitance in the double layer region [56].

The determination of the real surface area by the double layer concept implies the application of the modified Gouy–Chapman theory [57]. It is considered that the reciprocal of the measured capacity is

$$RC^{-1} = C_{inner}^{-1} + C_{outer}^{-1} \tag{12.1}$$

where
C_{inner}^{-1} and C_{outer}^{-1} are the reciprocals of the inner compact capacity and the diffuse layers capacity, respectively
R is the roughness factor

In this case, it is useful to apply the so-called Parsons–Zobel plot [58] that shows the dependence of the reciprocal measured differential capacity on the reciprocal of the calculated diffuse layer capacity at a constant charge density covering wide spectra of electrolyte concentrations. The slope of this plot must be a straight line and equal to the R when we have a non-adsorbing electrolyte. However, it has been observed that at a low electrolyte concentration the linearity is lost and the slope becomes lower [57] (electrolyte concentrations below 0.01 M).

In the determination of R by the Valette–Hamelin method [59] the inner layer capacity, C_{inner}, is calculated for the different values of the charge densities, σ, at constant electrolyte concentration, postulating that the C_{inner} vs. σ curve has to be monotonically close to a zero charge potential value. This condition is obtained by adjusting the R value.

According to the Gouy–Chapman theory, the thickness of the diffuse layer, δ^{-1}, depends on the electrolyte concentration according to

$$\delta^{-1} = \left(\frac{8\pi e N_A C^0 Z^2}{\varepsilon k T} \right)^{-1/2} \tag{12.2}$$

where
   C^0 is the electrolyte concentration
   ε is the dielectric constant of the solvent
   N_A is the Avogadro number
   Z is the atomic number of the particle and the rest of the symbols have their usual meaning

As expected the diffuse layer is wider if the electrolyte concentration is lower. The dependence of the roughness factor on the diffuse layer thickness in the aqueous solutions is strongly dependent on the solvent dielectric constant, and this parameter should influence the determined R [57].

The thickness of the diffuse layer influences the roughness factor determination and the R increases with the diffuse layer thickness. This is not expected for a primary current distribution. However, different irregularities in the surface at different points of the catalysts can be responsible for these conclusions. Such behavior would imply that the diffuse layer corresponds to the geometric area of the electrode and does not reproduce the irregularities of the real surface. These considerations are true if the height of the irregularities at the surface is of the order of the diffuse layer thickness or bigger (tertiary current distribution). To solve this problem, we must know by the STM, the roughness structure of the catalyst electrodes.

The use of potentiodynamic triangular- or square-wave programs on the quasi-noble surfaces as substrates also influences the surface roughness of the deposited metal layer. When titanium is used as the substrate in diluted solutions of the hexachloroplatinic acid, spherical platinum microparticles can be grown embedded into the simultaneously formed titanium oxide films [60]. The real surface of the platinum may be easily controlled by the number of potentiodynamic cycles, and roughness factors exceeding 100 were obtained. Within the potential region of both the hydrogen and the oxygen underpotential deposition, in both acidic and alkaline solutions, the electrode displayed an excellent electrochemical response that was characteristic of the smooth polycrystalline platinum, unlike the highly dispersed platinum electrodes produced by the other methods. In that paper [60], the titanium electrode was polished mechanically and immersed into an acidic solution with the hexachloroplatinic acid and immediately cycled between a fixed initial potential of 0 V and a final potential that was progressively increased in steps of 0.25 V up to a maximum of 1.75 V vs. NHE at the medium scan rates. The sweep was repeated, usually 20 times, until a repetitive profile of the hydrogen underpotential peaks was obtained.

The thickness of the TiO_2 layer as well as that of the other quasi-noble metal layers can be determined from the linear dependence of the consumed charge on the electrode potential. The same can be performed under potentiostatic polarization conditions.

An experiment was conducted by us [61] to prepare anode electrodes for cathodic protection in a gas turbine underwater recirculation system, mainly for the protection of the condenser. The procedure that allows the platinum deposition without the interference of the TiO_2 layers or the surface roughness has to be the following. The starting potential (0 V vs. NHE) has to be selected precisely at the onset of the hydrogen evolution. At this initial value, the platinum is deposited under diffusion control and simultaneously the TiO_2 is slowly formed by the spontaneous chemical corrosion. However at 0 V, the oxide formation was cathodically inhibited and concurrently, the first platinum nuclei were cathodically deposited on a bare titanium surface, since the experiment was performed using the continuous flow cell. According to the data available in the literature, the formed oxide layers may become 1 nm thick during the first day of exposure [62]. After the first polarization cycle, within the smallest potential window of 0.05 V, a new TiO_2 layer was electrochemically formed and its thickness increased during the further potential sweeps. This procedure was accomplished by Mentus [60], in which the platinum nuclei appear and grow primarily at the tips of the scratches that appear during the mechanical polishing of titanium. In the early stages of potential cycling, when the potential window is well within the region of the platinum deposition under diffusion control, the growth of the grains was favored more than the TiO_2 growth. During the electroplating procedure, it is possible to limit the growth of the real surface area by recording the height of the hydrogen underpotential-deposited peaks, since they increase with the number of cycles. The final TiO_2 thickness was 6 nm [63] and was limited by the extensive oxygen evolution on the electrodeposited platinum.

On the other hand, if the electrodeposit is not smooth enough for the prefixed technical purposes, it can be achieved by cyclic polarization that produces the condition of the plating under an alternating current that is known to act as the metal deposit leveler [64]. The platinum appears to be well adhered to the support and cannot be stripped mechanically by rubbing with a filter paper. This is the result of the simultaneous platinum growth and the thickening of the TiO_2 layer. One may assume that at this stage, the platinum particles will be raised in the form of irregular islands or lines that are captured within the compact TiO_2 layers. During subsequent cyclic polarizations, they continue to develop as small beads, probably in a uniform way along the catalyst.

This methodology has been applied to study electrocatalysis [60], since its high electrocatalytic effectiveness was demonstrated for the reduction of oxygen (was comparable to those obtained from 20% Pt/Vulcan catalyst) and bromide ion adsorption/oxidation. Moreover, the large overpotential developed for the production of oxygen was quite proper for the case of the cathodic protection [61]. However, investigations were particularly interesting for the performance of the catalyzed titanium mesh electrodes that were proposed for the methanol fuel cells instead of the carbon-supported catalyst. The open structure is likely to mitigate the elimination of the CO_2 bubbles developing during the cell operation [62,65].

12.2.2.2 Mechanism of the Oxidation Process on Metal Surfaces

The metal chemical and the electrochemical oxidation processes are extensively used in the technology for the protection of materials against corrosion and the production of ceramics and catalysts. The initial stages of oxidation (including oxygen adsorption and incorporation, oxide islands nucleation, and growth into a continuous film) have been actively studied and are now understood relatively well. The kinetics of thick (>200 Å) oxide films approximately follows the parabolic law. The Wagner model of oxidation, in which diffusion across the oxide film is the rate limiting process, can describe electrochemical corrosion, at least qualitatively. The situation is much more complicated in the case of thin (<100 Å) films. The kinetic laws in this region cannot be well-integrated and only approximations for the oxide growth rate as a function of the electrode potential or time can be obtained. This makes the experimental verification of the theory based only on the measurements of the oxidation kinetics difficult. However, the use of tracer atoms (both metal and oxygen) provides valuable information on the transport mechanism (e.g., migrating or diffusing species) in the oxidation process and can be a useful tool for the experimental test of theory.

The tracing of the oxygen species is usually performed by oxidation first in $^{16}O_2$ and then $^{18}O_2$ enriched gas in the gas/solid interface. Thin ($\sim$20 Å) oxide film growths on single crystal metals at moderate temperatures can be studied by oxygen isotope labeling combined with high-resolution microscopy. In the case of liquid/solid and electrochemical interfaces there is another complication that arises from the existence of an electric field and the presence of an electrolyte solution with adsorbable anions and solvents.

12.2.2.2.1 Electrooxidation Process of Metal Electrodes

The process of surface electrooxidation is an electrochemical situation in which the electrons are lost in the outer layer of the metal/solution interface. During the process, the metal atom undertakes the formation of either a metallic cation or a surface oxide. The electron loss goes with an increase in the electron density by the oxidizing agent. The formation of a soluble species on non-noble metals containing oxygenated molecules or the formation of noble metal oxides is probably one of the most studied topics in this field.

In this section, we are going to briefly summarize the mechanism of the oxide formation and later outline the adsorption of the organic precursors that inhibit metal corrosion. Two types of oxygen-containing species can be formed: surface and bulk oxides. The latter is typical of the non-noble metal oxidation and usually forms a three-dimensional new phase. This new phase has its own catalytic characteristic since it strictly obeys different physical and chemical properties. In comparison with the noble metal oxides, it is thick and hydrated with several monolayers of thickness, with different electronic and magnetic properties with respect to that of the surface. In most of the cases, the inner structure begins to age and loses its water by forming a compact and good conductor or semiconductor depending on the case.

It must be mentioned that many different methods of oxide formation can be found in the literature but most of them involve chemical or electrochemical procedures [66–73]. In the case of the gas phase environment, the oxygen molecule undergoes a surface dissociation and later the atomic chemisorbed oxygen suffers a charge transfer from the metal, creating a pair of species with opposite charges. In the case of the solid/liquid interface, especially in the electrochemical environment, the water molecule expends its oxidative discharge by forming an oxygen-containing species. These species at low potentials are the adsorbed hydroxyl species that experience a place-exchange mechanism with the metal [73,74]. Then, an electronic charge transfer occurs with a loss of protons with the formation of the surface oxide compounds. The aging and restructuring usually occurs with the formation of a three-dimensional new phase with successive electron transfers. It is worth noting that each pathway can change its order depending on the nature of the metal or on when an alloy is to be prepared. In all cases, the surface or bulk roughness can be developed because of the mismatch between the metal and the new oxide phases. A scheme is presented here in order to elucidate the formation of the roughness due to electrooxidation.

The kinetics of the oxide film growth was mostly studied in noble metals. Gilroy and Conway [75], Gilroy [76], and Biegler and Woods [77] were the first to study the oxide film growth platinum. The formation of thick oxide films was demonstrated and studied by Shibata [78] and later by Vassilyev and Gromyko [79]. In the analysis of the oxide film growth at high positive potentials Shibata [80] showed that a distinguishable oxide phase—maybe a bulk oxide film—can be formed on the platinum. This phase may be reduced by a cathodic current peak separate from that of the quasi two-dimensional oxide film and appearing at less positive potentials, near or overlapping the hydrogen adsorption/desorption system. This state of the oxide had a much higher resistivity [80] than that of the quasi two-dimensional film.

12.2.2.2.2 Electrooxidation Process of Noble Metals

In general and based on former works by many authors, it can be concluded [22,81–88] that the reduction of the quasi two-dimensional oxide film in a negative-going potential sweep leads to the deposition of the metal atoms on the outer most layer of the thicker oxide film. Moreover, the thick film implied that the oxide would become sandwiched between the bulk metal and a film of newly deposited metal atoms on the outside of the remaining oxide, resulting from the reduction of the latter.

In some cases, it was found that the electrode preparation was extremely important—grounding with emery paper, mirror-finish mechanical polishing, alumina or diamond-polished—surfaces can exhibit different results (some authors found that the surface finishing defines the appearance of the hydrated oxides. On the other hand, potentiostatic or potentiodynamic conditions give different results as shown before [22,87,88]. However, it can be stated that the potentiostatic conditions at the noble electrodes that were thermally annealed at approximately 800°C–1000°C give the best reproducibility of the oxide film growth behavior. It has also been found that it is possible to return to the initial cyclic voltammetric performance by repetitive and prolonged cycling between the solvent stability limits, independent of the state of the noble metal's surface. This performance is the result of an electrochemical activation that cannot be due to the dissolution and redeposition of the platinum since it occurs with undiminished efficiency when cycling is conducted in an extended and stirred large volume of solution. The same was found for the ruthenium [89], palladium [85], and iridium [90].

The general performance of the surface oxide layers that are at a distance from the initial, reversible region of the oxide film formation correspond to the hydroxyl electrosorption as an underpotential process. At this level the potential (and/or the time) for the oxide formation is inversely proportional to the potential required for the film electroreduction. This behavior reflects the increasing stability of the films formed at higher positive potentials (in some cases due to the ageing effects) and applies to both the reconstructed quasi two-dimensional hydroxyl or the atomic oxygen monolayer film formed between 1.1 and 1.4 V and to the more extended quasi three-dimensional phase oxide films formed as separately distinguishable species in reduction, at substantially higher potentials (from 1.7 to 2.3 V). This kind of performance is contrary to what is expected from the thermodynamics of the formation and the reduction of different oxidation state species. The performance on the noble metals is completely different and corresponds to an irreversible process with different pathways [91] of parallel and distinct processes. This situation is the result of the place-exchange reconstruction process [92] shown in Figure 12.1, which is connected to the conversion of the two dimension films to three dimension oxide phases.

In the case of a potentiodynamic thin oxide growth, it involves at least two distinct stages that are connected with two types of oxides, namely, α compact oxide and β hydrous oxide [67,83,85,88]. In the first few oxide growth cycles, only the compact α-oxide film (assumed in the case of noble metals to be MO and MO_2 below 1.4 V) is formed, according to the following reactions:

$$M_{(s)} + 2OH^-_{(aq)} \rightarrow MO_{(s)} + H_2O_{(l)} + 2e^-_{(M)} \qquad \text{(IVa)}$$

or

$$M_{(s)} + H_2O_{(l)} \rightarrow MO_{(s)} + 2H^+_{(aq)} + 2e^-_{(M)} \qquad \text{(IVb)}$$

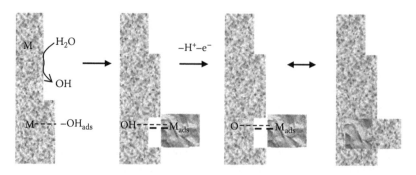

FIGURE 12.1 Scheme for electrochemical metal oxidation in the solid/liquid interface.

and

$$MO_{(s)} + 2OH^-_{(aq)} \rightarrow MO_{2(s)} + H_2O_{(l)} + 2e^-_{(M)} \tag{Va}$$

or

$$MO_{(s)} + H_2O_{(l)} \rightarrow MO_{2(s)} + 2H^+_{(aq)} + 2e^-_{(M)} \tag{Vb}$$

Figure 12.2 shows the case of the application of a square-wave symmetric routine within the platinum oxide stability region at medium frequencies in a strong acid solution. The development of the two types of α-oxides (α_1 and α_2) for different program times can be seen clearly. No hydrous

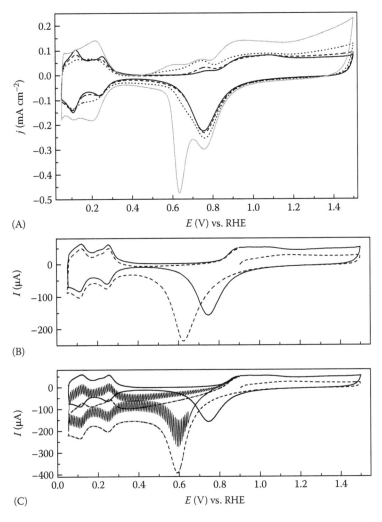

FIGURE 12.2 (A) Potentiodynamic profiles for smooth polycrystalline platinum in 0.5 M sulfuric acid solution at 0.10 V s^{-1} after a symmetric square-wave program between 1.0 and 1.5 V at 10 Hz. The time for application of the program was varied: 5 min (continuous line), 10 min (dashed line), 15 min (dotted line), and 20 min (gray lines); (B) 100 triangular cycles between 1.50 and 1.70 V at 1 V s^{-1}; (C) 100 triangular cycles between 1.50 and 1.90 V at 0.2 V s^{-1}.

β-oxide appeared for application times lower than 10 min. With this application, it is also possible to see the early stages of the platinum hydroxide formation in the anodic incursion.

The formation of β-oxides is also associated with the increase in the surface roughness of platinum, which is clearly seen in the hydrogen electrosorption region. The hydrogen adsorption/desorption system is fully altered, that is, the strongly bound hydrogen is favored.

It is always interesting to plot the effect of the chronoamperometric transient at a certain prefixed potential (Figure 12.3) and to check the stability of the modified layer with a chronopotentiometric transient (Figure 12.4) by the open circuit potential response. The break of the β-oxide growth on the platinum is observed after 250 s (Figure 12.4) when the current density becomes stable (Figure 12.3).

According to the experimental results, the two reduction peaks increase in intensity with the number of cycles. During these cycles, little generation of the β-oxide occurs, however, considering the case of Equations 12.4 and 12.5 and the case of noble metals, we have one monolayer of MO 420 μC cm^{-2} [93] and the same quantity for the MO_2 formation. Then the maximum charge

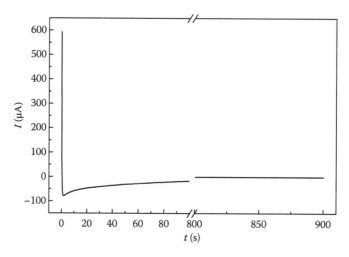

FIGURE 12.3 Chronoamperometric plots at 0.70 V for smooth polycrystalline platinum in 0.5 M sulfuric acid solution after applying 100 triangular cycles between 1.50 and 1.90 V at 0.2 V s^{-1}.

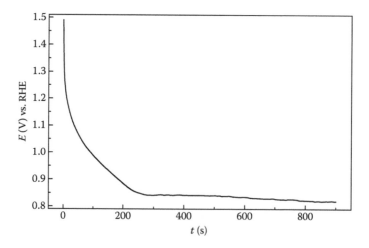

FIGURE 12.4 Open circuit potential transients for smooth polycrystalline platinum in 0.5 M sulfuric acid solution after applying 100 triangular cycles between 1.50 and 1.90 V at 0.2 V s^{-1}.

observed in the entire oxidation can be used to calculate the number of monolayers of each oxide. However, to achieve the results shown in Figure 12.2, the type of oxide that grows should be ascertained according to the number of cycles. In the case of noble metals, the reduction of approximately two monolayers of MO_2 to MO at 1.1 V is always observed [22,67,82,87,88,90]. For extremely outsized potential cycling, some regions of the compact α-oxide film must undergo some structural changes by allowing better internal access of solution ions and water, and consequently stabilizing the film in its hydrated form:

$$MO_{(s)} + nH_2O_{(l)} + 2OH^-_{(aq)} \rightarrow \{MO_2 \cdot (n+1)H_2O\}_{(s)} + 2e^- \tag{VIa}$$

and

$$MO_{(s)} + nH_2O_{(l)} \rightarrow \{MO_2 \cdot (n)H_2O\}_{(s)} \tag{VIb}$$

If the pH of the interface is sufficiently acidic, the breakdown of the α-oxide film will occur. This fact takes place at large potentials where the oxygen evolution reaction produces protons at the interface. Moreover, β-oxide layers are usually formed at the defects or at the "free" sites where residual α-oxides remain unreduced. It is suggested that this process proceeds further into the α-oxide layer, transforming almost the entire region into a β-oxide film, with continuous oxide growth by the subsequent potential cycles. The thickness of the MO_2 layer at the base of each β-oxide strands is depicted as the same as that of the MO_2 layer in the α-oxide film, where the charge remains constant at sufficiently high cycling number. For growth cycles after a critical value, both an increase in the surface coverage of β-oxide and a thickening of the oxide film are suggested to occur and the formation of strands of β-oxide within the compact α-oxide film arose. Under this model as the β-oxide film begins to form, the α-oxide charge density lessens very much. When a peak current density remains constant, the related number of cycles involved in the process is indicative of the formation of islands of MO with uniform thickness. We can assume that the β-oxides are not formed for the cycling values lower than the critical value, and from this value a β-oxide film formation occurs by the transformation of islands of α-MO layer. Therefore, the relative magnitude of the MO reduction charge density would reflect the fraction of the surface that is still occupied by the MO layer and the β-oxide surface coverage. However, one form of obtaining the fraction of β-oxide produced on the metal or on the α-oxide is by performing a subtracting experience—one covering the potential range usually involved in the α-oxide and the other covering a more positive anodic limit extended to lower reduction potentials. The extent of the increase of the MO formation-reduction charge densities with the fraction of the β-oxide film reduced in each cycle can be envisaged for the metal.

Very different results can be obtained in the case of a potentiostatic or potentiodynamic program within the hydrogen evolution potential region. For example, Figure 12.5 shows the anodic potentiodynamic profile of the platinum/acid aqueous solution interface after 60 min potential holding at −1.5 V.

In this case, the situation is very different since a complete reduction of the metal occurs with the formation of low-coordinated metallic atoms projected toward the outer layer of the interface. The model of conventional platinum electrochemistry suggests that there are two limiting types of surface metal atoms involved, which display significantly different actions. The overall reaction scheme may be presented as follows [94]:

$$M_{(high\ c)} + H_2O_{(l)} \rightarrow MOH_{(high\ c)} + H^+_{(aq)} + e^-_{(M)} \tag{VII}$$
$$\downarrow$$
$$M_{(low\ c)} + nH_2O_{(l)} \rightarrow M(OH)^{Z-}_{n(high\ c)} + nH^+_{(aq)} + (n-Z)e^-_{(M)} \tag{VIII}$$

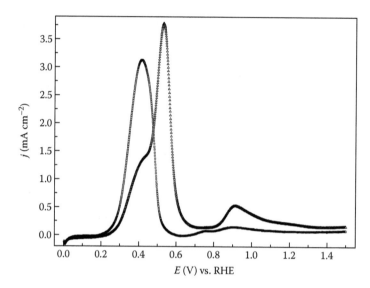

FIGURE 12.5 Current vs. potential anodic profile for polycrystalline platinum after 5 (open squares) and 15 min (open triangles) potential holding at −1.0 V in 1 M sulfuric acid solution at room temperature.

$M_{(high c)}$ denotes a high-coordination number surface metal atom. This is the most stable and usually the predominant state at the electrode surface and gives rise to the conventional cyclic voltammetry response. These well-embedded surface metal atoms remain in the lattice surface where, on oxidation, they lose electrons and coordinate the hydroxyl species, by Equation VIII.

With the application of a potentiostatic or potentiodynamic program, a large amount of electrical energy that comes into the lattice (as in [94,95]) produces the surface and subsurface diffusion and migration of the metal atoms. The result is the appearance of low-coordination number metals, $M_{(low c)}$, with a much higher surface energy, which come outside the outer layer of the interface. These high-energy surface metal atoms produced in Equation 12.9 constitute an intrinsically metastable surface (subsurface) state.

A metal adatom can coordinate several water molecules or hydroxyl species. Such a reaction will occur at unusually low potentials (the process being referred to as pre-monolayer oxidation [95,96]) due to the high activity of one of the main components, $M_{(low c)}$, in the couple. In the case of platinum, the product of this reaction is assumed to be a hydrous or β oxide that is assumed to be some type of platinate species, e.g., $[Pt(OH)_6]^{-2}$. On the other hand, the oxidation of high-coordinated atoms commences just above 0.8 V and the monolayer oxide deposit thus formed results in the deactivation of the surface. The surface atoms in the low-coordinated state are assumed great, as there is some occupancy of this state even in the case of conventional platinum. Oxidation of these atoms commences at approximately 0.7 V and for many of the electrocatalytic oxidation reactions with conventional platinum this is often the region of maximum activity. In many instances (when large cathodic potentials are applied), an anodic response was observed in the positive sweep at approximately 0.9 V (or even at 1.1 V). This is assumed to be due to the presence of high-energy surface atoms, $Pt_{(low c)}$, and the occupancy of the state is assumed to be quite low compared to conventional platinum. However, three independent processes demonstrate that there is a significant transition in this region with the multilayer hydrous oxide films on the platinum in the acid component that usually starts to undergo reduction on the negative sweep at approximately 0.6 V in Figure 12.3.

According to the present model, there are at least four states of the surface platinum atoms, which are represented in terms of their increasing energy. These states may not be associated with specific arrangements of the platinum atoms. They may simply be due to the different degrees of dispersion or lattice coordination (and hence different energies and reactivities) of the metal atoms at the interface.

When the treatments are intense in terms of time or cathodic potentials, platinum adatoms of a highly dispersed form are oxidized at even lower potentials than may be contained in the coordinated electrolyte, e.g., HSO_4^-. The dispersed platinum is the main species that is responsible for the E/pH shift in this electrode system because of a pH decrease at the interface and the counter-electrode platinum dissolution. The need to investigate, and understand, the behavior of rough and disordered metal surfaces was highlighted before by many authors, that is, defect systems are generally the most active from a catalytic viewpoint. The higher energy states of platinum are unstable and are usually not highly occupied. The interchange between these states may occur and, the activity of the interfacial atoms may be significantly greater during the course of an electro-catalytic reaction since they are invariably accompanied by the release of energy at the interface. Structural techniques have not evolved to the stage where minute changes in structure (or disorder) at the active sites at the surfaces and the interfaces can be readily detected, especially when these changes occur during the course of a reaction.

Further, the performance of single-crystal plane platinum electrodes in an aqueous media is also important, as the effect of the crystallographic orientation can be ascertained by using highly oriented surfaces. Such surfaces are regarded as being much better defined and more restrictively treated from a theoretical viewpoint than that of the polycrystalline surfaces. Such surfaces are now employed for research concerned with the structure, adsorption performance, and electrocata-lytic properties of the metal/solution interfaces. However, despite their apparent simplicity, the behavior of the platinum single crystal surfaces is surprisingly complex. For example, the potentio-dynamic performance of Pt(111) electrodes in an acid solution is still a matter of debate. According to various authors, it is still not clear whether the species responsible for the anomalous voltam-metric features, e.g., the well-known butterfly peaks, are due to the reaction of hydrogen, an oxygen-like species, a nickel oxide or a combination of these. These systems are important as they belong to a group of few where severe thermal pretreatment is widely used prior to recording the voltammetric responses for platinum.

As is outlined here, active metal atoms at the defect surface sites are much more likely to undergo oxidation at low potentials. A specific example of the similarity between the single crystal plane area and the data obtained from the polycrystalline platinum is the effect of thermal pretreatment on the response for the Pt(111) electrodes [97]. Prior to heating, sharp peaks due to reversible electro-chemical transitions were observed in the hydrogen region at approximately 0.11 and 0.26 V that correspond to the reaction of weakly and strongly bound hydrogen on the defect sites. However, after the heating process (at 1350°C) these hydrogen peaks were virtually absent—a major revers-ible response appeared between 0.3 and 0.5 V, with sharp peaks at approximately 0.45 V corrob-orating most published results. After the Pt(111) heating process, an additional, less reversible, response was also observed at approximately 0.7 V and the monolayer oxidation response in the range 0.8–1.2 V was dramatically reduced. The latter result obviously raises the possibility that much of the surface of the Pt(111) electrode undergoes oxidation in a rather reversible manner at potentials lower than 0.8 V. Reversible oxidation responses at low potentials were also discussed in the case of polycrystalline palladium electrodes in acid solution [98].

One of the basic assumptions, when working with single crystals, is the virtual absence of surface defects or irregularities. However, this may be far more difficult to achieve because the STM images show that it depends on the preparation method and on the part of the crystal we work on. Wagner and Ross [99] have claimed that a metallic single crystal consists of a mosaic of islands composed of crystallites with diameters in the order of 1 mm, showing that the minimization of the surface defects is not possible in these systems. The STM images of single crystal surfaces [100,101] frequently show the presence of ledges or steps. Such sites function well as points for the poorly stabilized surface metal atoms that, due to thermal promotion at room temperature, may attain the adatom state. Somorjai [102] pointed out that the mobile metal species could not be imaged at the room temperature. Evidently, the presence of the adatoms on a well-ordered single crystal surface cannot be discounted. The basic assumption here is that the atomic diffusion and migration occurs, and that, ideally, no disorder exists at the surface in the final state. However, it seems improbable that all the surface atoms can find a suitable high coordination site in the outermost layer, that is, whether the ideal monoatomic state can be achieved in practice. It is assumed here that where severe thermal pretreatment is involved, active platinum atoms exist at the interface, and that these are the species responsible for the unusual electrochemical responses at low potentials. The involvement of a hydroxyl radical is discounted for energetic reasons but a hydroxyl platinum species may be formed by the coordination of the hydroxide ions by the cations generated by the oxidation of the adatoms.

12.2.2.2.3 Oxidation and Electrooxidation Processes of Non-Noble Metals in the Gas Phase

Some of the transition metal–oxide systems have become a subject of intensive research in the last two decades. The relation between the parabolic oxidation kinetics and the predominating point defect in the oxide was verified. To discuss the high-temperature oxidation mechanism of non-noble metals it is appropriate to start with a brief survey of some of the literature on the point defect dependent properties of, for example, nickel oxide.

The point defect structure of nickel oxide has been investigated with various experimental techniques that include electrical conductivity [103], thermo-gravimetry [104], coulombimetric measurements [105], tracer diffusion measurements [106], and high-temperature oxidation [107]. It has been concluded that nickel oxide has to be a metal deficient p-type semiconductor, with nickel vacancies as the predominating defect. The oxygen pressure dependences of the different point-defect governed processes have been reported and interpreted to reflect that the vacancies may be singly or doubly charged. The non-stoichiometry in the nickel oxide is rather low and the concentration of the nickel vacancies at $1000°C$ in 1 atm oxygen is approximately 10^{-3} to 10^{-4} atomic fraction. Owing to the low concentration of the native defects, doping easily influences the defect structure of the nickel oxide [103]. Hence, the sensitivity toward the impurities is probably one important reason for the scatter among the values of the non-stoichiometry reported in the literature. To get reliable data on the native defect structure of the nickel oxide, caution must be shown with respect to the selection of the materials and the specimen treatment. Tracer studies are important in determining the transport mechanisms in the nickel oxide, both at high and low temperatures [106,108]. Atkinson and Taylor [108] carried out one of the most comprehensive investigations. The diffusion profiling of ^{63}Ni by secondary ion mass spectroscopy (SIMS) was conducted as a function of the annealing temperature from $500°C$ to $1400°C$. Through these measurements, Atkinson and Taylor [108] distinguished between the bulk, dislocation, and grain boundary diffusions and deduced semiempirical relations for the diffusivity of the transport mechanism. They also determined the approximate thickness, δ, of the grain boundary for the nickel oxide to be 700 nm. The measurements of the oxygen-pressure dependence for the grain boundary diffusion of the nickel indicate that the transport occurs according to a vacancy mechanism and that the singly charged vacancies predominate as the carriers [108].

Dubois and Monty [109] determined the self-diffusion of the oxygen in the nickel oxide between $1100°C$ and $1600°C$ by tracer diffusion measurements of $^{18}O_2$. When compared with the self-diffusion of nickel, the lattice diffusivity of oxygen is lower by more than five orders of magnitude in the temperature region to which most investigations of the oxidation mechanisms refer. The activation energy for the oxygen diffusion in the nickel oxide is 530 kJ mol^{-1}. There is some uncertainty as to the nature of the oxygen point defect but modeling indicates that the interstitials are the preferred defects [110]. Atkinson et al. [111] measured the oxygen diffusivity along the grain boundaries and found that the rate is much slower than the Ni grain-boundary diffusion. In fact, it was comparable with the nickel diffusivity in the nickel oxide lattice with an activation energy of 240 kJ mol^{-1}.

The high-temperature oxidation mechanism for the pure nickel can be classified based on the temperature range of operation. At the higher temperatures, the bulk diffusion determines the oxidation rate, while the short-circuit transport mechanisms occur at lower temperatures [112]. The physical distinction in the temperature is a consequence of the difference in the activation energy for the transport mechanisms. The temperature range in which the change in the growth mechanism occurs depends on factors that affect the grain size of the oxide, the purity of the nickel, and the surface preparation. Since the non-stoichiometry is rather low, the impurities that dissolve in the nickel oxide change the native defect's concentration. According to the point defect chemistry, ions with a higher valence than that of the Ni(II) increase the concentration of the metal vacancies and, in turn, the oxidation rate, whereas ions with a lower valence than that of the Ni(II) should, ideally, decrease the oxidation rate. The oxidation of the nickel in the temperature range above 1100°C is parabolic. The parabolic rate constant, k_p, is proportional to the surrounding oxygen pressure [113] with partial pressure dependences of the 1/4 to the 1/6 exponents. As for the other defect dependent properties, they are attributed to the growth governed by the lattice diffusion via singly or doubly charged nickel vacancies [113]. The activation energies reported for the oxidation at these temperatures are rather consistent and in the order of 220–250 kJ mol^{-1}. The oxide scales that are grown above 1100°C consist of columnar grains that extend through the entire scale. These oxides are essentially dense except for some microporosity at the oxide–metal interface. In a rather extensive investigation of the nickel oxidation from 600°C to 1200°C, Graham et al. [114] worked on the sub-parabolic kinetics and compared the changes in the instantaneous parabolic rate constant and oxide morphology with time for different surface pretreatments (annealing procedures prior to the oxidation). The oxidation rate was markedly influenced by the surface treatments of the specimen. At 600°C, differences in the oxidation rate constants were observed as high as four orders of magnitude. This was attributed to the varying amounts of the short circuit paths, such as grain boundaries and dislocations.

In the relative low temperature regime (below 700°C) a fine-grained surface structure with whiskers is formed. However, for increasing temperatures (above 900°C) ridges are often observed at the surface. Finally above 1100°C facetted nickel oxide grains dominate the surface morphology. The important morphological feature of the thermally grown nickel oxide cross-sections, both for the oxidation of the metal and, particularly, when oxidizing the Ni-rich alloys is the so-called duplex scale structure. In this structure the oxide scales can be divided into an outer region consisting of the columnar grains and an inner region consisting of the equally axed grains. The inner layer grows by the inward transport of oxygen. Since the oxygen diffusivity in the nickel oxide is much lower than the self-diffusion of the nickel, lattice diffusion cannot explain the growth of the inner fraction of the scale [108]. The occurrence of oxygen dissociation across the cavities during the outward nickel diffusion is responsible for the opening of the microfissures due to the anomalous diffusivity along the grain boundaries, and in the bulk phase by the transport though the micropores [115,116]. The thickness of the inner layer can be related to the purity of the material [117].

12.2.2.2.4 Passivation, Transpassivation, and Stability of Passive Layers of Non-Noble Metals

The electrochemical oxidation of the nickel is of special interest since it is a typical passivation metal in which very thin passive oxide films of a few nm thickness on the surface can cover the substrate metals efficiently. The passive oxide layer on the nickel was studied by Sikora and Mac Donald [118] who claimed that the passive film consisted of the inner nickel oxide of a barrier layer and an outer $Ni(OH)_2$ porous or hydrated layer, in which the inner layer behaves as a p-type oxide with a cation vacancy. Oblonsky and Devine measured the surface enhanced Raman spectra of the nickel passivized in a neutral borate solution and estimated the amorphous $Ni(OH)_2$ in the passive potential region and the NiOOH in the higher transpassive region [119]. Further, the passive films formed in the acidic and neutral solutions were assumed as partially hydrated nickel oxide [120,121]. The anodic film formed in the alkaline solution was assumed to be $Ni(OH)_2$ in the passive region and NiOOH in the high potentials close to the oxygen evolution. The films in the alkaline solution were supposed to be much thicker than those in the neutral solution [122]. Both passive layers formed in the neutral and the alkaline solutions behave as a p-type semiconductor from the Mott–Schottky plot [123].

The thickness of the passive film can be measured by ellipsometry. On one hand, some authors have found that the nickel oxide thickness in the neutral borate solution is 0.7–1.5 nm, depending on the anodic potential [124]. On the other hand, in acidic solution, the thickness was restricted to the high anodic dissolution that produces a roughened surface. Ohtsuka et al. estimated the thickness and optical properties of the nickel passive oxide film in an acidic sulfate solution by the reflectance measurement of polarized lights [125], that is, 1.5–2.0 nm. Besides, other authors have measured the thickness by using ellipsometry combined with the reflectance of the initial passive film in an acidic sulfate solution [126] by finding a complex refractive index that is dependent on the potential and the oxidation time. Recently, the thickness of the passive film on a roughening surface was estimated from ellipsometric parameters during the cathodic reduction that follows anodic passivation. It was found that the thickness of the passive film was about 1.4–1.7 nm and that the cathodic polarization introduced hydration of the passive film [127]. During the anodic passivation followed by the cathodic reduction, the roughness increases with the dissolution of the nickel as indicated by the gradual decrease of the reflectance. However, the ellipsometric parameters, Ψ (arctangent of the relative amplitude) and Δ (relative phase retardation), are relatively insensitive to the roughness increase. From the change of both the optic parameters, the thickness of the passive oxide film was estimated by assuming that the refractive index $n = 2.3$. The thickness covers between 1.4 and 1.7 nm in the passive region from 0.8 to 1.4 V vs. RHE. The cathodic reduction at a constant potential changes the oxide film to a new oxide with a lower refractive index ($n = 1.7$) accompanied by the thickening of about 30% more in the initial stage of the reduction for 30 s. According to these authors [127] the potential change from the passive region to the cathodic hydrogen evolution region may initially cause the hydration of the passive oxide to $Ni(OH)_2$, and during the latter stage of reduction, the hydrated nickel oxide gradually dissolves.

It is also important to consider the electrochemical oxidation of the nickel in the alkaline solutions, primarily, because of its application in rechargeable alkaline batteries [128]. Numerous studies show that during the anodic oxidation of the nickel in the alkaline solutions under galvanostatic [129], potentiodynamic [7], and voltammetric [130] conditions, the initial reaction is the formation of $Ni(OH)_2$. Some other authors [131] found that the film formed in the phosphate buffer is constant in structure during the growth and at high potentials the formed $Ni(OH)_2$ film experienced dehydration and decreased in thickness. In the borate buffer solutions, the nature of the film was considered most likely to be NiO [132]. Other studies [133] stated that the oxide film layer formed under these conditions must be a mixed layer consisting of NiO and Ni_3O_4. Ord et al. [134] have investigated the nature of the anodic films on the nickel in a neutral electrolyte by the use of alternating anodic and cathodic galvanostatic cycles and concluded that the oxide film formed on the first anodic cycle is NiO. De Souza et al. [128], using coupled ellipsometry, have suggested that the first layer of NiO is covered after the positive scan with a thick film of β-$Ni(OH)_2$. At potentials above a critical value, the β-$Ni(OH)_2$ film is converted to the Ni(III) oxide phase identified by ellipsometry as β-NiOOH [135]. The conversion of the β-$Ni(OH)_2$ to the β-NiOOH is a complex process [135], where as the potential increases, the oxygen begins to evolve and soon becomes the predominant reaction. With reverse polarization, the β-NiOOH is reduced back to the β-$Ni(OH)_2$ that cannot be reduced further to nickel even in the hydrogen evolution region [136]. If the polarization in the anodic direction is reapplied, the electrode behaves quiet differently than in the first anodic polarization. This is attributed to the fact that the nickel oxides have already covered the electrode during the second polarization. For the study of the initial nickel oxide formation, it was therefore, necessary to mechanically polish and cathodically reduce the nickel prior to each measurement [129]. Since the imposed current density, concentration of the electrolyte, temperature, and solution pH are expected to play important roles in determining the properties of the passive film, a systematic investigation on the effect of these parameters on the growth of the oxide films on nickel is necessary.

Nickel exhibits excellent corrosion resistance in aqueous aggressive environments that are attributed to the production of stable passive films [137]. Using coulometry, electron diffraction, and x-ray emission spectroscopy, the passive layer structure of the nickel in the sulfate solutions of various pH from 2.0 to 8.4 was studied [138]. These authors found that the passive film was composed of NiO, and that its physical and chemical properties were independent of the potential and the pH. Nickel passivity was also investigated in the pH 8.39 borate buffer solutions by the polarization potential decays [133]. They found that the passive film has a duplex structure consisting of NiO and Ni_3O_4. Sato and Kudo [124] studied the passivity of the nickel in the pH 8.4 solutions by the ellipsometric and coulometric methods, and found that the NiO is the only film present in the passive layer. The nickel in the alkaline solutions was analyzed by the reflectance and ellipsometric techniques [139] with the finding that the passive film consisted of the dehydrated $Ni(OH)_2$. Chao and Szklarska-Smialowska [131] have found that the film formed in phosphate buffer is $Ni(OH)_2$. However, other authors [134] investigated the nature of the anodic films in neutral solutions by alternate anodic and cathodic galvanostatic cycles with the result that the oxide film formed on the first anodic cycle was NiO. The growth of the oxide layers on the nickel in the alkaline solutions with cyclic voltammetry [140] produces a thin layer of $NiO \cdot nH_2O$ and with increasing anodic potential, it transforms into $Ni(OH)_2$. At higher pHs in borate and phosphate solutions by potentiostatic techniques, the view proposed by Okuyama and Haruyama [133] that the film in a passive region consists of $Ni(OH)_2$ in the outer region and NiO at the inner layer was confirmed.

The effect of the halides on the breakdown of the passivity of nickel is very important to understand the stability of the passive layer in a strong aggressive environment. In a study of alkaline solutions by potentiodynamic and galvanostatic techniques, it was found that the critical Cl^-/OH^- ratio required for breakdown increases at high hydroxyl concentrations [141]. The breakdown of the nickel passivity [142] was considered galvanostatically in dilute alkaline solutions in the presence of chloride anions from 0.0 to 0.6 M, and it was concluded that the electrode was activated by the desorption of the passivating oxygen by the chloride anions. The influence of the chloride, bromide, and iodide as aggressive anions [143] was also analyzed by the potentiostatic polarization of the nickel in the KOH and K_2CO_3 solutions. The addition of lower concentrations of aggressive anions has no effect on the mechanism of nickel passivity, while higher concentrations accelerate the dissolution of the nickel in both the active and passive regions. MacDougall and Graham [144] found that the passivity breakdown of the nickel occurs on an oxide surface having a small number of local defects. The main role of the aggressive anions in initiating pitting corrosion is to hinder the oxide repassivation at the local sites in the passive oxide rather than to the direct breakdown or thinning of the passive film. Nishimura [145] proposed that the passivity breakdown occurs in three stages: pit incubation time, pit nucleation, and pit growth. The time required for a transition from the pit nucleation to the pit growth was found to be a linear function of the thickness of the barrier layer next to the metal irrespective of the anion species and the solution pH.

In the case of the nickel alloys, the stability of the passive layer is a problem. The alloys depend on the oxide films or the passive layers for corrosion resistance and are susceptible to crevice corrosion. The conventional mechanism for crevice corrosion assumes that the sole cause for the localized attack is related to compositional aspects such as the acidification or the migration of the aggressive ions into the crevice solution [146]. These solution composition changes can cause the breakdown of the passive film and promote the acceleration and the autocatalysis of the crevice corrosion. In some cases, the classic theory does not explain the crevice corrosion where no acidification or chloride ion build up occurs [147].

The crevice corrosion occurred even at constant pH and zero chloride concentration, and was shown to be caused by the ohmic drop placed by the local electrode potential existing on the crevice wall in the active peak region of the polarization curve [148]. This can involve an increase in the size of the active peak or the formation of one if none exists in the bulk-solution polarization curve. Nevertheless, most researchers have reasoned that the ohmic drop is a consequence, rather than the cause, of the crevice corrosion. Thus, the actual importance of the *IR* voltage was not recognized until the works of Pickering [149] during the past decade. He demonstrated its operation and primary importance under the conditions of constant pH and zero chloride for the systems with active:passive transitions in their polarization curves. These studies also showed that the gas bubbles were not a necessary factor although, as shown earlier [150], they could promote susceptibility by increasing the magnitude of the *IR* voltage. In practice, the crevice corrosion can occur in the presence of an oxidant like oxygen or during anodic protection. The metal dissolves inside the crevice and the anodic current flows through the crevice electrolyte to the outer surface where the oxidant is reduced. The resulting *IR* voltage translates into an electrode potential on the crevice wall.

It is important to study the potential distribution inside the crevice and its relation to the polarization curve in order to obtain a better understanding of the mechanism by which the crevice corrosion occurs. Factors that influence both the potential distribution and the polarization curve include the metal environment reaction products and the crevice geometry as well as the well-known promoters of crevice corrosion: acidification and chloride ion build-up within the crevice. When excess oxygen entered the crevice via flushing air-saturated solution through it, the corrosion rate decreased which was consistent with the proposed need to keep the cathodic reaction outside the crevice, separate from the anodic reaction, for a stable crevice corrosion to occur [151].

In the case of crevice corrosion or for paint adhesion improvements, the phosphating of the metal surfaces is widely used. The extensive reviews on the various aspects of phosphating and phosphate coatings are given in the literature [152–155]. The phosphating process involves the dissolution of a base metal in an acidic solution of soluble primary phosphates, with the subsequent hydrolysis of these phosphates and the precipitation of the insoluble tertiary compounds. The phosphating baths for the mild steel and zinc usually contain ferrous, zinc, and manganese cations. Hence, the tertiary phosphates of these cations are the most common constituents of the phosphate coatings.

The characterization of the phosphate coatings involves determination of the porosity. Among the usual methods, we have those based on the electrochemical and corrosion measurements that assume that the magnitude of the anodic or the cathodic currents is related to the metal surface exposed inside the pores. Machu [156] measured the time for attaining a steady state current during the anodic polarization in 0.5 M Na_2SO_4 and estimated the pore area in the range of 0.145% of the total surface. Zurilla and Hospadaruk [157] estimated a pore surface area from the magnitude of the cathodic current in the 0.01 M NaOH solution. The pores can be revealed by the cementation of the copper in a dilute $CuSO_4$ solution or by corroding the phosphated steel in an aggressive solution with a color indicator. Cheever [158] recommended a solution containing 4% NaCl and 3% $K_3[Fe(CN)_6]$. Kwiatkowski et al. [159] evaluated the phosphate coatings by impedance measurements in 0.5 M sodium phosphate buffer of pH 7.0. The solutions used in the above tests are rather aggressive toward both the base metal and the phosphate coating. Hence, upon immersion the pore size and its number may increase and lead to an overestimation of the results. The occurrence of extensive corrosion can be inferred from a shift of the voltammetric curves for the zinc phosphated steel during cycling in 5% NaCl or from the potential and the pH dependence of the estimated porosity [160]. Other authors have reported that the porosity estimated in 0.01 M NaOH rose with the increasing cathodic polarization, while the porosity measured in pH 7.0 rose with the decreasing pH [161]. These data show that a considerable deterioration of the phosphate coatings can occur during the measurements, resulting in erroneous estimations of the coating porosity. Electrochemical impedance spectroscopy was used [161] as a non-aggressive test solution for the evaluation of the phosphate coatings in 5×10^{-4} M Na_2HPO_4 solutions on electrodeposited zinc 12% nickel. The ratios of the charge transfer resistance and of the interfacial capacitance for the phosphated and non-phosphated materials can be used for the quality evaluation. However, they cannot be assigned to real porosity of the phosphate coatings.

12.2.2.2.5   Inhibition of Non-Noble Metal Dissolution

The study of the interaction of the nickel with the substances that exhibit corrosion inhibitory properties offers the possibility for modifying the characteristics of the surface and making it more resistant to an aggressive environment [162,163]. Unfortunately, many common corrosion inhibitors used in corrosion inhibition in aqueous solutions are health hazards [164]. Many proven inhibitors with toxic properties include the aromatic and nitrogen containing heterocyclic compounds that find applications in pickling processes and in the oil and gas industry [165]. Some papers have dealt with the inhibition properties of the amino acids in strong and environments, since these compounds are nontoxic and easy to produce in purities greater than 99%. The corrosion potentials and the corrosion current densities can be determined by extrapolating the cathodic Tafel regions at the corrosion potentials [166]. Also the percentage inhibition efficiency, ε, can be calculated using the following formula:

$$\varepsilon = \frac{(j_{corr,o} - j_{corr})}{j_{corr,o}} \tag{12.3}$$

where $j_{corr,o}$ and j_{corr} are the corrosion current densities for the uninhibited and the inhibited surfaces, respectively.

On the other hand, it is possible to observe two current peaks in the anodic region in the current–potential curves of pure nickel in H_2SO_4 [167]. Most of the amino acids also gave the two peaks in the sulfuric acid solutions, but for some amino acids, the first peak or even both of the peaks are not visible due to corrosion inhibition. In the case of leucine, glutamic acid, and lysine, good inhibition results are obtained and for the lysine the highest efficiency of all the compounds can be obtained. When the pH increases the inhibition effect is decreased, however, the results showed that the amino acids are effective in acidic media. When the pH value increases, the passivation peaks become smaller and the corrosion rates increase.

It is known that the presence of carbon monoxide in the aqueous solutions inhibits the nickel corrosion over a certain potential range [168], this being the inhibition process explained by the CO-adsorption on the nickel active sites [169]. The interaction of some unsaturated molecules with the solid electrode surfaces is the result of adsorption and/or electro-adsorption processes. The interaction of the carbon monoxide, the double bonded species (ethylene and allyl alcohol), and the triple bonded species (acetylene and propargyl alcohol) produced a strong inhibition of nickel electrodissolution. The potentiodynamic profile of nickel in sulfuric acid is shown in Figure 12.6A and in the presence of carbon monoxide, ethylene, and acetylene it changes according to Figure 12.6B through D. Besides the important features appearing from the voltammetric results, the positive potential shift of the open circuit potential is a direct index of the corrosion inhibition behavior (Table 12.1).

Based on the accepted mechanism for nickel electrodissolution [170] and the potentiodynamic expressions of the electrocatalytic reactions, the mathematically simulated complex voltammetric contours were deconvoluted to obtain each anodic stripping peak. The values of k'_o (formal electrochemical rate constants) were also calculated using the Runge–Kutta method [171]. The voltammetric profile of the Ni/H_2SO_4 interface is deconvoluted in two peaks, namely, I and II (Figure 12.6A). The two peaks are related to the surface oxidation of nickel according to reactions (IX) to (XI). The peak I ($k'_o = 5.7 \times 10^{-2}$ cm s^{-1}) obeys the following reactions:

$$Ni_{(s)} + H_2O_{(l)} \rightarrow [NiOH]_{(ads)} + H^+_{(aq)} + e^-_{(Ni)} \tag{IX}$$

$$[NiOH]_{(ads)} \rightarrow [NiOH]^+_{(ads)} + e^-_{(Ni)} \tag{X}$$

and its chemical dissolutions to the $Ni(OH)_2$ formation. On the other hand, peak II ($k'_o = 1.2 \times 10^{-4}$ cm s^{-1}) is caused by

$$Ni(OH)_{2(s)} + Ni_{(s)} \rightarrow 2[NiOH]^+_{(ads)} + 2e^-_{(Ni)} \tag{XI}$$

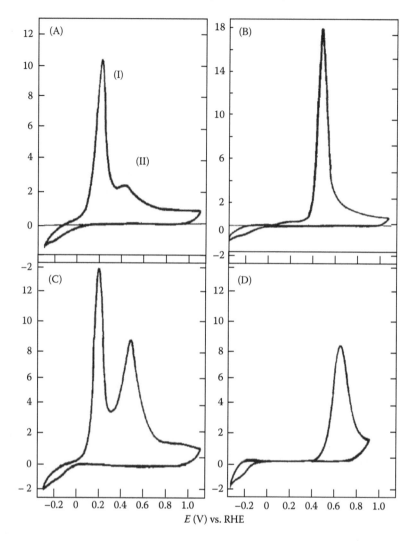

FIGURE 12.6   (A) Cyclic voltammogram for nickel in 0.5 M H_2SO_4 run at 0.1 V s^{-1}; (B) saturated carbon monoxide; (C) saturated ethylene; (D) saturated acetylene.

TABLE 12.1
Open Circuit Potentials ($E_{open\ circuit}$) for Nickel in 0.5 M Sulfuric Acid at Room Temperature

Organic Inhibitor	$E_{open\ circuit}$/V vs. RHE
Without any inhibitor	−0.038
Saturated carbon monoxide	0.018
Saturated ethylene	0.018
0.10 M Allyl alcohol	0.110
Saturated acetylene	0.288
0.10 M Propargyl alcohol	0.278

$$[NiOH]^+_{(ads)} + [NiOH]_{(ads)} \rightarrow Ni(OH)_{2(s)} + Ni^{2+}_{(aq)} + e^-_{(Ni)} \qquad \text{(XII)}$$

and its chemical aging is attributed to NiO. The stable $Ni(OH)_2$ species causes the early stages of the overall passivation process that is associated to an almost constant current observed after 0.6 V up to 1.0 V. The latter process competes with the chemical dissolution of the $Ni(OH)_2$, which also yields soluble nickel ions.

In the presence of carbon monoxide, the potentiodynamic response develops three peaks at $k'_o = 5.0 \times 10^{-2}$, 3.1×10^{-5}, and 8.3×10^{-5} cm s^{-1}, and an obscure peak between I and II can also be seen. On the other hand, in the presence of ethylene, we can find 4.5×10^{-2}, 1.4×10^{-7}, and 7.0×10^{-7} cm s^{-1} and, finally, in the presence of acetylene, 1.0×10^{-10} and 2.9×10^{-11} cm s^{-1}. The use of allyl or propargyl alcohol renders similar results to ethylene or acetylene with an easier handling, because of their liquid forms, besides the possibility of changing their concentrations.

The change in the nature of the ligand from water to, for example, carbon monoxide produces the subsequent reactions:

$$2Ni[H_2O]_{(ads)} + CO_{(sat)} \rightarrow Ni[H_2O]_{(ads)} + Ni[CO]_{(ads)} + H_2O_{(l)} \qquad \text{(XIII)}$$

Whether the oxidative or single nonoxidative desorption of the carbon monoxide occurs from the polycrystalline nickel surface was studied by differential electrochemical mass spectrometry (DEMS) [172]:

$$Ni[CO]_{(ads)} + 2H_2O_{(l)} \rightarrow Ni(OH)_{2(s)} + CO_{(sat)} + 2H^+_{(aq)} + 2e^-_{(Ni)} \qquad \text{(XIV)}$$

or

$$Ni[CO]_{(ads)} + 3H_2O_{(l)} \rightarrow Ni(OH)_{2(s)} + CO_{2(g)} + 3H^+_{(aq)} + 3e^-_{(Ni)} \qquad \text{(XV)}$$

The experimental results suggest the following sequence: first, a carbon monoxide and water species adsorption on nickel, second, the electrodissolution of nickel is inhibited by the adsorbed carbon monoxide, third, the anodic stripping of the adsorbate occurs at the rising branch of the current vs. potential profile, and this is finally followed by the oxidation of the underlying nickel layers. Thus, the carbon monoxide mass-current intensity signal reaches its maximum before the voltammetric peak that is almost totally determined by the bulk electrodissolution–precipitation process of nickel. A small portion of the remaining adsorbate is oxidized to carbon dioxide induced by the co-adsorbed hydroxyl present at the nickel interface.

The adsorption of CO was also used to protect the Ni(111) single crystal surface during the transfer from the UHV into the alkaline electrolytes [173]. These studies show that the CO is strongly adsorbed on the nickel and that the nickel electrodissolution–precipitation voltammetric peak shifts to more positive potentials in the presence of CO.

Hori and Murata [174], working at pH 6.8 in a phosphoric buffer electrolyte, proposed the electrooxidation of the protecting layer of the adsorbed CO to CO_2 followed by the electrooxidation of the Ni electrode. This was deduced from the values of the potential of the CO electrooxidation on the platinum and the rhodium electrodes and the standard equilibrium potential of the CO electrooxidation reaction. For solutions at pH values between 1 and 3 with sulfuric and perchloric electrolytes [175,176], it was proposed that the CO removal consisted of the stripping of an adsorption layer that includes the CO that was competitively adsorbed with water. This suggestion was because the Ni oxidation peak in the CO-containing acid solutions appeared only to be shifted in comparison with that in the pure acid solutions [167].

12.2.2.2.6 Study of Metal Corrosion by In Situ *Techniques*

Metal corrosion in the electrochemical environment can occur by homogeneous attacks in aqueous environments, localized heterogeneous corrosion, or stress corrosion cracking. The performance of a particular material is often controlled by the protective qualities of the formed surface oxide. The modern surface-analytical techniques can provide useful information regarding the nature and composition of the passive oxide films, leading to a better understanding of the causes of corrosion at an atomic scale. These techniques also provide useful information regarding the oxide composition and the transport processes during the oxide growth. Over the years, efforts have focused on the nature of passivity, its breakdown by aggressive ions, and the mass transport processes during the oxide growth on each metal or alloy.

Secondary mass spectrometry (SIMS), XPS and Auger spectroscopy are suitable to study the thick oxide films, as the nature and composition of them are strongly dependent on the electrode potential and the current distribution of the corrosion process. Thus, the local chemical composition can define the mass transport (metal ions, anions, water, etc.) effects to corrosion and then, the dissolution rate too.

Passivity with the nickel can be readily achieved in a wide variety of solutions over a large range of pH. A highly buffered neutral solution is not necessary to achieve good passivity. Different studies can be cited here; however, it seems that the passive film is entirely the NiO, of approximately 1 nm thickness [177], which is not affected by air exposure. Electron diffraction studies on the Ni(111) showed that the epitaxial growth of NiO is maintained during the anodic charging in the passive potential region, but is lost in the oxygen evolution region, where a porous, sponge-type oxide is formed. ^{18}O SIMS has been used to identify changes in the course of the oxide film during passivation [178]. The films formed in the ^{18}O-containing borate solution have been exposed to pH 1.0 containing ionic force solutions of the sulfate without the ^{18}O. The percentage of ^{18}O decreases with time suggesting that there are breakdown and repair events occurring within the oxide, which lead eventually to its complete reformation in the non-^{18}O electrolyte. The passive current is a monitor of the defect character of the film.

With respect to the passivity breakdown in the chloride-containing solution, there is considerable disagreement as to whether the chloride ion is incorporated into the film as a precursor for the initiation of the pitting corrosion. Diverse spectroscopic methods have been used to determine the presence of the chloride ion in the films formed on nickel. The passive films formed on the nickel in large concentrations of chloride are found to contain up to 4% chloride ion, whereas the films formed on iron do not incorporate the chloride ion even when formed in a chloride-containing solution [179]. The presence of the included chloride in the oxide films changes the open-circuit breakdown performance of nickel. Experiments have shown that the incorporated chloride is not a precursor for pit initiation and that the induction time to pitting in pH 4.0 sulfate-containing solution is longer when the chloride is incorporated in the film [180].

In the case of iron, the oxide film is crystalline consisting of an inner layer of magnetite (Fe_3O_4) and an outer layer of maghemite (γ-Fe_2O_3) that are cation deficient [181]. In other theories [182,183], the oxide film was considered as an amorphous gel with multiple oxygen and hydroxyl bonds. The use of the SIMS allows the detection of the hydrogen-containing species and the hydroxyl ions within the passive film. The absence of these species confirmed the first theory. The iron oxide film is removed by ion sputtering and the FeO^+ and $FeOH^+$ signals in SIMS experiments are measured through the depth of the film. The profile for the passive oxide film is very similar to that for the "dry" Fe_2O_3 standard. The use of conversion electron (back-scattered) Mossbauer spectroscopy (CEMS) is much more surface-sensitive than the conventional transmission technique and has been used to examine the passive oxide films on iron. These data show that the film consists of small particle size γ-Fe_2O_3 oxide [184].

The stability of the passive film has been checked also in aqueous solution and in ultrahigh vacuum conditions. In the case of iron, the unstable films thicken up to a constant thickness of

1.8 nm, whereas for Fe–Cr alloys, the thickness steadily diverges as the potential becomes less anodic. An explanation consistent with this behavior is that the instability of passive Fe–Cr films in air reflects the ability of the films to adapt to changing environmental conditions. The presence of exchangeable hydroxyl would be consistent with the decrease in ^{18}O enrichment observed by SIMS. A continuous increase in the concentration of the exchangeable hydroxyl would explain the steady divergence of the oxide thickness with the decreasing potential. In this model [185], the passive film on Fe–Cr alloys is considered as a flexible and labile layer. The ability of the passive film to accommodate to the environment may be a factor in the superior corrosion resistance of Fe–Cr alloys. On the contrary, the inflexible γ-Fe_2O_3/Fe_3O_4 layer formed on iron would break down under dramatic environmental conditions rather than rearrange itself to respond to the new circumstances.

REFERENCES

1. M.R. Tarasevich and B.N. Efremov, in: *Electrodes of Conductive Metallic Oxides* (S. Trasatti, Ed.) Part A, Elsevier, Amsterdam, the Netherlands (1980) p. 221.
2. S. Trasatti and G. Lodi, in: *Electrodes of Conductive Metallic Oxides* (S. Trasatti, Ed.) Part B, Elsevier, Amsterdam, the Netherlands (1981) p. 521.
3. C. Piovano and S. Trasatti, *J. Electroanal. Chem.* **180** (1984) 171.
4. V.V. Shalaginov, I.D. Belova, Y.E. Roginskaya, and D.M. Shub, *Elektrokhimiya* **14** (1978) 1708.
5. R. Garavagha, C.M. Mari, S. Trasatti, and C. De Asmurdes, *Surf. Technol.* **19** (1983) 197.
6. R. Boggio, A. Carugati, G. Lodi, and S. Trasatti, *J. Appl. Electrochem.* **15** (1985) 335.
7. R.S. Schrebler Guzmán, J.R. Vilche, and A.J. Arvia, *J. Electrochem. Soc.* **125** (1978) 1578.
8. H. Gomez Meier, J.R. Vilche, and A.J. Arvia, *J. Electroanal. Chem.* **134** (1982) 251.
9. H. Gomez Meier, J.R. Vilche, and A.J. Arvia, *J. Electroanal. Chem.* **138** (1982) 367.
10. O.A. Albano, J.O. Zerbino, J.R. Vilche, and A.J. Arvia, *Electrochim. Acta* **31** (1986) 1403.
11. L.D. Burke and O.J. Murphy, *J. Electroanal. Chem.* **109** (1980) 373.
12. T. Kessler, A. Visintín, M.R. de Chialvo, W.E. Triaca, and A.J. Arvia, *J. Electroanal. Chem.* **261** (1989) 315.
13. A.J. Bard, R. Parsons, and J. Jordan, Eds., *Standard Potentials in Aqueous Solution*, Marcel Dekker, New York (1985).
14. R.S. Schrebler Guzman, J.R. Vilche, and A.J. Arvia, *Electrochim. Acta* **24** (1979) 395.
15. R.S. Schrebler Guzman, J.R. Vilche, and A.J. Arvia, *J. Appl. Electrochem.* **11** (1981) 551.
16. G. Kreysa and B. Hakansson, *J. Electroanal. Chem.* **201** (1986) 61.
17. A. Baiker, *Faraday Discuss. Chem. Soc.* **87** (1989) 239.
18. G.A. Tsirlina, A. Petrii, and N.S. Kopylova, *Elektrokhimiya* **26** (1990) 1059.
19. F. Pearlstein, in: *Modern Electroplating* (F.A. Lowenheim, Ed.) John Wiley, New York (1974).
20. M. Schlesinger, in: *Electroless Deposition on Metals and Alloys, Proceedings*, Vol. 88–12, The Electrochemical Society, Pennington, New York (1988) p. 93.
21. J.J. Podestá, R.C.V. Piatti, A.J. Arvia, P. Edkunge, K. Júttner, and G. Kreysa, *Int. J. Hydrogen Energy* **17** (1992) 9.
22. A.C. Chialvo, W.E. Triaca, and A.J. Arvia, *J. Electroanal. Chem.* **146** (1983) **93**.
23. A. Visintín, A.C. Chialvo, W.E. Triaca, and A.J. Arvia, *J. Electroanal. Chem.* **225** (1987) 227.
24. T. Kessler, M.R.G. de Chialvo, A. Visintin, W.E. Triaca, and A.J. Arvia, *J. Electroanal. Chem.* **261** (1989) 315.
25. T. Kessler, W.E. Triaca, and A.J. Arvia, *J. Appl. Electrochem.* **23** (1993) 655.
26. T. Kessler, W.E. Triaca, and A.J. Arvia, *J. Appl. Electrochem.* **24** (1994) 310.
27. J. Haenen, W. Visscher, and E. Barendrecht, *Electrochim. Acta* **11** (1966) 1541.
28. L.D. Burke and E.J.M. O'Sullivan, *J. Electroanal. Chem.* **117** (1981) 155.
29. R. Boggio, A. Carugati, and S. Trasatti, *J. Appl. Electrochem.* **17** (1987) 828.
30. R. Sonnenfeld and P.K. Hansma, *Science* **232** (1986) 211.
31. F.R. Wagner and P.N. Ross, *J. Electroanal. Chem.* **150** (1983) 181.
32. H. Kono, S. Kobayashi, H. Takahashi, and M. Nagayama, *Corros. Sci.* **22** (1982) 913.
33. P. Zelenay, M.A. Habib, and J.O'M. Bockris, *Langmuir* **2** (1986) 393.
34. A.T. Hubbard, R.M. Ishikawa, and J. Katekaru, *J. Electroanal. Chem.* **86** (1978) 271.

35. J. Zerbino, A.M. Castro Luna, C.F. Zinola, E. Méndez, and M.E. Martins, *J. Electroanal. Chem.* **521** (2002) 168.
36. L. Blum, H.D. Abruña, S. White, J.B. Cordon, C.L. Barges, M.C. Samant, and O.R. Melroy, *J. Chem. Phys.* **85** (1986) 6732.
37. B. Trake, R. Sonnenfeld, J. Schneir, and P.K. Hansama, *Surf. Sci.* **181** (1987) 92.
38. J. Schneir, R. Sonnefeld, P.K. Hansma, and J. Tersoff, *Phys. Rev. B* **34** (1986) 4979.
39. L. Vázquez, J.M. Gómez-Rodríguez, J. Gómez-Herrero, A.M. Baró, N. García, J.C. Canullo, and A.J. Arvia, *Surf. Sci.* **181** (1987) 98.
40. F.-R. Fan and A.J. Bard, *J. Electrochem. Soc.* **136** (1989) 166.
41. M. Szklarczyk and J.O'M. Bockris, *J. Electrochem. Soc.* **137** (1990) 452.
42. J. Wintterlin, J. Wiecher, H. Brune, T. Gritsch, H. Hofer, and R.J. Behn, *Phys. Rev. Lett.* **62** (1989) 59.
43. M. Szklarczyk, L. Minevski, and J.O'M. Bockris, *J. Electroanal. Chem.* **289** (1990) 279.
44. S. Real, M. Urquidi-MacDonald, and D.D. MacDonald, *J. Electrochem. Soc.* **135** (1988) 1633.
45. R.C. Bhardwaj, A. Gonzalez-Martin, and J.O'M. Bockris, *J. Electroanal. Chem.* **307** (1991) 195.
46. S.-L. Yau, F.-R. Fan, T.P. Moffat, and A.J. Bard, *J. Phys. Chem.* **98** (1994) 5493.
47. P. Marcus and E. Protopopoff, in: *Electrochemical Surface Science of Hydrogen Adsorption and Absorption* (G. Jerkiewicz and P. Marcus, Eds.) Vol. 97–16, The Electrochemical Society Proceedings Series, Pennington, NJ (1997) p. 211.
48. P. Marcus, J. Oudar, and I. Olefjord, *J. Microsc. Spectrosc. Electron* **4** (1979) 63.
49. V. Maurice, W. Yang, and P. Marcus, *J. Electrochem. Soc.* **141** (1994) 3016.
50. B.J. Cruickshank, D.D. Sneddon, and A.A. Gewirth, *Surf. Sci. Lett.* **281**(1993) L308.
51. M.C. Asensio, M.J. Ashwin, A.L.D. Kilcoyne, D.P. Woodruff, A.W. Robinson, T. Linder, J.S. Somers, D.E. Ricken, and A.M. Bradshaw, *Surf. Sci.* **236** (1990) 1.
52. H.C. Zeng, R.A. McFarlane, and K.A.R. Mitchell, *Surf. Sci.* **208** (1989) L7.
53. H. Richter and U. Gerhardt, *Phys. Rev. Lett.* **51** (1983) 1570.
54. M.D. Pritzker and T.Z. Fahidy, *Electrochim. Acta* **31** (1992) 103.
55. D.P. Barkley, R.H. Müller, and C.W. Tobias, *J. Electrochem. Soc.* **136** (1989) 2207.
56. G. Valette, *J. Electroanal. Chem.* **260** (1989) 425.
57. G. Jarzqbek and Z. Borkowska, *Electrochim. Acta* **42** (1997) 2915.
58. R. Parsons and F.G.R. Zobel, *J. Electroanal. Chem.* **9** (1965) 333.
59. G. Valette and A. Hamelin, *J. Electroanal. Chem.* **45** (1973) 301.
60. S.V. Mentus, *Electrochim. Acta* **50** (2005) 3609.
61. M. Ohanian, V. Diaz, S. Martínez, and C.F. Zinola, Technical reports for Thermic Generation "Central Batlle" Cathodic Protection Project (2006) UDELAR.
62. V.V. Andreeva, *Corrosion* **20** (1964) 35.
63. J. Pjescić, S. Mentus, and N. Blagojević, *Mater. Corros.* **53** (2002) 44.
64. M.D. Maksimović and K. Popov, *J. Serb. Chem. Soc.* **64** (1999) 317.
65. C. Lim, K. Scott, R.G. Allen, and S. Roy, *J. Appl. Electrochem.* **34** (2004) 929.
66. N. Cabrera and N.F. Mott, *Rep. Prog. Phys.* **12** (1948–1949) 163.
67. L.D. Burke, in: *Electrodes of Conductive Metallic Oxides* (S. Trasatti, Ed.) Elsevier Science, New York (1980) pp. 141–181.
68. K. Hauffe, *Oxidation of Metals*, Plenum Press, New York (1965).
69. A. Damjanović and V.I. Birss, *J. Electrochem. Soc.* **130** (1983) 1688.
70. R. Woods, in: *Electroanalytical Chemistry* (A.J. Bard, Ed.) Vol. 9, Marcel Dekker, New York (1977) pp. 27–90.
71. B.E. Conway, *Prog. Surf. Sci.* **49** (1995) 331.
72. G. Jerkiewicz and J.J. Borodzinski, *J. Chem. Soc. Faraday Trans.* **90** (1994) 3669.
73. K.J. Vetter and J.W. Schulze, *J. Electroanal. Chem.* **34** (1972) 131; K.J. Vetter and J.W. Schulze, *J. Electroanal. Chem.* **34** (1972) 141.
74. H. Angerstein-Kozlowska, B.E. Conway, and W.B. Sharp, *J. Electroanal. Chem.* **43** (1973) 9.
75. D. Gilroy and B.E. Conway, *Can. J. Chem.* **46** (1968) 875.
76. D. Gilroy, *J. Electroanal. Chem.* **71** (1976) 257.
77. T. Biegler and R. Woods, *J. Electroanal. Chem.* **20** (1969) 73.
78. S. Shibata, *J. Electroanal. Chem.* **89** (1978) 37.
79. Y.B. Vassilyev, V.S. Bagotzkii, and O.A. Khazova, *J. Electroanal. Chem.* **181** (1984) 219.
80. S. Shibata, *Electrochim. Acta* **22** (1977) 175.
81. L.D. Burke and M.B.C. Roche, *J. Electroanal. Chem.* **137** (1982) 175.

82. L.D. Burke, in: *Modern Aspects of Electrochemistry* (R.E. White, J.O'M. Bockris, and B.E. Conway, Eds.) Vol. 18, Plenum, New York (1986) Ch. 4.

83. S.G. Roscoe and B.E. Conway, *J. Electroanal. Chem.* **224** (1987) 163.

84. T.C. Liu and B.E. Conway, *Langmuir* **6** (1990) 268.

85. A.J. Zhang, M. Gaur, and V.I. Birss, *J. Electroanal. Chem.* **389** (1995) 149.

86. V.I. Birss and G.A. Wright, *Electrochim. Acta* **27** (1982) 1.

87. A.J. Zhang, V.I. Birss, and P. Vanýsek, *J. Electroanal. Chem.* **378** (1994) 63.

88. E. Custidiano, A.C. Chialvo, and A.J. Arvia, *J. Electroanal. Chem.* **196** (1985) 423.

89. S. Hadzi-Jordanov, B.E. Conway, H. Angerstein-Kozlowska, and M. Vuković, *J. Electrochem. Soc.* **125** (1978) 1471.

90. J. Mozota and B.E. Conway, *J. Chem. Soc. Faraday Trans.* **178** (1982) 1717.

91. B.E. Conway, H. Angerstein-Kozlowska, and F.C. Ho, *J. Vac. Sci. Technol.* **14** (1977) 351.

92. A.K.N. Reddy, M. Genshaw, and J.O'M. Bockris, *J. Chem. Phys.* **48** (1968) 671.

93. T. Solomun, *J. Electroanal. Chem.* **255** (1988) 163.

94. L.D. Burke and L.M. Hurley, *Electrochim. Acta* **44** (1999) 3451.

95. V. Diaz and C.F. Zinola, *J. Colloid Int. Sci.* **313** (2007) 232.

96. L.D. Burke and P.F. Nugent, *Gold Bull.* **30** (1997) 43.

97. J. Clavilier, K. El Achi, M. Petit, A. Rodes, and M.A. Zamakhchari, *J. Electroanal. Chem.* **295** (1990) 333.

98. L.D. Burke and L.C. Nagle, *J. Electroanal. Chem.* **461** (1999) 52.

99. F.T. Wagner and P.N. Ross, *J. Electroanal. Chem.* **150** (1983) 141.

100. K. Itaya, S. Sugawara, K. Sashikata, and N. Furuya, *J. Vac. Sci. Technol. A* **8** (1990) 515.

101. R.J. Nichols, O.M. Magnussen, J. Hotlos, T. Twomey, R.J. Behm, and D.M. Kolb, *J. Electroanal. Chem.* **290** (1990) 21.

102. G.A. Somorjai, *Chem. Rev.* **96** (1996) 1223.

103. N.L. Peterson, *Solid State Ionics* **12** (1984) 201.

104. R. Haugsrud and T. Norby, *Solid State Ionics* **111** (1998) 323.

105. H.-G. Sockel and H. Schmalzried, *Ber. Bunsenges Physik. Chem.* **72** (1968) 745.

106. E.G. Moya, G. Deyme, and F. Moya, *Scr. Metall. Mater.* **24** (1990) 2447.

107. T. Karakasidis and M. Meyer, *Phys. Rev. B: Condens. Matter* **55** (1997) 13853.

108. A. Atkinson and R.I. Taylor, *Phil. Mag. A* **43** (1981) 979.

109. C. Dubois, C. Monty, and J. Philibert, *Phil. Mag. A* **46** (1982) 419.

110. D.M. Duffy and P.W. Tasker, *Phil. Mag.* **54** (1986) 759.

111. A. Atkinson, F.C.W. Pummery, and C. Monty, in: *Transport in Non Stoichiometric Compounds* (G. Simkovich and V.S. Stubican, Eds.) Plenum Press, New York (1985).

112. C.K. Kim and L.W. Hobbs, *Oxid. Met.* **45** (1996) 247.

113. R. Haugsrud, *Corros. Sci.* **45** (2003) 211.

114. M.J. Graham, R.J. Hussey, and M. Cohen, *J. Electrochem. Soc.* **119** (1973) 1523.

115. A. Atkinson and D.W. Smart, *J. Electrochem. Soc.* **135** (1988) 2886.

116. P. Kofstad, *Oxid. Met.* **24** (1985) 26.

117. J.S. Sheasby and D.S. Cox, *Oxid. Met.* **37** (1992) 373.

118. E. Sikora and D.D. Mac Donald, *Electrochim. Acta* **48** (2002) 69.

119. L.J. Oblonsky and T.M. Devine, *J. Electrochem. Soc.* **142** (1995) 3677.

120. B. Mac Dougall, D.F. Mitchel, and M.J. Graham, *Corrosion* **38** (1982) 85.

121. Y. Kang and W.-K. Paik, *Surf. Sci.* **182** (1987) 257.

122. J.O. Zerbino, C. De Pauli, D. Posadas, and A.J. Arvia, *J. Electroanal. Chem.* **330** (1992) 675.

123. S. Maximovitch, *Electrochim. Acta* **41** (1996) 2761.

124. N. Sato and K. Kudo, *Electrochim. Acta* **19** (1974) 461.

125. K.E. Heusler and T. Ohtsuka, *Surf. Sci.* **101** (1980) 194.

126. S.C. Tjong, *Mater. Res. Bull.* **17** (1982) 1297.

127. M. Iida and T. Ohtsuka, *Corros. Sci.* **49** (2007) 1408.

128. L.M.M. De Souza, F.P. Kong, F.R. McLarnon, and R.H. Muller, *Electrochim. Acta* **42** (1997) 125.

129. J.F. Wolf, L.-S.R. Yeh, and A. Damjanović, *Electrochim. Acta* **26** (1981) 811.

130. A. Seghiouer, J. Chevalet, A. Barhoun, and F. Lantelme, *J. Electroanal. Chem.* **442** (1998) 113.

131. C. Chao and Z. Szklarska-Smialowska, *Surf. Sci.* **96** (1980) 426.

132. B. MacDougall, D.F. Mitchell, and M.J. Graham, *J. Electrochem. Soc.* **127** (1980) 1248.

133. M. Okuyama and S. Haruyama, *Corros. Sci.* **14** (1974) 1.

134. J.L. Ord, J.C. Clayton, and D.J. DeSmet, *J. Electrochem. Soc.* **124** (1977) 1754.

135. M.A. Hopper and J.L. Ord, *J. Electrochem. Soc.* **120** (1973) 183.
136. E.E. Abd El Aal, *Corros. Sci.* **45** (2003) 759.
137. D.D.N. Singh and M.K. Banerjee, *Corros. NACE* **42** (1986) 156.
138. B. MacDougall and M. Cohen, *J. Electrochem. Soc.* **123** (1976) 191.
139. W. Paik and Z. Szklarska-Smialowska, *Surf. Sci.* **96** (1980) 401.
140. W. Vissker and E. Barendrecht, *Surf. Sci.* **135** (1983) 436.
141. J. Postlethwaite, *Electrochim. Acta* **12**(1967) 333.
142. K. Schwabe and R. Padeglia, *Werkst. Korros.* **5** (1962) 281.
143. M.S. Abd El Aal and A.H. Osman, *Corros. NACE* **36** (1980) 591.
144. B. MacDougall and M.J. Graham, *Electrochim. Acta* **27** (1982) 1093.
145. R. Nishimura, *Corros. NACE* **43** (1987) 486.
146. H.H. Uhlig and R.W. Revie, *Corrosion and Corrosion Control*, 3rd edn., John Wiley & Sons, New York (1985).
147. K. Cho and H.W. Pickering, *J. Electrochem. Soc.* **138** (1991) L56.
148. B.A. Shaw, P.J. Moran, and P.O. Gartland, *Corros. Sci.* **32** (1991) 707.
149. H.W. Pickering, *Corros. Sci.* **29** (1989) 325.
150. H.W. Pickering and R.P. Frankenthal, *J. Electrochem. Soc.* **119** (1972) 1297.
151. M.I. Abdulsalam and H.W. Pickering, *Corros. Sci.* **41** (1999) 351.
152. W. Machu, *Die Phosphatierung*, Verlag Chemie, Weinnharin (1950).
153. I.I. Khain, *Theory and Practice of Phosphating of Metals*, Izd. "Khimia", Leningrad, Russia (1973).
154. G. Lorin, *Phosphating of Metals*, Finishing Publications Ltd., Middlesex, England (1974).
155. D.B. Freeman, *Phosphating and Metal Pre-Treatment*, Woodhead-Faulkner, Cambridge, U.K. (1986).
156. W. Machu, *Korros. Metall.* **20**(6) (1944) 12.
157. R.W. Zurilla and V. Hospadaruk, SAE Technical Paper Series No. 780 (1978) 187.
158. G.D. Cheever, *J. Paint Technol.* **41** (1969) 259.
159. L. Kwiatkowski, A. Sadkowski, A. Kozlowski, and J. Flis, *Proceedings of the Australasian Corrosion Association Conference* **28**, Vol. 2, Perth Western, Australia (1988) pp. 7–12.
160. K. Kiss and M. Coll-Palagos, *Corrosion* **43** (1987) 8.
161. J. Flis, Y. Tobiyama, K. Mochizuki, and C. Shiga, *Corros. Sci.* **39** (1997) 1757.
162. J.I. Bregmann, *Corrosion Inhibitors*, Mac Millan, Ed., New York (1963) p. 1.
163. N. Hackerman, *Langmuir* **3** (1987) 922.
164. E.W. Flick, *Corrosion Inhibitors*, Development Science, Noyes Data Corporation, Park Ridge, NJ (1987) p. 68.
165. G. Wranglen, *An Introduction to Corrosion and Protection of Metals*, New York (1985) p. 173.
166. A.A. Aksüt and S. Bilgiç, *Corros. Sci.* **33** (1992) 379.
167. S.G. Real, J.R. Vilche, and A.J. Arvia, *Corros. Sci.* **20** (1980) 563.
168. D.W. McKee and M.S. Pak, *J. Electrochem. Soc.* **116** (1969) 516.
169. C.F. Zinola and A.M. Castro Luna, *Corros. Sci.* **37** (1995) 1919.
170. J.R. Vilche and A.J. Arvia, *Corros. Sci.* **15** (1975) 419.
171. A.D. Booth, *Numerical Methods*, 3rd edn., Butterworths & Co. Publishers Ltd., London, U.K. (1966).
172. C.F. Zinola, E.J. Vasini, U. Mueller, H. Baltruschat, and A.J. Arvia, *J. Electroanal. Chem.* **415** (1996) 165.
173. K. Wang, G.S. Chottiner, and D.A. Scherson, *J. Phys. Chem.* **97** (1993) 10108.
174. Y. Hori and A. Murata, *Electrochim. Acta* **35** (1990) 1777.
175. A.M. Castro Luna and A.J. Arvia, *J. Appl. Electrochem.* **21** (1991) 435.
176. C.F. Zinola, A.M. Castro Luna, and A.J. Arvia, *J. Appl. Electrochem.* **26** (1996) 325.
177. B. MacDougall, D.F. Mitchell, and M.J. Graham, *Corrosion* **38** (1982) 85.
178. B. MacDougall, D.F. Mitchell, and M.J. Graham, *J. Electrochem. Soc.* **132** (1985) 2895.
179. R. Goetz, B. MacDougall, and M.J. Graham, *Electrochim. Acta* **31** (1986) 1299.
180. B. MacDougall and M.J. Graham, *J. Electrochem. Soc.* **131** (1984) 727.
181. M.J. Graham, *Corros. Sci.* **37** (1995) 1377.
182. D.F. Mitchell and M.J. Graham, *J. Electrochem. Soc.* **133** (1986) 936.
183. M.J. Graham, J.A. Bardwell, R. Goetz, D.F. Mitchell, and B. Mac Dougall, *Corros. Sci.* **31** (1990) 139.
184. J.A. Bardwell, G.I. Sproule, D.F. Mitchell, B. Mac Dougall, and M.J. Graham, *J. Chem. Soc. Faraday Trans.* **87** (1991) 1011.
185. G. Okamoto, *Corros. Sci.* **13** (1973) 471.

13 Principles of Electrochemical Engineering

C. Fernando Zinola

CONTENTS

13.1 ELECTROCATALYSTS: THE HEART OF AN ELECTROCHEMICAL REACTOR

13.1.1 INTRODUCTORY CONCEPTS

The term electrochemical reactor is a modification of the chemical engineering term, chemical reactor, reflecting the electrochemical process. The theories in electrochemical engineering focus on the mass, heat, and charge transport at the electrodes and the arrangement of devices. We can define electrochemical engineering as the proper arrangement of the anode, cathode, electrolyte, diaphragm, and peripherals in order to produce electrochemical reactions. However, not every device is suitable for an efficient electrochemical reaction. We need to design it well. This means that we have to define the electrochemical and the engineering factors to make it efficient from all points of view. The overall chemical (catalytic) reaction results from at least two electrochemical (electrocatalytic) reactions that occur as a result of the electron transfer at the anode and the cathode either through an external electric power device or through an external motor or load. The reactants in the electroreduction reaction are the electrons, while they are one of the products in the electrooxidation reaction. The circuit is completed by the ions or the charged macromolecules that are transported through the electrolyte.

The common parameters of a good chemical reactor design are large chemical yields, high selectivity, low energy consumption, and large power production [1,2]. Reactors with good characteristics can be obtained from a proper selection of electrode materials and a good evaluation of the reactor geometry, the type of reactor, and an appropriate inclusion of the peripherals [1–3]. The electrochemical reactor differs from the heterogeneous reactor in that its performance mainly depends on the choice of the electrode material. Since most of the electrochemical reactions can be catalytic, the electrocatalyst can be considered as the "heart of the reactor." This book will cover the electrocatalyst as a whole including the electrode composition, morphology, local electrode potential, and current distribution at the catalyst.

From the electrochemical engineering point of view, the electrocatalyst design depends on the purpose of the electrochemical reactor, gas electrosynthesis, organic synthesis, batteries or supercapacitors, metal electrodeposition, and the fuel cells.

13.1.2 Electrocatalyst Selection

13.1.2.1 Knowledge of the Kinetics and Mechanisms of the Selected Electrochemical Reaction at the Considered Catalyst

The largest exchange current density, j_o, of the reaction has to be selected, if possible, since economic limitations are always prevalent in scaled-up engineering. However, with the development of nanodispersed substrates and carbon-supported metal catalysts, this limitation becomes a secondary consideration. At this point, it is important to say that most of the reported values of j_o usually refer to simple reactions on pure metal substrates using different shapes of electrode designs in a certain and single electrolyte. Thus, the measurement of the real j_o value at select industrial conditions of the electrochemical reactor has to be performed; that is, experimental measurements cannot be avoided [4,5].

Not only is the value of j_o important in electrocatalysis but also the experimental Tafel slope at the operating electrode potential. As expected in an electrocatalytic process, this complex heterogeneous reaction exhibits at least one intermediate (reactant or product) adsorbed species. Therefore, a single or simple Tafel slope for the entire process is not expected, but rather surface coverage and electrolyte composition potential dependent Tafel slopes within the whole potential domain are expected. Instead of calculating the most proper academic Tafel slope, the experimental current vs. potential curve is required for the selected electrocatalysts [4,6].

Corrosion of the material used is another factor that limits the selection of the electrocatalyst. The electrochemical corrosion of pure noble metals is not as important as in the case of binary or ternary alloys in strong acid or alkaline solutions, since these catalysts are widely used in electrochemical reactors. In the case of anodic bulk electrolysis, noble metal alloys used in electrocatalysis mainly contain noble metal oxides to make the oxidation mechanism more favorable for complete electron transfer. The corrosion problem that occurs from this type of catalyst is the auto-corrosion of the electrode surface instead of the electrode/electrolyte solution interface degradation. The problem of corrosion is considered in detail in Chapter 22.

Since the electrocatalytic reaction implies the existence of an adsorbed species as an intermediate, reactant, or product, the direct interaction with the electrode surface has to be considered first. In this sense, the kinetics of the formation and the stability of the adsorbate are of great importance and may be the determining step for the final value of j_o. The slow adsorption kinetics in the case of a reactive adsorbate will make the reaction at the electrocatalyst not fast enough to become operative. However, the same situation can occur in the case of an adsorbed product with a slow desorption kinetics. The most problematic situation can arise due to the stability of an adsorbed intermediate on the surface, which is the rate-determining step of the whole process. In the case of an anodic process, the species' desorption can be aided by the presence of a metal oxide on the surface. An interesting example of stable and efficient anodes is the dimensionally stable electrodes (*DSE*) used in brine electrolysis for the production of caustic soda and chlorine [7]. These electrodes consist of a titanium substrate coated with a mixture of tin and ruthenium oxides of approximately 1–10 μm of thickness. The porous and cracked structure of these catalysts is sufficiently revealing to offer an effective area of approximately 200 times that of the geometric area. Moreover, this fissuring structure is able to metallize titanium to gain some crystallites into the electrolyte solution, so that the triple composition, that is, the metallic titanium, the tin oxide, and the ruthenium oxide, is effective. The effect of titanium is to obtain a corrosion-resistant surface that is not active below 40 V either in a strong acidic or alkaline solution, whereas the two oxides, in similar proportions, are the real oxidants of the catalyst (in some cases the tin oxide is more active and in many cases the ruthenium oxide is more active).

In the case of cathodic reactions, the presence of metal hydrides can assist the reduction process. Sometimes this situation is not expected. Thus, in the case of chromium electrodeposition, the formation of CrH_2 and CrH_3 appears before the deposition of metallic chromium [4]. The presence of these hydrides produces an unexpected evolution of molecular hydrogen, which is not the desired process. Therefore, when possible it is important to first check the mechanism. With regard to the problem of electrocatalytic reactions, there is a close relationship between the presence of the adsorbed intermediate and the real effectiveness of the entire process. This is a major problem. Hence, the electrochemically active accessible electrode area has to be calculated precisely. Thus, with the aid of STM images, it was possible to calculate the real surface area and corroborate it with the calculation made from the integration of the medium sweep voltammetric curves of the substrate in the base electrolyte. Moreover, spectroscopic techniques also help us to get a complete "picture" of the electrode composition at each operating potential window. In the case of electrocatalytic reactions, the specificity of the surface toward a particular reaction, or more properly toward changing the kinetics of a particular reaction, requires a further analysis. The development of large active area electrodes compatible with lowest overpotentials for the reaction plays a fundamental role in the preparation of the technical electrode of the reactor. The construction of these electrodes was explained earlier in this book.

13.2 ELECTRODE POTENTIAL AND CURRENT DISTRIBUTIONS

13.2.1 Evaluation of the Most Possible Complete Current Distribution at the Electrocatalyst

The first treatment of the current distribution was applied for the case of a metal deposition [8,9] for the purpose of preparing a homogeneous metal layer on the substrate. It was observed that the thickness was completely heterogeneous as the rate of the electrodeposition was much higher on the edges of the electrode surface. The result was unsatisfactory and required an analysis of the current distribution to be performed.

On the other hand, the selectivity of the electrochemical deposition of the metal on the substrate must be 100% of the current efficiency, with no interference from the other metal deposition processes. Therefore, the potential distribution needs to be presented for any serious electrochemical reactor study and the electrocatalyst selection problem. The major problem of current distribution depends on the type of the process that controls the entire reaction rate, such as charge transfer, ohmic contributions, or mass transport to or from the electrode. Many parameters have to be evaluated in the course of an electrochemical process to obtain the desired uniform potential and current distributions. One of the conditions that has to be fulfilled is the continuity equation for the current density vector, $\vec{j}$:

$$\text{div}\,\vec{j} = 0 \tag{13.1}$$

In the case of a mass transport–controlled reaction, we have to make the complete transformation of (13.1) into

$$\text{div}(\chi \vec{\nabla}\Phi) + \text{div}\left(F \sum_i Z_i D_i \vec{\nabla} C_i \right) - \text{div}\left(F\vec{v} \sum_i Z_i C_i \right) = 0 \tag{13.2}$$

where

χ is the entire electrolyte and the anode's and cathode's specific conductivities

Φ is the electric potential difference between each electronic and ionic conductor

F is the Faraday constant

Z_i, D_i, C_i are the oxidation state, diffusion coefficient, and local concentration of the i-component, respectively

$\vec{v}$ is the hydrodynamic velocity vector

Occasionally, we can have a chemical reaction process coupled with the electrochemical reaction. In this case, Equation 13.2 changes to

$$\text{div}(\chi \vec{\nabla}\Phi) + \text{div}\left(F \sum_i Z_i D_i \vec{\nabla} C_i \right) - \text{div}\left(F\vec{v} \sum_i Z_i C_i \right) + \sum_{r=1}^n v_r = 0 \tag{13.3}$$

where v_r is r-sim chemical reaction rate.

Since we are approaching a problem of current and potential distributions, we have to consider the Gauss–Ostrogradsky theorem for the electric field, $\vec{E}$:

$$\text{div } \vec{E} = \nabla^2\Phi = \frac{\rho}{\varepsilon} = \frac{\sum_i FZ_i C_i}{\varepsilon} \tag{13.4}$$

where

ρ is the charge volumetric density of the electrochemical device

ε is the electrolyte permittivity

In the absence of concentration gradients, that is, for the primary and secondary current distributions, we can use the Laplace equation when there is no charge accumulation (bulk solution):

$$\text{div } \vec{E} = \nabla^2\Phi = 0 \tag{13.5}$$

For example, in the case of an aqueous electrolyte, $\varepsilon = 78.3$, and considering that the second variation of the electric potential at a distance close to the bulk of the solution (spatial coordinate $\to \infty$) is one-thousand of a volt, $\nabla^2\Phi \sim 10^{-3}$ V cm^{-2}, we can estimate that $\sum Z_i C_i^o = 10^{-8}$ mol cm^{-3}, which is a negligible value. On the other hand, at the electrode surface (spatial coordinate $\to 0$) we can have $\nabla^2\Phi \sim 1$ V cm^{-2} rendering $\sum Z_i C_i^o = 10^{-3}$ mol cm^{-3}, a value that cannot be neglected. Thus the electroneutrality of the interface is not always real.

When the mass transport process is rate determining, and in the presence of a supporting electrolyte, a convective-diffusion equation for this tertiary current distribution must be solved:

$$\frac{\partial C_i}{\partial t} = D_i \nabla^2 C_i + \vec{v} \cdot \vec{\nabla} C_i \tag{13.6}$$

13.2.2 PRIMARY CURRENT DISTRIBUTION

This is the simplest model of an electrocatalyst system where the single energy dissipation is caused by the ohmic drop of the electrolyte, with no influence of the charge transfer in the electrochemical reaction. Thus, fast electrochemical reactions occur at current densities that are far from the limiting current density. The partial differential equation governing the potential distribution in the solution can be derived from the Laplace Equation 13.5. This equation also governs the conduction of heat in solids, steady-state diffusion, and electrostatic fields. The electric potential immediately adjacent to the electrocatalyst is modeled as a constant potential surface, and the current density is proportional to its gradient:

$$\vec{j} = -\chi \vec{\nabla} \Phi \tag{13.7}$$

The solution for this equation is the same as that of the constant potential boundary conditions (Dirichlet's problem) and was solved not only for electrostatic fields but also for heat fluxes and concentration gradients (chemical potentials) [10]. The primary potential distribution between two infinitely parallel electrodes is simply obtained by a double integration of the Laplace equation 13.5 with constant potential boundary conditions (see Figure 13.3). The solution gives the potential field in the electrolyte solution and considering that the current and the electric potential are orthogonal, the direct evaluation of one function from the other is obtained from Equation 13.7.

13.2.2.1 Primary Current Distribution in Parallel Wire Electrodes

The primary current distribution equations were first derived by Kasper [11–13] for simple geometries of uniform distributions, such as infinite parallel plane plates, infinite height cylindrical surfaces, and concentric spheres (Figure 13.1). Further, some of these parameters were approached using the secondary current distribution.

The direct integration of the Laplace equation in the case of a one-dimensional Cartesian problem renders a linear function of distance between the two electrodes, when the current density is the same for the two electrodes.

In the case of Figure 13.1a, we can say that there are only electric potential variations in the radial direction with the following current density:

$$\vec{j} = -\chi \vec{\nabla} \Phi = -\chi \left(\frac{\partial \Phi}{\partial r} \right) \tag{13.8}$$

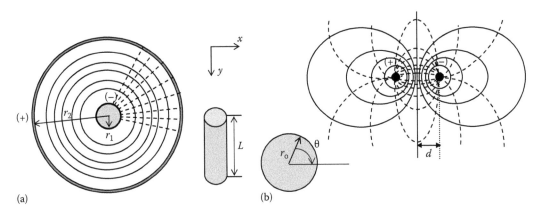

(a)   (b)

FIGURE 13.1 Equipotential surfaces (full circumferences) and equiflux lines (dashed lines) between (a) two concentric cylinders and (b) two parallel infinite lines.

From the definition of the current density,

$$I = \oiint_{r \cdot \hat{n}} \vec{j} \cdot d\vec{s} = \oint_{r} \vec{j} \cdot d\vec{r} = 2\pi(r_2 - r_1)Lj \tag{13.9}$$

By a simple integration, we can obtain

$$\int_{\Phi_1}^{\Phi_2} d\Phi = -\frac{I}{2\pi L\chi} \int_{r_1}^{r_2} \frac{dr}{r} \quad \text{and} \quad \Phi_2 - \Phi_1 = -\frac{I}{2\pi L\chi} \ln\frac{r_2}{r_1} \tag{13.10}$$

This expression demonstrates that the electric potential changes logarithmically with the radial distance of the cylindrical electrocatalyst. When the current flows from the outer to the inner cylinder, the electric potential difference is negative in this spatial system; the current density on the outer cylinder is lower than that of the inner because of the larger area. On the other hand, the electric potential difference is positive from the left side to the right side of the cylindrical system.

Since we have a primary current distribution, we can obtain the resistance of the electrolyte simply by applying Ohm's law:

$$R = \frac{1}{2\pi L\chi} \ln\frac{r_2}{r_1} \tag{13.11}$$

In the case of an electrocatalyst element, as shown in Figure 13.1b, we can say that there are only electric potential variations along the x axis; thus, the electric potential across the electrolyte, Φ, will be

$$\Phi = E_{(cathode-ref)} + \frac{(E_{(anode-ref)} - E_{(cathode-ref)})x}{d} \tag{13.12}$$

and the current density will be of the form

$$\vec{j} = -\chi\vec{\nabla}\Phi = -\chi\left(\frac{E_{(anode-ref)} - E_{(cathode-ref)}}{d}\right) \tag{13.13}$$

and both the magnitudes are strongly dependent on the actual values of the electrode potentials at the cathode and the anode.

These two-dimensional simulations can be reduced to a one-dimensional problem using the "image method," that is, the conversion of the electric field into an equivalent field, which is easier to calculate, especially for points, lines, planes, and cylinders either for the electric or the magnetic fields. In the case of Figure 13.1b, we have the following solution:

$$\Phi = \frac{V_{app}}{2 \ln\dfrac{d+u}{d-u}} \ln\left[\frac{2(d+u)(d+r\cos\theta) + r^2 + u^2 - d^2}{2(d-u)(d+r\cos\theta) + r^2 + u^2 - d^2}\right] \tag{13.14}$$

where
   V_{app} is the applied electric potential between the two electrodes
   r is the distance from the center of each cylindrical electrode to a certain point in the bulk of the solution
   u is $u = \sqrt{d^2 - r_o^2}$

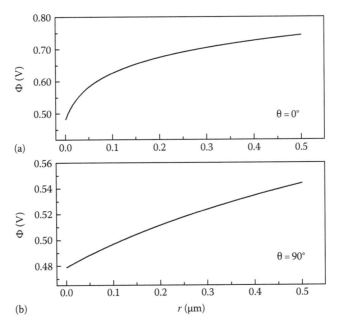

FIGURE 13.2 Primary electric potential distribution, Φ vs. r, on parallel wire electrodes for (a) $\theta = 0°$ and (b) $\theta = 90°$ with $V_{app} = 1$ V, $d = 10$ μm, and $r_o = 1$ μm.

It is evident that the equipotential in the electrolyte is cylindrical but not coaxial with the conductors. It is clear that the electric potential varies from V_{app} from a distance r in the electrolyte and specifically depends on the geometric parameters, d and r_o, but also with θ. For example, for $V_{app} = 1$ V, $d = 10$ μm and $r_o = 1$ μm. The variation of Φ with r is plotted in Figure 13.2.

It is clear that the maximum change in the electric potential in the solution occurs for $\theta = 0°$ and $180°$, since a direct interaction between the ions and the electrocatalyst takes place.

Considering Ohm's law (Equation 13.8), the current density at the catalyst surface is proportional to the potential gradient, so by differentiating (13.13) and taking its value at $r = r_o$,

$$\vec{j} = -\chi \vec{\nabla} \Phi = \frac{r_o u \chi V_{app}}{(d+u)(d-u)(d+r_o \cos \theta) 2 \ln \dfrac{(d+u)}{(d-u)}} \tag{13.15}$$

For example, for $V_{app} = 1$ V, $\chi = 1$ S cm^{-1}, and for $d = 10$ μm, the variation of $\vec{j}$ with θ at $r = r_o = 1$ μm is plotted in Figure 13.3.

The maximum value of the current density lies at the opposite side of the electrocatalyst, considering it is positioned as in Figure 13.1, that is, at the nearest distance of the closest approach.

Since we have a primary current distribution, we can obtain the resistance of the electrolyte simply from Ohm's law. If we assume that the wires have length L, we obtain

$$R = \frac{1}{2\pi L \chi} \ln \frac{(d+u)}{(d-u)} \tag{13.16}$$

13.2.2.2 Primary Current Distribution in Parallel Plate Electrodes

Some of the electric potential distributions in a two-dimensional set up can be expressed in infinite series terms, when other mathematical algorithms do not work. Special series terms of Bessel functions and Legendre polynomials are easily evaluated nowadays with the help of computer

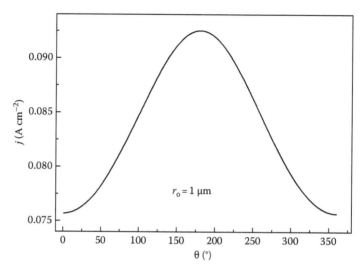

FIGURE 13.3 Primary current density distribution, j vs. θ, on parallel wire electrodes for $V_{app} = 1$ V, $\chi = 1$ S cm^{-1}, $d = 10$ μm, and $r_o = 1$ μm.

calculations using modern software [14,15]. We can consider, as an example, the simple electrochemical cell of Figure 13.4. The two-dimensional Laplace equation will be

$$\nabla^2\Phi = \frac{\partial^2\Phi}{\partial x^2} + \frac{\partial^2\Phi}{\partial y^2} = 0 \tag{13.17}$$

The electron-conducting surface is equipotential, since there is no potential variation on the surface ($\chi \to \infty$). On the other hand, on an insulating catalyst, it is equiflux since the current lines cannot get inside the surface ($\chi = 0$).

Two different situations arise from this cell: $H \gg L \wedge W \gg L$. The boundary conditions are as follows:

$$\frac{\partial\Phi}{\partial y} = 0 \quad L/2 < |x| < W/2 \tag{13.18a}$$

$$\frac{\partial\Phi}{\partial x} = 0 \quad y = 0 \, |x| = W/2 \tag{13.18b}$$

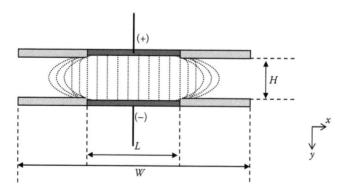

FIGURE 13.4 Current distribution for the primary approach in an electrochemical reactor of two parallel plate electrode geometries. Black and gray rectangles are the electrocatalyst and insulator surfaces, respectively.

The potential distribution in this case can be solved as Fredholm integrals [1,15]:

$$\Phi\left(\frac{2x}{L},\frac{2y}{L}\right) = \Phi^0 + \int_{-1}^{1} \left\{ \ln\left[\left(\frac{2x}{L}-\lambda\right)^2 + \left(\frac{2y}{L}\right)^2\right] \left[\frac{L}{4\pi m}\left(m\frac{\partial\Phi}{\partial x}-V_{app}+\Phi^0\right)\right.\right.$$

$$\left.\left. +\frac{L}{4\pi m}\int_{-1}^{1} f(\lambda)\ln\left(\frac{2x}{L}-\lambda\right)^2 d\lambda\right]\right\} d\lambda \tag{13.19}$$

with $m \equiv \chi(\partial\Phi/\partial j)$, and the other condition is $H \ll L$ (semi-infinite electrochemical reactor). The boundary conditions are as follows:

$$\frac{\partial\Phi}{\partial y} = 0 \quad x < 0 \wedge y = 0 \vee y = H \tag{13.20}$$

In this case, the Laplace solution is solved as a complex function [1,15]:

$$\Phi\left(\frac{\pi x}{2H},\frac{\pi y}{2H}\right) = \Phi\left(y=\frac{H}{2}\right)$$

$$+ \int_0^{+\infty} \left\{ \text{Re}\left\{\ln\text{th}\left[\hat{i}\left(\frac{\pi x}{2H}-\lambda\right)-\left(\frac{\pi y}{2H}\right)\right]\right\} \left[\begin{array}{c}\frac{H}{2\pi m}\left(m\frac{\partial\Phi}{\partial x}-V_{app}+\Phi\left(y=\frac{H}{2}\right)\right) \\ +\frac{H}{2\pi m}\int_0^{+\infty} f'(\lambda)\text{Re}\left\{\ln\text{th}\left[\hat{i}\left(\frac{\pi x}{2H}-\lambda\right)-\left(\frac{\pi y}{2H}\right)\right]\right\} d\lambda \end{array}\right] \right\} d\lambda \tag{13.21}$$

with $\hat{i} = \sqrt{-1}$

If we consider other initial conditions such as

(a) Symmetric translation of the scale potential

$$\Phi(x,H) = 0 \tag{13.22a}$$

(b) Negligible incidence of the electric potential at the counter electrode

$$\frac{\partial\Phi(x,H)}{\partial y} = 0 \tag{13.22b}$$

(c) Absence of the electric potential at the insulators

$$\frac{\partial\Phi(0,y)}{\partial x} = \frac{\partial\Phi(L,y)}{\partial x} = 0 \tag{13.22c}$$

then the expression for Φ is of a Fourier series expression:

$$\Phi(x,y) = \frac{1}{2}a_o(y) + \sum_{n=1}^{\infty} a_n(y)\cos\left(\frac{n\pi x}{L}\right) \tag{13.23}$$

By the principles of sectioning, we can place an insulator along the line at $x = H/2$, so we can reduce the form of Equation 13.24 to

$$\Phi(x, y) = \frac{4V_{app}}{\pi} \sum_{n=1}^{\infty} \frac{1}{n} \exp\left(\frac{-n\pi y}{H}\right) \sin\left(\frac{n\pi x}{L}\right) \tag{13.24}$$

Considering Ohm's law, we can calculate the current distribution along the electrode, therefore differentiating (13.24) with respect to y

$$j(x) = -\chi \left[\frac{\partial \Phi(x, y)}{\partial y}\right] = \frac{4\chi V_{app}}{H} \sum_{n=1}^{\infty} \exp\left(\frac{-n\pi y}{H}\right) \sin\left(\frac{n\pi x}{L}\right) \tag{13.25a}$$

and taking the value at the electrode surface, $y = 0$:

$$j(x) = -\chi \left[\frac{\partial \Phi(x, y)}{\partial y}\right]_{y=0} = \frac{4\chi V_{app}}{H} \sum_{n=1}^{\infty} \sin\left(\frac{n\pi x}{L}\right) \tag{13.25b}$$

Comparing the analytical expressions of Equations 13.24 and 13.25, we see that the electric potential converges with certain values at the electrodes surface, but the current distribution never converges, so an alternate solution is required. We can derive a solution to a primary current distribution based on conjugated functions [14,16]. In this case we obtain

$$j(x) = \frac{2\chi V_{app}}{H \sin\left(\frac{\pi x}{L}\right)} \tag{13.26}$$

13.2.3 Primary vs. Secondary Current Distributions

The primary current distribution model is an idealized scheme that gives us mathematical simplicity but cannot be justified when it is not possible to set constant electric potential boundary conditions to solve the Laplace equation.

A characteristic of the primary distribution, in general, is that it is less uniform than the secondary distribution for a given electrode geometry and the electrochemical cell device. There is only one exception that arises from the concentric cylindrical electrode system depicted in Figure 13.2a, where both the primary and the secondary current distributions are uniform in the case of the forced convective hydrodynamics (rotating electrodes).

The nonuniform nature of the primary distribution arises from the abrupt change in the electrolyte resistance in the neighborhood of the electrocatalyst. Since the surface roughness of the catalyst is much larger than the geometric area, especially in technical electrodes used in industry, the theory behind planar electrodes does not apply here. The minimum working of the auxiliary electrode separation distance gives less electric resistance and more favorable points for the current flow. However, there are exceptions to this rule—when there is a curvature at the catalyst surface or any intersection at the insulators (Figure 13.5).

Since there is no current flow at the insulator and the primary distribution predicts a normal contribution to the equipotential surfaces (intersecting with right angles to the insulating walls), the current tends to concentrate near the convex zones as shown in Figure 13.5. At the very near edge of the catalyst, the geometry defines a sharp cone or pyramid infinitesimal zone, making the electric potential differentiation discontinuous and the local current density infinite but with a finite electric current value. The different cases of electrode–insulator geometric configurations in a primary

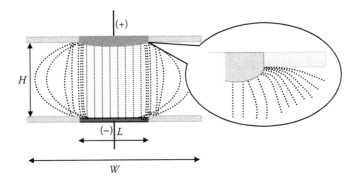

FIGURE 13.5 Current distribution of primary equiflux lines for two parallel plate electrode geometries. The working electrode has a convex curvature near the insulators, making the equiflux lines condensed at the convex zones. Black and gray rectangles are the electrocatalyst and insulator surfaces, respectively.

current distribution are explained in detail elsewhere [9,14]. The equipotential lines are parallel to the electrocatalyst surface; however, near the insulator the equipotential line has to be perpendicular to it. This fast change of direction resembles a jagged convex spike that concentrates the current lines at the intersection as shown in the figure. Surface defects and roughness are observed in technical volumetric catalysts and are far from the smooth ideal surface, resulting in an undesirable nonuniform current distribution. Considering the problem of a primary current distribution on a heterogeneous catalyst, the nonuniformity arises from excessive, defective populations of equipotential surfaces in a given collection of coordinates with respect to the bulk of the electrolyte. If we know the real resistance or the conductance current dependence, that is, solving Equation 13.26, the problem is then worked out:

$$\vec{j} = -\chi \vec{\nabla} \Phi \quad \text{with } \chi = f(\hat{x}, \hat{y}, \hat{z}) \quad \text{or} \quad \chi = f(\vec{j}) \tag{13.27}$$

However, the problem is not so simple since these are variable-dependent parameters in the equations that always lead to a numerical solution.

Another way to solve the problem is to consider the surface roughness of the catalyst. The surface roughness was described in a separate section of this chapter as being more difficult to characterize than a simple convex curvature. If we can describe it as a periodical perimeter of a given geometric assignation, it will be possible to characterize it. If there is a non-periodical surface distribution, then it will be very difficult to solve.

The topography of any solid object can be considered as regular (with smooth surface domains) or irregular (weak and strong disordered surfaces) [23,24]. The former can be described adequately by Euclidean geometry, but in the case of disordered surfaces Fractal geometry has to be used. The latter treats the disorders as an intrinsic rather than a perturbative phenomenon, since the fractal dimension of the object is involved between the Euclidean dimension and the topological dimension. The problem is not so difficult to comprehend as under scale transformations we can cover isotropic (self-similar fractals) or anisotropic disorders (self-affine fractals). Contrary to the self-similar fractal scale invariance at all lengths, the self-affine fractals have different local fractal dimensions depending on the selected direction of the characterization [25].

When the primary distribution does not illustrate the current or electric potential distribution well, an additional resistance, that is, the charge transfer electrode resistance, has to be considered. In such cases, we need to account for the electrode kinetics, and the secondary current and potential distributions emerge from the models. For industrial purposes the porous or tortuous electrocatalyst has to be considered as a dynamic system. This means that its porosity shape and density besides the surface roughness and the real geometric area changes all the time. This point makes us think that it will not be possible to predict what happens with these types of surfaces. Maybe this is true, but experience on working with this type of electrode makes us to analyze the situation more carefully. The effect of the supplementary resistance due to the charge transfer leads to an obvious breakdown as this additional resistor governs the process. We can think that the electrolyte resistance is a small constant value and the second new resistance is the current-dependent resistor. First, we can think of a smooth surface as the electrocatalyst, and then we can try to approach toward the reality.

In these cases, the electric potential near the surface of the electrocatalyst is governed by the Butler–Volmer equation of a pure charge transfer process:

$$j(q) = j_o \left\{ \exp\left(\frac{\overleftarrow{\alpha}F\eta(q)}{RT}\right) - \exp\left(\frac{-\overrightarrow{\alpha}F\eta(q)}{RT}\right) \right\} \tag{13.28}$$

where
   q is the collection of spatial coordinates
   $\vec{\alpha}$ is the charge transfer coefficient for the anodic and cathodic processes

A new variable is defined for the electrode kinetics at the interface, the overpotential η. This is a convenient variable that is dependent on the electric potential for a variable current circulation condition and for one without current.

$$\eta(j) \equiv \Delta\Phi(j) - \Delta\Phi(j=0) \tag{13.29}$$

The Butler–Volmer expression is not described here as it is described in all fundamental textbooks on electrodics, but instead the smaller expression is given here:

$$\lim_{\eta \to 0} j(q) = j_o \frac{(\overleftarrow{\alpha} + \overrightarrow{\alpha})F\eta(q)}{RT} \tag{13.30}$$

Large polarizations usually account for less than 1% of errors:

$$\overleftarrow{j}(q) = j_o \exp\left(\frac{\overleftarrow{\alpha}F\eta(q)}{RT}\right) \quad \text{for the Tafel expression of the electrooxidation current} \tag{13.31a}$$

$$\overrightarrow{j}(q) = j_o \exp\left(\frac{\overrightarrow{\alpha}F\eta(q)}{RT}\right) \quad \text{for the Tafel expression of the electroreduction current} \tag{13.31b}$$

13.2.3.1 Wagner Number

In spite of considering simple geometries to derive the electric potential and the current potential that were analytically solved from the Laplace equation, the secondary distributions always requires a numerical integration because of the current–electric potential-dependent boundary conditions. For example, in the case of the electrochemical reactor shown in Figure 13.5, the numerical solution for the secondary distribution of the current is generally presented as a plot of j_x/j vs. x/L [17,18], considering j_x to be the local current density along the x axis for a given y location. Also from the literature, the expression of j_x/j is found as a function of the Wagner number, Wa [9], defined as

$$Wa \equiv \left(\frac{\partial\eta}{\partial j}\right)\frac{\chi}{L} \tag{13.32}$$

Wa gives the ratio between the charge transfer resistance and the electrolyte ohmic resistance. This dimensionless value can be taken as a criteria or measure of the current uniformity at the electrocatalyst. When *Wa* → 0, the primary current contribution prevails. It is clear that the geometric characteristic distance *L* can be enlarged enough to simplify the analysis to a primary distribution. However, we should not forget that the variation of η with *j* contains several parameters. In the simplest case of a small polarization change, we have for an elementary electron transfer

$$Wa \equiv \frac{\chi RT}{nFj_oL} \tag{13.33}$$

Since both η (ohmic- and small polarization-charge transfer) vary linearly with the current density (Equations 13.34 and 13.35), its effect cancels each other and *Wa* remains independent of the current:

$$\eta = \frac{RT}{j_o(\bar{\alpha} + \vec{\alpha})F} \overset{\leftrightarrow}{j} \tag{13.34}$$

$$\eta = \frac{L}{\chi} \overset{\leftrightarrow}{j} \tag{13.35}$$

Again a primary current distribution can be obtained for small values of electrolyte conductivities, χ, or better with large values of exchange current densities, j_o. The latter is of interest in electrocatalysis, because by simply changing the nature or the composition of the electrocatalyst we can achieve small values of *Wa*, and then a uniform primary current distribution. For the case of the electrocatalytic agents used in the industry, the expression of the Equation 13.33 is not useful at all. By introducing large polarizations, we have

$$Wa \equiv \frac{\chi RT}{\alpha FL} \frac{1}{j} \tag{13.36}$$

since

$$\left(\frac{\partial j}{\partial \eta}\right) = \frac{j_o \vec{\alpha} F}{RT} \exp\left(\frac{\vec{\alpha} F\eta}{RT}\right) = \frac{\vec{j} \vec{\alpha} F}{RT} \tag{13.37}$$

In this case, the ohmic η varies linearly with the current density, whereas the large polarization charge transfer η varies logarithmically with the current density (Equations 13.35 and 13.38), and *Wa* becomes inversely proportional to the current density, as shown in Equation 13.36:

$$\eta = \frac{RT}{\vec{\alpha} F} \ln \frac{\vec{j}}{j_o} = \frac{RT}{\vec{\alpha} F} \ln \vec{j} - \frac{RT}{\vec{\alpha} F} \ln j_o \tag{13.38}$$

This expression is also very clear; for increasing the magnitudes of $\vec{j}$, we progressively obtain an uneven current distribution of the primary form *Wa* → 0.

Let us examine the changes of *Wa* with temperature. When we operate in the Tafel region, *Wa* shows little dependence since both χ and the Tafel slope are direct and linearly proportional. In the case of operating in the small polarization zone, j_o increases exponentially with temperature, leading to a strong temperature-dependent *Wa* number, which reveals a primary distribution at elevated temperatures.

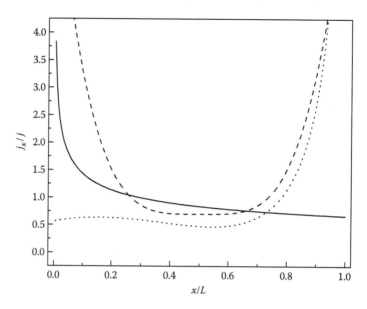

FIGURE 13.6 Current distribution on a plane electrode catalyst for primary distribution (dashed line), secondary distribution (dotted line), and tertiary distribution (continuous line).

Sometimes it is not possible or convenient to lower the values of Wa, so we have to play with all the adjustable parameters of Equation 13.37. In the case of the electrocatalyst and for an electrochemical reactor scale-up problem, Wa has to be of an appreciable magnitude; that is, we can change the electrode size (not geometry) but Wa should remain constant. The problem with real electrochemical reactors is that concentration gradients (especially for long-time uses) are inevitable, and thus the shape of the j_x/j vs. x/L plot is useful (Figure 13.6).

In the industry, we always apply a constant external electric potential in order to achieve a steady-state electric current. This process will last some time, but we expect a nearly constant current. Therefore, as the resistor representing the ohmic losses becomes negligible with respect to the current-dependent resistor of charge transfer, the current distribution becomes uniform only when this "kinetic" resistor gives rise to a constant value, that is, for a steady-state condition. For the secondary current distribution, this means $Wa \to \infty$. If Wa is not sufficiently large, it is better to use a primary current distribution. The main difference of the secondary current distribution is that it has a finite number at the edges of the electrode, that is, at $x = 0$ and $x = L$. Figure 13.7 shows the shape of the j_x/j vs. x/L plot for different Wa numbers. It is clear that $Wa = 0$ leads to a primary current distribution, and that the most uniform current profile is that of the highest Wa number (the most "pure" secondary current distribution).

13.2.4 CURRENT AND POTENTIAL DISTRIBUTIONS IN ROUGH AND POROUS ELECTROCATALYSTS

True ordered surfaces can be observed at the terrace of ideal single crystals, with practical corrugations less than 0.1 nm that are introduced by the electronic charge density of metal atoms. Thus, the surface ideal ordered domains can only be found in a small portion of the surface due to the induced surface relaxations and reconstructions [23].

In the case of rough surfaces, stochastic models often describe their topography with self-resemblance over a different range of scales. The common description of highly disordered surfaces is by means of the dynamic scaling theory [26]. The simple consideration of the theory is that growth proceeds in a single direction normal to the flat surface length, L, and increases with heights h_i. Thus, the instantaneous height can be described by a single function of x and t. The instantaneous

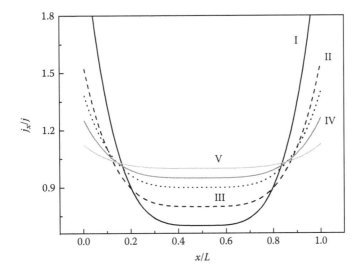

FIGURE 13.7 Current distribution (j_x/j vs. x/L) on a plane catalyst for a secondary distribution with different experimental conditions: (I) $Wa = 0$; (II) $Wa = 0.10$; (III) $Wa = 0.20$; and (IV) $Wa = 0.40$; (V) $Wa = 0.80$.

surface width, $w(x, t)$, is the measure of the roughness and is defined by the root mean square of h_i fluctuations:

$$w(x, t) = \sqrt{\frac{1}{N} \sum (h(x_i) - \tilde{h})^2} \tag{13.39}$$

where
   N is the number of surface sites
   $\tilde{h}$ is the average height normal to the surface x direction

In this model, the dependence of the width on t and L is considered as

$$w(L, h) \propto L^{\alpha} f(x) \tag{13.40}$$

where the following properties account for $f(x)$:

$$\lim_{x \to \infty} f(x) = \text{constant} \quad \lim_{x \to 0} f(x) = x^{\alpha/\gamma} \tag{13.41}$$

At $t \to 0$, $w(L, 0) \propto t^{\beta}$ where the superscript β is the roughness growth kinetics exponent along the x direction. After a certain time, a steady-state contour is reached, $w(L, \infty) \propto L^{\alpha}$, or in other words, a scale invariant fractal is obtained (self-affined fractal). The exponent α is the surface roughness (only when it is measured at scale lengths shorter than the interface).

Many factors can affect the rate of the electrocatalytic reaction, but the use of large-area rough electrodes makes almost no changes in the conceivable electrochemical reactor. Surface areas of more than 10^4 cm^2 cm^{-3} of the catalyst are attainable. This type of electrode is commonly used in fuel cells and in industrial gas-evolving electrodes. In this case, the matrix is electronically conductive but the electrode reactions take place on a highly dispersed catalyst area all over the substrate. There is a particular distribution of reaction rates due to the diffusion of reactants into the pores and the changes in the electrolyte resistances therein.

We have treated the catalyst like a one- or two-dimensional geometric device, but in reality the technical electrodes used in the industry are more likely to be of a volumetric three-dimensional

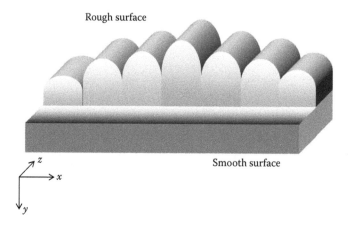

FIGURE 13.8 Real periodic electrocatalyst: smooth or rough electrode surfaces. A smooth catalyst exhibits no preferential growth along a single axis, while the rough periodic surface shows a (x, y) plane preferential crystallization.

system. The modeling of the volumetric electrocatalyst can be easy when the surface roughness has two properties: isotropy and absence of porosity. Figure 13.8 depicts the situation already described in the text.

13.2.4.1 Preferential Directions in Surface Roughness

In the example shown in Figure 13.8, we can see that the preferential surface roughness occurs along the (x, y) plane. This type of periodical behavior can be modeled by a mathematical expression for crystal growth, $G(x, y)$. On a first approximation, we can consider that along x the growth of the catalyst is constant so that

$$G(\vec{y}) = \sum_{n=1}^{\infty} \exp\left(-B(y) \cdot (\vec{y} - an)^2\right) sen^2(C(y) \cdot (\vec{y} - an)) \tag{13.42}$$

where a is the spatial constant parameter, which is the center of growth in the $\vec{y}$ direction (Figure 13.9a). If there is another component growth along x (as in Figure 13.8), we need to add another parameter to the series.

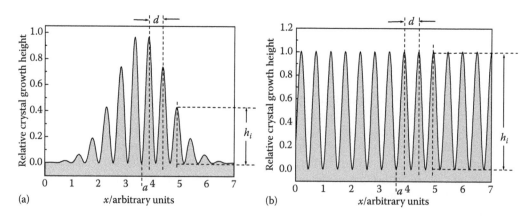

FIGURE 13.9 (a) Illustration of a catalyst periodic crystal growth of a rough surface. d denotes the separation between two columns of a crystal periodic growth of heights h_i with respect to a center a. (b) Particular situation of $B(y) = 0$, where the only changing parameter is h_i.

$B(y)$ symbolizes a parametric value proportional to the number of "columns" (or "peaks") of the crystalline growth and to some extent to the height (h_i) with respect to a hypothetical center value of $x = a$. The larger the value of $B(y)$, the lesser the number of "peaks." For middle values, the number of "peaks" is close to two and of the same height, positioned by a. It is possible to prove that when $B(y)$ approaches zero each "column" acquires nearly the same height. This is the case for platinum group metal depositions (columnar growth) that can be achieved by the roughening of smooth platinum wire by the application of a square-wave potential program. In this case, the modeling gives a growth function as shown in Figure 13.9b.

The application of periodic electric potentials to the electrodes in contact with the electrolytes produces morphological changes that depend on the upper and lower potential limits (E_u and E_l), the frequency (f), and the time-symmetry of the program [27–29]. In the case of a noble metal in a strong acid or alkaline solution, the effect is assigned to a pulse plating electrodissolution/deposition on different metal sites [30–32]. These methods can produce considerable modifications on electrodeposits such as morphology, adherence, uniformity, etc.

In this case, a similar crystal growth can be found as depicted in Figure 13.9b. The following expression denotes the type of growth of platinum group metals:

$$G(\vec{y}) = \sum_{n=1}^{\infty} sen^2(C(y) \cdot (\vec{y} - an)) \tag{13.43}$$

$C(y)$ is a parametric number that configures the separation (d) of the columns and the number of them all over the catalyst. The larger the $C(y)$, the higher the number of columns and the lower the value of d. It is also possible to show that for small values of $C(y)$, h_i also gets small and decreases the number of "peaks." It seems then, that the crystal growth over the catalyst at short times of metal deposition is strongly dependent on $C(y)$. For $C(y) = 0$ and $\forall B(y)$, all the columns are able to reach the height of the two sharp "twin peaks" and for $B(y) \to 0$ we obtain two wide peaks.

For small values of $B(y)$ and $C(y)$, we have two small but wide columns of the same height, whereas for large values of $B(y)$ and $C(y)$ we have two high columns of two short columns of the same height. When both parameters tend to infinity (at long-term deposition times), there are no "peaks" for the electrocatalyst growth; that is, a uniform and smooth two-dimensional growth occurs.

One interesting case for the rough electrocatalyst is when we are working at the limiting current density, that is, when we have to use the tertiary current distribution. Since the mathematics for mass transfer effects are very complex, we can get some help from semiempirical equations. Suppose we are working with an electrode surface with a roughness profile similar to that in Figure 13.8. Considering that at the most convex region, the current density is the largest of all, we can say that at the bottom of the pore we have the lowest concentration for the i-species. Thus, considering Equation 13.25a, we can propose the following variation of the concentration along y-axis with a constant x-component:

$$C_i(y) = C_i^{y=0} \exp(-Ay) \quad \text{with} \quad A \equiv \frac{\alpha K_M(1 - \varepsilon)}{|\vec{v}|} \tag{13.44}$$

where
$C_i^{y=0}$ is the concentration of the i-species at the bottom of the pore
α is the real surface area (calculated from STM images)
$|\vec{v}|$ is the module of the linear hydrodynamic rate for the transport of the i-species from the top to the bottom of the pore
ε is the porosity factor (cylindrical, conic, hemisphere, etc.)
K_M is the mass transport coefficient for the i-species

Since the current density is a function of the concentration gradient, we can say

$$j(y) = -Z_i FD_i \left[\frac{\partial C_i(y)}{\partial y} \right] = \frac{\alpha K_M (1 - \varepsilon) Z_i FD_i C_i^{y=0}}{|\vec{v}|} \exp\left(-\frac{\alpha K_M (1 - \varepsilon)}{|\vec{v}|} y \right) \tag{13.45}$$

where D_i is the diffusion coefficient for i-species along the pore, thus $K_M = (D_i/\delta)$, where δ is the thickness of the diffusion layer.

From

$$j(y) = -\chi \left[\frac{\partial \Phi(y)}{\partial y} \right] \tag{13.46}$$

we can say

$$\frac{Z_i FD_i C_i^{y=0} \alpha K_M (1 - \varepsilon)}{|\vec{v}|} \exp\left(-\frac{\alpha K_M (1 - \varepsilon)}{|\vec{v}|} y \right) = -\chi \left[\frac{\partial \Phi(y)}{\partial y} \right] \tag{13.47}$$

or

$$d\Phi(y) = -\left(\frac{\alpha Z_i FD_i^{\circ 2} C_i^{y=0} (1 - \varepsilon)}{\chi |\vec{v}(y)| \delta(y)} \right) \exp\left(-\frac{\alpha D_i^{\circ} (1 - \varepsilon)}{\delta(y) |\vec{v}(y)|} y \right) dy \tag{13.48}$$

The only two parameters that depend on the y-axis are $\delta(y)$ and $\vec{v}(y)$. The only parameter that can be considered as independent of y (within a certain error) is $\vec{v}(y)$; that is, we can consider a linear rate of transport for i along the y-axis. Therefore, considering as a first approach, we can say that the direct integration of Equation 13.48 between $y = 0$ (bottom of the pore) and y turns into

$$\Phi(y) = \Phi^0 - \left(\frac{Z_i FD_i^{\circ} C_i^{y=0}}{\chi} \right) \left\{ 1 - \exp\left(-\frac{\alpha D_i^{\circ} (1 - \varepsilon)}{\delta(y) |\vec{v}(y)|} y \right) \right\} \tag{13.49}$$

At the special case of $y = 0$, we have $\Phi(y) = \Phi^0$ constant.

Another interesting point from Equation 13.49 is the magnitude of the electrolyte conductivity inside the pore. χ depends on the function of the bulk electrolyte conductivity [15], χ^0, and the ε porosity factor in the case of hemispherical particles (silver, copper, and gold, for example) where ε is less than 0.45:

$$\chi = \chi^0 \left(\frac{2\varepsilon}{3 - \varepsilon} \right) \tag{13.50}$$

13.2.4.2 Agnesi Curves and Trochoid and Cycloid Surfaces

Another interesting case arising from the fast deposition of metal is when the crystal growth develops a single peak at the center of the substrate surface. This type of growth can be mathematically depicted as the *Agnesi curve*:

$$G(\vec{y}) = \sum_{n=1}^{\infty} \frac{B(y, n)^3}{B(y, n)^2 + (y - an)^2} \tag{13.51}$$

$B(y, n)$ symbolizes a parametric value proportional to the number of n-peaks for the crystalline growth, and a is a spatial constant parameter (the center of the growth along y).

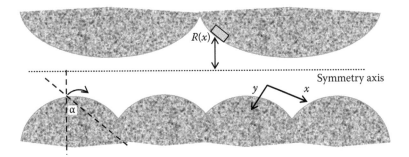

FIGURE 13.10   Approximation to the real curve catalyst. $R(x)$ denotes the electrodic infinitesimal element from which the angle α from the normal axis (of the catalyst surface) to the symmetry axis is defined.

The problem with these surfaces is that they do not exhibit a simple square or rectangular type of geometry, and we have to characterize the surface by means of the curvilinear coordinates, such as the example shown in Figure 13.10.

In the numerical simulation of an electrochemical system, it is advantageous to perform the simulation on the smallest domain to better describe the behavior of the cell. Taking symmetry as a criterion often reduces the problem of electric potential distribution. The division of the entire domain into smaller sub-domains enables to obtain a simpler function of linear or parabolic expressions. Then, by successive iterations the values from the approximating functions can be improved.

When we are approaching a nonuniform and sinusoidal, or "valley and mountain," catalyst profile, we need other spatial coordinates. This is the case of the cycloid curves, described by a point in a circumference, which rolls without slipping along a rectilinear line. Mathematically speaking, the representation of the crystal growth in the catalyst is the one of a trochoid (shortened cycloid) as depicted in Figure 13.11. The equations in a parametric form are

$$x(t) = a(t - \lambda \sin t)$$
$$y(t) = a(1 - \lambda \cos t)$$
(13.52)

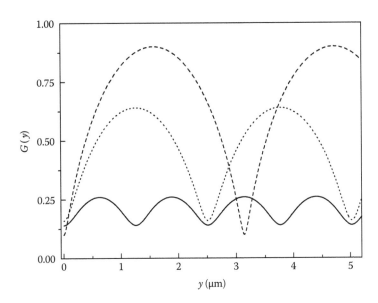

FIGURE 13.11   (a) Illustration of the growth of a catalyst periodic crystal (trochoid). a denotes the circumference radius (separation between two columns of the periodic crystal); $\lambda = 0.3$, $a = 0.2$ (continuous line); $\lambda = 0.6$, $a = 0.4$ (dotted line); and $\lambda = 0.8$, $a = 0.5$ (dashed line).

where

   a is the radius of the circumference

   λ is the type of the cycloid ($\lambda < 1$ is the lengthened cycloid and when $\lambda > 1$ we have a shortened cycloid or trochoid)

The length of one cycle L is

$$L = a \int\limits_{0}^{2\pi} \sqrt{1 + \lambda^2 - 2\lambda \cos t}\, dt \tag{13.53}$$

and the area under each cycle, S, is

$$S = \pi a^2 (2 + \lambda^2) \tag{13.54}$$

13.2.4.3   Lighthill–Acrivos Generalized Expression: Acrivos Number

Suppose now our surface is curvilinear such as that shown in Figure 13.10. The generalized expression for the current density as a function of the spatial coordinates is [1,15] given by the Lighthill–Acrivos generalized expression for axis-symmetric surfaces:

$$j\lim(x) = 0.55 Z_i F C_i^o D_i^{2/3} \frac{\left(R(x) \left[\dfrac{\partial v_x}{\partial y} \right]_{y=0} \right)^{1/2}}{\left[\int\limits_{0}^{x} \sqrt{R\left(x, \left[\dfrac{\partial v_x}{\partial y} \right]_{y=0} \right)}\, dx \right]^{1/3}} \tag{13.55}$$

To solve this equation first, we need to know the geometry of the catalyst surface and the convective component, $[\partial v_x / \partial y]_{y=0}$. In this sense, Table 13.1 shows the formulas for this hydrodynamic rate differentiation as a function of different electrode geometries.

   The use of this theory has a real advantage; as we do not need Newtonian fluids, it is applicable for large viscosity liquids, but the requisite is a laminar regime. Particularly, in the case of free convection, an Acrivos number (Ac) is defined.

   In this model, we ignored the activity of the catalyst itself. However, this is not true. We need to consider both the current density and the potential in the electrolyte (as we did before) and in the rough or porous catalyst. For the mass balance of the rough catalyst, we have to consider all the

---

TABLE 13.1
Dependence of Local Rate Change for Different Electrode Geometries

Electrode Geometry	$\left[\dfrac{\partial v_x}{\partial y} \right]_{y=0}$	Nomenclature
Rotating disk system	$0.51 \omega r \left[\dfrac{\omega}{\nu} \right]^{1/2}$	$r \equiv$ Radial distance from the center of the disk, $\omega \equiv$ angular rotation rate, $\nu \equiv$ cinematic viscosity
Single plane plate	$0.332 U_o \left[\dfrac{U_o}{\nu} \right]^{1/2}$	$U_o \equiv$ Limiting hydrodynamic rate
Parallel plane plates	$6 \dfrac{\bar{v}}{H}$	$\bar{v} \equiv$ Mean hydrodynamic rate, $H \equiv$ plate separation distance
Cylindrical surface	$-4 \dfrac{U_o}{D}$	$D \equiv$ Cylinder diameter

mass transport processes and the electrocatalytic reaction. The variation of the i-species concentration will occur in the presence of a supporting electrolyte:

$$\frac{\partial C_i}{\partial t} = D_i \nabla^2 C_i + \vec{v} \cdot \vec{\nabla} C_i + \frac{a_i}{nF} \operatorname{div} \vec{j}_i \tag{13.56}$$

where
 a_i is the real surface area of the catalyst
 n is the number of electrons transferred during the entire electrocatalytic reaction
 j_i is the current density developed by the reaction, since changes in i also occur due to the reactions at the catalysts

This current density is dependent on the distance, y, from the top to the bottom of the pores or from the top of the "mountains" to the bottom of the "valleys" at the rough surfaces. Therefore, we can consider that we have a mass-transport potential-dependent current density for an electrooxidation reaction:

$$\vec{j}(x,y) = j_0 \frac{C_i(x,y)}{C_i^o} \exp\left(\frac{\tilde{\alpha} F \Delta\Phi(x,y)}{RT}\right) \tag{13.57}$$

where $\Delta\Phi$ is the potential difference between the catalyst and the solution. In the case of a convective-free process, considering that we are within the thickness of the diffusion layer

$$\frac{\partial C_i}{\partial t} = D_i \left(\frac{\partial^2 C_i}{\partial x^2} + \frac{\partial^2 C_i}{\partial y^2}\right) + \frac{a_i}{nF}\left(\frac{\partial \vec{j}_i}{\partial x} + \frac{\partial \vec{j}_i}{\partial y}\right) \tag{13.58}$$

We can consider for simplicity that at the steady-state of the process, mass transport and current generation occurs only along the y-axis:

$$D_i\left(\frac{\partial^2 C_i}{\partial y^2}\right) = \frac{a_i}{nF}\left(\frac{\partial \vec{j}_i}{\partial y}\right) \tag{13.59}$$

Taking the first derivative of Equation 13.57 with respect to y,

$$\frac{d\vec{j}}{dy} = \left(\frac{C_i}{C_i^o}\right) \frac{\tilde{\alpha} F j_0}{RT} \exp\left(\frac{\tilde{\alpha} F \Delta\Phi}{RT}\right) \frac{d\Delta\Phi}{dy} \tag{13.60}$$

$$\frac{d\vec{j}}{dy} = \frac{\tilde{\alpha} F \vec{j}}{RT} \frac{d\Delta\Phi}{dy} \quad \text{or} \quad \frac{d(\ln \vec{j})}{dy} = \frac{\tilde{\alpha} F}{RT} \frac{d\Delta\Phi}{dy} \tag{13.61}$$

The solution for this equation implies the knowledge of the shape of the functions $\Delta\Phi(y)$ and $j(y)$. We can consider that $y = 0$ at the top of the pore ("mountain" of the rough surface) and $y = L$ at the bottom of it ("valley" of the rough layer). We can then, simplify the tertiary behavior to a primary one at the top of the pore so that the current obeys Ohm's law:

$$j(y) = -\chi \left[\frac{\partial \Delta\Phi(y)}{\partial y}\right] \tag{13.62}$$

Then substituting (13.61) into (13.60) at $y \to 0$:

$$\vec{j} = -\frac{\chi RT}{\overleftarrow{\alpha} F} \left(\frac{\partial \Delta \Phi}{\partial y} \right)_{y \to 0} \tag{13.63}$$

Then, it is clear that the conductivity at the surface of the catalyst is lower than that of the bulk of the electrolyte, χ.

Going back to (13.59), we can integrate the equation considering the following boundary conditions:

1. At $y = 0$ $j(y = 0) = -\chi^{\circ} \left(\frac{\partial \Delta \Phi}{\partial y} \right)_{y \to 0}$ and $C_i(y = 0) = C_i^{\circ}$ (13.64a)

2. At $y = L$ $\left(\frac{\partial C_i}{\partial y} \right)_{y \to L} = 0$ and $j(y = L) = 0$ (13.64b)

The derivation of Equation 13.60 offers

$$\frac{d^2 \vec{j}}{dy^2} = \frac{d\vec{j}}{dy} \left[\frac{\vec{j}^{\,2} L \vec{\alpha} F}{\chi RT} - \frac{\vec{j} L \vec{\alpha} F}{\chi^{\circ} RT} + \frac{d}{dy} \ln \left(\frac{C_i}{C_i^{\circ}} \right) \right] \tag{13.65}$$

This equation is nonlinear, and it can only be solved by numerical approaches.

13.2.5 Triphasic Interfaces

13.2.5.1 Quasi-Cylindrical Pores in an Electrocatalyst

Two types of porous electrodes can be considered: two- and three-phase systems, where the latter is the special case of a triphasic interface in fuel cells, where the gas, liquid, and solid coexist. In the former, the liquid reactant is dissolved in the electrolyte and transported to the active sites of the electrocatalyst. In each case, we can consider uniform, parallel, cylindrical, or conical pores that are topped at the bottom by the metal substrate and at the top by the electrolyte [19,20].

In the first case, we consider that the reactant and the product concentrations are high enough so that there are no mass transport limitations and that the conductivity inside the pore is large enough to avoid any potential drop. Let us examine the case of a single cylindrical pore (Figure 13.12). The number of pores per square centimeter is $1/4r^2$ for parallel and cylindrical arrays.

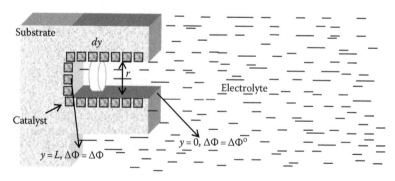

FIGURE 13.12 Quasi-cylindrical pore in an electrocatalyst for the analysis of current distribution in a two-phase system. An infinitesimal cylinder element of variable height dy and constant radius r is used for the calculations.

The expression of the current density for an electrooxidation reaction under Tafel conditions in one-dimension is

$$\bar{j}(y) = j_o \exp\left(\frac{\bar{\alpha}F\Delta\Phi(y)}{RT}\right) \tag{13.66}$$

Thus, the local current at the cylinder is

$$dI(y) = 2\pi r j_o \exp\left(\frac{\bar{\alpha}F\Delta\Phi(y)}{RT}\right) dy \tag{13.67}$$

where r is the constant radius of the cylindrical pore. The change in the potential from y to $y+dy$ can be simply given by Ohm's law:

$$d\Delta\Phi(y) = \frac{I(y)dy}{\chi\pi r^2} \tag{13.68}$$

where
  χ is the conductivity of the electrolyte inside the pore
  $I(y)$ is the local current generated in the pore from $y=0$ to $y=y$

Taking the second derivative with y,

$$d^2\Delta\Phi(y) = \frac{dI(y)dy}{\chi\pi r^2} \tag{13.69}$$

and substituting Equation 13.67, we get

$$\frac{d^2\Delta\Phi(y)}{dy^2} = \frac{2j_o}{\chi r} \exp\left(\frac{\bar{\alpha}F\Delta\Phi(y)}{RT}\right) \tag{13.70}$$

This differential equation can be solved with the aid of the following initial and boundary conditions:

$$\{y = 0 \quad \Delta\Phi(y) = \Delta\Phi^0 \quad I(y = 0) = \chi\pi r^2\left(\frac{\partial\Delta\Phi}{\partial y}\right)_{y=0} = 0$$

$$\{y = L \quad \Delta\Phi(y) = \Delta\Phi \quad I(y = L) = \chi\pi r^2\left(\frac{\partial\Delta\Phi}{\partial y}\right)_{y=L} \tag{13.71}$$

The solution to Equation 13.70 is simple when $\Delta\Phi^0$ is larger than 0.12 V. We can obtain a relation between the current distribution and the total current of the form [21,22].

$$\frac{I(y)}{I} = \frac{\tan\left[\left(\frac{\sqrt{a}}{2}\right)\exp\left(\frac{z_o}{2}\frac{y}{L}\right)\right]}{\tan\left[\left(\frac{\sqrt{a}}{2}\right)\exp\left(\frac{z_o}{2}\right)\right]} \tag{13.72}$$

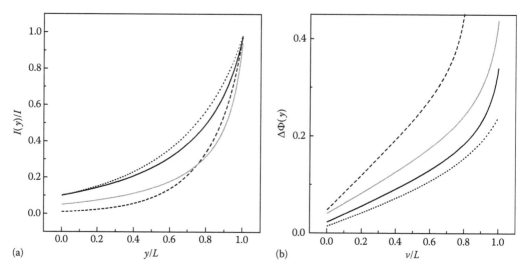

FIGURE 13.13 (a) Current distribution and (b) potential difference relationships as a function of the relative distance to the bottom of a cylindrical pore in an electrocatalyst. $z_0 = \Delta\Phi_0 F/2RT$ – value = 1.947 (dotted line), 2.92 (continuous line), 3.89 (gray line), 5.84 (dashed line). $\chi = 0.1$ S cm^{-1}, $j_0 = 10^{-6}$ A cm^{-2}, $r = 0.1$ μm.

where $a = \dfrac{2j_0 FL^2}{\chi rRT}$, $z_0 = \dfrac{\Delta\Phi_0 F}{2RT}$

with the total current given by

$$I = \chi\pi r^2 \frac{2RT}{L} \frac{\sqrt{a}}{L} \exp\left(\frac{z_0}{2}\right) \tan\left(\frac{\sqrt{a}}{2} \exp\left(\frac{z_0}{2}\right)\right) \tag{13.73}$$

The potential difference in the cylindrical pore is then (Figure 13.13) [21,22]:

$$\Delta\Phi(y) = \Delta\Phi^0 - \frac{4RT}{F} \ln \frac{1}{\tan\left[\dfrac{\sqrt{a}}{2} \exp\left(\dfrac{z_0}{2}\right)\dfrac{y}{L}\right]} \tag{13.74}$$

REFERENCES

1. F. Coeuret and A. Storck, *Eléments de Genie Electrochimique*, Lavoisier edition., Paris, France (1984).
2. D. J. Pickett, *Electrochemical Reactor Design*, Elsevier Science, Amsterdam, the Netherlands (1977).
3. E. Heitz and G. Kreysa, *Extended Version of a DECHEMA Experimental Course*, V.C.H., Weinheim, Germany (1986).
4. E. Raub and K. Müller, *Fundamentals of Metal Deposition*, Elsevier Publishing Co., Amsterdam, the Netherlands, London, U.K., New York (1967), Chap. 3, pp. 59–105, Chap. 4, pp. 124–136.
5. C. F. Zinola, Electrosulphometry in bright chromium baths, technical reports of the Galvanoplasty Section at Impresora Uruguaya Colombino S.A., Montevideo, Uruguay (1987).
6. C. F. Zinola and A. Colombino, Silver electroless deposition on commercial iron, technical reports of the Galvanoplasty Section at IUC S.A. Montevideo, Uruguay (1988).
7. L. J. J. Janssen, *Electrochim. Acta* **19** (1975) 257.
8. R. H. Rousselot, *Repartition du potential et du courant dans le électrolytes*, Dunod, Paris, France (1959).
9. N. Ibl, Current distribution, in E. B. Yeager, J.O'M. Bockris, B. E. Conway, S. Sarangapani (eds.), *Comprehensive Treatise of Electrochemistry*, Plenum Press, New York (1983), Vol. 6 Chap. 4.
10. H. S. Carlsaw and J. C. Jaeger, *Conduction of Heat in Solids*, Oxford Clarendon Press, New York (1959).

11. Ch. Kasper, *Trans. Electrochem. Soc.* **77** (1940) 353; Ch. Kasper, *Trans. Electrochem. Soc.* **77** (1940) 365.
12. Ch. Kasper, *Trans. Electrochem. Soc.* **78** (1940) 131; Ch. Kasper, *Trans. Electrochem. Soc.* **78** (1940) 147.
13. Ch. Kasper, *Trans. Electrochem. Soc.* **82** (1940) 153.
14. G. Prentice, *Electrochemical Engineering Principles*, Prentice Hall Inc., Upper Saddle River, NJ, 07458 (1991) Chap. 7, pp. 177–226.
15. T. Z. Fahidy, *Principles of Electrochemical Reactor Analysis*, Elsevier Science, Amsterdam, the Netherlands (1985).
16. P. J. Schneider, *Conduction Heat Transfer*, Addison-Wesley editions, Cambridge, U.K. (1955), pp. 121–128.
17. C. Wagner, *J. Electrochem. Soc.* **98** (1951) 116.
18. W. R. Parrish and J. Newman, *J. Electrochem. Soc.* **117** (1970) 43.
19. S. Srinivasan, H. D. Hurwitz, and J.O'M. Bockris, *J. Chem. Phys.* **46** (1967) 3108.
20. E. Justi and A. W. Winsel, *Kalte-Verbrennung*, Franz Steiner Verlag, Wiesbaden, Germany (1962).
21. J.O'M. Bockris and S. Srinivasan, *Fuel Cells: Their Electrochemistry*, Mc. Graw Hill Book Company, New York (1969), Chap. 5, pp. 230–288.
22. S. Srinivasan and H. D. Hurwitz, *Electrochim. Acta* **12** (1967) 495.
23. A. J. Arvia and R. C. Salvarezza, *Electrochim. Acta* **39** (1994) 1481.
24. P. Pfeifer, M. Obert, in *The Fractal Approach to the Heterogeneous Chemistry* (D. Avenir, Ed.), John Wiley & Sons, New York (1989), p. 11.
25. B. B. Mandelbrot, *The Fractal Geometry of Nature*, W. H. Freeman, New York (1982).
26. F. Family, *Physica* A **168** (1990) 561.
27. W. E. Triaca and A. J. Arvia, *J. Appl. Electrochem.* **20** (1990) 347.
28. L. D. Burke and M. B. C. Roche, *J. Electroanal. Chem.* **186** (1985) 139.
29. A. Chialvo, W. E. Triaca, and A. J. Arvia, *J. Electroanal. Chem.* **146** (1983) 93.
30. A. R. Despic and K. I. Popov, *J. Appl. Electrochem.* **1** (1971) 275.
31. N. Ibl, *Surf. Technol.* **10** (1980) 81.
32. C. Clerc and D. Landolt, *J. Appl. Electrochem.* **17** (1987) 1144.

14 Energy Balances and Hydrodynamic Conditions in the Electrochemical Reactor

C. Fernando Zinola

CONTENTS

14.1 DISCUSSION ON MASS BALANCES AND ENERGY EFFICIENCIES IN AN ELECTROCHEMICAL REACTOR

The optimization of an electrochemical reactor calls for a full description of the process to accomplish the specific objective of the mass and the energy balances together with heat transfer considerations and thermodynamic and enthalpy changes that are related to the unit cell and the whole stack [1,2]. A full description of the kinetics of both processes, the electric properties of the cell components, and the hydrodynamic aspects of the entire cell is also required.

The usual procedures for the conception of electrochemical reactors arise from the mass conservation laws and the hydrodynamic structure of the device. In fact, four types of balances can be considered: energy, charge, mass, and linear movement quantity. Since the reactor must include the anodic and the cathodic reactions, it is possible to make a complete balance for the mass. The temperature also governs the stability of a chemical reactor, but in the case of an electrochemical device, the charge involved in the entire process has to be considered first [3–5].

14.1.1 ENERGY BALANCE

For a given electrochemical reaction, the influence of the counter process has to be considered as a balance in the resulting chemical process. If φ_E and φ_S are the fluxes of the reactive substance at the entrance and the exit of the reactor, respectively, the first principle of thermodynamics will make an internal energy balance, U, as follows [4–6]:

$$\left(\frac{\partial U}{\partial n_E}\right) dn_E + P_E \left(\frac{\partial V}{\partial n_E}\right) dn_E + \dot{P} \, dt = \left(\frac{\partial U}{\partial n_S}\right) dn_S + P_S \left(\frac{\partial V}{\partial n_S}\right) dn_S + dU \tag{14.1}$$

where

     U is the electric, mechanic, and calorific energy
     $\dot{P}$ is the total power transferred to the ambient
     $(\partial V/\partial n)$ is the molar volume of the species
     dU is the amount of energy leaving the reactor and accumulated inside the cell

Similarly,

$$\left\{ \left(\frac{\partial U}{\partial n_E}\right) + P_E \left(\frac{\partial V}{\partial n_E}\right) \right\} \varphi_E + \dot{P} = \left\{ \left(\frac{\partial U}{\partial n_S}\right) + P_S \left(\frac{\partial V}{\partial n_S}\right) \right\} \varphi_S + \frac{dU}{dt} \tag{14.2}$$

Remembering the definition of enthalpy $H \equiv U + PV$, we can introduce it as the partial molar properties of the reactive substance at the entrance and exit of the reactor.

$$\left(\frac{\partial H}{\partial n_E}\right) \varphi_E + \dot{P} = \left(\frac{\partial H}{\partial n_S}\right) \varphi_S + \frac{dU}{dt} \tag{14.3}$$

If we define H as the enthalpy of the reactor and its contents as above, and in the case of an electrochemical reactor of constant volume, we can transform (14.3) into

$$\left(\frac{\partial H}{\partial n_E}\right) \varphi_E + \dot{P} = \left(\frac{\partial H}{\partial n_S}\right) \varphi_S + \frac{dH}{dt} - V \frac{dP}{dt} \tag{14.4}$$

14.1.2 ENERGY BALANCE FOR A CLOSED ELECTROCHEMICAL REACTOR UNDER GALVANOSTATIC AND ISOBARIC CONDITIONS

Equation 14.4 under constant current density and external pressure reduces to [6]

$$\dot{P} = \frac{dH}{dt} \tag{14.5}$$

since there is no difference in the molar fluxes for the reactant species at the start and end of the process. However, it has to be noted that the total enthalpy, H, refers to the reactor device, H_R, plus that of its electroactive, $\sum n_i h_i$, and inert components, $n_I h_I$:

$$\dot{P} = \frac{dH_R}{dt} + n_I \frac{dh_I}{dt} + \sum_i n_i \frac{dh_i}{dt} + \sum_i h_i \frac{dn_i}{dt} \tag{14.6}$$

On the other hand, considering the series for enthalpy [7]:

$$dH = \left(\frac{\partial H}{\partial V}\right) dV + \left(\frac{\partial H}{\partial P}\right) dP + \left(\frac{\partial H}{\partial T}\right) dT + \sum_i \left(\frac{\partial H}{\partial n_i}\right) dn_i \tag{14.7}$$

The following expressions for the change in enthalpy and power have to be considered in the case of a closed and isobaric reactor:

$$dH = \sum_k \left(\frac{\partial H_k}{\partial T}\right) dT = \left(m_R c_{p,R} + m_I c_{p,I} + \sum_i m_i c_{p,i}\right) dT \tag{14.8}$$

$$\dot{P} = \left(m_R c_{p,R} + m_I c_{p,I} + \sum_i m_i c_{p,i}\right) \frac{dT}{dt} + \sum_i h_i \frac{dn_i}{dt} \tag{14.8a}$$

By taking into account the mass balance based on Faraday's law extended to electrode kinetics [8]

$$\frac{jA\upsilon\varepsilon}{nF} = \sum_i \frac{dn_i}{dt} \tag{14.9}$$

where
  A is the electrode area
  ε is the current efficiency
  υ is the stoichiometric number (negative for reactants and positive for products)
  n is the total number of electrons consumed (or produced) during the process

Thus,

$$\dot{P} = \left(m_R c_{p,R} + m_I c_{p,I} + \sum_i m_i c_{p,i}\right) \frac{dT}{dt} + \frac{jA\varepsilon\Delta H}{nF} \tag{14.10}$$

where ΔH is the reaction enthalpy, $\Delta H = \sum_i \upsilon_i h_i$.
  On the other hand, the energetic power transferred to the outside of the reactor can be decomposed into two terms, the calorific and electric powers:

$$\dot{P} = jAE_j + \dot{Q} \tag{14.11}$$

where
  E_j is the electric potential at the current density j
  $\dot{Q}$ is the calorific power

In the case of pure electric power, it is well known that E_j can be developed into various components (overpotentials) that are dependent on j [9]:

$$E_j = E_{j=0} + a + b \log j + \frac{RT}{nF} \ln\left(1 - \frac{j}{j_{lim}}\right) + jAR_R \tag{14.12}$$

where
  a and b are the Tafel coefficients of the activation overpotential term
  j_{lim} is the mass transport limiting the current density
  R_R is the reactor linear resistance

Thus, the pure electric power $\dot{E}$ will be

$$\dot{E} = jAE_{j=0} + jAa + bAj \log j + \frac{RTAj}{nF} \ln\left(1 - \frac{j}{j_{\lim}}\right) + j^2 A^2 R_R \qquad (14.13)$$

In the case of $\dot{Q}$, we have to introduce the thermodynamic properties, ΔG as $-nFE_j$, and considering that $\Delta G = \Delta H - T\Delta S$, we can obtain the calorific power as

$$\dot{Q} = \left(m_R c_{p,R} + m_I c_{p,I} + \sum_i m_i c_{p,i}\right)\frac{dT}{dt} + \frac{jA\varepsilon T\Delta S}{nF}$$

$$- jAE_{j=0} + jAa + bAj \log j + \frac{RTAj}{nF} \ln\left(1 - \frac{j}{j_{\lim}}\right) + j^2 A^2 R_R \qquad (14.14)$$

In general, $\dot{Q}$ can be interpreted as the electric power due to the irreversibility of the processes minus the power absorbed by the reaction ΔS and those due to the reactant mixture and the body of the reactor. This power is of simple resolution in the case of galvanostatic conditions since the rate of the reaction is imposed from the outside. On the contrary, the solution to the equations is rather difficult due to the strong potential dependence on the current density (Equation 14.13). This power can be measured through the heat transport coefficient between the electrolyte and the outside or the refrigerant. However, in the case of isothermal electrochemical reactors, the flow of the power according to Equation 14.14 reduces to

$$\dot{Q} = \frac{jA\varepsilon T\Delta S}{nF} - jAE_{j=0} + jAa + bAj \log j + \frac{RTAj}{nF} \ln\left(1 - \frac{j}{j_{\lim}}\right) + j^2 A^2 R_R \qquad (14.15)$$

14.1.3 ENERGY BALANCE FOR AN OPEN ELECTROCHEMICAL REACTOR

Equation 14.4 resembles the conservation and mass balance for a hypothetical reactor:

$$\left(\frac{\partial H}{\partial n_E}\right)\varphi_E + \dot{P} = \left(\frac{\partial H}{\partial n_S}\right)\varphi_S + \frac{dH}{dt} - V\frac{dP}{dt} \qquad (14.4)$$

In the case of an open reactor, the equation reduces to

$$\left(\frac{\partial H}{\partial n_E}\right)\varphi_E + \dot{P} = \left(\frac{\partial H}{\partial n_S}\right)\varphi_S \qquad (14.16)$$

First, we will need to consider the different molar fluxes and molar enthalpies for the inert and electroactive species as in Equation 14.5:

$$\dot{P} = \left(\frac{\partial H_I}{\partial n_S}\right)\varphi_{IS} - \left(\frac{\partial H_I}{\partial n_E}\right)\varphi_{IE} + \sum_i \left(\frac{\partial H_i}{\partial n_S}\right)\varphi_{iS} - \left(\frac{\partial H_i}{\partial n_E}\right)\varphi_{iE} \qquad (14.17)$$

The temperature variation of the partial molar enthalpy at a constant specific heat will be

$$\left(\frac{\partial H_i}{\partial n_S}\right) = \left(\frac{\partial H_i}{\partial n_E}\right) + c_{p,i}(T_S - T_E) \qquad (14.18)$$

where T_S and T_E are the temperatures at the entrance and exit of the electrochemical reactor, respectively.

The energetic balance has different components depending on the nature of the species, but only the electroactive species renders a non-zero current density:

$$\begin{cases} \varphi_{iS} = \dfrac{\varepsilon_i j A \nu_i}{nF} + \varphi_{iE} \\ \varphi_{IS} = \varphi_{IE} \end{cases} \tag{14.19}$$

Considering expressions (14.19) and (14.18) and substituting them into (14.17) we have

$$\dot{P} = \sum_{i,I} \varphi_{kE} c_{p,k}(T_S - T_E) + \frac{\varepsilon_i j A}{nF} \Delta H(T_S) \tag{14.20}$$

Considering the same treatment as above, we can calculate $\dot{Q}$ with a similar physical meaning as in a closed electrochemical reactor:

$$\dot{Q} = \frac{jA\varepsilon T_S \Delta S(T_S)}{nF} + jAa + bAj \log j + \frac{RTAj}{nF} \ln\left(1 - \frac{j}{j_{\lim}}\right) + j^2 A^2 R_R - \sum_{i,I} \varphi_{kE} c_{p,k}(T_S - T_E) \tag{14.21}$$

This is of particular importance in the case of an adiabatic reactor, where the exchange of heat with the outside is zero, $\dot{Q} = 0$. Thus, in this case it follows

$$T_s \left[\sum_{i,I} \varphi_{kE} c_{p,k} + \frac{jA\varepsilon \Delta S(T_S)}{nF} \right] - \sum_{i,I} \varphi_{kE} c_{p,k} T_E = jA(a + b \log j) + \frac{RTAj}{nF} \ln\left(1 - \frac{j}{j_{\lim}}\right) + j^2 A^2 R_R \tag{14.22}$$

The right-hand side of Equation 14.22 is truly positive since it involves all the processes irreversibilities (always positive overpotentials), and then in the left-hand side the addition of the flux capacities has to be larger than the ΔS contribution. For the calculation of the operation temperature, T_S, in a continuous stirring reactor, we can distinguish two cases: (1) $\Delta S > 0$ as in the case of few electrolytic systems (water–alkaline electrolysis). We can calculate the value of T_S by the intersection of the plot of the right-hand term with the left-hand term as a function of T. (2) $\Delta S < 0$ as in most of the cases of electrolytic systems, it is possible to conduct the same plot but in this case, it is likely that the left-hand side of Equation 14.22 would be negative. Thus, a larger slope in the representation is expected with some negative currents (at the entrance temperature in the reactor). We can also determine the operation current, I, at T_S as

$$I = \frac{nF \varphi_{iE} c_{p,i}}{\varepsilon \Delta S(T_S)} \tag{14.23}$$

14.1.4 Chemical or Secondary Reactions

In most of the cases, the main electrochemical or electrocatalytic reaction is coupled with homogeneous chemical reactions that can be defined either in the bulk of the solution or in a Nernstian heterogeneous film. Whether one or the other takes place, the result is the modification of the kinetic constant of the main process. The topic is so general that we can only discuss some examples. In the case of electrosynthesis of inorganic compounds such as chlorate, the secondary formation of chloride from the saline electrolyte can also be demonstrated. The efficiency of the process evidently is reduced considerably, especially at high temperatures and from the formation of molecular chlorine; the subsequent chemical formation of the hypochlorite is accomplished. The same problem appears in the complex field of organic electrochemistry, especially in the case of non-polar solvents where the formation of free radicals occurs very often. When the free radicals reach the electrode surface, the formation of neutral species takes place mainly by the formation of adsorbed films. These films probably polymerize into a multilayer that changes the electronic properties of the catalyst. In some situations, the process is conceived as associated with a chemical reactant, such as in the oxidation of anthracene to anthraquinone with chromic acid on a lead oxide anode.

The diversity of the (electro)chemical reactions coupled with a main process can be classified as

1. *CE* mechanism, where the first process is the chemical formation of the real reactant or intermediate that is activated on the electrode forming the stable product.
2. *EC* mechanism, where the main process produces an intermediate or side-product that can be chemically transformed into the stable product.
3. *ECE* mechanism, where a complex process occurs after the first main electrochemical reaction. This is likely when the product of interest is (photo)chemically decomposed into another that is electrochemically active (example, electrooxidation of oleic acid with acetic acid as solvent and two products, mono and diacetooxyoleate acids, are formed).

Accordingly, the concentration profile of the processes changes with respect to the type of mechanism and to the rate determining specific constant, k (from 10^{-5} to 10^{10} s^{-1}). In the case of industrial electrochemistry, the optimized conditions of work imply the minimization of loss, according to the side reactions. This is a consequence of the selectivity condition needed in the case of an electrochemical reactor. In a general treatment the theoretical model of the reactor is based on mass conservation laws with the corresponding electrochemical kinetics (coupled or not to side reactions). For example, the *EC* mechanism can be treated as follows:

$$Cr_2O_7^{=}{}_{(aq)} + 14H^+ + 6e^- \rightarrow 2Cr^{3+}{}_{(aq)} + 7H_2O \tag{I}$$

$$Cr^{3+}{}_{(aq)} + 3H_2O \rightarrow Cr(OH)_{3(s)} + 3H^+{}_{(aq)} \tag{II}$$

We are going to consider, for simplicity, that only the diffusion of the species takes place (immobile electrolyte), using a high chemical rate constant (path II). Thus, the conservation equations, for $Cr_2O_7^{=}$ and Cr^{3+}, are for a semi-infinite diffusion process on a planar electrode:

$$\frac{\partial C_i(y, t)}{\partial t} = D_i \frac{\partial^2 C_i(y, t)}{\partial y^2} \quad \text{where } i \text{ denotes } Cr_2O_7^{=} \tag{14.24}$$

$$\frac{\partial C_j(y, t)}{\partial t} = D_j \frac{\partial^2 C_j(y, t)}{\partial y^2} - k\, C_j(y, t) \quad \text{where } j \text{ denotes } Cr^{3+} \tag{14.25}$$

The initial and contour conditions for i and j are

$$C_i(y, t = 0) = C_i^o \quad C_j(y, t = 0) = 0 \tag{14.26}$$

and

$$C_i(y \rightarrow \infty, t) = C_i^o \quad C_j(y \rightarrow \infty, t) = 0 \tag{14.27}$$

One simplified treatment is for working under galvanostatic conditions. Thus, both the diffusion current density components can be equal to that imposed by the equipment, j:

$$nFD_i\left[\frac{\partial C_i(y,t)}{\partial y}\right]_{y=0} = nFD_j\left[\frac{\partial C_j(y,t)}{\partial y}\right]_{y=0} = j \tag{14.28}$$

The solution of Equations 14.24 and 14.25 is of the Laplace transform as expected. Thus, in the case of (14.24), the transformation leads to

$$sC_i(y,s) = D_i\frac{\partial^2 C_i(y,s)}{\partial y^2} \tag{14.29}$$

with the integration form as

$$C_i(y,s) = \frac{C_i^o}{s} + A(s)\exp\left[-\left(\frac{s}{D_i}\right)^{1/2}y\right] \tag{14.30}$$

The value of $A(s)$ is obtained from the initial and contour conditions (14.28):

$$j = nFsD_i\left[\frac{\partial C_i(y,s)}{\partial y}\right]_{y=0} \tag{14.31}$$

Therefore,

$$j = -nFsD_iA(s)\left(\frac{s}{D_i}\right)^{1/2} \tag{14.32}$$

Then, obtaining $A(s)$ and substituting it into (14.30), we have

$$C_i(y,s) = \frac{C_i^o}{s} - \frac{j}{nFs^{3/2}D_i^{1/2}}\exp\left[-\left(\frac{s}{D_i}\right)^{1/2}y\right] \tag{14.33}$$

After making the anti-transformed expression of Equation (14.33) we can obtain the concentration profile dependence with space and time:

$$C_i(y,t) = C_i^o - \frac{j}{nFD_i}\left\{2\sqrt{\frac{D_it}{\pi}}\exp\left[-\frac{y^2}{4D_it}\right] - (y)\mathrm{erfc}\left[-\frac{y}{2\sqrt{D_it}}\right]\right\} \tag{14.34}$$

It is of particular interest the value of the concentration of the dichromate at the electrode surface, thus taking in Equation 14.34 $y = 0$ we will have

$$C_i(0,t) = C_i^o - \frac{2jt^{1/2}}{nF(\pi D_i)^{1/2}} \tag{14.35}$$

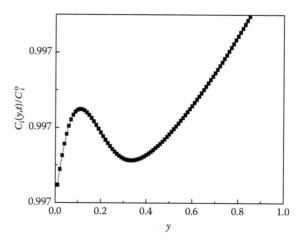

FIGURE 14.1   Variation of $C_i(y,t)/C_i^o$ as a function of the distance y for a constant time $\tau = 0.1$ s, $j = 0.5$ A cm^{-2}, $D_i = 10^{-5}$ cm^2 s^{-1}.

From this expression, it is evident that there is some condition where the surface concentration of the reactant is zero. This condition is temporal and for a galvanostatic statement is known as the transition time, τ. Thus, from Equation 14.35 we have

$$\tau = \frac{(nC_i^o F)^2 \pi D_i}{4j^2} \tag{14.36}$$

In Figure 14.1, we plot the rationalized expression of the concentration profile $C_i(y,t)/C_i^o$ vs. the y coordinate (distance from the electrode to the bulk of the electrolyte) considering $j = 0.5$ A cm^{-2}, $D_i = 10^{-5}$ cm^2 s^{-1}, $\tau = 1$ s. In this situation, we can observe a local saturation condition near the electrode surface, and then a fast increase in the concentration profile is observed, which reaches the maximum value (1) at y values that depend on the physicochemical constants of the system.

In the case of the Cr^{3+}, the chemical conversion of the treatment is nearly the same:

$$\frac{\partial C_j(y,t)}{\partial t} = D_j \frac{\partial^2 C_j(y,t)}{\partial y^2} - kC_j(y,t) \tag{14.25}$$

and it is transformed into

$$sC_j(y,s) = D_j \frac{\partial^2 C_j(y,s)}{\partial y^2} - kC_j(y,s) \tag{14.37}$$

Considering conditions in (14.26) and through the integration at $y = 0$:

$$C_j(0,s) = \frac{j}{nFs(k+s)^{1/2}D_j^{1/2}} \tag{14.38}$$

Finally, the anti-transformed expression will be

$$C_j(0,t) = \frac{j}{nF(k)^{1/2}D_j^{1/2}} \operatorname{erf}[kt]^{1/2} \tag{14.39}$$

Equations 14.39 and 14.25 can be used to formulate the expression of the potential obeying the situation of the first electrochemical reaction. Since it is reversible, we can consider

$$E(t) = E^\circ + \frac{RT}{nF} \ln \frac{C_i(0,t)}{C_j(0,t)}$$

$$E(t) = E^\circ + \frac{RT}{2nF} \ln \frac{D_i}{D_j} + \frac{RT}{nF} \ln \left(\sqrt{\frac{\tau}{t}} - 1 \right) + \frac{RT}{nF} \ln \left(\sqrt{\frac{4}{\pi}} \frac{k^{1/2}t}{\mathrm{erf}\{[kt]^{1/2}\}} \right)$$

(14.40)

For this expression, we can state the limiting conditions more easily in order to develop and understand. First, for $\sqrt{kt} < 0.1$ we can say that $\mathrm{erf}\left(\sqrt{kt}\right) \approx (2/\sqrt{\pi})\sqrt{kt}$ and the process occurs as no coupled chemical reactions occur. On the other hand, when $\sqrt{kt} > 2$, the error function reaches unity, leading to an extremely high contribution of the last term in the Equation 14.39. All this resemble a large chemical reaction that occurs and leads to an inverse process (oxidation of Cr^{3+} in this case).

14.2 ELECTROCATALYST DESIGN AND HYDRODYNAMIC CONDITIONS

14.2.1 ELECTROCATALYST FAILURE FACTORS

Electrocatalysis can modify the composition of the electrode surfaces and the nature of the electrolytic products. The perchlorate decomposition (cathodic production of chloride) on platinum catalysts is one of the examples [57] and the IrO_2 decomposition during the sodium chlorate production [58]. The electropolymerization of the organic substances is critically dependent on the type of the electronic/ionic conductors, electrolyte characteristics, and the electrolysis resident time of the monomer [59].

The failure of electrocatalysts is attributed to the mechanical, chemical, electrochemical, and metallurgical factors. However, the hydrodynamic aspects (liquid impact from the flow-jet reactors) cause the erosion of the electrodes, especially in the case of the non-noble metals. For lead-based alloys, the catalyst failure occurs in the case of the triphasic systems or substrate/gas systems [70]. Some other breakdown factors are due to the short-circuiting because of the preferential growth of the crystallites on the catalyst or due to the chemical dissolution and the redeposition of the metal substrates on defined sites of both the electrodes surfaces. In these cases, the use of the surface tension additives, the large organic inhibitors, and/or the brighteners is recommended. Moreover, the addition of the surfactants to the bath electrolyte is another interesting solution to this problem as they reduce the number of preferential sites for the crystal growth, inhibit surface frictions, and increase wettability and surface conductivity values [57].

The selection of the anodes and the cathodes for the industrial processes depends on the electrocatalytic properties of the different materials, such as corrosion stability, undemanding preparation, solvent decomposition potential, and basically current (electrolyzers) or energy (galvanic cells) efficiencies. Therefore, the catalysts for the technical purposes need to be capable of continuous operation for a longtime (say more than 6–12 months depending on the type of the electrochemical reactor). Further they must be capable of large current densities with low overpotentials (appropriate Tafel slopes for large j_o values) and minimization of the power density consumptions, long-term corrosion resistances, and large-surface real area/reactor volume ratios. The last requisite was achieved from the results obtained on the noble-metal supported catalysts some years before [8–15]. Roughness factors of 10^4 were achieved for platinum and platinum group metals and the easy preparation methods have been published elsewhere [16–18]. As it was stated above, the electrode design strongly depends on the type of the electrochemical reactor in which the catalyst is to be used. In the case of metal, inorganic or organic electrosynthesis, the electrodes can be grouped as follows:

1. Stable electrocatalyst for long-term performances that are typical for metal deposition and electrocrystallization and reduction reactions with the adsorbed intermediates such as cuprous ions. For example, most noble metals such as platinum and rhodium [19,20].
2. Electrocatalyst for gaseous reactants [21–23] where the electrooxidation reaction arises from the gas as a reactant. For example, dissolved molecular hydrogen or carbon monoxide oxidation and the electroreduction of molecular oxygen or carbon dioxide.
3. Active electrocatalyst forming soluble species whose utilization is likely for the electro-formation of organometallic compounds or low-chained polymers. One typical example is the industrial preparation of tetra-alkyl-lead through the electrooxidation of the Grignard reactant on bed-leached anodes [24,25].
4. Noble metal anodes passivated with a few microns of noble metal oxides. Typical for electroorganic oxidations and the most used catalyst is platinum or a titanium substrate covered with a layer of platinum oxide (the α-PtO_2 or the high-oxidation PtO_3 with β-PtO_2 oxide) [26,27]. Sometimes the pyrolytic-carbon graphite and the vitreous carbon surface also can act as a good catalyst for the organic reactions such as the oxidation of unsaturated hydrocarbons. These types of electrocatalysts are stable upon the application of high anodic potentials because of the presence of titanium (stable up to 40 V). In some cases, a high-oxidized nickel oxide supported on titanium can avoid the cost of employing a platinum catalyst. This layer of nickel oxide can be prepared using the Watts bath using large current densities, that is, 2–10 A cm^{-2} [8].

Most of the electrochemical reactors fail due to different attacks on the electrocatalysts, where the anodes are attacked faster than the cathodes (electrochemical corrosion, mechanical fissures due to electrodissolution, or bubble formation and evolutions, etc.) [43]. In new technologies, the use of the anode, membrane, or cathode assemblies solves this problem. In the case of the solid polymer electrolytes, the anode and the cathode catalysts are integrated to the membrane promoting the mechanical and electrochemical stability of the device [44,45]. This new technology replaces the problem of the diaphragm-based electrochemical industry that was established in the beginning of the twentieth century [46].

The mechanical failure is the result of a large number of reasons, but it is generally complicated in the case of long-term electrolysis, difficult hydrodynamic devices, or non-noble catalysts [47,48]. For example, when soft anodes are used the use of mechanical strengtheners and conductive additives is required. When the lead and lead alloy anodes are used, they are frequently exposed to fatigue corrosion in the presence of strong concentrated electrolytes [50–52]. The addition of ferrous-alkaline elements such as barium and calcium produces an increase in the substrate hardness. Former technologies, such as those of the chlorine-alkali industries, usually employ carbon anodes. The fissure and cracking of their surfaces were evident after 3 months of continuous use. The replacement of the carbon conical anodes was the most applicable solution. Nowadays, the carbon electrodes are replaced by titanium and Ti–RuO_2 or Ti–RuO_2–SnO_2 electrodes that are most expensive but exhibit good binding and mechanical properties [61]. In this case, after several years, the degradation of the Ti–RuO_2–SnO_2 substrate is observed by the formation of TiO_2 and RuO_4 and the elimination of SnO_2 due to chemical dissolution. It is also interesting to observe the cheaper electrodes such as coated titanium that suffer erosion of the surface layer and occurrence of fissures with the exhibition of the back substrate. The problem is solved by the addition of surface metals that produce the coalescence of the titanium grains again [53,54]. The most known folder is that of the ferrous metals and their alloys. Many cases can be enumerated, but we can summarize by saying that the stress corrosion cracking and erosion are the most important failures. The solution to this problem is simple; we have to add some more noble metals and produce a ferrous alloy of chromium and/or nickel. The selection of a proper chemical composition to the electrocatalyst can solve most of the corrosion properties [53].

The most interesting cases come from the stable *DSA* electrodes. Titanium substrates coated with IrO_2 and/or PtO_2 are sometimes unstable when the electrolysis products attack the oxide coatings or the chloride-pitting corrosion competes with the anodic reaction of the electrocatalytic process. This pitting corrosion occurs when the chloride concentration exceeds 150 g dm^{-3}. However, in the case of RuO_2-coated electrodes, the chemical corrosion occurs for a chloride solution concentration of lower than 100 g dm^{-3} [49]. For long-term experiments or most properly for industrial processes, the formation of non-conducting ruthenium oxide layers is clear [55]. The solution for the coating degradation problem is the addition of some certain impurities that improve the adhesion of the oxide layers. However, in the case of chlorine-evolution reaction, the chemical erosion of the *DSA* catalyst is evident, since an increase in the anodic potential is an indication of a chemical corrosion process. An electrochemical reactor design with the electrode plates with rather small holes improves the fast removal of the anodic gases that are produced by the catalysts [56,57]. These types of anodes are also sensible to selective corrosion, especially in the case of alloys in fuel cells, that is, for RuO_2 and ruthenium metal surface compounds [58].

In the case of soft anodes such as lead and lead alloys, the acid chemical corrosion for strong ionic force media can be prevented by the addition of nickel and cobalt cations to produce a mechanically more stable catalyst [59]. Ferrous metals and steels are also attractive for use in electrochemical reactors. The passivation of the catalytic surface of nickel, iron, and cobalt at low pH or the chemical hydride formation due to the simultaneous hydrogen evolution reactions (for electroplating) are obstacles for the metal electrodeposition process because of the less active sites that are available for the continuous process. pH control is an easy solution to avoid these problems and further it avoids the anodic dissolution of the metals in the strong acid solution [60–62]. The increase of the electric resistance of anode is also important in the case of long-term experiences, since the applied effective potential has to be continuously increased. However, pitting corrosion simultaneously produces an increase of surface conductivity due to the formation of micro-channels through which the electrolyte can flow. This effect is more noticeable when we are working with alloys as anode selective passivation or intergalvanic corrosion is inevitable, especially in the case of strong mineral acids [63–65]. The solution to this problem arises from not using a binary or ternary alloy or diminishes the experimental condition that is proper to electrochemical or chemical corrosion. For the case of general corrosion, there is no general rule for this problem, and thus *in situ* field tests are required.

Metallurgical factors are also mandatory for the preparation of an electrocatalyst. The occurrence of bulk oxidation of the alloy due to local thermal corrosion or by intergrain corrosion produces the inherent change in their physical characteristics [49]. The total absence of this process leads to a mechanical stable electrode, especially in the case of industrial applications. Sometimes internal cracks are caused by the inner diffusion of metal atoms producing phase transformations, preferentially crystallographic tendencies, or different grain dimensions [55]. The consequences of these facts are predictive for an electrochemical reactor using anodic alloys. When these facts are impossible to minimize, we can also observe during long-term experiments the changes in the electric properties of the catalyst, the appearance of inter-grain boundary layers causing electrolyte or gas inter-diffusions, intergalvanic progressive corrosion, and short-circuiting between grains (and the concomitant decrease in the number of surface active sites) [66,67]. In the case of alloys, it is better to minimize the proportion of the "problematic" component. For example for titanium substrates with PtO_2 and IrO_2 compounds, we can prevent problems by avoiding large proportions of the IrO_2 species, especially, when using chloride-containing solutions [58,68].

The changes in the anode composition and morphology are also responsible for the erratic response of the current density vs. potential difference tendency and the decrease in the electrocatalytic activity. Thermo-corrosion and thermal stress failure, stress corrosion cracking during gas evolution, and overcritical local pressure conditions can cause extensive pitting and large ohmic voltage drops.

The cathodes are similarly corroded (chemically or electrochemically) as the anodes but metal-lurgical factors affect to a lesser degree, that is, the transformation is gradual [69]. However selective attack, such as from electrolyte impurities, often takes place and the production of a homogeneous surface layer of sub-products or crystal metallic impurities produce micro-cracks reducing the durability [70]. In addition, microbial factors produce cathode failure especially in the electrochemical treatment of wastewaters [71].

As it has been stated before, spinels and spinel-type electrodes of nickel and iron are able to catalyze the solvent electrochemical decomposition. The production of hydrogen from strong alkaline solutions is of extreme importance taking seawater as a "mother prime." After long-term electrolysis, a mixture of FeO_3/NiO_2 spinel increases its nickel molar fraction value [19]. The control of the pH and the local temperature is of extreme importance to work out this problem. However, when the reactor is not properly working, it is possible to reduce the medium-higher oxidation states of NiO or Fe_2O_3 by cyclic electroreduction [72].

Many more cases that are interesting can be introduced here but because of problems of space, we will stop with the above.

14.2.2 Hydrodynamic Conditions and Gas Evolution

Electrochemical processes often involve gas evolution reactions. Formation and evolution of bubbles are most important in the case of the electrocatalyst since they can block the active sites from further reactions.

The first step of the process seems to be the nucleation of the bubbles that occur in a relatively short time before reaching the full-growth range. This process strongly depends on the interfacial forces working at the electrocatalyst. These physical forces are defined by the surface roughness of the catalyst and the type of the substrate and its chemical nature, for example, the type of material alloying metals, metal inclusions, etc. In addition, the solution composition has to be considered, that is, the presence of adsorbable anions, emulsions, the formation of suspensions, etc. However, the effective bubble radius is not only a compromise between the ionic and electronic conductors but also the consequence of the surface properties, since the detachment from it is actually defined by the morphology of the electrocatalyst.

The second step is the growth of the bubbles that mostly produces the insulation of the surface for successive acts of the reaction. The third and fourth steps, the detachment and uplift of the bubbles really imply a stirring effect near the electrocatalyst's surface. These two processes are interesting to electrochemical engineering, since when properly used it can take advantage as an *in situ* solution stirrer. Thus, it seems that the formation of the bubbles contributes to the increase in the rate of the electrocatalytic reaction but sometimes can lead to a blocking effect when the bubbles are larger than expected. The bubbles establish a nonconducting phase leading to a tortuous lane for the equiflux lines near the electrode with the development of an amplified local electric resistance (Figure 14.2).

Let us explain what happens with the formation of a bubble. When the electrocatalytic reaction occurs with the production of a gaseous product, it dissolves in the electrolyte until it reaches saturation. It is transported from the electronic conductor to the ionic conductor only by convective diffusion. When the solution concentration exceeds "supersaturation," we are able to activate the nucleation sites to the bubble formation. This condition depends on the morphology of the surface and the kind of electrolyte and its viscosity. The growth of the bubble induces a microconvective flow on the electrolyte, pushing in various radial directions each bubble from an ideal center of the surface ("active site"). When each bubble attains a certain size, the buoyancy exceeds its adhesion and the bubble leaves the surface producing a drag flow.

Therefore, we can say that four main processes can be described in the evolution of gases: *nucleation* (the bubble starts at the surface sites where preexisting gas rests, like in holes or kinks, or any defect of the crystal), *bubble growth* (which is controlled by the diffusion of dissolved gas from

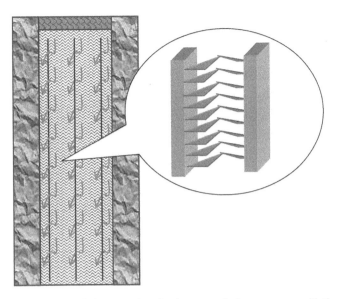

FIGURE 14.2   Composed structural electrocatalyst for the gas evolution processes with the flow pattern of gas evolving electrodes. The bubble takeoff promoter is detailed in the inset.

the surrounding liquid), *detachment* (which depends mostly on the surface topology of the catalyst), and the *departure* of the bubbles (bubble uplift with the concomitant stirring effect; that is, mass and heat transfer).

The nucleation process starts at defined active centers of the surface called the nucleation sites. Even in the case of an electrolyte supersaturated with gas the activation of the nucleation site is required. It depends not only on the nature of the gas and the electrolyte but also on the interfacial tension and number of substrate neighbors on the electrocatalyst. In summary, the existence of a gas–electrolyte interface besides the electrolyte–solid interface is required. Some authors explained this process as nucleate boiling [73] and others [74] as the super saturation of the electrolyte with the gas, developing different equations for the current density. Another important factor to consider is the number of surface active sites available for the bubble formation and the geometry of the bubble. Moreover, the surface roughness is certainly a factor to be considered for the stability of the nucleation site. The ageing of the nucleation site also has to be considered since it results in the loss of activity after a long time.

The bubble growth is initially controlled by the surface morphology; that is, a certain small area is attributed to a single hemispherical or spherical bubble as shown in Figure 14.3.

14.2.2.1 Mass Transport during Bubble Growth and Evolution

The mass transfer from the hemispherical bubble of time-dependent radius $R(t)$ can be characterized by the mass transfer coefficient, $K_M(x, t)$:

$$K_M(x, t) = \frac{D_i}{\delta(x, t)} \tag{14.41}$$

Since Sherwood number, Sh, is defined as

$$Sh(x, t) = \frac{K_M(x, t)x}{D_i} \tag{14.42}$$

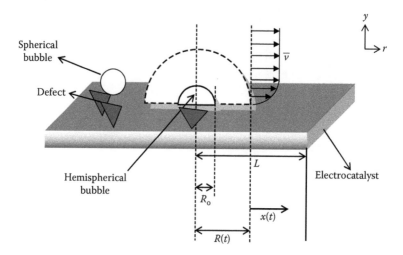

FIGURE 14.3 Bubble growth on a micro-defect site of an electrocatalyst.

We can consider that if the flow along the micro-defect is laminar and the convection flow between the bubble and the electrolyte behaves as a plug-flow reactor, we can say that [75]

$$Sh(x, t) = \sqrt{\frac{Re(x, t)Sc}{\pi}} \qquad (14.43)$$

where Sc is the Schmidt number defined as $Sc = (\nu/D_i)$ and $Re(x, t)$ is the length- and time-dependent Reynolds number defined as; $Re(x, t) = x((dV(x, t)/dt)/A\nu)$ where A is the micro-defect electrode area, ν the cinematic viscosity, $(dV(x, t))/dt$ the volumetric flow rate of the bubble and x is the length coordinate in the flow direction which reaches the largest value, R, that is, the break off diameter.

From Equations 14.42 and 14.43, we can say that

$$K_M(x, t) = \sqrt{\frac{D_i}{xA\pi} \frac{dV(x, t)}{dt}} \qquad (14.44)$$

The term $(dV(x, t)/Adt)$ is really $(dR(x, t)/dt)$ the bubble growth rate at $r = R$, but it is also interesting to obtain an equation for the electrolyte rate of micro-convection, that is, at a distance $x(t) = r - R(t)$. These expressions were originally used considering the continuity equation for the heat transfer in nucleate boiling [76].

14.2.2.1.1 Fourier's Number
If a Fourier number for mass transfer, Fo', is introduced with analogy to transient heat transfer as constant, we will have [78]

$$Fo' \equiv \frac{D_i t}{R^2} \qquad (14.45)$$

Then,

$$\frac{dR}{dt} = \frac{1}{2} \left(\frac{D_i}{Fo'} \right)^{1/2} t^{-1/2} \qquad (14.46)$$

Then, substituting (14.46) into (14.44) we have

$$K_M(x, t) = \sqrt{\frac{D_i^{3/2}}{2x\pi} \left(\frac{1}{tFo'}\right)^{1/2}} \tag{14.47}$$

However, from the definition of the Sherwood number (14.43)

$$Sh(x, t) \equiv \sqrt{\frac{xD_i^{1/2}}{2\pi} \left(\frac{1}{tFo'}\right)^{1/2}} \tag{14.48}$$

The Fourier number Fo' can be calculated from the break off radius of the bubble, R_b, at the residence limit time, t_b:

$$\frac{R_b^2}{t_b} \equiv \frac{D_b}{Fo'} \tag{14.49}$$

Then by substituting into (14.48), we can obtain an x-dependent Sherwood number for the bubble break off:

$$Sh(x) \equiv \sqrt{\frac{x}{2\pi} \left(\frac{R_b^2}{t_b^2}\right)^{1/2}} \tag{14.50a}$$

or

$$Sh(x) \equiv \sqrt{\frac{xR_b}{2\pi t_b}} \tag{14.50b}$$

On the other hand, the Reynolds number was defined as

$$Re(x, t) = \frac{x}{v} \frac{d(V/A)(x, t)}{dt} \tag{14.51}$$

As stated above, the ratio between the volumetric flow rate and the surface center active site for the evolved gas is the time-dependent bubble growth distance. For this ratio we have

$$\frac{d(V/A)(x, t)}{dt} = \frac{\pi 4 R_b^3}{3t_b} \frac{1}{\pi L^2} \tag{14.52}$$

where L is the perimeter of the active site area. Thus, the Reynolds number for the bubble break off is

$$Re(x) = \frac{4xR_b^3}{3vL^2 t_b} \tag{14.53}$$

As shown by Ibl and Venczel [77] we can define a degree of the surface coverage by the nucleated bubbles, θ, as

$$\theta \equiv \frac{\pi R^2}{\pi L^2} \tag{14.54}$$

and particularly from the bubble break off

$$\theta \equiv \left(\frac{R_b}{L}\right)^2 \qquad (14.55)$$

We can reorient the dimensionless numbers:

$$Sh(x) \equiv \sqrt{\frac{xL\theta^{1/2}}{2\pi t_b}} \qquad (14.56)$$

and

$$Re(x) = \frac{4x\theta R_b}{3vt_b} \qquad (14.57)$$

In other references, the mean Sherwood number is also presented as a double integration with time and the radial coordinate [76] for the special case of $\theta < 0.5$:

$$Sh = 2.343(ReSc)^{1/2}\theta^{1/4}(1 - \theta^{1/2})^{1/2} \qquad (14.58)$$

It has been explained before that during the gas evolution in electrolysis, it dissolves into the electrolyte within the diffusion layer. Because of the gases' low solubility and small diffusion coefficients, the diffusion layer becomes saturated with gases and then the nucleation starts. Due to the concentration gradients, we can also say that the diffusion flux for the bubble, J_b, is [79]

$$J_b(r, \vartheta) = D_b \left(\frac{\partial C_b}{\partial r}\right) + \frac{D_b}{r} \left(\frac{\partial C_b}{\partial \vartheta}\right) + \frac{D_b}{r \sin \vartheta} \left(\frac{\partial C_b}{\partial \varphi}\right) \qquad (14.59)$$

where
D_b is the diffusivity of the bubble
C_b is the local concentration at $r = R$ (radius of the bubble)
r, φ, and θ are the radial, angular, and polar coordinates, respectively (Figure 14.4)

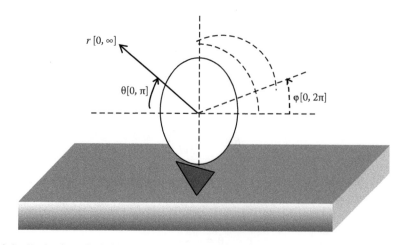

FIGURE 14.4   Evaluation of a bubble growth on an electrocatalyst site by spherical coordinates.

14.2.2.1.2 Influence of the Electric Charge on Bubble Growth

The evolution and growth of the bubble as a function of the spatial coordinates for a perfect spherical bubble can be only described by r and φ (only when the bubble is not an ellipsoid). Thus, taking the flux definition in Equation 14.59 of the first Fick's law, we have

$$\frac{\rho_b}{PM}\left(\frac{\partial R}{\partial t}\right) = D_b\left(\frac{\partial C_b}{\partial r}\right)_{r=R} + \frac{D_b}{R\sin\vartheta}\left(\frac{\partial C_b}{\partial \varphi}\right) \tag{14.60}$$

According to [79], two kinds of bubble local concentrations in the electrolyte can be distinguished, one arising from the electrolysis and the other that is dissolved in the bulk of the solution coming from the initial gas dissolution. Moreover, we can also consider that the change in the radius of the bubble simultaneously affects the change with the contact angle, so

$$\left(\frac{\partial R}{\partial t}\right) = \frac{2D_b PM}{\rho_b}\left(\frac{\partial C_b}{\partial r}\right)_{r=R} \tag{14.61}$$

and

$$\left(\frac{\partial R}{\partial t}\right) = \frac{2D_b PM}{\rho_b}\left[\left(\frac{\partial C_s}{\partial r}\right) + \left(\frac{\partial C_e}{\partial r}\right)\right]_{r=R} \tag{14.62}$$

where C_s and C_e, are the initial bulk dissolved gas and local electrochemical formed gas concentrations, respectively. If we consider that the concentration gradients are uniform around the electrocatalyst with no local defect sites, we can say that

$$C_e = \frac{Q}{nF4\pi r^2} \tag{14.63}$$

where
  n is the number of electrons per mol for the gas formation
  F is the Faraday's constant
  Q is the accumulated charge at the surface

$$\left(\frac{\partial C_e}{\partial r}\right)_{r=R} = \frac{Q}{nF2\pi R^3} \tag{14.64}$$

where the charge Q is only dependent on time.

On the other hand, the change in the dissolved gas concentration is directly the value of the gradient at $r = R$, and the radial convective-diffusion toward the bulk of the electrolyte because of the bubble growth [80]:

$$\left(\frac{\partial C_s}{\partial r}\right)_{r=R} = \frac{C_b^\circ - C_{b,r=R}}{R} + \frac{C_b^\circ - C_{b,r=R}}{\sqrt{\pi/3 D_b t}} \tag{14.65}$$

Substituting (14.65) and (14.64) into (14.62)

$$\left(\frac{\partial R}{\partial t}\right) = \frac{2D_b PM(C_b^\circ - C_{b,r=R})}{\rho_b}\left(\frac{1}{R} + \frac{1}{\sqrt{\pi/3 D_b t}}\right) + \frac{Q}{nF2\pi R^3} \tag{14.66}$$

14.2.2.1.3   Jakob's Number

The use of a dimensionless Jakob Number, Ja, is generally considered for large concentration changes:

$$Ja \equiv \frac{(C_b^o - C_{b,r=R})PM}{\rho_b} \tag{14.67}$$

Thus, after introducing Ja

$$dR = 2D_b Ja \left(\frac{1}{R} + \frac{1}{\sqrt{\pi/3D_b t}}\right) dt + \frac{Q\,dt}{nF2\pi R^3} \tag{14.68}$$

but considering the integral expression of Q

$$Q = \int I\,dt \tag{14.69}$$

Therefore,

$$R(t) = \int_{t=0}^{t=t} 2D_b Ja \left(\frac{1}{R} + \frac{1}{\sqrt{\pi/3D_b t}}\right) dt + \int_{t=0}^{t=t} \left(\frac{1}{nF2\pi R^3} \int_{t=0}^{t=t} I\,dt\right) dt \tag{14.70}$$

Two interesting cases arise from the Equation 14.70. The first one is the case when there is no current applied to the system or the current is very little with respect to the gas supersaturation component:

$$R(t) = \int_{t=0}^{t=t} \frac{2D_b Ja}{R}\,dt + 2Ja\sqrt{\frac{3D_b}{\pi t}}\,dt = \frac{4D_b Ja\sqrt{t}}{\sqrt{D_b}} + 4Ja\sqrt{\frac{3D_b t}{\pi}} \tag{14.71}$$

Then after considering that the bubble mean radius R is $(D_b t)^{1/2}$, we obtain

$$R(t) = 2\sqrt{D_b t}\left[\sqrt{2Ja} + 2Ja\sqrt{\frac{3}{\pi}}\right] \tag{14.72}$$

Second, the case when the current intensity is constant, then, the Equation 14.70 becomes

$$R(t) = 2\sqrt{D_b t}\left[\sqrt{2Ja} + 2Ja\sqrt{\frac{3}{\pi}}\right] + \int_{0}^{t} \left(\frac{It}{nF2\pi R^3}\right) dt \tag{14.73}$$

but we still have the problem of the time-dependent value of R^3. In the case of current values much larger than the corresponding values of the gas super saturation in the electrolyte, we simply have for stationary currents:

$$R(t) = \int_{0}^{t} \left(\frac{It}{nF2\pi R^3}\right) dt = \left(\frac{3It}{nF2\pi}\right)^{1/3} \tag{14.74}$$

The general case was earlier treated [79] as a convergent series expansion of the analytical pseudo-solution from the differential equation:

$$\frac{\partial R(t)}{\partial t} = 2D_b Ja \left(\frac{1}{R} + \frac{1}{\sqrt{\pi/3 D_b t}} \right) + \frac{Q}{nF2\pi R^2} \tag{14.75}$$

The proposed solution is of the form

$$\frac{\partial R(t)}{\partial t} = \frac{a}{R} + \frac{b}{\sqrt{t}} + \frac{c}{R^2} \tag{14.76}$$

where $a \equiv 2D_b Ja$, $b \equiv Ja\sqrt{3D_b/\pi}$, $c \equiv Q/(nF2\pi)$. They also established a dimensionless equation of the form

$$\frac{\partial S}{\partial \tau} = \frac{1}{S^2} + \frac{\alpha}{\sqrt{\tau}} + \frac{1}{S} \tag{14.77}$$

where $S \equiv R(a/c)$, $\tau \equiv t(a^3/c^2)$, and $\alpha \equiv (b/a^{1/2})$.

Since the two limiting cases of the absence of gas supersaturation and negligible current values led to $R(t) \propto t^{1/3}$ and $R(t) \propto t^{1/2}$, the following series is proposed:

$$S(\tau) = \tau^{1/3} \sum_{n=0}^{\infty} a_n \tau^{1/6} \tag{14.78}$$

Introducing the series (14.78) into the differential equation (14.77), we can obtain the values of a_n:

$$a_0 = 1/3, \quad a_1 = 6\alpha/7, \quad a_2 = 9/4 - 26.72\alpha^2 \tag{14.79}$$

Apparently, the series converges faster than expected and the first five terms are the most important of all. The dimensionless radii of convergence are much larger than the dimensionless maximal time that can be considered experimentally. This means that by simply applying this equation, we cannot model all the experimental situations; that is, the effect of the water vapor inside the gas bubble and the time-dependent current intensity during the electrolysis has to be also considered. When the relative humidity inside the bubble is large, the partial pressure of the gas in the bubble takes the value of the environmental pressure minus that of the saturation vapor pressure. This effect changes the molar volume occupied by the spherical gas bubble. On the other hand, galvanostatic experiments are the only ones that can be considered using the equations stated above. However, most of the experiments are performed by potentiostatic or potentiodynamic methodologies, where the current time-dependent processes have to be evaluated either in the accumulated charge density or in the current intensity during the electrochemical process.

The change in the electrolyte conductivity due to supersaturation is another problem that has been treated since many years ago. Kreysa and Kuhn have considered the gas voidage, ε_g, in the diffusion layer in a review paper [80]. A comparison of the relative conductivity, χ, expression considering the value of ε_g, where the expressions by Maxwell and Bruggemann are the most recommended [81,82]. The Maxwell equation

$$\frac{\chi}{\chi_e} = \frac{1 - \varepsilon_g}{1 + \varepsilon_g/2} \tag{14.80}$$

and the Bruggemann relationships

$$\frac{\chi}{\chi_e} = (1 - \varepsilon_g)^{3/2} \tag{14.81}$$

consider the base electrolyte conductivity, χ_e, without the presence of the gas and the bubbles.

However, all the equations require the knowledge of the gas voidage that depends on the volumetric gas flow rate, φ so the value of ε_g depends on the surface gas velocity, v_g:

$$\frac{v_g^s}{v_g} = \varepsilon_g \tag{14.82}$$

where v_g^s is the surface flow velocity.

On the other hand, an attempt to consider how the gas velocity changes with the value of ε_g was first evaluated by Nicklin [83]:

$$v_g = v_g^s + v_l^s + v_s \tag{14.83}$$

where
v_g^s and v_l^s are the surface gas and liquid velocities, respectively
v_s is the rise velocity of the neighbor layer due to the bubbles-swarm rise

The latter requires the calculation of the rise velocity due to the single bubble rise. Considering (14.82), we have

$$\frac{1}{1 + (v_l^s + v_s)/v_g^s} = \varepsilon_g \tag{14.84}$$

14.2.2.2   Growth of the Bubble

The growth of the bubble induces a micro-convective flow on the electrolyte that pushes each bubble from an ideal center of the surface ("active site") in various radial directions (Figure 14.5). When each bubble attains a certain size, the buoyancy exceeds its adhesion and the bubble leaves the surface producing a drag flow. However, in the electrochemical cell, the radial and the azimuthal directions produces the opposite drag flow to the platinum surface displaying the so-called surface pressure effect.

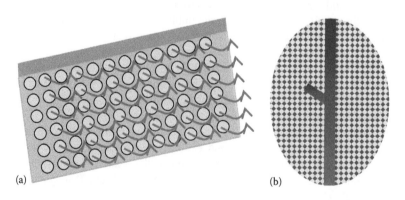

FIGURE 14.5 Structural hole-type patterns for the electrocatalyst used in gas evolution processes. (a) The diameters of the holes have to be larger than the mean bubble diameter to avoid gas clogging. Flow gas evolution in red arrows; (b) electrocatalyst patterns for gas evolving substrates with different mesh sectional areas.

Another important factor to be considered is the number of surface active sites available for the bubble formation and the geometry of the bubble. Moreover, the surface roughness is often considered as the certainty factor for the stability of the nucleation site. The ageing of the nucleation site, a factor expected for gas-producing electrodes, also has to be considered since it results in the loss of activity after a long time. The growth of the bubble is initially controlled by the surface morphology, that is, a certain small area is attributed to a single hemispherical or spherical bubble.

A balance of the electrical and mechanical forces contributes toward the stability of the bubble:

$$\mathrm{d}F = \mathrm{d}F_E - \mathrm{d}F_M = Q\,\mathrm{d}E + E\,\mathrm{d}Q - \frac{PM}{nF}\frac{\partial v}{\partial t}\,\mathrm{d}Q = \left(E - \frac{PM}{nF}\frac{\partial v}{\partial t}\right)\mathrm{d}Q + Q\,\mathrm{d}E \qquad (14.85)$$

where
  E is the gradient of the electrode potential
  v is the bubble growth rate
  r, φ, and θ are the radial, angular, and polar coordinates of the spherical bubble

The temporal change in the acceleration of the bubble is neglected due to conservative laws. On the other hand, it is possible to work under potentiostatic conditions:

$$\mathrm{d}F = E\,\mathrm{d}Q - \frac{PM}{nF}\frac{\partial v}{\partial t}\,\mathrm{d}Q = \left(E - \frac{PM}{nF}\frac{\partial v}{\partial t}\right)\mathrm{d}Q \qquad (14.86)$$

In the case of working under non-potentiostatic conditions, the gradient of the electrode potential in the spherical coordinates will be

$$\nabla E = \frac{\partial E}{\partial r} + \frac{1}{r}\left(\frac{\partial E}{\partial \theta}\right) + \frac{1}{r\sin\theta}\left(\frac{\partial E}{\partial \varphi}\right) \qquad (14.87)$$

but the increase in the bubble growth can be solely characterized by r and θ, since the φ is the resulting three-dimensional component of the growth, so

$$\nabla E = \frac{\partial E}{\partial r} + \frac{1}{r}\left(\frac{\partial E}{\partial \theta}\right) \qquad (14.88)$$

Since we are considering surface pressures, Equation 14.86 changes to

$$p = \frac{\mathrm{d}F}{\mathrm{d}A} = \left(E - \frac{PM}{nF}\frac{\partial v}{\partial t}\right)\sigma \qquad (14.89)$$

where
  A is the active surface area
  σ is the charge density on the electrode

The rate of the bubble growth has to be evaluated per surface active site for the evolved gas, which is the time-dependent bubble growth distance:

$$\frac{\mathrm{d}(v/A)}{\mathrm{d}t} = \frac{\partial^2 R}{\partial t^2}\frac{1}{\pi L^2} \qquad (14.90)$$

where L is the perimeter of the active site area.

However, the balance in the pressure will be simply taken by the second derivate of the radius:

$$p_b = \frac{dF_b}{dA} = \left(E - \frac{PM}{nF} \frac{\partial^2 R}{\partial t^2} \right) \sigma \tag{14.91}$$

The second derivate of the time-dependent radius will be

$$\frac{\partial^2 R(t)}{\partial t^2} = \frac{-2D_b PM}{\rho_b} \left(\frac{(R - (C_b^o - C_{b,r=R}))}{R^2} + \left\{ \frac{1 - (C_b^o - C_{b,r=R})(\pi/3D_b t)^{-1}}{\sqrt{\pi/3D_b t}} \right\} \right) - \frac{R\left(\frac{dQ}{dt}\right) - 3Q}{2nF\pi R^4} \tag{14.92}$$

Finally, the pressure caused by the potentiostatically-generated bubble will be

$$p_b = \frac{dQ}{dA} \left(E + \left\{ \begin{array}{c} \frac{2D_b PM^2}{nF\rho_b} \left(\frac{1}{R} + \frac{1}{\sqrt{\pi/3D_b t}} - (C_b^o - C_{b,r=R}) \left(\frac{1}{R^2} + \frac{1}{(\pi/3D_b t)^{-3/2}} \right) \right) \\ -\frac{PM}{2n^2 F^2 \pi R^4} \left(R\left(\frac{dQ}{dt}\right) - 3Q \right) \end{array} \right\} \right) \tag{14.93}$$

14.2.2.3 Effects of Mass and Heat Transfer on Bubble Growth

The effect of mass and heat transfer associated with the problem of the bubble evolution is also very interesting. In the case of mass transport, we can assume that at a given current density, or flux, N number of the adsorbed bubbles can be formed. At a given time t_R, the diameter reaches a critical value r_t, after which it breaks off. We assume that the fresh electrolyte arrives at a similar concentration of that of the bulk, especially when the convective flux is large enough. We also consider that the mass transport is rate determining, so when the new electrolyte arrives it rapidly converts into the product. Under this consideration, it follows the second Fick's law:

$$\frac{\partial C_{b,r=R}}{\partial t} = D_b \left(\frac{\partial^2 C_{b,r=R}}{\partial r^2} \right) \tag{14.94}$$

where the critical radius of the bubble, r_t is equal to the maximum radius R. In the case of a planar electrode, the solution is that of the Cottrell Equation:

$$j = nF \sqrt{\frac{D_b}{\pi t_R}} C_{b,r=R} \tag{14.95}$$

It is possible to consider that the duration of the bubble growth is similar to the period of non-stationary diffusion of the ion b. Initially the effective electrode section observed by the ion is πR^2, but it decreases because of the progressive growth of the new bubble. As usual, it is more practical to evaluate the average current density as the temporal integration of the current density:

$$j = \frac{nFC_{b,r=R}}{t_R} \int_0^{t_R} \sqrt{\frac{D_b}{\pi t}} \, dt \tag{14.96}$$

Since we are providing an average current density, $\langle j \rangle$, we also need to rationalize the value of the effective electrode area (A_R):

$$\langle j \rangle = \frac{N\pi R^2}{A_R} \frac{nFC_{b,r=R}}{t_R} \int_0^{t_R} \sqrt{\frac{D_b}{\pi t}}\, dt \qquad (14.97)$$

Thus,

$$\langle j \rangle = \frac{2N}{A_R} \frac{C_{b,r=R}nF}{R^2} \sqrt{\frac{\pi D_b}{t_R}} \qquad (14.98)$$

When the number of bubbles is identical to the whole electrode area, the equation converts into

$$\langle j \rangle = \frac{2nFC_{b,r=R}\sqrt{D_b}}{\sqrt{t_R \pi}} \qquad (14.99)$$

In this case, the coverage by the bubbles is complete and they can attain the radius R irrespective of the current density. The dependence of the radius of the bubble with time was followed in the case of vertical microelectrodes in an unstirred solution in Ref. [84]. The electrochemical evolution of the hydrogen on the platinum in 1 M sulfuric solution can be related by the following equation that correlates the radius of the bubble with the current that is exclusively related to the gas evolution:

$$r = k j_R^{1/3} \sqrt{t_R} \qquad (14.100)$$

where j_R is the current density for the critical radius of the bubble (gas evolution), R and $k = 1.5$ 10^{-2} cm$^{5/3}$ A$^{-1/3}$s$^{-1/2}$ for $t_R = 1$ s between $j_R = 0.04$ and 0.08 A cm^{-2}. Substituting t_R from (14.100) into (14.99), we can obtain:

$$\langle j \rangle = \frac{2nFkC_{b,r=R}\sqrt{D_b}j_R^{1/3}}{R\sqrt{\pi}} \qquad (14.101)$$

In the case of incomplete coverage by the bubbles, they only occupy a small fraction of the surface. We can calculate [85,86] the molar volume of the gas, $\bar{V}$, in this case using the critical value of a spherical bubble:

$$\bar{V} = \frac{4\pi R_F^3}{3} \qquad (14.102)$$

If this bubble is considered in a small fraction, N_B, of the surface area, A_E,

$$\bar{V} = \frac{4\pi R_F^3 N_B}{3A_E} \qquad (14.103)$$

According to the literature [86], the time t_F at which the radius of the bubble R_F at the critical point can only be attained when no bulk diffusion takes place. If this situation occurs, it is unlikely that the value of R_F would be identical to that of R. Therefore, we can write

$$\bar{V} = \frac{4\pi R_F^3 N_B}{3A_E} \qquad (14.104)$$

And in terms of time values

$$\frac{\overline{V}j_F}{nF} = \frac{4\pi R_F^3 N_B}{3A_E t_F}$$

(14.105)

This expression can help us to solve the relation obtained in (14.98). We can introduce Equation 14.104 as the volume per time and area to the current density, then

$$\langle j \rangle = \left(\frac{4\pi R_F^3 N_B}{3A_E t_F} \right) \frac{3}{2} \frac{C_{b,r=R} nF}{R^2} \sqrt{\frac{\pi D_b}{t_R}}$$

(14.106)

Using the expression derived by Glas and Westwater [84], we can obtain

$$\frac{j_R^{2/3}}{k} = \frac{2}{3} \left(\frac{\sqrt{\pi D_b} n \langle j \rangle}{D_b C_{b,r=R} nF \overline{V}} \right)$$

(14.107)

We can compare both the expressions (14.107) and (14.104) at the negligible and full coverage by the bubbles, respectively. The main difference is the exponent in the current density related to the gas evolution, j_R that can be experimentally confirmed in [87].

In the case of heat transfer, we can propose a similar model as above. If the area of the critical bubble is πR^2, the time variation in the temperature will be

$$\frac{\partial T}{\partial t} = \alpha \frac{\partial^2 T}{\partial x^2}$$

(14.108)

where α is the temperature diffusivity (cm^2 s). The change in the temperature in the bulk of the solution is the initial value T° ($x \to \infty$), then we can calculate the flow of heat, φ, ($J\ cm^{-2}\ s^{-1}$), as

$$\varphi = \lambda \frac{\partial T}{\sqrt{\alpha\pi} \partial t^{1/2}}$$

(14.109)

where λ is the heat conductivity ($J\ K^{-1}\ cm^{-1}\ s^{-1}$).

Similar equations can be derived for heat transfer during the bubble evolution to those of mass transfer. For a negligible coverage, we will have

$$\frac{j_R^{2/3}}{k} = \frac{2}{3} \frac{nF}{\overline{V}\sqrt{t_R}}$$

(14.110)

which is analogous to the mass transfer limiting (2/3 exponent) situation of small coverages of the nucleating bubbles.

In the case of an almost full coverage by the bubbles

$$\frac{k j_R^{1/3}}{R} = \frac{1}{2\sqrt{t_R}}$$

(14.111)

When deciding the case of the small or large coverage by the bubbles on the electrode surface difficulty arises. Data in the literature indicate that full coverages by the bubbles appear with transfer of heat as the limiting condition, whereas negligible coverage is observed when the mass transfer is the rate determining process [88].

14.2.2.4   Electrode Optimization and Electrochemical Reactor Design

The optimization of an electrochemical reactor calls for a full description of the process to accomplish the specific objective. The problem of the optimization of the electrocatalyst is of real importance in most of the recently developed technical electrodes that were prepared without detailed studies. It must be borne in mind that the strong experimental conditions in which the large electrical currents and large ionic forces of the electrolytes prevail change the morphology and the composition of the catalyst.

The noble metal electrodes are largely employed in electrocatalysis due to their high stability at extremely positive potentials and as some of them are very sensitive toward catalytic performances. The paper by Woods [28] points out that platinum has been so far the model metal for all electrocatalytic reactions. However, the conventional electrochemistry of platinum and their α-oxides is not sufficient to describe the most interesting possibilities of this metal toward electrocatalysis. In spite of the low oxidation onset potential for platinum and gold (0.8 and 1.3 V, for platinum and gold in acid, respectively), the preparation of thermally or electrochemically treated noble metal oxides (such as RuO_2 or RuO_2/IrO_2) as more active substrates is the basis for modern electrochemical technology [29]. However, Burke [30–33] has found anomalous electrochemical behavior of the platinum oxides after thermal and/or electrochemical treatments. The hydrated platinum oxide films are known to be difficult to reduce in the alkaline solutions, but not in acid. These "understandable" film formations were practically ignored in spite of some papers on them by Birss and Arvia [19,34–36]. These studies indirectly focused on the mechanism of the compact and hydrated film growths of α-platinum oxide and β-platinum oxide, respectively. Since then it is believed that the α-oxide growth occurs according to a place-exchange mechanism [37,38] and it thickens via a large-field growth model [39,40], and there is another large thickness β-oxide film that can grow at extremely anodic conditions. However, it has been found that by applying a cathodic treatment it is possible to completely reduce a first β-hydroxide and oxide formation that decomposes faster to an α-oxide layer at the first anodic incursion [41]. Figure 14.6 shows some of these interesting results, including those for symmetric periodic signals (triangular, sinusoidal, or square wave) but with larger relative contributions. The anodic peak III appears for constant applied potentials ranging from 0 to -3 V and is the most interesting due to its compromised catalytic

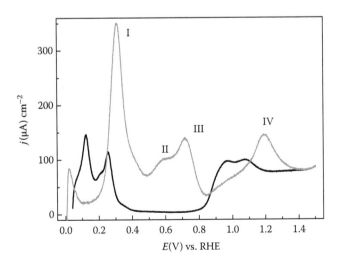

FIGURE 14.6   Cyclic voltammetric responses of polycrystalline platinum in 1 M sulfuric acid solution at 0.10 V s^{-1} after applying a constant cathodic potential V_{app} for 5 min. Repetitive response (continuous line), $V_{app} = -1.5$ V (gray line).

position. The potential at which all the electrooxidations of methanol, formic acid, formaldehyde, and reduced carbon dioxide lies ca. this peak (0.7 V).

The effect of the cathodic treatments was studied, though with limited detail, in the sixties as it is usually applied to clean surfaces. The voltammetric profiles are more interesting [41,42]. Figure 14.6 shows the sudden increase in the current (peak I) just after the potential scan is reversed which is attributed to the traces of molecular hydrogen oxidation. However, peak II is rather difficult to analyze since at least two features are involved during its formation. It partially disappears after stirring the electrolyte, so it might involve the oxidation of the soluble components together with the surface oxidation of some species arising from the cathodization. Considering that the peak II persists on the surface by approximately 30% after rotating the electrode at 500 rpm and the "normal" hydrogen adatom response (between 0.05 and 0.40 V) is totally observed during the first positive going potential scan, the formation of a new hydrogen state such as the subsurface hydrogen or the hydrogen adsorbed on a new active platinum center is likely. However, it is well known that the hydrogen evolution reaction at large negative potentials produces the increase of the pHs' interface on the noble metals. Thus, the alkalinization during the application of the cathodic treatment would produce molecular hydrogen from the water and not from the protons, and is proposed as the soluble component under peak II. The total diffusion of the base species to the bulk of the solution is completed after holding the potential at 0.05 V for more than 45 min. After screening the cyclic voltammetric responses of the Pt(111) at different pHs, it is possible to see the same large anodic's (and its cathodic counter peak) contribution in the 3.5–5.0 pH range. Moreover, when polarizing a Pt(111) surface at −1.5 V for 5 min in the supporting electrolyte, and after holding the potential at 0.05 V for 15 min, the first positive going potential sweep showed a similar voltammetric profile to that of the cathodically treated polycrystalline platinum in the same conditions. Moreover, the voltammetric features are similar to those obtained on the unperturbed Pt(111) at pH $= 4.0$, but with lower intensities in the anodic peaks.

The problem of the electrocatalyst response may be more important than we can understand here. The consequence of obtaining large electrode areas produces too many problems besides the expected positive catalytic effect. One of the problems is the short-circuiting between the anode and the cathode because of the large dissolution of the anodes or the increase in the cathode area due to metallic deposition. Another problem arises from the unstable areas that yield variable current densities, that is, not the electrochemical rates of the regime for the desired process. The only way to avoid this situation is by working under galvanostatic conditions. However, the electrode potential taken into consideration will definitely change to the values of another electrochemical reaction— mostly an undesirable reaction—that can occur.

Coming back to our case of a large cathodic polarization, we have to say that this can commonly happen in the case of the galvanoplastic processes. Thus, we can propose explanations with a different approach that globally justify the results observed in Figure 14.6. The procedure would certainly involve two main forces on the metal electrode: a large electric field created by the high electronic population on the platinum at very negative cathodic potentials and a strong convective flux by molecular hydrogen evolution from the platinum toward the electrolyte [41]. The mass transport process results in an excess of surface pressure exerted by the hydrogen bubbles on (or also in) the platinum that brings up the release of the crystal lattice upon hydrogen inclusion. The result is the formation of subsurface hydrogen arising from the proton electroreduction on the new platinum atoms at the top layers. The existence of the "subsurface hydrogen" on the platinum has been earlier suggested by Wieckowski [42], who proposed that the hydrogen-(subsurface)-platinum species has to be electronically different to the normal adsorbates. The relaxation of the crystal lattice has to occur after the surface expansion that produces the platinum restructuration with the hydrogen beneath it. If this is so, the lower-coordinated platinum species will be formed for which the interaction with the water and/or hydrogen species will be fast enough to produce new features. The atom rearrangement to the (111) planes is explained as being the thermodynamically most favored surface. The results performed above with the polycrystalline platinum and the Pt(111) at different pHs clearly demonstrate that the cathodic treatment produces a surface rearrangement to the (111) triangular sites.

REFERENCES

1. R. H. Rousselot, *Repartition du potential et du courant dans le électrolytes*, Dunod, Paris (1959).
2. N. Ibl, Current distribution, in E. B. Yeager, J.O'M. Bockris, B. E. Conway, and S. Sarangapani (Eds.) *Comprehensive Treatise of Electrochemistry*, Plenum Press, New York (1983), Vol. 6, Chap. 4.
3. H. S. Carlsaw and J. C. Jaeger, *Conduction of Heat in Solids*, Oxford Clarendon Press, Oxford, U.K. (1959).
4. Ch. Kasper, *Trans. Electrochem. Soc.* **77** (1940) 353; Ch. Kasper, *Trans. Electrochem. Soc.* **77** (1940) 365.
5. Ch. Kasper, *Trans. Electrochem. Soc.* **78** (1940) 131; Ch. Kasper, *Trans. Electrochem. Soc.* **78** (1940) 147.
6. Ch. Kasper, *Trans. Electrochem. Soc.* **82** (1940) 153.
7. G. Prentice, *Electrochemical Engineering Principles*, Prentice Hall Inc., Upper Saddle River, NJ (1991), Chap. 7, pp. 177–226.
8. T. Z. Fahidy, *Principles of Electrochemical Reactor Analysis*, Elsevier Science, Amsterdam, the Netherlands (1985).
9. P. J. Schneider, *Conduction Heat Transfer*, Addison-Wesley Editions, Cambridge, U.K. (1955), pp. 121–128.
10. C. Wagner, *J. Electrochem. Soc.* **98** (1951) 116.
11. W. R. Parrish and J. Newman, *J. Electrochem. Soc.* **117** (1970) 43.
12. S. Srinivasan, H. D. Hurwitz, and J.O'M. Bockris, *J. Chem. Phys.* **46** (1967) 3108.
13. E. Justi and A. W. Winsel, *Kalte-Verbrennung*, Franz Steiner Verlag, Wiesbaden, Germany (1962).
14. J.O'M. Bockris and S. Srinivasan, *Fuel Cells: Their Electrochemistry*, McGraw Hill Book Company, New York (1969), Chap. 5, pp. 230–288.
15. S. Srinivasan and H. D. Hurwitz, *Electrochim. Acta* **12** (1967) 495.
16. C. Clerc and D. Landolt, *J. Appl. Electrochem.* **17** (1987) 1144.
17. L. D. Burke and M. B. C. Roche, *J. Electroanal. Chem.* **186** (1985) 139.
18. A. J. Arvia, J. Canullo, E. Custidiano, C. Perdriel, and W. E. Triaca, *Electrochim. Acta* **31** (1986) 1359.
19. W. E. Triaca, T. Kessler, J. C. Canullo, and A. J. Arvia, *J. Electrochem. Soc.* **134** (1987) 1165.
20. N. R. Tacconi, J. Zerbino, M. E. Folquer, and A. J. Arvia, *J. Electroanal. Chem.* **85** (1977) 213.
21. A. J. Arvia, *Isr. J. Chem.* **18** (1979) 89.
22. A. C. Chialvo, W. E. Triaca, and A. J. Arvia, *J. Electroanal. Chem.* **146** (1983) 93.
23. A. C. Chialvo, W. E. Triaca, and A. J. Arvia, *J. Electroanal. Chem.* **171** (1984) 303.
24. T. Biegler, *J. Electrochem. Soc.* **116** (1969) 1131.
25. S. Shibata and M. P. Sumino, *Electrochim. Acta* **16** (1971) 1511.
26. S. Motoo and N. Furuya, *J. Electroanal. Chem.* **172** (1984) 339; S. Motoo and N. Furuya, *J. Electroanal. Chem.* **197** (1986) 209.
27. B. Bittins-Cattaneo, E. Santos, W. Vielstich, and U. Linke, *Electrochim. Acta* **33** (1988) 1499.
28. R. Woods, in A. J. Bard (Ed.) *Electroanalytical Chemistry*, Marcel Dekker, New York (1976), Vol. 9, pp. 1–162.
29. J. C. F. Boodts and S. Trasatti, *J. Appl. Electrochem.* **19** (1989) 255.
30. L. D. Burke, M. E. G. Lyons, in R. E. White, J.O'M. Bockris, and B. E. Conway (Eds.) *Modern Aspects of Electrochemistry*, Plenum, New York (1986), Vol. 38, pp. 169–248.
31. L. D. Burke and L. M. Hurley, *Electrochim. Acta* **44** (1999) 3451.
32. L. D. Burke, J. A. Collins, and M. A. Murphy, *J. Solid State Electrochem.* **4** (1999) 34.
33. L. D. Burke, J. A. Collins, M. A. Horgan, L. M. Hurley, and A. P. O'Mullane, *Electrochim. Acta* **45** (2000) 4127.
34. V. I. Birss and M. Goledzinowski, *J. Electroanal Chem.* **351** (1993) 227.
35. M. Farebrother, M. Goledzinowski, G. Thomas, and V. I. Birss, *J. Electroanal. Chem.* **297** (1991) 435.
36. A. Visintín, W. E. Triaca, and A. J. Arvia, *J. Electroanal. Chem.* **284** (1990) 465.
37. H. Angerstein-Kozlowska, B. E. Conway, and W. B. A. Sharp, *J. Electroanal. Chem.* **43** (1973) 9.
38. B. V. Tilak, B. E. Conway, and H. Angerstein-Kozlowska, *J. Electroanal. Chem.* **48** (1973) 1.
39. D. Gilroy and B. E. Conway, *Can. J. Chem.* **46** (1968) 875.
40. K. J. Vetter and J. W. Schulze, *J. Electroanal. Chem.* **34** (1972) 141.
41. V. Díaz and C. F. Zinola, *J. Colloid Interface Sci.* **313** (2007) 232.
42. A. Wieckowski, *J. Chim. Phys.* **88** (1991) 1247.
43. K. Mandhar and D. Pletcher, *J. Appl. Electrochem.* **9** (1979) 707.

44. P. L. Schilardi, S. L. Marchiano, R. C. Salvarezza, A. Hernández-Creus, and A. J. Arvia, *J. Electroanal. Chem.* **431** (1997) 81; P. Carro, S. Ambrosolio, S. L. Marchiano, A. Hernández-Creus, R. C. Salvarezza, and A. J. Arvia, *J. Electroanal. Chem.* **396** (1995) 183.
45. C. I. Elsner and S. L. Marchiano, *J. Appl. Electrochem.* **12** (1982) 735.
46. L. J. J. Janssen and J. G. Hoogland, *Electrochim. Acta* **15** (1970) 1013; L. J. J. Janssen and J. G. Hoogland, *Electrochim. Acta* **24** (1979) 11.
47. M. Ya. Fioshin, *Elektrokhimiya* **13** (1977) 3.
48. D. E. Danly and C. R. Campbell, in N. L. Weinberg and B. V. Tilak (Eds.) *Techniques of Chemistry*, John Wiley & Sons, Inc., New York (1982) Vol. 5, pp. 283–339.
49. M. I. Ismail, F. Hine, and H. Vogt, in M. I. Ismail (Ed.) *Electrochemical Reactors; Their Science and Technology Part A: Fundamentals, Electrolysers, Batteries and Fuel Cells*, Elsevier, Amsterdam, the Netherlands (1989) Chap. 6, pp. 114–144.
50. E. F. Sverdrup, D. H. Archer, and A. D. Glaser, *Am. Chem. Soc. Division of Fuel Cells Chemistry.* **11** (3) (Preprints 1967) 229–239.
51. M. I. Ismail, T. Z. Fahidy, H. Vogt, and H. Wendt, *Prog. Proc. Eng.* **19** (1981) 381.
52. M. I. Ismail, N. P. White, and S. Gupta, *Plat. Surf. Finish.* Dec. **67** (1980) 59.
53. J. Fleck, *Chem. Eng. Technol.* **43** (1971) 173.
54. L. M. Elina, V. M. Gitneva, V. I. Bystrov, and N. M. Shimygui, *Elektrokhimiya* **10** (1974) 163.
55. T. Loucka, *J. Appl. Electrochem.* **7** (1977) 211.
56. K. I. Fukuda, C. Iwakuda, and H. Tamura, *Electrochim. Acta* **24** (1979) 367.
57. D. M. Smyth, *J. Electrochem. Soc.* **113** (1966) 1271.
58. J. L. Weininger and R. R. Russell, *J. Electrochem. Soc.* **125** (1978) 1482.
59. T. N. Andersen, D. L. Adamson, and K. J. Richards, *Metall. Trans.* **5** (1974) 1345.
60. T. Takahashi, M. I. Ismail, and T. Z. Fahidy, *Electrochim. Acta.* **26** (1981) 1727.
61. M. I. Ismail and T. Z. Fahidy, *Can. J. Chem.* **57** (1979) 734.
62. M. I. Ismail and T. Z. Fahidy, *Can. J. Chem.* **58** (1980) 505.
63. G. Kissel, F. Kulesa, C. R. Davidson, and S. Srinivasan, *Proc. Symp. Water Electrolysis (Electrochem. Soc.)* **78**(4) (1978) 218–235.
64. Z. Abdel Hady and J. Pagetti, *J. Appl. Electrochem.* **6** (1976) 333.
65. R. D. Cowling and H. E. Hintermann, *J. Electrochem. Soc.* **118** (1971) 1912.
66. G. R. Wallwork, *Corrosion (Australia)* **4** (1979) 7.
67. M. A. Streicher, *Corrosion* **30** (1977) 77.
68. M. O. Coulter, *Modern Chlor-Alkali Technology*, Section 9. 7.2., Ellis Horwood Publ., Chichester, U.K. (1980).
69. P. R. Vassie and A. C. C. Tseung, *Electrochim. Acta* **20** (1975) 759.
70. W. Hofman, P. Wehr, and H. J. Engell, *Werkst. Korros.* **17** (1966) 227.
71. J. E. Smith, *Tribol. Int.* October **9** (1976) 225.
72. M. R. Tarasevich, G. L. Zakharkin, and A. M. Khutornoi, *Zh. Fiz. Khim.* **51** (1977) 2625.
73. S. van Stralen and R. Cole, *Boiling Phenomena*, Hemisphere Publ. Corp., Washington, DC (1980), Vol. 1.
74. S. Shibata, *Electrochim. Acta* **23** (1978) 619.
75. K. Stephan and H. Vogt, *Electrochim. Acta* **24** (1979) 11.
76. N. Zuber, *Int. J. Heat Mass Transfer* **6** (1963) 75.
77. N. Ibl and J. Venczel, *Metalloberfläche* **24** (1970) 365.
78. H. F. A. Verhaart, R. M. de Jonge, and S. J. D. van Stralen, *Int. J. Heat Mass Transfer* **23** (1980) 293.
79. P. S. Epstein and M. S. Plesset, *J. Chem. Phys.* **18** (1950) 1505.
80. G. Kreysa and M. Kuhn, *J. Appl. Electrochem.* **15** (1995) 517.
81. J. C. Maxwell, *A Treatise on Electricity and Magnetism*, 2nd edn. Clarendon Press, Oxford, U.K. (1881), Vol. 1.
82. D. A. G. Bruggemann, *Ann. Phys.* **24** (1935) 636.
83. D. J. Nicklin, *Chem. Eng. Sci.* **17** (1962) 693.
84. J. P. Glas and J. W. Westwater, *Int. J. Heat Mass Transfer* **7** (1964) 1427.
85. R. Darby and M. S. Haque, *Chem. Eng. Sci.* **28** (1973) 1129.
86. I. Rousar and V. Cezner, *Electrochim. Acta* **20** (1975) 289.
87. R. B. Mc. Mullin, K. L. Miles, and F. N. Ruehlen, *J. Electrochem. Soc.* **118** (1971) 1582.
88. J. Newman, *Electrochemical Systems*, Prentice Hall, New York (1973).

15 Current Density and Electrode Potential under Different Operating Conditions on Smooth and Rough Surfaces

C. Fernando Zinola

CONTENTS

15.1 GROWTH OF SURFACE ROUGHNESS DURING ELECTRODEPOSITION

Smooth surfaces can develop roughness or even porosity during metal electrodeposition since it is a very slow process that is controlled by many factors including applied potential, electrolyte composition, substrate morphology, and temperature [1,2]. The development of a smooth metal surface during electrodeposition concerns the amount of deposited metal, morphology of the surface, compactness of the deposit, current efficiency, and cell-applied potential. Excessive roughening can lead to short-circuiting between cathode–anode systems in technologically operating cells. Roughening can affect the mechanical stability of the anodes, thereby causing corrosion and pitting with crevice formation, and finally leading to the concentration of flux at their edges.

One of the most used theories for metal growth considers an infinitesimal perturbation on the surface that becomes unstable and continues to grow all along the surface [3,4]. Some authors [4] considered diffusion-controlled, steady-state galvanostatic deposition and used linear stability analysis to determine the critical size above which a planar surface becomes unstable. Their results were limited since the migration of the reacting species and kinetic polarization could not be ignored. It is expected that the kinetic growth would also be controlled by the applied potential in the cell at a defined geometry. Barkey et al. [1] extended these previous treatments including migration transport. However, diffusion and migration were only weakly coupled in their model since it was assumed that separate Laplace equations could be used for the concentration dependence and potential dependence within the boundary layer. It was only through the boundary conditions that the concentration and potential were linked. The ideas of mass transfer and electron transport at the interface led to the work of Pritzker and Fahidy [6]. Their theories are based on single-electrode models in which metal ions are assumed to be deposited on a cathode separated from an ionic conductor by a diffusion layer. Electrode kinetics is described by a Butler–Volmer expression containing both the reduction and oxidation directions, allowing both cathodic deposition and anodic dissolution to be considered simultaneously. The results of the analysis are displayed conveniently in the form of diagrams in the perturbation wavelength vs. applied voltage plane showing the regions where the planar interface is stable or unstable.

Solid state physics theory can predict the growth of a nucleus on a surface by just taking the growth rate of a metal deposit as v and the wavelength of the planar wave displacement on the surface as $2\pi/k$, where k denotes the corresponding wavenumber. The measurements of roughness can be acquired from the v vs. k^2 curve [3]. The interpretation of the curve is impeded as it responds to the entire electrochemical cell behavior and not to the cathode alone. All v vs. k^2 curves in electrodeposition appear to increase first, pass through a greatest value of v, and then decrease, all before crossing zero. The basic instability in electrodeposition stems from the fact that metal ions diffuse across the cell from the anode to the cathode [5,6]. Due to this, the metal ion concentration in the electrolyte increases as it is transported away from the cathode. This problem can be addressed by current distribution as explained in Chapter 13; that is, the presence of a curvature at the cathode weakens the rate of the metal growth at a crest. Another effect of the interelectrode distance can be used to increase the rate of the metal growth, but it destabilizes the system.

15.1.1 Current and Potential Distribution during Electrodeposition: Smooth and Rough Growth

The electrochemical system under consideration involves the electrodeposition on a cathode from an aqueous solution

$$M^{Z+} + Ze^- \rightarrow M \tag{I}$$

and the electrodissolution of a metal at the opposite anode

$$M \rightarrow M^{Z+} + Ze^- \tag{II}$$

that causes the displacement from the position of a planar electrode moving at its steady-state rate v.

The coordinate system on a planar electrode is that of a (x, z) plane along the surface and the y-axis is directly along the outward normal. Within a distance δ from the surface, transport of all species in the electrolyte occurs by diffusion and migration:

$$\frac{\partial C_i}{\partial t} = D_i \nabla^2 C_i + \frac{Z_i F D_i}{RT} \vec{\nabla}(C_i \vec{\nabla}\Phi) \tag{15.1}$$

where
   Φ is the electric potential difference between each pair of electronic and ionic conductors
   F is the Faraday constant
   Z_i, D_i, and C_i are the oxidation state, diffusion coefficient, and local concentration of the
       i-component, that is, M^{Z+}, respectively

The electric potential near the surface of the electrocatalyst is governed by the expression of the mass transport modified Butler–Volmer equation:

$$j_i(x, y, s) = \frac{C_i(x, y, s)}{C_i^{\circ}} j_0 \left\{ \exp\left(\frac{\tilde{\alpha} F \eta(x, y, s)}{RT}\right) - \exp\left(\frac{-\tilde{\alpha} F \eta(x, y, s)}{RT}\right) \right\} \qquad (15.2)$$

where
   (x, y, s) is the collection of the time-dependent spatial coordinates
   $\tilde{\alpha}$ is the charge transfer coefficient for the anodic and cathodic processes
   η is the overpotential
   C_i° is the bulk solution concentration of M^{Z+}

In the model referred to in [4], the metal is grown from the surface, s, which is a time-dependent function of the plane (x, z), producing a decrease in the boundary layer, δ, with time. Then, expression (15.1) has to be modified due to the curvature function or the surface development, which also changes the current density expression of the cathode reaction:

$$\frac{\rho}{PM}\left(\frac{\partial \vec{\delta}}{\partial t} + \vec{v}\right) = -\frac{\vec{j_i}}{Z_i F} \qquad (15.3)$$

where
   PM is the molecular weight
   ρ is the volumetric density of the electrodeposited metal M

The variation of i-species concentration in the mass transport equation will be

$$\frac{\partial C_i}{\partial t} = D_i \nabla^2 C_i + \frac{Z_i F D_i}{RT} \vec{\nabla}(C_i \vec{\nabla} \Phi) + \frac{\operatorname{div} \vec{j_i}}{Z_i F} \qquad (15.4)$$

or

$$\frac{\partial C_i}{\partial t} = D_i \nabla^2 C_i + \frac{Z_i F D_i}{RT} \vec{\nabla}(C_i \vec{\nabla} \Phi) - \frac{\rho}{PM} \operatorname{div}\left(\frac{\partial \vec{\delta}}{\partial t} + \vec{v}\right) \qquad (15.5)$$

We are considering that steady-state conditions account in any case; that is, no drastic changes occur on the electrode with time:

$$\frac{\partial C_{i,j}}{\partial t} = 0 \qquad (15.6)$$

As in all cases, we are going to consider that within the boundary layer, we have a Laplacian field:

$$\nabla^2 \Phi = 0 \qquad (15.7)$$

However, for these conditions, two cases also have to be evaluated:

1. A nonreactive species, j, for which neither a surface reaction occurs

$$\text{div } \vec{v} = 0 \tag{15.8a}$$

nor changes in the boundary layer with time can be detectable:

$$\frac{\partial \delta}{\partial t} = 0 \tag{15.8b}$$

Then, it follows

$$\frac{Z_j F D_j}{RT} \vec{\nabla} C_j \vec{\nabla} \Phi = -D_j \nabla^2 C_j \tag{15.9}$$

2. A reacting species, i, for which both the diffusion and migration (within the boundary layer) flux components, j_i, are balanced with the electrocatalytic reaction rate calculated by the module of the current density, $|j_i|$:

$$\frac{|\vec{j}_i|}{Z_i F} = -\vec{j}_i = -\left(\frac{Z_i F D_i C_i \vec{\nabla} \Phi}{RT} + D_i \vec{\nabla} C_i \right) \tag{15.10a}$$

$$|\vec{j}_i| = -\frac{Z_i^2 F^2 D_i C_i \vec{\nabla} \Phi}{RT} - Z_i F D_i \vec{\nabla} C_i \tag{15.10b}$$

For the surface growth and the subsequent variation in the boundary layer, the physical change of the electrocatalyst can be easily evaluated with the help of the current distributions. In general, the primary distribution can be used as Ohm's law:

$$\vec{j}_i = -\chi_{\text{int}} \vec{\nabla} \Phi \tag{15.11}$$

where χ_{int} is the bulk conductivity of the interface
However, near the electrocatalyst, we have

$$\vec{j}_i = -\chi_{\text{surface}} \vec{\nabla} \Phi \tag{15.12}$$

whereas the only component, far the surface, is that of y (normal to the electrode):

$$j_i|_y = -\chi_{\text{bulk}} \frac{\partial \Phi}{\partial y} \tag{15.13}$$

According to previous deductions [4,6], we need the mathematical expression of the surface and its curvature, as later explained in this chapter, to solve the above equations. We are going to obtain the concentration profile, the current density, and the electrode potential by considering either a null curvature or a rough surface with a defined profile. In each case, we are going to consider that there is no electrochemical activity along the z-axis or, in other words, that the charge transfer occurs only along the x-axis and the mass transfer along the y-axis. Thus, the overpotential $\eta_i(x, y, s)$ is defined by

$$\eta_i(x, y, s) = \Delta\Phi_{j_i}(x, y, s) - \Delta\Phi^o_{j=0,i} - \frac{RT}{Z_iF} \ln \frac{C_i(x = 0, y = 0, s = 0)}{C_i^o} \tag{15.14}$$

and in the case of an irreversible reaction at the catalyst, a generalized (mass transfer modified) Tafel expression can be considered:

$$j_i(x, y, s) = \frac{j_o C_i(x, y, s)}{C_i^o} \exp\left(\frac{\vec{\alpha}F\eta_i(x, y, s)}{RT}\right) \tag{15.15a}$$

$$j_i(x, y, s) = \frac{j_o C_i(x, y, s)}{C_i^o} \exp\left(\frac{\vec{\alpha}F\Delta\Phi_{j_i}(x, y, s)}{RT} - \frac{\vec{\alpha}F\Delta\Phi^o_{j=0i}}{RT} - \frac{\vec{\alpha}}{Z_i} \ln \frac{C_i(x = 0, y = 0, s = 0)}{C_i^o}\right) \tag{15.15b}$$

First, we can consider that the current density far from the catalyst is characterized by an electric field normal to the surface, from the "quasi-primary" distribution, by Equations 15.10b and 15.13:

$$j_i\big|_y = -\chi_{\text{bulk}} \frac{\partial\Delta\Phi(x, y, s)}{\partial y} = -\frac{Z_i^2 F^2 D_i C_i(x, y, s)}{RT}\left(\frac{\partial\Delta\Phi(x, y, s)}{\partial y}\right) - Z_iFD_i\left(\frac{\partial C_i(x, y, s)}{\partial y}\right) \tag{15.16}$$

By combining (15.15a) and (15.16), we have

$$-\frac{Z_i^2 F^2 D_i C_i(x, y, s)}{RT}\left(\frac{\partial\Delta\Phi(x, y, s)}{\partial y}\right) - Z_iFD_i\left(\frac{\partial C_i(x, y, s)}{\partial y}\right)$$

$$= \frac{j_o C_i(x, y, s)}{C_i^o}\left\{\exp\left[\frac{\vec{\alpha}F\left(\Delta\Phi(x, y, s) - \Delta\Phi^o_{j=0}\right)}{RT}\right] - \left(\frac{C_i(x, y, s)}{C_i^o}\right)^{\frac{\vec{\alpha}}{Z_i}}\right\} \tag{15.17}$$

By reordering terms, we obtain

$$-\frac{Z_i^2 F^2 D_i C_i(x, y, s)}{RT}\left(\frac{\partial\Delta\Phi(x, y, s)}{\partial y}\right) - \frac{j_o C_i(x, y, s)}{C_i^o}\exp\left(\frac{\vec{\alpha}F\left(\Delta\Phi(x, y, s) - \Delta\Phi^o_{j=0}\right)}{RT}\right)$$

$$= -j_o\left(\frac{C_i(x, y, s)}{C_i^o}\right)^{\frac{\vec{\alpha}}{Z_i}+1} + Z_iFD_i\left(\frac{\partial C_i(x, y, s)}{\partial y}\right) \tag{15.18}$$

There is no simple and direct solution of the differential equation of $\Delta\Phi$ and C_i with x, y, and s (or δ). However, we can consider Equation 15.16 for different cases of $\partial\Delta\Phi/\partial y$ as a primary distribution for a smooth and a rough surface growth.

15.1.1.1 Smooth Metal Surface Growth under Negligible Overpotentials

When $\eta(x, y) \to 0$, we will have a zero value for $\partial \Delta \Phi / \partial y$ and also for the current density. However, if we consider that the concentration at the point $(x, y) = (0,0)$ is also changing with the same function in the mass transfer Equation 15.14, we can calculate for almost zero values of $\eta(x, y)$:

$$\frac{\partial \Delta \Phi(x, y)}{\partial y} = \frac{RT}{Z_i F C_i(x, y)} \frac{\partial C_i(x, y)}{\partial y} \tag{15.19}$$

Then, by combining terms, we can obtain the mean value of $C_i(x, y)$:

$$\frac{RT \chi_{\text{int}}}{Z_i^2 F^2 D_i} = \overline{C_i(x, y)} \tag{15.20}$$

With this mean value of $C_i(x, y)$, we can obtain the mean value of $\Delta \Phi(x, y)$:

$$\overline{\Delta \Phi(x, y)} = \Delta \Phi^{\circ}_{j=0,i} + \frac{RT}{Z_i F} \ln \frac{\overline{C_i(x, y)}}{C_i^{\circ}} = \Delta \Phi^{\circ}_{j=0,i} + \frac{RT}{Z_i F} \ln \left(\frac{RT \chi_{\text{int}}}{C_i^{\circ} Z_i^2 F^2 D_i} \right) \tag{15.21}$$

Obviously, the mean current density along y has to be zero according to Equation 15.16:

$$\overline{j_i}|_y = -\frac{Z_i^2 F^2 D_i C_i}{RT} \left(\frac{\partial \overline{\Delta \Phi(x, y)}}{\partial y} \right) - Z_i F D_i \left(\frac{\partial \overline{C_i(x, y)}}{\partial y} \right) = 0 \tag{15.22}$$

This is a hypothetical case in which the influence of $\eta_i(x, y)$ is negligible.

15.1.1.2 Smooth Metal Surface Growth in the Presence of Nonzero Overpotentials

When $\eta(x, y) \neq 0$, we will have from Equation 15.14

$$\Delta \Phi(x, y) = \Delta \Phi^{\circ}_{j=0, i} + \frac{RT}{Z_i F} \ln \frac{C_i(x = 0, \ y = 0)}{C_i^{\circ}} + \eta_i(x, y) \tag{15.23}$$

but from Equation 15.15a

$$\eta_i(x, y) = \frac{RT}{\vec{\alpha} F} \ln \left[\frac{j_i(x, y) C_i^{\circ}}{C_i(x, y) j_0} \right] \tag{15.24}$$

Then, by substituting into Equation 15.14, we obtain

$$\Delta \Phi(x, y) = \Delta \Phi^{\circ}_{j=0, i} + \frac{RT}{Z_i F} \ln \frac{C_i(x = 0, y = 0)}{C_i^{\circ}} + \frac{RT}{\vec{\alpha} F} \ln \left[\frac{j_i(x, y)}{j_0} \right] - \frac{RT}{\vec{\alpha} F} \ln \left[\frac{C_i(x, y)}{C_i^{\circ}} \right] \tag{15.25}$$

We need the first derivate of $\Delta \Phi$ with y:

$$\frac{\partial \Delta \Phi(x, y)}{\partial y} = \frac{RT}{\vec{\alpha} F} \left(\frac{\partial \ln j_i(x, y)}{\partial y} \right) - \frac{RT}{\vec{\alpha} F} \left(\frac{\partial \ln C_i(x, y)}{\partial y} \right) \tag{15.26}$$

If we consider within a certain region of potentials that a linear relationship between $j_i|_y$ and $(\partial \Delta \Phi(x, y))/\partial y$ is obeyed in spite of considering mass transports, the current density far from the electrocatalyst can be calculated as

$$j_i|_y = -\chi_{\text{bulk}} \frac{\partial \Delta \Phi(x, y)}{\partial y} = \frac{\chi_{\text{bulk}} RT}{\bar{\alpha} F} \left\{ \left(\frac{\partial \ln C_i(x, y)}{\partial y} \right) - \left(\frac{\partial \ln j_i(x, y)}{\partial y} \right) \right\} \quad (15.27)$$

On the other hand, when can combine Equations 15.16 and 15.26, we we will obtain

$$j_i|_y = -\frac{Z_i^2 F D_i C_i(x, y)}{\bar{\alpha}} \left(\frac{\partial \ln j_i(x, y)}{\partial y} \right) + Z_i F D_i \left(\frac{Z_i}{\alpha} - 1 \right) \left(\frac{\partial C_i(x, y)}{\partial y} \right) \quad (15.28)$$

If the $j_i(x, y)$ dependence far from the surface is similar to the current density along the y-axis, $j_i|_y$ we can obtain a simpler version of (15.28) not shown here due to want of space.

Pritzker and Fahidy [6] considered the solution of Equation 15.9 under galvanostatic conditions as an unknown function of (x, y, δ). In this case, we are going to simplify Equations 15.26 through 15.28 by considering transport number, t_i, in the addition term of the diffusion and migration current density, $j_{\text{mig},i}$:

$$\frac{j_i(x, y)}{Z_i F} = -D_i \left(\frac{\partial C_i(x, y)}{\partial y} \right) + \frac{t_i j_{\text{mig},i}}{Z_i F} \quad (15.29)$$

with

$$j_{\text{mig},i} = -\frac{Z_i^2 F^2 D_i C_i(x, y)}{RT} \left(\frac{\partial \Delta \Phi(x, y)}{\partial y} \right) \quad (15.30)$$

The corrected version of the diffusion current density in (15.29) using t_i will result in

$$j_i(x, y) = \frac{-Z_i F D_i}{(1 - t_i)} \left(\frac{\partial C_i(x, y)}{\partial y} \right) \quad (15.31)$$

By using the above consideration, we can reduce the complexity of Equations 15.16 through 15.29 to a quasi-diffusion process (Equation 15.31). Thus, taking the first derivation of (15.31),

$$\left(\frac{\partial j_i(x, y)}{\partial y} \right) = \frac{-Z_i F D_i}{(1 - t_i)} \left(\frac{\partial^2 C_i(x, y)}{\partial y^2} \right) \quad (15.32)$$

and when using Equation 15.27b, we need the first differentiation of $\ln [j_i (x, y)]$:

$$\frac{\partial \Delta \Phi(x, y)}{\partial y} = \frac{RT}{\bar{\alpha} F j_i(x, y)} \left(\frac{\partial j_i(x, y)}{\partial y} \right) - \frac{RT}{\bar{\alpha} F} \left(\frac{\partial \ln C_i(x, y)}{\partial y} \right) \quad (15.33)$$

and using (15.32),

$$\frac{\partial \Delta \Phi(x, y)}{\partial y} = \frac{-Z_i F D_i}{(1 - t_i) j_i(x, y)} \left(\frac{\partial^2 C_i(x, y)}{\partial y^2} \right) - \frac{RT}{\bar{\alpha} F} \left(\frac{\partial \ln C_i(x, y)}{\partial y} \right) \quad (15.34)$$

Since we are considering a quasi-primary distribution for the current density, we can use Equation 15.13b and then

$$j_i|_y = \frac{\chi_{bulk} Z_i F D_i}{(1 - t_i) j_i|_y} \left(\frac{\partial^2 C_i(x, y)}{\partial y^2} \right) + \frac{\chi_{bulk} RT}{\vec{\alpha} F} \left(\frac{\partial \ln C_i(x, y)}{\partial y} \right) \tag{15.35}$$

The expression above is that of $j_i|_y$ with a null change in the curvature of the surface. However, we have to evaluate what happened with the changes along x. In this sense, the current density is a homogeneous function (of n grade) of two variables (x and y) using Euler's theorem:

$$n j_i(x, y) = x \left[\frac{\partial j_i(x, y)}{\partial x} \right] + y \left[\frac{\partial j_i(x, y)}{\partial y} \right] \tag{15.36}$$

Nevertheless, the variation of current along the x-axis can be taken as negligible compared with that along the y-axis since the mass transport occurs along y and we consider, at first thought, that there is no change in the curvature. Therefore, the (x, y)-dependent current density can be considered as a simple y-axis contribution. It has to be considered that the electrocatalytic reaction occurs along this axis at the electrode surface ($y = 0$).

We have almost two great cases to evaluate when $\eta_i(x, y)$ occurs; the current is under galvanostatic conditions or the current is variable. For the latter, only the case of a constant overpotential is considered. Apart from having the case of a smooth surface, a periodic trochoid surface is also solved and generalized with the help of [4] to non-periodic surfaces using the linear stability analysis.

15.2 GALVANOSTATIC CONDITIONS AND LARGE OVERPOTENTIALS

15.2.1 GALVANOSTATIC CONDITIONS WITH NEGLIGIBLE CURRENT DENSITIES (POOR PERFORMANCE CATALYSTS)

If we can say that $j_i(x, y)$ is constant and far from its limiting value, the differential equation will change to

$$\frac{\chi_{bulk} Z_i F D_i}{(1 - t_i) j_i|_y} \left(\frac{\partial^2 C_i(x, y)}{\partial y^2} \right) + \frac{\chi_{bulk} RT}{\vec{\alpha} F} \left(\frac{\partial \ln C_i(x, y)}{\partial y} \right) = j_i|_y \cong 0 \tag{15.37}$$

We can rewrite the equation as

$$\left(\frac{\partial^2 C_i(x, y)}{\partial y^2} \right) + \frac{\beta}{\alpha} \left(\frac{\partial \ln C_i(x, y)}{\partial y} \right) = 0 \tag{15.38}$$

with

$$\alpha \equiv \frac{\chi_{bulk} Z_i F D_i}{(1 - t_i) j_i|_y} \tag{15.39a}$$

$$\beta \equiv \frac{\chi_{bulk} RT}{\vec{\alpha} F} \tag{15.39b}$$

In the case of a second-order differential equation, either we can reduce the order by a change of variables or, if we know one solution $C_1(x, y)$, we can obtain the other solution $C_2(x, y)$ by the corollary of the Liouville formula [7]. Another solution can be assayed by changing variables as given in Appendix A. The solutions of a linear homogeneous differential equation can be arranged in a Wronskian formula, and when this is not zero, the Liouville formula can be used:

$$C_2(x, y) = AC_1(x, y) \int \frac{\exp\left[-\int \frac{\beta}{\alpha C_1(x, y)} dy\right]}{(C_1(x, y))^2} dy \qquad (15.40)$$

Then,

$$C_2(x, y) = A\left[\frac{\beta(C_1(x, y))^2(1 - 2\ln C_1(x, y))}{4\alpha}\right] \qquad (15.41)$$

We can propose for $C_1(x, y)$ the solution given in Ref. [4]:

$$C_1(x, y) = \frac{RT\chi_{\text{bulk}}}{(Z_iF)^2D_i} + \left(C_1^o - \frac{RT\chi_{\text{bulk}}}{(Z_iF)^2D_i}\right)\exp\left[-\frac{Z_iFj_i(\delta - y)}{RT\chi_{\text{bulk}}}\right] \qquad (15.42)$$

where
  j_i is the local and constant current density
  δ is the migration–diffusion layer

For simplification purposes, we are going to define

$$\lambda \equiv \frac{RT\chi_{\text{bulk}}}{(Z_iF)^2D_i} \qquad (15.43a)$$

$$\varepsilon \equiv \frac{Z_iFj_i}{RT\chi_{\text{bulk}}} \qquad (15.43b)$$

Therefore, the other solution will be

$$C_2(x, y) = A\left[\frac{\beta(\lambda + (C_1^o - \lambda)\exp[-\varepsilon(\delta - y)])^2(1 - 2\ln\{\lambda + (C_1^o - \lambda)\exp[-\varepsilon(\delta - y)]\})}{4\alpha}\right] \qquad (15.44)$$

These two solutions are extremely important, so the analysis of each one at the electrode surface, $(x, y) = (0, 0)$ gives rise to

$$C_1(0, 0) = \lambda + (C_1^o - \lambda)\exp[-\varepsilon\delta] \qquad (15.45a)$$

$$C_2(0, 0) = A\left[\frac{\beta(\lambda + (C_1^o - \lambda)\exp[-\varepsilon\delta])^2(1 - 2\ln\{\lambda + (C_1^o - \lambda)\exp[-\varepsilon\delta]\})}{4\alpha}\right] \qquad (15.45b)$$

The following graphs show the dependence of both solutions on the changes in y at a constant x for $\chi_{\text{bulk}} = 0.1$ S cm^{-1} and $j_i = 1$ A cm^{-2} at 298 K and $C_1^o = 10^{-4}$ mol cm^{-3}. We will then have

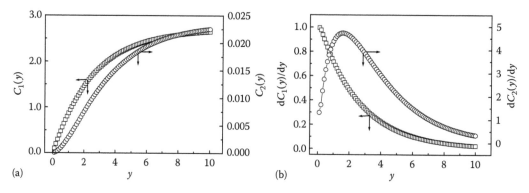

FIGURE 15.1 (a) The dependence of both concentration profiles, $C_1(x, y)$ ($\square$) and $C_2(x, y)$ ($\circ$), upon the distance, y, normal to the electrode surface; (b) the first derivative of each concentration profile. Galvanostatic conditions under negligible current densities.

$\varepsilon = 390.1$ and $\lambda = 2.7 \times 10^{-3}$ for a migration–diffusion layer of $\delta = 10^{-4}$ cm with $D_i = 10^{-5}$ cm^2 s^{-1}. Using $t_i = 0.5$ (for any ion except proton and hydroxyl) and $\alpha = 0.2$, with a normal value of $\bar{\alpha} = 0.5$, we will have $\beta = 0.005$. For simplicity, we are going to consider that $A = 1$.

Figure 15.1a clearly shows that the solution given by $C_1(x, y)$ changes sharper than that given by $C_2(x, y)$ involving various orders of lower concentration values. However, they are not qualitatively similar as expected since $C_2(x, y)$ shows an inflection point at distances near the electrode surface. This is better shown by the derivative expression of the curves in Figure 15.1b.

Taking the first derivation of $C_1(x, y)$ with respect to y and the second derivation, we are able to substitute into Equation 15.35. After considering $(Z_iF/\bar{\alpha}) \ll \exp[(Z_iFj_i(\delta - y))/RT\chi_{\text{bulk}}]$, it will be $(1 - t_i)\chi_{\text{bulk}}R^2T^2 \cong Z_i^3F^3D_i \exp[-(Z_iFj_i(\delta - y))/RT\chi_{\text{bulk}}]$. The expression for $j_i(y)$ is

$$j_i \cong - \frac{RT\chi_{\text{bulk}}}{Z_iF(\delta - y)} \ln\left[\frac{(1 - t_i)\chi_{\text{bulk}}R^2T^2}{Z_i^3F^3D_i}\right] \qquad (15.46)$$

which shows a reciprocal behavior with the distance inside the boundary layer.

On the other hand, since we have a quasi-primary current distribution

$$j_i(y) = -\chi_{\text{int}} \frac{\partial \Delta\Phi(x, y)}{\partial y} \approx - \frac{RT\chi_{\text{bulk}}}{Z_iF(\delta - y)} \ln\left[\frac{(1 - t_i)\chi_{\text{bulk}}R^2T^2}{Z_i^3F^3D_i}\right] \qquad (15.47a)$$

After a first integration, we obtain $\Delta\Phi(y)$

$$\Delta\Phi(y) = \Delta\Phi^\circ + \ln\left[\frac{(1 - t_i)\chi_{\text{bulk}}R^2T^2}{Z_i^3F^3D_i}\right] \frac{RT\chi_{\text{bulk}} \ln[(\delta - y)]}{Z_iF} \qquad (15.47b)$$

We can plot both, the current density and the electrode potential, for the same parameters of Figure 15.1. We can see, as expected, that $\Delta\Phi(y)$ has its minimum value (the more negative value) at the largest $j_i(y)$. On the other hand, it is also important to see that this high value is observed at the nearest distance to δ. The constancy in $j_i(y)$ near the expected zero value from a certain y is obvious since it was the first condition for working, that is, galvanostatic with negligible values (Figure 15.2).

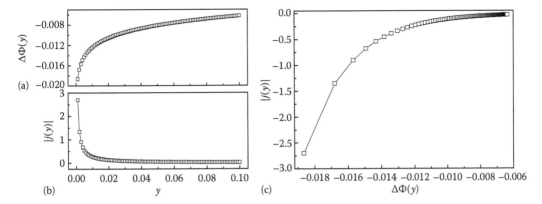

FIGURE 15.2 The dependence of (a) $\Delta\Phi(y)$ and (b) $j_i(y)$ upon the distance, y, normal to the electrode surface at galvanostatic conditions with negligible current densities. (c) $j_i(y)$ vs. $\Delta\Phi(y)$.

15.2.2 Galvanostatic Conditions with Large Current Densities (Good Performance Catalysts)

In this case, we cannot neglect the current densities from the differential equation since we are near the limiting current value:

$$\frac{\chi_{\text{bulk}}Z_iFD_i}{(1-t_i)j_i|_y}\left(\frac{\partial^2 C_i(x,y)}{\partial y^2}\right) + \frac{\chi_{\text{bulk}}RT}{\vec{\alpha}F}\left(\frac{\partial \ln C_i(x,y)}{\partial y}\right) - j_i|_y = 0 \qquad (15.48)$$

We can solve this equation by reducing the order of the quadratic expression:

$$\alpha \equiv \frac{\chi_{\text{bulk}}Z_iFD_i}{(1-t_i)j_i|_y} \qquad (15.49a)$$

$$\beta \equiv \frac{\chi_{\text{bulk}}RT}{\vec{\alpha}F} \qquad (15.49b)$$

$$\gamma \equiv j_i|_y \qquad (15.49c)$$

We can define a new function:

$$z_i(x,y) \equiv \frac{\left(\dfrac{\partial C_i(x,y)}{\partial y}\right)}{C_i(x,y)} \qquad (15.50)$$

and taking its first derivate,

$$\frac{\partial z_i(x,y)}{\partial y} = \frac{C_i(x,y)\left(\dfrac{\partial^2 C_i(x,y)}{\partial y^2}\right) - \left(\dfrac{\partial C_i(x,y)}{\partial y}\right)^2}{C_i^2(x,y)} \qquad (15.51)$$

By substituting $z_i(x,y)$ and its first derivate into (15.48) and (15.50), we have

$$\alpha e^{\int z_i(x,y)dy}\frac{\partial z_i(x,y)}{\partial y} + \alpha z_i^2(x,y) + \beta z_i(x,y) - \gamma = 0 \qquad (15.52)$$

This equation needs to be solved by reducing the quadratic expression by means of the Riccati differential equation (see Chapter 2) and then making it exact through an appropriate integrating factor, $\mu(y)$. The problem is the exponential term that multiplies the first derivate, which is a function affecting the solution of (15.54).

We can try the *Picard succession approximation method*, where the initial conditions of the equation are more important. If y_0 and $z_0(x, y_0)$ are the initial conditions and the differential equation is written as $z'(x, y) = f(y, z)$, the equation can be written as

$$z_i(x, y) = z_0(x, y_0) + \int_{y_0}^{y} f(y, z) dy \tag{15.53}$$

In this case, the differential equation given in (15.54) is rewritten as

$$\frac{\partial z_i(x, y)}{\partial y} = e^{-\int z_i(x, y) dy} \left[-z_i^2(x, y) - \frac{\beta}{\alpha} z_i(x, y) + \frac{\gamma}{\alpha} \right] \tag{15.54}$$

This differential equation has to obey the initial conditions of the equation. Thus, at $y = 0$ and considering Equation 15.50, we have

$$
\begin{cases}
z_0(0, 0) = Ae^{C_i y} \\[2mm]
z_1(x, y) = Ae^{C_i y} + e^{-\frac{A}{C_i}e^{C_i y}} \left[-Ae^{C_i y} + \frac{\beta}{\alpha} - C_i \right] + \dfrac{\frac{\gamma}{\alpha} \mathrm{Ei}\left(-\frac{A}{C_i} e^{C_i y} \right)}{C_i} \\[4mm]
z_2(x, y) = e^{\int C_i dy} + \int_0^y e^{-\int z_1(x, y) dy} \left[-(z_1(x, y))^2 - \frac{\beta}{\alpha} z_1(x, y) + \frac{\gamma}{\alpha} \right] dy
\end{cases}
\tag{15.55}
$$

where $\mathrm{Ei}(y)$ is the *exponential integral function* as explained in Equation 15.A.10. The series of solutions, $z_0, z_1, z_2, \ldots$, converges to the desired solution into a defined domain that contains the point $(0, 0)$. All the solutions have to exist and be continuous according to *Cauchy's theorem*. Anyway, we really need the concentration profile $C_i(x, y)$ instead of $z_i(x, y)$, so we use Equation 15.52b for first two terms of the series.

Some parts of the solution in the concentration profile are imaginary, so first we need to consider the definition of the exponential integral function:

$$\mathrm{Ei}\left(-\frac{A}{C_i} e^{C_i y} \right) = - \int_{\left(-\frac{A}{C_i} e^{C_i y} \right)}^{\infty} \frac{e^{-u} du}{u} \tag{15.56}$$

$\mathrm{Ei}(y)$ is closely related to the *incomplete Gamma function*, $\Gamma(0, y)$:

$$\Gamma(0, y) = -\mathrm{Ei}(-y) + \frac{1}{2} \left[\ln(-y) - \ln\left(-\frac{1}{y} \right) \right] - \ln y \tag{15.57}$$

The "complete" *Gamma function* $\Gamma(y)$

$$\Gamma(y) = \int_0^{\infty} t^{y-1} e^{-t} dt \quad \text{for } y > 0 \tag{15.58}$$

can be generalized to the incomplete Gamma function $\Gamma(a, y)$ such that $\Gamma(a) = \Gamma(a, 0)$. This "upper" incomplete Gamma function is given by

$$\Gamma(a, y) = \int_{y}^{\infty} t^{a-1} e^{-t} dt \qquad (15.59)$$

Therefore, for real y values,

$$\Gamma(0, y) = \begin{cases} -\text{Ei}(-y) - i\pi & \text{for } y < 0 \\ -\text{Ei}(-y) & \text{for } y > 0 \end{cases} \qquad (15.60)$$

After considering all the definitions above, $C_i(x, y)$ for real numbers is

$$C_i(x, y) = \exp\left\langle \frac{A}{C_i} e^{C_i y} - \frac{\left| \gamma A e^{(C_i y + 1)} \right|}{\alpha (C_i)^2} + e^{\left(-\frac{A}{C_i} e^{C_i y} \right)} \left[2 \left(C_i - \frac{\beta}{\alpha} \right) y + 1 \right] + \left(1 - \frac{\beta}{\alpha C_i} \right) \Gamma \left(0, -\frac{A}{C_i} e^{C_i y} \right) \right\rangle$$

$$(15.61)$$

Figure 15.3 illustrates the real components of $C_i(x, y)$ that has a singular point at $-(A/C_i)e^{C_i y} = 0$. We use here the values of Figure 15.1 again; thus, $\alpha = 0.193$; $\beta = 0.005$; $\gamma = 1$.
If we plot each part of $C_i(x, y)$, that is, the first term of the series, the single exponential, double exponential, and incomplete Gamma function terms, all of them show a smooth ascending and descending (in the case of both single exponential terms) behavior with the distance. However, the final exponential that makes us change from $z_i(x, y)$ to $C_i(x, y)$ produces an abrupt increase in the

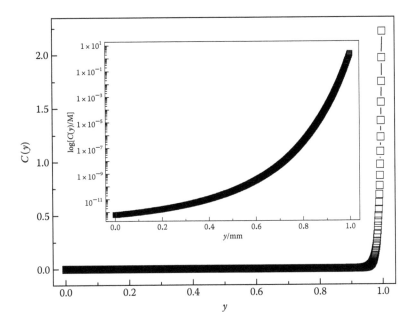

FIGURE 15.3 The dependence of the concentration profile, $C(x, y)$, upon the characteristic distance, y, normal to the electrode surface at a constant x for a good performance electrocatalyst. The inset shows the smoothness of the $C(x, y)$ for a logarithmic function.

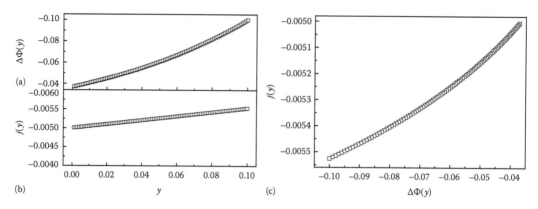

FIGURE 15.4 The dependence of (a) $\Delta\Phi(y)$ and (b) $j_i(y)$ upon the distance, y, normal to the electrode surface at galvanostatic conditions with negligible current densities. (c) $j_i(y)$ vs. $\Delta\Phi(y)$.

contour toward the distance with a positive concavity. This is better seen in the inset of the Figure 15.3 as smooth logarithm performance.

According to (15.49), we have to evaluate the first differentiation of $\ln C_i(x, y)$ and the second differentiation of $C_i(x, y)$ with respect to y. We can approach it after neglecting the $\mathrm{Ei}(x)$ function, and by substitutions into (15.48), we obtain the real part of $j_i(y)$:

$$j_i(y) \cong \frac{\chi_{\mathrm{bulk}} RTA e^{C_i y}}{\vec{\alpha} F} + \sqrt{\frac{\chi_{\mathrm{bulk}} Z_i F D_i A C_i e^{C_i y} e^{\left(\frac{A}{C_i} e^{C_i y}\right)}}{(1 - t_i)}} \tag{15.62}$$

On the other hand, the electrode potential obeys the "quasi-primary" current distribution:

$$\Delta\Phi(y) \cong \Delta\Phi^\circ - \frac{RTA}{\vec{\alpha} F C_i} \left(e^{C_i y} - 1\right) - \sqrt{\frac{\chi_{\mathrm{bulk}} Z_i F D_i A}{(1 - t_i) C_i}} e^{\left(-\frac{A}{2 C_i} e^{C_i y}\right)} \sqrt{e^{\left(C_i + \frac{A}{C_i} e^{C_i y}\right)}} \, \mathrm{Ei}\left(\frac{A}{2 C_i} e^{C_i y}\right) \tag{15.63}$$

Thus, Equations 15.62 and 15.63 represent the rough approximation of $\Delta\Phi(y)$ and $j_i(y)$ and can be plotted as a function of y in Figure 15.4 for the same data as in Figure 15.1. Both $\Delta\Phi(y)$ and $j_i(y)$ dependences are expected but the latter shows a quasi-limiting behavior not predicted by the condition of "quasi-primary" current distribution. The variation in $j_i(y)$ has to be zero if we consider all the components given by the Picard's method of successive approximation. Thus, the plots in Figure 15.4 have to be taken with care.

15.3 NON-GALVANOSTATIC CONDITIONS AND CONSTANT OVERPOTENTIALS

In some cases, the electrodeposition process can be performed under non-galvanostatic conditions; however, the mathematics is very difficult to solve. However, Equation 15.35 can be changed considering the "quasi-primary" distribution of current as

$$j_i\big|_y = \frac{\chi_{\mathrm{bulk}} Z_i F D_i}{(1 - t_i) j_i\big|_y} \left(\frac{\partial^2 C_i(x, y)}{\partial y^2}\right) + \frac{\chi_{\mathrm{bulk}} RT}{\vec{\alpha} F} \left(\frac{\partial \ln C_i(x, y)}{\partial y}\right) \tag{15.35}$$

so, using Equation 15.13, we obtain

$$-\chi_{bulk}\frac{\partial\Delta\Phi(x,y)}{\partial y} = -\frac{Z_iFD_i}{(1-t_i)}\left(\frac{\partial^2 C_i(x,y)}{\partial\Delta\Phi(x,y)\partial y}\right) + \frac{\chi_{bulk}RT}{\bar{\alpha}F}\left(\frac{\partial\ln C_i(x,y)}{\partial y}\right) \qquad (15.64)$$

The first term on the right-hand side of the equation is a mixed derivation, so the order of the variables operated with respect the derivation does not matter. Thus, when we use Equation 15.23 the first derivation of $\Delta\Phi(x,y)$ with $C_i(x,y)$ gives

$$\frac{\partial\Delta\Phi(x,y)}{\partial C_i(x,y)} = \frac{RT}{Z_iFC_i(x,y)} \qquad (15.65)$$

and substituting into (15.64), we obtain

$$\frac{\partial\Delta\Phi(x,y)}{\partial y} = \frac{(Z_iF)^2 D_i C_i(x,y)}{\chi_{bulk}RT(1-t_i)}\left(\frac{\partial C_i(x,y)}{\partial y}\right) - \frac{RT}{\bar{\alpha}F}\left(\frac{\partial\ln C_i(x,y)}{\partial y}\right) \qquad (15.66)$$

Since the overpotential is constant, we can use Equation 15.19:

$$\left[\frac{1}{Z_i}+\frac{1}{\bar{\alpha}}\right]\frac{RT}{F}\left(\frac{\partial\ln C_i(x,y)}{\partial y}\right) - \frac{(Z_iF)^2 D_i}{2\chi_{bulk}RT(1-t_i)}\left(\frac{\partial[C_i(x,y)]^2}{\partial y}\right) = 0 \qquad (15.67)$$

We can solve this equation by considering the first differentiation coefficients from (15.67) and defining the following parameters:

$$a \equiv \left(\frac{1}{Z_i}+\frac{1}{\bar{\alpha}}\right)\chi_{bulk}(RT)^2(1-t_i) \qquad (15.68a)$$

$$b \equiv (Z_i)^2 D_i F^3 \qquad (15.68b)$$

$$f \equiv \chi_{bulk}RTF(1-t_i) \qquad (15.68c)$$

We can propose the solution by reducing the order of the quadratic expression above. We can define a new function as

$$z_i(x,y) \equiv [C_i(x,y)]^2 \qquad (15.69)$$

and then,

$$\frac{\partial z_i(x,y)}{\partial y} = 2C_i(x,y)\left(\frac{\partial C_i(x,y)}{\partial y}\right) \qquad (15.70)$$

By substituting $z_i(x,y)$ and its first derivate into (15.70), we obtain

$$a\ln[C_i(x,y)]^2 - b[C_i(x,y)]^2 = 2f(x,y) \qquad (15.71)$$

We can expand $\ln[C_i(x,y)]^2$ according to a power series function for $C_i(x,y)>0$ and when we consider only the first two terms, $C_i(x,y)$ renders:

$$C_i(x,y) = \frac{b}{a}\sin^{-1}\left|\exp\left[\frac{2f}{a}(x,y)\right]\right| \qquad (15.72)$$

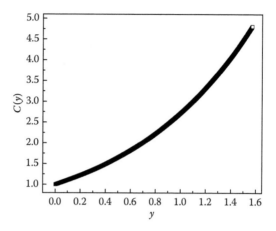

FIGURE 15.5 The dependence of the concentration profile, $C(x, y)$, upon the characteristic distance, y, normal to the electrode surface at a constant x under a constant overpotential.

The following graphs (Figure 15.5) show the dependence of $C_i(x, y)$ on the changes in y at a constant x for $\chi_{bulk} = 0.1$ S cm^{-1} at 298 K. We consider here $Z_i = 1$ and a normal value of $\vec{\alpha} = 0.5$. In addition, normal parameters for a diffusion-migration system are computed $D_i = 10^{-5}$ cm^2 s^{-1} and $t_i = 0.5$ (for any ion except proton and hydroxyl). Thus, $b/a = 10^{-4}$ and $f/a = 13$.

After a similar process, $j_i(y)$ and $\Delta\Phi(y)$ are

$$j_i(y) = \frac{\chi_{bulk} Z_i F D_i 4 i b f^2 e^{\frac{2fy}{a}}}{(1 - t_i) a^3 \left(e^{\frac{4fy}{a}} - 1\right)^{3/2}} \left[1 - \frac{2\chi_{bulk} R T f e^{\frac{2fy}{a}}}{a\vec{\alpha} F \sin^{-1}\left(e^{\frac{2fy}{a}}\right)\sqrt{1 - e^{\frac{4fy}{a}}}}\right]^{-1} \tag{15.73}$$

$$\Delta\Phi(y) = \Delta\Phi^\circ + \frac{Z_i F D_i 2bf}{(1 - t_i)a^2}\left\{\tan^{-1}\left[\sqrt{e^{\frac{4fy}{a}} - 1}\right] + \frac{1}{\sqrt{e^{\frac{4fy}{a}} - 1}}\right\} \tag{15.74}$$

At constant overpotentials, the evolution of the current density is approximately linear with the distance as shown above (Figure 15.6).

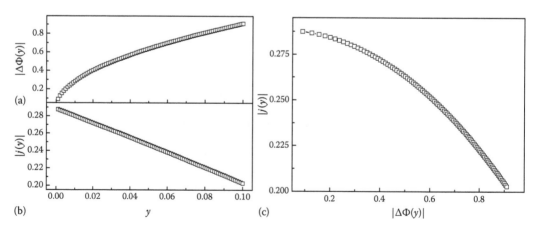

FIGURE 15.6 The dependence of (a) $|\Delta\Phi(y)|$ and (b) $|j_i(y)|$ upon the distance, y, normal to the electrode surface at non-galvanostatic conditions and constant overpotentials. (c) $|j_i(y)|$ vs. $|\Delta\Phi(y)|$.

15.4 METAL ROUGH GROWTH IN THE ABSENCE AND PRESENCE OF OVERPOTENTIALS

15.4.1 METAL ROUGHNESS AT REGULAR AND PERIODIC SURFACES

When we approach a regular and quasi-sinusoidal, commonly known as "valley and mountain" surface profile, we need to integrate it as a first-type curvilinear in other coordinates. Suppose we have a concentration profile in the (x, y) plane: $C_i(x, y)$ in a conexing region, s, over a segment, K, of a plane curve, the curvilinear integral of the first type is the following function:

$$\int_{(K)} C_i(x, y) ds \qquad (15.75)$$

This segment, K, is known as *integrating road*, which, when parameterized, (in two or three dimensions) allows us to calculate the curvilinear integral as a normal Riemman's definite integral. If the integrating road is written in a parameteric form

$$\begin{cases} x = x(t) \\ y = y(t) \\ z = z(t) \end{cases} \qquad (15.76)$$

then the integral will be

$$\int_{(K)} C_i(x, y, z) ds = \int_{t_o}^{t} C_i(x(t), y(t), z(t)) \sqrt{[x'(t)]^2 + [y'(t)]^2 + [z'(t)]^2} dt \qquad (15.77)$$

where t_o and t are the parameter values at the beginning and end of the integrating road, with $t_o < t$.

15.4.1.1 Cycloid Surfaces

The cycloids are described by a point in a circumference that rolls without slipping along a rectilinear line. Mathematically speaking, the representation of the crystal growth in the catalyst is trochoid (shortened cycloid) as depicted in Chapter 13. The equations given in a parameteric form in a conexing plane are

$$\begin{aligned} x(t) &= \alpha(t - \lambda \sin t) \\ y(t) &= \alpha(1 - \lambda \cos t) \end{aligned} \qquad (15.78)$$

where
   α is the radius of the circumference
   λ defines the type of cycloid ($\lambda < 1$ is the lengthened cycloid, and when $\lambda > 1$ we have a shortened cycloid or trochoid) (Figure 15.7)

We can calculate the length of the segment in the plane curve, L, using the derivates of $x(t)$ and $y(t)$:

$$\begin{aligned} x'(t) &= \alpha(1 - \lambda \cos t) \\ y'(t) &= \alpha\lambda \sin t \end{aligned} \qquad (15.79)$$

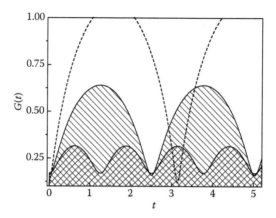

FIGURE 15.7   The growth of a periodic surface as a trochoid. Examples: $\lambda = 0.3$; $\alpha = 0.2$ (continuous line, striped area); $\lambda = 0.6$; $\alpha = 0.4$ (continuous line, checkered area); $\lambda = 0.8$; and $\alpha = 0.5$ (dashed line, striped area).

$$L = \int_{t_o}^{t} \alpha \sqrt{(1 - \lambda \cos t)^2 + \lambda^2 \sin^2 t} \; dt \tag{15.80}$$

$$L = \left[\frac{\alpha \left(2\sqrt{\lambda^2 - 2\lambda \cos t + 1}\right) E\left(\dfrac{4\lambda}{(\lambda - 1)^2}, \dfrac{t}{2}\right)}{\sqrt{\dfrac{\lambda^2 - 2(\lambda + 1)\cos t}{(\lambda - 1)^2}}} \right]_{t_o}^{t} \tag{15.81}$$

where $E(x, m)$ is the *elliptic integral of the second kind* (see Appendix C).

   We can also calculate the length of one cycle of the trochoid:

$$L = \alpha \int_{0}^{\pi} \sqrt{1 + \lambda^2 - 2\lambda \cos t} \; dt \tag{15.82}$$

In this case, the elliptic integral of the second kind is converted to a complete integral:

$$L = \frac{2\alpha(\lambda + 1)}{\sqrt{\dfrac{\lambda^2 + 2(\lambda + 1)}{(\lambda - 1)^2}}} E\left(\frac{4\lambda}{(\lambda - 1)^2}, \frac{\pi}{2}\right) \tag{15.83}$$

   On the other hand, the area under each cycle, S, will be

$$S = \pi \alpha^2 (2 + \lambda^2) \tag{15.84}$$

with a radius of curvature, r_{curv}, calculated as

$$r_{curv} = \alpha \frac{\left(\sqrt{1 + \lambda^2 - 2\lambda \cos t}\right)^3}{\lambda(\cos t - \lambda)} \tag{15.85}$$

The curvature, κ, of the trochoid can be defined as

$$\kappa = -\frac{\partial^2 S}{\partial x^2} - \frac{\partial^2 S}{\partial y^2} \tag{15.86}$$

with

$$S = \pi \alpha^2 \frac{\left(1 + \lambda^2 - 2\lambda \cos t\right)^3}{\lambda^2 (\cos t - \lambda)^2} \tag{15.87}$$

The parameterization given in (15.78) gives rise to the following identity:

$$\kappa = -\left(\frac{1}{\alpha\lambda \sin t} + \frac{1}{\alpha\lambda \cos t}\right)\frac{d^2 S}{dt^2} = -\frac{2}{\alpha\lambda}\frac{d^2 S}{dt^2}\left(\frac{\sin t + \cos t}{\sin 2t}\right) \tag{15.88}$$

Then, by taking both derivations of the surface area, we can obtain the curvature:

$$\kappa = \frac{4\pi\alpha(\sin t + \cos t)}{3\lambda^3 \sin 2t (\cos t - \lambda)^2} *$$
$$*\left\{(1 + \lambda^2 - 2\lambda \cos t)^2 \left\{4 + \cot t + [\lambda\cos t + 1 - 2\lambda^2 - \lambda \sin t]\lambda^2 + (1 - 2\lambda^2 + \lambda\cos t)\right\} - 2\lambda\cot t\right\} \tag{15.89}$$

15.4.1.1.1 Constant Overpotentials

Now, we need to know $C_i(x, y)$ on an electrode surface under non-galvanostatic conditions and constant overpotentials. The differential equation for the "quasi-primary" current distribution is that of Equation 15.35, which is reconstructed as

$$a\left(\frac{\partial C_i(x, y)}{\partial y}\right) - b[C_i(x, y)]^2\left(\frac{\partial C_i(x, y)}{\partial y}\right) - fC_i(x, y) = 0 \tag{15.90}$$

with

$$a \equiv \left(\frac{1}{Z_i} + \frac{1}{\bar{\alpha}}\right)\chi_{\text{bulk}}(RT)^2(1 - t_i) \tag{15.91a}$$

$$b \equiv (Z_i)^2 D_i F^3 \tag{15.91b}$$

$$f \equiv \chi_{\text{bulk}} RTF(1 - t_i) \tag{15.91c}$$

This expression was transformed to reduce the order of the quadratic expression by defining a new function $z_i(x, y) = [C_i(x, y)]^2$ that is converted using the parameteric functions of Equation 15.78:

$$z_i(t) = \alpha \int \left(\frac{a - bz_i(t)}{2fz_i(t)}\right)\left(\sqrt{1 - 2\lambda \cos t}\right)dt \tag{15.92}$$

In this case, the solution after some algebra and integration is $[z_i(t)]_1$

$$[z_i(t)]_1 = \frac{-\alpha b\sqrt{(1 - 2\lambda)(1 - 2\lambda \cos t)}E\left(\frac{4\lambda}{(\lambda - 1)^2}, \frac{t}{2}\right)}{2f\sqrt{1 - 2\lambda \cos t}} \tag{15.93}$$

We can also simplify the solution given in (15.93) using the Picard successive approximation method when we know the initial conditions. If t_o and $z_o(t_o)$ are the initial conditions and the differential equation is written as $z'(t) = f(t, z)$, the equation can be written as

$$z_i(t) = z_o(t_o) + \int_{t_o}^{t} f(t, z)dt \tag{15.94}$$

In this case, the differential equation given in (15.90) is rewritten and after the resolution

$$z_1(t) = A\sqrt{C_i} + \left(\frac{2fA\sqrt{C_i}}{a - bA\sqrt{C_i}}\right)\left[-\frac{2}{3}tE_{1/3}\left(\frac{2}{3}At^{3/2}\right)\right] \tag{15.95}$$

where $Ei_n(t)$ is the *exponential integral function* with an exponential factor, n, equal to $1/3$.

The method implies the finding of subsequent functions $z_k(t)$ by iteration that is substituted into each other in the second term of the right-hand side of (15.94). The series of solutions, $z_0, z_1, z_2, \ldots$, converges to the desired solution into a defined domain that contains the point $(0, 0)$. The solution of $C_i(x, y)$ needs the back parameterization of the trochoid:

$$t = \cos^{-1}\left[\frac{1}{\lambda} - \frac{y}{\alpha\lambda}\right] \tag{15.96}$$

The $Ei_n(t)$ can be approached as shown above with the incomplete Gamma function. Thus, the concentration profile along y will be for the firsts two terms:

$$\begin{cases} C_0(0) = A^{1/2}C_i^{1/4} \\ C_1(y) = \sqrt{A\sqrt{C_i} - \left(\frac{4fA\sqrt{C_i}}{3(a - bA\sqrt{C_i})}\right)\cos^{-1}\left(\frac{\alpha - y}{\alpha\lambda}\right)\Gamma\left\{0, \frac{2A}{3}\left[\cos^{-1}\left(\frac{\alpha - y}{\alpha\lambda}\right)\right]^{3/2}\right\}} \end{cases} \tag{15.97}$$

Figure 15.8 shows a positive concavity similar to the same situation of constant overpotential on a smooth and planar electrocatalyst. The difference arises in the early stages (at low distances) since

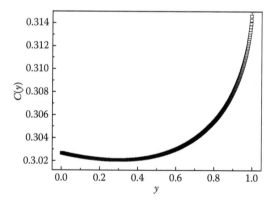

FIGURE 15.8 The second term of the concentration profile for a trochoid surface catalyst with $\lambda = 0.8$ and $\alpha = 0.5$ under non-galvanostatic conditions and constant overpotentials, $C_i = 0.01$ mol cm^{-3}. For simplicity, we consider $A \equiv 1$. According to the text, the selected parameters are $b/a = 10^{-4}$ and $f/a = 13$.

the changes are softer than those found in the smooth surface; more properly, a quasi-uniform distribution is observed. It sharply increases at medium distances; however, this rise of the concentration is much lower in absolute values compared with a planar surface. This effect is an indication of a uniform current distribution and will probably be more effective when we consider a tertiary current distribution.

Now, we have to make an approach to the current density expression. Since $C_i(x,y)$ is given by Equation 15.97 and $\Gamma(y) = \int_0^\infty t^{y-1} e^{-t} dt = \Gamma(0,y)$, its derivation [8] is

$$\frac{d\Gamma(y)}{dy} = \int_0^\infty \frac{dt^{y-1}}{dy} e^{-t} dt = \int_0^\infty t^{y-1} e^{-t} \ln(t) dt \qquad (15.98)$$

Then, after the subsequent change of variables, chain rules, and some algebra, we have

$$\frac{dC_1(y)}{dy} = \frac{\left(A\sqrt{C_i} - [C_1(y)]^2\right)\left(-\frac{3}{4}\Gamma\left\{0, \frac{2A}{3}\left[\cos^{-1}\left(\frac{\alpha-y}{\alpha\lambda}\right)\right]^{3/2}\right\} - \frac{1}{2}\right)}{C_1(y)\cos^{-1}\left(\frac{\alpha-y}{\alpha\lambda}\right)\sqrt{\left[2 - \left(\frac{\alpha-y}{\alpha\lambda}\right)\right]\left(\frac{\alpha-y}{\alpha\lambda}\right)}} \qquad (15.99)$$

Since the overpotential is constant, we can use Equation 15.19. Then, after the substitution of (15.99) as the principal value of $C_i(x,y)$

$$j_i(y) \cong -\left(\frac{\chi_{bulk} RT}{Z_i F}\right)\frac{A\sqrt{C_i}}{A\sqrt{C_i} - \left(\frac{4fA\sqrt{C_i}}{3(a - bA\sqrt{C_i})}\right)\left[\cos^{-1}\left(\frac{\alpha-y}{\alpha\lambda}\right)\right]^2\sqrt{\left[2 - \left(\frac{\alpha-y}{\alpha\lambda}\right)\right]\left(\frac{\alpha-y}{\alpha\lambda}\right)}} \qquad (15.100)$$

Thus, $\Delta\Phi(y)$ will be the first integration of $j_i(y)$:

$$\Delta\Phi(y) \cong \Delta\Phi^\circ - \left(\frac{RT}{Z_i F}\right)\int \frac{A\sqrt{C_i}}{A\sqrt{C_i} - \left(\frac{4fA\sqrt{C_i}}{3(a - bA\sqrt{C_i})}\right)\left[\cos^{-1}\left(\frac{\alpha-y}{\alpha\lambda}\right)\right]^2\sqrt{\left[2 - \left(\frac{\alpha-y}{\alpha\lambda}\right)\right]\left(\frac{\alpha-y}{\alpha\lambda}\right)}} dy \qquad (15.101)$$

Considering that in most of the cases $\alpha \approx \alpha\lambda$, we can take $\sqrt{(2 - (1 - y/\alpha)(1 - y/\alpha)} = (1 - y/\alpha)$. Then, under these adjusted conditions

$$\Delta\Phi(y) \cong \Delta\Phi^\circ - \alpha\left(\frac{RT}{Z_i F}\right)\ln y \qquad (15.102)$$

The constancy in $\eta(y)$ is only obeyed after certain distance from the surface, y, where $j_i(y)$ sharply increases (Figure 15.9).

15.4.1.1.2 Negligible Overpotentials

When $\eta(y) \to 0$ we will have zero values for $\partial\Delta\Phi/\partial y$ and also for $C_i(y)$ and $j_i(y)$ as explained above. If we consider that the concentration at the point $(x,y) = (0,0)$ is also changing with the same

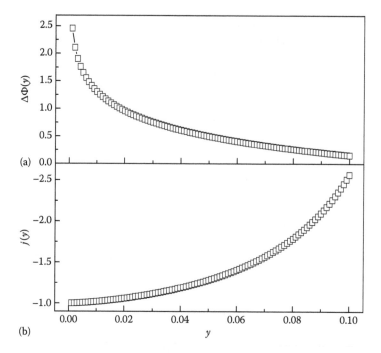

FIGURE 15.9 The dependence of (a) $|\Delta\Phi(y)|$ and (b) $|j_i(y)|$ upon the distance, y, normal to the electrode surface at constant overpotentials. We consider a trochoid surface with $\lambda = 0.8$ and $\alpha = 0.5$ with $C_i = 0.01$ mol cm^{-3}, $\Delta\Phi^\circ = -1$ V, $b/a = 10^{-4}$, and $f/a = 13$. For simplicity, we consider $A \equiv 1$.

function in the mass transfer Equation 15.14, we can calculate $\Delta\Phi(y)$ in the case of a trochoid using the parameterization given above, $dy = \alpha\lambda \cos t\, dt$:

$$\frac{d\Delta\Phi(t)}{dt} = \frac{RT}{Z_i F}\frac{d(\ln C_i(t))}{dt} \tag{15.103}$$

and then by combining terms, we can obtain the mean value of $C_i(t)$:

$$\frac{RT\chi_{\text{bulk}}}{Z_i^2 F^2 D_i} = \overline{C_i(t)} \tag{15.104}$$

With the mean value, we can obtain both the mean electrode potential $\overline{\Delta\Phi(t)}$

$$\overline{\Delta\Phi(t)} = \Delta\Phi^\circ_{j=0,i} + \frac{RT}{Z_i F}\ln\left(\frac{RT\chi_{\text{bulk}}}{C_i^\circ Z_i^2 F^2 D_i}\right) \tag{15.105}$$

and the mean current density along the y axis that has to be zero due to steady-state conditions in Equation 15.16:

$$\bar{j}_i = -\left(\frac{Z_i F D_i}{\alpha\lambda \cos t}\right)\left\{\frac{Z_i F C_i}{RT}\left(\frac{d\overline{\Delta\Phi(t)}}{dt}\right) + \left(\frac{d\overline{C_i(t)}}{dt}\right)\right\} = 0 \tag{15.106}$$

15.4.1.1.3   Galvanostatic Conditions with Large Current Densities
The new form of (15.37) under the trochoid parameterization for the catalyst will be

$$\frac{\chi_{bulk}Z_iFD_i}{(1-t_i)j_i|_y\alpha\lambda\cos t}\left(\frac{d^2C_i(t)}{dt^2}\right)+\frac{\chi_{bulk}RT}{\tilde{\alpha}F\alpha\lambda\sin t}\left(\frac{d\ln C_i(t)}{dt}\right)-j_i|_y=0 \tag{15.107}$$

Using the following abbreviations,

$$a\equiv\frac{\chi_{bulk}Z_iFD_i}{\alpha\lambda(1-t_i)j_i|_y} \tag{15.108a}$$

$$b\equiv\frac{\chi_{bulk}RT}{\tilde{\alpha}F\alpha\lambda} \tag{15.108b}$$

$$\gamma\equiv j_i|_y \tag{15.108c}$$

we can obtain

$$\frac{a}{\cos t}\left(\frac{d^2C_i(t)}{dt^2}\right)+\frac{b}{\sin tC_i(t)}\left(\frac{dC_i(t)}{dt}\right)-\gamma=0 \tag{15.109}$$

We can solve (15.109) by reducing the order of the quadratic differential equation. We can define a new function as

$$z_i(t)\equiv[C_i(t)]^{-1}\left(\frac{dC_i(t)}{dt}\right) \tag{15.110}$$

By substituting $z_i(t)$ and its first derivate into (15.109), we have

$$\frac{dz_i}{dt}=-z_i^2-\frac{b}{a}\cot tz_i^2+\frac{\gamma}{a}\cos te^{-\int z_idt} \tag{15.111}$$

We can try the Picard approximation method, where the initial conditions of the equation are definite in the proposed solution. Using $t_o=0$ and $z_o(0)=0$ since $C_i(0)=C_i^o$ the bulk solution concentration is $(dC_i^o/dt)=0$. Thus,

$$z_1=z_o+\int_0^t\left(\frac{\gamma}{a}\cos t\exp\left[-\int z_odt\right]\right)dt=\int_0^t\left(\frac{\gamma}{a}\cos t\right)dt=\frac{\gamma}{a}\sin t \tag{15.112}$$

The third term of the series will be

$$z_2=\frac{\gamma}{a}\sin t-\int_{t_o}^t\left(\frac{\gamma}{a}\sin t\right)^2dt-\int_{t_o}^t\frac{b\gamma^2\sin^2 t}{a^3\tan t}dt+\frac{\gamma}{a}\int_{t_o}^t\cos t\exp\left[\frac{\gamma}{a}(\cos t-1)\right]dt \tag{15.113}$$

Considering $t_o=0$ and after the complete integration of terms, we obtain

$$z_2=\frac{\gamma}{a}\sin t-\left(\frac{\gamma}{2a}\right)^2(t-\cos t\sin t)+\frac{b\gamma^2(\cos^2 t-1)}{2a^3}+\frac{\gamma(\gamma\cos t+a\sin t)ECi\left[\frac{\gamma}{a}(\cos t-1)\right]}{a^2+\gamma^2}$$

$$\tag{15.114}$$

where ECi the exponential cosinoidal integral which is mainly approached by series. It is defined as

$$\text{ECi}(y) = -\int_{y}^{\infty} \cos u \cdot e^{\cos u} du \qquad (15.115)$$

Since we really need $C_i(t)$ instead of $z_i(t)$, after neglecting the ECi function, we obtain

$$C_i(t) = \exp\left[-\frac{2\gamma}{a}(\cos t - 1) - \left(\frac{\gamma}{4a}\right)^2 (2t^2 + \cos 2t) + \frac{b\gamma^2(\sin 2t - 2t)}{8a^3} + \cdots\right] \qquad (15.116)$$

Figure 15.10a shows the concentration profile with $\chi_{\text{bulk}} = 0.1$ S cm^{-1} at 298 K with $Z_i = 1$ and $\vec{\alpha} = 0.5$. Also $D_i = 10^{-5}$ cm^2 s^{-1} and $t_i = 0.5$ are taken into account. We are going to consider $C_i = 1$ mol cm^{-3} and $\gamma = 1$ A cm^{-2}. To avoid further calculations, we use $A \equiv -1$. Thus, $a = 0.193$; $b = 0.005$; and $\gamma = 1$. As in the above case, we are going to pursue the following parameterization for the trochoid with $\lambda = 0.8$ and $\alpha = 0.5$. Surprisingly in this case, the large and constant current density applied to the electrocatalyst produces a maximum, but considering that the surface is a periodic cycloid, this can be expected.

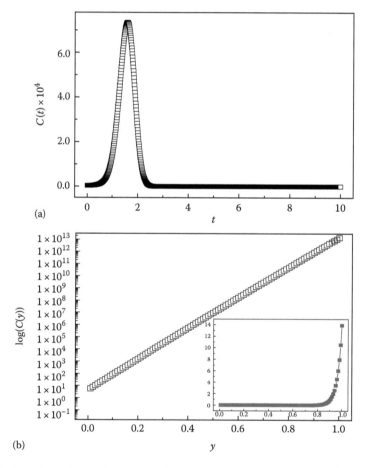

(a)

(b)

FIGURE 15.10 The dependence of the concentration profile (a) $C(t)$ upon the characteristic temporal magnitude, t, and (b) log $C(y)$ against the distance normal, y to the surface electrocatalyst parameterized for a trochoid with $\lambda = 0.8$ and $\alpha = 0.5$. The inset shows the normal $C(y)$ profile.

Anyhow, the back parameterization (Equation 15.116) with y makes us change to

$$C_i(y) = \exp\left[\begin{array}{l} \frac{2\gamma(\lambda - 1)}{a\lambda} + \frac{2\gamma}{a\alpha\lambda}y - \left(\frac{\gamma}{4a}\right)^2 2\left(\cos^{-1}\left[\frac{\alpha - y}{\alpha\lambda}\right]\right)^2 \\ -\left(\frac{\gamma}{4a}\right)^2\left\{\cos\left[\frac{\alpha - y}{\alpha\lambda}\right] - \sin^2\left(\cos^{-1}\left[\frac{\alpha - y}{\alpha\lambda}\right]\right)\right\} \\ +\frac{b\gamma^2}{4a^3}\left\{\left[\frac{\alpha - y}{\alpha\lambda}\right]\sin\left(\cos^{-1}\left[\frac{\alpha - y}{\alpha\lambda}\right]\right) - 2\cos^{-1}\left[\frac{\alpha - y}{\alpha\lambda}\right]\right\} + \cdots \end{array}\right]$$ (15.117)

considering $\cos 2t = \cos^2 t - \sin^2 t$ and $\sin 2t = 2\sin t \cos t$.

Figure 15.10b shows the logarithm contour of the $C_i(y)$ profile against the distance y and in the inset the normal $C_i(y)$ plot.

Now, $j_i(y)$ can be deduced from the mass balances and using the following abbreviations:

$$\xi \equiv \frac{\chi_{\text{bulk}}Z_iFD_i}{(1 - t_i)}$$ (15.118a)

$$\zeta \equiv \frac{\chi_{\text{bulk}}RT}{\bar{\alpha}F}$$ (15.118b)

$$-[j_i(y)]^2 + \zeta\left(\frac{\mathrm{d}\ln C_i(y)}{\mathrm{d}y}\right)j_i(y) + \xi\left(\frac{\mathrm{d}^2 C_i(y)}{\mathrm{d}y^2}\right) = 0$$ (15.119)

with $C_i(y)$ given by (15.117).

We are going to use a limited expression of $C_i(y)$ also with the approximation $\alpha \approx \alpha\lambda$:

$$j_i(y) \cong \left(\frac{\gamma}{a\alpha^2}\right)\zeta + \sqrt{\left(\frac{\gamma}{a\alpha^2}\right)^2 + \xi\left(\frac{2\gamma}{a\alpha^2}\right)^2}$$ (15.120)

After some substitutions by considering $t_i \approx 0.5$, the square root simplifies. The initial condition obeys similar concentration constancy at C_i^0 to $(j_i|_y)$, and so after some algebra,

$$j_i|_y \cong \frac{\alpha Z_i F^2 D_i C_i^0 \bar{\alpha}}{\lambda(1 - t_i)RT\chi_{\text{bulk}}}$$ (15.121)

which clearly shows the constancy and independence of y (galvanostatic conditions).

On the other hand, $\Delta\Phi(y)$ will be the integration in the ohmic drop:

$$\Delta\Phi(y) = \Delta\Phi^\circ - \frac{\alpha Z_i F^2 D_i C_i^0 \bar{\alpha}}{\lambda(1 - t_i)RT(\chi_{\text{bulk}})^2}y$$ (15.122)

When we are working with a constant current density, and when the distribution is of the first type, the drop of the electrode potential is linear.

15.5 METAL ROUGHNESS AT NON-REGULAR SURFACES

15.5.1 PERTURBATION THEORY AND LINEAR STABILITY THEOREM

In the case of generalizing the rough surface growth to a non-regular surface, a generalized methodology can be used. Thus, a linear stability analysis has been carried out in Refs. [6,9].

As always, one introduces a small and infinitesimal perturbation to the catalyst surface starting with a perfectly planar surface. The perturbations can be performed on either the electrode potential, the concentration profile, or even the current density. In the case of the electrode potential $\Phi(x, y, s)$ with $s = s(x, y, t)$, the situation can be constant or a variable overpotential, $\eta(x, y, s)$. In any case, since s is a function of (x, y, t), the perturbation has to be of the Euler form:

$$\Phi[x, y, s(x, y, t)] = \Phi^\circ(y) + \Phi_{\text{pert}}(y) e^{[i(k_x x + k_y y)]} e^{[\omega t]} \tag{15.123}$$

where we are assuming that the electric perturbation occurs mainly along the y-axis and the surface growth is dependent on both spatial components, x and y, with the surface growth dependent on time, t. As Euler predicted, k_i are the spatial amplitudes along x or y directions and ω is the temporal frequency. $\Phi^\circ(y)$ is the electrode potential value subject to the initial condition, and $\Phi_{\text{pert}}(y)$ is the perturbation infinitesimal potential variation dependent on the boundary conditions. Thus, $\Phi_{\text{pert}}(y)$ has to obey:

$$\nabla^2 \Phi = 0 \tag{15.7}$$

$$|\vec{j}_i| = -\frac{Z_i^2 F^2 D_i C_i \vec{\nabla} \Phi}{RT} - Z_i F D_i \vec{\nabla} C_i \tag{15.10b}$$

$$j_i|_y = -\chi_{\text{bulk}} \left(\frac{d\Phi}{dy} \right) \tag{15.13}$$

The substitutions of $\Phi_{\text{pert}}(y)$ into the Laplacian field (15.7) produce a linearized homogeneous differential equation of constant coefficients:

$$\frac{d^2 \Phi_{\text{pert}}(y)}{dy^2} - \left(k_x^2 + k_y^2 \right) \Phi_{\text{pert}}(y) = 0 \tag{15.124}$$

The typical solution of the homogeneous equation is

$$\Phi_{\text{pert}}(y) = \Phi_1^\circ(y) e^{-\left(k_x^2 + k_y^2 \right) y} + \Phi_2^\circ(y) e^{\left(k_x^2 + k_y^2 \right) y} \tag{15.125}$$

where $\Phi_i^\circ(y)$ is the electrode potentials defined by the boundary conditions (15.13) and $\Phi^\circ(y)$ at the initial condition. The expression is then transformed into

$$\Phi_{\text{pert}}(y) = 2\Phi_1^\circ \left[e^{-\left(k_x^2 + k_y^2 \right) y} + e^{\left(k_x^2 + k_y^2 \right) y} \right] = 2\Phi_1^\circ(y) \cosh[k(\delta - y)] \tag{15.126}$$

Since Equation 15.13 is considered exactly valid within the boundary layer δ, but far away from that place, the current density is zero (or as more precisely, the first derivate of the electrode potential is null).

$$j_i(y = \delta) = 0 = -\chi_{\text{bulk}} \frac{d\Phi}{dy}\bigg|_{y=\delta} \tag{15.127}$$

Using this equation, we can calculate the $j[x, y, s(x, y, t)]$ and then using (15.10b), $C_i[x, y, s(x, y, t)]$. The latter is clearly shown in Ref. [4] under galvanostatic conditions using C_i° as the bulk solution concentration and j_{pert} as the perturbation coefficient for the current density.

$$C_{i\,\text{pert}}(y) = C_1^\circ(y) e^{-\varsigma_1 y} + C_2^\circ(y) e^{\varsigma_2 y} - \frac{2Z_i F \Phi_1^\circ}{RT} \left[C_i^\circ - \frac{RT\chi_{\text{bulk}}}{Z_i^2 F^2 D_i} \right] e^{\left[-\frac{Z_i F j_{\text{pert}}}{RT\chi_{\text{bulk}}} (\delta - y) \right]} \cosh[k(\delta - y)] \tag{15.128}$$

with

$$k \equiv \sqrt{k_x^2 + k_y^2};$$ (15.129a)

$$\varsigma_{1,2} \equiv \frac{1}{2}\left[\frac{Z_iFj_{\text{pert}}}{RT\chi_{\text{bulk}}} \pm \sqrt{\left(\frac{Z_iFj_{\text{pert}}}{RT\chi_{\text{bulk}}}\right)^2 + 4k^2}\right]$$ (15.129b)

$$C_{i\,\text{pert}}(\delta) = C_1^0 e^{-\varsigma_1\delta} + C_2^0 e^{\varsigma_2\delta} - \frac{2Z_iF\Phi_1^0}{RT}\left[C_i^0 - \frac{RT\chi_{\text{bulk}}}{Z_i^2F^2D_i}\right]$$ (15.130)

The perturbed expression of the current density is much more complex and the limiting conditions at $y=0$ and $y=\delta$ are obligated to obtain a potable expression.

Anyway, it is desirable to the dispersion relationship coefficients that determine the stability of the surface. This expresses a linear and nontrivial dependence of ω on k and can be determined from the following equation system:

$$C_1^0 e^{-\varsigma_1\delta} + C_2^0 e^{\varsigma_2\delta} - \frac{Z_iF}{RT\chi_{\text{bulk}}k\sinh[k\delta]}\left[C_i^0 - \frac{RT\chi_{\text{bulk}}}{Z_i^2F^2D_i}\right]j^0 = 0$$ (15.131)

$$C_1^0\left(\varsigma_1 - \frac{Z_iFj_i^0}{RT\chi_{\text{bulk}}}\right) + C_2^0\left(\varsigma_2 - \frac{Z_iFj_i^0}{RT\chi_{\text{bulk}}}\right) = 0$$ (15.132)

$$\omega(C_1^0 + C_2^0)\left\{\frac{\vec{\alpha}j^0}{C^0} - \frac{j_i^0}{Z_iC^0}\left(\frac{C^0}{C_i^0}\right)^{\vec{\alpha}}\vec{\alpha}e^{-\left[\frac{\vec{\alpha}F\eta_i}{RT}\right]}\right\}$$

$$-\left(\begin{array}{l}\left\{\omega j_i^0 + \frac{\omega Z_iFj_i^0}{2RT\chi_{\text{bulk}}k\sinh[k\delta]}\right\}\left\{\frac{\vec{\alpha}j^0}{C^0} - \frac{j_i^0}{Z_iC^0}\left(\frac{C^0}{C_i^0}\right)^{\vec{\alpha}}\vec{\alpha}e^{-\left[\frac{\vec{\alpha}F\eta_i}{RT}\right]}\right\}* \\ \\ \left\langle\left[C_i^0 - \frac{RT\chi_{\text{bulk}}}{Z_i^2F^2D_i}\right]e^{\left[-\left(\frac{Z_iFj_{\text{pert}}}{RT\chi_{\text{bulk}}}\pm k\right)\delta\right]} - \frac{j_i^0}{Z_iC^0}\left(\frac{C^0}{C_i^0}\right)^{\vec{\alpha}}\frac{\vec{\alpha}F}{RT}e^{-\left[\frac{\vec{\alpha}F\eta_i}{RT}\right]}\right\rangle\cosh[k\delta] \\ \\ +\frac{1}{Z_iF\rho}\left\{\frac{j_i^0\vec{\alpha}j^0}{C^0}\left(\frac{dC_i}{dy}\right)_{y=0} + j_i^0\left(\frac{C^0}{C_i^0}\right)^{\vec{\alpha}}\left[\frac{j^0}{Z_i^2FD_iC^0}\right]\vec{\alpha}e^{-\left[\frac{\vec{\alpha}F\eta_i}{RT}\right]}\right\}\end{array}\right) = 0$$ (15.133)

where
j^0 and C^0 are the unperturbed coefficients for the current density and concentration profile at $y=0$, respectively
j_i^0 and C_i^0 are the exchange current density and bulk solution concentration for i-species, respectively
ρ is the bulk density of the deposited metal on the catalyst

A determinant built with (15.131) through (15.133) gives rise, after making it zero, to the values of coefficients C_1^0, C_2^0, and j^0. This equation, according to [6], leads to the dependence of ω on k:

$$\omega = \frac{-A\left(\varsigma_1 e^{\varsigma_1\delta} - \varsigma_2 e^{-\varsigma_2\delta}\right)}{Z_iF\rho B}$$ (15.134)

where

$$A \equiv \left\{ \frac{j_i^o \bar{\alpha} j^o}{C^o} \left(\frac{dC_i}{dy} \right)_{y=0} + j_i^o \left(\frac{C^o}{C_i^o} \right)^{\bar{\alpha}} \left[\frac{j^o}{Z_i^2 F D_i C^o} \right] \bar{\alpha} e^{-\left[\frac{\bar{\alpha} F \eta_i}{RT} \right]} \right\} \qquad (15.135a)$$

and

$$B \equiv \left(\varsigma_1 e^{\varsigma_1 \delta} - \varsigma_2 e^{-\varsigma_2 \delta} \right) \left[1 - \frac{C}{\chi_{\text{bulk}} k \sinh k\delta} \right] + \frac{(\varsigma_2 - \varsigma_1) Z_i F D}{RT \chi_{\text{bulk}} k \sinh k\delta} \left[C_i^o - \frac{RT \chi_{\text{bulk}}}{Z_i^2 F^2 D_i} \right] \qquad (15.135b)$$

where

$$C \equiv - \frac{D Z_i F}{2RT} \left\langle \left[C_i^o - \frac{RT \chi_{\text{bulk}}}{Z_i^2 F^2 D_i} \right] e^{\left[- \left(\frac{Z_i F j_{\text{pert}}}{RT \chi_{\text{bulk}}} \pm k \right) \delta \right]} - \frac{j_i^o}{Z_i C^o} \left(\frac{C^o}{C_i^o} \right)^{\bar{\alpha}} \frac{\bar{\alpha} F}{RT} e^{-\left[\frac{\bar{\alpha} F \eta_i}{RT} \right]} \right\rangle \cosh [k\delta]$$

$$(15.136a)$$

and

$$D \equiv \left\{ \frac{\bar{\alpha} j^o}{C^o} - \frac{j_i^o}{Z_i C^o} \left(\frac{C^o}{C_i^o} \right)^{\bar{\alpha}} \bar{\alpha} e^{-\left[\frac{\bar{\alpha} F \eta_i}{RT} \right]} \right\} \qquad (15.136b)$$

According to the authors who developed the above equations, B is always positive and independent of the sign of the current density flowing on the electrocatalyst. Thus, whether the real part of ω is negative or not (stability or instability of the surface) depends exclusively on A since $\varsigma_2 < \varsigma_1$. Hence, A has to be positive in order to keep up the surface stability; that is, this is strongly dependent on the sign of $j^o (dC_i/dy)_{y=0}$. This term is always positive since under cathodic conditions, the current density and the slope of the concentration profile are both negative. A further analysis of the behavior of the concentration profile slope shows the largest concentration in any case at $y = \delta$.

Moreover, the authors decided to include the surface energy contribution to the total overpotential as

$$\eta_{\text{surf}} = \frac{\varepsilon \chi}{\rho Z_i F} \qquad (15.137)$$

where ε and χ are the surface energy and surface conductivity of the electrocatalyst during electrodeposition, respectively. Both terms have to be dependent on the current density (or even on the surface concentration of the i-species). Thus, including this term, the value of A changes to

$$A' \equiv \left\{ \frac{j_i^o \bar{\alpha} j^o}{C^o} \left(\frac{dC_i}{dy} \right)_{y=0} + j_i^o \left(\frac{C^o}{C_i^o} \right)^{\bar{\alpha}} \left[\frac{j^o}{Z_i^2 F D_i C^o} + \frac{\varepsilon k^2}{\rho Z_i RT} \right] \bar{\alpha} e^{-\left[\frac{\bar{\alpha} F \eta_i}{RT} \right]} \right\} \qquad (15.138)$$

This new parameter is now able to compensate the instability produced by the first term through the surface energy that is always negative. The definition of the situation arises in the third term that comprises the modified expression of the Tafel behavior due to ohmic drop and mass transport effects:

$$\frac{j_i^o}{Z_i C^o} \left(\frac{C^o}{C_i^o} \right)^{\bar{\alpha}} \frac{\bar{\alpha} F}{RT} e^{-\left[\frac{\bar{\alpha} F \eta_i}{RT} \right]} \qquad (15.139)$$

The sign of the expression strongly depends on that of j_i^o.

The treatment by Pritzker and Fahidy [6] also involves anodic dissolution of the metal by a generalized expression of the activation overpotential by a mass transport modified Butler–Volmer equation. In both cases, anodic dissolution and cathodic deposition and stability and instability are possible. Some other studies have been presented with fewer terms that contribute to the surface stability such as those in Refs. [1,10] but with some erroneous conclusions. Anyhow, the limitations of Fahidy's work [6] arise from the fact that only a linear perturbation is considered, and so it can be applied only to the early stages of the instability.

APPENDIX A: RESOLUTION OF THE DIFFERENTIAL EQUATION (15.37) BY CHANGING VARIABLES

There is another way to solve the differential equation (15.37). We can reduce the order of the differential equation by an appropriate change of variables. Thus, (15.37) can be solved by writing the differential equation as

$$\left(\frac{\partial^2 C_i(x,y)}{\partial y^2}\right) + \frac{\beta}{\alpha C_i(x,y)}\left(\frac{\partial C_i(x,y)}{\partial y}\right) = 0 \qquad (15.A.1)$$

We can define a new function as

$$p_i(x,y) \equiv \frac{\partial C_i(x,y)}{\partial y} \qquad (15.A.2)$$

Then,

$$\frac{\partial^2 C_i(x,y)}{\partial y^2} = p_i(x,y)\frac{\partial p_i(x,y)}{\partial y} \qquad (15.A.3)$$

By substituting $p_i(x,y)$ and its first derivative into (15.A.2), we have

$$\frac{\partial p_i(x,y)}{\partial y} + \frac{\beta}{\alpha C_i(x,y)} = 0 \qquad (15.A.4a)$$

Or,

$$\frac{\partial p_i(x,y)}{\partial y} = -\frac{\beta}{\alpha C_i(x,y)} \qquad (15.A.4b)$$

The solution is

$$\int dp_i(x,y) = -\frac{\beta}{\alpha}\int\frac{dy}{C_i(x,y)} \qquad (15.A.5a)$$

Or,

$$p_i(x,y) = -cte\frac{\beta}{\alpha}\ln\ C_i(x,y) \qquad (15.A.5b)$$

However, by exchanging the variables,

$$\frac{dC_i(x, y)}{dy} = -cte\,\frac{\beta}{\alpha}\,\ln C_i(x, y) \tag{15.A.6}$$

and by simple integration of the function

$$\frac{dC_i(x, y)}{\ln C_i(x, y)} = -cte\,\frac{\beta}{\alpha}dy \tag{15.A.7}$$

Since the integration has a discontinuity at some points, the solution has to be

$$\text{li}\,C_i(x, y) = -cte\,\frac{\beta}{\alpha}(y - \delta) \tag{15.A.8}$$

where
δ is the migration–diffusion boundary layer
$\text{li}(x, y)$ is the *logarithmic integral function*, which is defined by $\text{li}\,C_i(x, y) = \int_0^{C_i(x, y)} (dt/\ln t)$

It is important to note that the function has a branch-cut discontinuity in the complex $C_i(x, y)$ plane running from $-\infty$ to 1 [11] (see Appendix B).
The function $\text{li}(x, y)$ also obeys:

$$\text{li}\,C_i(x, y) = \text{Ei}(\ln C_i(x, y)) \tag{15.A.9}$$

where $\text{Ei}(x, y)$ is the expression of the *exponential integral function* for $C_i(x, y)$. However, the exponential integral cannot act as the inverse function; we can only compute each point of the concentration profile.
The definition of $\text{Ei}(y)$ is given as follows:

$$\text{Ei}(y) = \int_1^{\infty} \frac{e^{-ty}dt}{t} = \int_y^{\infty} \frac{e^{-u}du}{u} \tag{15.A.10}$$

where $\text{Ei}(1) = 1.89511\ldots$ and the real root of the $\text{Ei}(y)$ function is given by [12] $\ln \mu$ where $\mu = 1.451369\ldots$ and is known as Soldner's constant [13] for which $\text{li}(y)$ has to be zero.
The Puiseux series of $\text{Ei}(y)$ along the positive real-axis is given by

$$\text{Ei}(y) = \gamma + \ln y + y + \frac{1}{4}y^2 + \frac{1}{18}y^3 + \frac{1}{96}y^4 + \cdots \tag{15.A.11}$$

where γ is the Euler–Mascheroni constant (Equation 15.B.6) and the denominators of the coefficients are given by $n \cdot n!$ [14]. For computing expression (15.A.9), we can either use the Ramanujan formula [15] for $C_i(x, y)$ at constant x values

$$\text{li}\,C_i(x, y) = \gamma + \ln \ln C_i(x, y) + \sum_{k=1}^{\infty} \frac{\ln C_i(x, y)^k}{k!k} \tag{15.A.12}$$

or we can use Equation 15.A.9:

$$\text{li } C_i(x, y) = \text{Ei}(\ln C_i(x, y)) =$$

$$\ln \gamma + \ln[\ln C_i(x, y)] + \ln C_i(x, y) + \frac{1}{4}(\ln C_i(x, y))^2 + \frac{1}{18}(\ln C_i(x, y))^3 + \cdots \qquad (15.A.13)$$

However, the best way to obtain the concentration profile would be the use of the inverse function of Ei(ln$C_i(x, y)$). According to Pecina [16], by delimiting some intervals of definitions for the variable y, it is possible to obtain the inverse function of the exponential integral function, $\text{Ei}^{-1}(y)$. Besides, a rough approximation for Ei(y) is possible [17] and then it will be for the inverse function by means of an empirical formula valid for $x > 1.6$, such as

$$\text{Ei}(y) = -\int_y^{\infty} \frac{e^{-u} du}{u} \approx \frac{e^{-y}}{(y + 2)(y - d)} \qquad (15.A.14)$$

with

$$d = \frac{16}{(y^2 - 4y + 84)} \qquad (15.A.15)$$

A suitable way of obtaining an analytical approach for $\text{Ei}^{-1}(y)$ is the use of the series expansion in Chebyshev polynomials. If $f(y)$ is a function defined in the region $-1 < y < 1$, the series will be [18]

$$f(y) = \sum_{n=0}^{\infty} a_n(y) T_n(y) \qquad (15.A.16a)$$

with

$$a_n(y) = \frac{\varepsilon_n}{\pi} \int_{-1}^{1} \frac{f(y) T_n(y)}{(1 - y^2)^{1/2}} dy \qquad (15.A.16b)$$

From the definition of Ei(y), it can be clearly seen that

$$\frac{d\text{Ei}^{-1}(y)}{dy} = \text{Ei}^{-1}(y) \exp\left[-\text{Ei}^{-1}(y)\right] \qquad (15.A.17)$$

Expanding $\exp(-\text{Ei}^{-1}(y)) \approx 1 - \text{Ei}^{-1}(y)$ and using the limiting condition $\lim_{y \to -\infty} \text{Ei}^{-1}(y) = 0$ Pecina [16] arrived at the asymptotic inverse function, $\text{Ei}^{-1}(y)$,

$$\text{Ei}^{-1}(y) = \frac{e^y}{1 + e^y} \qquad (15.A.18)$$

in the $-40 < y < 0$ domain.

On the other hand, the asymptotic formula for the $-\infty < y < -40$ region is

$$\text{Ei}^{-1}(y) = e^{(y-\gamma)} \qquad (15.A.19)$$

whereas for $y \to \infty$, we can demonstrate:

$$\mathrm{Ei}^{-1}(y) = \frac{e^y}{y} \tag{15.A.20}$$

The problem is that the positive values of y are normally used, and so we have to use approximations [16]. No asymptotic formula has been obtained yet for $0 < y < 10$, but at large values,

$$\mathrm{Ei}^{-1}(y) = \begin{cases} \ln[y(\ln y)] & \text{for } 10 < y < 10^6 \\ \ln\{y[\ln\langle y(\ln y)\rangle]\} & \text{for } 10^6 < y < 10^{75} \end{cases} \tag{15.A.21}$$

We consider again Equation 15.A.9 but plot $C_i(x, y)$ vs. $(y - \delta)$ from $\mathrm{Ei}^{-1}(\ln C_i(x, y))$ for $10 < y < 10^6$ (Equations 15.A.16a and 15.A.21):

$$\mathrm{Ei}^{-1}[\mathrm{Ei}(\ln C_i(x, y))] = \ln C_i(x, y) = \mathrm{Ei}^{-1}\left[-\frac{\beta}{\alpha}(y - \delta)\right] \tag{15.A.22}$$

Then,

$$\ln C_i(x, y) = \ln\left[-\frac{\beta}{\alpha}(y - \delta)\left(\ln\left|\frac{\beta}{\alpha}(y - \delta)\right|\right)\right] \tag{15.A.23a}$$

or

$$C_i(x, y) = -\frac{\beta}{\alpha}(y - \delta)\left(\ln\left|\frac{\beta}{\alpha}(y - \delta)\right|\right) \tag{15.A.23b}$$

Again using $t_i = 0.5$, the value of α is 0.2 and with a normal value of $\bar{\alpha} = 0.5$, we will have $\beta = 0.005$. Using $\delta = 10^{-4}$ cm, we will have the plot of $C_i(x, y)$ with y in the same units (cm) but between 10 and 10^6. The approximation at $y < 10$ gives rise to negative values of $C_i(x, y)$ as shown in Figure 15.A.1. Therefore, at low concentrations, it is possible to use the solutions given in Equations 15.42 and 15.44.

To obtain the electrode potential and current density, we take the first and the second derivations of Equation 15.A.23b to obtain the desired current density:

$$\frac{\partial C_i(x, y)}{\partial y} = -\frac{\beta}{\alpha}\left(\ln\left|\frac{\beta}{\alpha}(y - \delta)\right| - 1\right) \tag{15.A.24a}$$

$$\frac{1}{C_i(x, y)}\frac{\partial C_i(x, y)}{\partial y} = \frac{\partial \ln C_i(x, y)}{\partial y} = \frac{-\left(\ln\left|\frac{\beta}{\alpha}(y - \delta)\right| - 1\right)}{(y - \delta)\left(\ln\left|\frac{\beta}{\alpha}(y - \delta)\right|\right)} \tag{15.A.24b}$$

$$\frac{\partial^2 C_i(x, y)}{\partial y^2} = -\frac{\left(\frac{\beta}{\alpha}\right)}{|(y - \delta)|} \tag{15.A.24c}$$

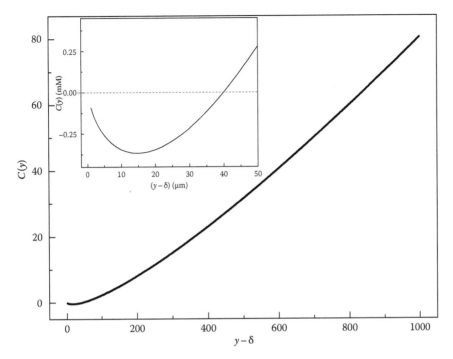

FIGURE 15.A.1 The dependence of the concentration profile, $C(x, y)$ upon the characteristic distance, $y - \delta$, normal to the electrode surface at a constant x. Inset shows the negative values of $C(x, y)$ obtained with the inverse exponential function at low $y - \delta$ values as expected. Galvanostatic conditions.

According to Equation 15.35 and after the substitution of the corresponding derivates,

$$j_i(y) + \frac{\chi_{\text{bulk}}\beta Z_i F D_i}{(1 - t_i)\alpha|(y - \delta)|j_i(y)} = -\frac{\chi_{\text{bulk}}RT}{\vec{\alpha}F(y - \delta)} \frac{\left(\ln\left|\frac{\beta}{\alpha}(y - \delta)\right| - 1\right)}{\left(\ln\left|\frac{\beta}{\alpha}(y - \delta)\right|\right)}$$

(15.A.25a)

and taking some approximations

$$(1 - t_i)\alpha|(y - \delta)|j_i^2(y) + \frac{\chi_{\text{bulk}}RT(1 - t_i)\alpha j_i(y)}{\vec{\alpha}F} + \chi_{\text{bulk}}\beta Z_i F D_i = 0$$

(15.A.25b)

the solutions of the second order equation are

$$j_i(y) = \frac{\chi_{\text{bulk}}RT}{\vec{\alpha}2F|(y - \delta)|} \pm \sqrt{\frac{(\chi_{\text{bulk}}RT)^2}{\vec{\alpha}^2 4F^2(y - \delta)^2} - \frac{\chi_{\text{bulk}}\beta Z_i F D_i}{(1 - t_i)\alpha|(y - \delta)|}}$$

(15.A.25c)

By making some approximations and considering the real solution for $y < \delta$,

$$j_i(y) = \frac{\chi_{\text{bulk}}RT}{\vec{\alpha}F|(y - \delta)|}$$

(15.A.25d)

On the other hand, the primary current distribution leads us to consider again

$$j_i(y) = -\chi_{\text{bulk}} \frac{\partial \Delta\Phi(x, y)}{\partial y} = \frac{\chi_{\text{bulk}} RT}{|\vec{\alpha}F|(y - \delta)|} \tag{15.A.26a}$$

so

$$\Delta\Phi(y) = \Delta\Phi^\circ - \int \frac{RT}{|\vec{\alpha}F|(y - \delta)|} dy \tag{15.A.26b}$$

Then,

$$\Delta\Phi(y) = \Delta\Phi^\circ - \frac{RT}{\vec{\alpha}F} \ln|(y - \delta)| \tag{15.A.26c}$$

The plots of $j_i(y)$ and $\Delta\Phi(y)$ as a function of y exhibit nearly the same behavior as those observed in Figure 15.2, when the same magnitudes of the parameters are used, so we avoid its representation.

APPENDIX B: LOGARITHMIC INTEGRAL FUNCTION

$\text{li}(x, y)$ is the logarithmic integral function and is defined as

$$\text{li}\, C_i(x, y) = \int_0^{C_i(x, y)} \frac{dt}{\ln t} \tag{15.B.1}$$

It is important to note that the function has a branch cut discontinuity in the complex $C_i(x, y)$ plane running from $-\infty$ to 1 [11]. Therefore, we can say that

$$\text{li}\, C_i(x, y) = \begin{cases} \displaystyle\int_0^{C_i(x, y)} \frac{dt}{\ln t} & \text{for } 0 < C_i(x, y) < 1 \\[2ex] \displaystyle\lim_{\varepsilon \to 0} \left[\int_0^{1-\varepsilon} \frac{dt}{\ln t} + \int_{1-\varepsilon}^{C_i(x, y)} \frac{dt}{\ln t} \right] & \text{for } C_i(x, y) > 1 \end{cases} \tag{15.B.2}$$

There is a unique positive number, $\mu = 1.451369\ldots$, known as Soldner's constant [13] for which $\text{li}(x)$ has to be zero. Thus, the logarithmic integral can also be written for $x > \mu$ as

$$\text{li}\, C_i(x, y) = \int_\mu^{C_i(x, y)} \frac{dt}{\ln t} \tag{15.B.3}$$

Some interesting values arise:

$$\text{li}(0) = 0, \tag{15.B.4a}$$

$$\text{li}(\mu) = 0, \tag{15.B.4b}$$

and also

$$\mathrm{li}(1) = -\infty \tag{15.B.4c}$$

Ramanujan independently discovered [15] that

$$\mathrm{li}(x) = \gamma + \ln\ln(x) + \sum_{k=1}^{\infty} \frac{\ln x^k}{k!k} \tag{15.B.5}$$

where γ is the Euler–Mascheroni constant $(0.57721566\ldots)$ in Ref. [15, pp. 3 and 11]. The Euler–Mascheroni constant is defined as the limit of the sequence:

$$\gamma \equiv \lim_{n\to\infty} \left(\sum_{k=1}^{n} \frac{1}{k} - \ln n \right) \tag{15.B.6}$$

In fact, the definition of $\mathrm{li}(x)$ for real x values gives rise to the plot given in Figure 15.B.1 and for the entire region with the discontinuity in Figure 15.B.2.

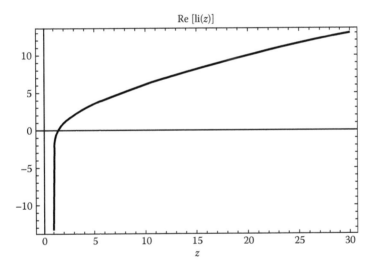

FIGURE 15.B.1 Plot of the real part of the logarithm integral function, $\mathrm{Re}[\mathrm{li}(z)]$ vs. z, from Soldner's constant.

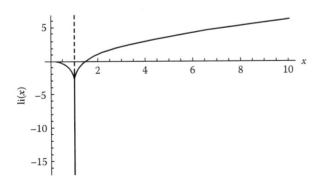

FIGURE 15.B.2 Plot of the real part of the logarithm integral function, $\mathrm{Re}[\mathrm{li}(z)]$ vs. z.

APPENDIX C: ELLIPTIC INTEGRAL FUNCTION

The integrals of the type

$$\int R\left(x,\ \sqrt{ax^3 + bx^2 + cx + d}\,\right)dx \tag{15.C.1}$$

cannot be expressed as elemental functions and they are known as *elliptics*.

They can be reduced after a series of transformations into elemental functions to three types of integrals [19]:

$$\int \frac{1}{\sqrt{(1 - t^2)(1 - k^2 t^2)}}\,dt \tag{15.C.2}$$

$$\int \frac{t^2}{\sqrt{(1 - t^2)(1 - k^2 t^2)}}\,dt \tag{15.C.3}$$

$$\int \frac{1}{(1 + ht^2)\sqrt{(1 - t^2)(1 - k^2 t^2)}}\,dt \tag{15.C.4}$$

with $0 < k^2 < 1$, k is known as the *elliptic module*.

With the substitution $t = \sin \varphi$ $(0 < \varphi < \pi/2)$, the above integrals can be expressed as the Legendre–Jacobi formulae:

$$\int \frac{1}{\sqrt{(1 - k^2 \sin^2 \varphi)}}\,d\varphi \tag{15.C.5}$$

$$\int \sqrt{(1 - k^2 \sin^2 \varphi)}\,d\varphi \tag{15.C.6}$$

$$\int \frac{1}{(1 + h\sin^2 \varphi)\sqrt{(1 - k^2 \sin^2 \varphi)}}\,d\varphi \tag{15.C.7}$$

where (15.C.5) through (15.C.7) are the *elliptic integral functions* of the *first*, *second*, and *third types*.

When the above integrals are defined between 0 and φ, they are known as incomplete integrals of the same types:

$$F(k, \varphi) \equiv \int_0^{\varphi} \frac{1}{\sqrt{1 - k^2 \sin^2 \varphi}}\,d\varphi \tag{15.C.8}$$

$$E(k, \varphi) \equiv \int_0^{\varphi} \sqrt{1 - k^2 \sin^2 \varphi}\,d\varphi \tag{15.C.9}$$

$$\prod(h, k, \varphi) \equiv \int_0^{\varphi} \frac{1}{(1 + h\sin^2 \varphi)\sqrt{1 - k^2 \sin^2 \varphi}}\,d\varphi \tag{15.C.10}$$

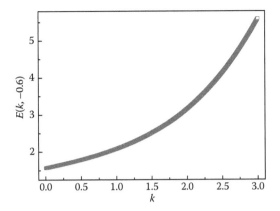

FIGURE 15.C.1   Plot of the incomplete elliptic integral function of the second type, $E[k, -0.6]$ vs. k (elliptic module), from a series approximation with $|\varphi| = 0.6$.

When $\varphi = \pi/2$, the integrals of the first and second types are considered as complete elliptic integrals. The complete elliptic integral of the second type, illustrated above as a function of k, is defined by

$$E(k) \equiv E\left(\frac{\pi}{2}, k\right) = \frac{\pi}{2}\left\{1 - \sum_{n=1}^{\infty}\left[\frac{(2n-1)!!}{2n!!}\right]^2 \frac{k^{2n}}{2n-1}\right\} \qquad (15.C.11)$$

or

$$E(k) = \frac{\pi}{2}{}_2F_1\left(-\frac{1}{2}, \frac{1}{2}; 1; k^2\right) \qquad (15.C.12)$$

where ${}_2F_1(a, b; c; x)$ is the hypergeometric function.

For example, $E(k, -0.6)$ can be developed in series of the form:

$$E(k, -0.6) = \frac{\pi}{2} + \frac{\pi k}{8} + \frac{3\pi k^2}{128} + \frac{5\pi k^3}{512} + \frac{175\pi k^4}{32768} + \cdots \qquad (15.C.13)$$

The solution is plotted in Figure 15.C.1.

REFERENCES

1. D. P. Barkey, R. H. Muller, and C. W. Tobias, *J. Electrochem. Soc.* **136** (1989) 2207.
2. J. de Bruyn, *Phys. Rev. E* **56** (1997) 3326.
3. W. W. Mullins and R. F. Sekerka, *J. Appl. Phys.* **34** (1963) 323.
4. T. Okada, *J. Electrochem. Soc.* **132** (1985) 537.
5. G. Kahanda, X.-Q. Zou, R. Farrel, and P.-Z. Wong, *Phys. Rev. Lett.* **68** (1992) 3741.
6. M. Pritzker and T. Fahidy, *Electrochim. Acta* **37** (1992) 103.
7. I. Bronshtein and K. Semendiaev, *Mathematics Handbook for Engineers and Students*, 2nd edn., Moscow, Russia: Mir Editorial (1973).
8. M. Abramowitz and C. A. Stegun (Eds.), Gamma (factorial) function and incomplete gamma function §6.1 and 6.5, in *Handbook of Mathematical Functions with Formulas, Graphs, and Mathematical Tables*, 9th printing, New York: Dover, pp. 255–258 and 260–263 (1972).
9. Q. BuAli, L. E. Johns, and R. Narayanan, *Electrochim. Acta* **51** (2006) 2881.

10. R. Aogaki, K. Kitazawa, Y. Kose, and K. Fuecki, *Electrochim. Acta* **25** (1980) 965.
11. M. Abramowitz and I. A. Stegun (Eds.), *Handbook of Mathematical Functions with Formulas, Graphs and Tables*, 9th printing, New York: Dover, p. 879 (1972).
12. S. R. Finch, Euler–Gompertz constant (§6.2), in *Mathematical Constants*, Cambridge, England: Cambridge University Press, pp. 423–428 (2003).
13. N. J. A. Sloane, The on-line encyclopedia of integer sequences, Sequences A0021624074, A069284 and A070769.
14. D. J. Mundfrom, *Eur. J. Comb.* **15** (1994) 555.
15. N. Nielsen, Theorie des Integrallograrithmus und Verwandter Transzendenten, Part II in *Die Gammafunktion*, New York: Chelsea (1965).
16. P. Pecina, *Czech. Acad. Sci.* **37** (1986) 8.
17. J. Zsakǎ, *J. Therm. Anal. Calorim.* **8** (1975) 593.
18. L. Y. Luke, *Mathematical Functions and Their Approximations*, Russian Translation, Moscow: Mir Editorial (1980).
19. J. Spanier and K. B. Oldham, The complete elliptic integrals K(p) and E(p) and the incomplete elliptic integrals F(p,φ) and E(p,φ), Chs. 61 and 62, in *An Atlas of Functions*, Washington, DC: Hemisphere, pp. 609–633 (1987).

16 The Electrochemical Fuel Cell Reactor

C. Fernando Zinola

CONTENTS

16.1 INTRODUCTORY REMARKS

The design of the electrochemical reactor, in the case of a fuel cell, is not yet totally solved as classical heterogeneous chemical reactors do not meet the requirements of the triphasic interface anode and the cathode binary system. Some papers [1–3] have considered the problem at the cathode and at the anode independently. However, the electrocatalytic reactions on both the electrodes produce a single chemical reaction, which is the chemical outlet of the energy conversion process.

We are going to study the case of a smooth carbon-paste electrode plate with channels, as those used for fuel cells. The problem of surface roughness in the case of gas diffusion electrodes was treated in Chapter 15.

Let us first consider that each channel has a semi-infinite extension and that we are treating the problem far away from the edges. Since each channel has an inter-distance length, defined as a, which is much less than the total length of the channel, defined as L, we have to consider that the electrolyte is under a limiting layer condition. We are evaluating the problem as a rectilinear and uniform displacement of the fluid along the x-axis. The Navier–Stokes equation for the fluid hydrodynamic current in the absence of the gravitational forces can serve for evaluation purposes:

$$\frac{\partial \vec{v}}{\partial t} + (\vec{v}\,\nabla)\vec{v} = -\frac{1}{\rho}\operatorname{grad} p + \vec{v}\,\nabla^2\vec{v} + \left(\frac{\varsigma}{\rho} + \frac{\nu}{3}\right)\operatorname{grad}\operatorname{div}\vec{v} \qquad (16.1)$$

where
- ρ is the linear density
- ν is the kinematic viscosity
- ς is the second viscosity
- p is the mean surface pressure of the electrolyte on the catalyst
- $\vec{v}$ is the hydrodynamic rate of the electrolyte in the fuel cell reactor

In the case of having a stationary fluid electrolyte,

$$(\vec{v}\,\nabla)\vec{v} = -\frac{1}{\rho}\operatorname{grad} p + \vec{v}\,\nabla^2\vec{v} + \left(\frac{\varsigma}{\rho} + \frac{\nu}{3}\right)\operatorname{grad} \operatorname{div} \vec{v} \tag{16.2}$$

In most of the cases, the stationary fluid is of a Newtonian type, so $\operatorname{div} \vec{v} = 0$:

$$(\vec{v}\,\nabla)\vec{v} = -\frac{1}{\rho}\operatorname{grad} p + \vec{v}\,\nabla^2\vec{v} \tag{16.3}$$

We assume that an approximation of the Re of the fluid in the channel is $\mathrm{Re} \equiv U_o L/\nu \ll 1$, since under this condition we can take that $(\vec{v}\,\nabla)\vec{v} \ll \vec{v}\,\nabla^2\vec{v}$. The meaning of U_o is the same as in hydrodynamics (limiting characteristic rate of the fluid). Therefore,

$$\vec{v}\,\nabla^2\vec{v} - \frac{1}{\rho}\operatorname{grad} p = 0 \tag{16.4}$$

When Re values are large, we can consider that the viscosity coefficient affects only the portion of the fluid near the surface of the electrocatalyst (*limiting layer* and *thickness*). One of the useful conditions is the adherence situation, that is, the rate of the fluid at the surface is zero. On the other hand, the rate of the fluid very far from the surface depends on the geometric construction of the channel and the real value of U_o. This limiting layer is going to be thinner when the Re values are high enough to compress it to the surface of the electrocatalyst. Thus, the only pressure component to be considered in Equation 16.3 is that which is along the x-axis (direction of electrolyte flow). Then, for a laminar flow, we can derive

$$v_x \frac{\partial v_x}{\partial x} + v_y \frac{\partial v_y}{\partial y} - \nu \frac{\partial^2 v_x}{\partial y^2} = -\frac{1}{\rho}\frac{dp}{dx} \tag{16.5}$$

where $\vec{v}$ and p are the rate and pressure, respectively, at the external limit of the limiting layer.

As we are interested in what happens within the limiting layer, the condition of Equation 16.4 applies:

$$\nu \frac{\partial^2 v_x}{\partial y^2} = \frac{1}{\rho}\frac{dp}{dx} \tag{16.6}$$

If the catalyst surface is smooth and the electrolyte can be considered as an uncompressible fluid, the tangential surface tension produced by the electrolyte is of the form

$$\gamma = 0.332\sqrt{\frac{\rho\eta U_o^2}{x}} = \frac{0.332\rho U_o^2}{\sqrt{Re_x}} \tag{16.7}$$

with $Re_x \equiv U_o x/\nu$, which is very important to evaluate the transition between the laminar and the turbulent flows at the limiting layer (Figure 16.1).

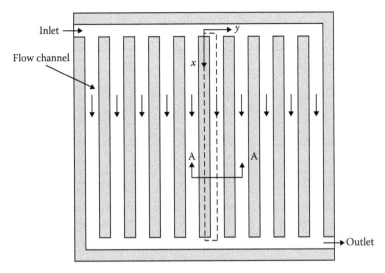

FIGURE 16.1 Bipolar plate without codes or edges for the flow rate of reactants on carbon paste electrodes. The spatial coordinates are indicated as the flux hydrodynamic directions.

16.2 SOLUTION FOR A TWO-DIMENSIONAL FUEL CELL SYSTEM WITH SMOOTH ELECTRODES

Let us consider a two-dimensional system for the fuel cell similar to a plug flow reactor model.

Since mass transfer must be rate determining, first, we consider the problem of convective diffusion of the *i*-species moving toward the surface. The steady-state conditions must be fulfilled so we can say that in the presence of an effective supporting electrolyte

$$\frac{\partial C}{\partial t} + \vec{v} \cdot \vec{\nabla} C = D\nabla^2 C \tag{16.8}$$

and under steady state

$$\vec{v} \cdot \vec{\nabla} C = D\nabla^2 C \tag{16.9}$$

Therefore, the diffusion convective equation in two dimensions is

$$v_x \frac{\partial C_i}{\partial x} + v_y \frac{\partial C_i}{\partial y} = D_i \left(\frac{\partial^2 C_i}{\partial x^2} + \frac{\partial^2 C_i}{\partial y^2} \right) \tag{16.10}$$

In the case of a fuel cell, we can also say that $D_i(\partial^2 C_i/\partial x^2) = 0$ since there is a uniform distribution along the flux direction (x) normal to y (Figure 16.2):

$$v_x \frac{\partial C_i}{\partial x} + v_y \frac{\partial C_i}{\partial y} = D_i \frac{\partial^2 C_i}{\partial y^2} \tag{16.11}$$

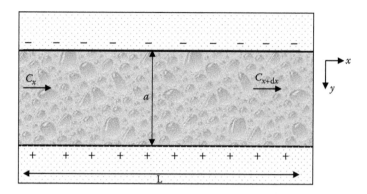

FIGURE 16.2 One-dimensional reactor scheme with the inlet and outlet reactant concentrations as a function of the penetration (along x channel). The parameters a and L depict the one-dimensional cross section and the characteristic channel distance, respectively.

We use the following initial (fast electrode kinetics at the catalyst) and boundary conditions (negligible influence in the bulk of the electrolyte and no surface diffusion by the i-species):

$$
\begin{aligned}
\overbrace{\lim_{y \to 0} C_i = 0} \\
\lim_{y \to \infty} C_i = C_i^{\circ} \\
\left[\frac{\partial^2 C_i}{\partial x^2} \right]_{x=0} = 0
\end{aligned}
\tag{16.12}
$$

The concentration profile is of the known form:

$$
C_i(x) = 0.678 \left(\frac{v}{D_i} \right)^{1/3} C_i^{\circ} \int_0^A \exp\left(\frac{-0.22v}{D_i} \lambda \right) d\lambda
\tag{16.13}
$$

where
     A is an integration limit $A = 1/2(U_o/v)^{1/2}(y/\sqrt{x})$
     D_i is the diffusion coefficient of the i-particle
     U_o is the limiting hydrodynamic rate
     v is the kinematic viscosity
     λ is a variable that joins the spatial coordinates (y, x)

The integration gives rise to the concentration profile along x for a defined y:

$$
C_i(x) = -3.08 C_i^{\circ} \left(\frac{D_i}{v} \right)^{2/3} \left[\exp\left(\frac{-0.11v^{1/2}U_o^{1/2}}{D_i} \frac{y}{\sqrt{x}} \right) - 1 \right]
\tag{16.14a}
$$

or

$$
C_i(x) = -3.08 C_i^{\circ} Sc^{2/3} \left[\exp\left(\frac{-0.11v^{1/2}U_o^{1/2}}{D_i} \frac{y}{\sqrt{x}} \right) - 1 \right]
\tag{16.14b}
$$

where Sc is the Schmidt number.

We are interested in the slopes of the concentration profile along y:

$$\frac{dC_i(x)}{dy} = \frac{0.339 C_i^{\circ} U_o^{1/2}}{D_i^{1/3} v^{1/6} \sqrt{x}} \exp\left(\frac{-0.11 v^{1/2} U_o^{1/2}}{D_i} \frac{y}{\sqrt{x}}\right) \tag{16.15}$$

Since the current density is defined as

$$j(x,y) = Z_i FD_i \left[\frac{\partial C_i(x,y)}{\partial y}\right]_{y=0} \tag{16.16}$$

we take the value at the electrode surface as $y = 0$:

$$\left(\frac{dC_i(x)}{dy}\right)_{y=0} = \frac{0.339 C_i^{\circ} U_o^{1/2}}{D_i^{1/3} v^{1/6} \sqrt{x}} \tag{16.17}$$

Then, the expression of the current density for the reactor under limiting conditions is

$$\lim_{y \to 0} j(x) = Z_i FD_i \left(\frac{\partial C_i}{\partial y}\right)_{y=0} = \frac{0.339 Z_i FD_i C_i^{\circ}}{\sqrt{x}} \left(\frac{v}{D_i}\right)^{1/3} \left(\frac{U_o}{v}\right)^{1/2} \tag{16.18}$$

$$j(x) = \frac{0.339 Z_i FD_i C_i^{\circ}}{\sqrt{xL}} Re^{1/2} Sc^{1/3} \tag{16.19}$$

where L is the characteristic distance of each channel.

It has to be noticed that the thickness of the hydrodynamic layer will be dependent on the square root of the distance along the catalyst surface:

$$\delta(x) = \frac{\sqrt{x}}{0.339} \left(\frac{D_i}{v}\right)^{1/3} \left(\frac{v}{U_o}\right)^{1/2} \tag{16.20}$$

This expression is very important since it shows the asymmetric thickness on the catalyst from $x = 0$ to $x = L$, even on a smooth plane electrode.

It is important to show the complete formulation of the current density with x and y (current density distribution), especially for the case of a two-dimensional conception of the reactor:

$$j_y(x) = Z_i FD_i \left(\frac{\partial C_i}{\partial y}\right) = \frac{0.339 C_i^{\circ}}{\sqrt{x}} \left(\frac{v}{D_i}\right)^{1/3} \left(\frac{U_o}{v}\right)^{1/2} \exp\left(\frac{-0.22 y}{\sqrt{x D_i A^3}}\right) \tag{16.21}$$

In the case of a plug flow reactor in one dimension, the expression of the concentration profile (mass balance) is a function of the current density:

$$\left(\frac{\partial C_i}{\partial x}\right) = -\frac{j_y(x)a}{Z_i F \varphi} \tag{16.22}$$

where
φ is the volumetric molar flow rate of the reactive entrance, which is considered constant
a is the separation between the plates (also constant)

We are proposing a similar expression for a two-dimensional reactor, but with a (x, y) dependent current density:

$$j_y(x) = \frac{Z_i F D_i^{2/3} 0.339 C_i^\circ U_0^{1/2}}{\nu^{1/6}\sqrt{x}} \exp\left(\frac{-0.11\nu^{1/2}U_0^{1/2}}{D_i} \frac{y}{\sqrt{x}}\right) \qquad (16.23)$$

Then, using Equation 16.22, we have a complete expression for the concentration profile:

$$\left(\frac{\partial C_i}{\partial x}\right) = -\frac{j_y(x)a}{Z_i F \varphi} = -\frac{0.339 a D_i^{2/3} C_i^\circ U_0^{1/2}}{\varphi\nu^{1/6}\sqrt{x}} \exp\left(\frac{-0.11\nu^{1/2}U_0^{1/2}}{D_i} \frac{y}{\sqrt{x}}\right) \qquad (16.24)$$

We are interested in this case only in the spatial coordinate at which there is a reactive hydrodynamic direction, x:

$$dC_i = -\frac{0.339 a D_i^{2/3} C_i^\circ U_0^{1/2}}{\varphi\nu^{1/6}} \int_0^L \frac{1}{\sqrt{x}} \exp\left(\frac{-0.11\nu^{1/2}U_0^{1/2}}{D_i} \frac{y}{\sqrt{x}}\right) dx \qquad (16.25)$$

This integral is solved by using the following variable change:

$$\frac{1}{\sqrt{x}} \equiv t, \quad dt = -2/3 x^{-3/2} dx \qquad (16.26)$$

Thus, substituting into (16.25) and defining as a the collection of the geometric and physicochemical parameters, we have

$$-3/2 \int_0^{1/\sqrt{L}} \frac{e^{-at}}{t^2} dt \quad \text{with } a \equiv \frac{0.11(\nu U_0)^{1/2} y}{D_i} \qquad (16.27)$$

From the *part integration theorem*, we have

$$-3/2 \int_0^{1/\sqrt{L}} \frac{e^{-at}}{t^2} dt = -3/2 \left[-\frac{e^{-at}}{t} + a \int_0^{1/\sqrt{L}} \frac{e^{-at}}{t} dt\right] \qquad (16.28)$$

The integral inside the parenthesis in Equation 16.28 is known as the exponential-integral function $Ei(x)$:

$$Ei(x) \equiv \int_{-\infty}^{x} \frac{e^{-t}}{t} dt \qquad (16.29)$$

When $x > 0$, the function diverges for $t \to 0$. In this case, the value of $Ei(x)$ is the main part of the undefined integral. Thus,

$$\int_{-\infty}^{x} \frac{e^{-at}}{at} dt = C + \ln|ax| - ax + \frac{(ax)^2}{2} - \frac{(ax)^3}{3!} + \cdots + \frac{(ax)^n}{n \cdot n!} \qquad (16.30)$$

with $C \equiv 0.5772$ being the Euler's constant or C number.

Therefore, applying the addition limiting integral theorem and considering the definition of the gamma function, we have

$$a^2 \int_{-\infty}^{1/\sqrt{L}} \frac{e^{-at}}{at} dt = a^2 \int_{-\infty}^{0} \frac{e^{-at}}{at} dt + a^2 \int_{0}^{1/\sqrt{L}} \frac{e^{-at}}{at} dt = -a^2 + a^2 \int_{0}^{1/\sqrt{L}} \frac{e^{-at}}{at} dt \qquad (16.31)$$

Taking the limiting property for the discontinuity at $t=0$ in the non-eigen integral,

$$\int_{0}^{1/\sqrt{L}} \frac{e^{-at}}{at} d(at) = \lim_{\varepsilon \to 0} \int_{\varepsilon}^{1/\sqrt{L}} \frac{e^{-at}}{at} d(at) = -\ln(\sqrt{L}) \qquad (16.32)$$

Substituting (16.32) and (16.31) into (16.28),

$$-3/2 \int_{0}^{1/\sqrt{L}} \frac{e^{-at}}{t^2} dt = \frac{3}{2} \left[\frac{e^{-at}}{t} + \ln \sqrt{L} \right] \qquad (16.33)$$

and changing back the variables for large values of x

$$\int_{0}^{L} \frac{e^{-a/\sqrt{x}}}{\sqrt{x}} dx = \frac{3}{2} \left[\sqrt{x} \exp\left(-\frac{a}{\sqrt{x}} \right) \right] + \frac{3}{2} \ln \sqrt{L} \qquad (16.34)$$

Finally, the difference between the inlet and outlet concentration of the i-species will be

$$C_i^{out} - C_i^{in} = -\frac{0.509 a D_i^{2/3} C_i^o U_o^{1/2}}{\varphi \nu^{1/6}} \left[\sqrt{x} \exp\left(-\frac{0.11 \nu^{1/2} U_o^{1/2} y}{D_i \sqrt{x}} \right) + \ln \sqrt{L} \right] \qquad (16.35)$$

We can write the change in the concentration with dimensionless numbers:

$$\Delta C_i = -\frac{0.509 a D_i C_i^o Sc^{1/3} Re^{1/2}}{\varphi L^{1/2}} \left[\sqrt{x} \exp\left(-\frac{0.11 Sc Re^{1/2}}{L^{1/2}} \frac{y}{\sqrt{x}} \right) + \ln \sqrt{L} \right] \qquad (16.36)$$

where Re and Sc are the Reynolds and Schmidt numbers, respectively (Figure 16.3).

All the concentration profiles exhibit a zero change in ΔC_i between $0 < (x/L)^{1/2} < 1$, because of the competition between the progress of the electrochemical reaction along y and the direction of the inlet flux along x. However, there are some important points to be noted. Those curves in which the increase in D_i or in a is observed show the lowest ΔC_i null value in the abscissa (those with the highest conversion of the reactive). Besides, it is also important to say that the increase in φ produces no changes in the ΔC_i null value at the abscissa (i.e., at $(x/L)^{1/2} = 0.05$). On the other hand, the most perceptible feature of this characteristic is the increase in U_o, that is, at $(x/L)^{1/2} = 0.15$.

The most interesting and useful situation is that of the increase in the diffusion coefficient of the particle. However, this is a property that can be enhanced only by using electro-dispersed electrodes for diffusing gases. These porous structures are discussed in the following sections. The separation between the anode and the cathode also enhances the conversion efficiency, probably because of the minimization of the interaction between the anode and the cathode diffusion layers. No interesting results are obtained when the molar volumetric flux and the limiting hydrodynamic rate are increased.

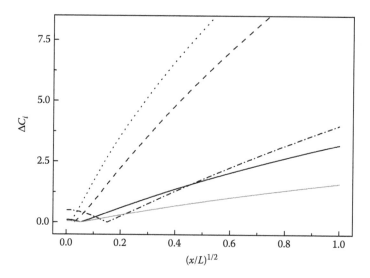

FIGURE 16.3 Absolute difference between the outlet and inlet concentrations, ΔC_i, as a function of square root of x/L. Arbitrary values of $a = 0.01$ cm, $D_i = 10^{-6}$ cm^2 s^{-1}, $C_i^{\circ} = 0.01$ mol cm s^{-3}, and $U_{\mathrm{o}} = 1$ cm s^{-1}, $\varphi = 1$ mol cm^{-3} (solid line), for $a = 0.1$ cm (dashed line); $D_i = 10^{-5}$ cm^2 s^{-1} (dotted line); $U_{\mathrm{o}} = 10$ cm s^{-1} (dash-dotted line); and $\varphi = 10$ mol cm^{-3} (grey solid line).

Let us examine the situation where the maximum reaction rate is reached, that is, at the catalyst surface ($y = 0$). In this case, we can also use the local Re number, Re_x, with x instead of L.

$$\Delta C_i\big|_{y=0} = -\frac{0.509aD_iC_i^{\circ}Sc^{1/3}Re^{1/2}}{\varphi}\left(\sqrt{\frac{x}{L}}+\frac{\ln\sqrt{L}}{\sqrt{L}}\right) \qquad (16.37)$$

Figure 16.4 shows the simplest model of the concentration profiles at the catalyst surface where the reaction occurs.

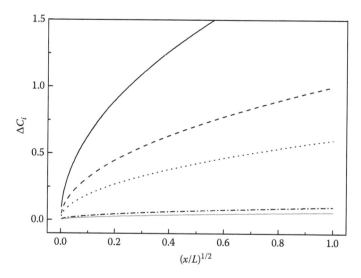

FIGURE 16.4 Outlet and inlet concentration absolute difference, ΔC_i, as a function of $\sqrt{x/L}$ for $y = 0$. Arbitrary values of $a = 0.1$ cm, $D_i = 10^{-5}$ cm^2 s^{-1}, $C_i^{\circ} = 0.01$ mol cm s^{-3}, $U_{\mathrm{o}} = 10$ cm s^{-1}, and $\varphi = 10$ mol cm^{-3} (solid line), for $a = 0.01$ cm (dashed line); $D_i = 10^{-6}$ cm^2 s^{-1} (dotted line); $U_{\mathrm{o}} = 1$ cm s^{-1} (dash-dotted line); and $\varphi = 1$ mol cm^{-3} (grey solid line).

When we decrease the distance between the anode and the cathode (a) both the diffusion layers interact to produce minimal changes in ΔC_i. The same dramatic diminution in the concentration change is produced when D_i decreases, that is, most of the electrocatalytic reactions in the fuel cells are mass-transport determining. Little variations are observed in ΔC_i for changes in φ and U_o.

16.3 GAS DIFFUSION ELECTRODE ASSEMBLY

16.3.1 CONICAL PORE DIFFUSION ELECTRODE

In the case of a three-phase electrocatalytic system, such as those of fuel cells, the cylindrical pore model is not applicable since it does not consider the problem of the partial pressure of the reactant in the gas phase. In this case, an equilibrium between the gaseous pressure inside the pore (which tends to force the electrolyte out of it) and the capillary forces of the electrolyte (which tend to flood the electrolyte away from the pore) must occur. This is known as the stable meniscus condition:

$$P = \frac{2\gamma}{r} \cos \theta \tag{16.38}$$

where
  P is the partial pressure of the gaseous reactant
  γ is the surface tension of the electrolyte over the solid catalyst
  θ is the contact angle of the electrolyte with the catalyst surface

If the pores have a mean radii less than the critical radius r of Equation 16.38, the electrolyte floods the pores; on reversing the conditions, the pores get dry. One way to overcome this problem is the use of a quasi-conic pore system.

Figure 16.5 shows the model proposed by Justi and Winsel [4] where the base of the pore faces the gas phase and the closest zone near the top of the cone on the electrolyte side. The meniscus is flat and with a finite contact angle forms a thin-film electrolyte layer on the catalyst.

Srinivasan et al. [5], for the first time, used a model for a conic single pore in a non-wetting three-phase (gas reactant, aqueous electrolyte, and a solid catalyst on a metal substrate) fuel cell system.

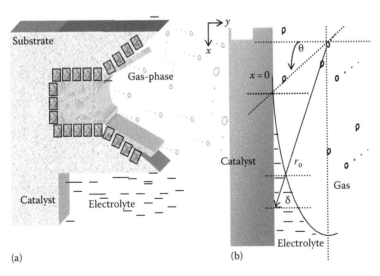

(a)   (b)

FIGURE 16.5 (a) Quasi-conical pore for an electrocatalyst in a three-phase system: gaseous reactant, liquid electrolyte, and solid electrocatalyst; (b) triphasic scheme with the angle of contact translated from the point $x = 0$ to θ; δ is the thickness of the diffusion layer of the electrolyte to the substrate at a radial distance, r_o.

The authors considered analytical and numerical solutions for the differential equations by using the three main overpotentials: charge transfer, mass transport, and ohmic drop. We discuss only the case of a two-dimensional system, where the current distribution varies with the radial and the azimuthal coordinates.

16.3.2 Current and Potential Distribution in a Conical Pore Electrode

If we consider again the equation of charge transfer, but for a mass transport–controlled process, the change in the current distribution at a radius r will be

$$dI(r, y) = 2\pi r j_0 \frac{C(r, y)}{C^o} \exp\left(\frac{\bar{\alpha} F \Delta\Phi(r, y)}{RT}\right) dy \qquad (16.39)$$

where
   r is the variable radius of the conical pore of length dy
   C^o is the concentration for the i-species at $y = 0$

Since we need to obtain the concentration of the reactant as a function of the distance along the pore and in the radial direction, we need to solve a Laplacian equation for the concentration:

$$\nabla^2 C_i(r, y) = 0 \qquad (16.40)$$

which is valid for $0 < y < L$ and $0 < r < r_o$ where r_o is the largest value of r.
   The total current density flowing through the pore will be

$$j_i(r, y) = Z_i F D_i \left[\frac{\partial C_i(r, y)}{\partial(r, y)}\right]_{y=L} \qquad (16.41)$$

To solve this equation, we need to set the following initial and boundary conditions:

$$
\begin{array}{ll}
C_i(r, y) = C_i^o & \text{for } y = 0 \quad \text{and} \quad 0 \leq r \leq r_o \\
\left(\dfrac{\partial C_i(r, y)}{\partial y}\right) = 0 & \text{for } y = L \quad \text{and} \quad 0 \leq r \leq r_o \\
dI = -Z_i F D_i 2\pi r_o dy \left(\dfrac{\partial C_i(r, y)}{\partial r}\right) & \text{for } r = r_o \quad \text{and} \quad 0 < y < L
\end{array}
\qquad (16.42)
$$

The first initial condition is the one that we have used in all the other models, that is, no reaction at the top of the pore (or at the top of the pore, there is no catalyst only the metal substrate). The second boundary condition means that the electrocatalytic reaction occurs completely, that is, all the gaseous reactant is consumed. The third condition denotes that the generated current in an infinitesimal dy element is due to the radial flux of the i-species toward the catalyst.
   The solution to Equations 16.39 and 16.41 is neither analytical nor numerical because the local potential difference is strongly dependent on y, and the surface concentration on both y and r. Two assumptions can account for this resolution:

1. All the electrocatalysts are prepared as columnar surfaces (a preferential direction along y, as explained in Chapter 15), that is, $d \ll h_i$; therefore, the potential or the current (depending on the electrochemical system) is generated mostly in the y direction. We can say

that $[\partial C_i / \partial r] = 0$ or the local overpotential for the charge transfer process is less than the concentration overpotential.

2. We can consider that the local potential difference is constant along the pore; thus, the solution to Equation 16.40 is possible.

However, an attempt to solve the differential equations can be a one-dimensional resolution by neglecting the radial diffusion. The value of the current can be deduced from (16.41):

$$dI(r, y) = \pi r_o^2 Z_i F D_i \left[\frac{\partial C_i^2(r, y)}{\partial y^2} \right] dy \tag{16.43}$$

Within an infinitesimal dy element, we can consider that the change in the electric current obeys Ohm's law:

$$d\Delta\Phi(r, y)\chi\pi r_o^2 = I(r, y)dy \tag{16.44a}$$

or

$$\left[\frac{\partial \Delta\Phi(r, y)}{\partial y} \right]\chi\pi r_o^2 = I(r, y) \tag{16.44b}$$

Thus, taking a second derivation with respect to y

$$dI_i(r, y) = \chi\pi r_o^2 \left[\frac{\partial^2 \Delta\Phi(r, y)}{\partial y^2} \right] dy \tag{16.45}$$

and combining (16.43) and (16.45), we have

$$\left[\frac{\partial^2 \Delta\Phi(r, y)}{\partial y^2} \right] = \left(\frac{Z_i F D_i}{\chi} \right) \left[\frac{\partial C_i^2(r, y)}{\partial y^2} \right] \tag{16.46}$$

This is a very important equation since it is clear that the second derivative of the electrode potential distribution is proportional to the second derivative of the surface concentration by the dissolved i-species. As always, we proceed to the set of initial and boundary conditions:

$$
\begin{array}{ll}
\overbrace{C_i(r, y) = C_i^o} & \text{for } y = 0 \\[2mm]
\left(\dfrac{\partial C_i(r, y)}{\partial y} \right) = 0 & \text{for } y = L \\[2mm]
I = -\pi r_o^2 Z_i F D_i dy \left(\dfrac{\partial C_i(r, y)}{\partial y} \right) & \text{for } y = 0
\end{array}
\tag{16.47}
$$

Since we also need to obtain a current distribution, we can combine Equation 16.39 with 16.45:

$$\left[\frac{\partial^2 \Delta\Phi(r, y)}{\partial y^2} \right] = \frac{2j_o}{\chi r_o} \left(\frac{C(r, y)}{C^o} \right) \exp \left(\frac{\tilde{\alpha} F \Delta\Phi(r, y)}{RT} \right) \tag{16.48}$$

The methods for solving Equation 16.48 are found in the literature [5,7]. We are not going to discuss further since we are especially interested in the case of wetting porous gas-diffusion electrodes with respect to the thin-film model.

16.3.3 Current and Potential Distribution in a Cylindrical Pore Electrode

Suppose that our thin-film layer covers cylindrically the pore as shown in Figure 16.6. We are going to consider that three types of polarizations occur obeying

$$j_i(r, y) = j_0 \left(\frac{C_i(r, y)}{C_i^o} \right) \exp \left(\frac{\bar{\alpha} F \Delta \Phi(r, y)}{RT} \right) \tag{16.49}$$

The wetting perimeter renders the electric current, $I(r, y)$, at the metal contact surface, $2\pi r_2$:

$$dI(r, y) = 2\pi r_2 dy \, j_0 \left(\frac{C_i(r, y)}{C_i^o} \right) \exp \left(\frac{\bar{\alpha} F \Delta \Phi(r, y)}{RT} \right) \tag{16.50}$$

Since we are taking into account all the polarization processes, the gas-diffusion electrode must obey a concentration gradient, which shows the following change in $I(r, y)$:

$$dI(r, y) = -2\pi r \, dy D_i Z_i F \left(\frac{\partial C_i(r, y)}{\partial r} \right) \tag{16.51}$$

$$\frac{dI(r, y)}{dy} \int_{r_1}^{r_2} \frac{dr}{r} = -2\pi D_i Z_i F \int_{C_i^o}^{C_i} dC_i(r, y) \tag{16.52}$$

since $C_i = C_i^o$ for $r = r_1$

$$C_i(r, y) = C_i^o + \frac{1}{2\pi D_i Z_i F} \left(\frac{dI(r, y)}{dy} \right) \ln \frac{r_1}{r_2} \tag{16.53}$$

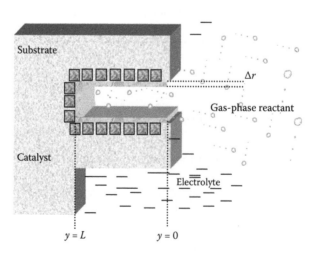

FIGURE 16.6 Schematic representation of the porous gas-diffusion catalyst using the thin-layer model. The thickness of the film is $\Delta r = r_2 - r_1$; the top and bottom of the pore are located at $y = 0$ and L, respectively.

and substituting in (16.50)

$$\left(\frac{\mathrm{d}I(r,y)}{\mathrm{d}y}\right) = 2\pi r_2 j_0 \left(1 + \frac{1}{2\pi D_i Z_i F} \ln \frac{r_1}{r_2} \left(\frac{\mathrm{d}I(r,y)}{\mathrm{d}y}\right)\right) \exp\left(\frac{\tilde{\alpha}F\Delta\Phi(r,y)}{RT}\right) \qquad (16.54)$$

On the other hand, the change in $\Delta\Phi(r,y)$ within the thin-film layer model obeys Ohm's law, for the ring cylindrical area of the electrolyte wetting the pore:

$$\left(\frac{\mathrm{d}\Delta\Phi(r,y)}{\mathrm{d}y}\right) = \frac{I(r,y)}{\chi\pi\left(r_2^2 - r_1^2\right)} \qquad (16.55)$$

If we take the derivative expression for $I(r,y)$ in (16.55), we will obtain

$$\left(\frac{\mathrm{d}^2\Delta\Phi(r,y)}{\mathrm{d}y^2}\right) = \frac{1}{\chi\pi\left(r_2^2 - r_1^2\right)} \left(\frac{\mathrm{d}I(r,y)}{\mathrm{d}y}\right) \qquad (16.56)$$

and substituting in (16.54)

$$\left(\frac{\mathrm{d}^2\Delta\Phi(r,y)}{\mathrm{d}y^2}\right) = \frac{2j_0 \exp\left(\dfrac{\tilde{\alpha}F\Delta\Phi(r,y)}{RT}\right)}{\chi\left(r_2^2 - r_1^2\right)\left[1 - \dfrac{r_2 j_0}{D_i Z_i F C_i^\circ}\left(\ln\dfrac{r_1}{r_2}\right)\exp\left(\dfrac{\tilde{\alpha}F\Delta\Phi(r,y)}{RT}\right)\right]} \qquad (16.57)$$

If we change the variables to

$$\Psi \equiv \frac{\tilde{\alpha}F\Delta\Phi(r,y)}{RT} \wedge \xi \equiv \frac{y}{L} \qquad (16.58)$$

we can simplify the resolution of the differential equation to

$$\left(\frac{\mathrm{d}^2\Psi}{\mathrm{d}\xi^2}\right) = \frac{A\exp(\Psi)}{1 + B\exp(\Psi)} \qquad (16.59)$$

where

$$A \equiv \frac{FL^2 j_0}{RT\chi\left(r_2^2 - r_1^2\right)} \wedge B \equiv \frac{r_2 j_0}{D_i Z_i F C_i^\circ}\left(\ln\frac{r_2}{r_1}\right) \qquad (16.60)$$

But we can make expression (16.59) straightforward by considering the following property:

$$\left(\frac{\mathrm{d}^2\Psi}{\mathrm{d}\xi^2}\right) = \frac{1}{2}\frac{\mathrm{d}}{\mathrm{d}\xi}\left(\frac{\mathrm{d}\Psi}{\mathrm{d}\xi}\right)^2 \qquad (16.61)$$

Thus, only a single integration is required now:

$$\left(\frac{\mathrm{d}\Psi}{\mathrm{d}\xi}\right)^2 = \int \frac{2A\exp(\Psi)}{1 + B\exp[\Psi]}\,\mathrm{d}\xi \qquad (16.62)$$

The resolution of this equation leads to two possible situations [6]:

1. $B \exp(\Psi) \gg 1$, that is, mass transport limitations (concentration overpotential in the pore). This happens when j_o and Δr are large (fast kinetic reaction and no thin-film model) or when D_i and C_i^o are small (slow diffusion process and very dilute solutions):

$$\left(\frac{d\Psi}{d\xi}\right)^2 = \frac{2A}{B}(\Psi - \Psi_o) \quad \text{where} \quad \Psi_o \equiv \frac{\bar{\alpha}F\Delta\Phi_o(r,y)}{RT} \tag{16.63}$$

Then,

$$(\Psi - \Psi_o) = \frac{A}{2B}\xi^2 \tag{16.64}$$

This means that the electric potential variation is proportional to the distance from the top to the bottom of the pore:

$$\Delta\Phi(r,y) = \Delta\Phi_o + \frac{D_i Z_i F C_i^o}{2\bar{\alpha}\chi r_2\left(r_2^2 - r_1^2\right)\ln\frac{r_2}{r_1}} y^2 \tag{16.65}$$

On the other hand, the current in Equation 16.55 as a function of y is

$$I(r,y) = \chi\pi\left(r_2^2 - r_1^2\right)\left(\frac{d\Delta\Phi(r,y)}{dy}\right) = \frac{D_i Z_i F C_i^o \pi}{\overleftarrow{\alpha} r_2 \, \ln\frac{r_2}{r_1}} y \tag{16.66}$$

2. $B \exp(\Psi) \ll 1$, that is, there is an absence of mass transport polarization in the pore. In this case, we cannot consider j_o and Δr as large, but the values of D_i and C_i^o are large. Thus, by considering Equations 16.59 and 16.61, we will have to solve

$$\left(\frac{d\Psi}{d\xi}\right)^2 = \int 2A \exp(\Psi)d\xi \tag{16.67}$$

but if we take the boundary condition $(d\Psi/d\xi) = 0$ for $\Delta\Phi = \Delta\Phi^o$, we can perform the first integration:

$$\left(\frac{d\Psi}{d\xi}\right)^2 = 2A(\exp\Psi - \exp\Psi_o) = 2A\exp\Psi_o(\exp(\Psi - \Psi_o) - 1) \tag{16.68}$$

where Ψ_o is equivalent to $\Delta\Phi^o$.

We can solve Equation 16.68 by changing the variable to

$$u^2 \equiv \exp(\Psi - \Psi_o) - 1 \tag{16.69a}$$

With this new variable, we have

$$du \equiv \frac{d\Psi \exp(\Psi - \Psi_o)}{\sqrt{\exp(\Psi - \Psi_o) - 1}} = \frac{d\Psi(1 + u^2)}{u} \tag{16.69b}$$

After substituting into (16.68) both Equations 16.69a and 16.69b

$$\left(\frac{du}{d\xi}\right)\frac{1}{(1+u^2)} = \sqrt{2A\exp\Psi_o} \tag{16.70}$$

where the solution is

$$\int\frac{1}{(1+u^2)}\,du = \tan^{-1}u = \sqrt{2A\exp\Psi_o}\int d\xi \tag{16.71}$$

and finally taking the integration limits we obtain

$$[\exp\Psi - 1]^{1/2} = [\exp\Psi_o - 1]^{1/2} + \tan(\sqrt{2A\exp\Psi_o}\,\xi) \tag{16.72}$$

For simplicity, we consider that $\exp\Psi_o \to 1$. Thus,

$$\exp\Psi = 1 + \tan^2(\sqrt{2A\exp\Psi_o}\,\xi) \tag{16.73}$$

Recombining the arbitrary defined parameters, Ψ and ξ, the variation of the electric potential, $\Delta\Phi$, will be

$$\Delta\Phi(y) = \Delta\Phi^o + \frac{2RT}{\overline{\alpha}F}\ln\sqrt{1 + \tan^2\left[\sqrt{\frac{2Fj_o\exp(\overline{\alpha}F\Delta\Phi^o/RT)}{RT\chi(r_2^2 - r_1^2)}}y\right]} \tag{16.74a}$$

or

$$\Delta\Phi(y) = \Delta\Phi^o + \frac{2RT}{\overline{\alpha}F}\ln\sec\left[\sqrt{\frac{2Fj_o\exp(\overline{\alpha}F\Delta\Phi^o/RT)}{RT\chi(r_2^2 - r_1^2)}}y\right] \tag{16.74b}$$

where $\Delta\Phi^o$ is the value of the electric potential at $y=0$.

For the calculation of the current, we need the first derivative of the $\Delta\Phi$ distribution with respect to y:

$$\left(\frac{d\Delta\Phi(y)}{dy}\right) = \sqrt{\frac{4Fj_o\exp(\overline{\alpha}F\Delta\Phi^o/RT)}{RT\chi(r_2^2 - r_1^2)}}\tan\left[\sqrt{\frac{2Fj_o\exp(\overline{\alpha}F\Delta\Phi^o/RT)}{RT\chi(r_2^2 - r_1^2)}}y\right] \tag{16.75}$$

Finally, the current distribution, $I(y)$, along y will be

$$I(y) = \sqrt{\frac{4\pi^2 Fj_o\chi(r_2^2 - r_1^2)\exp(\overline{\alpha}F\Delta\Phi^o/RT)}{RT}}\tan\left[\sqrt{\frac{2Fj_o\exp(\overline{\alpha}F\Delta\Phi^o/RT)}{RT\chi(r_2^2 - r_1^2)}}y\right] \tag{16.76}$$

Figures 16.7 and 16.8 show the electric potential profile for the conditions explained in the legends. The influence of $\Delta\Phi^o$ on the electric potential distribution is not very critical, but the values of j_o and χ strongly affect the absolute values of $\Delta\Phi$ but not its linear dependence with y.

This extreme simplification makes a strong dependence on relative distance and a largely nonuniform current distribution profile that sharply depends on j_o and $\Delta\Phi^o$. It is common to perform

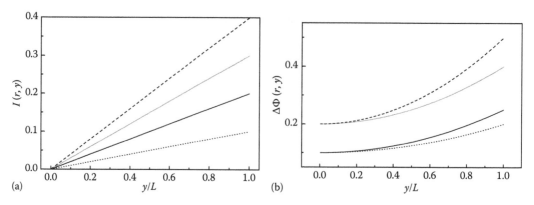

FIGURE 16.7 (a) Current and (b) potential difference profiles as a function of the relative pore penetration distance of a wetted thin-layer cylindrical pore electrocatalyst. Mass transport limitations: $\Delta\Phi_o = 0.10$ V, $r_2 = 100\,\mu m, r_1 = 1\,\mu m$ (dotted line), $\Delta\Phi_o = 0.10$ V, $r_2 = 10\,\mu m, r_1 = 1\,\mu m$ (continuous line), $\Delta\Phi_o = 0.20$ V, $r_2 = 100\,\mu m, r_1 = 10\,\mu m$ (grey line), $\Delta\Phi_o = 0.20$ V, $r_2 = 10\,\mu m$, and $r_1 = 10\,\mu m$ (dashed line). $\chi = 0.1\,\mathrm{S\,cm^{-1}}$, $j_o = 10^{-6}\,\mathrm{A\,cm^{-2}}$, $D_i = 10^{-5}\,\mathrm{cm^2\,s^{-1}}$, and $C_i^o = 10^{-4}\,\mathrm{mol\,cm^{-3}}$.

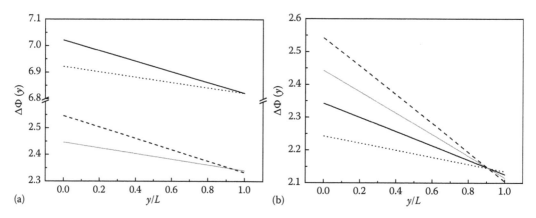

FIGURE 16.8 Potential difference profiles as a function of the relative pore penetration distance in a thin-layer cylindrical pore electrocatalyst. Only the charge transfer process and the ohmic drop are considered. (a) $r_2 = 100\,\mu m$, $r_1 = 10\,\mu m$, $\chi = 1\,\mathrm{S\,cm^{-1}}, j_o = 10^{-3}\,\mathrm{A\,cm^{-2}}$. $\Delta\Phi_o = 0.10$ V (grey line), $\Delta\Phi_o = 0.20$ V (dashed line). $r_2 = 100\,\mu m$, $r_1 = 10\,\mu m$, $\chi = 0.1\,\mathrm{S\,cm^{-1}}, j_o = 10^{-6}\,\mathrm{A\,cm^{-2}}$. $\Delta\Phi_o = 0.10$ V (dotted line), $\Delta\Phi_o = 0.20$ V (continuous line). (b) $r_2 = 10\,\mu m$, $r_1 = 1\,\mu m$, $\chi = 0.1\,\mathrm{S\,cm^{-1}}, j_o = 10^{-6}\,\mathrm{A\,cm^{-2}}$, $\Delta\Phi_o = 0.10$ V (dotted line), $\Delta\Phi_o = 0.20$ V (continuous line), $\Delta\Phi_o = 0.30$ V (grey line), $\Delta\Phi_o = 0.40$ V (dashed line).

the current vs. potential curves as Tafel plots to see the kinetic parameters in the wetted cylindrical pore along the catalyst.

In the case of the current distribution, for the charge transfer and ohmic limitations within a wetted cylindrical pore, the complex expression in (16.76) can be reduced in the case of small activation overpotentials (or large values of electrolyte conductivity within the pores at large $r_2 - r_1$ differences).

$$\tan\left[\sqrt{\frac{2Fj_o}{RT\chi(r_2^2 - r_1^2)}}\exp\left(\frac{\overline{\alpha}F\Delta\Phi^o}{RT}\right)y\right] \approx \sqrt{\frac{2Fj_o}{RT\chi(r_2^2 - r_1^2)}}\exp\left(\frac{\overline{\alpha}F\Delta\Phi^o}{RT}\right)y \qquad (16.77)$$

So for small polarizations, we have

$$I(y) = \frac{2\sqrt{2}\pi F j_0}{RT} \exp\left(\frac{2\bar{\alpha}F\Delta\Phi^\circ}{RT}\right)y \tag{16.78}$$

Under this condition, it follows that

$$\sec\left[\sqrt{\frac{2Fj_0 \exp(\bar{\alpha}F\Delta\Phi^\circ/RT)}{RT\chi(r_2^2 - r_1^2)}}y\right] \to 1 \tag{16.79}$$

Then, we can say that

$$\Delta\Phi \approx \Delta\Phi^\circ \tag{16.80}$$

Substituting Equation 16.80 into (16.78), we will have

$$\ln I(y) = \ln\left(\frac{2\sqrt{2}\pi F j_0 y}{RT}\right) + \frac{2\bar{\alpha}F\Delta\Phi}{RT} \tag{16.81}$$

This expression allows us to obtain different electrochemical features.

Tafel lines, $\Delta\Phi$ vs. $\ln I(y)$, give constant slopes at small polarizations, but the theoretical values of the relative exchange current densities (from the ordinate) are dependent on y. For large polarizations, we need numerical solutions to Equations 16.59 or 16.62 with fixed values of r_2, r_1, j_0, D_i, C_i°, and χ. Figure 16.9 shows some of the results.

We will not illustrate the polarization curves because a more generalized treatment and a more complete expression of the current vs. potential profile comprising the entire mass transfer–modified Butler–Volmer equation are available in the literature [7]. We only discuss here the most determining features of the curves that are deduced from our detailed comments presented above:

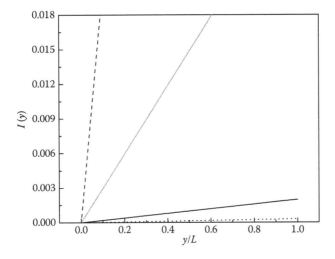

FIGURE 16.9 Current profiles as a function of the relative pore penetration distance in a thin-layer cylindrical pore electrocatalyst. Only small polarizations and high-conductivity electrolytes are considered. $j_0 = 10^{-5}$ A cm^{-2}. $\Delta\Phi_o = 0.10$ V (grey line), $\Delta\Phi_o = 0.15$ V (dashed line), $j_0 = 10^{-7}$ A cm^{-2}. $\Delta\Phi_o = 0.10$ V (dotted line), $\Delta\Phi_o = 0.15$ V (continuous line).

1. The initial response of the polarization curves is that of a combined charge transfer-ohmic behavior, and the influence of the reactant diffusion along the pore can be neglected.
2. The first portion of the curve depends on the magnitudes of Δr, D_i, C_i^o, and χ. The higher the D_i and C_i^o (reactant solubility), the more extended is the Tafel region. The lower the Δr and χ, the longer is the Tafel region of the curve.
3. When $B \equiv (r_2 j_o / D_i Z_i F C_i^o)(\ln(r_2/r_1))$ is less than 10^{-4}, no limiting current density is reached at $\Delta\Phi - \Delta\Phi^o$ lower than 0.7 V.
4. As accepted for the thin-layer models, the lower the value of Δr the higher is the current density value. Similar expressions apply for larger values of D_i, C_i^o, and r_2.
5. As expected, we arrive at larger current density values by a comparison between the thin-layer model and a single pore scheme.
6. Oversimplifications of the mathematical treatment usually lead to nonuniform current distributions, but in this case, since the values of D_i and C_i^o are not so high, the current profile is nonuniform for low overpotentials. When we are at large overpotentials, the situation is totally opposite, since we approach the limiting current.

REFERENCES

1. F. Coeret and A. Storck, *Elements de Génie Electrochimique*, Tec at Doc, Paris, France (1984).
2. A.A. Kulikovsky, *Electrochem. Commun.* **5** (2003) 530.
3. R. Chen, T.S. Zhao, K.T. Jeng, and C.W. Chen, *J. Power Sources* **152** (2005) 122.
4. E. Justi and A.W. Winsel, *Kalte-Verbrennung*, Franz Steiner Verlag, Wiesbaden, Germany (1962).
5. S. Srinivasan, H.D. Hurwitz, and J.O'M. Bockris, *J. Chem. Phys.* **46** (1967) 3108.
6. J.O'M. Bockris and S. Srinivasan, *Fuel Cells: Their Electrochemistry*, McGraw Hill Book Company, New York (1969), Chapter 5, pp. 230–288.
7. S. Srinivasan and H.D. Hurwitz, *Electrochim. Acta* **12** (1967) 495.

17 Diffusion-Convective Systems in Electrochemical Reactors

C. Fernando Zinola

CONTENTS

17.1 BULK ELECTROLYZERS: GENERALITIES

The fluid hydrodynamics and geometric configuration of the electrochemical reactor are key to understand the mixed processes that occur in a system. Though the specific geometry of the electrocatalysts is important, the mass transfer can be determined solely by fluid hydrodynamics [1].

One of the most commonly used electrochemical reactors for synthesis or metal deposition is the perfect agitated reactor, where a disc turbine in a vessel with deflectors generates the flux [2]. The measurements of the fluxes and discharge currents from the impeller are of fundamental importance since they define the dead volume of the vessel (especially in the case of viscous fluids) [3,4]. The problem of turbulence is also important, and has been studied elsewhere [5,6].

Multiple impellers have been mainly restricted to the *Rushton turbines*, inclined pallet turbines with up or down fluxes, and their combinations [7]. The flux patterns are now measured by the *Doppler* laser velocimeter (LDV) in systems with a single impeller [8] and double impellers [9]. Figure 17.1 shows the hydrodynamic fluxes of a single impeller with a double lace. It is important to notice that in the steady state, the flux varies between the two patterns periodically, the frequency being linearly dependent on the rotating rate of the impeller [9].

Rutherford et al. [10], by using LDV, studied the hydrodynamic characteristics of the agitated systems in the *Rushton double turbine*, where the height of the electrolyte, H, is equal to the tank diameter, D. They detected four unstable and three stable flux patterns. The stable flux patterns were named as parallel, mixed, and divergent (Figure 17.2).

The flux patterns are stable when the clarity, C, is larger than $0.2H$ and the clearance between the impellers, IC, is larger than $0.385H$. To prevail over the flux patterns, C has to be higher than $0.17H$ and IC less than $0.385H$. To maintain the divergent fluxes, C has to be less than $0.15H$ and IC larger than $0.385H$.

The power consumption of an impeller in a perfectly agitated tank of a Newtonian-type liquid can be calculated with the *power number*, P [11]. This dimensionless number is conceived as a

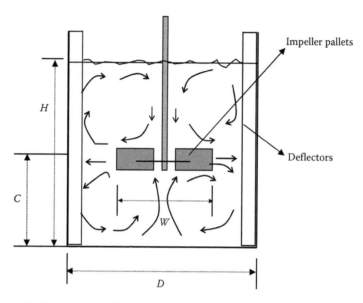

FIGURE 17.1 Standard geometric configuration for monophasic systems under the turbulent regime with a radial flux impeller.

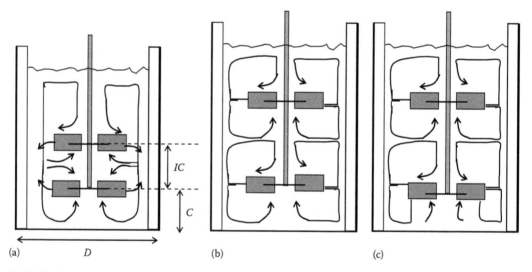

FIGURE 17.2 Standard flux patterns for a double disk turbine system. (a) Mixed flux, (b) parallel flux, and (c) divergent fluxes.

compromised between the geometric parameters of the reactor and the hydrodynamic conditions generated during the process. After a certain rotation rate, the gravitational force can be neglected:

$$P = (Fr, \text{geom}, Re) \qquad (17.1)$$

where Fr and Re are the Froude and the Reynolds numbers, respectively.

$$Fr = U^{o2}/gL \qquad (17.2)$$

where

U° is the characteristic rate

L the linear dimension of the tank

This number gives a correlation between the inertial and gravitational forces, defined by g, inside the fluid.

$$Re = U°L/v \qquad (17.3)$$

where v is the kinematic viscosity. This equation clearly shows the balance between the inertial and frictional forces inside the flux. For example, in the case of an incompressible fluid flowing inside a cylindrical tube of diameter, D, the value of L has to be D, but $U°$ equals the mean speed in the tube, $4\varphi/\pi D^2$, where φ is the volumetric flux in $m^3\ s^{-1}$.

The complexity of the integration of the geometric factors, such as those of the impellers and the clearness region, define that when $Re > 2 \times 10^4$, the turbulent flux, P, is constant. This is important because if P is dependent on the rotating rate, then the construction of the electrochemical reactor will become uneconomical.

Calderbank and Moo-Young [13] have derived the following formula:

$$P = 160WL(D - L)/D^3 \qquad (17.4)$$

where W is the width of the impeller pallets.

In this case, the height of the electrolyte, H, and that of the reactor, T, Equation 17.4 is valid for $0.57 < L/W < 1.33$; $2 < D/W < 6$; $1.5 < T/D < 3$; $0.6 < C/D < 1.6$. In this case, $0.67 < H/T < 1.33$ and $J/T = 0.1$, where J is the width of the deflectors.

Many other factors have to be considered for the characterization of a turbine in a chemical or electrochemical reactor. First, the impeller pumping capacity, defined as the liquid flow, is obtained from the revolution volume of the impeller. In addition it is also considered here the circulation flux, conceived as the fluid flowable to drag by the circulation laces generated by the impellers. The renovation time—the time that the entire electrolyte contained in the vessel remains before being drawn across the impeller—has to be also considered. The circulation time is the time that taken by the electrolyte in the reactor to circulate along all the circulation laces (flux pattern of the impeller). Finally, the index of the turbulence is simply the ratio between the mean fluctuant speed in the entire reactor volume from the edge of the impeller.

17.1.1 BOUNDARY LAYER CONDITION

When the fluid is limited by a solid surface, we can substitute the adherence condition (null speed) with the non-penetration of the fluid in the electrocatalyst or the walls of the reactor. Thus, the fluid flow will have a nonzero tangential component that does not obey the adherence condition. The description of the electrolyte movement shows a surface discontinuity at which the fluid rate changes from a nonzero value, v_t, to a negligible value. However, this is not the case since instead of a surface discontinuity, there is a very thin layer, δ, which is much lower than the length of the electrodes, the radius of the pipe, or the hydraulic diameter of the tube. Through this layer, the rate evolves as a continuous function from v_t to the outside of the *limiting layer* until a zero value is attained on the wall of the electrode. Inside this *limiting layer*, known as the *boundary layer*, the effect of viscosity and heat conduction is important (Figure 17.3).

This part of the chapter deals with the effects of viscosity on an electrolyte flowing in the electrochemical reactor in two dimensions. The boundary layers appear on the surface of bodies in viscous flow because the fluid seems to "stick" to the electrocatalyst's surface. As we have described above, right at the surface, the flow has zero speed, and this fluid transfers a linear momentum to the adjacent layers through the action of dynamic viscosity. Therefore, a thin fluid

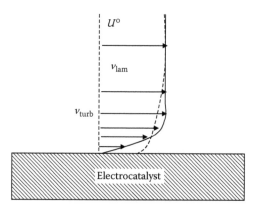

FIGURE 17.3   Velocity contours of the laminar (continuous lines) and the turbulent (dashed lines) boundary layers.

layer with a velocity lower than that of the outer flow develops. The condition where the flow at the surface has no relative motion is called the "no slip condition."

The boundary layer concept is attributed to Ludwig Prandtl (1874–1953). His manuscript, published in 1904, formed the basis for the future work on skin friction, heat transfer, and fluid separation. He later made original contributions to finite wing theory and compressibility effects. Theodore von Kármán and Max Munk were among his many famous students.

In a laminar flow, the fluid moves in smooth layers or films. There is relatively little mixing and, consequently, the velocity gradients are small and the shear stresses are low. The thickness of the laminar boundary layer increases with the distance from the start of the boundary layer and decreases with Re. For the laminar boundary layer, the average width of the diffusion layer, δ, is approximately $Re^{-1/2}$, but it is $Re^{-1/5}$ for a turbulent flow.

However, in the case of industrial electrocatalysts, the electrode is rough and porous so the fluid is sheared across the electrode's surface. Instabilities develop and eventually the flow transitions into turbulent motion. The turbulent boundary layer flow is characterized by unsteady mixing due to the eddy currents that occur at different distances to the electrode surface. This causes a larger shear stress at the electrode, a denser speed profile, and a greater boundary layer thickness. The electrode or tank wall shear stress is higher because the rate gradient near the wall is greater. This is because of the more effective mixing associated with turbulent flow. However, the lower velocity fluid is also transported outward resulting in a larger distance to the edge of the layer.

17.1.2 BOUNDARY LAYER THICKNESS

The boundary layer thickness, δ, is defined as the distance that is required for the flow to almost reach U°. We might take an arbitrary number (say 99%) to define practically what we mean by "nearly," but certain other definitions are used for convenience. The displacement and the momentum thicknesses are alternative measures of the boundary layer thickness and are used in the calculation of various boundary layer assets.

The displacement thickness is defined by considering the total mass flow through the boundary layer. This mass flow is the same as if the boundary layer were completely at rest, with a thickness, δ':

$$\delta' = \int_0^{+\infty} \left(1 - \frac{v(y)}{U^\circ}\right) dy \tag{17.5}$$

where $v(y)$ is the rate of fluid flow along the y-axis that is normal to the electrode's surface.

For the laminar boundary layers, δ' is about $1/3$ of the distance to the edge of the boundary layer, δ. The momentum thickness, ϕ, is defined similarly, using the momentum flux rather than the mass flux:

$$\phi = \int\limits_{0}^{+\infty} \frac{v(y)}{U^\circ}\left(1 - \frac{v(y)}{U^\circ}\right)dy \tag{17.6}$$

For the laminar boundary layers, δ tends to be about an order of magnitude greater than ϕ. The ratio of δ' to ϕ is termed the shape factor, H:

$$H = \delta'/\phi \tag{17.7}$$

17.1.3 Boundary Layer Equations

Newton's law applied to a fluid electrolyte in two-dimensions is

$$F = (\rho\, dx\, dy)\frac{dv}{dt} \tag{17.8}$$

where the mass $= (\rho\, dx\, dy)$. The force balance in the x-direction will be

$$F = dx\, dy\left(\frac{d\tau}{dy} - \frac{dp}{dx}\right) \tag{17.9}$$

This yields directly

$$\mu v_y - p_x = \rho(u_t + uu_x + vu_y) \tag{17.10}$$

This boundary layer equation is combined with the continuity equation and the condition of the constant static pressure through the boundary layer, $dp/dy = 0$, to obtain the three equations of the unknowns: p, u, and v. As long as the curvature radius of the catalyst is much larger than δ and the local velocities are not too large, there is a constant static pressure.

In the case of a steady state flow, Equation 17.10 reduces to

$$\mu v_y - p_x = \rho(uu_x + vu_y) \tag{17.11a}$$

After rewriting, this becomes

$$vv_y - \frac{p_x}{\rho} = uu_x + vu_y \tag{17.11b}$$

One approach to the solution of this equation is to assume that the pressure does not vary with y, and so it is specified by the external velocity distribution. v is computed from the continuity equation, leaving a differential equation in u to be integrated. However, this holds only for the laminar flow since turbulent boundary layers are inherently unsteady. The subsequent sections deal with the solutions of these equations in more detail.

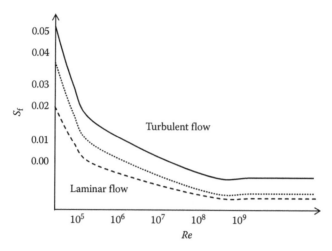

FIGURE 17.4   Surface friction drag coefficient vs. Reynolds number: transition location from the turbulent to the laminar flow.

17.1.4   VISCOUS DRAG

17.1.4.1   Surface Friction

The shearing stresses at the surface of a body produce a surface friction drag. We define the surface friction coefficient, S_f, by

$$S_f = \tau/A_\infty \tag{17.12}$$

where τ is the shear stress that is related to the dynamic viscosity, μ, by

$$\tau = \mu \frac{du}{dy} \tag{17.13}$$

S_f is related to the drag coefficient by S_D (skin friction) Figure 17.4:

$$S_D = S_f(A_{wetted}/A_{ref}) \tag{17.14}$$

where
    A_{wetted} is the area "wetted" by the fluids
    A_{ref} is the reference area used to define the drag coefficient

This expression applies to a flat plate. When the body has a thickness, the local velocities on the surface may be higher than the free-stream speed, and the skin friction is increased.
    We usually write

$$S_D = k\, S_f(A_{wetted}/A_{ref}) \tag{17.15}$$

where k is a "form factor" that depends on the shape of the body.
    The skin friction coefficient varies with Re, the Mach number, M, (when it involves incompressible fluids of local speeds less than the sonic speed, $M \ll 1$), and the character of the boundary layer. The momentum transferred between the air and the body surface appears as a velocity deficit in the viscous wake behind the body.

The plot below shows how Re and the location of the transition from the laminar flow to the turbulent flow affect the skin friction coefficient.

From the basic boundary layer theory combined with experimental fits, the following results are obtained.

For laminar boundary layers on smooth electrodes,

$$S_f = \frac{1.328}{Re^{1/2}} \tag{17.16}$$

For boundary layers of completely turbulent, smooth electrodes,

$$S_f = \frac{0.874}{(\log Re)^{2.58}} \tag{17.17}$$

17.1.5 PRESSURE DRAG

In addition to direct skin or film frictions, the presence of a boundary layer creates a pressure or forms a drag on the electrode. This does not appear in the case of a smooth electrode as the pressure always acts normal to the drag direction. In an adverse pressure gradient, the skin-friction drag is reduced, but the pressure drag is increased. This increase in the pressure drag compensates for some of the reduction in the skin friction. The combined drag may be estimated by a handy expression derived by Squire and Young, and gives amazingly good estimates of the total profile drag:

$$S_D = \left(2\phi U^{o\,\frac{H_{te}+5}{2}}\right)_{upper} + \left(2\phi U^{o\,\frac{H_{te}+5}{2}}\right)_{lower} \tag{17.18}$$

where ϕ is non-dimensionalized by the chord length and the velocity outside the boundary layer at the trailing edge, U^o, is normalized by the free stream U. H_{te} is the shape factor of the boundary layer at the trailing edge. Note that when U^o is 1.0, the drag is just twice the momentum thickness on the upper and the lower surfaces.

The presence of an adverse pressure gradient causes a deceleration of the fluid decreasing the efficiency of the conversion in an electrochemical reactor. Just as when one fluid starts climbing, the pressure hill with a low speed rolls backward after sometime (Figure 17.5).

This picture explains why the flow does not separate as readily at a higher Re. In this case, the velocity profile is "fuller" with the high external velocities extending down and closer to

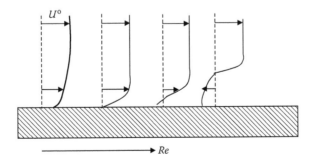

FIGURE 17.5 Presence of an adverse pressure gradient causes a deceleration of the fluid (increasing Re numbers).

the surface. The turbulent boundary layers also have a greater velocity near the surface and are therefore better able to handle the adverse pressure gradients.

Since the speed near the surface in a laminar boundary layer has a lower velocity than its turbulent counterpart, the laminar boundary layer is more likely to separate. When this occurs, the laminar boundary layer leaves the surface and usually undergoes a transition to a turbulent flow away from the surface. This process takes place over a certain distance that is inversely related to Re, but if it happens quickly enough, the flow may reattach as a turbulent boundary layer and continue along the surface. To compute when the separation will occur, we can solve the Navier–Stokes equations or apply one of the several separation criteria to the solutions of the boundary layer equations.

17.1.6 Conservation Laws

To derive the equations of motion for the fluid particles, we rely on various conservation principles. These principles are entirely intuitive. They are a statement of the fact that the rate of change of mass, the momentum, or the energy in a certain volume is equal to the rate at which it enters the borders of the volume in addition to the rate at which it is created inside. The first two of these parameters will be used extensively here:

$$\frac{d}{dt}\left(\int_V \rho \, dV\right) = -\int_S \rho(\vec{v} \cdot \vec{n})dS \tag{17.19}$$

This is a continuity equation, given in terms of the volumetric density, ρ, and the velocity of mass transfer, v, over a volume, V, and a surface, S, with a normal versor, n. In addition, similar equations that obey for the linear momentum, p, and energy, e, are

$$\frac{d}{dt}\left(\int_V \rho \vec{v} dV\right) = -\int_S [\rho \vec{v}(\vec{v} \cdot \vec{n}) + (\vec{p} \cdot \vec{n})]dS + \left(\int_V \rho \vec{f} dV\right) \tag{17.20}$$

$$\frac{d}{dt}\left(\int_V \rho\left(\vec{e} + \frac{v^2}{2}\right)dV\right) = -\int_S \left[\rho\left(\vec{e} + \frac{v^2}{2}\right)(\vec{v} \cdot \vec{n}) + (\vec{p}n \cdot \vec{v})\right]dS \tag{17.21}$$

These expressions are combined with the divergence theorem and by the fact that they can hold over arbitrary volumes to obtain the differential form of the equations:

$$\int_V (\vec{\nabla} \cdot \vec{F})dV = \int_S (\vec{F} \cdot \vec{n})dS \tag{17.22}$$

$$\frac{d\rho}{dt} + (\vec{\nabla} \cdot \rho\vec{v}) = 0 \tag{17.23}$$

$$\frac{d\vec{v}}{dt} + (\vec{v} \cdot \vec{\nabla})\vec{v} + \frac{\nabla\vec{p}}{\rho} = \vec{f} \tag{17.24}$$

We can use the momentum theorem by itself to obtain useful results. In this example, we apply the momentum theorem to relate the force on a body (catalyst particle) to the properties of the flow that is some distance away from the particle. This technique is useful in wind tunnel tests and is the basis of several fundamental theorems that are related to the lift and the induced drag of wings or, in our

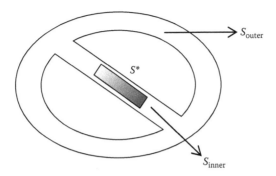

FIGURE 17.6 Control volume by a single surface, divided into three regions: the outer surface, S_{outer}, the inner surface, S_{inner}, and the pieces of the surface connecting the two, S^*.

case, for the electrolyte or the gas (reactive or product) flowing over a particle of a catalyst. Usually in industrial electrocatalysis, the electrocatalyst prepared by an active metal that is spread all over the substrate is porous and has a high roughness.

We take the control volume shown below, bounded by the single surface, S, which we divide into three parts: the outer surface, S_{outer}, the inner surface, S_{inner}, and the pieces of the surface connecting the two, S^* (Figure 17.6). We can write the integral form of the momentum equation for a steady flow with no body forces as

$$\frac{d \int_V \rho \vec{v} dV}{dt} = - \int_S [\rho \vec{v}(\vec{v} \cdot \vec{n}) + (\vec{p} \cdot \vec{n})]dS = 0 \qquad (17.25)$$

The linear momentum is zero since we are working with steady flows. Deriving for the three pieces of the surface,

$$\int_{S_{inner}} [\rho \vec{v}(\vec{v} \cdot \vec{n}) + (\vec{p} \cdot \vec{n})]dS + \int_{S_{outer}} [\rho \vec{v}(\vec{v} \cdot \vec{n}) + (\vec{p} \cdot \vec{n})]dS = 0 \qquad (17.26)$$

If the inner part of the particle at the electrocatalyst is shrunk until it touches the substrate, the electrolyte or fluid flow is tangential to it and then, $\vec{v} \cdot \vec{n} = 0$. Then, the force developed by the body, F, will be

$$\int_{S_{inner}} [(\vec{p} \cdot \vec{n})]dS = F \qquad (17.27)$$

and then

$$F = - \int_{S_{outer}} [\rho \vec{v}(\vec{v} \cdot \vec{n}) + (\vec{p} \cdot \vec{n})]dS \qquad (17.28)$$

Note that the contribution from the part of the surface connecting S_{inner} and S_{outer} to the integrals is zero because, as the two pieces of S^* are made close together; the unit normals point in the opposite directions while p and V are equal.

17.2 FLOW ELECTROLYZERS: GENERALITIES

An alternate method for the perfectly agitated electrolysis involves the flow of the solution that is to be electrolyzed to be continuous through a porous working electrode with a large surface area. The flow electrolytic methods are highly efficient and produce conveniently fast conversions of huge amounts of solution that are treated. The flow methods are used in industrial applications (e.g., removal of metals such as copper from waste streams) and are applied broadly to inorganic electrosyntheses, separations, and electroanalyses. The flow electrolytic cell contains a large area working electrode that is prepared by screens of fine mesh of metal or a cradle of conductive material. If the two-cell compartment is not necessary as in metal electrodeposition, the counter electrode can be interleaved with the working electrode and "isolated" from it by simple separators. In the case of industrial applications, as in chlorine-soda divided cells, more complex structures are required, including separators such as porous glass, ceramics, or ion-exchange membranes and a careful placement of the counter and reference electrodes is needed to minimize the ohmic drops. The cells are designed to show high conversions with a minimum length of electrode and a maximum flow of velocities [14].

First, we have to consider a flow on a rough and porous electrode of length L (cm) and a cross-sectional area A (cm^2) immersed in a torrent of electrolyte with a volumetric flow rate φ (cm^3 s^{-1}). The *linear flow rate* of the electrolyte stream, v (cm s^{-1}), is given by

$$\vec{v} = \hat{v}\frac{d\varphi}{dA} \tag{17.29}$$

where $\hat{v}$ is the unitary vector.

We can suppose that the electrocatalytic reaction is

$$\text{Ox}_{(aq)}^{+Z} + ne^- \rightarrow \text{Red}_{(aq)}^{+(Z-n)} \tag{I}$$

that is assumed to occur with 100% faradaic efficiency. The inlet concentration of Ox is $C_{Ox}(\text{in})$ and $C_{Red}(\text{in})$ is assumed zero. At the outlet, the concentrations are $C_{Ox}(\text{out})$ and $C_{Red}(\text{out})$. Then, the mass balance will be

$$C_{Ox}(\text{out}) = C_{Ox}(\text{in}) - \frac{jA}{nF\varphi} \tag{17.30}$$

The overall conversion from Ox^{+Z} to Red$^{+(Z-n)}$, Θ, when the applied current density is j, at the electrocatalyst is $j \ln F$ (mol cm^{-2} s^{-1}):

$$\Theta = 1 - \frac{C_{Ox}(\text{out})}{C_{Ox}(\text{in})} = \frac{jA}{nF\varphi C_{Ox}(\text{in})} \tag{17.31}$$

Under mass control, we can say that

$$j_{Ox}(x, t) = Z_{Ox}FD_{Ox}\frac{\partial C_{Ox}(x, t)}{\partial x} \tag{17.32}$$

and that under the limiting conditions it is (the current is no longer dependent on time)

$$j_{Ox}(x) = Z_{Ox}FD_{Ox}\frac{C_{Ox}(x)}{\delta} \tag{17.33}$$

When considering the electrocatalytic reaction under a stationary regime, the following constant slope of the concentration profile along L is obeyed:

$$\frac{\partial C_{Ox}(x)}{\partial x} = -\frac{j_{Ox}(x)L}{nF\varphi} \tag{17.34}$$

Then, after combining both the equations,

$$\frac{\partial C_{Ox}(x)}{\partial x} = -\frac{LZ_{Ox}D_{Ox}}{n\varphi}\frac{C_{Ox}(x)}{\delta} \tag{17.35}$$

Reordering the terms and integrating,

$$\int \frac{dC_{Ox}(x)}{C_{Ox}(x)} = -\frac{LZ_{Ox}D_{Ox}}{\delta n\varphi}\int dx \tag{17.36}$$

$$\ln\left[\frac{C_{Ox}(x)}{C_{Ox}(in)}\right] = -\frac{LZ_{Ox}D_{Ox}}{\delta n\varphi}x \tag{17.37}$$

or

$$C_{Ox}(x) = C_{Ox}(in)\exp\left[-\frac{LZ_{Ox}D_{Ox}}{n\varphi\delta}x\right] \tag{17.38}$$

Then, the current density along x is determined after Equation 17.33:

$$j_{Ox}(x) = \frac{Z_{Ox}FD_{Ox}C_{Ox}(in)}{\delta}\exp\left[-\frac{LZ_{Ox}D_{Ox}}{n\varphi\delta}x\right] \tag{17.39}$$

The local current density, $j_{Ox}(x)$, is highest at the front face of the electrode, and it decreases exponentially with x. The conversion factor Θ is independent of the initial concentration of $Ox_{(aq)}^{+Z}$:

$$\Theta = 1 - \frac{C_{Ox}(out)}{C_{Ox}(in)} = 1 - \exp\left[-\frac{LZ_{Ox}D_{Ox}}{n\varphi\delta}x\right] \tag{17.40}$$

By defining a mass transfer coefficient, K_M, as the ratio between D_{Ox} and δ, we can use its empirical form:

$$K_M = \frac{D_{Ox}}{\delta} = a\varphi^m \tag{17.41}$$

where $m = 1/3$–$1/2$ for a laminar flow and $3/4$–1 for a turbulent flow and the value of a depends on the geometric construction of the system. Then, after substitutions,

$$\Theta = 1 - \exp\left[-\frac{LZ_{Ox}a\varphi^{m-1}}{n}x\right] \tag{17.42}$$

In the case of the oxidation state and the number of electrons having the same values, we simply enclose

$$\Theta = 1 - \exp[-La\varphi^{m-1}x] \tag{17.43}$$

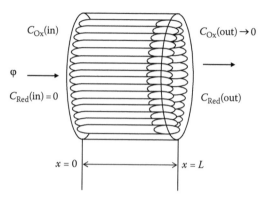

FIGURE 17.7 Scheme of a portion of the electrocatalyst in a flow electrolytic cell.

Thus, the conversion efficiency increases with the decreasing flow velocity and, the increasing specific area and the length of the electrode. From Equation 17.38, it can be seen that the concentration of Ox^{+z} varies exponentially with the distance along the electrode.

However, the porosity terms at the electrocatalyst's surface also has to be considered. According to Figure 17.7, the frontal surface area with open pores is A_{pores}, and then the porosity, p (in analogy to the roughness factor), is defined as

$$p = A_{pores}/A \tag{17.44}$$

Due to the porosity term, the linear rate for the electrolyte flow, v, that is related to φ increases upon entering the electrode to an interstitial speed, w, and generally to a turbulent flow regime:

$$\vec{w} = \vec{v}/p = \varphi/Ap \tag{17.45}$$

Then, the evaluation of the velocity profile along the entire length of the pore is

$$|\vec{w}| = \frac{1}{p}\frac{dx}{dt} \tag{17.46}$$

The most important factor that needs to be ascertained is the correct length of the pores in the electrocatalyst at which the total conversion of the substance is achieved. Therefore, if we integrate Equation 17.46 between 0 and L and substitute into Equation 17.43, we will have to reach the maximum Θ:

$$\ln(1 - \Theta) = -La\varphi^{m-1}wpt \tag{17.47}$$

In the case of a laminar flow with $m = 1/2$, the value of L will be

$$L = -\frac{\ln(1 - \Theta)\varphi^{1/2}}{awpt} \tag{17.48}$$

It is clear that the increase in the porosity factor produces a decrease in the required value of L to reach a maximum value of Θ (Figure 17.8). From the interstitial speed, w, we can obtain the residence

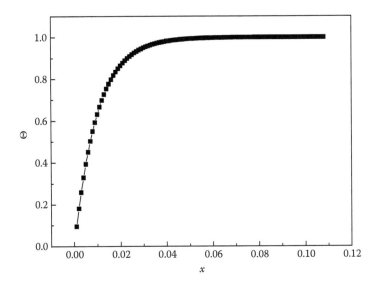

FIGURE 17.8 Hypothetical plot between the conversion factor, Θ, as a function of the distance, x, along which the electrolyte flows at a constant volumetric flow, φ.

time, τ, that is, the time taken by the electrolyte to flow through the electrode. This is defined clearly in most common chemical engineering books as [15,16]:

$$\tau = \frac{Lp}{w} = -\frac{\ln(1-\Theta)p\varphi^{1/2}}{awL} \tag{17.49}$$

A complete analysis of this is very difficult to achieve, since w and φ are both dependent on time. Therefore, we can use the value of $K_{M,Ox}$ as

$$\tau = -\frac{\ln(1-\Theta)pK_{M,Ox}}{a^2wL} \tag{17.50}$$

To achieve the desired value of Θ, an increase in porosity produces larger values of τ. In addition, the increase of w produces a lower τ. In order to attain an appropriate Θ, better magnitudes of $K_{M,Ox}$ are needed (that is adequate φ values).

These equations become more complex when the limitations due to the charge transfer kinetics arise, and other overpotentials, such as that of the bubble formation/break-off and the crystal deposition/reconstructions have to be considered. In this chapter, some typical cases were considered, however, the resolution of the other equations (Nernst–Planck equations) with their continuity and the corresponding Navier–Stokes expressions have to be evaluated carefully for each real situation.

REFERENCES

1. G.B. Tatterson, *Fluid Mixing and Gas Dispersion in Agitated Tanks*, McGraw-Hill, New York, 1991.
2. V.V. Ranade and J.B. Joshi, *Chem. Eng. Res. Des.* **68** (1990) 19.
3. K. Rutherford, K.C. Lee, S.M.S. Mahomoudi, and M. Yianneskis, *Chem. Eng. Res. Des.* **74** (1996) 369.
4. M. Schafer, M. Hofken, and F. Durst, *Chem. Eng. Res. Des.* **75** (1997) 729.
5. E.S. Wernersson and C. Trägardh, *Chem. Eng. J.* **70** (1998) 37.
6. S. Baldi and M. Yianneskis, *Chem. Eng. Sci.* **59** (2004) 2659.
7. P.R. Gogate, A.A.C.M. Beenackers, and A.B. Pandit, *Biochem. Eng. J.* **6** (2000) 109.

8. D. García-Cortés, E. Ferrer, and E. Barbera, *Chem. Eng. Res. Des.* **79** (2001) 269.
9. C. Galletti, E. Brunazzi, M. Yianneskis, and A. Paglianti, *Chem. Eng. Sci.* **58** (2003) 3859.
10. K. Rutherford, K.C. Lee, S.M.S. Mahmoudi, and M. Yianneskis, Hydrodynamics characteristics of dual Rushton impeller stirred vessels. *AIChE J.* **42** (1996) 332–346.
11. Y.Q. Cui, R.G.J.M. van der Lans, and K.Ch.A.M. Luyben, *Chem. Eng. Sci.* **51** (1996) 2631.
12. J.H. Rushton, E.W. Costich, and H.J. Everett, *Chem. Eng. Prog.* **46** (1950) 395.
13. P.H. Calderbank and M.B. Moo-Young, *Trans. Inst. Chem. Eng.* **39** (1961) 337.
14. A.N. Strohl and D.J. Curran, *Anal. Chem.* **51** (1979) 353.
15. D. Himmelbau and J. Riggs, *Basic Principles and Calculations in Chemical Engineering,* 7th edn., Prentice Hall Int., Upper Saddle River, NJ, 2003.
16. O. Levenspiel, *Chemical Reaction Engineering*, 3rd edn., John Wiley & Sons, Hoboken, NJ, 1998, Part I, Chapter 4, Part II, Chapters 11 and 15.

18 Electrocatalytic Reactor Design

Shriram Santhanagopalan and Ralph E. White

CONTENTS

18.1 INTRODUCTION

Reactor design is a crucial part of plant engineering in any chemical industry. The chemical reactor constitutes the core of the plant, and other parts of the system are designed so as to enable a maximum yield from the reactor. For example, the heat exchangers are designed to provide the ambient temperature for maximizing the rate of the reaction, the flow ducts are designed to ensure maximum conversion based on the reaction rate, separation devices operate at conditions that maximize selectivity, and the valve designs ensure that the pressure within the reactor is ideal for the reaction to proceed at a reasonable rate. Hence the design of a chemical reactor is of atmost importance in ensuring optimum performance of the plant. At the same time, constraints exist in terms of practical limits, for example, when designing packed beds when a specific pressure may be ideal for the reaction, this may result in channeling effects; or the packing size may not be practical. Another typical example is the case when a higher temperature favors the reaction rate, whereas the decomposition temperature of the product may be the limiting factor. The distribution of temperature along the reactor is another aspect of concern in catalytic reactors.

The promotion of catalytic reactions by the application of an electrochemical potential is now widely recognized as a viable industrial practice. The reduction of oxygen, the dimerization of acetonitrile, the synthesis of sorbitol, the detection and the monitoring of pollutants and contaminants, and the hydrogenation of organic compounds are a few examples of commercial processes employing electrocatalysis. In some cases, the electrocatalytic route presents specific advantages over other alternate paths because of product selectivity or the ease of operation. Figure 18.1 presents a flow chart illustrating the key role played by the reactor in a plant synthesizing sorbitol from fructose using a platinum/rhodium catalyst loaded on a membrane. This process offers several advantages—for example, the electrocatalytic route is commercially viable due to operability at

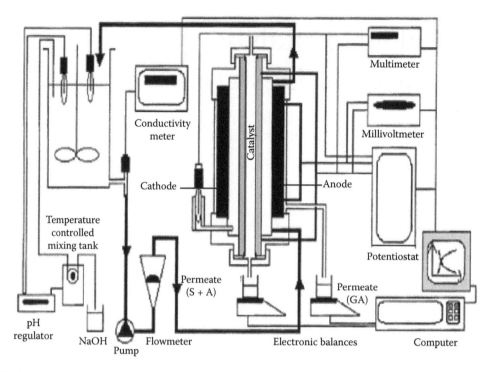

FIGURE 18.1 An electrocatalytic plant for synthesizing sorbitol (S), mannitol (M), and gluconic acid (GA).

ambient temperatures and atmospheric pressure; an alternate route for the synthesis of sorbitol via organic reactions involves temperatures as high as 130°C–160°C and pressures well above 50 bar. Also, in the electrocatalytic process, the oxidation products (sorbitol and mannitol) as well as the reduction product (gluconic acid) are obtained simultaneously, using the same feed (fructose) from the cathode and anode compartments of the reactor, respectively.

18.2 PRINCIPLES OF ELECTROCATALYTIC REACTOR DESIGN

The design of an electrocatalytic reactor draws most principles from conventional reactor design. In addition to the challenges one faces in catalytic reactor design, which does not involve an electric field, an electrocatalytic reaction involves an additional parameter as discussed earlier: the voltage. The extent to which the reaction is promoted depends largely on the activation of the catalyst surface by the electrochemical potential. The catalytic activity for an electrochemical reaction depends on the potential in addition to factors such as temperature and pH, which makes the optimization of the operating conditions more challenging than for a conventional reactor. However, the availability of an additional parameter can be leveraged to our advantage by appropriate design, to enhance the selectivity and/or yield of the product of interest.

Catalytic reactor design is usually classified into three different phases: the mechanistic design, the catalyst design, and the reactor design. The first phase of design is centered on the reaction mechanism: the adsorption of the reactants or products on the catalyst particles is a commonly studied problem at this phase. The dependence of the catalytic activity or the isotherms on the temperature and the competitive adsorption of multiple species have been studied extensively. Competing reaction pathways present another example of design at the microscopic phase: the design of a catalyst with activities favoring the pathway that results in the highest yield for the desired products is a commonly encountered objective. As one may realize, most of this phase focuses on the kinetic aspects of the reactor design. This chapter will discuss some aspects of design at this phase; however, a few other aspects like catalyst design by molecular level simulations are beyond the scope of this chapter. The second aspect of designing a catalytic reactor is the design of the catalyst particle in itself. Typical examples of problems that one focuses at this phase include the effectiveness of the catalyst particles, the diffusion of the reactants through the porous catalyst particles, and the optimization of the surface area versus the increase in temperature due to rapid reactivity. In the context of an electrocatalytic reactor, a significant potential drop across the catalyst pellet leads to ohmic losses, and hence must be minimized. The distribution of current density on the surface of the particles depends on the shape and the size of the particle. Sharp edges on the particles typically resulting from the cracking of particles due to pressure or thermal stresses lead to a leakage of the electrical flux resulting in suboptimal performance of the reactor. Most aspects at the second phase are based on the geometry and the dimensional stability of the catalyst pellets. A third phase of design focuses on the reactor at the macro-level. A typical example is balance between the packing fraction and the pressure profile across the reactor: a larger packing fraction results in more available area for reaction on the surface of the particles until a threshold is reached, when the flow rate of the reactants is reduced by a buildup of pressure within the reactor. The channeling of the electrolyte due to excess buildup of pressure leads to the isolation of the reaction zones leading to lower yields. Potential distribution across the reactor largely governs the selectivity of the process; hence, ensuring that the potential distribution across the reactor is fairly uniform is critical for the performance of the reactor. Also, one must ensure that the overpotential loss caused by lower conductivity or poor contact between particles across the bed is minimized. Ensuring a uniform distribution of the temperature (both rapid removal of heat during an exothermal reaction as well as supplying adequate heat for maintaining a viable rate of reaction) and adequate mass transport constitutes commonly encountered problems addressed at this phase of reactor design. Figure 18.2 summarizes the various aspects of reactor design discussed earlier.

The following sections focus on each of these phases in more detail. In order to elaborate on the different aspects of reactor design, examples are included to emphasize design problems at each level.

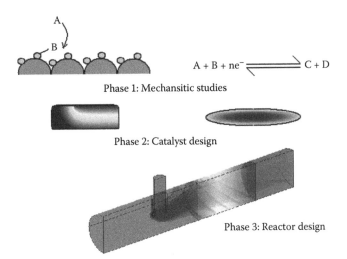

FIGURE 18.2 Various phases of catalytic reactor design: the design of an electrocatalytic reactor begins with the investigation of the reaction mechanism, extends to factors like temperature and potential distribution within a pellet, and then to the design of the actual reactor.

18.3 MECHANISTIC DESIGN

Mechanistic design is used to identify the limiting factors in a reaction sequence. For example, any catalytic reaction usually involves the following steps: the transfer of the reacting species to the surface of the catalyst, adsorption on the active sites, reaction, the desorption of the products, and the subsequent transfer of the products to the bulk of the electrolyte. One or more of these steps is the limiting factor in ensuring appropriate yield of the desired product. An estimate of the transport and reaction coefficients obtained, by comparing experimental data with a kinetic model provides insight into the limiting step for a reaction. Another interesting aspect is the prevalence of competing reaction pathways. Usually, several mechanisms are put forward for an electrochemical reaction in practice. These pathways may actually coexist. A mechanistic model for such a scenario helps identify the energetically favored path under a given set of operating conditions.

Let us consider the oxygen reduction reaction (ORR) that occurs in the cathode of the polymer electrolyte membrane fuel cell (PEMFC), in an acidic environment. Although a variety of ORR mechanisms have been proposed, the four-electron pathway is primarily used to characterize the behavior of this reaction at a platinum electrode or a glassy carbon electrode coated with a platinum-based catalyst. The overall reaction is given by

$$O_2 + 4H^+ + 4e^- \rightarrow 2H_2O \tag{I}$$

A rotating disc electrode setup is commonly used in mechanistic studies for reactions of this type. When the potential at the working electrode is changed in small sequential steps from 0 to 1.0 V versus standard hydrogen electrode (SHE) and a logarithmic plot of the measured reaction current versus the applied potential is made, a change in the Tafel slopes is observed at about 0.85 V (see inset of Figure 18.3). In order to explain the change in the slope of the current–potential curve, the following two-step mechanism is proposed for the ORR:

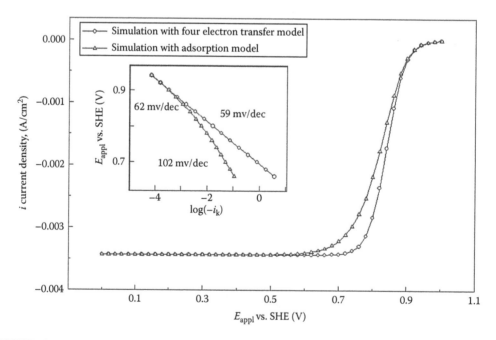

FIGURE 18.3 Polarization curves comparing the one-step electrochemical reaction model with the adsorption model for the ORR in acidic media: note that the adsorption model captures the change in the slope of the current–voltage curve as observed in experiments.

$$O_{2,sur} \leftrightarrow O_{2,ads} \tag{II}$$

$$O_{2,ads} + 4H^+ + 4e^- \longrightarrow 2H_2O \tag{III}$$

In the first step [Reaction (II)], the oxygen molecules adsorb on to the surface of the catalyst and during the second step, the actual electrochemical reduction reaction takes place between the adsorbed oxygen molecules and the protons. Let us now derive the kinetic expression relating the current to the overpotential for this adsorption mechanism.

The following assumptions are made in the process of this derivation: there are no double-layer effects and the adsorption process is fast; the Langmuir isotherm is applicable; the adsorbed oxygen at the electrode surface $c_{O_2,ads}$ occupies only one monolayer on the electrode surface and its concentration is proportional to the fractional coverage (θ) of the adsorbed oxygen at the electrode surface; the fractional coverage of oxygen on the surface of the electrode in all the situations encountered in this work is much smaller than one; the concentration of the solvent (water) is constant; the protons do not undergo an adsorption process.

The adsorption mechanism consists of an adsorption step (II), and a charge-transfer step (Reaction III) as shown in reaction mechanism (II). The adsorption step shown in Reaction (II) is a chemical reaction rather than an electrochemical reaction; however, in order to derive the kinetic expression for mechanism (II), it is convenient to express the rate equation in a form similar to that of an electrochemical reaction:

$$i_a = F \frac{dc_{O_2,ads}}{dt} = -\left[kc_{O_{2,0}}(1 - \theta) - k''\theta \right] \tag{18.1}$$

where
   i_a is the rate of adsorption of oxygen at the electrode surface
   θ is the fraction of the electrode surface covered by the adsorbed oxygen

The parameter is not measured at equilibrium value and hence cannot be obtained from the adsorption isotherms. In order to obtain an expression for the current density for the mechanism proposed in Reaction (II), we eliminate the variable θ between the rate expressions for the individual steps as shown in the following steps.

The first step is to relate the rate of adsorption to the surface coverage at equilibrium. When the adsorption process is in equilibrium with the solution immediately adjacent to the electrode surface, the adsorption rate i_a should be 0 (since the rate of adsorption equals the rate of desorption, the net change in $c_{O_2,ads}$ is 0):

$$0 = kc_{O_{2,0}}(1 - \theta_0) - k''\theta_0 \tag{18.2}$$

where θ_0 is the surface coverage under equilibrium conditions. Solving for k'', we have

$$k'' = \frac{kc_{O_{2,0}}(1 - \theta_0)}{\theta_0} \tag{18.3}$$

Substituting Equation 18.3 into Equation 18.1, we have

$$i_a = -\left[kc_{O_{2,0}}(1 - \theta) - \frac{kc_{O_{2,0}}(1 - \theta_0)}{\theta_0}\theta \right] = kc_{O_{2,0}} \frac{\theta - \theta_0}{\theta_0} \tag{18.4}$$

If we define the rate constant for the adsorption reaction ($i'_{0,a}$) as follows

$$i'_{0,a} = kc_{O_2,0} \tag{18.5}$$

we have

$$i_a = i'_{0,a} \frac{\theta - \theta_0}{\theta_0} \tag{18.6}$$

For the charge-transfer step occurring at the surface of the electrode as shown in Reaction (III), the current density can be expressed according to the Butler–Volmer equation:

$$i_1 = k''_{a1} c^2_{H_2O} \exp\left(\frac{\alpha_a F}{RT} V\right) - k'_{c1} c_{O_2,ads} c^4_{H^+} \exp\left(-\frac{\alpha_c F}{RT} V\right) \tag{18.7}$$

The electrode potential V is given by

$$V = \Phi_{met} - \Phi_0 \tag{18.8}$$

where both Φ_{met} and Φ_0 are measured with respect to the same reference electrode. Since the adsorption reaction occurs only at that fraction of the electrode covered by the active sites (θ), and the reverse reaction occurs only at the surface not covered by oxygen, we need to update Equation (18.7) by multiplying the anodic part with $(1 - \theta)$ and the cathodic part by θ:

$$i_1 = (1 - \theta) k''_{a1} c^2_{H_2O} \exp\left(\frac{\alpha_a F}{RT} V\right) - \theta k'_{c1} c_{O_2,ads} c^4_{H^+} \exp\left(-\frac{\alpha_c F}{RT} V\right) \tag{18.9}$$

The concentration of the adsorbed oxygen ($c_{O_2,ads}$) can be expressed in terms of θ. Since it is assumed that the adsorbed oxygen occupies a monolayer at the surface of the electrode, $c_{O_2,ads}$ is linearly related to θ, that is, $c_{O_2,ads}$ is related to θ by the following expression:

$$c_{O_2,ads} = k_{ads} \theta \tag{18.10}$$

where k_{ads} is a proportionality constant. The rate expression Equation 18.9 can be rewritten in terms of θ as follows:

$$i_1 = k'_{a1}(1 - \theta) \exp\left(\frac{\alpha_a F}{RT} V\right) - k_{c1} \theta^2 c^4_{H^+} \exp\left(-\frac{\alpha_c F}{RT} V\right) \tag{18.11}$$

where k'_{a1} and k'_{c1} are the new rate constants given by

$$k'_{a1} = k''_{a1} c^2_{H_2O} \tag{18.12}$$

$$k_{c1} = k_{ads} k'_{c1} \tag{18.13}$$

When the electrode is in equilibrium with the solution adjacent to the electrode surface, the net current density i_1 should be 0 in Equation 18.11:

$$0 = k'_{a1}(1 - \theta_0) \exp\left(\frac{\alpha_a F}{RT} V'_{0,1}\right) - k_{c1} \theta_0^2 c^4_{H^+} \exp\left(-\frac{\alpha_c F}{RT} V'_{0,1}\right) \tag{18.14}$$

where
$\quad \theta_0$ is the surface coverage at equilibrium
$\quad V'_{0,1}$ is the corresponding electrode potential

Using Equation 18.14 to define the equilibrium exchange current density $i'_{0,1}$, we have

$$i'_{0,1} = k'_{a1}(1 - \theta_0) \exp\left(\frac{\alpha_a F}{RT} V'_{0,1}\right) = k_{c1} \theta_0^2 c_{H^+}^4 \exp\left(-\frac{\alpha_c F}{RT} V'_{0,1}\right) \tag{18.15}$$

Rewriting Equation 18.11 in terms of $i'_{0,1}$, we have

$$i_1 = i'_{0,1}\left[\frac{1 - \theta}{1 - \theta_0} \exp\left(\frac{\alpha_a F}{RT}\eta\right) - \left(\frac{\theta}{\theta_0}\right)^2 \exp\left(-\frac{\alpha_c F}{RT}\eta\right)\right] \tag{18.16}$$

where

$$i'_{0,1} = k'^{\alpha_c/4}_{a1} k^{\alpha_a/4}_{c1} c^{\alpha_a}_{H^+} (1 - \theta_0)^{\alpha_c/4} \theta_0^{\alpha_a/2} \tag{18.17}$$

$$V'_{0,1} = \frac{RT}{4F} \ln\left(\frac{k_{c1}}{k'_{a1}}\right) + \frac{RT}{4F} \ln\left(\frac{\theta_0^2 c_{H^+}^4}{1 - \theta_0}\right) \tag{18.18}$$

$$\eta = V - V'_{0,1} \tag{18.19}$$

Since we have assumed that there are no double-layer effects and that the adsorption of oxygen to the electrode surface is fast, at steady state, the rate of the adsorption step should be the same as the charge-transfer step and the same as the total current density across the cell (i):

$$i = i_a = i_1 \tag{18.20}$$

This relationship can be used to eliminate θ. From Equation 18.16, solving for θ we have

$$\theta = \theta_0 \sqrt{\frac{\exp\left(\dfrac{\alpha_a F}{RT}\eta\right) - \dfrac{i_1}{i'_{0,1}}}{\exp\left(-\dfrac{\alpha_c F}{RT}\eta\right)}} \tag{18.21}$$

Substituting Equation 18.20 and the expression for θ (i.e., Equation 18.21) into the rate expression for the adsorption step (i.e., Equation 18.6), we have

$$i = i'_{0,a}\left(\sqrt{\frac{\exp\left(\dfrac{\alpha_a F}{RT}\eta\right) - \dfrac{i}{i'_{0,1}}}{\exp\left(-\dfrac{\alpha_c F}{RT}\eta\right)}} - 1\right) \tag{18.22}$$

Solving for the current density across the cell (i) gives

$$i = \frac{1}{2}\left(-2i'_{0,a} - X + \sqrt{X^2 + 4i'_{0,a}X + 4i'_{0,1}X \exp\left(\frac{\alpha_a F}{RT}\eta\right)}\right) \tag{18.23}$$

where

$$X = \frac{i_{0,a}'^2}{i_{0,1}' \exp\left(-\dfrac{\alpha_c F}{RT}\eta\right)} \tag{18.24}$$

Since the surface coverage at all times was assumed to be much smaller than 1, the following approximation was made in arriving at Equations 18.21 through 18.24:

$$\frac{1-\theta}{1-\theta_0} \approx 1 \tag{18.25}$$

In Equations 18.23 and 18.24, η is given by Equation 18.19, $i_{0,a}'$ is given by Equation 18.5, and $i_{0,1}'$ is given by Equation 18.17.

Thus we have an expression relating the current density (i) to the cell potential (V) in terms of the equilibrium surface coverage (θ_0) and the concentration of oxygen at the surface of the electrode ($c_{O_2,0}$). To complete the derivation, Langmuir's isotherm for chemisorption is introduced to relate the equilibrium surface coverage to the concentration of the adsorbed species:

$$\theta_0 = \frac{c_{O_2,0} \exp\left(-\dfrac{\Delta G^0}{RT}\right)}{1 + c_{O_2,0} \exp\left(-\dfrac{\Delta G^0}{RT}\right)} \tag{18.26}$$

where ΔG^0 is the Gibbs free energy change for the adsorption process. Since we have assumed that $\theta \ll 1$ under all conditions, this implies that the surface adsorption is not energetically favorable or in other words, the ΔG^0 value for this reaction is a large positive value. Hence, we have

$$\theta_0 \approx c_{O_2,0} \exp\left(-\frac{\Delta G^0}{RT}\right) \tag{18.27}$$

Equations 18.5 and 18.17 used to calculate $i_{0,a}'$ and $i_{0,1}'$ contain terms that are dependent on the concentration at the surface of the electrode. It is convenient to define these quantities in terms of the exchange current densities defined at reference conditions:

$$i_{a,\text{ref}} = k c_{O_2,\text{ref}} \tag{18.28}$$

$$i_{0,\text{ref}} = k_{a1}'^{\alpha_c/4} k_{c1}^{\alpha_a/4} c_{H^+,\text{ref}}^{\alpha_a} (1-\theta_{0,\text{ref}})^{\alpha_c/4} \theta_{0,\text{ref}}^{\alpha_a/2} \tag{18.29}$$

Here, $\theta_{0,\text{ref}}$ is the fractional coverage of oxygen with respect to a reference solution that is practically chosen to be 0.5 M H_2SO_4 solution saturated with oxygen. Since the exponent in Equation 18.27 is constant at a given temperature, we have

$$\frac{\theta_0}{\theta_{0,\text{ref}}} \approx \frac{c_0}{c_{0,\text{ref}}} \tag{18.30}$$

Therefore, Equations 18.5 and 18.17 become

$$i_{0,a}' = i_{a,\text{ref}} \frac{c_{O_2,0}}{c_{O_2,\text{ref}}} \tag{18.31}$$

$$i'_{0,1} = i_{0,\text{ref}} \left(\frac{c_{H^+}}{c_{H^+,\text{ref}}} \right)^{\alpha_a} \left(\frac{c_{O_2,0}}{c_{O_2,\text{ref}}} \right)^{\alpha_a/2} \tag{18.32}$$

Once again, in obtaining Equation 18.32, the approximation shown in Equation 18.25 is used.

The expression for overpotential in terms of the reference concentrations is given by Equations 18.17 through 18.19:

$$\eta = (\Phi_{\text{met}} - \Phi_{\text{RE}}) - (\Phi_0 - \Phi_{\text{RE}}) - U_{\text{ref}} + \frac{RT}{nF} \ln \left(\left(\frac{c_{O_2,0}}{c_{O_2,\text{ref}}} \right)^{\frac{\alpha_a}{2}} \left(\frac{c_{H^+,0}}{c_{H^+,\text{ref}}} \right)^{\alpha_a} \right) \tag{18.33}$$

Substituting Equation 18.8 into Equation 18.33, we have

$$\eta = V - U_{\text{ref}} + \frac{RT}{nF} \ln \left(\left(\frac{c_{O_2,0}}{c_{O_2,\text{ref}}} \right)^{\frac{\alpha_a}{2}} \left(\frac{c_{H^+,0}}{c_{H^+,\text{ref}}} \right)^{\alpha_a} \right) \tag{18.34}$$

where U_{ref} is the equilibrium potential with respect to SHE.

Thus, the final expression relating the current density (i) to the cell potential (V) is given by

$$i = i_{a,\text{ref}} \frac{c_{O_2,0}}{c_{O_2,\text{ref}}} \left(\frac{-X_{\text{ref}} + \sqrt{X_{\text{ref}}^2 + 4 \exp\left(-\frac{\alpha_c F}{RT} \eta \right) \left(X_{\text{ref}} + \exp\left(\frac{\alpha_a F}{RT} \eta \right) \right)}}{2 \exp\left(-\frac{\alpha_c F}{RT} \eta \right)} - 1 \right) \tag{18.35}$$

$$X_{\text{ref}} = \frac{i_{a,\text{ref}} \dfrac{c_{O_2,0}}{c_{O_2,\text{ref}}}}{i_{0,\text{ref}} \left(\dfrac{c_{H^+}}{c_{H^+,\text{ref}}} \right)^{\alpha_a} \left(\dfrac{c_{O_2,0}}{c_{O_2,\text{ref}}} \right)^{\alpha_a/2}} \tag{18.36}$$

Equations 18.34 through 18.36 are more convenient to use since the exchange current densities and the overpotential are evaluated at the reference concentrations as opposed to the surface concentrations. An immediate implication of using reference concentrations is that the exchange current density values can be obtained experimentally at a particular reference condition and the values at other conditions can be calculated readily using Equations 18.31 and 18.32. The exchange current density of the adsorption step, $i'_{0,a}$, is linearly proportional to the concentration of oxygen adjacent to the surface of the electrode $c_{O_2,0}$. The exchange current density for a one-step mechanism involving no adsorption step as intermediate has the following form for the exchange current density (i_0):

$$i_0 = i_{0,\text{ref}} \left(\frac{c_{H^+}}{c_{H^+,\text{ref}}} \right) \left(\frac{c_{O_2,0}}{c_{O_2,\text{ref}}} \right)^{3.0} \tag{18.37}$$

when $\alpha_a = 1.0$. Note that the exchange current density for the charge-transfer step ($i'_{0,1}$) in Reaction (III) has almost the same form as i_0 in Equation 18.37 except that the exponent of the concentration of oxygen is different, which is due to the difference in the reaction mechanism; and the overpotential η defined in Equation 18.33 is similar to that for a one-step charge-transfer reaction (Reaction I) but with a different exponent for the oxygen concentration term.

The actual mechanism for oxygen reduction may involve more complicated steps and usually includes competitive adsorption of multiple species and multiple reaction pathways. Figure 18.4 shows the different competing pathways for the ORR. Under these circumstances, a generalized extension of the model presented earlier can be used, provided the experimental parameters such as

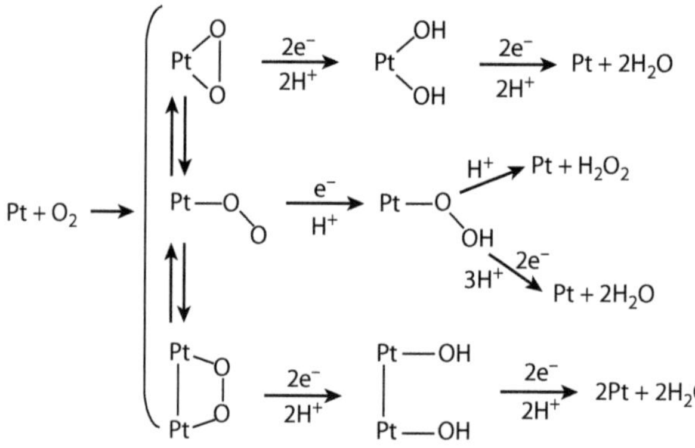

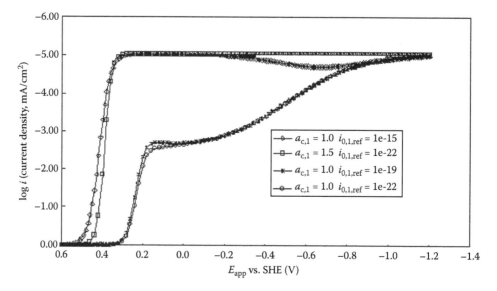

FIGURE 18.4 Various competing pathways for the ORR on a platinum catalyst surface.

FIGURE 18.5 Polarization curves for competing reaction pathways for the ORR. The first two reactions shown in Figure 18.4 were simulated; no regeneration of peroxide was considered in the simulations.

adsorption isotherms are readily available. The solution of the set of equations for the more complicated scenario usually involves intense numerical computations. Figure 18.5 shows a set of polarization curves for different sets of kinetic parameters. The first two reaction pathways shown in Figure 18.4 were considered. The exchange current density for the second reaction was fixed at 10^{-12} mA cm^{-2} and the transfer coefficient for this reaction was held constant at 1.2. As shown in Figure 18.5, for cases when the exchange current densities for the competing reaction producing peroxide are comparable, the limiting current density sets in earlier when the transfer coefficients for the two reactions are not very different. However, this trend is reversed when the transfer coefficient for the first reaction is increased to 1.5.

18.4 CATALYST DESIGN

Once the mechanism of the reaction of interest has been investigated and the optimal operating conditions have been arrived at, the next step is to design a suitable geometry for the catalyst. Several important aspects go into the design of the catalyst: the surface area of the catalyst is directly proportional to the extent of reaction in many cases. Hence it is intuitive to use as small a particle size as possible; however, such a choice may introduce handling issues for the catalyst. In an exothermic catalytic reactor, it is desirable to keep the reaction temperatures under control, while maximizing the reaction rate. Minimizing any ohmic drop within the pellet ensues as an additional constraint in the presence of an electric field to ensure uniform current distribution within the reactor. Fick's second law gives the material balance equation that governs the concentration distribution:

$$\frac{\partial c}{\partial t} = \nabla \cdot (D \nabla c) \tag{18.38}$$

In addition to the material balance Equation 18.38 that one normally encounters in design problems involving chemical reactions with no charge transfer, for the case of an electrochemical reaction, one must consider a charge balance equation that governs the distribution of the electric potential across the particle or the pores. For the particle in itself, such a conservation equation is usually presented in the form of Ohm's law given as follows:

$$\nabla \cdot i = -\sigma_e \nabla^2 \phi \tag{18.39}$$

where
  i is the applied current density
  σ_e is the electronic conductivity across the particle
  $\nabla \phi$ signifies the potential drop across the domain

In the representation given earlier, isotropic behavior of the properties is assumed. This is a good design requirement for the aspect ratios used in current industrial practice. In some cases, a composite pellet may be used, wherein an inactive core may be coated with a layer of the catalyst. Under these circumstances, the regime of interest is restricted to the thin active layer.

Let us consider the potential and concentration distribution across a cylindrical catalyst pellet. The geometry is set up as shown in Figure 18.6. The electrolyte flows along the axial direction and the charge-transfer reaction takes place on the surface of the catalyst pellet.

For this case, the Equations 18.38 and 18.39 can be rewritten in the dimensionless form. The dimensionless concentration C is then given as follows:

$$\frac{\partial C}{\partial \tau} = \frac{\partial^2 C}{\partial X^2} + \frac{1}{Y} \frac{\partial}{\partial Y} \left(Y \frac{\partial C}{\partial Y} \right) \tag{18.40}$$

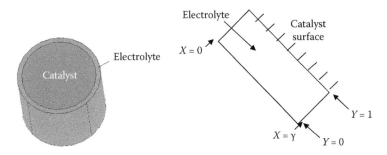

FIGURE 18.6 Concentration and potential distribution within the electrolyte undergoing an electrochemical reaction on a cylindrical catalyst pellet: the definition of the geometric coordinates.

where Y is the radial direction and X is the axial direction as shown in Figure 18.6. Similarly, the potential distribution is governed by the following equation:

$$\frac{\partial^2 E}{\partial X^2} + \frac{1}{Y}\frac{\partial}{\partial Y}\left(Y\frac{\partial E}{\partial Y}\right) = 0 \tag{18.41}$$

The initial and boundary conditions for Equations 18.40 and 18.41 are described next.

The initial (i.e., at $\tau = 0$) dimensionless concentration is set equal to 1. The concentration at the point of entry ($X = 0$), is related to the bulk concentration by means of a mass-transfer coefficient. The resulting equation in the dimensionless form is as follows:

$$\left.\frac{\partial C}{\partial Y}\right|_{X=0,Y} = \frac{4}{\pi}(C-1) \tag{18.42}$$

At the exit, as well as at distances far away from the surface, the concentration is uniform:

$$\left.\frac{\partial C}{\partial X}\right|_{X=\gamma} = 0, \quad \left.\frac{\partial C}{\partial Y}\right|_{Y=0} = 0 \tag{18.43}$$

On the surface of the catalyst particle, a Butler–Volmer type electrochemical reaction takes place. The flux of the ions equals the rate of reaction as given by a jump material balance at the interface:

$$\left.\frac{\partial C}{\partial Y}\right|_{Y=1} = \sqrt{\sigma}I^* \tag{18.44}$$

where I^* is the dimensionless applied current density. The current density is coupled with the potential according to the following kinetic expression:

$$I^* = \Lambda\left[C_W e^{-\alpha E_W} - \left(\frac{1+\xi-\xi C_W}{\xi}\right)e^{(1-\alpha)E_W}\right] \tag{18.45}$$

In this expression, α represents the transfer coefficient (assumed to be equal to 0.5) and ξ is the dimensionless initial concentration (in this example equal to 1.0). The symbols C_W and E_W represent the concentration and the potential at the catalyst wall, respectively. The boundary conditions for the potential include a reference value at $X = 0$, and Ohm's law at the catalyst wall. These expressions along with the continuity of potential at distances far from the reaction site are represented as follows:

$$E|_{X=0} = \ln\xi - \sigma\tau \quad \left.\frac{\partial E}{\partial X}\right|_{X=\gamma} = 0 \tag{18.46a,b}$$

$$\left.\frac{\partial E}{\partial Y}\right|_{Y=0} = 0 \quad \left.\frac{\partial E}{\partial Y}\right|_{Y=1} = \left(\frac{\theta}{\gamma^2}\right)I^* \tag{18.47a,b}$$

These set of equations present a simple albeit interesting case study. The parameter Λ is the ratio between the resistance to mass transport and the resistance from the reaction kinetics. At the limiting case, when the reaction is extremely slow, Λ approaches a value close to 0. Similarly, for the case when the mass-transfer resistance reaches extremely large values, the value for this parameter approaches infinity. The parameter θ represents the ratio between the ohmic resistance and the mass-transfer resistance. Hence it compares the contributions of the gradient in the potential versus the gradient in concentration toward the movement of the ions. At very high conductivity values, this parameter approaches values close to 0. However, if the electrolyte (or the particle) is not highly conductive, a significant part of the resistance arises from the ohmic drop and the parameter θ assumes values of comparable magnitude. The parameter σ is the ratio between the diffusion time and the time for the potential sweep. It compares the effect of processes of varying time constants coupled with one another, on the movement of the ions. In mechanistic studies, choosing the appropriate value for this parameter is critical to interpret experimental results, for example, from voltammetric curves. In order to understand fast reactions, sufficiently low sweep rates must be used. The parameter σ can in such cases be used as a good measure for the choice of experimental conditions. At the same time, models can be simplified, for example, if the diffusion time lengths are of much larger timescales in comparison to the time regime of interest, a pseudo-steady-state approximation can be used to decouple the mass-transfer equations. The parameter γ is the ratio between the diameter of the reaction regime and the length along the axial surface. This is a geometric parameter and is important at the design phase of the catalyst pellets. Several limiting cases can be obtained by studying the dependence of the concentration and/or potential profiles on the values for this parameter (γ). In the conventional chemical reactor design process, this parameter is used to study the influence of axial dispersion on the reaction rate. For smaller values of γ, the concentration or potential may be assumed to be uniform along the radius. This case corresponds to the case of thin film diffusion along the surface of the catalyst pellet. While it simplifies the rigors of the mathematical solution for the above set of equations by eliminating any dependence on the Y coordinate, the physical significance of this process is denoted as the effective reaction zone from the surface of the pellet, that is, this parameter characterizes how far out from the surface of the catalyst pellet are the gradients in the concentration and the potential significant. Thus for the case of an electrolyte in which the ions move fast enough, the packing fraction for the catalyst pellets can be reduced accordingly. However, for the case with slow kinetics or if the movement of ions to and/or from the surface is much slower, then one must implement as thin a reaction zone as practically feasible.

Figure 18.7 presents the conversion and potential profiles in the vicinity of the pellet. As observed, the potential levels off at the boundary far from the reacting surface (i.e., for any given X, close to $Y = 0$ the cell voltage reaches a constant value). Also, as the electrolyte flows toward the end of the reaction zone, the gradients in the voltage taper off. Essentially, at the exit point, the concentration distribution is influenced solely by the flow profile, if any. The concentration of the species generated at the surface of the catalyst pellet shows the maximum gradient at the surface. As the ions move farther into the electrolyte away from the reacting surface, driven by the concentration gradients as

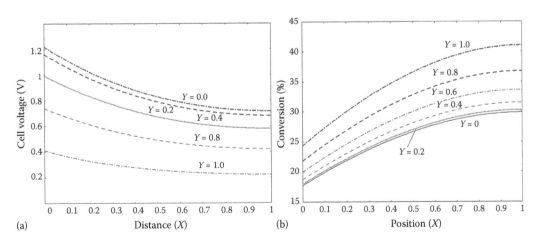

(a)   (b)

FIGURE 18.7 Potential and concentration distribution in the vicinity of a cylindrical catalyst pellet: $Y = 1.0$ corresponds to the surface of the pellet and $Y = 0.0$ corresponds to the bulk electrolyte. $X = 0$ is the initial point of contact of the incoming stream to the catalyst and $X = 1$ is the point of exit.

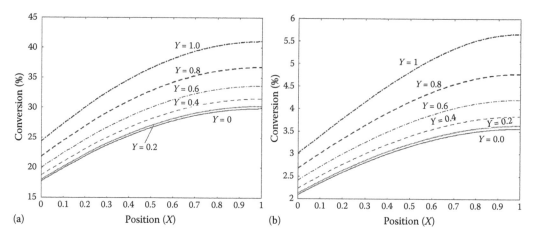

FIGURE 18.8 Comparison of conversions for two cases: (a) ohmic drop in the overpotential is comparable to the mass-transfer polarization effects and (b) ohmic resistance is far greater than the mass-transfer resistance.

well as the potential, the flux is reduced gradually due to the limited zone of influence for the potential. Figure 18.8 compares two cases with different conductivity values for the electrolyte. In Figure 18.8a, the value for θ is set to 0.01, corresponding to an excellent conductivity for the electrolyte. As a result of the enhanced transport of ions, the reaction is driven forward, leading to a theoretical maximum of about 42% conversion. In Figure 18.8b, the parameter θ is set to 400, corresponding to a poor conductor. As a result, for this case, the conversion is about 5%–6%. Figure 18.9 presents the electric field superimposed with the concentration gradients for the case discussed in Figure 18.8b. The

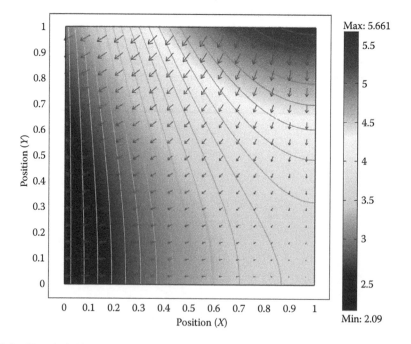

FIGURE 18.9 Electric field and concentration gradients for an electrochemical reaction at a catalyst surface: the contours indicate the isopotential surfaces in the electrolyte and the arrow marks indicate the flux of the species generated at the catalyst surface (i.e., $Y = 1$). The scale indicates the conversion. Parameters correspond to the case shown in Figure 18.8b.

contours indicate the isopotential surfaces and the arrows indicate the flux of the species generated. The color scale corresponds to the conversion shown in Figure 18.8b.

As observed for this case, the potential contours remain vertical for most part indicating that the potential gradient does not aid the reaction on the surface of the pellet. Toward the point of exit, the increase in the concentration gradients allows for the buildup of a stronger field.

The analysis presented earlier is readily extendable to the case of a parallel plate or plug flow reactor, where the walls are coated with the catalytic material and the electrolyte flows between the electrodes.

18.5 EFFECT OF TEMPERATURE

Thermal effects constitute a significant portion of the study devoted to catalysis. This is true of electrochemical reactions as well. In general the reaction rate constants, diffusion coefficients, and conductivities all exhibit Arrhenius-type dependence on temperature, and as a rule of the thumb, for every 10°C rise in temperature, most reaction rates are doubled. Hence, temperature effects must be incorporated into the parameter values. Fourier's law governs the distribution of temperature. For the example with the cylindrical catalyst pellet described in the previous section, the equation corresponding to the energy balance can be written in the dimensionless form as follows:

$$\frac{\partial T}{\partial \tau} = \frac{\partial^2 T}{\partial X^2} + \frac{1}{Y} \frac{\partial}{\partial Y} \left(Y \frac{\partial T}{\partial Y} \right) \tag{18.48}$$

For a closer understanding, let us consider the catalyst pellet from the previous example. At the surface of this pellet, an exothermic reaction takes place according to Equation 18.45 leading to the generation of heat on the surface of the pellet (i.e., at $Y = 1.0$). If the radius of the pellet is δ, then we have for the boundary conditions

$$\left. \frac{\partial T}{\partial Y} \right|_{Y=1} = KI^* \tag{18.49}$$

where K is the ratio between the heat generated due to the reaction and the heat utilized to heat the catalyst pellet. At the center of the pellet, symmetry constrains are employed. If we assume that the temperature distribution along the axial direction is negligible or consider a thin slice along the radial direction, the boundaries at $X = 0$ and $X = 1$ correspond to that of an infinite cylindrical rod. Under these assumptions, we have

$$\left. \frac{\partial T}{\partial Y} \right|_{Y=1+\delta} = 0 \quad \left. \frac{\partial T}{\partial X} \right|_{X=0} = 0 \quad \left. \frac{\partial T}{\partial X} \right|_{X=1} = 0 \tag{18.50}$$

The initial dimensionless temperature is set equal to 1.0. Figure 18.10 shows the distribution of the temperature for the case where the Biot number (K) and the dimensionless radius of the catalyst are both set equal to 1.0. Since we did assume that the variations along the axial direction are negligible, the temperature is constant along X. The reaction temperature heats up about 45% of the catalyst pellet to over 1.5 times the initial temperature. This implies that if the initial reaction were to start at room temperature (300 K), the heat generated due to the reaction will raise to about 45°C, and the rate of reaction predicted will be enhanced by about 40% compared to the isothermal predictions. Once again, limitations in terms of conductivity of the electrolyte prevent the reaction rates and hence the conversion is not doubled with every 10°C rise in the local temperature.

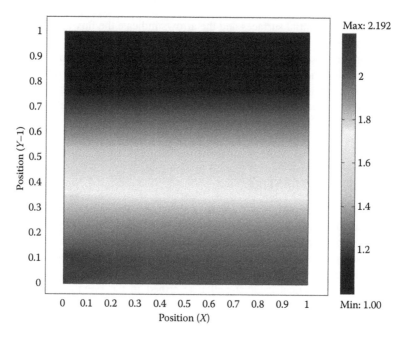

FIGURE 18.10 Temperature distribution within the catalyst pellet: the changes along the axial direction are assumed to be negligible.

18.6 DOUBLE-LAYER EFFECTS

Double-layer effects are significant in cases where the reactions happen within a short time frame: for example, the carbon-based electrical double-layer capacitors employ a carbon cloth loaded with nickel particles as the catalyst. Under these circumstances, the equation for the electrode potential is modified to include the transient effects as follows:

$$\frac{\partial E}{\partial \tau} = \frac{\partial^2 E}{\partial X^2}$$

$$\frac{di}{dX} = C\frac{\partial E}{\partial \tau}$$

(18.51a,b)

where C is the double-layer capacitance. Under constant current conditions, the boundary conditions are given by

$$\left.\frac{\partial E}{\partial X}\right|_{X=0} = -iR \quad \left.\frac{\partial E}{\partial X}\right|_{X=1} = 0$$

(18.52)

where the reaction takes place at the surface of the catalyst ($X=0$) and bulk conditions prevail at $X=1$. There exists a closed form analytical solution for the potential distribution in this case:

$$\frac{E}{iR} = \tau + \frac{2}{\pi^2}\sum_{k=1}^{\infty}\frac{(-1)^k}{k^2}\cos\left[k\pi(1-X)\right]\left[1 - \exp(-k^2\pi^2\tau)\right]$$

(18.53)

Figure 18.11a shows the evolution of the voltage across the capacitor with time. The potential at the surface of the catalyst changes significantly over the double-layer buildup time. Figure 18.11b shows the current distribution within the reactor at the corresponding instants of time: at very short

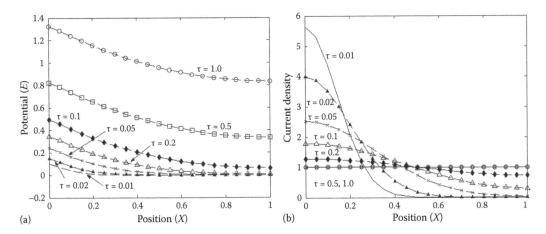

FIGURE 18.11 Transient potential and current distributions within a parallel plate reactor in the presence of a double layer: (a) Distribution of the potential near the surface of a spherical catalyst pellet at short times (b) Distribution of the transient current density near the surface of a spherical catalyst pellet.

periods of time, the current spikes at the surface where the reaction takes place. With the evolution of the double layer, the current densities become progressively uniform until it becomes flat after a critical time constant.

18.7 REACTOR DESIGN

The previous examples considered a regular geometry. Often, catalytic reactors employ a porous catalytic bed and the electrolyte flows through the mesh. A typical example for this type of reactor is the one employed for the regeneration of chlorine. Often hydrochloric acid is evolved as a by-product in plants manufacturing polyurethanes, fluorocarbons, pigments, and dyes, where chlorine is used as a promoting agent. The recovery of the chlorine gas is carried out by the electrochemical oxidation of hydrochloric acid on a catalyst frame made of copper, using, for example, the Uhde process. There exist several commercial processes involving other catalysts as well as thermochemical methods for the recovery of chlorine. Let us consider such a porous flow through a reactor, in which chlorine is produced. A simple electrochemical model can be put forward for this case of a porous catalytic bed. Figure 18.12 illustrates such a reactor.

The mesh is usually made conductive to reduce drop in the potential across the bed and to ensure a uniform current distribution. The current collector ensures adequate contact with the external power supply, across the length of the reactor.

The porous bed reactor presents several challenges: the geometry is not regular like in the previous example with the catalyst pellet; the current is carried both by the solution as well as the catalyst mesh; the reaction may deposit products that reduce the porosity for the flow of electrolyte or may corrode the mesh, resulting in an excessive flow of electrolyte, essentially reducing the residence time. In order to address the concerns over the geometry, a framework that treats the reactor as a superposition of two continuous phases (i.e., the electrolyte and the mesh) was developed by Newman and Tobias. This theoretical framework is referred to as the porous electrode theory. All the variables of interest are then reformulated into quantities averaged over a macroscopic volume that is sufficiently representative of the domain; for example, the concentration within the reference volume, even though it might change from the surface of a particular pore to the bulk, will be represented by a volume-averaged concentration term, that shows the mean profile for the change in the concentration of the species of interest from the surface of a reference pore that has a volume-averaged geometry to the bulk. Similarly, the current or potential across the electrolyte

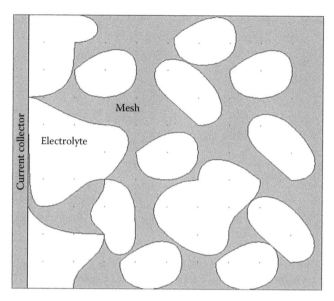

FIGURE 18.12 Porous catalytic bed: the catalyst is loaded into the mesh and the electrolyte flows across the reactor through the pores. The mesh can be a catalytic membrane, a flow-through reactor, or a zeolite bed.

will be represented by a volume-averaged quantity that has a profile in a volume-averaged sense. Such a volume-averaged current density for the solid phase can be related to a corresponding potential as follows:

$$i_1 = -\kappa_1 \frac{d\phi_1}{dx} \qquad (18.54)$$

A similar expression can be used to relate the current in the solution phase to a corresponding solution phase potential:

$$i_2 = -\kappa_2 \frac{d\phi_2}{dx} \qquad (18.55)$$

The total current density is then the sum of the current carried by the mesh and that carried by the electrolyte:

$$I = i_1 + i_2 \qquad (18.56)$$

The flux is related to the potential by a Butler–Volmer type kinetic expression:

$$\frac{di_1}{dx} = -ai_0 \left\{ \frac{c_O}{c_O^*} \exp[\beta(\phi_1 - \phi_2)] - \frac{c_R}{c_R^*} \exp[-\beta(\phi_1 - \phi_2)] \right\} \qquad (18.57)$$

The concentrations for species O and R can be obtained using material balances similar to Equation 18.40. For simplicity, in this example, we use Faraday's law and a flux expression with a mass-transfer coefficient:

$$\frac{di_1}{dx} = \frac{-nFaN_i}{s_i}$$

$$\text{where } N_i = k_{m,i}\left(c_i^0 - c_i\right) \qquad (18.58)$$

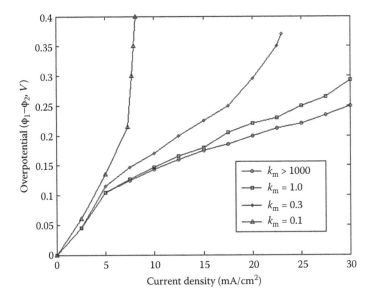

FIGURE 18.13 Polarization curves for a porous catalyst bed: The limiting currents are reduced with a reduction in the mass-transfer coefficient.

The stoichiometric coefficient is positive for the species O and negative for R. Figure 18.13 shows the polarization curves for this example at various rates of mass transfer. The mass-transfer coefficient was varied over three orders of magnitude. Note that the change in the polarization curves is not as much between the cases where $k_m = 1$ and 1000, as it is for the cases where $k_m = 1$ and below. This signifies a limiting value for mass transport, below which the overpotential rises significantly. Under these circumstances, mass transport is promoted by agitation, for example. Another means to enhance mass transport is to increase the porosity and allow for a greater convectional velocity. Note that even if the initial overpotential values are not high, as the current density is increased, there is a drastic rise in the overpotential. This indicates that if one opts for the higher current densities, this fact results in smaller residence times before the limiting current is reached.

Figure 18.13 can also be used to define an effectiveness factor, Φ, for the porous bed reactor as

$$\Phi = \text{total obtainable current density/maximum limiting current density} \qquad (18.59)$$

The effectiveness factor then is closer to 1.0 for lower current densities and appears to reduce exponentially with increase in the value of the current density. Figure 18.14 shows the flux across the bed at the beginning of the reaction. The position corresponding to 0 represents the current collector and that corresponding to 1 represents the edge of the bed facing the bulk electrolyte. The negative sign on the ordinate is indicative of the direction of the flux. During the first few seconds, the reaction proceeds at a higher rate at the current collector surface resulting in a larger concentration of the product accumulating near the surface of the current collector. As a result, the flux is driven to larger values at the end facing the bulk electrolyte. As shown in Figure 18.13, enhancing the mass-transfer coefficient reduces differences in the flux across the bed. As a result, more uniform reaction rates are observed; this ensures maximum utility of the entire surface area of the catalyst, across the thickness of the bed. By comparing the profiles in Figures 18.13 and 18.14, it is observed that a small change in the flux has a significant impact on the overpotential at higher values of the voltage.

Figure 18.15 shows the reaction rate distribution across the reactor bed. As observed during the initial few seconds of the reaction, the flux follows a nonuniform profile similar to that in

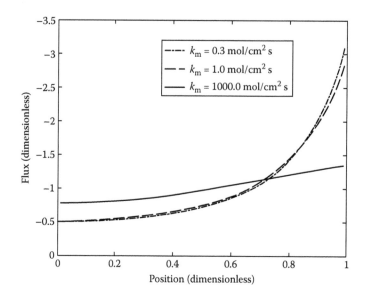

FIGURE 18.14   Effect of the rate of mass transfer on the flux across a porous catalytic bed.

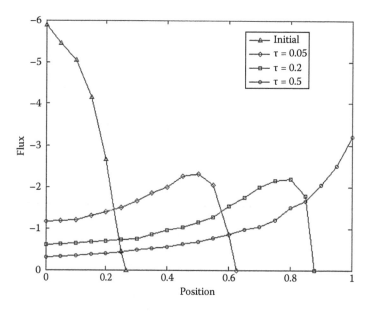

FIGURE 18.15   The movement of the reaction front across a catalytic bed with increase in the residence time: the point of maximum flux moves from the end of the bed facing the bulk solution toward the end facing the current collector and is maximum at the surface of the current collector toward the end of the residence time.

Figure 18.14. This results in a reduction of mass-transfer at that end of the bed close to the bulk solution, which in turn dampens the local mass-transfer coefficient. As this process continues, there builds up a point across the profile of the reactor, where the flux is maximum. Toward the current collector, more and more of the product is produced and as a result, the flux values are lower. Toward the bulk, the mass transfer is sluggish and thus on either side of the point of maximum, the flux values are lower. This point of maximum reaction front moves toward the current collector surface with increase in the residence time (τ) and toward the end of the residence time, the electrolyte is depleted of the reactant, slowing down the reaction at the current collector surface. As a result, the value of the flux is maximum on the surface of the current collector, at the end of the pass.

One may vary the ionic and/or electronic conductivity values as shown in the earlier example with the catalyst pellet to compare the contributions from mass-transfer resistance and the migration resistance toward the polarization curve.

18.8 PRESSURE DROP IN A PACKED-BED REACTOR

Another significant aspect in the design of reactors with a packed bed is the pressure drop across the reactor. The pressure drop determines the flow rate through the reactor. The buildup of excess pressure within the reactor results in choking or channeling within the bed resulting in suboptimal performance. The pressure drop inside a packed catalytic bed is governed by the Ergun equation:

$$\frac{dP}{dW} = -\frac{G}{\rho g_c D_P}\left(\frac{1}{\phi^3}\right)\left[\frac{150(1-\phi)\mu}{D_P} + 1.75G\right]\frac{4}{\pi D_c^2 \rho_c} \tag{18.60}$$

This is an empirical equation relating the change in the pressure drop across the bed as a function of the weight of the catalyst (W), the flow velocity (G), densities of the electrolyte, and the catalyst particle (ρ and ρ_c), as well as the diameter of the pellets and that of the bed (D_P and D_c). The parameter g_c is a conversion factor, and ϕ is the porosity of the bed. The right hand side of Equation 18.60 has only two parameters that vary with pressure if the fluid is compressible: the density of the electrolyte (ρ) and its viscosity (μ). However, these parameters do not change as much for liquids as for gaseous reactants—especially for aqueous electrolytes. Hence the pressure dependence on packing can be simplified for incompressible fluids as follows:

$$\frac{dP}{dW} = -\overline{\beta} \tag{18.61}$$

where $\overline{\beta}$ is a constant calculated from Equation 18.60.

Let us consider the epoxidation of ethylene. The reactions are given by

$$2C_2H_4 + \frac{5}{2}O_2 \xrightarrow{Ag} C_2H_4O + 2CO + 2H_2O \tag{IV}$$

$$CO + \frac{1}{2}O_2 \rightarrow CO_2 \tag{V}$$

This reaction takes place in an electrocatalytic reactor containing particles of silver as the catalyst. The individual concentrations are monitored using material balance equations as discussed in the earlier sections. If the flow rate is fixed, Reaction (IV) or (V) can be used to deduce the pressure drop across the reactor. Besides, the temperature profile is defined by an energy balance (see Equations 18.48 through 18.50). The objective in this case is to maintain the concentration of CO at the exit stream to be less than 0.4%. For good processability, the exit stream pressure should be about 1.5 atm. This example calculates the optimal length of catalyst bed required to achieve a concentration to match this amount of conversion, while maintaining an exit pressure at 1.55 atm, allowing room for some deviations in the exit stream pressure.

The concentration profiles for the reactants are shown in Figure 18.16. At about 1100 cm³, the exit stream concentration of CO drops to 0.004 times that in the inlet. The temperature and pressure profiles are given in Figure 18.17. The inlet stream is preheated to 550 K. This value of temperature rises up until about 700 K due to the exothermal nature of the reaction. After the initial 400 cm³ the reaction rate reaches a maximum and then tapers off, leading to a drop in the temperature due to heat losses to the surrounding packing.

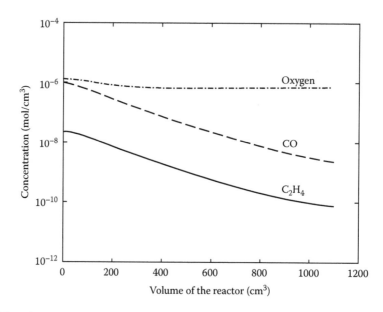

FIGURE 18.16 Distribution of reactants inside a packed-bed nonisothermal reactor during the electrochemical reduction of ethylene.

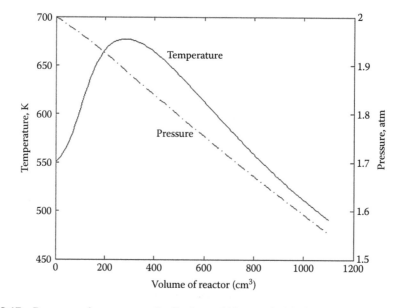

FIGURE 18.17 Pressure and temperature distributions within a packed-bed reactor.

The heat transfer to the catalyst material in the packing leads to differences in reaction rates along the length of the reactor. Since the exchange current densities for the oxidation of CO depend exponentially on the temperature, corresponding conversions of CO is impacted by the temperature variations. This effect is captured in Figure 18.18, which shows the distribution of CO along the radial direction of the catalyst pellets, which in this example were assumed to be perfect spheres. Profiles close to the entrance show steep gradients in the concentration profiles, toward

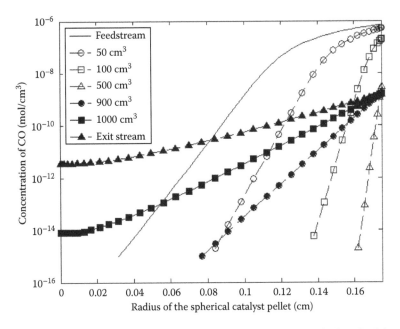

FIGURE 18.18 Concentration profiles of CO within the catalyst pellet across the length of the reactor.

about a third of the bed length down the stream, temperature effects promote the conversion at an exponential rate: as a result, the concentration of CO is reduced by an order of magnitude (see Figure 18.16). As a result, the concentration of CO drops to 0, rapidly near the surface of the catalyst pellets. Further down the length, near the exit, the temperature levels off, due to heat loss to the surroundings, and as a result, the concentration profile increases while maintaining a mild gradient.

Note that in Figure 18.17, the pressure follows a linear profile across the length of the reactor according to Equation 18.61. For a higher packing fraction, this profile may be nonlinear. Under such circumstances, the pressure profile is either approximated using a polynomial profile or calculated using the rigorous Equation 18.60.

18.9 SUMMARY

The successful design of an efficient electrocatalytic reactor relies on basic principles of chemical engineering and electrochemistry. The role of the designer ranges from choosing reaction pathways that can maintain yields at commercially viable levels, appropriate choice of catalysts, and adequate design of the flow fields. Since power consumption plays a major portion of the operating costs for any electrochemical reactor, the selection of a good effectiveness factor becomes critical. The kinetic aspects of a catalytic electrochemical reaction were investigated: design challenges when facing competing reaction pathways were discussed. The design of the catalyst and that of the reactor draws inspiration from conventional plant design: however, specific attention must be paid to the distribution of the electric field as well as the current density, which in turn determine the flux distribution for the reaction. In this vein, an electrochemical engineer is faced with challenges to combine the reaction engineering aspects with mechanical aspects like allowing for adequate heat transfer and pressure distribution. Once the optimal operating conditions for the reactor are determined using the principles illustrated earlier, other aspects such as the design of flow fields are usually adapted from conventional flow design tools.

LIST OF VARIABLES

c_i	concentration of species i, mol/cm^3
$c_{i,0}$	concentration of species i at equilibrium with the solution adjacent to the surface of electrode, mol/cm^3
$c_{i,\text{ads}}$	transient concentration of species i at electrode surface, mol/cm^3
$c_{i,\text{ref}}$	concentration of species i at reference concentrations, mol/cm^3
D_i	diffusion coefficient of species i, cm^2/s
E	dimensionless potential
E_{app}	applied potential between working electrode and reference electrode, V
F	Faraday's constant, 96,487 C/mol
G	mass flow velocity across the packed bed, kg/min
ΔG^0	free energy change for adsorption process, J/mol
I	total current density, A/cm^2
i_0	exchange current density at concentrations adjacent to the electrode surface, A/cm^2
i_1	current density in the solid phase of the porous catalyst, A/cm^2
i_2	current density in the electrolyte phase of the porous catalyst, A/cm^2
i_a	current density of adsorption step, A/cm^2
$i_{0,\text{ref}}$	exchange current density of the charge-transfer reaction at reference concentrations, A/cm^2
$i_{a,\text{ref}}$	exchange current density of the adsorption step at reference concentrations, A/cm^2
k, k''	potential independent rate constants for the adsorption step, A cm/mol
k'_{a1}, k_{c1}	potential independent rate constants
k_{ads}	proportionality constant (see Equation 18.13), mol/cm^3
$k_{m,i}$	mass-transport coefficient for species i
N	number of electrons transferred in a reaction
O_2,ads	O_2 absorbed at the surface of the electrode
O_2,sur	O_2 in the solution phase adjacent to the surface of the electrode
R	universal gas constant, 8.3143 J/mol K
s_i	stoichiometric coefficient of species i in a reaction
T	temperature, K
U_{ref}	open-circuit potential of the reaction at the reference concentrations relative to a standard reference electrode of a given kind, V
V	potential of the working electrode, V
X	dimensionless position coordinate
Y	dimensionless position coordinate
z_i	charge on species i

LIST OF GREEK SYMBOLS USED

α_a, α_c	anodic and cathodic transfer coefficient for the charge-transfer reaction
γ_i	exponent in the composition dependence of the exchange current density
η	overpotential of a reaction corrected for ohmic drop in the solution measured with respect to the reference electrode of a given kind containing a solution at the reference concentrations, V
θ	fractional coverage of the electrode surface by oxygen
θ_0	equilibrium fractional surface coverage of oxygen with respect to the concentration of the solution adjacent to the electrode surface
$\theta_{0,\text{ref}}$	relative surface coverage of oxygen with respect to reference concentration at equilibrium
ν	kinematic viscosity, cm^2/s
ξ	dimensionless position coordinate
ρ_0	density of the pure solvent, kg/cm^3
Φ	potential in the solution within the diffusion layer, V

Φ_0 potential in the solution adjacent to the electrode surface, V
Φ_{RE} potential of the reference electrode at the experimental conditions, V
Φ_{met} potential of working electrode, V
τ dimensionless time

FURTHER READINGS

1. T. Fahidy, *Principles of Electrochemical Reactor Analysis*, Elsevier Science Ltd., Amsterdam, the Netherlands, 1985.
2. D. Pletcher and F. Walsh, *Industrial Electrochemistry*, 2nd edn., Springer Verlag, Berlin, Germany, 1990.
3. J.S. Newman and K.E. Thomas-Alyea, *Electrochemical Systems*, 3rd edn., Wiley-Interscience, Hoboken, NJ, 2004.
4. A.J. Bard and L.R. Faulkner, *Electrochemical Methods: Fundamentals and Applications*, John Wiley & Sons, New York, 1980.
5. J. O'M Bockris, A.K.N. Reddy, and M.E. Gamboa-Aldeco, *Modern Electrochemistry*, Volume 1: *Ionics* and Volume 2A: *Electrodics*, 2nd edn., Springer Verlag, Berlin, Germany, 2006.
6. R.S. Nicholson and I. Shain, *Anal. Chem.* **36** (1964) 706.
7. P.K. Adanuvor, R.E. White, and S. Lorimer, *J. Electrochem. Soc.* **134** (1987) 625.
8. S.E. Lorimer, A mathematical model of the current—Potential characteristics for the bromine/bromide ion electrochemical system, Masters thesis, Department of Chemical Engineering, Texas A&M University, College Station, TX, 1982.
9. C.G. Vayenas, P.G. Debenedetti, and Iannis Yentekakis, L.L. Hegedus, *Ind. Eng. Chem. Fundam.* **24** (1985) 316.
10. A.M. Couper and D. Pletcher, *Chem. Rev.* **90** (1990) 837.
11. F. Walsh and G. Reade, *Analyst* **119** (1994) 791.
12. R. Reddy and R.G. Reddy, *Electrochim. Acta* **53** (2007) 575.
13. Y. Owobi-Andely, K. Fiaty, P. Laurent, and C. Bardot, *Catalysis Today* **57** (2000) 173.
14. A. Varma and M. Morbidelli, *Mathematical Methods in Chemical Engineering*, Oxford University Press, New York, 1997.
15. H. Scott Fogler, *Elements of Chemical Reaction Engineering*, 3rd edn., Prentice Hall PTR, New York, 1998.
16. J.S. Dunning, D.N. Bennion, and J. Newman, *J. Electrochem. Soc.* **118** (1971) 1251.
17. J.B. Rawlings and J.G. Ekerdt, *Chemical Reactor Analysis and Design Fundamentals*, 2nd edn., Nob Hill Publishing LLC, Madison, WI, 2004.
18. R.B. Bird, W. Stewart, and E.N. Lightfoot, *Transport Phenomena*, John Wiley & Sons, New York, 1960.
19. J.W. Weidner and P.S. Fedkiw, *J. Electrochem. Soc.* **138** (1991) 2514.
20. C.M. Villa and T.W. Chapman, *Ind. Eng. Chem. Res.* **34** (10) (1995) 3445.
21. J. Larminie and A. Dicks, *Fuel Cell Systems Explained*, 2nd edn., John Wiley & Sons, New York, 2003.
22. R.F. Probstein, *Physicochemical Hydrodynamics: An Introduction*, 2nd edn., Wiley-Interscience, Hoboken, NJ, 2003.

19 Electrocatalysis of Electroless Plating

Z. Jusys and A. Vaškelis

CONTENTS

19.1 INTRODUCTION TO ELECTROLESS PLATING

The chemical reduction of metal ions and its practical applications dates back to the medieval ages, for instance, the "purple of Cassius"—colloidal gold prepared via the reduction of Au(III) ions by Sn(II)—was described in the seventeenth century by the Dutch alchemist A. Cassius [1], enabling the preparation of a stable and long-lasting pigment for ruby-red tainting of glass and ceramics. The textbook "silver mirror" reaction (the reduction of an ammonia complex of silver ions by lactose), discovered by Liebig [2], was formerly used to prepare high-quality mercury-free mirrors. Nowadays, the development of electroless metal plating has been stimulated by technological progress that offers the ability to form thin metal films on dielectric materials (plastics, ceramics, glasses) and semiconductors, thus allowing to considerably reduce metal consumption and to uniquely combine materials of very different physical and chemical properties. Furthermore, the chemical reduction of metal ions to fabricate nanometer-sized particles (monometallic, bimetallic, etc.) is an important method, which is used to prepare high-surface-area catalysts, for example, for fuel cell applications [3].

Electroless Ni–P plating, introduced by Brenner and Riddell [4], has opened a new trend in metal coating formation—the deposition of metals and alloys without applying an external current, in contrast to a variety of known electrochemical deposition processes. Due to unique properties, electrolessly deposited metals and alloys have found wide applications in such high-tech industries as high-density information storage (soft, hard, longitudinal, and perpendicular recording media for hard disks) and reading devices of a high resolution (layered multimetallic composites with a giant magnetoresistance for hard disk reading heads), or the fabrication of printed circuits (through-hole plating on multilayered wiring boards), as well as for decorative purposes (the metalizing of plastics, ceramics, and other dielectric materials, corrosion protection) [5]. This stimulated both applied and fundamental research of electroless plating processes for better understanding the mechanism of occurring reactions and their fine control. A great number of such studies are summarized in recent reviews devoted to practical and theoretical aspects of electroless plating processes [6–12]. However, mechanistic studies are not numerous, and a lot of crucial questions regarding electroless metal deposition still remain under discussion. One of the recent reviews on electroless plating claims: "...electroless plating is still considered by many to be 'black magic', with some mysterious elements playing a critical role in ensuring the successful operation of the process. Clearly, much more work is needed to advance solid, scientific understanding if the technique of electroless plating is to be accepted with greater confidence" [11].

19.2 THEORETICAL CONSIDERATIONS

Several mechanistic proposals were suggested to explain electroless deposition processes, which are discussed in detail in recent reviews [2–12]. In summary, metal ions are supposed to be reduced by either some intermediate formed during the oxidation of the reducing agent (atomic hydrogen [1] or hydride ion [13]) or the electron transferred from the reducing agent to metal ions through the metal surface (Figure 19.1A). A specific feature of electroless plating is that the reduction of metal ions by the reducing agent to metal occurs preferably on the substrate surface, the latter being the catalyst for further oxidation of the reducing agent and metal deposition. Therefore, electroless deposition processes are often defined as autocatalytic.

The mixed-potential concept, originally proposed for interpreting corrosion phenomena [14], was successfully applied to the analysis of electroless deposition processes and other electrocatalytic redox systems [15–18]. According to the electrochemical interpretation of electroless plating, the electrons are transferred from the reducing agent to the metal ion via the metal surface (Figure 19.1A). In general, electroless plating solutions contain metal ions (Me^{n+}), complexing (L) and reducing (Red) agents as well as various (buffering, stabilizing, accelerating) additives. According to the mixed-potential theory, the overall process of electroless plating

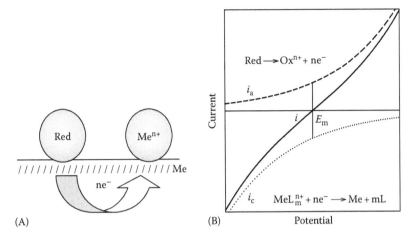

FIGURE 19.1 (A) The electron transfer scheme from the reducing agent to the metal ion through the electrode surface; (B) schematic diagram of the mixed potential concept for electroless plating: i, net current; i_a, anodic partial current (19.2); i_c, cathodic partial current (19.3); E_m, mixed potential.

$$MeL_m^{n+} + Red \rightarrow Me + mL + Ox^{n+} \tag{19.1}$$

results from the superposition of anodic

$$Red \rightarrow Ox^{n+} + ne^- \tag{19.2}$$

and cathodic

$$MeL_m^{n+} + ne^- \rightarrow Me + mL \tag{19.3}$$

partial reactions occurring simultaneously at the metal catalyst surface (Figure 19.1B). The electrode attains the mixed potential (E_m) under open-circuit conditions. The net current measured due to either negative or positive potential shift from E_m is always an algebraic sum of partial reactions (19.2) and (19.3) (solid line in Figure 19.1B).

19.3 HYDROGEN-CONTAINING REDUCING AGENTS

Despite a great variety of electroless metal deposition processes, there are certain distinct common features valid for many different reducing agents used for electroless plating, such as the following: (1) reducing agents (RH) used for electroless deposition usually contain hydrogen bonded to the corresponding atom (C–H in formaldehyde, P–H in hypophosphite, B–H in tetrahydroborate, N–H in hydrazine); (2) electroless plating processes are accompanied by hydrogen evolution upon the oxidation of the reducing agent; (3) metals known to be platable electrolessly are often also effective catalysts in hydrogenation–dehydrogenation reactions (Ni, Co, Pd, Cu, Ag, Au, etc.); (4) additives used as stabilizers (e.g., thiourea, mercapto compounds) are poisons for hydrogenation–dehydrogenation reactions; (5) the reduction of metal does not occur in the solution bulk as can be expected from the potentials of half-cell reactions.

The so-called unified mechanism for electroless plating systems involving hydrogen-containing reducing agents have been proposed [19], taking into account similarities of electroless deposition discussed earlier. In this mechanism, anodic half-reaction (19.2) is assumed to consist of several steps, viz.,

1. Metal-catalyzed dehydrogenation (dissociation of the R–H bond upon adsorption on the surface of the depositing metal)

$$R - H \rightarrow R + H \qquad\qquad (19.4)$$

2. Oxidation of the intermediate formed

$$R + OH^- \rightarrow ROH + e^- \qquad\qquad (19.5)$$

3. Depending on the nature of the catalyst, either the oxidation of hydrogen atoms formed

$$H + OH^- \rightarrow H_2O + e^- \qquad\qquad (19.6)$$

or their recombination

$$H + H \rightarrow H_2 \qquad\qquad (19.7)$$

Electrons for the reduction of metal ions to metal and other cathodic processes are supplied by reactions (19.5) and (19.6).

Possible cathodic reactions according to [19] are:

1. Reduction of metal ions

$$Me^{n+} + ne^- \rightarrow Me \qquad\qquad (19.8)$$

2. Reduction of protons from water

$$2H_2O + 2e^- \rightarrow H_2 + 2OH^- \qquad\qquad (19.9)$$

3. In some cases, the reduction of the reducing agent itself, for example, hypophosphite to phosphorus or tetrahydroborate to boron, and their co-deposition with metal (the formation of MeP or MeB alloys)

Thus, the unified mechanism involves both chemical (19.4) and electrochemical (19.5 through 19.9) contributions and, therefore, could be assigned to an electrocatalytic process by nature (the catalytic dissociation of the bond, new bond formation, charge, and mass transfer). This reaction sequence explains why only some metals (catalysts for hydrogen bond dissociation/formation) can be deposited by electroless plating as well as autocatalysis upon their deposition (preferable reduction of metal ions on the catalyst surface without appreciable reduction in the solution bulk). The mechanism proposed in [19] is in agreement with the mixed-potential interpretation of electroless plating, assuming the coupling of anodic (19.2) and cathodic (19.3) partial reactions, the rate of the overall electroless metal deposition process (19.1) being determined by a slow catalytic dehydrogenation step in the anodic oxidation of the reducing agent (19.4).

Density functional calculations of dimethylamine borane, formaldehyde, and hypophosphite oxidation on Ag surfaces indicate initial addition of OH^- to form a five-coordinated intermediate species resulting in the destabilization of the hydrogen bond (see [20] and references cited therein). The isotopic molecules containing either R–H or R–D bonds are involved in elementary reaction in which the R–H or R–D bond is broken in the rate-determining step (19.4); a decrease in reaction rate due to isotopic H/D substitution (the vibrational frequency of corresponding harmonic oscillator) might be expected [21], resulting in a primary kinetic isotope effect. Maximum values of primary kinetic H/D isotope effect predicted from the vibrational frequencies are substantial, for example, at 25°C for the C–H bond, k_H/k_D is ca. 7 and for the O–H bond 10.5 [21,22]. The H/D substitution in water can cause a solvent isotope effect [23,24], which may be due to the difference in pK_W (auto-protolysis constant at 25°C for H_2O is 10^{-14} and for D_2O 1.5×10^{-15})—D_2O is a weaker acid than H_2O. If the O–H bond is broken in the rate-determining step, a primary kinetic isotope effect contributes to the solvent isotope effect. Kinetic isotope effects, therefore, can provide indispensable mechanistic information on the details of the reaction transition state.

Summarizing the mixed-potential interpretation of electroless plating discussed above, it should be noted that the crucial requirement is the simultaneous occurrence of both anodic and cathodic half-reactions in the common potential region (see Figure 19.1B). The redox potential for the hydrogen-containing reducing agents could be varied to a certain extent by changing the solution pH. In addition, the redox potential for the cathodic half-reaction can be varied not only by the solution pH but also by chelating the metal ions binding them into complexes, and defined by the equilibrium constant of the complex. Importantly, the concentration of noncomplexed (free) metal ions must be low enough (regarding the corresponding solubility product) to avoid the precipitation of metal hydroxide. A variety of complexing agents that were found to ensure these parameters for electroless copper deposition using formaldehyde as a reducing agent are summarized in our previous publications ([25,26] and references cited therein).

19.4 METAL ION COMPLEXES AS REDUCING AGENTS

A variety of multivalent metal ions at lower oxidation states, for example, Cr^{2+}, Ti^{3+}, V^{2+}, Cu^+, Sn^{2+}, Fe^{2+}, are reducing agents capable of reducing other metal ions to the metallic state. However, the use of these redox couples, Cr^{2+}–Cr^{3+}, Ti^{3+}–Ti^{4+}, V^{2+}–V^{3+}, Cu^+–Cu^{2+}, Sn^{2+}–Sn^{4+}, Fe^{2+}–Fe^{3+}, in electroless plating systems is rather limited, because of the absence of an appreciable catalytic effect of metal in most of these processes, although the reduction reaction is possible even in the absence of the catalyst, so that metal particles are formed in the solution volume. A rather different behavior of a metal ion redox couple was observed for the Co^{2+}–Co^{3+} system. Co(II) complexes with ammonia and amines were found to be efficient reducing agents for silver and copper electroless deposition at a low extent of metal ion reduction in the solution [27–34].

Based on the standard potentials of redox couples such as Co(III)/Co(II), Cu(II)/Cu, and Ag(I)/Ag ($E^0_{Co^{3+}/Co^{2+}} = 1.82$ V, $E^0_{Cu^{2+}/Cu} = 0.345$ V, and $E^0_{Ag^+/Ag} = 0.799$ V, respectively) and on the thermodynamic analysis of equilibrium in the systems containing Co(II) and Cu(II) or Ag(I) ions chelated by corresponding amines (ethylenediamine [29], propylenediamine [30], diethylenetriamine [31]) or ammonia [29,32–34], the distribution of Co(III), Co(II), Cu(II), and Ag(I) among the complexes can be calculated as a function of solution pH, by using tabulated metal ion complex stability constants as well as ligand protonation constants, providing redox potential values for a corresponding couple in the wide range of pH values. A resulting pH dependence of the difference between the redox potentials of Co(II)/Co(III) complexes and metal ion complex/metal [Cu(II)/Cu or Ag(I)/Ag] implies that the reduction of Cu(II) or Ag(I) complexes to corresponding metal by Co(II) complexes is possible only if this difference (change in the free energy) is negative, which allows to successfully predict novel electroless plating systems and their operation conditions (complexing agent, the concentration of the components, solution pH, and temperature) [29,31–33]. Notably, in this type of electroless plating processes (metal ion complex reduction by metal ion complex), the redox potential of both partial reactions can be tuned sensitively by varying the concentration of a proper complexing agent and the solution pH, thus offering the ability to theoretically predict the new constrains, in contrast to a simple trial-and-error approach. Similarly, from estimating the redox potentials, the electroless copper plating was developed recently using Cu(II) and Fe(II) ethylenediamine and citrate complexes [35]. It should be noted, however, that Co(II)–En complexes are easily oxidized in the solution bulk by oxygen [36]. Therefore, the preparation of the solutions and using them require strictly anaerobic conditions.

Based on these theoretical considerations, practical electroless metal plating solutions of this type were developed, allowing the formation of more homogeneous metal layers [30,37,38] because of the absence of hydrogen evolution and its inclusion into the deposits using conventional reducing agents releasing hydrogen. Furthermore, it is possible to regenerate the reducing agent by the electrochemical or chemical reduction of Co(III) back to the initial Co(II) complex [29,39–41], in contrast to the irreversible transformation of the conventional reducing agents.

19.5   EXPERIMENTAL VERIFICATION OF ELECTROLESS PLATING MECHANISM

A principal evidence of the electrochemical mechanism is metal deposition in a two-compartment cell when one of the compartments contains only the metal ions and the other only the reducing agent— metal deposition occurs on one of the electrodes being driven by the oxidation of the reducing agent on the other electrode [42]. An experimental support for a mixed potential interpretation of Cu(II) reduction by formaldehyde was provided based on electrochemical measurements of copper electrode in formaldehyde and Cu(II) solutions [15–17], which stimulated an extensive use of the electrochemical techniques for studying electroless deposition processes. However, voltammetric measurements alone applied in electroless plating systems can provide only limited information on partial reaction rates, because the net current measured is an algebraic sum of both partial processes (see Figure 19.1B). Thus, detecting the partial reaction rates is necessary. The rate of cathodic half-reaction (3) can be found as a function of the potential by ex situ or in situ (electrochemical quartz crystal microbalance, EQCM) [25,26,39,40,43–51] weighing the deposit. The instantaneous rate of anodic partial reaction in a complete electroless plating bath can be measured by online mass spectrometry (differential electrochemical mass spectrometry, DEMS) as demonstrated for hypophosphite oxidation on Ni electrode [52,53] and formaldehyde oxidation on Cu, Ag, and Au electrodes [55–63].

Deuterium tracer allows the mechanism of electroless plating to be studied—there are distinct sources of hydrogen to originate from the reducing agent (19.4) or the solvent (19.9), respectively. The mass spectrometric analysis of the isotopic composition of the evolved gas using deuterium-labeled reducing agent or water shows the contribution of each reaction in the overall process and makes it possible to check the validity of the mechanism proposed. On the other hand, a recent understanding of electroless plating mechanism [19] is drawn on the basis of studies using deuterium labeling [64–69]. The kinetic isotope effect for H/D substitution in formaldehyde on the electroless copper deposition rate was found using quartz crystal microgravimetry [43].

This book presents an overview of systematic model studies of different types of reducing agents involving either both charge and mass transfer (hydrogen-containing reducing agents— hypophosphite and formaldehyde) or charge transfer only [metal ion complexes as reducing agents—Co(II) ethylenediamine complexes] upon their oxidation, addressing, in particular, mechanistic and kinetic aspects of complex electrocatalytic processes of the electroless deposition of nickel and copper. Modern online and in situ methods (electrochemical mass spectrometry and quartz crystal microgravimetry) using deuterium tracer combined with a controlled mass transport (flow-through thin-layer and wall-jet cells) were employed in the present work and allowed the kinetics and the mechanism of electroless plating (the dependence of partial reaction rates on the electrode potential, their mutual interaction, the rate-determining step of the process) to be studied. Overall, the data presented confirm the validity of electrochemical mechanism of electroless plating, demonstrate similarities and peculiarities of the processes studied, and provide a better understanding of these practically important processes.

In the following text, we will present the data on catalytic hypophosphite oxidation on nickel using deuterium tracer and online electrochemical mass spectrometry studies of partial reactions and their mutual interaction as a function of electrode potential, the modeling of the catalyst surface state upon the oxidation of hypophosphite, isotopic gas composition during electroless Ni–P alloy deposition under open-circuit conditions and at different electrode potentials. Further, the kinetic and mechanistic aspects of Cu(II) ion reduction by formaldehyde will be discussed, referring to individual half-reactions occurring in the complete bath monitored by electrochemical mass spectrometry (formaldehyde oxidation) and quartz crystal microgravimetry (copper deposition) as a function of the electrode potential, the kinetic H/D isotope effect in the overall process and in corresponding partial reactions, their mutual interaction, the extraction of partial currents in the complete electroless plating bath, complexing agent effects on the overall process and partial reactions. Finally, we will present the EQCM data in a novel type of electroless copper plating using Co(II) ions as reducing agent [the anodic oxidation of Co(II) on copper, the anodic dissolution of copper occurring in parallel to Co(II) oxidation, the reduction of Co(III) formed during Co(II) oxidation back to Co(II), the effect of the electrode potential and additives on partial reaction rates, mass transport phenomena]. In summary, despite different (in terms of metals and reducing agents) electroless metal deposition systems addressed to in this chapter, the kinetic and mechanistic data obtained are largely supporting the mixed potential interpretation of these electrocatalytic processes.

19.6 REDUCTION OF Ni(II) IONS BY HYPOPHOSPHITE STUDIED BY ONLINE MASS SPECTROMETRY

19.6.1 CATALYTIC OXIDATION OF HYPOPHOSPHITE ON NICKEL

Deuterium labeling can be used for assessing the mechanistic information of catalytic hypophosphite oxidation by water—earlier studies of the isotopic composition of hydrogen evolved (sampled off-line by collecting the evolved gas) during the catalytic oxidation of hypophosphite by water on Ni showed similar amounts of hydrogen originating from hypophosphite and water [64,65,69]. This fact was later interpreted as the evidence of the electrochemical mechanism of hypophosphite oxidation; according to this mechanism [19], the catalytic oxidation of hypophosphite on nickel in heavy water solution

$$H_2PO_2^- + D_2O \xrightarrow{Ni} HDPO_3^- + HD \qquad (19.10)$$

can be expressed as the sum of two partial reactions

$$H_2PO_2^- + D_2O \xrightarrow{Ni} HDPO_3^- + D^+ + H + e^- \qquad (19.11)$$

$$D^+ + e^- \xrightarrow{Ni} D \qquad (19.12)$$

The isotopic labeling of the source molecules should lead to the ratio H/D = 1 (D_2 mol% = 50) in the evolved gas under open-circuit conditions at equal rates of reactions (19.11) and (19.12) (equal partial currents i_{11} and i_{12}). Furthermore, as it follows from the mixed potential theory, the isotopic composition of the evolving gas could be varied by changing the electrode potential due to the change in the partial reaction rates.

The online mass spectrometric analysis of the evolving gas under open-circuit conditions and at different electrode potentials was carried out using nickel film sputter deposited onto a thin Teflon film as a working electrode, which was interfaced to the inlet of the mass spectrometer. Deuterium labeling allowed the rate of partial reactions (19.11) and (19.12) and the isotopic composition of the evolving gas to be monitored as a function of the electrode potential in parallel to faradaic current measurements, providing a solid evidence of the electrochemical mechanism of (electro) catalytic hypophosphite oxidation.

19.6.2 CATALYTIC HYPOPHOSPHITE OXIDATION UNDER OPEN-CIRCUIT CONDITIONS

The recombination of H and D atoms formed on the nickel surface as resulting from partial reactions (19.11) and (19.12) leads to the formation of three types of molecular hydrogen, that is, H_2, HD,

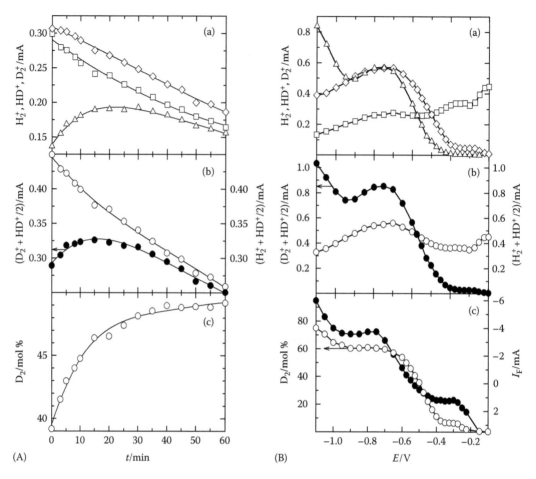

FIGURE 19.2 (A) Time dependence of H_2^+ ($\square$), HD^+ ($\diamond$), and D_2^+ ($\triangle$) ion current intensities (a), light ($\circ$) and heavy ($\bullet$) hydrogen formation rate (b), and deuterium content (c) in the gas evolved during the oxidation of hypophosphite on nickel catalyst under open-circuit conditions. (B) Potential dependence of H_2^+ ($\square$), HD^+ ($\diamond$), and D_2^+ ($\triangle$) ion current intensities (a), light ($\circ$) and heavy ($\bullet$) hydrogen formation rate (b), and deuterium content ($\circ$) and faradaic current ($\bullet$) (c). Solution contained (mol L^{-1}) CH$_3$COONa—0.15; NaH$_2$PO$_2$—0.19, $t = 80°C$, pH $= 6.0$, solvent—D$_2$O (pH adjusted by adding CD$_3$COOD). (From Jusys, Z. et al., *J. Electroanal. Chem.*, 325, 247, 1992.)

and D$_2$. Figure 19.2A(a) shows a temporal development of the H_2^+, HD^+, and D_2^+ ion current intensities for the gas evolved during catalytic hypophosphite oxidation on nickel catalyst under open-circuit conditions monitored using online mass spectrometry. H_2^+ and HD^+ intensities show a monotonic decrease with time, while D_2^+ intensity initially increases, passes a broad maximum, and then decreases in a similar manner as the H_2^+ and HD^+ ion current intensities. To deconvolute the kinetics of light and heavy hydrogen formation in partial reactions (19.11) and (19.12) from the mass spectra, containing H_2^+, HD^+, and D_2^+ contributions (Figure 19.2A(a)), half of the HD^+ intensity for each mass spectra was added to H_2^+ and HD^+ ion current intensities (Figure 19.2A(b)), respectively, yielding to corresponding hydrogen formation rates from hypophosphite and heavy water. Figure 19.2A(b) clearly shows considerable initial differences in hydrogen formation from hypophosphite and water under the catalytic oxidation of hypophosphite by water on nickel catalyst under open-circuit conditions. This may be due to additional cathodic reactions occurring in parallel to (19.12), for example, the reduction of oxy-species at the nickel surface and/or trace amounts of the dissolved oxygen, which are consuming the electrons released in the anodic half-reaction, thus diminishing the rate for proton discharge from water. The decrease in the hypophosphite oxidation rate with time (Figure 19.2A(b)) may be related to the decrease of catalytic activity of nickel [69].

The differences between hydrogen production rate from hypophosphite and water are reflected in the time dependence of deuterium content in the gas evolving during the catalytic oxidation of hypophosphite by heavy water on nickel catalyst (Figure 19.2A(c)), as calculated from the corresponding mass spectra using

$$D_2 \text{ mol\%} = [(\tfrac{1}{2}HD^+ + D_2^+)/(H_2^+ + HD^+ + D_2^+)] \cdot 100 \qquad (19.13)$$

The largest deviations from H/D = 1 (corresponding to deuterium content of 50 mol%) appear at the early stages of reaction. The deuterium content increases with time and reaches saturation at about the constant value of 50 mol% in about 60 min (Figure 19.2A(c)) in excellent agreement with the theoretical value H/D = 1 expected from simultaneously occurring partial reactions (19.11) and (19.12). Integral (over 60 min) deuterium content in the gas is about 46–47 mol%, which is comparable with the data reported in the literature for the gas mixtures sampled for a certain time [64,65]. However, at 10 times lower hypophosphite concentration, the deuterium content in evolving gas is approximately stable within the range 35 ± 2 mol% for 1 h (not shown, see Ref. [53]), as could be expected for a relatively large contribution of the above-mentioned side reactions at a lower rate of the process (10-fold lower hypophosphite concentration). These findings can explain the scattering in isotopic composition reported in the literature [64,65,69], considering the analysis of gas sampled under different reaction conditions (hypophosphite concentration, solution pH, reaction temperature, catalyst loading). Furthermore, online mass spectrometric analysis using the membrane inlet system allows to largely avoid metal-catalyzed isotope exchange between hydrogen gas and deuterium from water, which could lead to deuterium content in the gas higher than 50 mol% when sampling the gas evolving at large surface area catalyst [69], since during online analysis the evolved gas is instantly diffusing into the mass spectrometer via the membrane thus having a limited contact time with water and the catalyst surface. Notably, the homogeneous H/D exchange between hypophosphite and water that can occur at considerable rates at much lower pH values [70–72] can be neglected under the conditions applied.

19.6.3 DEPENDENCE OF PARTIAL REACTION RATES ON THE ELECTRODE POTENTIAL

As mentioned earlier, the electrochemical mechanism of catalytic hypophosphite oxidation [partial reactions (19.11) and (19.12)] implies the dependence of partial reaction rates on the electrode potential. To gain further mechanistic insight and to test the validity of the electrochemical mechanism, the catalytic hypophosphite oxidation in heavy water was performed at different constant nickel electrode potentials (versus standard hydrogen electrode, SHE) at 80°C temperature, simultaneously monitoring the isotopic composition of the evolved gas. Figure 19.2B(a) shows a plot of the ion current intensities for H_2^+, HD^+, and D_2^+ measured online at different nickel electrode potentials. The resulting mass spectra display a complex character as a function of the applied electrode potential. Notably, the molecular hydrogen ions having one or two deuterium atoms, which are originating from heavy water, show more pronounced potential dependence compared to H_2^+, which originates solely from hypophosphite. The rate of light and heavy hydrogen formation [corresponding to the rate of reactions (19.11) and (19.12)] as a function of the electrode potential may be assessed from the corresponding mass spectra by adding half of the HD^+ ion current intensity to H_2^+ and D_2^+ ion current intensities. The resulting dependences for hydrogen and deuterium formation from hypophosphite and water, respectively, on the Ni electrode potential are plotted in Figure 19.2B(b).

The influence of the electrode potential on the deuterium formation rate in hypophosphite solution demonstrates a typical electrochemical behavior (Figure 19.2B(b)). The rate of hydrogen evolution from hypophosphite [the rate of partial reaction (19.11)] is less affected by the electrode potential at pH = 6.0 (Figure 19.2B(b)). The resulting dependence of the isotopic composition of the gas evolved during hypophosphite oxidation on nickel electrode potential (Figure 19.2B(c)) shows a pronounced variation from ca. 60 to 10 mol% deuterium in the potential range from -0.6 to -0.4 V, approaching the plateau at 60 mol% up to -0.9 V and decreasing to 1–2 mol% at more positive potentials of -0.2 V. Figure 19.2B(c) shows a clear correlation of the deuterium content in the gas and the faradaic passed, which is driven mainly by the potential-dependent rate of proton discharge from water (Figure 19.2B(b)).

A decrease in solution pH from 6.0 to 4.5 enhances deuterium formation rates at corresponding potentials (Figure 19.3A(a)) as could be expected from the increase in concentration of D^+ and the rate of reaction (19.12). The hydrogen formation rate from hypophosphite at pH = 4.5 is about constant in the potential range -0.55 to -0.37 V; however, decreases gradually at potentials more

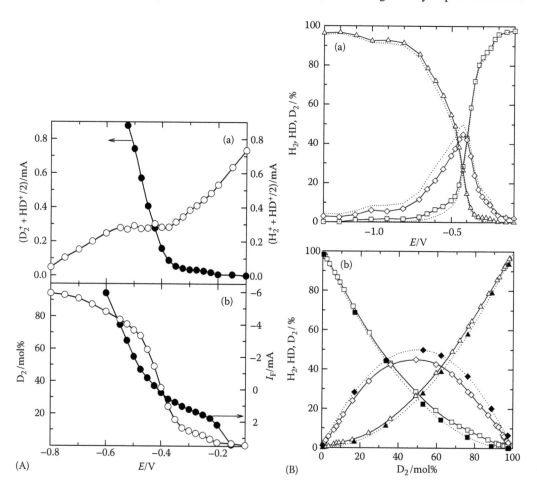

FIGURE 19.3 (A) Potential dependence of light ($\circ$) and heavy ($\bullet$) hydrogen formation rate (a), deuterium content and faradaic current (b) in the gas evolved during the oxidation of hypophosphite on nickel electrode potential. (B) Potential dependence of H_2 ($\square$), HD ($\diamond$), and D_2 ($\triangle$) content on the electrode potential (a) and isotopic composition of the gas (b) evolving during hypophosphite oxidation on nickel. Solution contained (mol L^{-1}) CH_3COONa—0.15; NaH_2PO_2—0.19; $t = 80°C$; pH = 4.5; solvent—D_2O (pH adjusted by adding CD_3COOD). Dotted lines—equilibrium H_2–HD–D_2 gas mixture values (calculated from Equation 19.14), black symbols in B(b)—the H_2 (squares) HD (diamonds) and D_2 (triangles) content in H_2–HD–D_2 gas mixtures produced by the reduction of equilibrium water mixtures (H_2O–HDO–D_2O) on uranium metal at 600°C. (From Jusys, Z. and Vaškelis, A., *Ber. Bunsenges. Phys. Chem.*, 101, 1865, 1997.)

negative of -0.55 V or increases steadily at positive potentials of -0.37 V (Figure 19.3A(a)). In the former case, the decrease in hypophosphite oxidation rate at low potentials could result from increasing cathodic deuterium evolution rate from water that is blocking the sites required for hypophosphite oxidation, while the increase in hypophosphite oxidation at more positive potentials can result from the increase in the catalyst area due to dissolution on nickel (see faradaic current in Figure 19.3A(b)).

Minor variations in hypophosphite oxidation rate on the electrode potential (Figures 19.2A(b) and 19.3A(a)) could be interpreted as the evidence of a non-electrochemical rate-determining step in the anodic oxidation of hypophosphite, for example, catalytic dehydrogenation [19], possibly, via hypophosphite radical intermediate formation, as was evidenced from the electron spin-trap resonance spectroscopy measurements [73], although this was not confirmed by in situ IR spectroscopy, apparently, due to a short lifetime of the radical intermediate [74].

A pronounced variation of the deuterium formation rate from water on the electrode potential superimposed with minor variations of hydrogen formation from hypophosphite (Figures 19.2A(b) and 19.3A(a)) implies a potential dependence of the isotopic composition of gas, evolved during the catalytic oxidation of hypophosphite on nickel electrode. Indeed, the isotopic composition of the evolving gas depends on the nickel electrode potential: deuterium prevails at potentials more negative than E_m while light hydrogen predominates at more positive potentials (deuterium content 1–2 mol% at -0.2 V) (Figures 19.2B(c) and 19.3A(b)). Correspondingly, the decrease in solution pH from 6.0 to 4.5 leads to higher values of deuterium content at more negative potentials due to increase in the rate of reaction (19.12). Correlation between the faradaic current and the deuterium content in the evolving gas (Figures 19.2B(c) and 19.3A(b)) provide an experimental evidence for the electrochemical mechanism of the process (19.10), which proceeds via the coupling of partial reactions (19.11) and (19.12).

19.7 MODELING OF THE NICKEL SURFACE STATE DURING ELECTROCATALYTIC HYPOPHOSPHITE OXIDATION

Online mass spectrometry data presented and discussed in the previous sections suggest that catalytic hypophosphite oxidation on nickel in D_2O solutions proceeds via the coupling of anodic (19.11) and cathodic (19.12) half-reactions at the catalyst surface. The classical mixed-potential theory for simultaneously occurring electrochemical partial reactions [14] presupposes the catalyst surface to be equally accessible for both anodic (19.11) and cathodic (19.12) half-reactions. Equilibrium mixtures of H_2, HD, and D_2 should be formed in this case due to the statistical recombination of H_{ad} and D_{ad} atoms randomly developed in half-reactions (19.11) and (19.12); for example, the catalytic oxidation of hypophosphite on nickel in D_2O solution under open-circuit conditions should result in the formation of gas containing equal amounts of hydrogen and deuterium ($H/D = 1$) with the distribution $H_2:HD:D_2 = 1:2:1$ (the probability of HD molecule formation is twice as high as for either H_2 or D_2 formation [75]). Therefore, to get further mechanistic insight, the distribution of H_2, HD, and D_2 species in the evolved gas was compared to the equilibrium values at the respective deuterium content [54].

Figure 19.3B(a) shows the H_2, HD, and D_2 product distribution for electrocatalytic oxidation of hypophosphite on nickel electrode over a wide potential range at pH $= 4.5$, which were fitted recently using a kinetic model [76]. Nonequilibrium mixtures of H_2, HD, and D_2 with lower HD content and, correspondingly, higher H_2 and D_2 content, than those predicted theoretically (dotted lines) are formed in the course of catalytic hypophosphite oxidation on nickel in D_2O solutions in a wide range of electrode potentials, according to online electrochemical mass spectrometry data (Figure 19.3B(a)). The largest deviations of $H_2\%$, HD%, and $D_2\%$ from the equilibrium values occur at the potentials more negative than an open-circuit potential (-0.4 V) up to ca. -1.0 V, and at more positive potentials than -0.3 V. At either negative (-1.3 V) or positive (-0.1 V) potential limit, the gas composition reaches the equilibrium values (Figure 19.3B(a)). The same product distribution data can be plotted versus the deuterium content in the gas evolved during hypophosphite oxidation (Figure 19.3B(b)) and compared with the equilibrium distribution of H_2, HD, and D_2 (black symbols) mixtures produced by the reduction of corresponding water mixtures (H_2O and D_2O) over uranium metal at 600°C. The data of Figure 19.3B(b) indicate that the content of H_2, HD, and D_2 in equilibrium gas mixtures obtained by the reduction of water vapor with different deuterium content on uranium metal is in good agreement (dotted line) with that calculated according to the following equation:

$$K = \frac{[HD]^2}{[H_2][D_2]} = 4 \tag{19.14}$$

Therefore, the analysis of the product distribution in the H_2, HD, and D_2 mixtures obtained during electrocatalytic hypophosphite oxidation on nickel electrode suggests that the hydride mechanism, assuming the release of hydride ion and instantaneous reaction with water, is unlikely due to HD content lower than the equilibrium values (hydride mechanism should lead to HD as a prevailing component [77]). Furthermore, this also puts to a question the electrochemical mechanism, according to which equilibrium H_2, HD, and D_2 mixtures must be formed due to the statistical recombination of H and D atoms for equally accessible electrode surface. To clarify this issue, computer simulations for the H_2, HD, and D_2 formed by the recombination of H and D atoms were performed.

Only the final product of the process (19.10), that is, molecular hydrogen (H_2, HD, and D_2), can be detected by online mass spectrometry. However, the catalyst microstructure may be reconstructed by the computer simulation, since the gas evolved "remembers" the surrounding hydrogen atoms in the environs: H_{ad} and D_{ad} atoms serve as peculiar markers of reactions (19.11) and (19.12) and thereby define on which site—either anodic or cathodic—they appear. The location of reactions (19.11) and (19.12) at specific site ensembles seems to be probable, taking into account a heterogeneity (structural, chemical, and energetic) of the real catalyst surface. This could be reconstructed from the H_2, HD, and D_2 distribution in the gas.

A simplified lattice-gas model has been used for the simulation of the microstructure of the nickel catalyst in a wide range of electrode potentials according to online mass spectrometry data. Possible distributions of H_{ad} and D_{ad} atoms at the catalyst surface, which result from the gas composition determined experimentally, were simulated using a home-written software. For this the following approximate model has been used: reactions (19.11) and (19.12) were supposed to occur at two types of the sites—anodic and cathodic, respectively. A two-dimensional ideal surface fully covered by H_{ad} and D_{ad} atoms with hexagonal packing (the so-called lattice gas) was examined. The following assumptions were made: (1) the possibility of recombination of any atom with every nearest neighbor (n–n) is equal, that is, the distance between each atom is the same (hexagonal packing of the elements); (2) interaction forces between H–H, H–D, and D–D atoms are equal; (3) the rate of irreversible simultaneous dimerization of the atoms is much higher than their diffusion rate—next nearest neighbor (n–n–n) interactions were not taken into account.

For the computer simulation, a two-dimensional matrix of 60×60 with a hexagonal packing of elements was randomly filled by 1 and 0 (corresponding to D_{ad} and H_{ad} atoms formed at the sites where appropriate reactions took place) with a deuterium content equal to that detected experimentally. A numerical value of the sum of n–n elements (0–6) was used to calculate the probability of H_2, HD, and D_2 formation for each element. The content of H_2, HD, and D_2 was calculated in the central part of the matrix of 30×30 elements to avoid the edge effects during both simultaneous irreversible recombination of the elements into dimmers and rearrangements of the matrix (alternatively, periodic boundary conditions could be applied). The calculated ratio $H_2\%:HD\%:D_2\%$ for the matrix randomly filled by H_{ad} and D_{ad} at D_2 mol% = 50.17 was found to be 25.02:49.61:25.37 confirming a proper statistical distribution of atoms in the initial matrix.

In the case when calculated H_2, HD, and D_2 composition was deviating from that observed experimentally, the distribution of elements in the matrix was rearranged for those elements that had no more than one n–n of the same kind. If there was no n–n of the same kind, the central element could randomly exchange with any of the six n–n with the probability of 0.5. And if the number of the n–n of the same kind was equal to 1, two rearrangements for the central element were possible with the same probability. Transformations of the matrix yielding to the clustering to the atoms of the same type were repeatedly performed until the calculated H_2%, HD%, and D_2% fitted the experimental values to an accuracy of ± 1%.

The computer simulated representative distributions of H and D atoms, yielding upon their statistical recombination to the formation of H_2, HD, and D_2 mixtures with distribution equal to that determined experimentally (Figure 19.3B(a and b)), are shown in (Figure 19.4). A good agreement with the experimental data was achieved by transforming (the number of transformations is included at each figure) the statistical distribution of H and D atoms toward the cluster formation of the atoms of the same time. This suggests that the deviations from equilibrium values (Figure 19.3B(a and b)), resulting in lower HD fraction and higher H_2 and D_2 fractions, could be explained

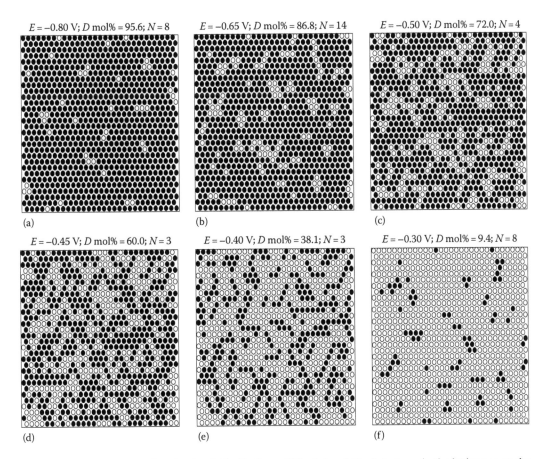

$E = -0.80$ V; D mol% = 95.6; $N = 8$ (a)
$E = -0.65$ V; D mol% = 86.8; $N = 14$ (b)
$E = -0.50$ V; D mol% = 72.0; $N = 4$ (c)
$E = -0.45$ V; D mol% = 60.0; $N = 3$ (d)
$E = -0.40$ V; D mol% = 38.1; $N = 3$ (e)
$E = -0.30$ V; D mol% = 9.4; $N = 8$ (f)

FIGURE 19.4 Representative simulated distributions of H_{ad} ($\circ$) and D_{ad} ($\bullet$) atoms in the lattice gas on the nickel surface developed in reactions (19.11) and (19.12), respectively, leading to the formation of the gas mixtures with H_2, HD, and D_2 content equal to that detected using online mass spectrometry (Figure 19.3B). N—the number of transformations starting from random distribution at a defined H/D ratio to achieve H_2, HD, and D_2 content equal to that found experimentally at corresponding potentials. (From Jusys, Z. and Vaškelis, A., *Ber. Bunsenges. Phys. Chem.*, 101, 1865, 1997.)

by the presence of ensembles of sites at the catalyst surface that are active either to anodic or cathodic half-reactions. Such catalytically active regions as the anode site ensembles (white sections), where reaction (19.11) predominantly occurs, and the cathode site ensembles (black sections), where reaction (19.12) proceeds, are clearly distinguishable on the transformed matrix. Figure 19.4(e) corresponds to the "snapshot" of the catalyst surface at the initial stages of hypophosphite oxidation under open-circuit conditions. A potential shift to more negative values leads to a gradual increase in the number and the area of cathode site ensembles due to increase in the number of electrons supplied from the external circuit for reaction (19.12), which causes a fragmentation and a diminution of catalytically active regions for hypophosphite oxidation (Figure 19.4a through d). This corresponds to the increase in the rate of reaction (19.12) and some decrease in the rate of reaction (19.11) at more negative potentials as found according to online mass spectrometry measurements (Figure 19.3A(a)). The catalytically active areas [site ensembles for anodic partial reaction (19.11)] are particularly distinct on the electrode surface at cathodic potentials from -0.65 to -0.8 V (Figure 19.4(a and b)) where deviations of $H_2\%$, HD%, and $D_2\%$ from equilibrium values are the largest (Figure 19.3B(a and b)). At potentials more negative -1.2 V, equilibrium H_2–HD–D_2 mixtures are formed and the corresponding distribution of H_{ad} and D_{ad} atoms in the matrix becomes random (not shown). The shift of the electrode potential to the values more positive than E_m suppression reaction (19.12) and causes the number as well as the size of cathode ensembles to be reduced, which, however, are still exhibiting a distinct clustering (Figure 19.4(f)). At more positive potential, a part of electrons generated in anodic half-reaction (19.11) area captured to the external circuit and are not used for reaction (19.12) (corresponding anodic current increase as well as decrease (19.12) and increase in the rate of reaction (19.11) is found at more positive potentials, see Figure 19.3A(a)). Finally, this results in the random distribution of H_{ad} and D_{ad} atoms at -0.2 V and the formation of equilibrium H_2–HD–D_2 mixtures (not shown). Notably, the real areas of the ensembles of sites active for the specific reaction could be even larger if the surface diffusion of hydrogen atoms is accounted.

The formation of nonequilibrium mixtures of H_2, HD, and D_2, with a lower HD content than that predicted from Equation 19.14, in the course of the catalytic oxidation of hypophosphite under open-circuit conditions and in the wide range of electrode potentials (-0.2 to -1.2 V) detected using online mass spectrometry, suggests that partial reactions occur at the specific surface site ensembles for anodic (19.11) and cathodic (19.12) half-reactions, respectively, on the catalyst surface, as follows from a nonrandom of H_{ad} and D_{ad} atoms. Similar conclusion was drawn using a linear source model for nonequilibrium H_2, HD, and D_2 mixture formation for catalytic hypophosphite oxidation on the nickel catalyst under open-circuit conditions in [69]. It should also be noted that the formation of intermediate hydride ion in the course of reaction (19.10) as a result of the P–H bond rupture seems to be less probable: according to the literature data [77], the reaction of CaH_2 with D_2O leads to the formation of nonequilibrium H_2–HD–D_2 mixtures with $H_2\%$:HD%:$D_2\% = 15:70:15$, that is, HD prevails in the evolved gas in this case as opposed to the HD content lower than the equilibrium values in the gas evolving during catalytic hypophosphite oxidation.

19.8 ELECTROLESS Ni–P DEPOSITION STUDIED BY ONLINE ELECTROCHEMICAL MASS SPECTROMETRY

19.8.1 ELECTROLESS Ni–P PLATING UNDER OPEN-CIRCUIT CONDITIONS

According to the electrochemical mechanism of electroless Ni–P plating, the electrons generated in the anodic half-reaction of hypophosphite oxidation (19.11) are involved in simultaneous cathodic reactions, that is, proton discharge from water (19.12), Ni(II) reduction

$$Ni^{2+} + 2e^- \rightarrow Ni \qquad (19.15)$$

and hypophosphite reduction to phosphorus

$$H_2PO_2^- + 4D^+ + 3e^- \rightarrow P + 2D_2O + H_2 \qquad (19.16)$$

It should be noted that such a stoichiometry of the phosphorus formation reaction during electroless Ni–P deposition was confirmed by using analytical methods to determine the amounts of Ni and P deposited, the overall amount of hypophosphite used, and the isotopic composition of the evolved gas, which allows to quantify the contribution of reaction (19.12) under these conditions [78,79]. Therefore, anodic partial current is equal to the sum of cathodic currents during electroless Ni–P plating under open-circuit conditions:

$$i_{11} = i_{12} + i_{15} + i_{16} \qquad (19.17)$$

Partial reaction (19.15) has no direct influence on the isotopic composition of the evolving gas, since hydrogen is not produced in the course of this reaction. However, a part of the electrons released in hypophosphite oxidation is consumed in reaction (19.15), rather than being used in the reduction of protons from water. Furthermore, a new light hydrogen source—cathodic reaction (19.16)—occurs alongside with anodic reaction (19.11). Assuming that the total amount of hypophosphite used per 1 mol of nickel deposited is about 2.5 mol ($i_{11} = 2.5i_{15}$) and the phosphorus content in the nickel deposit is 15–20 mol% under similar conditions [78], the contribution of reaction (19.16) to the total amount of light hydrogen formed is not exceeding 10%, and thus could be neglected for simplicity.

Since reaction (19.12) is the only source of deuterium in the course of electroless Ni–P deposition, the rate of reactions (19.11) and (19.12) can be studied by online mass spectrometry under open-circuit conditions as well as a function of the electrode potential similar to analogous measurements in Ni(II)-free hypophosphite solutions (see Sections 19.6.1 and 19.6.2). To avoid changes in the membrane permeability due to nickel deposition, each measurement has been carried out on a new specimen of the Ni-sputtered Teflon membrane.

Online mass spectrometry data listed in Table 19.1 show decrease in the deuterium content in the evolving gas from 36 mol% in Ni(II)-free solution to 5–10 mol% in the presence of Ni(II) under open-circuit conditions. This is in accordance with the literature data of mass spectrometric analysis of the isotopic composition of the gas evolved in various electroless plating solutions [64,65] and

TABLE 19.1
The Deuterium Content in the Evolving Gas and the Mixed Potential Values during Electroless Ni–P Deposition in Solutions (mol L^{-1}): CH_3COONa—0.15; NaH_2PO_2—0.01–0.19; $NiCl_2$— 0–0.08; $t = 80°C$; pH = 4.5; Solvent—D_2O (pH Adjusted by Adding CD_3COOD)

Exp. No.	$[H_2PO_2^-]$, mol L^{-1}	$[Ni^{2+}]$, mol L^{-1}	D_2, mol%	E_m, V
1.	0.19	—	36	−0.400
2.	0.19	0.02	8	−0.400
3.	0.19	0.04	6	−0.375
4.	0.19	0.08	5	−0.355
5.	0.10	0.08	5	−0.350
6.	0.05	0.08	5	−0.345
7.	0.02	0.08	6	−0.340
8.	0.01	0.08	10	−0.330

Source: Adapted from Jusys, Z., et al., *J. Electroanal. Chem.*, 307, 87, 1991.

previous studies in glycine solution [78,79], and can be simply explained by the occurrence of new cathode reaction(s) in parallel to (19.12) consuming the electrons released by hypophosphite, in agreement with the electrochemical mechanism.

The value of the mixed potential becomes more positive with decrease in hypophosphite concentration (Table 19.1) as could be expected according to the mixed potential theory due to the decrease in the rate of reaction (19.11) (partial current i_{11}). In the simplest case of independent electrochemical reactions, the occurrence of reactions (19.15) and (19.16) in parallel to (19.11) and (19.12) should shift the value of E_m to more positive values. However, there is no shift of E_m at low (0.02 mol L^{-1}) concentration of Ni(II) ions in comparison with Ni(II)-free solution (in both cases $E_m = -0.4$ V) although the deuterium content in the evolving gas significantly decreases from 36 to 8 mol%. This implies a mutual interaction of reactions (19.11 and 19.12) when occurring simultaneously at the catalyst surface. A plot of light and heavy hydrogen formation rate versus Ni(II) concentration (Figure 19.5A) shows some increase in the rate of reaction (19.11) and considerable decrease in the rate of reaction (19.12) when introducing Ni(II) into the solution, while further increase in Ni(II) concentration does not change remarkably the rate of both reactions (19.11) and (19.12). The dependence of reaction (19.11) rate on hypophosphite concentration shows transition from the first-order kinetics (up to 0.02 mol L^{-1}) to zero-order kinetics (over 0.10 mol L^{-1}) though the rate of reaction (19.12) is nearly constant (Figure 19.5B). This causes a relative increase in D$_2$ mol% in the evolving gas at low hypophosphite concentration under open-circuit conditions (Table 19.1).

19.8.2 DEPENDENCE OF PARTIAL REACTION RATES ON THE ELECTRODE POTENTIAL

The light and heavy hydrogen formation rate as a function of the electrode potential (Figure 19.5B(a)) can give a more detailed picture on how individual reaction rates (19.11 and 19.12), which are

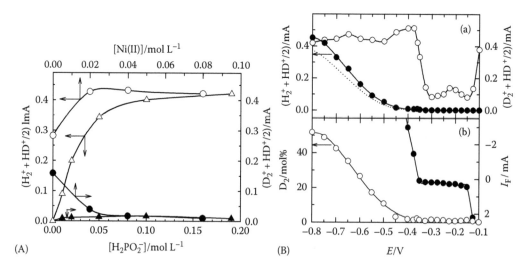

FIGURE 19.5 (A) Dependence of light ($\circ$ and $\triangle$) and heavy ($\bullet$ and $\blacktriangle$) hydrogen formation rate on Ni(II) (circles) concentration (at constant hypophosphite concentration), and hypophosphite (triangles) concentration (at constant Ni(II) concentration) during electroless nickel plating in D$_2$O solution under open-circuit conditions (for deuterium content in the evolved gas and corresponding mixed-potential values see Table 19.1). (B) Dependence of light ($\circ$) and heavy ($\bullet$) hydrogen formation rate (a), deuterium content in the evolving gas ($\circ$) and faradaic current ($\bullet$) (b) on nickel electrode potential. Solution contained (mol L^{-1}) NiCl$_2$ (un-hydrous)—0.08; CH$_3$COONa—0.15; NaH$_2$PO$_2$—0.19; t = 80°C; pH = 4.5; solvent—D$_2$O (pH adjusted by adding CD$_3$COOD). Dotted line in B(a)—deuterium evolution in the same solution without hypophosphite. (From Jusys, Z. et al., *J. Electroanal. Chem.*, 307, 87, 1991.)

reflected in the isotopic composition of the gas, formed during electroless Ni–P plating. The deuterium formation from water in reaction (19.12) is retarded by simultaneous occurrence of Ni(II) reduction in reaction (19.15), while hydrogen formation in hypophosphite oxidation [reaction (19.11)] is accelerated at potentials more negative than -0.35 V in the presence of Ni(II) ions (Figure 19.5B(a)), compared to Ni(II)-free hypophosphite solution (see Figure 19.3A(a)).

The rate of light hydrogen formation from hypophosphite is most markedly enhanced in the presence of Ni(II) (Figure 19.5B(a)), compared to Ni(II)-free hypophosphite (see Figure 19.3A(a)), close to E_m and occurs at about constant rate at more negative potentials (Figure 19.5B(a)), in contrast to the decay with potential in Ni(II)-free solution (Figure 19.3A(a)). This might be related to a continuous renewal of the catalyst surface due to Ni(II) reduction, which occurs in the expense of deuterium evolution from water in Ni(II)-free solution. The latter is blocking the sites required for the anodic partial reaction (19.11). The decrease in the rate of hypophosphite oxidation at more positive potentials in Ni(II)-containing electrolyte (-0.3 to -0.15 V) (Figure 19.5B(a)) could be related to the passivation of the Ni electrode [80], which is also seen in faradaic current decrease to low values in this potential region (Figure 19.5B(b)), while the electrode is reactivated at positive potentials of -0.15 V due to an increase in the catalyst active surface caused by the anodic dissolution of Ni.

The rate of cathodic partial reaction (19.12) for deuterium evolution from water in Ni(II)-containing solution (Figure 19.5B(a)) is largely suppressed compared to Ni(II)-free solution (Figure 19.3A(a)), which clearly indicates the competition of two [or three, when accounting hypophosphite reduction to phosphorus, reaction (19.16)] cathodic reactions occurring simultaneously. The presence or absence of hypophosphite in the solution does not change more significantly the rate of deuterium formation from water (Figure 19.5B(a)), indicating that Ni(II) reduction to Ni is the major cathodic reaction, while the reduction of water is just a side reaction in the presence of Ni(II) in the solution.

A large suppression of deuterium evolution from water with enhanced hydrogen evolution from hypophosphite at more negative potentials in the presence of Ni(II) (Figure 19.5B(a)) are causing corresponding variations in the potential dependence of deuterium content in the gas evolved (Figure 19.5B(b)), compared to Ni(II)-free hypophosphite solution (Figure 19.3A(b)), clearly indicating that the variations of the isotopic gas composition with the potential can be thoroughly explained in terms of simultaneously occurring competitive cathodic reactions in agreement with the electrochemical mechanism of electroless plating. These results can also help to correctly interpret the experimental and theoretical findings for electroless deposition of Ni–P alloy [80–83].

19.8.3 REDUCTION OF CU(II) IONS BY FORMALDEHYDE

Presuming the electrochemical mechanism electroless copper plating [19], namely, the catalytic reduction of Cu(II) ions by formaldehyde, the partial reactions occurring at equal rates under open-circuit conditions could be written (using the deuterium tracer to specify the origin of hydrogen) in a simplified form as follows:

$$D_2CO + 2OH^- \xrightarrow{Cu} DCOO^- + H_2O + D + e^- \tag{19.18}$$

$$Cu^{2+} + 2e^- \xrightarrow{Cu} Cu \tag{19.19}$$

The overall process of electroless plating in this case could be expressed as

$$Cu^{2+} + 2D_2CO + 4OH^- \xrightarrow{Cu} Cu + 2DCOO^- + 2H_2O + D_2 \tag{19.20}$$

Notably, formaldehyde in aqueous solutions is hydrated to methylene glycol [84] ($D_2C(OH)_2$ for the case of deuterated formaldehyde) with $pK = 13$, indicating that at $pH = 13$ methylene glycol is present at equal amounts with its anion [23,24]. Nevertheless, due to the absence of hydrogen exchange in the C–D bond, it is possible to convincingly distinguish if the hydrogen originates from the rupture of carbon–hydrogen bond. Furthermore, the difference in the formaldehyde oxidation reaction rate when using C–H or C–D formaldehyde could be expected if the carbon–hydrogen bond dissociation occurs in the rate-determining step [43].

It should also be noted that Cu(II) ions must be chelated in formaldehyde-containing electroless copper-plating baths operating in strongly alkaline medium to prevent Cu(OH)$_2$ formation and precipitation—the concentration of "free" (un-complexed) Cu(II) ions in the pH range 11–14 should not exceed 10^{-12}–10^{-18} mol/L, respectively [25,26]. Therefore, the proper Cu(II) complex must have the stability constant, which could ensure extremely low concentration of Cu(II) ions in the solution. On the other hand, too strong binding of Cu(II) ions into the complex can shift the overpotential for its reduction, and the essential requirement of the mixed potential theory for the overlap of potentials for both anodic and cathodic half-reactions will fail, that is, the electroless plating will not occur (see Figure 19.1B).

Online mass spectrometry studies using deuterium tracer showed the hydrogen to evolve entirely from the C–H (C–D) bond of formaldehyde during anodic oxidation on copper, silver, and gold electrodes [55–63]. This convincingly confirms the origin of the evolving hydrogen and thus the stoichiometry of the anodic partial reaction (19.18). Furthermore, comparative online mass spectrometric studies of the isotopic composition of the gas evolving during electroless copper deposition carried out using either -d$_2$ formaldehyde and -h$_2$ water or in the "mirror-image" system of -h$_2$ formaldehyde and -d$_2$ water, confirmed that all the gas (hydrogen or deuterium) originates solely from the C–H (C–D) bond of formaldehyde, respectively [59], which supports the assumptions based on the electrochemical mechanism of electroless copper plating. However, in [85] the gas evolved during electroless copper plating using CD$_2$O was found to contain more than 99% of HD, which could not be supported by the mechanism given earlier and, possibly, deals with the experimental artifacts as discussed in [59]. Moreover, the evolution of gas containing equal amounts of hydrogen originating from formaldehyde and water [85] contradicts the isotopic analysis of the gas evolved during anodic formaldehyde oxidation on copper, silver, and gold electrodes [55–63].

To further elucidate the kinetic and mechanistic aspects of electroless copper plating using formaldehyde as the reducing agent, the potential dependence of simultaneously occurring partial reactions needs to be studied. The electrochemical methods alone, as mentioned in Section 19.1, cannot provide such information since the net current is the sum of two partial reactions at each potential. The rates of specific partial reactions were monitored employing either online mass spectrometry for monitoring hydrogen evolution rate from formaldehyde [60,61], or in situ quartz crystal microgravimetry for monitoring the copper deposition rate as a function of the electrode potential, in parallel to the measurements of the net current [43–51]. Notably, the electroless copper plating consists only of two (anodic and cathodic) partial reactions and thus is less complex compared to Ni–P alloy electroless deposition as discussed in the previous section (one anodic and three cathodic partial reactions). Therefore, by employing hyphenated probing techniques, it is possible to determine the individual rates of simultaneously occurring partial reactions during electroless copper plating by measuring the rate of one of them (either via online mass spectrometry or in situ quartz crystal microbalance) and calculating the rate of the other from the difference with the net current. The kinetic isotope H/D effect will be evaluated for hydrogen exchange to deuterium in formaldehyde and water using these techniques to gain mechanistic information about partial reactions and the overall process.

19.9 FORMALDEHYDE OXIDATION DURING ELECTROLESS COPPER PLATING STUDIED BY ONLINE ELECTROCHEMICAL MASS SPECTROMETRY

DEMS allows the rate of the anodic half-reaction of formaldehyde oxidation to be measured online with a higher accuracy in comparison with cyclic voltammetry, since the contribution of the double-layer charging and the electrode surface oxidation/reduction features (for instance, oxide formation/reduction) [50], and the superposition of cathodic partial reaction can be avoided [60,61]. Additionally, the formaldehyde oxidation rate can also be monitored by DEMS under open-circuit conditions. However, quantitative mass spectrometric measurements during electroless plating using porous copper-sputtered electrodes are complicated because of the permeability changes of the porous DEMS electrodes as a result of metal deposition [52,59]. Therefore, a thin-layer flow-through DEMS cell (for details see [60]) with a solid Cu working electrode separated from a bare membrane inlet system by a thin electrolyte layer was employed to study the electroless copper deposition process.

The comparative DEMS measurements for formaldehyde (h_2- or d_2-) oxidation on copper electrode in alkaline (pH = 13) electroless copper-plating solution using light or heavy water were performed [61]. The DEMS data (cyclic voltammograms and mass spectrometric cyclic voltammograms, CVs (a) and MSCVs (b), respectively) obtained using a thin-layer flow-through cell in electroless copper-plating solution of four different isotopic compositions ($-h_2$ or $-d_2$ formaldehyde in $-h_2$ or $-d_2$ water) are shown in Figure 19.6A and B.

The corresponding CVs (Figure 19.6A(a) and B(a)) demonstrate a rather complex character due to the superposition of cathodic and anodic processes for Cu(II) reduction and formaldehyde oxidation, with the voltammetric features of copper electrode in alkaline solution (all potentials are referred to the SHE). These features could be better understood with the aid of simultaneously

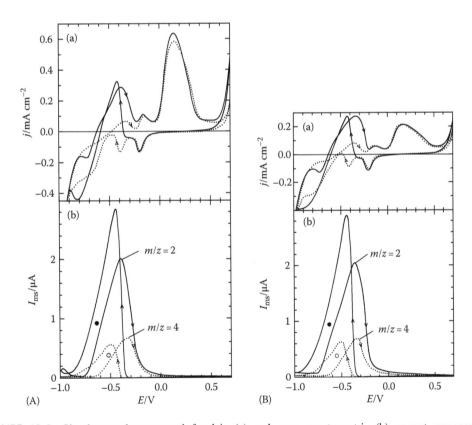

FIGURE 19.6 Simultaneously measured faradaic (a) and mass spectrometric (b) current response on the copper electrode potential in a thin-layer flow-through DEMS cell. Solution contained (mol L^{-1}) CuSO$_4$ (un-hydrous)—0.008; EDTA—0.01; H$_2$CO (solid lines) or D$_2$CO (dotted lines)—0.02; solvent H$_2$O (A) or D$_2$O (B). pH = 13 (adjusted by adding NaOH or NaOD, respectively); temperature 25°C. Potential sweep rate 5 mV s^{-1}. Electrolyte flow rate 30 μL s^{-1}. Circles indicate open-circuit conditions (filled—H$_2$ evolution, open—D$_2$ evolution). (From Jusys, Z. and Vaškelis, A., *Electrochim. Acta*, 42, 449, 1997.)

measured MSCVs for hydrogen evolution rate. The pseudo-mass-transport limited current at ca. -0.8 V could be attributed to the cathodic reduction of Cu(II) ion (19.19) occurs (cf. Figure 19.6A(a) and B(a)) since it is absent in Cu(II)-free formaldehyde solution [55,61]. Formaldehyde anodic oxidation to formate with hydrogen evolution proceeds within the potential region from -0.7 to -0.2 V in the positive-going potential scan and from -0.35 to -0.8 V in the negative-going scan as is evidenced by both CVs and MSCVs of Figure 19.6A and B. Cu_2O formation inhibits anodic formaldehyde oxidation and causes the anodic current peak at -0.15 V in the positive-going scan (Figure 19.6A(a) and B(a)). An increase in the anodic current in the electrode potential range from 0.0 to 0.4 V (Figure 19.6A(a) and B(a)) could be attributed to the formation of either oxy- or hydroxy-Cu(II) species and their dissolution because of chelation by ethylenediamine tetraacetic acid (EDTA) (the experimental evidence for that will be provided by EQCM measurements, see Section 19.5), as confirmed when comparing with the CVs in EDTA-free formaldehyde solution, which are exhibiting lower anodic current in this potential region (not shown, see Ref. [55]). Further potential increase causes a passivation of the electrode, which is followed by a breakthrough positive potential of 0.4 V. Importantly, the voltammetric features at more positive potentials of -0.3 V (Figure 19.6A(a) and B(a)) have no corresponding counterpart in the MSCVS for hydrogen evolution (Figure 19.6A(b) and B(b)).

Hydrogen evolution from H_2O onsets at ca. -1.0 V, while cathodic deuterium evolution in D_2O solutions occurs at potentials more negative than -1.0 V as evidenced formaldehyde-free in sodium hydroxide solutions of the same pH value [58]. Formaldehyde reduction to methanol proceeds also at the negative potential limit as detected using porous electrodes and in accordance with [57,58].

The reduction of Cu(II) oxy-species occurs at about -0.2 V in the negative-going potential scan (CVs in Figure 19.6A(a) and B(a)) [50] in agreement with electrochemical [86,87] microgravimetric [88,89] measurements in alkaline solutions. The reduction of Cu(I) oxide species in the negative-going potential scan appears at -0.4 V in deuterated formaldehyde solutions (dotted lines), in contrast to the absence of this peak for light formaldehyde (solid lines) due to the overlap of high anodic current with the reduction of Cu(I) oxide (Figure 19.6A(a) and B(a)). Anodic formaldehyde oxidation onsets immediately after the reduction of Cu_2O species in the negative-going potential scan (Figure 19.6A(b) and B(b)). The formaldehyde oxidation current, however, is overlapping with Cu(II) reduction current, not allowing to discriminate between these contributions from the net current (Figure 19.6A(a) and B(a)). The mass spectrometric analysis of the evolving gas shows either hydrogen or deuterium to evolve entirely from the C–H (C–D) bond of formaldehyde, respectively, during anodic formaldehyde oxidation, and under open-circuit conditions during electroless copper plating (Figure 19.6A(b) and B(b)), in agreement with the data reported in [55–63]. Furthermore, the rate of formaldehyde oxidation differs significantly for light and heavy formaldehyde, indicating the occurrence of the kinetic H/D isotope effect upon the C–H bond breaking in the rate-determining step. These differences are also reflected in the hydrogen evolution rate under open-circuit conditions (Figure 19.6A(b) and B(b)), implying corresponding variations in the electroless copper deposition rate. It should be noted, however, that the open-circuit potential in electroless copper plating is a mixed one, that is, it depends on the rate of both cathodic and anodic half-reactions. When diminishing the rate of the latter by H/D substitution in formaldehyde, the value of the open-circuit potential becomes more positive (cf. Figure 19.6A(b) and B(b)) in accordance with the mixed potential theory. Therefore, the electrode potential control is necessary when studying the kinetic H/D isotope effect in formaldehyde oxidation.

The dependences of the kinetic H/D isotope effect, calculated as the ratio of mass spectrometric currents for hydrogen and deuterium formation, on the copper electrode potential according to the MSCV data during the positive-going potential scan (Figure 19.6A(b) and B(b)) are shown in Figure 19.7A for several cases, viz., for H/D substitution in formaldehyde in electroless plating solution in H_2O (solid line, calculated from Figure 19.6A(b)) and D_2O (dotted line, calculated from Figure 19.6B(b)), and compared to the kinetic H/D isotope effect in Cu(II)-free EDTA-containing D_2O solution, as well as for H/D substitution in the solvent for the anodic oxidation

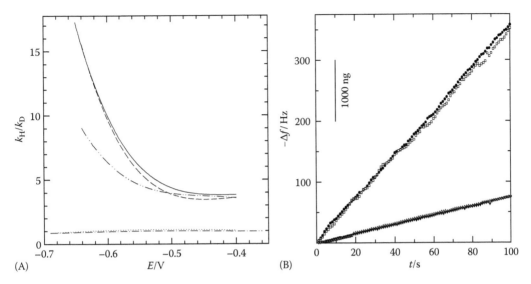

FIGURE 19.7 (A) Dependences of the kinetic isotope effect on copper electrode potential for H/D substitution in formaldehyde in light (—) and heavy water (- - -), and for H/D substitution in water for light (–·–·–) and heavy (·····) formaldehyde-containing electroless copper-plating solutions, according to the data of Figure 19.6A(b) and B(b). For comparison the kinetic isotope effect for H/D substitution in formaldehyde in D_2O solution in the absence of Cu(II) ions is included (–··–··–). (From Jusys, Z. and Vaškelis, A., *Electrochim. Acta*, 42, 449, 1997.) (B) The effect of H/D substitution in formaldehyde and water on the quartz crystal oscillation frequency during electroless copper plating. The solution contained (mol L^{-1}) $CuSO_4$—0.008; EDTA—0.01; formaldehyde—0.02; pH = 13 and t = 25°C. Systems assigned as follows: CH_2O–H_2O (●), CH_2O–D_2O (□), CD_2O–H_2O (△), and CD_2O–D_2O (▼). Open-circuit potential values (V): −0.62 (CH_2O–H_2O); −0.65 (CH_2O–D_2O); −0.56 (CD_2O–H_2O) and −0.57 (CD_2O–D_2O). (From Jusys, Z. et al., *Electrochim. Acta*, 43, 301, 1998.)

of -h_2 (dash-dotted line in Figure 19.7A) and -d_2 (dotted line in Figure 19.7A) formaldehyde [50]. The formaldehyde (either -h_2 or -d_2) anodic oxidation rate is nearly not affected if H_2O is replaced by D_2O in electroless copper-plating solution and k_H/k_D is 1 ± 0.2 within the potential range from −0.7 to −0.4 V (Figure 19.7A) in agreement with analogous studies in EDTA-free alkaline formaldehyde solutions [55]. The k_H/k_D value for H/D substitution in formaldehyde is strongly dependent on the electrode potential. The k_H/k_D reaches the highest value of 16 ± 1 at −0.65 V in electroless copper-plating solutions with either H_2O or D_2O as a solvent and drops to ca. 4 at −0.5 V and remains constant at 3.7 ± 0.4 up to −0.4 V. For comparison, k_H/k_D reaches the maximum value of 8 ± 1 at −0.65 V in Cu(II)-free formaldehyde solutions containing EDTA, then decreases to ca. 4 at −0.5 V and remains nearly constant up to −0.4 V. It should be noted here that the highest theoretical value of the primary kinetic H/D isotope effect is 6.9 at 25°C as could be found from the ratio of stretching frequencies of the C–H and C–D bonds [22]. It is comparable with the k_H/k_D value at −0.65 V in Cu(II)-free formaldehyde solution (Figure 19.7A, dash-double dotted line). Therefore, the scission of the C–H bond could be assumed as the rate-determining step in formaldehyde anodic oxidation. About twice higher k_H/k_D value in electroless plating solution (16 ± 1), compared to that detected in Cu(II)-free solutions and predicted theoretically could be due to increase in the surface catalytic activity as a result of continuous copper deposition, which occurs at a higher rate for non-deuterated formaldehyde (see Figure 19.7B) and thus is enhancing the rate of H_2 formation due to the mutual interaction of cathodic and anodic partial reaction.

The dependence of the k_H/k_D value for the H/D substitution in formaldehyde on the copper electrode potential could be interpreted, assuming that the methylene glycol anion is electroactive species in alkaline formaldehyde solutions [90]. At more negative potentials, a dissociative hydrogen-loss adsorption of methylene glycol anion with subsequent oxidation of the adsorbate could be presumed, whereas at higher potentials initial electron transfer and the subsequent decomposition of the adsorbate could be speculated upon. The value of kinetic isotope effect can also be affected by strongly adsorbed species—the catalytic dehydrogenation of formaldehyde (confirmed by catalytic hydrogen evolution from formaldehyde in the absence of Cu(II) ions in the solution under open-circuit conditions, see Refs. [55,57,58,63]), resulting in either adsorbed CO [58] or formate [63] poisoning species. The former effect can be excluded under the continuous renewal of the catalyst surface during electroless copper deposition, while the formate is produced at even larger amounts in the course of electroless copper plating. Therefore, the apparent increase in the kinetic H/D isotope effect at low potentials in the electroless plating bath can be due to increase in the surface catalytic activity as a result of copper deposition.

19.10 IN SITU ELECTROCHEMICAL QUARTZ CRYSTAL MICROGRAVIMETRY MEASUREMENTS OF COPPER DEPOSITION RATE

19.10.1 THE KINETIC H/D ISOTOPE EFFECT IN ELECTROLESS COPPER DEPOSITION UNDER OPEN-CIRCUIT CONDITIONS

EQCM makes it possible to measure the rate of reaction (19.3) in situ under open-circuit conditions and as a function of the electrode potential [25,26,43–47,51,60]. QCM study of the effect of H/D substitution in formaldehyde on the rate of electroless copper deposition suggests that the rupture of the C–H bond is the rate-determining step of the overall process [43]. Less attention, however, was paid to the detailed mechanism of reaction (19.19), which involves, apparently, the two-step reduction of cupric ions to Cu with the formation of intermediate Cu(I) species [91].

Therefore, the overall reaction for the reduction of Cu(II) ions (usually chelated by an anion of ethylenediamine tetracetic acid Y^{4-}) by formaldehyde—which exists dominantly in the form of methylene glycol and its anion—may be rewritten as follows (assuming that methylene glycol anion, and $CuY(OH)^{3-}$ [92] are reacting species in alkaline Cu(II)–EDTA solutions at pH 12–14):

$$CuY(OH)^{3-} + 2H_2C(OH)O^- + OH^- \rightarrow Cu + 2HCOO^- + 2H_2O + Y^{4-} + H_2 \quad (19.21)$$

Reaction (19.21) occurs at the catalyst surface as a result of the coupling of the oxidation of methylene glycol anion

$$2H_2C(OH)O^- + 2OH^- \rightarrow 2HCOO^- + 2H_2O + H_2 + 2e^- \quad (19.22)$$

and the reduction of Cu(II) ions

$$CuY(OH)^{3-} + 2e^- \rightarrow Cu + Y^{4-} + OH^- \quad (19.23)$$

The influence of H/D substitution in both formaldehyde and water on the rate of reaction (19.23) was studied under open-circuit conditions as well as in a wide electrode potential range.

EQCM measurements were carried out to investigate the influence of H/D substitution on the rate of electroless copper plating in both formaldehyde and water (Figure 19.7B). Here the unique properties of the quartz crystal microgravimetry (to in situ sense the mass change in the nanogram range) were employed to monitor the electroless copper deposition under open-circuit conditions. A linear decrease in the quartz crystal oscillation frequency corresponds to continuous electrode mass increase due to electroless copper deposition. One important thing is that the increase of mass is not affected by the used solvent (H_2O or D_2O, respectively); however, it is largely affected by

H/D substitution in formaldehyde. The differentiation of the overall mass increase rate and converting the quartz crystal oscillation frequency to the mass units results in the electroless copper deposition rate of ca. 36 ng s^{-1} in the light formaldehyde solution, and about 8 ng s^{-1} in the deuterated formaldehyde solution. The value of the kinetic H/D isotope effect (k_H/k_D) of ca. 4.5 can be found from the slope of the plots in Figure 19.7B for H/D substitution in formaldehyde during electroless plating in solutions of both light and heavy water. k_H/k_D is approximately 1 for H/D substitution in a solvent for either -h$_2$ or -d$_2$ formaldehyde. The kinetic isotope effect for H/D substitution in formaldehyde implies that the rupture of the C–H bond in formaldehyde is the rate-determining step of reaction (19.18) and of the overall process (19.20) in agreement with the data presented in a previous section and reported in [43]. The absence of the kinetic isotope effect for H/D substitution in the solvent suggests that the difference in deprotonation constants of methylene glycol [23,24] has no significant influence on the rate of reaction (19.18) in light and heavy water solutions.

However, according to the mixed potential theory, the value of the open-circuit potential varies when changing the rate of reaction (19.18) or reaction (19.19) due to H/D substitution in formaldehyde or water (for corresponding E_m values see Figure 19.8A(b) and B(b)). Therefore, the effect of H/D substitution on the rates of individual partial reactions must be studied under electrode potential control.

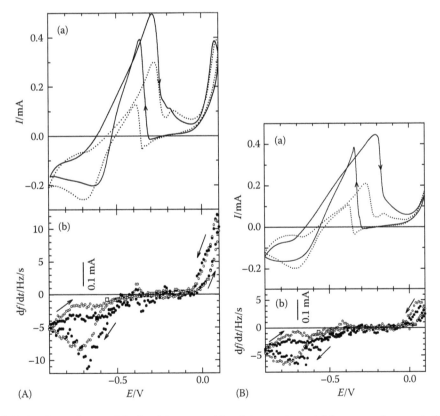

FIGURE 19.8 Dependences of the faradaic current (a) and the quartz crystal frequency change rate (b) on the electrode potential. The solution contained (mol L^{-1}) CuSO$_4$—0.008; EDTA—0.01; H$_2$CO [solid line in (a) and filled circles in (b)] or D$_2$CO [dotted line in (a) or empty circles in (b)]—0.02; pH = 13 and t = 25°C; solvent: H$_2$O (A) or D$_2$O (B). Squares indicate open-circuit conditions for H$_2$CO (■) and D$_2$CO (□)-containing electrolytes. Potential scan rate: 5 mV s^{-1}. (From Jusys, Z. et al., *Electrochim. Acta*, 43, 301, 1998.)

19.10.2 POTENTIAL-DEPENDENT EQCM STUDIES IN Cu(II)–CH$_2$O/CD$_2$O–H$_2$O/D$_2$O SOLUTIONS

Simultaneously measured CVs and differential EQCM curves in electroless copper-plating solutions of various levels of deuteration are shown in Figure 19.8A and B. EQCM allowed the rate of partial reaction (19.19) in a complete electroless plating bath to be studied as a function of the electrode potential (referred to SHE) in parallel to CV. (The latter alone does not allow to discriminate between formaldehyde oxidation and Cu(II) reduction currents.) The rate of Cu(II) reduction has been found to increase in the presence of formaldehyde in comparison with Cu(II)–EDTA solutions (Figure 19.8A(b) and B(b)). The highest enhancement of the Cu(II) reduction rate is achieved using -h$_2$ formaldehyde and water, the lowest is reached in the case of -d$_2$ formaldehyde and water in the potential range of formaldehyde oxidation. A higher noise level in differential EQCM data (Figure 19.8A(b) and B(b)) occurs due to the gas evolution.

At positive potentials of −0.1 V, the anodic copper dissolution onsets as could be seen from the decrease in the electrode mass (Figure 19.8A(b) and B(b)). The presence of formaldehyde does not cause a significant change in the rate of anodic copper dissolution at the positive potential limit. However, it does cause a marked (two- to threefold) increase in the rate of copper deposition in the potential range from −0.5 to −0.85 V as is evident from both CVs and differential EQCM data. The hysteresis in the cathodic current between negative-going and positive-going scan directions in the potential region from −0.6 to −0.9 V is followed by the corresponding electrode mass change. The rate of copper deposition increases during the negative potential scan at −0.5 to −0.85 V in D$_2$O solutions containing formaldehyde (Figure 19.8A and B). The highest rate of copper deposition is achieved at ca. −0.7 V in a negative-going scan and depends mostly on the solvent employed, viz., it reaches the frequency change rate of approximately −6 Hz s^{-1} (corresponding to the increase in mass rate 60 ng s^{-1}) in D$_2$O solution (Figure 19.8A(b) and B(b) empty circles) and about −8 Hz s^{-1} (mass increase rate 80 ng s^{-1}) in H$_2$O solution (Figure 19.8A(b) and B(b), filled circles). The deposition rate of copper in the reverse scan within the electrode potential region from −0.8 to −0.5 V diminishes with the deuteration of both water and formaldehyde as is seen from Figure 19.8A (b) and B(b). It is important to note that the ratio between the current passed and the frequency change rate is nearly constant (about 3.5 Hz s^{-1} per 0.1 mA) for the cathodic copper deposition in electroless copper-plating solutions in all cases, in spite of a higher noise in systems containing light formaldehyde (e.g., Figure 19.8A(b) and B(b), black circles); compared to that obtained in formaldehyde-free solution (not shown, see Ref. [25]). Therefore, the direct chemical reduction of Cu(II) or of intermediate Cu(I) species by formaldehyde or evolving hydrogen does not occur to a significant extent in the course of the cathodic reduction of cupric ions to Cu.

Electroless copper deposition rates under open-circuit conditions (shown as squares in Figure 19.8A(b) and B(b)), calculated from Figure 19.7B, demonstrate a satisfactory fit with differential EQCM data in the positive-going potential scan (Figure 19.8A(b) and B(b)). The kinetic isotope effect for H/D substitution in water is absent in electroless copper plating under open-circuit conditions (Figure 19.7B), though it occurs in the cathodic half-reaction at equal electrode potential values (Figure 19.8A(b) and B(b)). This might be caused by the shift of the open-circuit potential due to H/D substitution, and corresponds to the findings, that the Cu(II) reduction is mass transport limited reaction, while formaldehyde oxidation is a kinetically limited process [91], which is controlling the Cu(II) deposition rate under open-circuit conditions.

When comparing EQCM (Figure 19.8A(b) and B(b)) and DEMS (Figure 19.6A(b) and B(b)) data, one can find that the increase in Cu(II) reduction rate coincides with the onset of hydrogen formation resulting from formaldehyde oxidation. It is important to note that hydrogen evolves as a result of the rupture of the C–H bond of formaldehyde in reaction (19.18) in the potential region more positive than the cathodic proton discharge from water. Apparently, the simultaneous occurrence of hydrogen evolution from formaldehyde with depositing copper changes characteristics for Cu crystallization as supported by the capacitance measurements (extreme roughness of the depositing copper) [93]. The deposition of Cu_2O instead of copper metal when reducing Cu(II) ions by formaldehyde [94] or electrochemically [95,96] support the possibility of the formation of an oxygen-containing intermediate such as Cu_2O. Furthermore, a freshly depositing metal can increase the catalytic activity of the surface with respect to the initial dehydrogenation/oxidation of formaldehyde due to a preferable structure [37] and/or intermediate oxy-species formation [25,26,97].

19.10.3 EXTRACTION OF PARTIAL REACTION RATES DURING ELECTROLESS COPPER DEPOSITION

The data presented earlier demonstrate that the intrinsic rates of partial reactions (formaldehyde oxidation or Cu(II) reduction) could be monitored by applying an additional non-electrochemical probing technique (online mass spectrometry or in situ quartz crystal microgravimetry, respectively) in parallel to the net current measurements, which results from both contributions being their algebraic sum according to the mixed potential theory (see Figure 19.1B). Based on that, by detecting and quantifying the rate of one of the partial reactions, the rate of the second could be easily found from the difference of the net current and the measured partial current if only two reactions are occurring. This assumption is valid for electroless copper deposition using formaldehyde as a reducing agent [see reactions (19.21) through (19.23)]; however, it is not fully applicable for electroless Ni–P deposition by means of hypophosphite (three parallel cathodic reactions). Figure 19.9A shows an example of the extraction of partial currents during electroless copper plating as a function of the electrode potential from the DEMS data [26,60]. The hydrogen evolution during formaldehyde oxidation on Cu electrode in electroless copper-plating solution was measured online by mass spectrometry and converted to the partial formaldehyde oxidation current (Figure 19.9A, circles) using a calibration constant found for cathodic hydrogen evolution on the same electrode and under the same conditions in the absence of formaldehyde and Cu(II) ions. From the difference with the measured net current (Figure 19.9A, solid line), the partial current for Cu(II) reduction to copper metal can be calculated (Figure 19.9A, dotted line).

Similarly, partial reaction currents in electroless copper-plating solution can be extracted using electrochemical quartz crystal microgravimetry (EQCM) to in situ monitor the rate of copper deposition under open-circuit conditions and as a function of the electrode potential

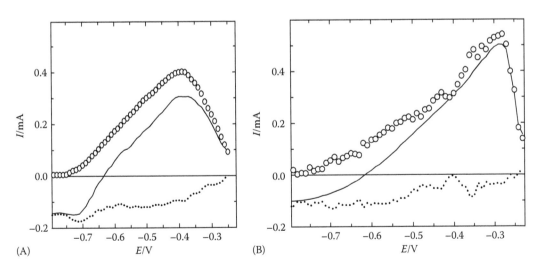

FIGURE 19.9 Extraction of the partial reaction rates for formaldehyde oxidation ($\circ$) ((A) measured by DEMS, Figure 19.6A) and copper Cu(II) ion reduction (dotted line) ((B) measured by EQCM, Figure 19.8A) as the difference with the net current (solid line) in the positive-going scan in electroless copper-plating solution (for details see captions of Figures 19.6A and 19.8A). (From Pauliukaitė, R. et al., *J. Appl. Electrochem.*, 36, 1261, 2006.)

[25,26,44,45,50,51]. Figure 19.9B shows the measured copper deposition rate converted to the partial Cu(II) reduction current (dotted line) using a calibration constant for cathodic Cu deposition in the absence of formaldehyde, and the partial formaldehyde oxidation current (circles), calculated as the difference between the net current (solid line) and partial Cu(II) reduction current (dotted line) [26]. Clearly, despite the differences in DEMS and EQCM cell designs, working electrodes, and electrolyte convection (thin-layer flow-cell used in DEMS measurements), the partial anodic and cathodic currents for electroless copper deposition, found that using these two techniques are in a good qualitative and quantitative agreement. Thus, from the data presented in Figure 19.9A and B, it could be concluded that applying one of these techniques (either DEMS or EQCM) a kinetic and mechanistic information about partial reactions of electroless deposition process can be obtained, as is demonstrated for the electroless copper-plating solution with EDTA as a complexing agent.

Notably, EDTA is the most widely used complexing agent for Cu(II) ions in alkaline electroless copper-plating solutions; however, EDTA is not the only possible complexing agent for applications in the electroless copper-plating baths (see Refs. [25,26] and references cited therein): other aminoacetic acids, such as hydroxyethyl ethylenediamine triacetic acid (HEDTA), diethylenetriamine pentaacetic acid (DTPA), trans-cyklohexane-1,2-diamine tetraacetic acid (CDTA), and triethylenetetramine hexacetic acid (TTHA) are known as alternative chelating agents, which bind Cu(II) ions strong enough allowing to prevent copper hydroxide formation (low concentration of free Cu(II) ions) in strongly alkaline solutions. In practical applications, N, N, N', N'-tetrakis-(2-hydroxypropyl)-ethylenediamine (Quadrol) and tartrates, both natural L(+)-tartrate (Rochelle salt) and synthetic DL($\mp$)-tartrate, are used frequently in electroless copper-plating baths. For some time, polyhydroxylic alcohols such as glycerol and sucrose were employed as Cu(II) complexing agent in electroless copper-plating solutions. Importantly, different complexing agents bind Cu(II) into the complexes of different stability constant, resulting in corresponding change of the equilibrium Cu–Cu(II) potential and the overpotential for Cu(II) complex reduction, thus changing the rate of the cathodic partial reaction and the overall process. Furthermore, formaldehyde anodic oxidation rate can also be affected because of copper deposition rate and the morphology of the deposit [37], as discussed earlier. Therefore, a systematic EQCM study of electroless copper-plating solutions using formaldehyde as reducing agent and the above-mentioned different complexing agents [25,26] aiming to subtract the partial currents of individual reactions as demonstrated in Figure 19.9B and to evaluate the electroless copper deposition rate under open-circuit conditions.

The major findings for the EQCM studies in these electroless copper solutions (for details see Refs. [25,26]) could be summarized as follows:

1. The equilibrium potential of copper in the solutions studied shifts to more negative values by 0.22 V going from nitrilotriacetic acid (NTA) and sucrose to TTHA, and the free Cu(II) ion concentration decreasing by almost four orders. The entire group of 11 Cu(II) complexes presents a sequence of solutions where Cu(II) complexation [free Cu(II) ion concentration] changes by more than seven orders of magnitude (resulting in corresponding shift of equilibrium potential).

2. No direct correlation between the stability constant of the complex and the rate of Cu(II) complex reduction, which could also be related to the different structures of complexes containing branched chelating molecules yielding to possible steric hindrances.

3. A pronounced increase in Cu(II) reduction rate in electroless copper-plating solution, compared to formaldehyde-free solutions, was found for EDTA-type aminocarboxylates used as complexing agents, namely, EDTA, HEDTA, DTPA, CDTA, TTHA, and amino-polycarboxylate (NTA), and hydroxypolyamine (Quadrol); however, is less affected for hydroxycarboxylates (tartrates) and polyhydroxylic compounds (glycerol and sucrose), indicating nonadditive effects, which could be tentatively explained by the influence of simultaneously occurring formaldehyde oxidation (hydrogen evolution) on the cathodic reaction intermediates (Cu(I) oxy-species, copper crystallization phenomena).

4. Formaldehyde oxidation rate in electroless plating solution is also higher compared to Cu(II)-free formaldehyde solution. In this case, a nonadditive could be equally explained by the higher catalytic activity of freshly deposited copper metal.

5. Additional factors contributing to a nonadditive phenomena were suggested reduction of Cu(II) ions not only to metal but, possibly, partly to Cu_2O (as shown for tartrate complex reduction in alkaline solution [98]), which can be responsible for periodic oscillations of the deposition rate in some electroless plating solutions [26]. Furthermore, sucrose and glycerol can exhibit the ability for Cu(II) reduction [26] in addition to reduction by formaldehyde, which makes it difficult to discriminate between the individual contributions. The adsorption of large molecules on the electrode surface [99] can also induce certain deviations in the microgravimetric behavior according to Faraday's law.

6. The highest rate of electroless copper deposition under open-circuit conditions is achieved at the most positive E_m values: both E_m and process rate values are determined by electrochemical characteristics of coupled partial reactions. Notably, Cu(II) reduction partial reaction is more sensitive to the nature of the complexing agent compared to anodic formaldehyde oxidation. The decrease in the rate of electroless copper deposition in solutions with E_m value becoming more negative corresponds to a negative shift of the Cu(II)/Cu potential, due to the increase in the pK value of the Cu(II) complexes, as well as due to kinetic and structural factors [37].

In summary, it could be concluded that a combination of electrochemical and microgravimetric probes for simultaneous sensing reaction rates provides important kinetic and mechanistic details on the ongoing processes during electroless copper deposition using formaldehyde as a reducing agent. However, even for these systems, the rate of anodic partial reaction, calculated as the difference between net current and partial cathodic current found independently from microgravimetry data, can be affected in specific systems as described earlier. The same is true for the DEMS measurements, which allow to quantify the rate of formaldehyde oxidation only. To overcome these uncertainties and limitations, simultaneous measurements of both partial reaction rates could be possible using a combined DEMS/EQCM setup introduced recently [100], which could offer simultaneous quantitative detection of both gaseous and solid phase formation during the electroless deposition.

19.11 ELECTROLESS COPPER PLATING USING CO(II) COMPLEXES AS REDUCING AGENT

The reduction of Cu(II)–ethylenediamine (En) complexes to copper by Co(II)–En species recently has been found to occur preferably at the catalyst surface, allowing the copper coatings to be deposited without applying an external current [27–30], similar to electroless copper deposition using formaldehyde as a reducing agent. Halide additives were found to increase significantly the rate of electroless copper plating [27].

These novel processes—the reduction of metal ions to metal by other metal ions [31–35]—are interesting from both practical and fundamental points of view. Toxic formaldehyde could be replaced in the solutions for electroless copper plating by the reducing agent of quite a different type, which is possible to recover electrochemically [39–41]. The oxidation of the conventional hydrogen-containing reducing agents for electroless metal plating (e.g., hypophosphite to phosphite oxidation of formaldehyde to formate) are more complicated electrocatalytic reactions, since the charge transfer is accompanied by the mass transfer, which is causing significant rearrangements of the reactant molecule (the rupture of C–H or P–H bonds and the formation of new bonds with oxygen atoms). A metal-catalyzed hydrogen atom abstraction is assumed to be the decisive step of the oxidation of conventional reducing agents. However, some different mechanism of catalytic reaction could be expected for the oxidation of the reducing agent, containing no hydrogen bond, such as Co(II) ions. Notably, the metal ions [both Co(II) and Cu(II)] must be chelated into stable complexes, ensuring low concentrations of free metal ions in the solution bulk, and thus not allowing a redox reaction to occur to an appreciable extent in the solution bulk. Moreover, the redox process (both partial reactions) must be catalyzed by the metal surface. Therefore, detailed model studies of the new type of autocatalytic reaction—copper(II) reduction by cobalt(II) complexes—are important in more broad aspects apart from the significance of one specific process. In particular, a general question could be asked—Are these processes fulfilling the mixed potential theory concepts discussed earlier? (Figure 19.1B) or are they having some different mechanistic background?

This work is aimed at EQCM studies allowing the instantaneous rate of either the deposition or the dissolution of copper to be measured in parallel to cyclic voltammetry, and a partial current of either Co(II) oxidation to Co(III) or Co(III) reduction to Co(II) to be extracted from the EQCM data. The effect of halide additives on partial reaction rates was studied. A wall-jet EQCM cell [39] was used to ensure a continuous mass transport to/from the electrode diffusion limitations [39–41]. Comparative EQCM measurements were also performed in a stagnant electrolyte.

19.11.1 OXIDATION OF Co(II)–En COMPLEXES ON COPPER ELECTRODE STUDIED BY EQCM

EQCM studies in Co(II)–En chloride solution of different pH carried out under stopped-flow (Figure 19.10A) and wall-jet conditions (Figure 19.10B) show increase in Co(II) oxidation rate

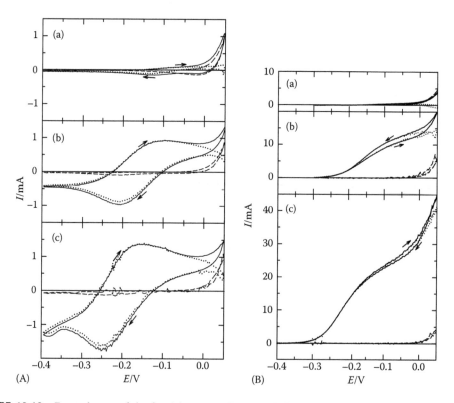

FIGURE 19.10 Dependences of the faradaic current [measured directly (—), calculated from EQCM data (- - - -), and their difference (· · · · · ·)] on the copper electrode potential under (A) stopped-flow conditions and (B) wall-jet conditions (electrolyte flow rate 0.9 mL s^{-1}). The solution contained (mol L^{-1}) CoCl$_2$—0.15; En—0.45; pH = 6 (a); 7 (b); 8 (c); t = 20°C. Potential scan rate 2 mV s^{-1}. (From Jusys, Z. and Stalnionis, G., *J. Electroanal. Chem.*, 431, 141, 1997.)

with solution pH [39]. Importantly, in situ microgravimetric measurements in parallel to simultan-eously recorded voltammetric response are necessary to independently monitor and quantify anodic dissolution of copper, which onsets at positive potentials of -0.05 V (the current for copper dissolution/deposition, plotted as dashed lines in Figure 19.10A and B, was calculated from the differential microgravimetric response using a calibration constant determined for Cu(II) reduction in Cu(II)–ethylenediamine solution at same pH values). The reduction of Cu(II)–En species formed as a result of anodic copper dissolution occurs also in this case during a reverse scan of the potential as evidence from cathodic current at ca. -0.25 V calculated from the increase in mass measured by EQCM. From the difference between the measured net current (solid lines in Figure 19.10A and B) and the partial current for copper dissolution/re-deposition, the partial reaction for Co(II) oxidation and Co(III) reduction can be calculated (dotted lines in Figure 19.10A and B). Cathodic current in the potential region from -0.12 to -0.3 V during the negative-going potential scan under stopped-flow conditions can be attributed to the reduction of Co(III) formed back to Co(II) (Figure 19.10A), since it is completely vanished under forced mass transport conditions (Figure 19.10B). The occurrence of the limiting Co(II) oxidation current in stagnant electrolyte is caused by the diffusion limitations of Co(II) species from the solution and due to slow diffusion of Co(III) species, formed due to Co(II) oxidation, resulting in the reduction of Co(III) back to Co(II) (Figure 19.10A). Both Co(II) transport to the electrode and the removal of Co(III) formed causes an increase in Co(II) oxidation current under wall-jet conditions and the absence of Co(III) reduction features (cf. Figure 19.10A and B). It should be noted that the electrochemical reduction of Co(III) ions back to Co(II) is an important issue, allowing to regenerate the reducing agent in the consumed bath for the repetitive usage, in contrast to irreversible transformation of hydrogen-containing reducing agents for electroless plating. Such recycling of the reducing agent clearly has a practical import-ance, since the Co(III) ions could be "recharged" to lower oxidation state (however, avoiding reduction to Co metal) back to Co(II) for multiple use in electroless plating process, where the electrons are released to reduce Cu(II) ions. In this case, "recharged" Co(II) ions could be accounted as the electron carrier for the "electroless" reduction of Cu(II) ions under open-circuit conditions.

In the absence of diffusion limitations, Co(II) oxidation rate increases with solution pH under wall-jet conditions in the manner similar to the rise in the concentration of $CoEn_3^{2+}$ species as a function of solution pH, calculated from the stability constants of corresponding complexes and the protonation of ethylenediamine [29,36]—their content at pH $= 6$ is negligible, about 25% at pH $= 7$, and over 80% at pH $= 8$ (see Figure 19.11A). From this correlation, a trivial conclusion could be drawn that $CoEn_3^{2+}$ species are directly responsible for Co(II) oxidation current, which could be expressed in a simplified form as

$$CoEn_3^{2+} \rightarrow CoEn_3^{3+} + e^- \tag{19.24}$$

However, Co(II) oxidation rate strongly depends not only on the solution pH value and ethyledia-mine concentration but also on the anion of the original Co(II) salt used. For instance, the EQCM studies in Co(II)–En sulfate solution show that Co(II) oxidation does not occur in the potential region studied (Figure 19.11B(a)). The KCl additive at 0.5 mmol L^{-1} concentration considerably increases the rate of Co(II) oxidation as to compare with the additive-free sulfate solution (Figure 19.11B(a and b)). Even higher rates of Co(II) oxidation is achieved for Br^- additive of the same 0.5 mmol L^{-1} concentration in sulfate solution (Figure 19.11B(c)). The acceleration of Co(II) oxidation by iodide additive occurs even at 0.005 mmol L^{-1} concentration (Figure 19.11B(d)) due to a strong adsorption of I^- ions. Therefore, the attribution of Co(II) oxidation to the prevailing Co(II)–En complex in this pH range appears to be oversimplified, since it is not accounting the role of the anion employed. While in the chloride solution (using Co(II) chloride) chloride anions can be involve in a mixed Co(II) complex formation together with En species in the solution bulk, at low concentrations of halide additives in sulfate solution this possibility could be neglected. A detailed study of the bromide additive concentration effect on Co(II) oxidation rate in sulfate solution under

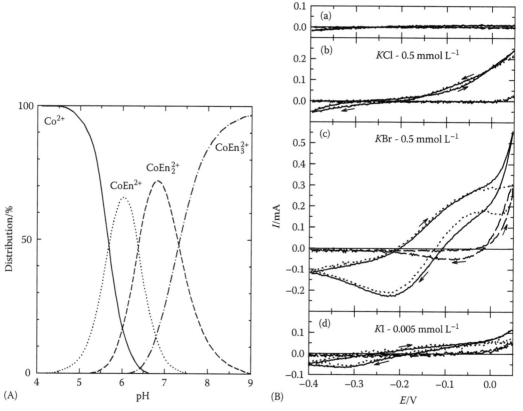

FIGURE 19.11 (A) Calculated distribution of Co(II)–En complexes as a function of solution pH. Solution contained (mol L^{-1}) Co(NO$_3$)$_2$—0,16; En—0, 64; 20°C. (From Norkus, E. et al., *Chemija* (Lithuania), 3, 21, 1995.) (B) Dependences of the faradaic current [measured directly (—), calculated from EQCM data (----), and their difference (·······)] on the copper electrode potential under stopped-flow conditions. The solution contained (mol L^{-1}) CoSO$_4$—0.15; En—0.45; pH = 7; t = 20°C; (a—additive-free; b, c, d—containing halide additives). Potential scan rate 2 mV s^{-1}.

wall-jet conditions exhibits a typical adsorption isotherm behavior with a nearly linear increase of Co(II) oxidation current up to 0.1 mM bromide concentration and hardly any further increase in the oxidation rate with bromide concentration increase up to 1.0 mM [101], indicating that adsorbed halide ions facilitate the electron transfer in Co(II)–En complex oxidation.

Similar EQCM measurements in the additive-free Cu(II)–En sulfate solution showed that Cu(II) reduction occurs at a low rate, while chloride additive of 0.5 mmol L^{-1} increases the rate of Cu(II) reduction ca. twice as compared with the additive-free sulfate solution. About sevenfold higher rate of Cu(II) reduction compared with additive-free sulfate solution and ca. fourfold higher in comparison with 0.5 mmol L^{-1} chloride additive is achieved in solution with KBr additive at the same concentration. Iodide additive at the same concentration increases the rate of Cu(II) reduction by the factor of approximately 20 compared with the additive-free sulfate solution. These data also clearly show that the halide ions even at low concentration facilitate Cu(II) ion reduction, which in ethylenediame solutions at pH > 5 exist mainly (>99%) as CuEn$_2^{2+}$ complex [29]. Accelerating effect of halide additives on Cu(II) reduction rate in sulfate solution corresponds to the increase in adsorption ability of halides. The acceleration of Cu(II) reduction rate might be explained by either the displacement of inhibiting species, for example, adsorbed ethylenediamine, or preventing from copper surface oxidation (for potential-pH phase diagrams of copper electrode in ethylenediamine solution see [96]) due to the adsorption of halide ions. A crucial role of chloride anion versus sulfate in Cu(II) electrochemical reduction was demonstrated recently by Monte Carlo simulations [102].

19.12 AUTOCATALYTIC REDUCTION OF CU(II) BY CO(II) COMPLEXES STUDIED BY EQCM

In the following text, we will present and discuss Cu(II)–En complex reduction and Co(II) complex oxidation in chloride solution at pH = 7 occurring separately (Figure 19.12A), and simultaneously in complete electroless plating bath (Figure 19.12B), employing the EQCM under stopped-flow (Figure 19.12A(a, c) and B(a)) and wall-jet conditions (Figure 19.12A(b, d) and B(b)).

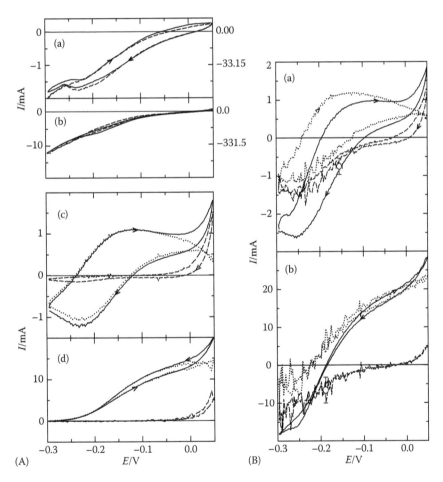

FIGURE 19.12 (A) Dependence of the faradaic current (—) and the frequency change rate (- - - -) on the copper electrode potential (a, b), and Co(II)–En oxidation rate (b, c) under stopped-flow (a, c) and wall-jet (b, d) conditions. (B) Extraction of the partial reaction currents in an electroless copper deposition bath under stopped-flow (a) and wall-jet (b) conditions. Faradaic current measured directly (—), calculated from EQCM data (- - - -), and their difference (······). The solution contained (mol L^{-1}) A: CuCl$_2$—0.05; En—0.15 (a, b); CoSO$_4$—0.15; En—0.45 (c, d). B: CuCl$_2$—0.05; CoCl$_2$—0.15; En—0.6. pH = 7; t = 20°C. Potential scan rate 2 mV s^{-1}. Electrolyte flow rate [in A(b), A(d), B(b) = 0.9 mL s^{-1}]. (From Jusys, Z. and Stalnionis, G., *J. Electroanal. Chem.*, 431, 141, 1997.)

Accounting $CuEn_2^{2+}$ as the predominant species in Cu(II)–En solutions at pH > 5 [29], the Cu(II)–En complex reduction can be expressed as

$$CuEn_2^{2+} + 2e^- \rightarrow Cu + 2En \qquad (19.25)$$

Figure 19.12 A(a and b) shows the response of both faradaic current and frequency change rate on the copper electrode potential in Cu(II)–En chloride solution under stopped-flow (a) and flow-through (b) conditions in a wall-jet EQCM cell. Increase in the rate of copper deposition due to electrolyte jet impinging at the electrode surface indicates diffusion limitations to occur in the course of Cu(II) discharge—the diffusion of both ethylenediamine formed at the electrode as a result of Cu(II)–En complex discharge into the bulk of the solution, and reacting Cu(II)–En species from the bulk to the electrode. In both stagnant electrolyte and under forced convection, the quartz crystal frequency change rate exhibits the same proportionality with the faradaic current allowing the calibration constant to be evaluated, which is further used to convert the frequency change rate directly to faradaic current for copper deposition/dissolution (see Figures 19.10, 19.11B, 19.12A(c, d), and 19.12B, dashed lines). EQCM measurements for Co(II)–En oxidation on copper electrode in the same potential region in stagnant electrolyte (Figure 19.12A(c)) and under forced convection (Figure 19.12A(d)) reproduce the features discussed in detail earlier (see Section 2.3.1). Importantly, a common potential region exists in which both Cu(II) reduction and Co(II) oxidation processes occur, as can be seen from the comparison of voltammetric Co(II) oxidation and Cu(II) reduction curves (Figure 19.12A). This potential range extends from ca. −0.1 to −0.25 V at pH = 7. Therefore, the coupling of these electrochemical processes resulting in autocatalytic copper(II) reduction by cobalt(II)—electroless copper deposition—could be predicted in the system containing copper, Co (II) and Cu(II) ethylenediamine complexes according to the mixed potential theory (Figure 19.1B). The mixed potential of copper in such a solution should attain a value in this potential range under open-circuit conditions—at equal rates of anodic and cathodic reactions.

The EQCM measurements in chloride-containing Co(II)–Cu(II)–En electroless plating solution were performed under stopped-flow (Figure 19.12B(a)) and wall-jet conditions (Figure 19.12B(b)) to extract the rates of partial reactions. The current for copper deposition/dissolution (dashed lines) was found from the quartz crystal oscillation frequency change rate as discussed earlier, while the current for Co(II) reduction/Co(III) oxidation (dotted lines) was found as the difference between the net current (solid lines) and the partial current for copper deposition/dissolution. The net current response to the electrode potential in stagnant electrolyte (solid line, Figure 19.12B(a)) exhibits a broad feature, as could be expected from the superposition of Cu(II)–En reduction (Figure 19.12A(a)) and Co(II)–En oxidation/Co(III) reduction (Figure 19.12A(b)). The partial current for copper deposition/dissolution (dashed line in Figure 19.12B(a)) is to the sum of Cu(II) reduction (Figure 19.12A(a)) and copper dissolution (Figure 19.12A(c)) in Cu(II)–En and Co(II)–En solutions, respectively. The partial current for Co(II)–En oxidation Co(III) reduction (dotted line, Figure 19.12B (a)), calculated as the difference between the faradaic current and the partial current for copper deposition/dissolution, is similar to that found in Co(II)–En oxidation/Co(III)–En reduction on copper electrode in a stagnant electrolyte (Figure 19.12A(c)). Importantly, the copper deposition rate under open-circuit conditions (open circle in Figure 19.12B(a)), converted to Cu(II) reduction current using a calibration constant found from Figure 19.12A(a), is similar to the Cu(II) reduction partial current at the potential equal to the E_m value under open circuit conditions. Furthermore, at this potential both partial currents for Cu(II) reduction and Co(II) oxidation occur at similar rates (ca. ±1 mA), supporting the validity of the mixed potential concept in this system.

To avoid the contributions of Co(III) reduction and to ensure a continuous delivery of Cu(II)–En and Co(II)–En species to the electrode surface, the EQCM measurements under wall-jet conditions were performed (Figure 19.12B(b)). A forced mass transport results in the increase in the rate of both partial reactions under wall-jet conditions (Figure 19.12B(b)) due to the transport of both reactants ($CuEn_2^{2+}$ and $CoEn_3^{2+}$) to the electrode and off-transport of the reaction products ($CoEn_3^{3+}$ and En) from the electrode. The current for copper deposition dissolution in the electroless plating bath (dashed line, Figure 19.12(b)) under forced convection is similar to the sum of currents for Cu(II) reduction in Cu(II)–En solution (dashed line, Figure 19.12A(a)) and copper dissolution in Co(II)–En solution oxidation (dashed line, Figure 19.12A(b)). The partial current for Co(II) oxidation is higher in electroless deposition solution compared to Co(II)–En (dotted lines, Figure 19.12A(d) and B(b)), which could be related to higher catalytic activity of freshly deposited copper in the former case. The E_m value for electroless copper deposition under open-circuit conditions upon enforced mass transport is shifted negatively, compared to the stagnant electrolyte (open circles, Figure 19.12B(a) and B(b)) due to some larger increase in Cu(II) reduction current under wall-jet conditions, compared to Co(II) oxidation, in a common potential range. Notably, the rate of copper deposition under open-circuit conditions is equal to the partial current for Cu(II) reduction at the potential equal to the E_m (Figure 19.12B(b)). Furthermore, at this potential, the rates of Cu(II) reduction and Co(II) oxidation are equal leading to the zero net current (Figure 19.12B(b)), as could be predicted from the theoretical considerations of a mixed potential theory (Figure 19.1B). Overall this provides a clear proof of a general validity of the mixed potential theory in this system, applying quite different reducing agents compared to conventional electroless plating solutions.

Notably, a variety of amines (ethylenediamine [27–29], propylenediamine [30], diethylenetriamine [31]) or ammonia [32–34] could be employed as complexing agents in metal ion/metal ion electroless plating systems. A correlation between the electroless deposition rate and predominate complexes as a function of solution pH was found in these systems [29,31]. It should be noted, however, that the predominant component in the solution is not necessarily the active species involved in the partial reaction and the overall process. Furthermore, halide ions at trace amounts were found to considerably increase the rate of electroless deposition of copper and silver and anodic Co(II) oxidation on these electrodes, indicating an enhanced electron transfer rate via the metal surface. Therefore, further kinetic studies applying in situ probing techniques are necessary for more detailed kinetic and mechanistic understanding of individual reactions and the overall process, as well as theoretical considerations accounting the anion effects.

19.13 SUMMARY AND OUTLOOK

The data presented and discussed in the present contribution underlines the necessity of applying modern electroanalytical techniques for online and in situ monitoring of the individual partial reactions of the electroless plating processes for a detailed kinetic and mechanistic understanding of these complex processes (otherwise not accessible). Based on these systematic model studies, general implications of the mixed potential theory were demonstrated to be valid for various electroless plating processes. Importantly, based on the theoretical considerations of the complex stability constants and their dependence on the solution pH, a new class of electroless plating processes employing metal ion complexes as the reducing agent had been developed. This clearly demonstrates the validity of the mixed potential theory not only to explain the basics of the already known processes but also to predict and develop novel systems. The variety of the electroless plating process extends also to the metallization of semiconductors where the displaced Si is operating as a reducing agent [103]. This further demonstrates that the basic theoretical considerations of the mixed potential theory are important for developing novel processes and materials to meet the challenges and calls from hi-tech industries.

REFERENCES

1. L.B. Hunt, *Gold Bull.* **9** (1976) 134.
2. J. Liebig, *Ann. Chem. Pharm.* **21** (1856) 132.
3. H. Bönnemann, W. Brijoux, R. Brinkmann, E. Dinjus, T. Joussen, and B. Korall, *Angew. Chem. Int. Ed. Engl.* **30** (1991) 1312.
4. A. Brenner and G.E. Riddell, *J. Res. Nat. Bur. Stand.* **37** (1946) 31.

5. N. Matsuoka, T. Osaka, and Y. Ito, Eds., *Electrochemical Technology. Innovation and New Developments*. Co-published by Kodansha Ltd., Tokyo, and Gordon and Breach Science Publishers S.A., Amsterdam, the Netherlands (1996).
6. A. Vaškelis and M. Šalkauskas, in *Plastics Finishing and Decoration*, Satas, D., Ed., Van Nostrand Reinhold Company Inc., New York, (1986) p. 287.
7. M. Šalkauskas and A. Vaškelis, *Electroless Deposition on Plastics*, Chemija, Leningrad, Russia (1985) (in Russian).
8. G.O. Mallory and J.B. Hajdu, *Electroless Plating: Fundamentals and Applications*, American Electroplaters and Surface Finishers Society, Inc., Orlando, FL (1990).
9. A. Vaškelis, in *Coatings Technology Handbook*, Satas, D., Ed., Marcel Dekker, New York (1991), p. 187.
10. V. Brusić, J. Horkans, and D.J. Barclay, in *Advances in Electrochemical Science and Engineering*, Vol. 1, Gerischer, H. and Tobias, Ch.W., Eds., VCH Verlagsgesellschaft, Weinheim, Germany (1990), p. 249.
11. Y. Okinaka and T. Osaka, in *Advances in Electrochemical Science and Engineering*, Vol. 3, Gerisher, H. and Tobias, Ch.W., Eds., VCH Verlagsgesellschaft, Weinheim, Germany (1994), p. 55.
12. E.J. O'Sulivan, in *Advances in Electrochemical Science and Engineering*, Vol. 7, Alkire, R.C. and Kolb, D.M. Eds., Wiley-VCH, Weinheim, Germany (2002), p. 225.
13. R.M. Lukes, *Plating* **51** (1964) 969.
14. C. Wagner and W. Traud, *Z. Elektrochem.* **44** (1938) 391 (in Germany).
15. M. Saito, *J. Metal Finish. Soc. Japan* **16** (1965) 300.
16. A. Vaškelis and M. Šalkauskas, *Lietuvos MA Darbai. Proc. Lit. Acad. Sci.* **B4(51)** (1967) 3 (in Russian).
17. M. Paunović, *Plat. Surf. Fin.* **55** (1968) 1161.
18. M. Spiro, *J. Chem. Soc. Faraday Trans.* **75** (1979) 1507.
19. J.E.A.M. Van den Meerakker, *J. Appl. Electrochem.* **11** (1981) 395.
20. T. Shimada, H. Nakai, and T. Homma, *J. Electrochem. Soc.* **154** (2007) D273.
21. L. Melander and W.H. Saunders, *Reaction Rates of Isotopic Molecules*, John Wiley & Sons, New York (1980).
22. K.A. Connors, *Chemical Kinetics: The Study of Reaction Rates in Solution*, VCH, New York (1990).
23. A. Calusaru, I. Crisan, and J. Kuta, *J. Electroanal. Chem.* **46** (1973) 51.
24. E. Norkus et al., *J. Chem. Res. (S)* (1998) 320.
25. Z. Jusys, R, Pauliukaitė, and A. Vaškelis, *Phys. Chem. Chem. Phys.* **1** (1999) 313.
26. R. Pauliukaitė, G. Stalnionis, Z. Jusys, and A. Vaškelis, *J. Appl. Electrochem.* **36** (2006) 1261.
27. A. Vaškelis, J. Jačiauskienė, and E. Norkus, *Chemija* (Lithuania) **3** (1995) 16.
28. A. Vaškelis et al., *Trans. Inst. Metal Finish.* **75** (1997) 1.
29. A. Vaškelis and E. Norkus, *Electrochim. Acta* **44** (1999) 3667.
30. A. Vaškelis, J. Jačiauskiené, A. Jagminiené, and E. Norkus, *Solid State Sci.* **4** (2002) 1299.
31. I. Stankevičienė, A. Vaškelis, A. Jagminienė, L. Tamašauskaitė-Tamašiūnaitė, and E. Norkus *Chemija* (Lithuania) **18** (2007) 8.
32. A. Vaškelis, E. Norkus, A. Jagminienė, J. Reklaitis, and L. Tamašauskaitė-Tamašiūnaitė, *Galvanotechnik* **91** (2000) 2129 (in German).
33. A. Vaškelis, E. Norkus, A. Jagminienė, J. Reklaitis, and L. Tamašauskaitė-Tamašiūnaitė, *Galvanotechnik* **91** (2000) 3395 (in German).
34. A. Vaškelis, A. Jagminienė, and L. Tamašauskaitė-Tamašiūnaitė, *J. Electroanal. Chem.* **521** (2002) 137.
35. M. Sone, K. Kobayakawa, M. Saitou, and Y. Sato, *Electrochim. Acta* **49** (2004) 233.
36. E. Norkus, A. Vaškelis, J. Reklaitis, and A. Grigucevičienė, *Chemija* (Lithuania), **3** (1995) 21.
37. A. Vaškelis, E. Norkus, I. Stalnionienė, and G. Stalnionis, *Electrochim. Acta* **49** (2004) 1613.
38. E. Norkus, A. Vaškelis, J. Jačiauskienė, I. Stalnionienė, and G. Stalnionis, *Electrochim. Acta* **51** (2006) 3495.
39. Z. Jusys and G. Stalnionis, *J. Electroanal. Chem.* **431** (1997) 141.
40. A. Vaškelis, G. Stalnionis, and Z. Jusys, *J. Electroanal. Chem.* **465** (1999) 142.
41. Z. Jusys and G. Stalnionis, *Electrochim. Acta* **45** (2000) 3675.
42. C.H. De Minjer, *Electrodep. Surf. Treat.* **3** (1975) 261.
43. R. Schumacher, J.J. Pesek, and O.R. Melroy, *J. Phys. Chem.* **89** (1985) 4338.
44. H. Wiese and K.G. Weil, *Ber. Bunsenges. Phys. Chem.* **91** (1987) 619.
45. B.J. Feldman and O.R. Melroy, *J. Electrochem. Soc.* **136** (1989) 640.
46. M. Matsuoka, J. Murai, and C. Iwakura, *J. Electrochem. Soc.* **139** (1992) 2446.
47. A.H. Gafin and S.W. Orchard, *J. Appl. Electrochem.* **22** (1992) 830.
48. A.H. Gafin and S.W. Orchard, *J. Electrochem. Soc.* **140** (1993) 3458.
49. L.M. Abrantes and J.P. Correia, *J. Electrochem. Soc.* **141** (1994) 2356.
50. Z. Jusys, G. Stalnionis, E. Juzeliūnas, and A. Vaškelis, *Electrochim. Acta* **43** (1998) 301.

51. A. Zouhou, H. Vergnes, and P. Duverneuil, *Microelectr. Eng.* **56** (2001) 177.
52. Z. Jusys, J. Liaukonis, and A. Vaškelis, *J. Electroanal. Chem.* **307** (1991) 87.
53. Z. Jusys, J. Liaukonis, and A. Vaškelis, *J. Electroanal. Chem.* **325** (1992) 247.
54. Z. Jusys and A. Vaškelis, *Ber. Bunsenges. Phys. Chem.* **101** (1997) 1865.
55. Z. Jusys, *J. Electroanal. Chem.* **375** (1994) 257.
56. H. Baltruschat et al., *Ber. Bunsenges. Phys. Chem.* **94** (1990) 996.
57. N.A. Anastasijević, H. Baltruschat, and J. Heitbaum, *Electrochim. Acta* **38** (1993) 1067.
58. Z. Jusys and A. Vaškelis, *J. Electroanal. Chem.* **335** (1992) 93.
59. Z. Jusys and A. Vaškelis, *Langmuir* **8** (1992) 1230.
60. A. Vaškelis and Z. Jusys, *Anal. Chim. Acta* **305** (1995) 227.
61. Z. Jusys and A. Vaškelis, *Electrochim. Acta* **42** (1997) 449.
62. R. Stadler, Z. Jusys, and H. Baltruschat, *Electrochim. Acta* **47** (2002) 4485.
63. M.V. ten Kortenaar, Z.I. Kolar, J.J.M. de Goeij, and G. Fren, *J. Electrochem. Soc.* **148** (2001) E327.
64. A.A. Sutiagina, K.M. Gorbunova, and M.P. Glazunov, *Zh. Fiz. Khim.* **37** (1963) 2022 (in Russian).
65. A.A. Sutiagina, K.M. Gorbunova, and M.P. Glazunov, *Zh. Fiz. Khim.* **37** (1963) 2214 (in Russian).
66. M.V. Ivanov, K.M. Gorbunova, A.A. Nikiforova, and V.P. Shcheredin, *Trans. Acad. Sci. USSR Chem. Ser.* **199** (1971) 1317 (in Russian).
67. K.M. Gorbunova, M.V. Ivanov, and V.P. Moiseev, *J. Electrochem. Soc.* **120** (1973) 613.
68. K.A. Holbrook and P.J. Twist, *J. Chem. Soc. A* **7** (1971) 890.
69. J.H. Marshall, *J. Electrochem. Soc.* **130** (1983) 369.
70. W.A. Jenkins and D.M. Yost, *J. Inorg. Nucl. Chem.* **11** (1959) 297.
71. A. Fratiello and E.W. Anderson, *J. Am. Chem. Soc.* **85** (1963) 519.
72. G.C. Ropper, T.E. Haas, and H.D. Gillma, *Inorg. Chem.* **9** (1970) 1049.
73. Y. Zeng, Y. Zheng, S. Yu, K. Chen, and S. Zhou, *Electrochem. Commun.* **4** (2002) 293.
74. L.M. Abrantes, M.C. Oliveira, J.P. Correia, A. Bewick, and M. Kalaji, *J. Chem. Soc. Faraday Trans.* **93** (1997) 1119.
75. K. Christmann, *Surf. Sci. Rep.* **9** (1988) 1.
76. Y. Zeng et al., *Chinese Chem. Lett.* **12** (2001) 641.
77. R.A. Sutula and J.B. Hunt, *J. Inorg. Nucl. Chem.* **31** (1969) 613.
78. Z. Jusys and J. Liaukonis, *Zh. Neorg. Khim.* **34** (1989) 337 (in Russian).
79. I. Genutienė, J. Lenkaitienė, Z. Jusys, and A. Luneckas, *J. Appl. Electrochem.* **26** (1996) 118.
80. Y. Zheng et al., *Chinese Chem. Lett.* **15** (2004) 1483.
81. Y.-S. Kim and H.-J. Sohn, *J. Electrochem. Soc.* **143** (1996) 505.
82. X. Yin, L. Hong, and B.-H. Chen, *J. Phys. Chem. B* **108** (2004) 10929.
83. G. Cui, H. Liu, G. Wu, J. Zhao, S. Song, and P. Kang Shen, *J. Phys. Chem. C* **112** (2008) 4601.
84. P. Greenzaid, Z. Luz, and D. Samuel, *J. Am. Chem. Soc.* **89** (1967) 749.
85. T. Ogura, M. Malcomson, and Q. Fernando, *Langmuir* **6** (1990) 1709.
86. A.M. Castro Luna, S.L. Marchiano, and A.J. Arvia, *J. Electroanal. Chem.* **59** (1975) 335.
87. L.D. Burke, M.J.G. Ahern, and T.G. Ryan, *J. Electrochem. Soc.* **137** (1990) 553.
88. U. Bertocci, *Electrochim. Acta* **49** (2004) 1831.
89. A. Survila, A. Survilienė, S. Kanapeckaitė, J. Büdienė, P. Kalinauskas, G. Stalnionis, and A. Sudavičius, *J. Electroanal. Chem.* **582** (2005) 221.
90. M. Beltowska-Brzezinska and J. Heitbaum, *J. Electroanal. Chem.* **183** (1985) 167.
91. P. Bindra, D. Light, and D. Rath, *IBM J. Res. Dev.* **28** (1984) 668.
92. E. Norkus, A. Vaškelis, and I. Žakaitė, *Talanta* **43** (1996) 465.
93. M. Wanner, H. Wiese, and K.G. Weil, *Ber. Bunsenges. Phys. Chem.* **92** (1988) 736.
94. A. Vaškelis et al., *Galvanotechnik* **86** (1995) 2114 (in Germany).
95. M. Izaki, Y. Kobayashi, and J.-I.S. Ohtomo, *J. Electrochem. Soc.* **153** (2006) C612.
96. S. Aksu and F.M. Doyle, *J. Electrochem. Soc.* **149** (2002) B340.
97. B. Lertanantawong, A.P. O'Mullane, W. Surareungchai, M. Somasundrum, L.D. Burke, and A.M. Bond, *Langmuir* **24** (2008) 2856.
98. S. Leopold, J.C. Arrayet, J.L. Bruneel, M. Herranen, J.-O. Carlsson, F. Argoul, and L. Servant, *J. Electrochem. Soc.* **150** (2003) C472.
99. H.-J. Yan, D. Wang, M.-J. Han, L.-J. Wan, and C.-L. Bai, *Langmuir* **20** (2003) 7360.
100. Z. Jusys, H. Massong, and H. Baltruschat, *J. Electrochem. Soc.* **146** (1999) 1093.
101. G. Stalnionis, PhD Thesis, Institute of Chemistry, Vilnius (1997) 478.
102. S. Harippriya and V.R. Subramanian, *J. Phys. Chem. B* **112** (2008) 4036.
103. C.P. daRosa, E. Iglesia, and R. Maboudian, *J. Electrochem. Soc.* **155** (2008) D244.

20 Fine Electrodeposition

P. C. Hsu, C. S. Lin, Y. Hwu, J. H. Je, and G. Margaritondo

CONTENTS

20.1   INTRODUCTION: ELECTRODEPOSITION AND OPEN ISSUES

The objective of this chapter is to illustrate the use of microradiology with coherent x-rays [1] to investigate open problems in electrodeposition. Metal electrodeposition is an old and widely exploited technique, one of the most frequently used for protective and decorative coating [2], semiconductor device fabrication processes, and other industrial tasks [3,4].

However, in spite of these widespread applications, the fundamental understanding of the electrodeposition process is still limited at the microscopic level. One should note that it is a very complicated process, sensitive to many parameters such as ionic concentration, local pH values, temperature, and conditions of the electrode surface. A major handicap in probing and understanding such a complex process has been the lack of experimental techniques fast enough for the dynamic effects associated to the many different parameters.

The negative consequences of this situation are not limited to fundamental issues. Indeed, industrial objectives such as coating homogeneity require a good understanding of the process and suitable monitoring techniques. The same conclusion applies to the growth of microstructures and nanostructures by electrodeposition, in particular on patterned substrates. What is really needed to tackle such problems is a new class of experimental techniques capable of establishing the links between the electrodeposition dynamics, the microscopic features, and the final quality of the products.

These problems significantly limit the expansion of electrodeposition in industry. This is particularly true for technologically advanced processes such as fabrication for microelectronics.

We specifically note in this latter case the link between the geometric features of the electro-deposition system and the electric field distribution near the electrode that influences the growth rate and the final product quality [5]. For example, dendrite growth is considered as a cause for failure of rechargeable batteries [6–9]. Thus, a good knowledge of dendrite growth is necessary, in particular for very small batteries with limited electrode separation [10,11].

All of these facts point again to the same conclusion: electrodeposition studies and the practical use of the corresponding technologies suffer from the lack of fast microprobes. In this context, the technique discussed here—microradiology with coherent x-rays [1]—can provide effective solutions for many of the open problems.

Before specifically dealing with coherent x-ray imaging, its foundations, and its advantages, we note that alternate experimental solutions were used to tackle these problems. Scanning electron microscopy (SEM) and transmission electron microscopy (TEM) probe the surface morphology and the overall microstructure of metal electrodeposits. However, they do not work in real time: they are used to analyze the final products after the end of the growth.

In situ observations with 10–20 micron lateral resolution can be performed with x-ray projection microscopy [12]. However, the probed system must be placed in a vacuum chamber, and the deposition process must be artificially slowed down: such limitations affect the applications of this probe. Likewise, other microscopy approaches have technical restrictions that, as we shall see, are not present for microradiology with coherent x-rays [1].

20.1.1 SOME EXAMPLES OF OPEN PROBLEMS

A complete or even extensive review of the problems that could be tackled by microradiology with coherent x-rays is beyond our scope. To give an idea of the potential impact, we briefly discuss here a few relevant cases that also provide the background for the later discussion of our experimental results.

As we shall see, microradiology with coherent x-rays is very effective for studying bubbles. This is important since the role of hydrogen and hydrogen bubbles in electrodeposition is relevant but still controversial.

Specifically, hydrogen formation was linked to the transition from compact to ramified deposits [13,14]. This notion became controversial after compact layer growth was observed in the presence of hydrogen evolution. Conversely, spongy electrodeposits were found to form without hydrogen evolution at high applied potentials from copper sulfate electrolytes. The link between hydrogen evolution and growth morphology is thus still unclear, and microradiology with coherent x-rays can elucidate it.

In addition to the role of bubbles, morphological transitions during electrodeposition present several interesting aspects that can be explored with microradiology. Striking morphological changes are often observed during ramified growth. These phenomena were first discovered by Hecker [15] and known as the "Hecker effect." The original case was a sudden change in the branching rate and deposit color at a certain anode–cathode distance. Garik et al. [16] described in detail transition phenomena in copper ramified growth in two-dimensional circular cells, and also observed other transition features affecting current and voltage.

The chemical aspect of the Hecker effect was analyzed by Melrose et al. [17]. Kuhn and Argoul [18,19] found that the growth morphology is very sensitive to small chemical perturbations, for example, pH changes, the presence of small amounts of oxygen, and minute quantities of alkali metal ions. Fleury et al. [20] studied the role of impurities and their dynamic behavior.

These facts stress both the importance of the transition phenomena and their complexity. Small changes in the growth parameters can strongly affect them, hence the need for probes that can show in detail the morphological changes and link them in real time to the parameter changes. This is the role that microradiology with coherent x-rays can play.

Finally, as another interesting example we would like to propose the already mentioned use of electrodeposition for the fabrication of batteries. Open pore structures with large surface areas lead to enhanced reactions and better efficiency. Ramified or porous structures are not difficult to synthesize, but it is difficult to control the microscopic morphology and achieve the desired features.

This optimization is not easy to obtain with techniques that were developed for uniform film growth. It requires instead suitable experimental probes to link the growth condition and the product features on the microscopic scale.

In the case of batteries, the key challenge is to better control the electrode–electrolyte interface and possibly to design new solid–solid or solid–liquid interfaces. This is very difficult without using the aforementioned experimental techniques to dynamically and locally probe the evolution of the electrode–electrolyte interfaces. Posthumous rather than *in situ* analysis is still widely used, thus missing key information. This is, therefore, another good example of the potential advantages offered by coherent x-ray microradiology [1].

20.1.2 INTRODUCTION TO COHERENT MICRORADIOLOGY

Microradiology with coherent x-rays [1,21–30] possesses all the necessary features to tackle the issues discussed in the previous sections. It can already deliver high-quality information on the scale below 1 μm, operating in real time with a resolution below 1 ms. This chapter briefly presents the essential features of this technique and then some of its recent applications to electrodeposition studies. We shall see, in particular, how it is able [1,21,31] to deliver fine morphological information and to correlate it with essential parameters like electrochemical potentials, pH values, ionic concentration, etc.

Conventional radiology is well known, primarily because of its medical applications. We are familiar with its strong points—such as the high penetration of x-rays into solids and liquids—as well as with its limitations [32,33]. In particular, the weak absorption that allows the high penetration limits the image contrast and therefore the space and time resolution. Such problems make conventional radiology unsuitable for studies of electrodeposition processes.

Quite recently, a new type of radiology was developed, under names such as "phase contrast radiology," "refractive index radiology," or "coherent radiology" [1,21–41]. The foundation is the use of image contrast mechanisms different from absorption contrast, the basis of conventional radiology.

All contrast mechanisms in radiology are related to the interactions between x-rays and the object; in turn, such interactions are characterized by the complex refractive index. The imaginary part of the refractive index corresponds to x-ray absorption, the foundation of conventional radiology. Coherent radiology corresponds instead to the real part and to phenomena like refraction, diffraction, and interference.

A simple and widely used contrast [24] mechanism of this kind is illustrated in Figure 20.1. Consider first Figure 20.1a, in which an x-ray beam passes through the tapered edge separating two different regions of the object. For the sake of simplification, assume that the two regions have similar x-ray absorption but (slightly) different x-ray refraction. As shown in Figure 20.1a, the corresponding beam deviation by the edge region produces a bright–dark fringe pair on the detector, and enhances the visibility of the edge.

The mechanism is somewhat similar to the interaction of visible light with the edge of a glass that makes it more visible than the glass itself. Figure 20.2 shows a practical example of edge enhancement in a microradiograph taken during an electrodeposition process. We see, in particular, the fringe pairs at the edges of gas bubbles developing in the solution during the deposition. In this way, it is possible to obtain high-quality images with excellent time and space resolution. There is, however, a catch in this argument. Figure 20.1b shows that a small

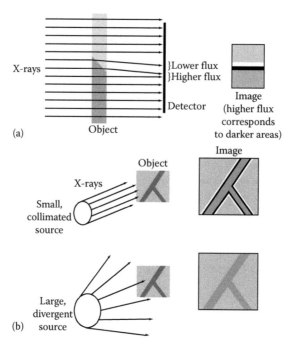

FIGURE 20.1   A widely used contrast mechanism for microradiology with coherent x-rays. (a) As a collimated x-ray beam passes through an object with two different regions and a tapered edge between them, the refraction-induced deviations create a bright–dark fringe pair at the detector that enhances the visibility of the edge. (b) The effect is visible if the x-ray beam is collimated and emitted by a small source; otherwise penumbra effects wash it out. This illustrates the need for spatial (or lateral) coherence.

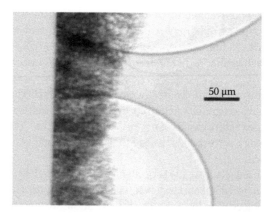

FIGURE 20.2   An example illustrating the practical application of the mechanism of Figure 20.1 to the study of a copper electrodeposition process. Note in particular the dark–bright fringe pair marking the edge of each gas bubbles.

source size emitting a collimated beam is required to see the refraction-related edge enhancement. If the source is too large and/or the emission is too divergent, penumbra-like effects wash out the edge enhancement.

A conventional x-ray source is large and divergent, therefore not suitable for this task. It could in principle be converted into a small-size source by using a screen with a pinhole. The negative side effect, however, is to waste most of the source emission, which makes it difficult or impossible to achieve high signal level together with high space and time resolution.

The solution is offered by modern synchrotron sources that are based on the emission of accelerated electrons in accelerators of the class known as "storage rings" [23,42]. This emission comes from a small source area (of the order of microns) and as an exceedingly small angular divergence.

Such properties correspond in fundamental optics [23] to a high level of "spatial coherence" or "lateral coherence." Thus, the implementation of the new type of radiology requires highly (spatially) coherent x-rays. This explains the name utilized for the technique: "coherent radiology."

Before the advent of synchrotron sources of the third generation [23], coherent x-rays were in practice unavailable. In recent years, coherent radiology has grown exponentially [1–21,41], with broad applications ranging from biology and medical research to a variety of materials problems. Several reviews are available [22,24,27] for a detailed description of the technique and of its potential and limitations. For our present scope, a simple picture like that of Figure 20.1 is quite sufficient.

One should also specifically note that coherent radiology is quite easy to use. It has no vacuum requirements as many other synchrotron techniques, and it delivers results rapidly with instrumentation that is simpler than most other synchrotron beamlines. It can be readily applied [1–21,41] to a variety of condensed-matter systems including liquids and mixed liquids and solids.

The first appearance [1,21] of coherent radiology in electrodeposition studies dates to 2002. As we shall see in the next section, the impact was immediately important. The unprecedented levels of spatial and time resolution made it quickly possible to observe some surprising and quite counter-intuitive phenomenon.

20.1.3   First Results: Building on Bubbles

The first electrodeposition studies [1,21] with coherent microradiology were performed on Zn coatings on copper electrodes. The real-time experiments were performed *in situ* using a specially designed mini electrochemical cell with Kapton films windows (see the following text). The current density was controlled using a copper rod covered with corrosion-resisting epoxy. This coating blocked electrical conduction everywhere except for a small window. The tests were performed at constant voltage bias or at constant current to investigate the effects related to the electric field and to the current density. This enabled us to verify that the final electrodeposited coating quality depends on such parameters and can be optimized by controlling them.

Figures 20.3 and 20.4 show the most surprising result of these first experiments. This is the observation of metal deposition on the rapidly evolving gas bubbles developed during electrodeposition. The phenomenon [42–47] has been hypothesized by some authors but its counter-intuitive character made it hard to believe without direct observation, which was impossible before the advent of coherent microradiology. The effects of gas bubbles on the coating quality were already empirically known [1,21,42–47]. During the coating process, water electrolysis at the cathode produces hydrogen bubbles. An excessive generation of such bubbles results in poor morphology and poor adhesion of the coating.

The coating quality [1,21,42–47] was in fact negatively affected in the presence of bubbles by the development of void defects. An intriguing hypothesis explained these defects as negative finger-prints of the bubbles, created by their dynamic coating with a metal overlayer. Such a mechanism occurred on a short timescale and for small bubbles: that is why, as we mentioned, it could not be easily observed before the advent of real-time coherent microradiology.

Figures 20.3 and 20.4 remove all questions about this surprising mechanism by directly showing that Zn layers do indeed grow on the gas bubbles. After the initial coating, the bubble disappears

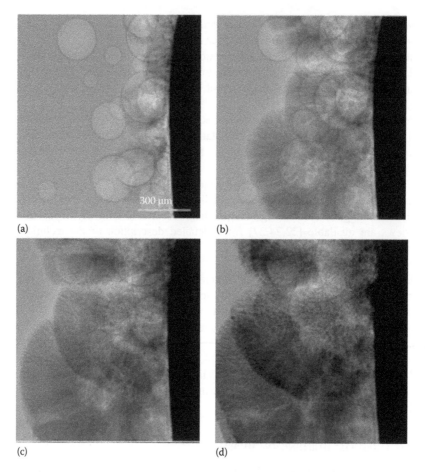

FIGURE 20.3 A series of microradiographs (Ref. [3]) taken in real time with coherent x-rays, showing the presence of bubbles during the electrodeposition of zinc and revealing the fact that deposition occurs dynamically on the bubble surfaces. Images (a) through (d) are different snapshots taken with high time resolution and illustrating the growth on bubbles at different times.

leaving a spherical void that constitutes a coating defect with negative impact on the quality. The phenomenon is clearly visible both in Figures 20.3 and 20.4, extracted from movie sequences. Figure 20.4 specifically reveals that the coating only occurs on a subset of the bubbles and appears related to their contact with the electrode.

20.2 CASE STUDIES

The initial success presented in the previous chapter paved the way to a series of systematic studies linking the final deposit, its structure, and other characteristics to relevant growth parameters, in view of possible optimization strategies. We discuss here some relevant case studies: Zn, Cu, and Ni coatings. Such studies elucidate in more detail the role of bubbles during the growth.

20.2.1 ZINC-RAMIFIED STRUCTURES

The main result of a more detailed investigation of Zn electrodeposition is that when intense hydrogen bubbling occurs at high potential or current density, the morphology of the ramified zinc deposit changes from dense-branching to fern-shaped dendrite [21]. The fern-shaped dendrite morphology results in part from the constricted growth due to hydrogen bubbles but also from the highly concentrated electric field.

This morphology was observed during the early stages of electrodeposition for both the potentiostatic and galvanostatic modes; however, the deposit plated in the galvanostatic mode densified via lateral growth during the later deposition stages. This indicates that potentiostatic plating is better than galvanostatic plating for fabricating fern-shaped deposits, which are, for example, ideal electrodes for Zn-air batteries due to the relatively large specific area.

20.2.1.1 Background to Zinc-Ramified Structures

We already mentioned that hydrogen bubble formation during metal electrodeposition is often considered a nuisance because of its negative impact on the quality of the deposit [48,49]. However, one can also take advantage of the bubbles as they can provide active sites for zinc electrocrystallization. Hydrogen bubbles can also serve as a dynamic template for fabricating self-supporting three-dimensional (3D) foams of metal deposit [50]. Our studies demonstrated that hydrogen bubbles directly affect the growth morphology and can make it possible to control the dendrite growth as required for certain important applications. This can offer the opportunity to exploit the presence of bubbles to obtain positive results.

Electrodeposition is used to fabricate ramified structures with high surface areas [51–55], in particular zinc structures. On the other hand, zinc is widely used in battery electrode technology. To enhance the battery efficiency, zinc electrodes with high surface areas are favored, and their fabrication is one of the objectives of electrodeposition research.

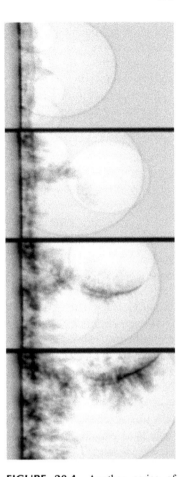

FIGURE 20.4 Another series of images illustrating the phenomenon in Figure 20.3.

A wide variety of ramified Zn deposit patterns can be obtained by varying the growth conditions [56–58]. For example, Lopez and Choi [59] studied zinc dendrite morphologies electroplated in nonaqueous formamide media. Their results indicate that the morphological features of dendrite, such as particle size, shape, and the degree of branching, markedly influence the physical, chemical, and electrochemical properties of the zinc electrodes. However, the detailed understanding of the electrochemistry and the physics of the ramified growth [60–64] was so far limited by the lack of suitable monitoring instruments.

Coherent microradiology with synchrotron x-rays enabled us to investigate in detail the hydrogen bubble formation process and the corresponding electrodeposition morphology in three dimensions and in a standard environment. We observed morphological transitions of the ramified zinc deposits—electrodeposited in both the galvanostatic and potentiostatic modes—that are clearly linked to the presence of hydrogen bubbles.

Specifically, we could quantitatively measure the hydrogen bubble formation and assess its dependence on the applied potential and the impact on the resulting morphology of the deposit [65]. The studies included the macroscopic analysis of the growth morphology, the removal of the deposit from the electrolyte for microanalysis, and calculations of the metal electrodeposition efficiency with qualitative information on the bubble formation.

The observed morphology changes from dense-branching [66,67] to fern-shaped dendrite in the potentiostatic mode. They are directly related to the increased hydrogen bubble formation.

The hydrogen bubbles also affected the ramified growth in galvanostatic electrodeposition. Lateral growth was found to occur during the later plating stages when the effect of the hydrogen bubbles decreases. Among the different morphologies, the fern-shaped dendrite has the highest surface area and, as we mentioned, can be of practical use in batteries—if its growth can be controlled as indicated by our results.

Previous real-time studies of ramified electrodeposited formations were performed in two-dimensional cells due to the depth of focus problem while using optical microscopy. The role of hydrogen bubbles could not be investigated in detail since the bubbles in two-dimensional electrodeposition cells markedly retard the development of ramified electrodeposits. On the contrary, our experimental approach enabled us to monitor hydrogen bubble evolution and the related Zn deposit morphological changes in 3D and with no restrictions on the electrodeposition geometry.

20.2.1.2 Experimental Procedure

Zinc was electrodeposited either in the potentiostatic (constant potential) mode or in the galvanostatic (constant current) mode at room temperature in an aqueous solution composed of 2.2 M $ZnCl_2$ and 4.8 M KCl without stirring. The electrolyte was prepared with analytical chemicals and deionized water, and deaerated with bubbling nitrogen prior to electroplating.

A three-electrode electrochemical cell was specifically designed for our *in situ* microradiography tests. The cell was machined in Teflon and had two windows with adjustable positions to control the solution thickness. A 5 mm thick solution was selected so as to avoid unnecessarily excessive x-ray absorption by the electrolyte while providing an adequate space for ramified growth. The distance between the cathode and the anode was adjusted to 10 mm.

The working electrode was made of copper rod, had a cross-sectional area of $0.5 \times 2 \text{ mm}^2$ and was embedded in corrosion-resisting epoxy. Before each run, the electrode was polished with 2400 grit emery paper, cleaned in a 5% sulfuric acid solution for 5 s, and finally rinsed with deionized water. The counter-electrode was an electrolytic zinc plate, and the reference electrode was a saturated calomel electrode (SCE) positioned close to the working electrode using a Luggin capillary. Electroplating was performed using an Autolab PGSTAT 30 potentiostat/galvanostat. All potentials were measured with respect to the SCE. During electrodeposition, microradiography with coherent x-rays was performed in real time using the 7B2 beamline of Pohang light source, Korea [68] and the 01A beamline of NSRRC, Taiwan.

In addition to *in situ* coherent microradiology observation, the electric field distribution across zinc anode and copper cathode was calculated with the ANSYS software based on the assumption that the electrolyte composition is uniform and the solution's electrical receptivity remains constant during electrodeposition.

20.2.1.3 Some Results on Zinc-Ramified Structures

It is well known that the common "well defined morphologies"—such as the diffusion limiting aggregation, the dense-branching morphology, and the dendrite morphology [56,57,66,67]—cannot be obtained in a two-dimensional electrodeposition cell when hydrogen bubbles are developed together with the ramified growth [69]. In fact, the deposition is retarded or jeopardized and spongy deposits are usually formed instead of a well-defined morphology. On the contrary, a well-defined morphology does often coexist with hydrogen bubbles in 3D environments [70].

Two-dimensional experiments with optical microscopy are also affected by other negative factors. Laser confocal microscopy has a relatively slow imaging speed and is not suitable for real-time electrodeposition studies. Furthermore, hydrogen bubbles interfere with optical microimaging making it difficult to observe their effects. Why, then, have two-dimensional studies so far dominated this domain? The answer is that they are considerably simpler than those in 3D since they are not affected by complications like convection and, indeed, hydrogen bubbles. It is time,

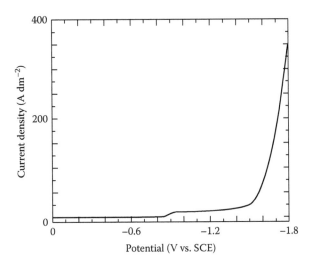

FIGURE 20.5 A typical cathodic polarization curve of zinc electrodeposition in the solution used for this study.

however, to move from two dimensions to 3D: coherent microradiology enabled us to do it by exploiting the high penetration of x-rays.

Before performing imaging experiments, we measured the cathodic polarization (IV) curve at a scan rate of 0.01 V s^{-1} (Figure 20.5) to determine the basic electrochemical parameters. Such a curve should not be considered a rigorous measurement of the electrochemical properties of the electrolyte because the changes in the electrode surface and morphology cannot be ignored. In our case, even during a fast IV scan the deposit on the cathode changed. However, the main characteristics of the electrolyte could still be qualitatively derived.

For example, an obvious plateau was observed at ~13 A dm^{-2} and identified as the "limiting current density" [70] determined by the maximum rate of mass transfer by diffusion. When the negative potential increased above this plateau region, the current density dramatically increased and intense hydrogen bubble formation dominated the electrochemical reaction, and the clear threshold, as seen in Figure 20.5, indicated the onset of water electrolysis.

A small number of hydrogen bubbles can be already observed when Zn is electrodeposited slightly below the limiting current density region. Figure 20.6 shows an example of galvanostatic deposit obtained at 13 A dm^{-2}. We saw a dense-branching structure growing linearly with the electrodeposition time. The formation of hydrogen bubbles did not appear to affect the ramified growth and the deposit became denser as the deposition continued. This is the typical growth morphology without the interference of the hydrogen bubbles.

Above the limiting current density region, the electrodeposition mechanism was dominated by the creation of hydrogen bubbles; simultaneously, a morphological transition was observed during the ramified growth. Sequential microradiographs of the deposit plated in the galvanostatic mode of 20 A dm^{-2} are shown in Figure 20.7.

In the early stages of electrodeposition, hydrogen bubbles, leaving a limited number of spots available for deposit nucleation and limited directions for deposit growth, covered most of the cathode surface. A fern-shaped dendrite was formed and consisted of a main stem, around which many side-branches were quite uniformly distributed. The fern-shaped structure grew outward within a short period of time (Figure 20.7a through c) producing a completely nucleated structure— before lateral growth prevailed at a rate significantly higher than that of the outward growth. The fern-shaped dendrite morphology gradually disappeared and a dense deposit formed with top-heavy characteristics (Figure 20.7d through h).

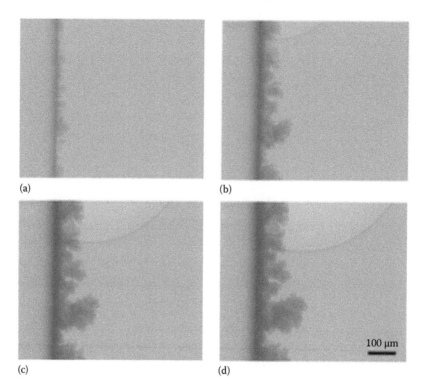

FIGURE 20.6 Sequential microradiographs of Zn dense-branching growth on copper in the galvanostatic mode at 13 A dm^{-2}, taken at different times during growth: (a) 99 s, (b) 255 s, (c) 411 s, and (d) 515 s.

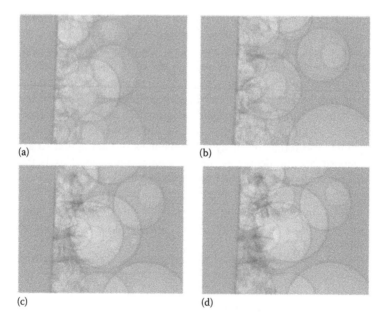

FIGURE 20.7 Sequence of microradiographs showing a zinc deposit on copper obtained with galvanostatic electrodeposition at 20 A dm^{-2}. The images correspond to the deposition times (a) 5.4 s, (b) 10.6 s, (c) 15.8 s, (d) 21 s.

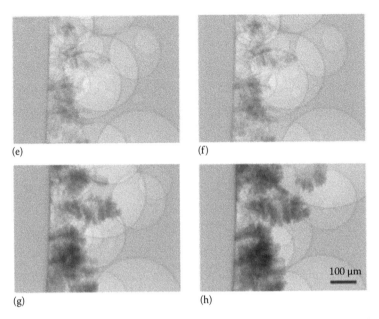

(e)   (f)

(g)   (h)

100 μm

FIGURE 20.7 (continued)   (e) 26.2 s, (f) 31.4 s, (g) 47 s, and (h) 151 s. Images (a–d) show that the deposited zinc forms fern-shaped dendrites with outward growth, whereas images (e–h) show dense-branching dendrites with lateral growth.

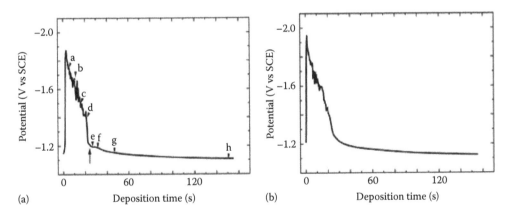

(a)   Deposition time (s)   (b)   Deposition time (s)

FIGURE 20.8   (a) Potential transient curve at 20 A dm^{-2}. The arrows labeled (a–h) correspond to the microradiographs in Figure 20.7. (b) Potential transient curve at 30 A dm^{-2}.

The electrostatic potential was measured *in situ* during the galvanostatic electroplating. Figure 20.8a shows a potential transient curve at 20 A dm^{-2}. The curve can be divided into three stages: when the current was switched on, the voltage magnitude immediately rose to a given value, then drastically dropped over a few seconds, and afterward decreased more gradually. These stages correspond to the transitions in the electrode conditions [71].

The first stage was associated with the double layer charging, when the conductivity of the electrolyte went to zero as the concentration of zinc ions became negligible at the cathode. After the maximum voltage magnitude, a linear decrease with the plating time was observed together with the appearance of fern-shaped dendrites. Such a decrease, occurring instead of the expected steady voltage, is caused by the evolution of the deposit.

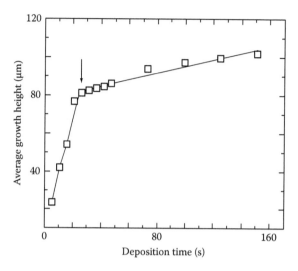

FIGURE 20.9 Average growth height versus deposition time for zinc galvanostatic electrodeposition at 20 A dm^{-2}.

Indeed, the ramified structure increases the cathodic surface area and therefore reduces the potential magnitude at constant current. Strong hydrogen bubble formation interfered with the potential measurements leading to voltage oscillation during the second stage. During the final stage, the voltage magnitude gradually decreased (the voltage changed approximately from -1.2 to -1.0 V) and hydrogen bubble creation nearly terminated. We found that the potential of the final stage in the potential transition curves was always -1.2 to -1.0 V in the region above limiting current density. Experiments performed with a current density of 30 A dm^{-2} yielded the same potential trend (Figure 20.8b). For all current densities, the last growth stage resulted in a dense branching morphology instead of an outward growth, developed by lateral growth of the fern-shaped dendrites.

Figure 20.9 shows the average growth height (measured from the substrate/deposit interface to the deposit surface) as a function of the electrodeposition time at 20 A dm^{-2}. The fern-shaped dendrites grew generally faster than dense-branching deposits regardless of the current density. The average growth height of the fern-shaped dendrites increased linearly with time when the outward growth prevailed. The growth rate of the fern-shaped dendrite decreased dramatically at the transitional point (the arrow in Figure 20.9), and a smaller growth rate was observed afterward for the dense-branching deposit. This rate change coincided with the slope change of the potential transient curve, marked by an arrow in Figure 20.8a.

Quite interesting were the Zn electrodeposition studies for potential magnitudes well above the limiting current density. Figure 20.10 shows an example obtained with an applied potential of -1.8 V. The hydrogen bubble effects dominate the deposition and fern-shaped dendrites are formed. Unlike what happens in the galvanostatic mode, the zinc deposit remained fern-shaped dendrite throughout the entire potentiostatic-mode deposition. The morphology transition from the fern-shaped dendrite with a major stem in the axial direction to the dense-branching observed in Figure 20.7 can thus be attributed to the bubble formation.

Figure 20.11 shows a schematic diagram for the formation of fern-shaped dendrites. Once an existing dendrite is surrounded by a couple of hydrogen bubbles, the concentration of zinc ions changes: a Zn^{2+}-depleted zone appears in the space between the two bubbles and a zone with the original Zn^{2+} concentration is found outside them. The Zn^{2+} concentration difference between these two zones can cause local convection [72,73], replenishing the zinc ions necessary for the growth of the main stem of the fern-shaped dendrites (Figure 20.11a). The creation of hydrogen bubbles suppresses the lateral growth of fern-shaped dendrites, whereas a fast tip growth is

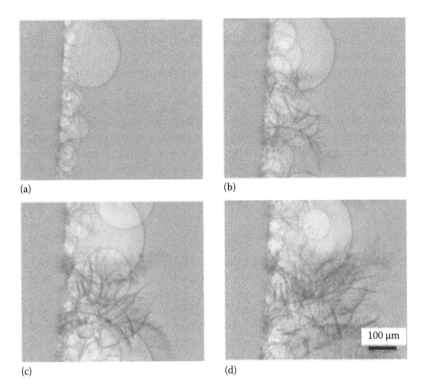

(a)   (b)

(c)   (d)

FIGURE 20.10   Sequential microradiographs of fern-shaped dendrite growth at −1.8 V, taken at different times: (a) 5.4 s, (b) 10.6 s, (c) 15.8 s, and (d) 21 s.

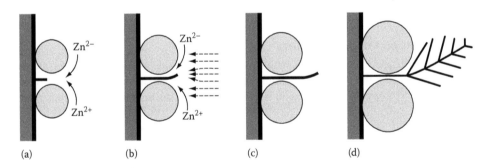

(a)   (b)   (c)   (d)

FIGURE 20.11   Schematic diagram showing the formation of fern-shaped dendrites. (a) The circles indicate hydrogen bubbles. (b) The dotted arrowhead lines indicate the electric field. (c and d) The main stem of the fern-shaped dendrite grows between two hydrogen bubbles.

promoted by the concentrated electric field (Figure 20.11b and c). When the main stem outgrows the surrounding hydrogen bubbles, its side-branches start to split, leading to the appearance of a fern-shaped dendrite morphology (Figure 20.11d).

In order to reach a better understanding of this phenomenon, we performed a simple theoretical electric field analysis. We found that the electric field is not only concentrated on the dendrite tip but also distributed on the dendrite stem. This causes the growth of a long dendrite stem and side-branches. In the case of galvanostatic plating, the lateral growth gradually dominates the overall ramified formation because the potential magnitude drops as the electrodeposition continues, reaching values that are too low for sufficient bubble formation and for the growth of fern-shaped dendrites. As a result, in the galvanostatic mode, the fern-shaped dendrites evolve to a more compact ramified deposit.

This is an important result for practical applications. The tendency to eventually produce densified growth makes the galvanostatic mode less suitable than potentiostatic plating for technologies requiring large surface areas—since fern-shaped dendrites have much larger surface areas than dense-branching structure. These interlaced dendrites, once collapsed after removal from the electrolyte, produce a porous structure.

Zinc electrodeposits composed of fern-shaped dendrites were shown to possess large electrochemically active surface areas [59], ideal for Zn-air batteries. Our results suggest that products of this type can be obtained by controlling the hydrogen bubble formation. Such a practical conclusion can lead to interesting applications in electrodeposition technology, transforming bubbles from negative factors to potentially useful ones. In general, our results stress once again the importance of real-time studies—made possible by coherent radiology—to fully understand electrodeposition and identify new ways to exploit its potential capabilities.

20.2.2 COPPER-RAMIFIED STRUCTURES

Copper electrodeposition is an important process in microelectronics technology. An unusual deposit is often formed at the edge of the micropatterns, typically a ramified structure [74]. This is an unwanted feature whose creation merits a careful study.

An excess amount of sulfuric acid is usually added to the copper electrodeposition solution to increase the bath conductivity [75]. We used coherent microradiology to monitor the ramified growth with and without sulfuric acid and observed striking and interesting differences.

20.2.2.1 Experimental Procedure

The electrolyte composition was 0.24 M copper sulfate; for the second part of the experiments, we added sulfuric acid. Tests were performed with 0.5–1.5 M of sulfuric acid obtaining the same type of morphology. The solutions were deaerated with bubbling nitrogen before experiments. The depositions were performed at ambient temperature without stirring. The rest of the procedure was similar to that used for Zn electrodeposition, with the necessary modifications.

20.2.2.2 Results on Copper Structures

In the first set of experiments, we investigated copper electrodeposition on micropatterns with different aspect ratios in the presence of sulfuric acid. Figure 20.12 shows the results for an applied potential of −0.2 V. The effect of hydrogen bubbles is evident: the bubbles block the microtrenches

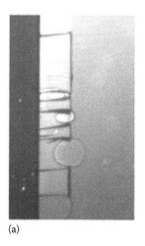
(a)

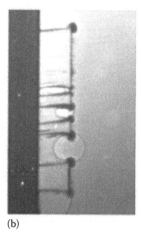
(b)

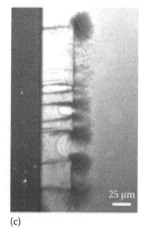
(c)

FIGURE 20.12 Copper plating on a micropattern at −0.2 V, showing the ramified structure growing on the micropattern edges. Images (a) through (c) illustrate different stages of the growth.

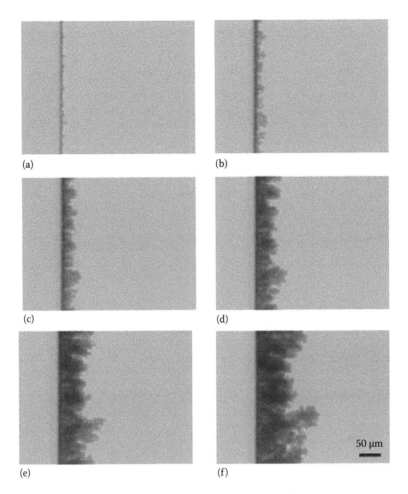

(a)   (b)

(c)   (d)

(e)   (f)

50 µm

FIGURE 20.13 Sequence of microradiographs showing copper-ramified structures growing on a copper substrate at an applied potential of −0.5 V from an electrolyte consisting of 0.24 M copper sulfate. The images were taken at the deposition times (a) 47 s, (b) 99 s, (c) 151 s, (d) 255 s, (e) 359 s, and (f) 515 s.

preventing copper penetration and filling. We also see the ramified deposit growing at the edges, although the applied potential magnitude is below the limiting current density region.

We then investigated copper electrodeposition on a flat substrate with and without sulfuric acid, both at the limiting current density region and above it. Figure 20.13 shows results without sulfuric acid for an applied potential of −0.5 V, at the limiting current density region. The ramified deposit evolved by repeated tip splitting and no main stem was observed. As the growth continued, this type of deposit became dense.

The growth evolution is quite different in Figure 20.14 that corresponds to an applied voltage of −0.9 eV, a magnitude well above the limiting current density. The morphology progressively changes as schematically shown in Figure 20.15: a needle-like dendrite is present in the first stage, with a main stem and several side-branches (Figure 20.15a). The second stage corresponds to tip splitting that starts from front end of needle-like dendrite and subsequent splits everywhere (Figure 20.15b and c). In the third stage, both tip thickening and tip splitting occur. Such changes are well correlated with the measured current density and can be explained with simple theoretical arguments. No hydrogen bubbles are observed in this case.

On the contrary, hydrogen bubbles are developed when sulfuric acid is present and their rate of creation increases with the applied voltage. Figure 20.16 shows an example for an applied potential

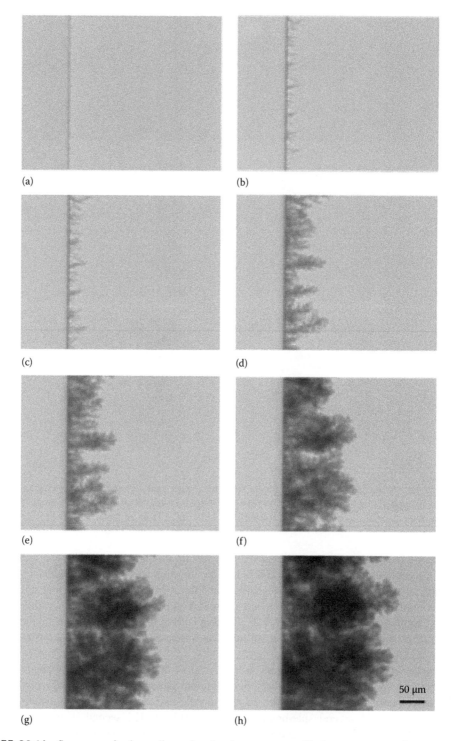

FIGURE 20.14 Sequence of microradiographs showing copper-ramified structures growing on a copper substrate at an applied potential of −0.9 V from the same electrolyte as in Figure 20.13. The images were taken at the deposition times (a) 21 s, (b) 31.4 s, (c) 36.6 s, (d) 73 s, (e) 99 s, (f) 203 s, (g) 359 s, and (h) 515 s.

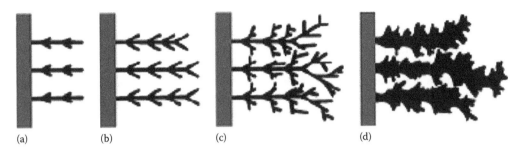

FIGURE 20.15 Schematic diagram showing three stages of the growth process of Figure 20.14: (a) needle-like dendrite formation, (b and c) the tip-splitting stage, and (d) the combined tip-thickening and tip-splitting stage.

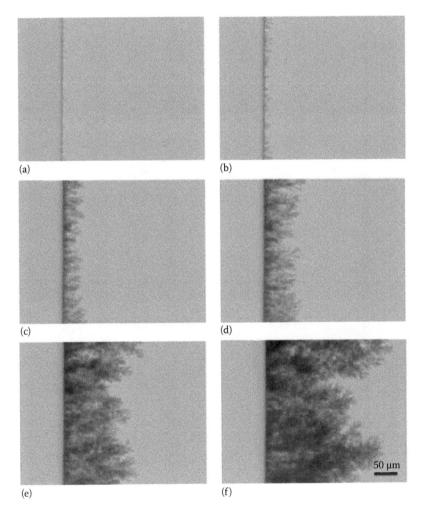

FIGURE 20.16 Sequence of microradiographs showing copper-ramified structures grown on a copper substrate at an applied potential −0.5 V from an electrolyte consisting of 0.24 M copper sulfate and 0.5 M sulfuric acid. The images were taken at the deposition times (a) 26.2 s, (b) 47 s, (c) 99 s, (d) 151 s, (e) 307 s, and (f) 515 s.

of −0.5 V. We initially see an anisotropic structure consisting of needle-like dendrites dispersed on cathode surface. In the second stage, we observe tip-splitting growth and eventually the morphology changes into a thin-branching structure with no main stems and thinner branch tips. The size of the branch tips decreases as the applied potential increases.

Measurements of the current density revealed a linear increase as the applied potential increases up to a certain level, followed by fluctuations. This phenomenon is related to the presence of hydrogen bubbles, and specifically to the bubble growth, detachment, coalescence, and collapse. This agrees with the results of Kahanda and Tomkiewicz [76] who explained the current density fluctuations in terms of the competition between the metal deposition and the hydrogen evolution.

In order to better assess the role of hydrogen bubbles, we injected diluted sulfuric acid (0.05 M) into a 0.24 M copper sulfate solution (pH value 2.41) during growth. Figure 20.17 illustrates the

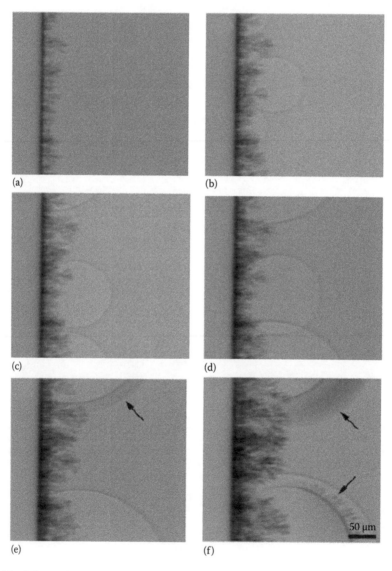

FIGURE 20.17 Microradiographs of the ramified Cu growth on a copper electrode at −0.7 V from a 0.24 M copper sulfate solution, showing the change of growth pattern after the injection of diluted sulfuric acid close to the cathode surface for 27 s, after 87 s electrodeposition. The images were taken at the deposition times (a) 60.9 s, (b) 96.6 s, (c) 102.9 s, (d) 123.9 s, (e) 144.9 s, and (f) 207.9 s.

results for growth at −0.7 V. Before injection, we see the deposition of a dense branching structure (Figure 20.17a). As the diluted sulfuric acid is injected, hydrogen bubbles are generated (Figure 20.17b) and the local pH drops to ∼1.9 whereas that of the bulk solution remains at ∼2.41. With continued injection, the hydrogen bubble generation becomes more intense (Figure 20.17c and d). After stopping the injection, the deposit grows with a dendrite morphology on the upper hydrogen bubbles (upper arrow in Figure 20.17e and f); however, the newly nucleated copper deposit on the lower hydrogen bubble exhibits a dense-branching morphology (lower arrow in Figure 20.17f). The original deposit on cathode surface still has a dense-branching morphology as the electrodeposition continues.

All these observations stress the importance of bubbles during growth. Their role cannot be neglected in the theoretical interpretations of the interplaying phenomena and in the search of optimal conditions to obtain high-quality copper deposits.

20.2.3 NICKEL OVERLAYERS

Nickel electrodeposition is often performed in the presence of boric acid; the corresponding mechanisms and the impact on the deposit quality are still controversial. Our studies demonstrate that the growth morphology is radically different with or without boric acid and that the mechanism is related once again to the formation of hydrogen bubbles.

20.2.3.1 Experimental Procedure

Two electroplating solutions were used in our experiments: the first one had 100 g/L of nickel sulfate, 10 g/L of nickel chloride, and no boric acid; the pH value was 4. The second (Watts bath) had 330 g/L of nickel sulfate, 45 g/L of nickel chloride, and 38 g/L of boric acid; the pH value was again 4. In each case, the solution was deaerated with bubbling nitrogen before the experiments and electrodeposition was performed, as in the previous cases, at ambient temperature without stirring.

20.2.3.2 Results on Nickel Overlayers

The basic findings obtained with real-time microradiology were the following. Nickel electrodeposition without boric acid is accompanied by strong hydrogen bubble evolution. The corresponding deposits are affected by internal stress and by crack formation. On the contrary, in the Watts bath the hydrogen bubble evolution was quite gentle and the resulting morphology was dense and uniform.

Figures 20.18 through 20.20 visually illustrate these findings. Figure 20.18 shows that in the absence of boric acid bubbles are present and the deposit is of rather poor quality. This is in sharp contrast with Figures 20.19 and 20.20 that correspond to electrodeposition in the Watts bath at two different applied voltages, −1 and −4 V. The quality improvement with respect to Figure 20.18 is quite striking. Although the boric acid does not altogether suppress the evolution of bubbles, it does strongly decrease it with the reduction of the cracks in the deposit.

20.3 CONCLUSIONS AND OUTLOOK

Both microradiology with coherent x-rays and its application to the study of electrodeposition phenomena are in their infancy. In particular, the potential limits in time and space resolution have not yet been reached. Nevertheless, this approach already yielded very important information, quite difficult to obtain otherwise. Considering the contrast mechanism, microradiology with coherent x-rays is particularly effective in the study of electrodeposition. Specifically, it can easily and dynamically image the substrate–overlayer system in great detail as well as the gas bubbles. This makes it specifically possible to directly monitor the role of the bubbles in a series of important systems. The results are in some cases surprising and in general very important for understanding the factors that influence the final deposit quality, and to optimize them.

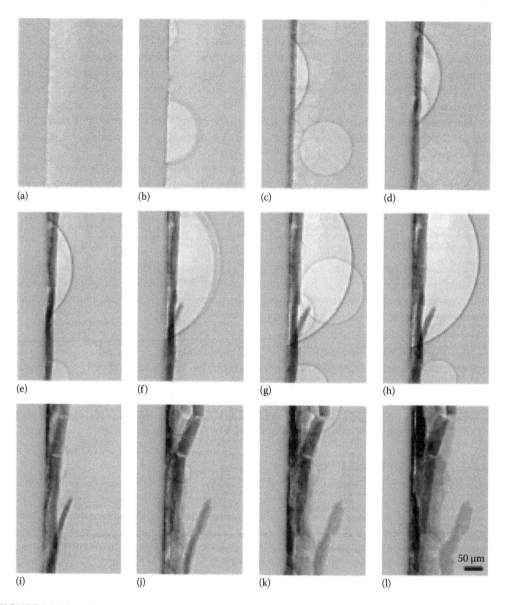

FIGURE 20.18   Nickel deposition on a copper substrate and fracture process stages at an applied potential of −1.6 V. The sequential microradiographs were taken at the deposition times (a) 21.6 s, (b) 22.8 s (note that the low contrast on hydrogen bubbles is due to the too rapid hydrogen bubble evolution), (c) 34.8 s, (d) 43.2 s, (e) 46.8 s, (f) 52.8 s, (g) 57.6 s, (h) 63.6 s, (i) 67.2 s, (j) 80.4 s, (k) 94.8 s, and (l) 108 s.

In addition to the role of bubbles, our technique can provide other kinds of valuable information. We specifically note the possibility to image the ionic concentration; see Ref. [2]. This creates another significant difference between our approach and more conventional studies: the corresponding information is not easy to obtain otherwise. We also note that the technique can be applied in real time to monitor electrodeposition on microscopic structures. This approach can be used, for example, to study the growth on microtrenches. Once again, comparable dynamic information would be quite difficult to obtain otherwise, and is very valuable for many practical applications. The near future of this developing field is therefore quite clear: our general approach can be applied to a variety of electrodeposition phenomena. The technique can be specifically used to optimize the

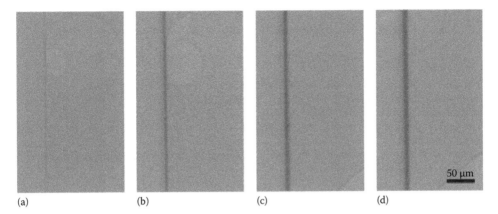

(a)   (b)   (c)   (d)

FIGURE 20.19 Sequential microradiographs of nickel deposition at -1 V in the presence of boric acid solution (Watts bath). Images (a) through (d) illustrate different stages of the growth.

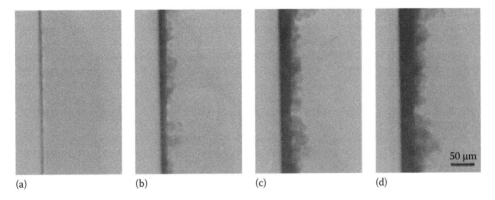

(a)   (b)   (c)   (d)

FIGURE 20.20 Sequential microradiographs of nickel deposition at -4 V. Images (a) through (d) illustrate different stages of the growth.

parameters in real time and to transfer this valuable information to researchers dealing with practical technological applications. In the future, we expect to use the new performances that are being developed in microradiology with coherent x-rays for this type of studies. Better time and space resolution are expected. Two other important elements are the introduction of quantitative analysis and the implementation of the technique in a microtomographic mode [8,11]. As far as this second development is concerned, notwithstanding the issues concerning the 3D tomographic reconstruction of phase images, spectacular progress has been already achieved.

REFERENCES

1. W. L. Tsai, P. C. Hsu, Y. Hwu, C. H. Chen, L. W. Chang, J. H. Je, H. M. Lin, A. Groso, and G. Margaritondo, *Nature* **417** (2002) 139, and the references therein.
2. M. Schlesinger and M. Paunovi, *Modern Electroplating*, Wiley, New York, 2000.
3. C. Steinbruchel and B. L. Chin, *Copper Interconnect Technology*, SPIE, Washington, DC, 2001.
4. M. Madou, *Fundamental of Microfabrication*, CRC Press, Boca Raton, FL, 1997.
5. J. O. Duković IBM, *J. Res. Dev.* **34** (1990) 693.
6. C. Chakkaravarthy, A. K. Abdul Waheed, and H. V. K. Udupa, *J. Power Sources* **6** (1981) 203.
7. H. Chang and C. Lim, *J. Power Sources* **66** (1997) 115.

8. C. Monroe and J. Newman, *J. Electrochem. Soc.* **150** (2003) A1377.
9. Y. F. Yuan, J. P. Tu, H. M. Wu, B. Zhang, X. H. Huang, and X. B. Zhao, *J. Electrochem. Soc.* **153** (2006) A1719.
10. N. Takami, T. Ohsaki, H. Hasabe, and M. Yamamoto, *J. Electrochem. Soc.* **149** (2002) A9.
11. L. Fu, *J. Phys. Conf.* **34** (2006) 700.
12. S. Rondot, O. Aaboubi, J. Cazaux, and A. Olivier, *J. Phys. Chem. B* **101** (1997) 6695.
13. H. Nissenson and H. Danneel, *Z. Elektrochem.* **9** (1903) 760.
14. V. P. Chalyi, *Chem. Abstr.* **47** (1953) 11047b.
15. N. Hecker, D. G. Grier, and L. M. Sander, in *Fractal Aspects of Materials*, R. B. Laibowitz, B. B. Mandelbrot, and D. E. Passoja (Eds.), Materials Research Society, University Park, TX, 1985.
16. P. Garik, D. Barkey, E. Ben-Jacob, E. Bochner, E. Bochner, N. Broxholm, B. Miller, B. Orr, and R. Zamir, *Phys. Rev. Lett.* **62** (1989) 2703.
17. J. R. Melrose, D. B. Hibbert, and R. C. Ball, *Phys. Rev. Lett.* **65** (1990) 3009.
18. A. Kuhn and F. Argoul, *J. Electroanal. Chem.* **371** (1994) 93.
19. A. Kuhn and F. Argoul, *Phys. Rev. Lett.* **49** (1994) 4298.
20. V. Fleury, M. Rosso, and J.-N. Chzazlviel, *Phys. Rev. A* **43** (1991) 6908.
21. W. L. Tsai, P. C. Hsu, Y. Hwu, C. H. Chen, L. W. Chang, J. H. Je, and G. Margaritondo, *Nuclear Instrum. Methods B* **199** (2003) 451.
22. Y. Hwu, J. H. Je, J. M. Yi, and G. Margaritondo, in *Highlights of Spectroscopies on Semiconductors and Nanostructures*, G. Guizzetti, L. C. Andreani, F. Marabelli, and M. Patrini (Eds.), Italian Physical Society, Bologna, Italy, 2007, p. 375.
23. G. Margaritondo, *Elements of Synchrotron Light for Biology, Chemistry, and Medical Research*, Oxford University Press, New York, 2002.
24. G. Margaritondo, Y. Hwu, and G. Tromba, in *Synchrotron Radiation: Fundamentals, Methodologies and Applications*, S. Mobilio and G. Vlaic (Eds.), Società Italiana di Fisica, Bologna, Italy, 2003, p. 25.
25. Y. Hwu, J. H. Je, and G. Margaritondo, *Chinese J. Phys.* **43** (2005) 285; *Nucl. Instrum. Methods* **A551** (2005) 108.
26. S.-K. Seol, A.-R. Pyun, Y. Hwu, G. Margaritondo, and J. H. Je, *Adv. Funct. Mater.* **15** (2005) 934.
27. G. Margaritondo, Y. Hwu, and J. H. Je, *Riv. Nuovo Cimento* **27** (2005) 1.
28. B. M. Weon, J. H. Je, Y. Hwu, and G. Margaritondo, *Int. J. Nanotechnol.* **3** (2006) 280.
29. J. M. Yi, J. H. Je, Y. S. Chu, Y. Zhong, Y. Hwu, and G. Margaritondo, *Appl. Phys. Lett.* **89** (2006) 074103.
30. E. Gallucci, K. Scrivener, A. Groso, M. Stampanoni, and G. Margaritondo, *Cement Concrete Res.* **37** (2007) 360.
31. P.-C. Hsu, PhD thesis, Department of Materials Science and Engineering, College of Engineering, National Taiwan University, Taipei, Taiwan (unpublished).
32. G. Margaritondo and R. Meuli, *Eur. Radiol.* **13** (2004) 2633.
33. R. Meuli, Y. Hwu, J. H. Je, and G. Margaritondo, *Eur. Radiol.* **14** (2004) 1550.
34. F. Arfelli, M. Assante, V. Bonvicini, A. Bravin, G. Cantatore, E. Castelli, L. Dalla Palma, M. Di Michiel, R. Longo, A. Olivo, S. Pani, D. Pontoni, P. Poropat, P. Prest, A. Rashevsky, G. Tromba, A. Vacchi, E. Vallazza, and F. Zanconati, *Phys. Med. Biol.* **43** (1998) 2845.
35. F. Arfelli, V. Bonvicini, A. Bravin, G. Cantatore, E. Castelli, L. Dalla Palma, M. Di Michiel, M. Fabrizioli, R. Longo, R. H. Menk, A. Olivo, S. Pani, D. Pontoni, P. Poropat, P. Prest, A. Rashevsky, M. Ratti, L. Rigon, G. Tromba, A. Vacchi, E. Vallazza, and F. Zanconati, *Radiology* **215** (2000) 286.
36. A. Snigirev, I. Snigireva, V. Kohn, S. Kuznetsov, and I. Schelokov, *Rev. Sci. Instrum.* **66** (1995) 5486.
37. D. Chapman, W. Thomlinson, R. E. Johnston, D. Washburn, E. Pisano, N. Gmur, Z. Zhong, R. Menk, F. Arfelli, and D. Sayers, *Phys. Med. Biol.* **42** (1997) 2015.
38. A. Pogany, D. Gao, and S. W. Wilkins, *Rev. Sci. Instrum.* **68** (1997) 2774.
39. S. W. Wilkins, T. E. Gureyev, D. Gao, A. Pogany, and A. W. Stevenson, *Nature* **384** (1996) 335.
40. K. A. Nugent, T. E. Gureyev, D. F. Cookson, D. Paganin, and Z. Barnea, *Phys. Rev. Lett.* **77** (1996) 2961.
41. P. Cloetens, M. Pateyron-Salomé, Y. Buffière, G. Peix, J. Baruchel, F. Peyrin, and M. Schlenker, *J. Appl. Phys.* **81** (1997) 5878.
42. G. Margaritondo, *Introduction to Synchrotron Radiation*, Oxford University Press, New York, 1988.
43. D. H. Coleman, B. Popov, and R. E. White, *J. Appl. Electrochem.* **28** (1998) 889.
44. B. Popov, G. Zheng, and R. E. White, *Corrosion* **51** (1995) 429.
45. G. Zheng, B. Popov, and R. E. White, *J. Electrochem. Soc.* **140** (1993) 3153.
46. M. Zamanzadeh, A. Allam, C. Kato, B. Ateya, and H. W. Pickering, *J. Electrochem. Soc.* **129** (1982) 285.
47. H. Yumoto, Y. Kinase, M. Ishihara, N. Baba, and K. Kamei, *J. Surf. Sci. Soc. Japan* **17** (1996) 43.
48. D. H. Coleman, B. N. Popov, and R. E. White, *J. Appl. Electrochem.* **28** (1998) 889.

49. M. Monev, L. M. Mirkova, I. Krastev, Hr. Tsvetkova, St. Rashkov, and W. Richtering, *J. Appl. Electrochem.* **28** (1998) 1107.
50. H. C. Shin and M. Liu, *Chem. Mater.* **16** (2004) 5460.
51. J. W. Long, L. R. Qadir, R. M. Stroud, and D. R. Rolison, *J. Phys. Chem. B* **105** (2001) 8712.
52. J. S. Sakamoto and B. Dunn, *J. Mater. Chem.* **12** (2002) 2859.
53. G. Q. Zhang, X. G. Zhang, and H. L. Li, *J. Solid State Electrochem.* **10** (2006) 955.
54. K. Kinoshita, X. Song, K. Kim, and M. Inaba, *J. Power Sources* **81–82** (1999) 170.
55. C. C. Yang and S. J. Lin, *J. Power Sources* **112** (2002) 174.
56. Y. Sawada, A. Dougherty, and J. P. Gollub, *Phys. Rev. Lett.* **56** (1986) 1260.
57. D. Grier, E. Jacob, R. Clarke, and L. M. Sander, *Phys. Rev. Lett.* **56** (1986) 1264.
58. P. P. Trigueros, J. Charet, F. Mas, and F. Sagues, *J. Electroanal. Chem.* **312** (1991) 219.
59. C. M. Lopez and K. S. Choi, *Langmuir* **22** (2006) 10625.
60. J.-N. Chazalviel, *Phys. Rev. A* **42** (1991) 7355.
61. V. Fleury, J.-N. Chazalviel, and M. Rosso, *Phys. Rev. Lett.* **68** (1992) 2492.
62. V. Fleury, J.-N. Chazalviel, and M. Rosso, *Phys. Rev. E* **48** (1993) 1279.
63. C. Livermore and P. Z. Wong, *Phys. Rev. Lett.* **72**, 3847 (1994).
64. D. P. Barkey, D. Watt, Z. Liu, and S. Raber, *J. Electrochem. Soc.* **141** (1994) 1206.
65. N. D. Nikolic, K. I. Popov, Lj. J. Pavlović, and M. G. Pavlović, *J. Electroanal. Chem.* **588** (2006) 88.
66. E. Ben-Jacob, G. Deutscher, P. Garik, N. D. Goldenfeld, and Y. Lareah, *Phys. Rev. Lett.* **57** (1986) 1903.
67. E. Ben-Jacob, P. Garik, and D. Geier, *Superlattices Microstruct.* **3** (1987) 599.
68. S. Baik, H. S. Kim, M. H. Jeong, C. S. Lee, J. H. Je, Y. Hwu, and G. Margaritondo, *Rev. Sci. Instrum.* **75** (2004) 4355.
69. G. L. Kahanda and M. Tomkiewicz, *J. Electrochem. Soc.* **136** (1989) 1497.
70. D. Pletcher and F. C. Walsh, *Industrial Electrochemistry*, 2nd edn., Chapman and Hall, London, U.K., 1993.
71. J. R. Bruyn, *Phys. Rev. E* **56** (1997) 3326.
72. L. J. J. Janssen and J. G. Hoogland, *Electrochim. Acta* **15** (1970) 1013.
73. F. Ajersch, D. Mathieu, and D. L. Piron, *Can. Metall. Q.* **24** (1985) 53.
74. S. C. Chang, J. M. Shieh, K. C. Lin, B. T. Dai, T. C. Wang, C. F. Chen, M. S. Feng, Y. H. Chen, and C. P. Lu, *J. Vac. Sci. Technol. B* **19** (2001) 767.
75. J. W. Dini, *Modern Electroplating*, Wiley, New York, 2000, p. 65.
76. G. L. M. K. S. Kahanda and M. Tomkiewicz, *J. Electrochem. Soc.* **136** (1989) 1497.

21 Electrode Modification: Application in Organic Electrocatalysis

N. Alonso-Vante and K. B. Kokoh

CONTENTS

21.1   INTRODUCTION

The electrochemical processes are determined by Faraday's law according to which the quantity of reagent converted electrochemically is proportional to the current that crosses the surface of the electrode and the residual current capacitance. However, the electrochemical reaction is essentially a heterogeneous process, and for thermodynamic and kinetic reasons, the reaction is possible in a certain domain of potential on a defined electrode surface. It comprises several elementary processes, namely, mass transport of the reactive species toward the electrode, adsorption on an active site, the exchange of electrons, possible chemical reactions, desorption, and then mass transport from the electrode toward the solution, which describe the global reaction, as depicted in Figure 21.1.

Electrocatalysis is the interfacing science between liquid-phase heterogeneous catalysis and electrochemistry. According to a general definition, electrocatalysis may be defined as the

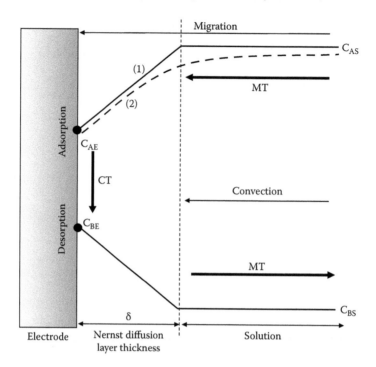

FIGURE 21.1 Schematic representation of an electrode–solution interface. MT, mass transport; CT, charge transfer; C_{AS} and C_{BS}, concentration of A and B in a bulk solution; C_{AE} and C_{BE}, concentration of A and B at the electrode surface.

acceleration of an electrode reaction by a substance that is not consumed in the overall reaction. The substance is mostly the electrode itself; however, the catalytic effects of other components (for instance, the solvent) may also be envisaged.[1]

If we restrict the discussion to the catalytic effect of the nature and the structure of the electrode material and consider systems with polar liquid components (for instance, water and aqueous solutions), it is relatively easy to find a link between electrocatalysis and liquid-phase heterogeneous catalysis. A liquid-phase heterogeneous catalytic system where an electrified interphase is formed between the solid and liquid phases should also be considered as an electrochemical system. Thus, the interphase may be envisaged as consisting of the surface regions of two phases in contact where the accumulation or the depletion of free charged components can occur, resulting in net charges on the phases. In such systems, charged components may or may not cross the interface between the two phases. On the basis of this condition, electrochemists talk about unpolarizable and polarizable systems. The thermodynamic description of an electrified solid–liquid interphase is similar to that of nonionic systems with one important difference, that is, the description requires the introduction of electrochemical parameters—the thermodynamic charge and the electrical potential difference between the solid phase considered and the reference electrode. This means that a heterogeneous catalytic system where an electrified interphase is formed cannot be unambiguously treated ignoring the electrochemical parameters.

Electrocatalysis is a heterogeneous process that involves the adsorption and the chemisorption of reactants or intermediates at the interface. These phenomena are encountered, for example, in fuel cells (FCs) or in organic electrosynthesis. The electrocatalytic activity of a given electrode for a certain reaction may be characterized by the current density at a chosen potential, which is proportional to the specific activity, when referred to the effective active surface. As shown in Figure 21.2, the role of a heterogeneous catalyst is to adsorb the electro-reactive species (reactant and intermediate) and transform it to another compound that can more readily undergo the desired

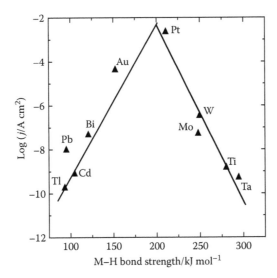

FIGURE 21.2 Volcano-shaped plot of the exchange current density for the hydrogen evolution reaction as a function of the metal–hydrogen bond energy.

chemical transformation. Therefore, if the heat of adsorption is low, the extent of adsorption will be very small. If the binding energy to the surface is too high, the adsorbed intermediate or its product may remain "attached" to the surface and effectively poison it. The "best" catalyst is thus one giving rise to an intermediate adsorption energy value.[2]

21.2 MULTI-ELECTRON CHARGE-TRANSFER REACTIONS

The need to convert chemical energy to electricity, efficiently and at low temperature, has increased the development of materials with electrocatalytic activity toward multi-electron charge transfer. Reactions of technical relevance are, for example, cathodic processes, such as the oxygen reduction reaction (ORR),[3–13] and anodic processes, such as small organics (R-OH, where $R = CH_3$-[3–12] or CH_3CH–[13–17]) and sugars.[18–20] These complex electrochemical processes are useful in low-temperature systems such as the direct methanol FC (DMFC) or biofuel cell systems.

21.2.1 OXYGEN REDUCTION REACTION

The marked irreversibility of this electrochemical process, which leads to excessive voltage losses at reasonable temperatures, has complicated mechanistic studies. Indeed, in aqueous solvents, molecular oxygen requires a strong interaction with the electrode surface for the reaction to proceed at a reasonable rate, with a number of elementary steps involving various reaction intermediates. The reduction can proceed either via the formation of water (four electrons) without the intermediate formation of hydrogen peroxide, or hydrogen peroxide (two electrons) as a product. Studies of the ORR in acid medium on platinum at rotating ring-disk electrodes[26] reported that the mechanism proposed by Damjanović et al.[27] appears suitable. A similar scheme adapted for alkaline medium was later reported by Wróblowa et al.[28] Furthermore, the ORR is also sensitive to the electrolyte nature (adsorbed anions),[26,29,30] to the structure of surface of the electrode materials,[30–34] and to the particle size.[29,30,33–35]

This electrochemical cathodic process is the complementary reaction, aiming at developing an energy-converting system with the anodic process based on an organic fuel such as glucose (see following sections). However, if the electrode material is based on platinum, the anodic process is affected by the nature of the fuel, other than hydrogen, inducing an anodic overpotential "η_a."

Efforts to reduce the overpotential at the cathode as well as the anode reactions are focused on novel material synthesis. A recent review on cluster-like materials, obtained via carbonyl chemical route, to enhance the selectivity and the reactivity of ORR has been reported.[36]

21.2.2 SUGAR OXIDATION REACTION

Glucose is an example of a reducing carbohydrate. It becomes a pyranose when dissolved in aqueous solution by hemiacetalization (a chemical reaction between the aldehydic group and the secondary alcohol in C5 position):

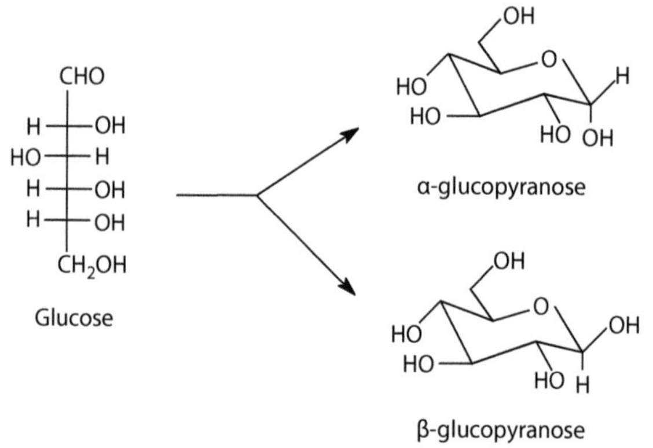

α-glucopyranose

β-glucopyranose

Glucose

$$(21.1)$$

Among these two anomeric forms, β-glucopyranose is described as the electroreactive species of glucose.[37,38] The oxidation of the carbon in the C1 position on a noble electrode material such as Pt or Au is representative of reducing sugars, which proceeds by dehydrogenation:

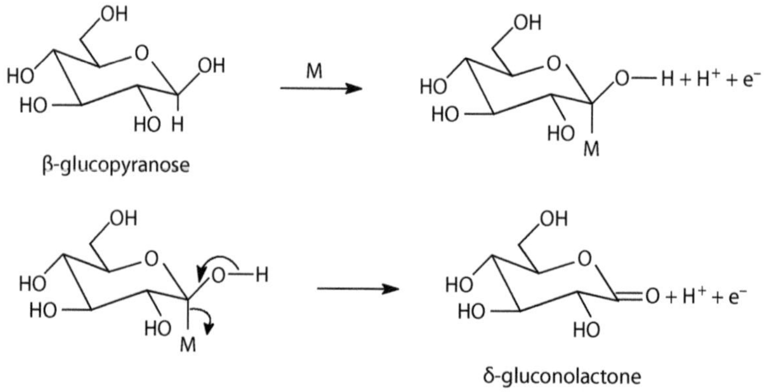

β-glucopyranose

δ-gluconolactone

$$(21.2)$$

The lactone undergoes hydrolysis in solution to form the corresponding aldonic acid, that is, here, gluconic acid, as follows:

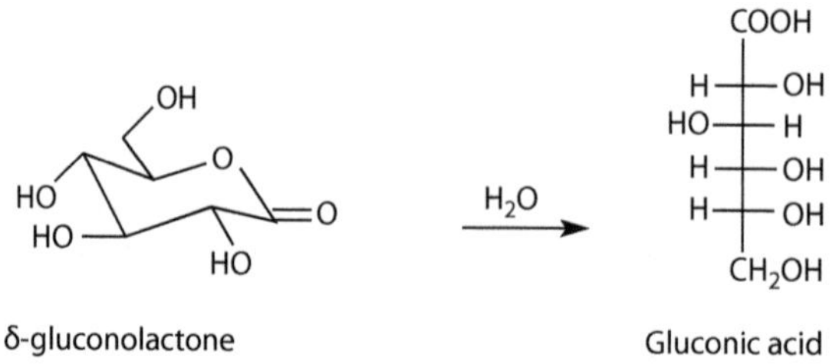

δ-gluconolactone

Gluconic acid

$$(21.3)$$

The electrocatalytic oxidation of the primary alcohol (in C6 position) is similar to the reaction mechanism that is well known for methanol. Therefore, the transformation of gluconic acid into glucaric acid can be written as follows:

$$
\text{Gluconic acid} + H_2O \longrightarrow \text{Glucaric acid} + 4H^+ + 4e^- \tag{21.4}
$$

21.3 ELECTRODE SURFACE MODIFICATION FOR ELECTROCATALYSIS

21.3.1 METAL ADATOMS: PLATINUM AND GOLD ELECTRODES

21.3.1.1 Characterization of the Electrode Surface

Cyclic voltammetry is a widely used electrochemical technique, which allows the investigation of the transient reactions occurring on the electrode surface when the potential applied to the electrode is varied linearly and repetitively at a constant sweep rate between two given suitable limits. The steady-state current–potential curves or voltammograms provide direct information as to the adsorption–desorption processes and allow estimating the catalytic properties of the electrode surface.

The voltammogram of a platinum electrode recorded at 25°C in 0.5 M H_2SO_4 between a lower limit ($E_C = 0.05$ V versus reference hydrogen electrode [RHE]) and an upper limit ($E_A = 1.5$ V versus RHE) displays three regions according to the applied potential (Figure 21.3). Region I at lower potentials is associated with the adsorption (reduction peaks)–desorption (oxidation peaks) of hydrogen, which is very sensitive to the crystallographic orientation of the Pt surface mainly constituted of (110), (100), and (111) faces. Region III at higher potentials reveals the adsorption (oxidation peaks)–desorption (reduction peak) of hydroxyl and oxygenated species at the electrode

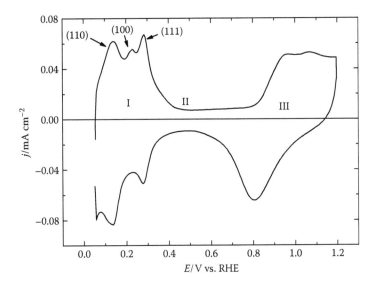

FIGURE 21.3 Voltammogram of a Pt electrode in 0.5 M H_2SO_4 at 0.050 V s^{-1} and 25°C.

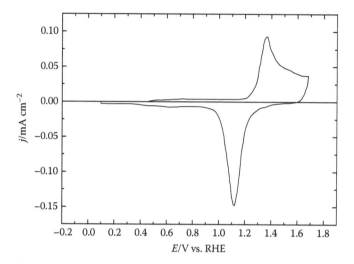

FIGURE 21.4 Voltammogram of an Au electrode in 0.5 M H_2SO_4 at 0.05 V s^{-1} and 25°C.

surface.[39–41] Between them the double-layer region II, where no electrochemical reaction occurs, is characterized by a small current associated with the charge and the discharge of the double-layer capacitance.

Considering now a gold electrode, the voltammogram recorded, under the same experimental conditions (0.5 M H_2SO_4, 25°C), is relatively simple, the hydrogen region having disappeared (Figure 21.4). The double-layer region is remarkably spread out in a potential interval from 0.05 to 1.3 V versus RHE. The oxygen region is characterized by the occurrence of several adsorption peaks and usually a single desorption peak at ca. 1.1 V versus RHE.

When the same experiments are performed in an alkaline medium, 0.1 M NaOH, the main features of both voltammograms remain unchanged. However, it is interesting to point out the existence of important currents in the double-layer region for both electrodes. These currents arise from faradaic processes associated with the oxygen region, which begin earlier in the alkaline medium, particularly for the gold electrode, where the onset of the adsorption of oxygenated species begins at 0.4 V versus RHE (Figure 21.4). Although the voltammetric investigation does not allow one to identify this kind of adsorbed species, it is reasonable to think that this current is related to earlier OH adsorption, as far as this process is supposed to be favored in alkaline medium. Therefore, this layer will necessarily affect the catalytic properties of a gold electrode surface in alkaline solution, as compared with its behavior in acid medium, and explain part of the results, which will be given below.

CO-stripping voltammetry is also used for the calculation of the true electrode surface area (Figure 21.5). CO adsorption on the electrode was carried out by bubbling CO gas for 2 min at 0.05 V versus RHE. Thereafter, CO was removed from the solution by bubbling nitrogen for 15 min.

The effective electroactive surface may be determined directly by the voltammogram in evaluating the quantity of electricity involved in the adsorption–desorption of hydrogen (region I). The quantity of electricity is obtained by the integration of the current–potential curves, because the potential varies linearly with time. The active surface area ($S = 0.56$ cm^2) is then calculated from the charge under the hydrogen underpotential-deposited (UPD) peaks, which needs 210 μC cm^{-2}, as follows:

$$Pt + H^+ + e^- \longrightarrow Pt—H_{ads} \tag{21.5}$$

The charge involved in the CO oxidation allowed the determination of the surface areas of platinum. Assuming that the monolayer of adsorbed CO needs 420 μC cm^{-2} for its oxidation to CO_2 on the Pt surface[42]

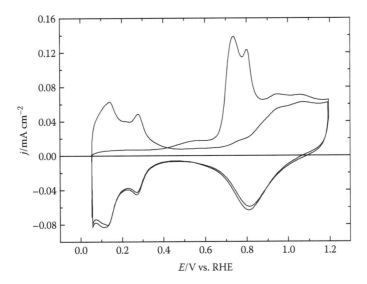

FIGURE 21.5 CO stripping on a Pt electrode in 0.5 M H_2SO_4 at 0.05 V s^{-1}. All the electrochemical experiments were carried out at 21°C ± 1°C. The electrolytic solution was deoxygenated by bubbling nitrogen during 15 min.

$$Pt + H_2O \longrightarrow Pt\text{—}OH_{ads} + H^+ + e^- \qquad (21.6)$$

$$Pt\text{—}CO_{ads} + Pt\text{—}OH_{ads} \longrightarrow CO_2 + 2Pt + H^+ + e^- \qquad (21.7)$$

The quantity of electricity involved in this reaction allowed determining the active platinum surface (0.52 cm^2). The surface composition of a catalyst can also be determined in alkaline medium using the oxide reaction formation at a given potential. Figure 21.6 shows a cyclic voltammogram of $Pt_{50}Au_{50}$ dispersed on carbon XC-72 Vulcan. The charge associated to reduce oxide species can be used to determine the surface composition.

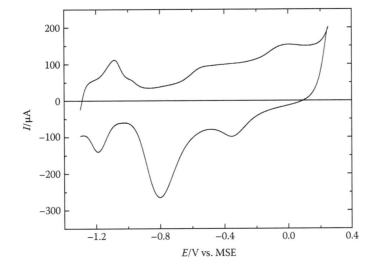

FIGURE 21.6 Voltammogram of $Au_{50}Pt_{50}$ nanoparticles deposited on a carbon-RDE recorded in 0.1 M NaOH at 0.05 V s^{-1} and 25°C.

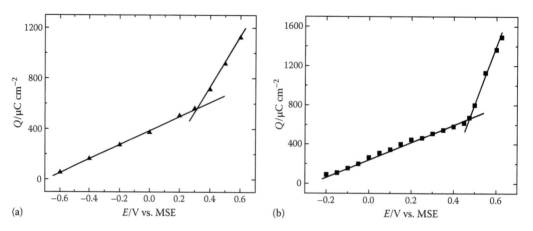

FIGURE 21.7   Quantity of electricity involved as a function of the upper potential limit of voltammograms recorded in alkaline medium 0.1 M NaOH and at 25°C. (a) on a Pt electrode and (b) on a Au electrode.

The peak around -0.38 V versus mercury/mercurous sulfate reference electrode (MSE), during the negative sweep, is associated to gold species whereas the one at -0.8 V versus MSE corresponds to platinum species. For pure catalysts, with an upper potential limit of $+0.25$ V versus MSE, the charge values of 493 and 543 $\mu C\ cm^{-2}$ were obtained for platinum and gold, respectively (Figure 21.7).[43,44] The atomic content of the Pt–Au nanoparticles can be deduced as follows:

$$x = \frac{S_{Au}}{S_{Au} + S_{Pt}} \times 100 \qquad (21.8)$$

where

  x represents the Au content

  S_{Au} and S_{Pt} are the electrode surface covered by gold and platinum oxides, respectively

The active surface area due to the surface composition of the catalyst was estimated as 0.5 cm^2, which represents 18.4% of the gold content in $Pt_{50}Au_{50}$.

21.3.1.2   Underpotential Deposition of Metal Ions

The modification of surface catalytic properties by the deposition of metal adatoms has been studied extensively in the field of electrocatalysis.[45–59] The main and common metals used as adatoms have low energy of adsorption (see Figure 21.2). Adatoms are considered as a means to enhance the catalytic activity of the electrode surface. Underpotential deposition is the process of the deposition of submonolayer amounts of metal ions (M^{z+}) on an inert foreign metal (M) support from a solution considered at more positive potentials than the corresponding M^{z+}/M equilibrium potential.

The electrocatalytic oxidation of organic molecules has been studied very extensively using various supports and adatoms. There are several proposed mechanisms in the literature to interpret the role of adatoms. It is assumed, for instance, that adatoms block the poison formation not leaving enough space for these reactions. According to another hypothesis, adatoms act as redox intermediates. Independently from the explanation, it is a fact that very often a significant enhancement of the catalytic activity can be observed and these results could provide very useful information for liquid-phase catalytic reactions.[60]

21.4 ELECTROSYNTHESIS IN AQUEOUS MEDIUM

21.4.1 ELECTROOXIDATION OF CARBOHYDRATES AND ALCOHOLS

As a first attempt in applying electrocatalysis to the valorization of carbohydrates, the regioselective electrooxidation of monosaccharides (glucose) and disaccharides such as sucrose and lactose was investigated. Corresponding acids or ketones keeping their initial skeleton are of interest in the pharmaceutical and detergent branches. Mono and dicarboxylic acid of disaccharides are important intermediates in the selective synthesis of ester and amides of carbohydrates, which can be used as biodegradable tensioactive compounds.

The oxidation reaction of different sugars was carried out in alkaline medium because

- They are not electroreactive in acid medium.
- The disaccharides such as sucrose undergo inversion reaction in acid medium implying the cleavage of the C–O–C glycosidic bond.
- The carbohydrates present a fragile structure so that the removal of the aqueous solution by evaporation on vacuum degrades the molecules. However, at the end of the oxidation process, the electrolyte is easily neutralized with a cation-exchange resin (Amberlite 200), which allows a lyophilization of the aqueous solution free from inorganic species.

21.4.1.1 Electrochemical Oxidation of Ascorbic Acid at a Platinum Electrode

Investigation was made into the transport characteristics of ascorbic acid (AA). Cyclic voltammograms of 10 mM AA at different scan rates were recorded at platinum (Figure 21.8) without stirring the solution with the rotating disk electrode (RDE).

If the electrooxidation of AA at the platinum electrode is controlled solely by the mass transfer process in the solution (Figure 21.9), the relationship between the maximum current density at the maximum potential should obey the Randles–Sevčik law

$$I_P = 0.4463nF\sqrt{\frac{nF}{RT}}SD^{\frac{1}{2}}C_{AA}\sqrt{v} \tag{21.9}$$

where D, S, and C_{AA} are the diffusion coefficient, the surface area of the electrode, and the concentration of AA in the solution, respectively.

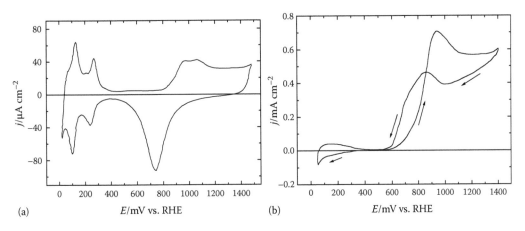

(a) E/mV vs. RHE   (b) E/mV vs. RHE

FIGURE 21.8 Cyclic voltammograms of Pt in supporting electrolyte 0.5 M H_2SO_4 (a) and in the presence of 10 mM AA (b) with a scan rate of 0.01 V s^{-1} at 20°C.

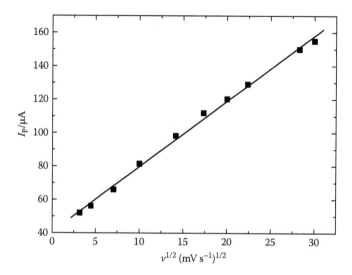

FIGURE 21.9   Plot of I_p versus $v^{1/2}$ for the electrooxidation of AA on a Pt electrode in 0.1 M H_2SO_4.

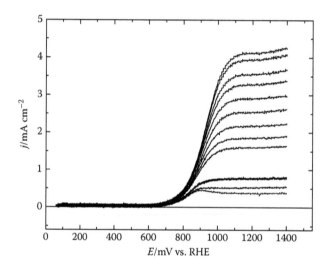

FIGURE 21.10   Current density–potential profiles at an RDE Pt electrode with 10 mM AA in 0.5 M H_2SO_4 at different values of rotation rate (49, 100, 490, 700, 900, 1200, 1600, 2000, 2500, 2900, 3600, and 4000 rpm) at 0.01 V s^{-1}.

    In order to determine the kinetic parameters of the reaction corrected from diffusion phenomenon, a voltammetric investigation was pursued with a rotating platinum ring disk electrode. Figure 21.10 represents current density–potential curves obtained for different rotation speeds and for the positive variations of potential. It appears that the current density of the rotating disk increases with the rotation rate.

    For reactions controlled by both activation and mass transfer, the total current density is expressed as follows and known as Koutecký–Levich plots:

$$\frac{1}{j} = \frac{1}{j_k} + \frac{1}{j_L} \qquad\qquad (21.10)$$

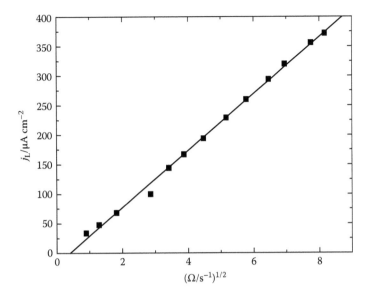

FIGURE 21.11   Levich plot of the main data in Figure 21.10 at $E = 1.2$ V versus RHE.

where

   j_L is the current density due to mass transfer (according to the hypothesis of Nernst double layer)
   j_k is the kinetic current density, that is, the oxidation current obtained in the absence of mass
      transfer

As j_L is proportional to $\Omega^{1/2}$ (Ω being the rotation rate), the current density of the disk is related to the rotation rate according to

$$\frac{1}{j} = \frac{1}{j_k} + \frac{1}{B\sqrt{\Omega}} \tag{21.11}$$

B is given by

$$B = 0.2nF(D_{AA})^{2/3}v^{-1/6}C_{AA} \tag{21.12}$$

where

   n is the number of electrons transferred per molecule of AA
   D_{AA} is the diffusion coefficient of AA
   C_{AA} is its concentration
   v is the kinematics viscosity of the solvent ($v_{H_2SO_4} = 1.009 \times 10^{-2}\,cm^{-2}\,s^{-1}$)[61]

Therefore, as shown by Levich, a plot of $1/j$ versus $\Omega^{-1/2}$ should be linear (Figure 21.11). Its slope allowed the calculation of the diffusion coefficient of AA for $6.5 \times 10^{-6}\,cm^2\,s^{-1}$.

21.4.1.2   Electrooxidation of Glucose

21.4.1.2.1   Polarimetric Measurements

The specific rotation angle $[\alpha]$ is related to the experimental rotation angle of light α through the so-called Biot and Savart law:

$$[\alpha] = \frac{100\alpha}{l \times C} \tag{21.13}$$

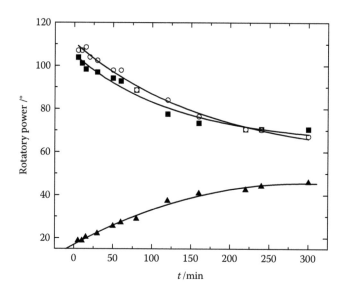

FIGURE 21.12   Evolution of the rotatory power for α-glucose (○), β-glucose (▲), and D-glucose (■) in acid medium (0.1 M HClO₄).

where

    l is the length of the polarimetric tube thermostated at 2°C, expressed in decimeter

    C is the concentration of the active substance, expressed in grams per 100 cm³

    $[\alpha]$ and α are expressed in degrees and decimal fractions of α degree

Different solutions of glucose were characterized by polarimetry, referring to the specific rotation angles of the two glucose anomers, namely, +112° for the α form and +18.7° for the β form. Measuring the specific rotation angle $[\alpha]$ of the solution is thus a way of determining the proportion of each form, and thus of determining which anomeric form prevails. For instance, the percentage of β form can be calculated by

$$\%\beta = \frac{[\alpha] - 112}{18.7 - 112} \times 100 \tag{21.14}$$

The evolution of the specific rotation angle in acid medium (pH 1) at 2°C is given in Figure 21.12 for the three solutions prepared from the initial "α," "β," and "commercial" D-glucose (Pro-analysis from Merck). In each case, the equilibrium is attained slowly. After 5 h, the β form, which amounts to about 71% of β-D-glucose in solution, exhibits the highest stability. At the same time, α and D-glucose form solutions containing only 52% and 56% of α-D-glucose, respectively.

In neutral medium (Figure 21.13), both α and β forms exhibit stability, but it should be noted that the β form appears to be more stable (95% in solution) and that the D-glucose leads to solutions where the α form is dominant (73% in solution).

The influence of the alkaline medium is interesting (Figure 21.14). It is remarkable that equilibrium is reached very quickly (first measurement at $t = 5$ min after glucose dissolution). Each solution contains ca. 80% of the β form, which is unstable in alkaline medium.

21.4.1.2.2   Electrochemical Measurements

Figure 21.15 represents the voltammograms of platinum in alkaline medium (0.1 M NaOH). In agreement with polarimetric experiments, indicating that the composition of the three solutions is almost the same in alkaline solution, the voltammograms are similar and correspond to those previously described for D-glucose under the same conditions. However, if we can conclude that

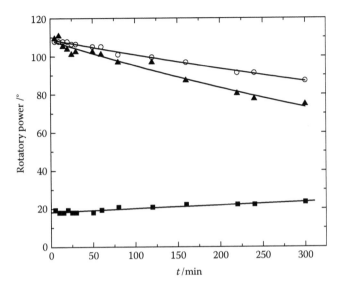

FIGURE 21.13 Evolution of the rotatory power for α-glucose (○), β-glucose (▲), and D-glucose (■) in acid water.

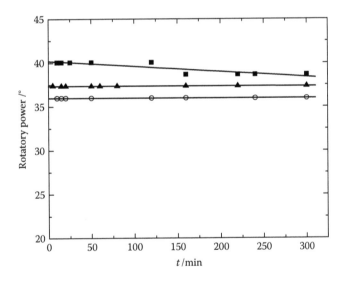

FIGURE 21.14 Evolution of the rotatory power for α-glucose (○), β-glucose (▲), and D-glucose (■) in alkaline medium (0.1 M NaOH).

β-glucose predominant in solution is reactive, it is more difficult to say the same for the α form that is always the minor component (ca. 20% in solution).

The situation is different in acid medium and the solutions prepared at low temperature retain α and β predominant forms long enough to obtain evidence for a possible difference in reactivity among the anomers.

The oxidation reaction of glucose at 0.3 V versus RHE was carried out during long-term electrolysis. The activity of the platinum electrode decreased quickly because of poisoning adsorbed intermediates, a pulse potential set at 1.4 V was added to the time-program in order to reactivate *in situ* the platinum surface.

Figure 21.16 shows the variations of the current intensity and the charge involved during the oxidation reaction of glucose on a platinum electrode. At the end of electrolysis, gluconate was

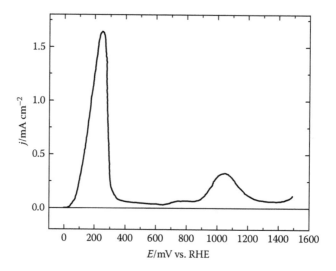

FIGURE 21.15 Anodic sweep curve of a Pt electrode in alkaline medium (0.1 M NaOH) recorded at 0.05 V s^{-1} and at 2°C, with 0.1 M D-glucose.

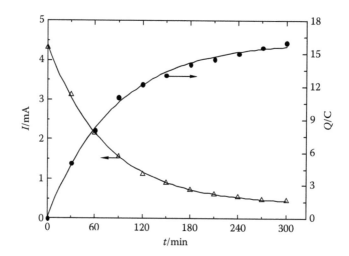

FIGURE 21.16 Evolution of the current intensity and the quantity of electricity involved during electrolysis of 0.1 M glucose in 0.1 M NaOH at 0.3 V versus RHE and 2°C.

detected by ionic chromatography as the main oxidation product of glucose, that is, the chemical transformation processes in alkaline medium (epimerization and interconversion) are nearly stopped at low temperature. The following reaction mechanism of the dehydrogenation of glucose is commonly proposed[38,61–63]:

$$Pt + H_2O + e^- \longrightarrow Pt\text{—}H_{ads} + HO^- \tag{21.15}$$

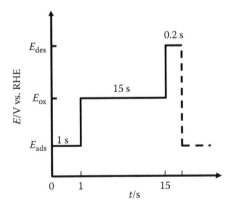

$$\delta\text{-gluconolactone} \qquad (21.17)$$

The first reaction shown produces δ-gluconolactone with + Pt + H₂O + e⁻

$$(21.18)$$

with product Gluconate (COO⁻, Na⁺).

The electrolytic solution was neutralized with a cation-exchange resin (Amberlite 200 from Sigma). The aqueous solution, free from inorganic ions, was then lyophilized, and the crystals obtained were trimethylsilylated for analysis on GC/MS*:

- Main fragments of δ-lactone m/e: 377; 361; 320; 271; 259; 243; 220; 204; 189; 169; 157; 147; 129
- Main fragments of glucose m/e: 393; 361; 317; 305; 271; 243; 217; 204; 191; 147; 129

As a result, β-glucose is the major configuration in alkaline solution at low temperature. Its most planar approach of the anomeric form of the molecule to the electrode surface facilitates its dehydrogenation at low potentials to δ-gluconolactone; the latter reaction product undergoes hydrolysis in the bulk solution to gluconate as shown by Largeaud et al. using *in situ* reflectance infrared spectroscopy.[64]

21.4.1.3 On Platinum Adatoms

Because of the poisoning phenomena occurring in long-term electrolysis and in order to maintain the electrode activity at a sufficient level, the oxidation reaction of carbohydrates was carried out using an optimized triple-pulse potential program (Figure 21.17). It consists in controlling the

FIGURE 21.17 Potential program used for long-time electrolysis of carbohydrates.

---

* Gas chromatography was equipped with a DB-5, 95% dimethyl 5% diphenyl polysiloxane–bounded capillary column. It was combined with a mass spectrometer (ITS 40 Finnigan Matt).

oxidation potential that, by itself, is a parameter of selectivity. The oxidation plateau was applied during the long time of the sequence in a potential range corresponding to the maximum of electrode activity ($E = 0.7$ V versus RHE). Then, clearing out the poisoning species reactivated the electrode surface by oxidation at higher potential during 0.2 s. The third pulse was set during 1 s in the hydrogen region, allowing the reduction of the surface and/or the *in situ* underpotential deposition of adatoms. This sequential unit program was repeated all along during the time of electrolysis.

Lead perchlorate and bismuth nitrate were added to the electrolytic solution at 5×10^{-5} and 10^{-5} M, respectively, to modify a platinum electrode, which has a real surface area of 40 cm^2. The electrolysis cell has two identical compartments separated by an ion-exchange membrane (Nafion® 423). A Pt$_{90}$Ir$_{10}$ sheet and the Hg/Hg$_2$SO$_4$, SO$_4$$^{2-}$$_{sat}$ electrode served as counter and reference electrodes, respectively.

The analysis of the reaction products was performed using ion-exchange liquid chromatography (Dionex-4500i). It works with a ternary gradient elution and includes an ion-exchange column (AG11 + AS11) and a double detection, that is, a conductimeter (for the ionizable compounds such as carboxylic acids) followed by a refractive index detector that is specifically dedicated to the D-glucose consumption.

According to the nature of the metal used as adatom to partially cover the platinum surface, the oxidation route is strongly modified as shown in Schemes 21.1 and 21.2.

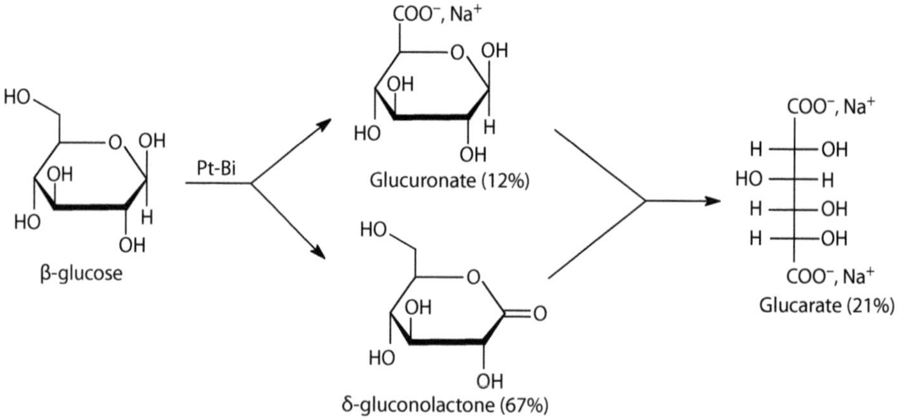

SCHEME 21.1   Oxidation of glucose at 0.7 V versus RHE in carbonate buffer on a Pt electrode modified by Bi adatoms.

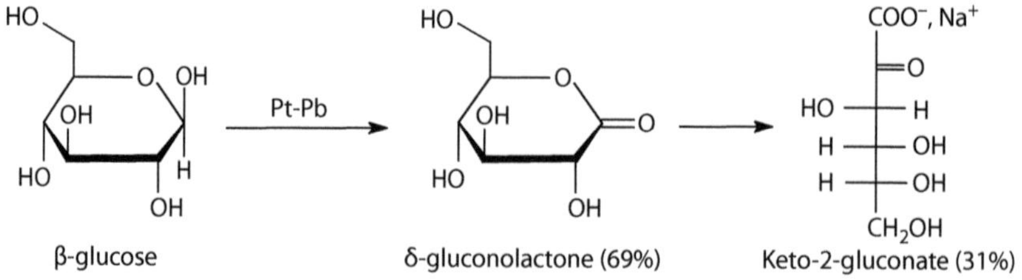

SCHEME 21.2   Oxidation of glucose at 0.7 V versus RHE in carbonate buffer on a Pt electrode modified by Pb adatoms.

21.4.1.4 On Gold Adatoms

As shown in different works,[41,65–71] the mechanism of glucose electrooxidation on platinum is similar to that on gold electrode. The primary and main reaction product remains δ-gluconolactone that is transformed into gluconate by hydrolysis.

In the presence of metal adatoms, a triple-pulse potential program shown in Figure 21.17 is also applied during the electrolysis of glucose on a gold electrode. The adsorption of adatoms (e.g., bismuth in Figure 21.18) occurs simultaneously with glucose dehydrogenation at the electrode surface. At higher potentials ($E > 0.4$ V versus RHE), the desorbed adatoms are quickly replaced by hydroxyl groups, which could later on contribute to the surface oxidation of the organic adsorbates. This leads us to postulate a reaction mechanism of glucose oxidation on Pt–Bi as follows:

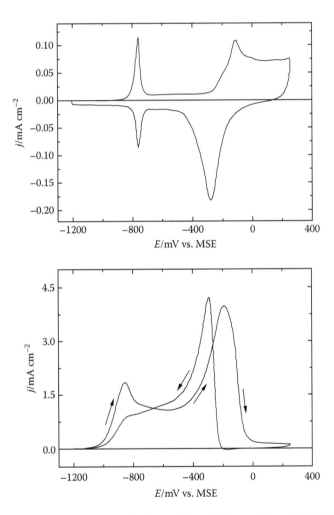

FIGURE 21.18 Voltammograms of a Au–Bi electrode in alkaline medium (0.1 M NaOH), recorded at 0.05 V s^{-1} and at 22°C.

- At ca. 0.4 V, the adsorption of the metal adatoms from Bi^{3+} in solution and that of glucose at the electrode surface first occurs:

$$Au + Bi^{3+} + 3e^- \rightarrow Au\text{–}Bi_{ads} \qquad (21.19)$$

$$Au + C_6H_{12}O_6 + HO^- \rightarrow Au(C_6H_{11}O_6)ads + H_2O + e^- \qquad (21.20)$$

At lower potentials, the desorption of glucose species can lead to glucono-δ-lactone

- For $0.5 < E_{ox} < 0.9$ V, the hydroxyls necessary for the oxidation of the electrode surface competes with Bi adatoms still adsorbed. Both adsorbates modify the organics' oxidation, such as gluconate (40%) leading to the formation of keto-2-gluconate (Scheme 21.3) that is the second compound produced in an appreciable amount (19%):

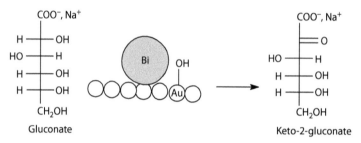

SCHEME 21.3 Conversion of gluconate to keto-2-gluconate.

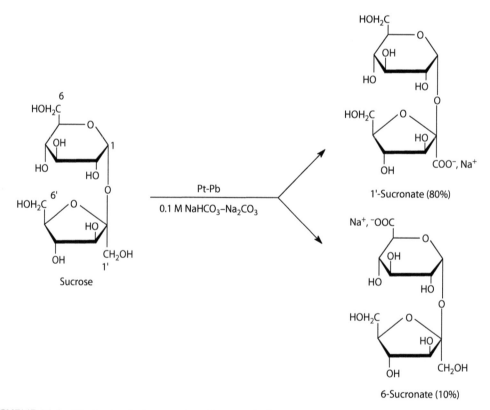

SCHEME 21.4 Electrochemical conversion of sucrose to 6-sucronate.

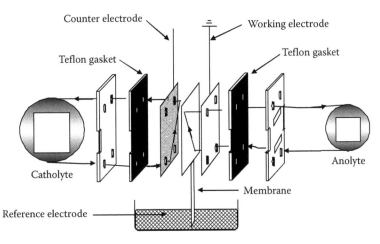

FIGURE 21.19 Schematic representation of the filter press cell used to carry out the electrooxidation of carbohydrates.

21.4.1.5 Electrooxidation of Sucrose

21.4.1.5.1 On Adatoms Modified Platinum Electrodes

The selective oxidation of sucrose is likely to lead to mono-, di-, and polycarboxylic acids (Scheme 21.4). These derivatives have a potential application in detergents, to replace phosphates, thanks to their properties of nontoxicity, biocompatibility, and biodegradability.

The electrolysis of sucrose was performed in a filter press cell (micro-flow cell, electro-cell AB) (Figure 21.19). The working electrode was platinum deposited electrochemically on a titanium plate. The counter electrode was a plate of stainless steel. The two compartments of the cell were separated by an ion-exchange membrane (Nafion 423). A part of this membrane immersed in a saturated potassium sulfate permitted to connect by capillarity the MSE. The electrolyte in the cell was circulated by an external peristaltic pump ($1 \text{ cm}^3 \text{ min}^{-1}$) and passed through a reservoir (100 cm^3).

Some attempts* in alkaline solution (0.1 M NaOH) have shown an important current decay caused by the participation of the hydroxyl species of the electrolyte to the formation of carboxylates. Accordingly, the oxidation of sucrose was carried out on a lead-modified platinum electrode, in $0.1 \text{ M Na}_2\text{CO}_3\text{–NaHCO}_3$ (Figure 21.20). During the positive variation of potential, two oxidation peaks of sucrose, A and B, are observed at 0.8 V versus RHE ($j = 0.8 \text{ mA cm}^{-2}$) and at 1.1 V versus RHE ($j = 0.44 \text{ mA cm}^{-2}$). During the negative variation of potential, another peak, C, is noticed at 0.7 V versus RHE ($j = 0.3 \text{ mA cm}^{-2}$) after the desorption of the oxygenated species from the electrode surface.

The electrolysis of the same sucrose solution was carried out under the same experimental conditions during 8.3 h with the potential program described in Figure 21.17. The oxidation pulse step was fixed at 0.74 V versus RHE. The quantity of electricity at each potential plateau was evaluated and presented in Figure 21.21. It can be noted that Q_{des} and Q_{ads} have an almost symmetrical profile with a small advantage for Q_{ads}. Q_{ads} is the quantity of electricity, which served to adsorb the organic molecule and eventually to reduce the electrode surface.

---

* The electrochemical instrumentation consisted of a Hewlett Packard HP 33120A Arbitrary Waveform Generator coupled to a Bank High Power Potentiostat (Wenking Model HP88). The potential program is synthesized with an HP BenchLink Software (HP BenchLink/Arb) under a Windows™ environment, and then it is transferred to the function generator. Data such as the applied potential, the current intensity, and the quantity of electricity were acquired by a PC microcomputer equipped with an AD/DA convertor (Keithley DAS 20-Viewdac).

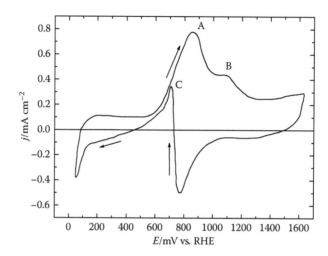

FIGURE 21.20 Voltammogram of a Pt–Pb electrode recorded in 0.1 M Na_2CO_3–$NaHCO_3$ + 10.2 mM sucrose at 0.05 V s^{-1}; $[Pb^{2+}] = 10^{-5}$ M.

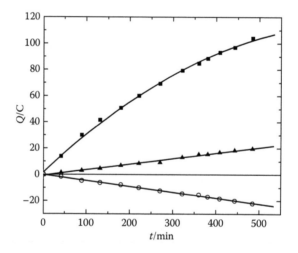

FIGURE 21.21 Variation of the quantity of electricity at each step potential of the time program applied during the oxidation of sucrose on a Pt–Pb electrode.

As for Q_{des}, it can be assumed that this quantity of electricity allowed clearing out of the electrode surface from the poisoning organic species. But if the active surface of platinum was covered by OH_{ads} at the desorption plateau (1.63 V versus RHE), after 8.3 h of electrolysis, $Q_{(OH)ads}$ would be estimated as follows:

$$Q_{(OH)ads} = n \times e \times (OH)_{ads} \times S_a \times \tau \tag{21.21}$$

where

  n is the electron number ($n = 1$)
  e is the electron elementary charge
  $(OH)_{ads}$ is the number of adsorbed hydroxyls, that is, the number of platinum sites per cm^2 (1.35×10^{15})
  S_a is the active surface of the working electrode (45 cm^2)
  τ is the number of cycles with t_{des} (0.2 s) used during the electrolysis ($\tau = 1852$), which gives $Q_{(OH)ads} = 17.34$ C

TABLE 21.1
Chemical Shifts of the ^{13}C NMR Spectrum of the Lyophilized Reaction Products Obtained on a Pt–Pb Electrode in Comparison with Sucrose as Reference

$\delta(C)$/ppm	C_1'	C_2'	C_3'	C_4'	C_5'	C_6'	C_1	C_2	C_3	C_4	C_5	C_6
Sucrose	63.3	104.4	77.4	75.0	82.2	63.4	92.9	72	73.6	70.2	73.3	61.1
1'-sucronic acid	183.2	106	78.6	77.7	81.5	62.9	93.7	72.7	74.5	70.8	73.4	61.6

This estimation is close to the experimental value ($Q_{des} = 19.7$ C). This means that the electro-oxidation of sucrose took place at 0.74 V versus RHE with nearly no poisoning species involved. The sole quantity of electricity, which served to oxidize sucrose, was thus Q_{ox}. At the end of electrolysis, this value was estimated to ca. 104 C, and the yield of the sucrose transformation was estimated close to 60%. The solution was sampled and neutralized on a cation-exchange resin (Amberlite 200, Sigma). The solutions obtained, free from inorganic cations (Na^+), were chromatographed again to assure the reproducibility of the last analysis.

The chemical shifts of the ^{13}C NMR spectrum recorded in D_2O with methanol as an internal reference and in the presence of NaOH in order to keep the pH in the alkaline range and to prevent the lactone formation, are given in Table 21.1.

The confirmation of the oxidation of the primary hydroxyl group located at C_1' to a carboxylic acid group is based on the two peaks (at 62 and 63 ppm) in the (CH_2OH) region and on the peak at 183 ppm in the "carbonyl region" of the spectrum. This is in accordance with the interpretation of Eyde et al.[72]

All these analytical results are in agreement with the experimental number of electrons involved during prolonged electrolyses of sucrose, which were found to be very close to 4. This confirms that the oxidation of one primary alcohol group leads to one carboxylic acid group (which needs four electrons per alcohol group), either the 1'mono sucronic acid (1'MSA) with a selectivity close to 80% or to the 6-mono sucronic acid (6-MSA) (with a selectivity close to 10%) on a UPD-Pb/Pt electrode[73,74]:

21.4.1.6 Electrooxidation of Lactose

Different voltammograms were recorded at various concentrations of metal adatoms (lead, bismuth, and thallium) in the presence of 10 mM lactose in 0.1 M Na_2CO_3–$NaHCO_3$. Their optimized concentrations gave evidences of the electrocatalytic effect by the ratio I_{Pt-M}/I_{Pt} for the current densities, with and without adatoms, versus electrode potential (Figure 21.22). In fact, the presence of metal adatoms at the platinum surface induces a shift of the lactose oxidation peaks toward lower potential and an increase in the current densities.

Using the previous information provided by the voltammetric measurements, the oxidation of 10 mM lactose was carried out in carbonate buffer and in the presence of lead adatoms ($[Pb^{2+}] = 5 \times 10^{-6}$ M). Electrolysis was carried out in a two-compartment cell (270 cm^3) for 3 h by applying the suitable triple pulse potential program repeatedly. The oxidation potential was set to 0.6 V versus RHE on platinum, which had an active surface area of 18 cm^2. After 3 h, the recorded quantity of electricity, $Q_{exp} = 15.1$ C, showed that the lactose conversion yield was 87% and 90% of selectivity in lactobionate was obtained[75,76]:

$$\text{(21.22)}$$

Lactose   Lactobionate

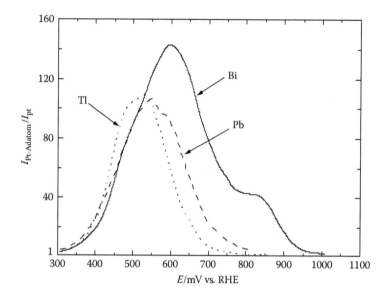

FIGURE 21.22   Electrocatalytic effect of metal adatoms on lactose oxidation in carbonate buffer during the positive potential going scan. (——) with 5.10^{-6} mol L^{-1} bismuth; (– – – –) with 5.10^{-6} mol L^{-1} lead; ($\cdots$) with 5.10^{-6} mol L^{-1} thallium.

TABLE 21.2

Chemical Shifts of the ^{13}C NMR Spectrum of the Lyophilized Reaction Product Obtained on a Pt–Pb Electrode in Comparison with Lactose Used as Reference

δ(C)/ppm	C_1'	C_2'	C_3'	C_4'	C_5'	C_6'	C_1	C_2	C_3	C_4	C_5	C_6
Lactose	103.0	70.0	81.1	68.9	84.7	61.0	91.94	75.6	71.7	90.6	72.9	62.0
Lactobionic acid	103.6	70.8	72,5	68.5	75.2	60.8	175.5	71,5	71.0	80.6	71.4	61.9

Treatments of the electrolytic solution were required to remove carbonates with an exchange resin (Amberlite 200). Then a dry and solid material was recovered by the lyophilization of water. Finally, the ^{13}C NMR spectrum of the reaction product was compared with that of lactose to determine the oxidized group (Table 21.2).

21.4.1.7   Cathodic Process: Carbonyl Compounds

Various methods have been used to produce selectively lactic acid or dimethyltartaric acid (or pinacol) by the reduction of pyruvic acid and its derivatives. The hydrogenation product (lactic acid) is a useful intermediate in the biological sector, food, and polymer industries.[77]

Cyclic voltammetry was used to check the degree of purity of the reaction medium (0.5 M H$_2$SO$_4$ or 0.5 M NaHCO$_3$–Na$_2$CO$_3$) and that of the electrode. In sulfuric acid (Figure 21.23), it can be seen during the positive potential scan, an oxidation peak at $E = -0.8$ V versus Hg/Hg$_2$SO$_4$/K$_2$SO$_4$ sat (MSE: $E_{MSE} = 0.650$ V versus RHE at pH $= 0$). This peak followed by a shoulder is attributed to PbSO$_4$ species that desorb irreversibly from the electrode surface at $E = -1.05$ versus MSE. A reduction wave of pyruvic acid begins in 0.5 H$_2$SO$_4$ at ca. -1.05 V versus MSE where lead sulfate desorbs from the electrode surface. In carbonate buffer (Figure 21.24), the electrode surface is oxidized at -1.02 V versus MSE to lead carbonate that was reduced reversibly during the negative potential sweep. In this medium, the reduction of pyruvic acid keeps the same shape, which starts at -1.65 V versus MSE. Hydrogen evolution was observed at the lead cathode from -1.4 to -2.2 V versus MSE

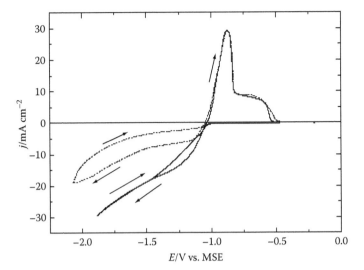

FIGURE 21.23   Voltammograms of a Pb electrode recorded at 0.05 V s^{-1}: (– – –) in 0.5 M H$_2$SO$_4$ supporting electrolyte alone and (—) in the presence of 0.1 M pyruvic acid.

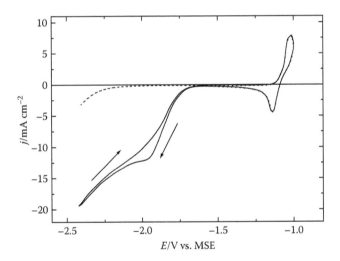

FIGURE 21.24   Voltammograms of a Pb electrode recorded at 0.05 V s^{-1}: (– – –) in 0.5 M Na$_2$CO$_3$ + 0.5 M NaHCO$_3$ supporting electrolyte alone and (—) in the presence of 0.1 M pyruvate.

according to the nature of the supporting electrolyte. Therefore, the production of the latter should be competitive with that of lactic acid and dimethyltartaric acid at more negative potentials.

A series of electrolysis of pyruvic acid were carried out at different fixed potentials in an undivided conventional three-electrode Pyrex cell (50 cm^3). In sulfuric acid, the electrohydrodimerization process was favored (69%) when increasing the concentration of pyruvic acid and the electrode potential:

$$2 \quad \underset{\text{H}_3\text{C} \quad \text{COOH}}{\overset{\text{O}}{\parallel}} + 2\text{H}^+ + 2e^- \xrightarrow[\text{Pb (–1.1 V vs. MSE)}]{0.5\,\text{M H}_2\text{SO}_4} \quad \text{HOOC} \underset{\text{H}_3\text{C} \quad \text{CH}_3}{\overset{\text{HO} \quad \text{OH}}{\vert\vert}} \text{COOH} \qquad (21.23)$$

Pyruvic acid   Dimethyltartaric acid

In carbonate buffer, the selectivity toward lactate was higher (90%) and seemed to be promoted when the electrode potential was set close to the hydrogen evolution reaction:

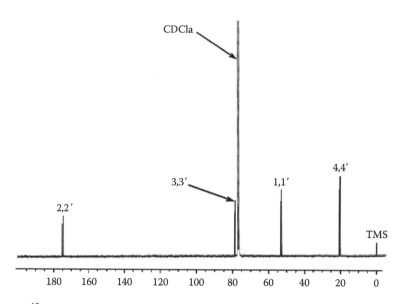

$$\qquad \qquad \qquad \qquad (21.24)$$

The analysis of the electrolyzed solutions by HPLC* allowed the determination of lactic and dimethyltartaric acids.[78–81] A chemical step of esterification is necessary to separate and isolate the reaction products. However, the aqueous electrolyte was removed at the end of electrolysis to recover dry organic compounds free from inorganic ions. The esterified corresponding products were separated by flash chromatography and then identified spectroscopically by GC–MS–MS[†], and ^{13}C NMR[‡] (Figure 21.25).

FIGURE 21.25   ^{13}C NMR spectrum of dimethylester of 2,3-dimethyltartaric acid in CDCl$_3$ obtained at the end of electrolysis of 0.1 M pyruvic acid at -1.1 V versus MSE on the lead cathode and after esterification. The internal reference is TMS.

---

* The analysis of the composition mixture was carried out by HPLC, which consisted of a pump (Knauer Pump 64) and two detectors settled online (UV Applied Biosystems 785A and refractometer Shodex RI-71). The partition was performed on an HPX-87H (300 mm × 7.8 mm, BioRad, 3.33 mM H$_2$SO$_4$ in water, 0.6 mL min^{-1}) and the quantitative analyses were carried out using the so-called external standard method. According to the used medium as supporting electrolyte, an ionic-exchange resin was necessary to neutralize it (for the sulfuric acid solution: Amberlite IRA-900 Cl and for the alkaline solutions: Amberlite IRA-400 Na). The aqueous solutions of the electrolyzed products, free from inorganic ions, were removed at 50°C under vacuum, and then dissolved in methanol (10 cm^3) for esterification using 2,2-dimethoxypropane (1 cm^3). The separation was performed on silica gel 15–40 μm (Silica Gel 60 from Sigma) with ethyl acetate/petroleum ether (60:40) as eluent.
[†] The different reaction products were identified by GC–MS–MS (EI: 70 eV, CI/NH$_3$ and CI/CH$_4$: 100 eV, 1200L Varian).
[‡] ^{13}C NMR: 300 MHz, CDCl$_3$, WP 200 SY Bruker spectrometer.

21.5 SUMMARY AND OUTLOOK

The basis of the surface modification of electrodes by adatoms has been delineated by selected examples with regard to the oxidation and reduction processes of small organic molecules in aqueous medium. The complex interplay of surface adatoms favors the electroactivity of a determined species, and delivers some insights as to the mechanism of such organic molecules in favor of a product with an added value.

The future progress of organic activation via electrocatalysis can be associated with the development of the following topics:

- In the case of platinum and gold, some incipient progress has already been achieved in finding the optimum alloy combination in the nanoscale domain to enhance the glucose oxidation, which is the anodic reaction in a biofuel cell system, concomitant with surface defects control.[25,82,83]
- Value added products, such as the reduction of oxalic acid to glyoxylic acid, are nowadays developed at an industrial scale (BASF—Badische Anilin und Soda-Fabriken). This process uses non-precious metals such as lead. This electrode material reduces carbon dioxide to formic acid in an aqueous medium and to oxalic acid in an organic solvent. This approach can be used to develop environmentally friendly value added products from abundant anthropogenic carbon dioxide.

ACKNOWLEDGMENT

The authors kindly acknowledge their students and collaborators for their contributions.

REFERENCES

1. S. Trasatti and R. Parsons, *Pure Appl. Chem.* **58** (1986) 437.
2. E. Gileadi, *Electrode Kinetics for Chemists, Chemical Engineers, and Materials Scientists*; VCH: Weinheim, Germany, 1993.
3. V. S. Bagotskii and Y. B. Vasil'ev, *Electrochim. Acta* **12** (1967) 1323.
4. B. D. McNicol, A. G. Chapman, and R. T. Short, *J. Appl. Electrochem.* **6** (1976) 221.
5. L. D. Burke and O. J. Murphy, *J. Electroanal. Chem.* **101** (1979) 351.
6. C. Lamy, J. M. Leger, and J. Clavilier, *J. Electroanal. Chem.* **135** (1982) 321.
7. T. Frelink, W. Visscher, and J. A. R. van Veen, *J. Electroanal. Chem.* **382** (1995) 65.
8. Z. D. Wei, L. L. Li, Y. H. Luo, C. Yan, C. X. Sun, G. Z. Yin, and P. K. Shen, *J. Phys. Chem. B* **110** (2006) 26055.
9. A. S. Arico, P. Creti, N. Giordano, V. Antonucci, P. L. Antonucci, and A. Chuvilin, *J. Appl. Electrochem.* **26** (1996) 959.
10. G. Bronoel, S. Besse, and N. Tassin, *Electrochim. Acta* **37** (1992) 1351.
11. T. Iwasita and W. Vielstich, *J. Electroanal. Chem.* **201** (1986) 403.
12. A. S. Lin, A. D. Kowalak, and W. E. O'Grady, *J. Power Sources* **58** (1996) 67.
13. H. Hitmi, E. M. Belgsir, J. M. Leger, C. Lamy, and R. O. Lezna, *Electrochim. Acta* **39** (1994) 407.
14. J. P. I. Souza, F. J. Botelho Rabelo, I. R. de Moraes, and F. C. Nart, *J. Electroanal. Chem.* **420** (1997) 17.
15. F. Delime, J.-M. Leger, and C. Lamy, *J. Appl. Electrochem.* **29** (1999) 1249.
16. D. T. Shieh and B. J. Hwang, *J. Electrochem. Soc.* **142** (1995) 816.
17. X. Xue, J. Ge, T. Tian, C. Liu, W. Xing, and T. Lu, *J. Power Sources* **172** (2007) 560.
18. I. G. Casella, M. Gatta, and M. Contursi, *J. Electroanal. Chem.* **561** (2004) 103.
19. S. Berchmans, H. Gomathi, and G. P. Rao, *J. Electroanal. Chem.* **394** (1995) 267.
20. T. R. I. Cataldi, A. Guerrieri, I. G. Casella, and E. Desimoni, *Electroanalysis* **7** (1995) 305.
21. H. S. Liu, C. J. Song, L. Zhang, J. J. Zhang, H. J. Wang, and D. P. Wilkinson, *J. Power Sources* **155** (2006) 95.
22. M. Baldauf and W. Preidel, *J. Power Sources* **84** (1999) 161.

23. P. Atanassov, C. Apblett, S. Banta, B. Susan, S. C. Barton, M. Cooney, B. Y. Liaw, S. Mukerjee, and S. D. Minteer, *Electrochem. Soc. Interface* **16** (2007) 28.
24. A. Habrioux, K. Servat, B. Kokoh, and N. Alonso-Vante, *ECS Trans.* **6** (2007) 9.
25. A. Habrioux, E. Sibert, K. Servat, W. Vogel, K. B. Kokoh, and N. Alonso-Vante, *J. Phys. Chem. B* **111** (2007) 10329.
26. K. L. Hsueh, E. R. Gonzalez, and S. Srinivasan, *Electrochim. Acta* **28** (1983) 691.
27. A. Damjanovic, M. A. Genshaw, and J. O'M. Bockris, *J. Chem. Phys.* **45** (1966) 4057.
28. H. S. Wroblowa, P. Yen Chi, and G. Razumney, *J. Electroanal. Chem.* **69** (1976) 195.
29. G. Tamizhmani, J. P. Dodelet, and D. Guay, *J. Electrochem. Soc.* **143** (1996) 18.
30. N. M. Markovic, T. J. Schmidt, V. Stamenkovic, and P. N. Ross, *Fuel Cells* **1** (2001) 105.
31. P. N. Ross, *J. Electrochem. Soc.* **126** (1979) 78.
32. P. Fischer and J. Heitbaum, *J. Electroanal. Chem.* **112** (1980) 231.
33. K. J. J. Mayrhofer, B. B. Blizanac, M. Arenz, V. R. Stamenkovic, P. N. Ross, and N. M. Markovic, *J. Phys. Chem. B* **109** (2005) 14433.
34. V. Komanicky, A. Menzel, and H. You, *J. Phys. Chem. B* **109** (2005) 23550.
35. K. Kinoshita, *J. Electrochem. Soc.* **137** (1990) 845.
36. N. Alonso-Vante, *Pure Appl. Chem.* **80** (2008) 2103.
37. J. Giner, L. Marincic, J. S. Soeldner, and C. K. Colton, *J. Electrochem. Soc.* **128** (1981) 2106.
38. F. Largeaud, K. B. Kokoh, B. Beden, and C. Lamy, *J. Electroanal. Chem.* **397** (1995) 261.
39. H. Angerstein-Kozlowska, *Surfaces, Cells, and Solutions for Kinetics Studies*; Plenum Press: New York, 1984.
40. B. E. Conway and L. Bai, *J. Electroanal. Chem.* **198** (1986) 149.
41. A. J. Appleby. Electrocatalysis. In *Modern Aspects of Electrochemistry*; B. E. Conway, and J. O'M. Bockris, Eds.; Plenum Press: New York, 1974.
42. D. Zurawski, L. Rice, M. Hourani, and A. Wieckowski, *J. Electroanal. Chem.* **230** (1987) 221.
43. A. N. Kahyaoglu. Oxydation électrocatalytique du glycérol sur le platine, l'or et leurs alliages binaires, PhD, Poitiers, France, 1981.
44. H. Möller and P. C. Pistorius, *J. Electroanal. Chem.* **570** (2004) 243.
45. R. R. Adzic and N. M. Markovic, *Electrochim. Acta* **30** (1985) 1473.
46. N. Furuya and S. Motoo, *J. Electroanal. Chem.* **98** (1979) 189.
47. R. Gomez and J. M. Feliu, *J. Electroanal. Chem.* **554–555** (2003) 145.
48. G. Kokkindis, J. M. Leger, and C. Lamy, *J. Electroanal. Chem.* **242** (1988) 221.
49. G. Kokkinidis, D. Sazou, and I. Moumtzis, *J. Electroanal. Chem.* **213** (1986) 135.
50. F. Matsumoto, M. Harada, N. Koura, and S. Uesugi, *Electrochem. Commun.* **5** (2003) 42.
51. S. Motoo and N. Furuya, *J. Electroanal. Chem.* **184** (1985) 303.
52. M. Sakamoto and K. Takamura, *Bioelectrochem. Bioenerg.* **9** (1982) 571.
53. M.-J. Shao, X.-K. Xing, and C.-C. Liu, *Bioelectrochem. Bioenerg.* **17** (1987) 59.
54. M. Shibata, N. Furuya, M. Watanabe, and S. Motoo, *J. Electroanal. Chem.* **263** (1989) 97.
55. P. C. C. Smits, B. F. M. Kuster, K. van der Wiele, and S. van der Baan, *Appl. Catal.* **33** (1987) 83.
56. M. Watanabe, M. Horiuchi, and S. Motoo, *J. Electroanal. Chem.* **250** (1988) 117.
57. C. Paul Wilde and M. Zhang, *Electrochim. Acta* **38** (1993) 2725.
58. Q.-H. Wu, S.-G. Sun, X.-Y. Xiao, Y.-Y. Yang, and Z.-Y. Zhou, *Electrochim. Acta* **45** (2000) 3683.
59. G. Kokkinidis and N. Xonoglou, *Bioelectrochem. Bioenerg.* **14** (1985) 375.
60. B. Beden, J.-M. Léger, and C. Lamy, *Electrocatalytic Oxidation of Oxygenated Aliphatic Organic Compounds at Noble Metal Electrodes*; Plenum Press: New York, 1992.
61. S. Ernst, C. H. Hamann, and J. Heitbaum, *Ber. Bunsen-Ges. Phys. Chem.* **84** (1980) 50.
62. H.-W. Lei, B. Wu, C.-S. Cha, and H. Kita, *J. Electroanal. Chem.* **382** (1995) 103.
63. L. H. E. Yei, B. Beden, and C. Lamy, *J. Electroanal. Chem.* **246** (1988) 349.
64. B. Beden, F. Largeaud, K. B. Kokoh, and C. Lamy, *Electrochim. Acta* **41** (1996) 701.
65. M. L. F. De Mele, H. A. Videla, and A. J. Arvia, *Bioelectrochem. Bioenerg.* **16** (1986) 213.
66. M. Fleischmann and Z. Q. Tian, *J. Electroanal. Chem.* **217** (1987) 385.
67. M. W. Hsiao, R. R. Adzic, and E. B. Yeager, *Electrochim. Acta* **37** (1992) 357.
68. M. W. Hsiao, R. R. Adzic, and E. B. Yeager, *J. Electrochem. Soc.* **143** (1996) 759.
69. K. B. Kokoh, J. M. Leger, B. Beden, H. Huser, and C. Lamy, *Electrochim. Acta* **37** (1992) 1909.
70. L. A. Larew and D. C. Johnson, *J. Electroanal. Chem.* **262** (1989) 167.
71. N. N. Nikolaeva, O. A. Khazova, and Y. B. Vassilyev, *Elektrokhimiya* **18** (1982) 1120.
72. L. A. Edye, G. V. Meehan, and G. N. Richards, *J. Carbohydr. Chem.* **10** (1991) 11.
73. P. Parpot, K. B. Kokoh, B. Beden, and C. Lamy, *Electrochim. Acta* **38** (1993) 1679.

74. P. Parpot, K. B. Kokoh, E. M. Belgsir, J. M. Léger, B. Beden, and C. Lamy, *J. Appl. Electrochem.* **27** (1997) 25.
75. H. Druliolle, K. B. Kokoh, and B. Beden, *Electrochim. Acta* **39** (1994) 2577.
76. H. Druliolle, K. B. Kokoh, F. Hahn, C. Lamy, and B. Beden, *J. Electroanal. Chem.* **426** (1997) 103.
77. M. Atobe and T. Nonaka, *Chem. Lett.* **24** (1995) 669.
78. K. Matsuda, M. Atobe, and T. Nonaka, *Chem. Lett.* **23** (1994) 1619.
79. C. Martin, H. Huser, K. Servat, and K. B. Kokoh, *Electrochim. Acta* **50** (2005) 2431.
80. C. Martin, H. Huser, K. Servat, and K. B. Kokoh, *Tetrahedron Lett.* **47** (2006) 3459.
81. C. Martin, K. Servat, H. Huser, and K. B. Kokoh, *J. Appl. Electrochem.* **36** (2006) 643.
82. A. Habrioux, K. Servat, S. Tingry, and K. B. Kokoh, *Electrochem. Commun.* **11** (2008) 111.
83. A. Habrioux, W. Vogel, M. Guinel, L. Guetaz, K. Servat, B. Kokoh, and N. Alonso-Vante, *Phys. Chem. Chem. Phys.* **11** (2009) 3573.

22 Fundamental Aspects of Corrosion of Metals and Semiconductors

N. Sato

CONTENTS

This chapter provides a fundamental context of materials corrosion in aqueous solution on the basis of the current state of corrosion science. The corrosion of metals and semiconductors involves anodic and cathodic charge-transfer processes across the material–solution interface and is hence under the control of the interfacial potential. We describe how semiconductor corrosion differs from metallic corrosion, emphasizing the importance of the interfacial potential and the electrode potential. Passivation occurs not only with metals but also with semiconductors, resulting in the protection of the materials. Passivity breakdown causes localized corrosion such as pitting and crevice corrosion. We discuss the stability of pitting corrosion and crevice corrosion with the criterion of the polishing and the active modes of metal dissolution, proposing a potential-dimension diagram for localized corrosion. Finally, we survey the influence of corrosion precipitate oxides on metallic corrosion and show that the ion-selective property and the semiconductive property of the oxides may affect the corrosion of underlying metals.

22.1 INTRODUCTION

Solid materials, in general, are more or less subject to corrosion in the environments where they stand, and materials corrosion is one of the most troublesome problems we have been frequently confronted with in the current industrialized world. In the past decades, corrosion science has steadily contributed to the understanding of materials corrosion and its prevention. Modern corrosion science of materials is rooted in the *local cell model* of metallic corrosion proposed by Evans [1] and in the *mixed electrode potential* concept of metallic corrosion proved by Wagner and Traud [2]. These two magnificent achievements have combined into what we call the *electrochemical theory of metallic corrosion*. It describes metallic corrosion as a coupled reaction of anodic metal dissolution and cathodic oxidant reduction. The electrochemical theory of corrosion can be applied not only to metals but also to other solid materials.

Solid materials may be classified into three categories according to the bonding characteristics prevailing in the solid: metallic solids, ionic solids, and covalent solids. In metallic solids, electrons bind metal ions into a crystalline lattice and are freely moving around all over the metal lattice, the metal ions hardly move though. For ionic solids, the crystal lattice is made of cations and anions with electrostatic attractive and repulsive forces, and the ions at the lattice sites are almost immobile under normal conditions. Covalent solids, whose lattice atoms are combined with each other by localized covalent electrons, apparently contain no ions and no delocalized electrons. Both ionic and covalent solids, if containing no lattice defects and impurities, are nonconductors for electrons because of the absence of free electrons in the solids. These nonconductive solids, however, turn to be semiconductors if foreign additives and/or lattice defects produce freely mobile electrons and electron vacancies (holes).

The corrosion of solid materials normally occurs in a variety of gaseous and liquid environments. The most common are atmospheric corrosion and aqueous corrosion. For convenience, we restrict ourselves to the environment of the aqueous solution, in which solid materials normally corrode first in the form of hydrated ions and then turn to be corrosion precipitates such as oxides and hydroxides. In this chapter, based on the current state of understanding of corrosion science, we discuss from a fundamental point of view the corrosion processes of solid materials mainly in terms of the energy-level concepts of reacting particles. For simplification, we deal with only inorganic solid materials excluding organic solids whose corrosion appears to be somewhat different from that of inorganic ones.

22.2 CORROSION PROCESSES

The corrosion of materials normally occurs across the interface between a solid material and an aqueous solution through the transfer processes of chemical particles such as ions and electrons. In electrochemistry, we define the transfer of positively charged particles from a solid into an aqueous solution as *anodic oxidation* and the transfer of negatively charged particles in the identical direction as *cathodic reduction*. In chemistry, what we call the reduction–oxidation (*redox*) reaction as the one that involves a complete transfer of electrons from the uppermost electron-donor orbital (HOMO) of a reductant particle to the uppermost electron-acceptor orbital (LUMO) of an oxidant particle. For a redox reaction to occur, then, the donor orbital of the reductant has to be sufficiently higher in energy than the acceptor orbital of the oxidant so that no hybridization between the donor orbital and the acceptor orbital may occur. The greater the energy difference between the donor orbital of the reductant and the acceptor orbital of the oxidant, the higher is the probability of the complete electron transfer.

In the case of metallic corrosion, the local cell model assumes that corrosion occurs as a combination of anodic metal oxidation and cathodic oxidant reduction. The anodic metal oxidation (dissolution) is a process of metal ion transfer across the metal–solution interface, in which the metal ions transfer from the metallic bonding state into the hydrated state in solution. We note that, before they transfer into the solution, the metal ions are ionized forming surface metal ions free from the metallic bonding electrons. The metal ion transfer is written as follows:

$$M_M^+ \rightarrow M_{aq}^+ \quad \text{anodic oxidation} \qquad (22.1)$$

where
 M_M^+ is the metal ion in the metallic bonding state
 M_{aq}^+ is the metal ion in the hydrated state

On the other hand, the cathodic oxidant reduction is an electron transfer process across the metal–solution interface, and electrons transfer from the Fermi level of the metal to the Fermi level of the redox reaction in solution, involving the reorganization of the hydrated structure of the redox particles.

$$Ox_{aq} + e_M^- \rightarrow Red\left(e_{redox}^-\right)_{aq} \quad \text{cathodic reduction} \qquad (22.2)$$

where

e_M^- is the electron in the metal

e_{redox}^- is the redox electron in the hydrated redox particles in the solution

The hydration of metal ions is a process that belongs to what we call the *acid–base reaction* in chemistry. The acid–base reaction involves no electron transfer between separate partners but a localized electron rearrangement to make up or break down a hybrid molecular orbital between an acid particle and a base particle, that is, the formation or the dissolution of a bonding molecular orbital due to the interaction between the frontier donor orbital of a particle (*Lewis base*) and the frontier acceptor orbital of another particle (*Lewis acid*). In order for an acid–base reaction to occur between a base particle and an acid particle, the electronic energy levels of the frontier orbitals for both acid and base particles are required to be close enough to each other for the orbital hybridization to prevail [3].

Let us consider, as examples for redox and acid–base reactions, water molecules, H_2O, in the chemical reactions with four different chemical species of metallic iron, Fe, chloride ions, Cl^-, ferrous ions, Fe^{2+}, and fluorine, F_2, in aqueous solution. In the cases of Fe and F_2, the uppermost frontier orbital level is sufficiently far away from that of H_2O, and hence the complete electron transfer takes place for these two species out of or into the water molecules. Water, H_2O, is an oxidant against metallic iron, Fe, and is a reductant against fluorine, F_2:

$$4H_2O + Fe \rightarrow Fe_{aq}^{2+} + 2OH_{aq}^- + H_{2(gas)} \qquad (22.3)$$

$$2H_2O + 2F_2 \rightarrow 4F_{aq}^- + 4H_{aq}^+ + O_{2(gas)} \qquad (22.4)$$

The addition of H_3O^+ ions in solution enhances Reaction 22.3, while the addition of OH^- ions boosts Reaction 22.4. In the case of Cl^- and Fe^{2+}, on the other hand, the electron energy level of the uppermost frontier orbital is very close to that of the frontier orbital of water molecules, H_2O, and hence the orbital hybridization, instead of the complete electron transfer, occurs between the frontier orbital of H_2O and that of each of the two species. The water molecule thus reacts as an acid with Cl^- forming hydrated chloride ion clusters [4], $Cl^- \cdot (H_2O)_n$, whereas it reacts as a base with Fe^{2+} forming hydrated ferrous ion clusters, $Fe^{2+} \cdot (H_2O)_6$:

$$nH_2O + Cl^- \rightarrow Cl^- \cdot (H_2O)_n \qquad (22.5)$$

$$6H_2O + Fe^{2+} \rightarrow Fe^{2+} \cdot (H_2O)_6 \qquad (22.6)$$

In general, various metal ions fall into two categories: *hard acids* such as Fe^{3+} and Al^{3+}, which are nonpolarizable, and *soft acids* such as Cu^+ and Ag^+, which are highly polarizable. There are several acids such as Fe^{2+} and Cu^{2+} on the borderline, and they act either as hard acids or as soft acids depending on the partners to associate with. It is convenient to divide bases also into two categories: *hard bases* such as Cl^-, OH^-, and SO_4^{2-}, which are nonpolarizable, and *soft bases* such as I^- and S^{2-}, which are polarizable. It is a useful generalization that hard acids prefer associating with hard bases, and soft acids prefer soft bases.

Chemical and electrochemical processes that cause materials corrosion usually involve both reduction–oxidations and acid–base reactions. The reduction–oxidation reaction is dependent on the electron energy level of the particles involved in the reaction, and hence managing the electrode potential of corroding materials may control the corrosion reaction. The acid–base reaction, on the other side, is determined by the HSAB characteristics (hard and soft acids and bases) of the particles involved in the reaction. It is mainly through the acid–base property that the environmental substances such as aggressive salts affect the corrosion of solid materials.

22.3 ELECTRONIC AND IONIC LEVELS

22.3.1 Electronic Levels

Electrons in a solid phase occupy the energy bands for electrons successively from the inner bands to the uppermost frontier bands. In the case of *ionic solids*, the bonding electrons occupy the atomic orbital of the anion only. The anion orbital, hence, normally forms the occupied uppermost donor band (*valence band*), whereas the orbital of the cation provides the unoccupied uppermost acceptor band (*conduction band*). For *metals* and *covalent solids*, the orbitals of the bonding partners overlap with each other, forming wide energy bands in which the bonding electrons are delocalized. In the case of metals such as sodium and aluminum, the frontier band is only partially filled with electrons, and hence the delocalized bonding electrons are freely mobile all over the bulk of the metals. In the case of covalent solids, the occupied frontier band (valence band) is entirely filled, and hence the bonding electrons therein are immobile. Beyond the valence band, there is an unoccupied band (the conduction band), and a forbidden band (the *band gap*) separates the conduction band from the valence band. A solid whose band gap is great falls into the category of insulators, while a solid with a narrow band gap such as silicon and gallium arsenide makes the category of semiconductors.

We normally define the energy level of electrons in a solid in terms of the *Fermi level*, ε_F, which is essentially equivalent to the electrochemical potential of electrons in the solid. In the case of metals, the Fermi level is equal to the highest occupied level of electrons in the partially filled frontier band. In the case of semiconductors of covalent and ionic solids, by contrast, the Fermi level is situated within the band gap where no electron levels are available except for localized ones. A semiconductor is either *n-type* or *p-type*, depending on its impurities and lattice defects. For *n*-type semiconductors, the Fermi level is located close to the conduction band edge, while it is located close to the valence band edge for *p*-type semiconductors. For examples, a zinc oxide containing indium as donor impurities is an *n*-type semiconductor, and a nickel oxide containing nickel ion vacancies, which accept electrons, makes a *p*-type semiconductor. In semiconductors, impurities and lattice defects that donate electrons introduce freely mobile electrons in the conduction band, and those that accept electrons leave mobile holes (electron vacancies) in the valence band. Both the conduction band electrons and the valence band holes contribute to electronic conduction in semiconductors.

Liquid water (hydrogen oxide) makes an electronic insulator in an amorphous condensed phase, and its electronic energy band structure is basically similar to that of metal oxides, the band edges being indistinct though due to its amorphous nature. The conduction band is made of hydrogen 1s orbitals and the valence band is made of oxygen 2p orbitals. The band gap between the conduction band and the valence band amounts to more than 8 eV [5] as shown in Figure 22.1 [6]. Within the band gap, there are localized electron levels attached to solute particles in aqueous solution. These

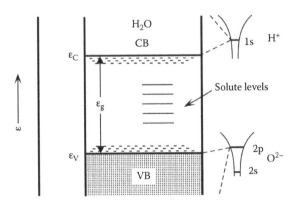

FIGURE 22.1 Electron energy bands of liquid water made of atomic orbitals of hydrogen and oxygen [5]: CB = conduction band, VB = valence band, $\varepsilon_C \geq -1.2$ eV, $\varepsilon_V \leq -9.3$ eV, $\varepsilon_g \geq 8$ eV.

localized electron levels are usually defined by what we call the redox electron level, that is, the Fermi level of redox electrons, at which level electrons are in equilibrium with a reductant–oxidant pair (redox pair) in aqueous solution:

$$RED(e^-_{redox}) \Leftrightarrow OX + e^-_{redox} \quad \varepsilon_{F(redox)} \tag{22.6a}$$

where

e^-_{redox} is the *redox electron*
$\varepsilon_{F(redox)}$ is the Fermi level of the redox electron

The Fermi level, $\varepsilon_{F(redox)}$, of the redox electron depends on the ratio of the concentration of the reductant to the oxidant:

$$\varepsilon_{F(redox)} = \frac{1}{2}(\varepsilon_{red} + \varepsilon_{ox}) + kT \ln \frac{N_{red}}{N_{ox}} = \varepsilon^0_{F(redox)} + kT \ln \frac{N_{red}}{N_{ox}} \tag{22.7}$$

where

ε_{red} is the *most probable donor level* of the reductant
ε_{ox} is the *most probable acceptor level* of the oxidant
$\varepsilon^0_{F(redox)}$ is the *standard* Fermi level
N_{red} and N_{ox} are the concentrations of the reductant and the oxidant, respectively

For ordinary redox couples in aqueous solution, the most probable acceptor level, ε_{ox}, is higher by about 3 eV than the most probable donor level, ε_{red}. The standard Fermi level, $\varepsilon^0_{F(redox)}$, of a redox couple is set just at the middle between the most probable acceptor level, ε_{ox}, of the oxidant and the most probable donor level, ε_{red}, of the reductant. The hydrogen–hydrogen-ion redox reaction is one of the basic reactions in electrochemistry and is called hydrogen electrode reaction:

$$H_{2(gas)} \Leftrightarrow 2H^+_{aq} + 2e^-_{nhe} \quad \varepsilon_{F(nhe)} \tag{22.8}$$

where

e^-_{nhe} is the redox electron of the *normal hydrogen electrode*
$\varepsilon_{F(nhe)}$ is the Fermi level of the redox electron of the normal hydrogen electrode: $\varepsilon_{F(nhe)} = -4.5$ eV

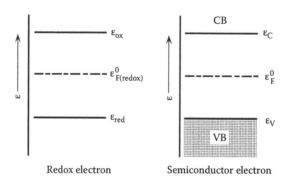

FIGURE 22.2 Electron energy levels for a standard pair of hydrated redox particles and for an intrinsic semiconductor: ε_{red} = the most probable electron level of oxidant, ε_{ox} = the most probable electron level of reductant, $\varepsilon^0_{F(redox)}$ = the standard Fermi level of redox electrons, ε^0_F = Fermi level of an intrinsic semiconductor, ε_V = valence band edge level, and ε_C = conduction band edge level.

Figure 22.2 shows a comparison of the electron energy-level diagram of an intrinsic solid semiconductor and that of an aqueous solution containing a redox species at the standard concentration. We see resemblance between the two diagrams in that the Fermi level is situated at the middle between the occupied level and the vacant level. We see further that the occupied level and the vacant level in semiconductors are delocalized energy bands in which electrons and holes can move around; whereas, for aqueous redox particles, they are localized levels producing no mobile electrons and hence giving no electronic conduction to aqueous solution. Aqueous redox solutions, however, are characterized by their ionic conduction due to the diffusion and the migration of hydrated redox ions, in contrast to semiconductors in which ions hardly move under normal conditions.

In electrochemistry, we define the zero level of electron energy as the electron energy situated at a position close to the solid or liquid surface but free from the image force of the solid or the liquid, namely, at the position of the *outer potential* ψ. We then describe the Fermi level, ε_F, of electrons in a solid as follows:

$$\varepsilon_F = \mu_e - e\chi = \alpha_e \tag{22.9}$$

where
  μ_e is the chemical potential of electrons in the solid or the liquid
  χ is the surface potential of the solid or the liquid
  α_e is called in electrochemistry the *real potential* of electrons

If we define the zero level at infinity as with physicists, the *electrochemical potential* η_e of electrons in a solid is given as

$$\eta_e = \mu_e - e\phi = \alpha_e - e\psi = \varepsilon_F - e\psi \tag{22.10}$$

where
  ϕ is the inner potential of the solid or the liquid
  ψ is the outer potential

The difference between the Fermi level and the electrochemical potential of electrons is thus equal to the outer potential term $e\psi$, which turns to be zero when the solid or liquid is free of electrostatic charge $\psi = 0$. Practically, thereby, we may assume that the Fermi level is essentially equal to the electrochemical potential of electrons.

22.3.2 IONIC LEVELS

The *unitary energy* of surface metal ions of a metallic solid, referred to the energy level of the gaseous metal ion at the position of the outer potential outside the solid metal, may be derived from the metal sublimation energy, the ionization energy of the metal atom, and the work function of the metal. This unitary energy, $\alpha^0_{M^+(s)}$, of the surface metal ions is essentially equivalent to the energy, $\alpha_{M^+(M)}$, of the metal ions in the solid metal [6]. We write the energy level, $\alpha_{M^+(s)}$, of surface metal ions as follows:

$$\alpha_{M^+(s)} = \alpha^0_{M^+(s)} + kT \ln x_k \tag{22.11}$$

where x_k is the fraction of surface kink sites: $x_k \approx 10^{-2}$ to 10^{-4}. The unitary energy $\alpha^0_{M^+(s)}$ of common metals amounts to several electron volts below zero: $\alpha^0_{M^+(s)} = -5.84$ eV for silver ions at the surface of metallic silver [6]. Similarly, the energy, $\alpha_{S_s^+}$, of surface ions of a covalent semiconductor is given as

$$\alpha_{S_s^+} = \alpha_{S_s^+}^0 - \varepsilon_{V_s} + kT \ln x_k \qquad (22.12)$$

where

$\alpha_{S_s^+}^0$ is the unitary energy of the surface ion

ε_{V_s} is the electron energy at the valence band edge

For germanium, we have $\alpha_{S_s^+}^0 = -6.8$ eV [6].

The energy level of an ion, A^+, in aqueous solution is defined by the energy required for the ion to transfer from the standard gaseous state into the hydrated state, namely, the hydration energy. The ion energy consists of the *unitary energy* (the standard energy) and the *communal energy*: the former is the one in the standard state and the latter depends on the concentration of the hydrated ion. We normally write the energy, $\alpha_{A_{aq}^+}$, of the hydrated ion in terms of the unitary one and the mixing one:

$$\alpha_{A_{aq}^+} = \alpha_{A_{aq}^+}^0 + kT \ln m_{A^+} \qquad (22.13)$$

where

$\alpha_{A_{aq}^+}^0$ is the standard level

m_{A^+} is the *molality* of the hydrated ion

According to [7], $\alpha_{A_{aq}^+}^0 = -11.28$ eV for H_{aq}^+, $\alpha_{A_{aq}^+}^0 = -4.95$ eV for Ag_{aq}^+, $\alpha_{A_{aq}^+}^0 = -44.85$ eV for Fe_{aq}^+, and $\alpha_{A_{aq}^+}^0 = 3.08$ eV for Cl_{aq}^-.

In aqueous solution, the energy level of hydrated protons determines the *acid–base* characteristics of the solution. A water molecule, H_2O, accepts a proton in its vacant proton level forming an acidic proton, H_3O^+, which we call the *acidic* or occupied *proton level*. Furthermore, a water molecule donates a proton out of its proton level leaving a basic proton vacancy OH^-, which we call the *basic* or vacant *proton level*. The difference in energy is 1.03 eV between the standard acidic proton level and the standard basic proton level. Figure 22.3 shows an energy diagram for the ionic dissociation of water molecules. The proton level is dependent on the proton concentration. The acidic proton level increases with acidic proton concentration and the basic vacant proton level increases with growing basic vacant proton concentration:

$$\alpha_{H_3O^+} = \alpha_{H_3O^+}^0 + kT \ln m_{H_3O^+} \qquad \alpha_{OH^-} = \alpha_{OH^-}^0 + kT \ln m_{OH^-}, \qquad (22.14)$$

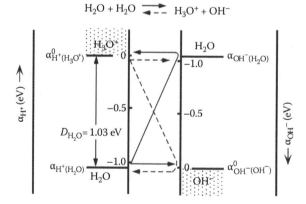

FIGURE 22.3 Energy levels of protons and proton vacancies in aqueous solution showing the ionic dissociation of water molecules: α_{H^+} = occupied proton level (donor), α_{OH^-} = vacant proton level (acceptor), and α^0 = the standard level.

where $\alpha^0_{H_3O^+}$ and $\alpha^0_{OH^-}$ are the standard acidic and basic proton levels, respectively. In an aqueous solution where the proton transfer is in equilibrium, the acidic proton level is equal to the basic vacant proton level on the same scale: $\alpha_{H_3O^+} = \alpha_{OH^-}$.

In general, the ionic level can be thermodynamically related to the electronic level through the ionization energy or electron affinity [6]. The ionic level may, hence, be put on the same energy scale as that of the electronic level as will be mentioned in the following sections, where the ionic level of metal ions in a metal electrode will be set equal to the Fermi level of the metal.

It is worth seeing some resemblance between the proton-level diagram of aqueous solutions and the electron-level diagram of semiconductors as shown in Figure 22.4. In the proton-level diagram, the ionic dissociation energy, 1.03 eV, of water molecules makes the energy gap between the standard acidic proton level, $\alpha^0_{H_3O^+}$, and the standard basic proton vacancy level, $\alpha^0_{OH^-}$. This energy gap may be compared with the band gap between the conduction band edge, ε_C, for electrons and the valence band edge, ε_V, for electron vacancies (holes) in semiconductors. Further, the concentration product, $m_{H_3O^+} \times m_{OH^-}$, of acidic protons and basic proton vacancies remains constant in aqueous solutions. Similarly, the concentration product, $n \times p$, of electrons and holes is also kept constant in a semiconductor. Furthermore, in the case of pure water, the proton level, $\alpha_{H_3O^+}$, is located at the midpoint between the standard level of acidic protons, $\alpha^0_{H_3O^+}$, and that of basic proton vacancies, $\alpha^0_{OH^-}$, where the concentration of acidic protons is equal to that of basic proton vacancies: $m_{H_3O^+} = m_{OH^-}$. An intrinsic semiconductor, likewise, situates its Fermi level, ε_F, at the midpoint between the conduction band, ε_C, and the valence band, ε_V, making the electron concentration equal to the hole one: $n = p$.

We now consider an aqueous solution containing foreign solutes. Increasing the acidic solute concentration shifts the proton level, $\alpha_{H_3O^+}$, upward in the acidic direction and hence increases the acidic proton concentration: $m_{H_3O^+} > m_{OH^-}$. Increasing the basic solute concentration, by contrast, shifts the proton level, $\alpha_{H_3O^+}$, downward in the basic direction resulting in an increase in the basic proton vacancy concentration; $m_{OH^-} > m_{H_3O^+}$. In the same way, with extrinsic semiconductors containing donor or acceptor impurities, increasing the donor concentration shifts the Fermi level, ε_F, upward to the conduction band edge, ε_C, hence increasing the electron concentration in the

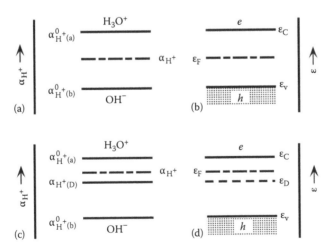

FIGURE 22.4 Energy diagrams for proton levels in aqueous solutions and for electron levels in semiconductors: (a) proton levels in pure water, (b) electron levels in an intrinsic semiconductor, (c) proton levels in an acidic solution, (d) electron levels in an n-type semiconductor: α_{H^+} = proton level, $\alpha^0_{H^+(a)}$ = the standard acidic proton level, $\alpha^0_{H^+(b)}$ = the standard basic proton level, $\alpha_{H^+(D)}$ = a proton donor level of solute, and ε_D = an electron donor level.

conduction band, $n > p$; whereas, increasing the acceptor concentration shifts the Fermi level ε_F, downward to the valence band edge, ε_V, hence increasing the hole concentration in the valence band, $p > n$.

22.3.3 FERMI LEVELS OF ELECTRODE REACTIONS

The electronic energy level and the ionic energy level can be arranged on the same energy-level scale [6]. For instance, if we equate the energy level, $\alpha_{M^+(M)}$, of metal ions in a metal to the Fermi level, ε_F, of the metal, we then obtain the energy level, $\alpha_{M^+(aq)}$, of the hydrated metal ions that can be determined by the Fermi level, $\varepsilon_{F\left(M/M^{2+}\right)}$, of the electrochemical metal dissolution in equilibrium:

$$M = M_{aq}^{2+} + 2e_{redox}^- \quad \varepsilon_{F\left(M/M_{aq}^{2+}\right)} \tag{22.15}$$

where
e_{redox}^- is the redox electron in equilibrium with the metal dissolution
$\varepsilon_{F\left(M/M_{aq}^{2+}\right)}$ is the Fermi level of the reaction

In this case of metal dissolution equilibrium, the Fermi level, $\varepsilon_{F\left(M/M^{2+}\right)}$, of the reaction is set equal to the ionic energy level of hydrated metal ions, that is, the real potential, $\alpha_{M^{2+}(s)}$, of hydrated metal ions on the same energy scale as that of electrons.

In general, all the reactions that involve transfer of electrons and/or ions in the electrode are called electrode reactions in electrochemistry. Any electrode reaction has its electron energy level, which we call the *Fermi level of the reaction*. In fact, the Fermi level of an electrode reaction is equivalent to what we call the equilibrium electrode potential of the reaction.

22.4 ELECTRODE POTENTIAL

22.4.1 ELECTRONIC ELECTRODE POTENTIAL

In electrochemistry, the electrode potential is defined by the electronic energy level in a solid electrode referred to the energy level of the standard gaseous electron just outside the surface of an electrolyte (aqueous solution) in which the electrode is immersed [6]:

$$E = -\frac{\varepsilon_{F(M)}}{e} = -\frac{\mu_{e(M)}}{e} + \Delta\phi_{M/S} + \chi_{S/G} = \frac{\Phi_{M/S/G}}{e} \tag{22.16}$$

where
E is the electrode potential
$\varepsilon_{F(M)}$ is the Fermi level of the electrode
$\mu_{e(M)}$ is the chemical potential of electrons in the electrode
$\Delta\phi_{M/S}$ is the interfacial potential between the electrode M and the solution S
$\chi_{S/G}$ is the surface potential of the electrolyte
$\Phi_{M/S/G}$ is the work function of the electrode–solution system
e is the elemental charge

The electrode potential as defined earlier is called the *absolute electrode potential*, and it is compared to the electrode potential referred to normal hydrogen electrode. The Fermi level of the normal hydrogen electrode has been estimated near -4.5 eV, and the normal hydrogen electrode potential is 4.5 V on the scale of the absolute electrode potential.

An electrode in equilibrium with a redox electron transfer reaction sets its Fermi level equal to the Fermi level of the redox electrons in the electrolytic solution. The electrode potential, then, determines the Fermi level, $\varepsilon_{F(redox)}$, of the redox electrons in the solution:

$$E = -\frac{\varepsilon_{F(M)}}{e} = -\frac{\varepsilon_{F(redox)}}{e} = -\frac{\mu_{e(redox)}}{e} + \chi_{S/G} = E_{redox} \qquad (22.17)$$

where E_{redox} denotes the redox potential of the electron transfer reaction.

22.4.2 IONIC ELECTRODE POTENTIAL

The electrode potential can also be defined by the energy level of an ion in a solid electrode. The energy level of an ion in a condensed phase is normally referred to the energy level of the ion in the standard gaseous state just outside the surface of the condensed phase. The gaseous ion level is related to the gaseous electron level through the ionization energy of the ion. Let us consider, as an example, a metal electrode in an aqueous solution. The energy level, $\alpha_{M^{z+}(M)}$, of the metal ion, M_M^{z+}, in the metal electrode, M, is then given as follows [6]:

$$\alpha_{M^{z+}(M)} = zeE + \mu_{M(M)} \qquad E = \frac{\alpha_{M^{z+}(M)}}{ze} - \frac{\mu_{M(M)}}{ze} \qquad (22.18)$$

where
   E is the electrode potential defined by the electron level of the metal electrode
   z is the ionic valence of the metal ion
   $\mu_{M(M)}$ is the energy required to form the solid metal from the standard gaseous metal ion and the standard gaseous electrons; $\mu_{M(M)}$ consists of the sublimation energy of the metal and the ionization energy of the metal atom

The ion level, $\alpha_{M^{z+}(M)}$, in the electrode, thus, depends on E. A metal–solution electrode, if in equilibrium with the metal ion transfer, has its metal ion level, $\alpha_{M^{z+}(M)}$, equal to the energy level, $\alpha_{M^{z+}(aq)}$, of the hydrated metal ion in the solution. The hydrated ion level hence determines the value of E:

$$E = \frac{\alpha_{M^{z+}(aq)}}{ze} - \frac{\mu_{M(M)}}{ze} \qquad (22.19)$$

where $\alpha_{M^{z+}(aq)}$ is a function of the concentration of hydrated metal ions.

22.4.3 INTERFACIAL POTENTIAL AT NONMETALLIC SOLID ELECTRODES

In the case of metallic electrodes, any change in the electrode potential, E, always occurs at the interfacial potential, $\Delta\phi_H$, between the metal and the solution. The amount of change in the electrode potential, then, is equal to that in the interfacial potential, which arises across what we call the Helmholtz layer or the electrical compact double layer. This is, however, not the case for nonmetallic solid electrodes such as ionic or covalent semiconductors, at which the interfacial potential usually remains constant irrespective of the electrode potential.

The interfacial potential is made up of what we call the interfacial charge. In aqueous solution, a solid electrode of ionic or covalent bonding has adsorbed water molecules on its interface, and they are normally hydroxylated forming two types of adsorbed hydroxyl groups, the *acid*-type and the *base*-type. These hydroxyl groups undergo dissociation or association and produce an interfacial charge, which normally is positive in acidic solution and negative in basic solution. The interfacial potential is thus determined by the association–dissociation of the interfacial hydroxyl groups. We learn that the interfacial potential is dependent on solution pH as shown in the following equation:

$$\Delta\phi_H = \Delta\phi_{iep} - 2.3kT(pH - pH_{iep}) \tag{22.20}$$

where $\Delta\phi_{iep}$ and pH_{iep} are the interfacial potential and the pH, respectively, at the *isoelectric point* where the interface has no electric charge. The linear relation between the interfacial potential and the solution pH has been observed with a number of nonmetallic electrodes [8].

22.4.4 SEMICONDUCTOR ELECTRODES

In semiconductor electrodes, we have a space charge layer in addition to an electrical compact double layer (Helmholz layer) at the electrode interface. The electrode potential, then, is the sum of the space charge potential, $\Delta\phi_{SC}$, and the interfacial potential, $\Delta\phi_H$:

$$\Delta\phi_{SC/aq} = \Delta\phi_{SC} + \Delta\phi_H \tag{22.21}$$

where $\Delta\phi_{SC/aq}$ is the total difference in the inner potential between the electrode body and the solution bulk. It is, in fact, the total difference, $\Delta\phi_{SC/aq}$, in potential that is a linear function of the electrode potential, E. As we mentioned in the foregoing text, the interfacial potential of nonmetallic electrodes is normally maintained constant, and hence the band edge levels, ε_C and ε_V, are pinned relative to the Fermi level of the solution (e.g., the Fermi level of the normal hydrogen redox electron). This is the state that we call the *band edge pinning*, and at this state any change in the electrode potential occurs in the space charge layer, $\Delta\phi_{SC}$, with the interfacial potential, $\Delta\phi_H$, remaining constant.

The electrode potential where the space charge potential becomes zero is called the *flat band potential*, E_{fb}. The space charge is positive at electrode potentials more positive (i.e., more anodic) than E_{fb}, and it is negative at electrode potentials less positive (i.e., more cathodic) than E_{fb}. In fact, the space charge potential, $\Delta\phi_{SC}$, is defined by the difference between the band edge level in the semiconductor interior and that at the semiconductor surface. Since the Fermi level is not allowed to move out of the band gap, the space charge potential always amounts to less than the band gap of the semiconductor.

When the surface Fermi level comes to one of the band edge levels (the valence band edge or conduction band edge), an interfacial charge arises breaking down the band edge pinning state, and the Fermi level is now pinned at the band edge level of the electrode. Under this *Fermi-level pinning state*, the interface is degenerated in the electron state density causing quasi-metallization of the electrode surface. With the degenerate interface, the semiconductor electrode behaves like a metal electrode and the interfacial potential then turns to be a linear function of the electrode potential. The quasi-metallized surface will occasionally play an important role in semiconductor electrodes. Figure 22.5 shows for an intrinsic semiconductor electrode the interfacial potential, $\Delta\phi_H$, and the space charge potential, $\Delta\phi_{SC}$, in relation to the electrode potential, E, that is, the Fermi level, ε_F. We see that, as the electrode potential shifts in the positive direction, the electrode interface changes from the cathodic degeneration where the electrode Fermi level is pinned at the conduction band edge, ε_C, through the band edge pinning state where the interfacial potential remains constant, to the anodic degeneration where the electrode Fermi level is pinned at the valence band edge, ε_V. Under normal conditions, semiconductor electrodes are in a state of band edge pinning.

Since the Fermi level, $_p\varepsilon_F$, of a *p*-type semiconductor electrode is inevitably lower than the Fermi level, $_n\varepsilon_F$, of an *n*-type electrode of the same semiconductor, the electrode potential of the *p*-type is always more positive than that of the *n*-type under the flat band condition. The difference in the flat band electrode potential between the *p*-type and the *n*-type electrode is nearly equivalent to the band gap of the semiconductor. It is an observed fact that the electrode potential of most

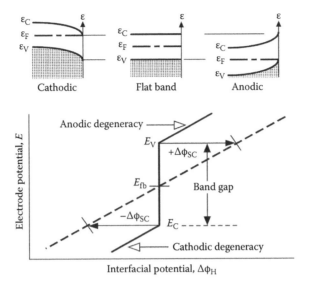

FIGURE 22.5 Electrode potential consisting of interfacial potential $\Delta\phi_H$ and space charge potential $\Delta\phi_{SC}$ for an intrinsic semiconductor: E_{fb} = flat band potential, E_V = valence band edge potential, and E_C = conduction band edge potential.

semiconductors in aqueous solution ranges within a few volts around the normal hydrogen electrode potential.

22.4.5 SEMICONDUCTOR PHOTO-ELECTRODES

Semiconductor electrodes whose band gap is relatively narrow receive photon energy and produce photoexcited electron–hole pairs in the space charge layer. The photoexcited electron–hole pair formation significantly increases the concentration of minority charge carriers (holes in the n-type), but influences little the concentration of majority carriers (electrons in the n-type). The photoexcited electrons and holes set their energy levels not at the electrode Fermi level, ε_F, but at what we call the *quasi-Fermi levels*, $_n\varepsilon_F^*$ and $_p\varepsilon_F^*$, respectively. The quasi-Fermi level for majority carriers is close to the electrode Fermi level, ε_F, but the quasi-Fermi level for minority carriers is far away from the electrode Fermi level.

Photoexcitation also introduces a photo-potential, $\Delta\varepsilon_{ph}$, which reduces the space charge potential and shifts the electrode potential toward the flat band potential, E_{fb}, of the semiconductor electrode. Furthermore, as mentioned earlier, the electrode Fermi level splits into two quasi-Fermi levels, $_n\varepsilon_F^*$ and $_p\varepsilon_F^*$, for electrons and holes in the space charge layer. Figure 22.6 shows the photoinduced Fermi-level splitting in the space charge layer for an n-type semiconductor electrode. We suggest that a photoexcited n-type semiconductor electrode with its interfacial quasi-Fermi level at $_p\varepsilon_F^*$ is energetically equivalent to a p-type electrode of the same semiconductor whose Fermi level, ε_F, is equal to $_p\varepsilon_F^*$. The electrode potential, however, differs from each other and the potential of the photoexcited n-type electrode is much less positive than the potential of the p-type electrode by a magnitude nearly equivalent to $_p\varepsilon_F^* - \varepsilon_F$ in the photoexcited n-type electrode.

It is noticed that, although the Fermi level, ε_F, of holes in the photoexcited n-type semiconductor bulk reflects the electrode potential, E, the energy level of holes directly taking part in reactions at the electrode interface is obviously far away from the Fermi level, ε_F, that we measure as the electrode potential. This situation makes it thermodynamically possible for certain reactions, which are unallowable with a semiconductor electrode in the dark, to occur on the same semiconductor electrode under the condition of photoexcitation [6].

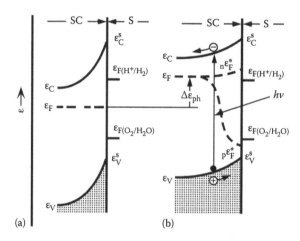

FIGURE 22.6 Electron energy diagrams for an n-type semiconductor electrode in the dark (a) and in the photoexcited state (b): ε^s = band edge level at the interface, $\varepsilon_{F(H^+/H_2)}$ = Fermi level of the normal hydrogen electrode reaction, $\varepsilon_{F(O_2/H_2O)}$ = Fermi level of the normal oxygen electrode reaction, $\Delta\varepsilon_{ph}$ = photo potential, $_p\varepsilon_F^*$ = quasi-Fermi level of photoexcited holes, and $_n\varepsilon_F^*$ = quasi-Fermi level of photoexcited electrons ($_n\varepsilon_F^* \approx \varepsilon_F$ for n-type semiconductors).

22.5 ELECTROCHEMICAL DISSOLUTION OF SOLID ELECTRODES

22.5.1 Anodic Dissolution of Metal Electrodes

Let us consider as an example a metallic iron electrode in an aqueous solution. The anodic metal dissolution is a process in which iron ions in the metallic bonding state in the metal phase transfer across the interfacial compact double layer (Helmholtz layer) into the hydrated state of the ions in the solution:

$$Fe_M^{2+} \rightarrow Fe_{aq}^{2+} \quad A = \alpha_{Fe^{2+}(M)} - \alpha_{Fe^{2+}(aq)} \qquad (22.22a,b)$$

where
   A is the affinity of the iron ion transfer
   $\alpha_{Fe^{2+}(M)}$ is the energy level of iron ions in the metallic bonding state
   $\alpha_{Fe^{2+}(aq)}$ is the energy level of iron ions in the hydrated state

This ionic transfer will occur if $\alpha_{Fe^{2+}(M)}$ is higher than $\alpha_{Fe^{2+}(aq)}$. The metal ion transfer is, in fact, a complicated process frequently involving hydrated anions as ligands coordinated with transferring metal ions. Electrons in the metal phase participate in forming and breaking the metallic bonding of iron ions at the metal surface, but during the ion transfer they do not transfer across the Helmholtz layer at the electrode interface where an intense electric field exists.

   The difference between $\alpha_{Fe^{2+}(M)}$ and $\alpha_{Fe^{2+}(aq)}$ corresponds to the affinity for the ion transfer, and it controls the ion transfer current. In electrochemical kinetics, the ion transfer current, i_d, is given by an exponential function of the interfacial potential, $\Delta\phi_H$, and hence of the electrode potential, E:

$$i_d = k_d \exp\left(\frac{2e\beta\Delta\phi_H}{kT}\right) = K \exp\left(\frac{2e\beta E}{kT}\right) \qquad (22.23)$$

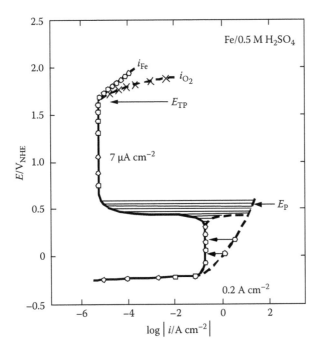

FIGURE 22.7 Polarization curves for the anodic dissolution and the passivation of metallic iron in 0.5 kmol m^{-3} sulfuric acid solution at 25°C [9,10]: i_{Fe} = anodic iron dissolution current, i_{O_2} = oxygen evolution current, E_P = passivation potential, and E_{TP} = trans-passivation potential.

The anodic metal dissolution current thus increases with the electrode potential up to a certain potential, called the *passivation potential*. As shown in Figure 22.7 for iron in acid solution [9,10], in the potential region more positive than the passivation potential, the metal passivates into almost no dissolution current due to the formation of a surface oxide film several nanometers thick, which we call the *passive film* [11]. In contrast to the *passive state* of the metal, the *active state* refers to the metal undergoing anodic dissolution at significant rates below the passivation potential.

22.5.2 Anodic Dissolution of Covalent Semiconductors

We consider, for instance, germanium in aqueous solution. The anodic dissolution of germanium occurs by either donating electrons into the conduction band or accepting holes out of the valence band of germanium as follows:

$$Ge \rightarrow Ge^{4+}_{aq} + 4e^-_{Ge} \quad \varepsilon_{F(d)} \tag{22.24}$$

$$Ge + h^+_{Ge} \rightarrow Ge^{4+}_{aq} \quad \varepsilon_{F(d)} \tag{22.25}$$

where
   e^-_{Ge} is the electron in the conduction band
   h^+_{Ge} is the hole in the valence band
   $\varepsilon_{F(d)}$ is the Fermi level of the dissolution reaction

We call these reactions an electron injecting dissolution and a hole emitting dissolution of germanium, respectively. The Fermi level of the reaction, $\varepsilon_{F(d)}$, relative to the Fermi level of the electrode determines which of the two reactions dominates over the other. If the band gap is comparatively narrow, we may assume that the two reactions occur simultaneously.

Let us consider the hole accepting dissolution and see the dissolution rate i_d:

$$i_d = k_d c_{Ge_s^{4+}} \exp\left(\frac{4e\beta\Delta\phi_H}{kT}\right) \tag{22.26}$$

where
$c_{Ge_s^{4+}}$ is the surface concentration of germanium ions
$\Delta\phi_H$ is the interfacial potential, which is constant due to the band edge pinning

Since $c_{Ge_s^{4+}}$ is one fourth the surface hole concentration, c_{h_s}, due to the electrical charge balance ($c_{Ge_s^{4+}} = 4c_{h_s}$), the surface ion concentration, $c_{Ge_s^{4+}}$, is described as a function of both the space charge potential, $\Delta\phi_{SC}$, and the hole concentration, c_h^0, in the bulk of the semiconductor:

$$i_d = K_d c_h^0 \exp\left(\frac{e\beta\Delta\phi_{SC}}{kT}\right) = K_d^0 c_h^0 \exp\left(\frac{e\beta E}{kT}\right) \tag{22.27}$$

where $K_d = k_d \exp (4e\beta\Delta\phi_H/kT)$ is constant under the band edge pinning condition. The anodic dissolution rate is much greater with the p-type than with the n-type electrode as shown in Figure 22.8 [12].

As with metals, semiconductors are also subject to passivation. Figure 22.9 shows the anodic dissolution and the passivation of n-type and p-type silicon electrodes in sodium hydroxide solution [13]. Silicon dissolves in basic solution in the form of soluble divalent silicon, $Si(OH)_{aq}^+$ or $Si(OH)_{2,aq}$, and passivates forming a silicon dioxide film.

22.5.3   ANODIC PHOTO-DISSOLUTION OF SEMICONDUCTORS

Let us consider an n-type semiconductor electrode, which due to the lack of holes is not subject to the anodic hole emitting dissolution in the dark. Photoexcitation, however, produces a significant concentration of holes in the valence band. These photoexcited holes migrate in a space charge layer

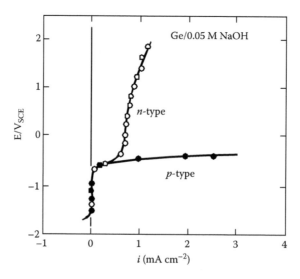

FIGURE 22.8 Polarization curves for anodic dissolution of n-type and p-type germanium electrodes in 0.05 kmol m^{-3} sodium hydroxide solution [12].

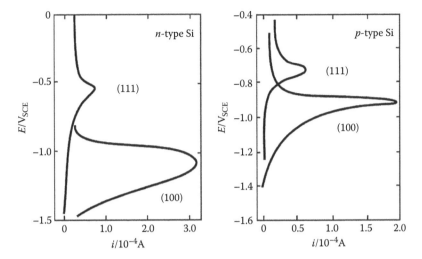

FIGURE 22.9 Polarization curves for the anodic dissolution and the passivation of n-type and p-type silicon electrodes in 5 kmol m^{-3} sodium hydroxide solution; (100) and (111) denote crystal faces; the potential scan rate is 10 mV s^{-1} at 60°C for n-type silicon and at 50°C for p-type silicon [13].

toward the interface, and they participate in the anodic hole emitting the dissolution of the semiconductor. As we saw in the foregoing section, photoexcitation provides photo-created holes at an interfacial quasi-Fermi level, $_p\varepsilon_F^*$, which is lower than the Fermi level, $_n\varepsilon_F$, of the n-type semiconductor electrode. The difference between $_p\varepsilon_F^*$ and $_n\varepsilon_F$ is, in fact, equivalent to the photo-energy absorbed by the semiconductor. We may then assume that the photoexcited n-type electrode is equivalent to a p-type electrode of the same semiconductor with the Fermi level, $_p\varepsilon_F$, equal to $_p\varepsilon_F^*$ of the photoexcited n-type electrode.

There is, however, a definite difference in the electrode potential between the photoexcited n-type electrode and the p-type electrode of the same semiconductor. The electrode potential of the former is less positive than that of the latter by a magnitude nearly equal to $\Delta\varepsilon = _n\varepsilon_F - _p\varepsilon_F$, which is the difference between the flat band potential of the n-type and that of the p-type. The anodic dissolution of the photoexcited n-type semiconductor electrode, as a result, will occur in the potential range less positive by about $\Delta\varepsilon$ than that for the p-type electrode of the same semiconductor. Figure 22.10 shows schematic polarization curves of the anodic hole emitting dissolution with an n-type electrode and with a p-type electrode of the same semiconductor under the dark and photoexcitation conditions. Such photoexcited dissolution was observed with n-type GaAs electrodes [14,15].

22.5.4 ANODIC AND CATHODIC DISSOLUTION OF COMPOUND SEMICONDUCTORS

Compound semiconductors may be subject to both anodic oxidative dissolution and cathodic reductive dissolution. Let us now consider a metal oxide, MO, in aqueous solution. The anodic dissolution will occur with the oxidation of oxide ions at the metal oxide surface producing gaseous oxygen. Concurrently, the metal ions transfer from the metal oxide across the oxide–solution interface forming hydrated metal ions in the aqueous solution:

$$O_{MO}^{2-} + 2h_{MO}^+ \rightarrow (1/2)O_{2(gas)} \tag{22.28}$$

$$M_{MO}^{2+} \rightarrow M_{aq}^{2+} \tag{22.29}$$

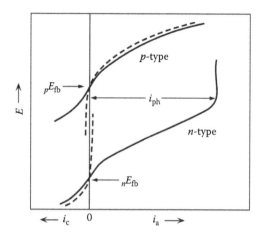

FIGURE 22.10 Schematic polarization curves for anodic hole-emitting dissolution of an n-type electrode and a p-type electrode of the same semiconductor in the dark and photoexcited conditions [14,15]: solid curve = photoexcited, dashed curve = dark, i_a = anodic dissolution current, i_c = cathodic current, i_{ph} = photoexcited dissolution current, and E_{fb} = flat band potential for n-type and p-type electrodes.

$$MO + 2h^+_{MO} \rightarrow M^{2+}_{aq} + (1/2)O_{2(gas)} \quad \varepsilon_{F(MO/O_2)} \tag{22.30}$$

where

     h^+_{MO} is the hole in the valence band of the metal oxide
     $\varepsilon_{F(MO/O_2)}$ is the Fermi level of the electrode reaction

One of the anodic oxidative dissolution of metal oxides frequently encountered in the corrosion of stainless steels is the transpassive dissolution of metallic chromium:

$$Cr_2O_3 + 5H_2O + 6h^+ \rightarrow 2CrO^{2-}_{4(aq)} + 10H^+_{aq} \tag{22.31}$$

where, instead of oxide ions, insoluble trivalent chromium oxide is oxidized to soluble hexavalent chromate ions. The cathodic dissolution, by contrast, will occur with the reduction of metal ions in the oxide producing metallic particles. The oxide ions, concurrently, transfer from the oxide across the oxide–solution interface forming water molecules in the aqueous solution:

$$M^{2+}_{MO} + 2e^-_{MO} \rightarrow M \tag{22.32}$$

$$O^{2-}_{MO} + 2H^+_{aq} \rightarrow H_2O_{aq} \tag{22.33}$$

$$MO + 2H^+_{aq} + 2e^-_{MO} \rightarrow M + H_2O_{aq} \quad \varepsilon_{F(M/MO)} \tag{22.34}$$

where

     e^-_{MO} is the electron in the conduction band of the metal oxide
     $\varepsilon_{F(M/MO)}$ is the Fermi level of the electrode reaction

One of the cathodic dissolution of metal oxides frequently encountered in the corrosion of metallic iron is the reductive dissolution of ferric oxide:

$$Fe_2O_3 + 6H^+_{aq} + 2e^- \rightarrow 2Fe^{2+}_{aq} + 3H_2O \tag{22.35}$$

where sparingly soluble ferric oxide is reduced to soluble ferrous ions.

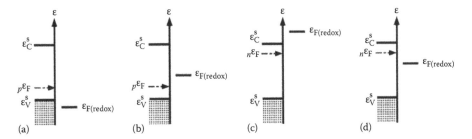

FIGURE 22.11 Energy diagrams for anodic and cathodic dissolution of compound semiconductor electrodes: (a) anodic dissolution impossible, (b) anodic dissolution possible, (c) cathodic dissolution impossible, and (d) cathodic dissolution possible [16]: $\varepsilon_{F(redox)}$ = Fermi level of anodic and cathodic semiconductor dissolution reactions.

From the thermodynamic view point, the anodic dissolution is allowed to occur only when the Fermi level, $\varepsilon_{F(MO/O_2)}$, of the oxide ion oxidation is higher than the Fermi level, $\varepsilon_{F(MO)}$, of the oxide electrode. The cathodic dissolution, by contrast, is allowed to occur only when the Fermi level, $\varepsilon_{F(M/MO)}$, of the metal ion reduction is lower than the Fermi level, $\varepsilon_{F(MO)}$, of the oxide electrode. Further, under normal conditions, semiconductor electrodes in aqueous solution are in the state of band edge pinning, in which state the band levels, ε_C and ε_V, are held constant relative to the electron level of the solution. Furthermore, the Fermi level of semiconductors is confined within the range of the band gap between ε_C and ε_V. With these situations, we hence see that only those anodic and cathodic dissolution reactions whose Fermi level is located within the band gap of the semiconductor electrodes are thermodynamically allowed to occur. In contrast, it is thermodynamically impossible for those dissolution reactions to occur whose Fermi level stands outside the band gap, that is, in the conduction band or in the valence band as shown in Figure 22.11 [16]. In aqueous solution, the anodic dissolution is thermodynamically possible, for instance, for ZnO, TiO$_2$, Cu$_2$O, GaP, and GaAs, whereas the cathodic dissolution is allowed to occur with ZnO, TiO$_2$, and CdS [16].

22.6 CORROSION OF IONIC SOLIDS

22.6.1 IONIC TRANSFER PROCESSES

For an ionic solid whose electronic band gap is so great that no mobile electrons and holes are available, the corrosion occurs through the transfer of cations and anions from the ionic bonding state into the state of hydrated ions in aqueous solution. Let us suppose an ionic solid, MO, consisting of cation M^{2+} and anion O^{2-}. The ionic transfer occurs across the solid–aqua-solution interface:

$$M_s^{2+} + nH_2O_{aq} \rightarrow M_{aq}^{2+}(H_2O)_n \quad \left(M_s^{2+} \rightarrow M_{aq}^{2+}\right) \quad A_{M^{2+}} \tag{22.36}$$

$$O_s^{2-} + 2H_{aq}^+ \rightarrow H_2O_{aq} \quad O_s^{2-} \rightarrow O_{aq}^{2-} \quad A_{O^{2-}} \tag{22.37}$$

where
   subscripts "s" and "aq" denote surface sites on the solid and hydrated ions in the solution
   $A_{M^{2+}}$ and $A_{O^{2-}}$ are the affinities of ionic transfer for the cation and anion, respectively

In the electrochemical sense, the cation transfer is an anodic reaction and the anion transfer is a cathodic reaction. The two ionic transfer processes make up an overall process:

$$MO + 2H_{aq}^+ \rightarrow M_{aq}^{2+} + H_2O_{aq} \tag{22.38}$$

When the two ionic transfers are simultaneously in equilibrium, the concentration of hydrated ions corresponds to the solubility product of the ionic solid in the solution.

The affinity of the ion transfer is given by the difference in the energy level of the transferring ions between the solid and the solution [6]:

$$A_{M^{2+}} = \alpha_{M^{2+}(s)} - \alpha_{M^{2+}(aq)} = \mu_{M^{2+}(s)} - \mu_{M^{2+}(aq)} + 2\Delta\phi_H \tag{22.39}$$

$$A_{O^{2-}} = \alpha_{O^{2-}(s)} + 2\alpha_{H^+(aq)} - \alpha_{H_2O(aq)} = \mu_{O^{2-}(s)} + 2\mu_{H^+(aq)} - \mu_{H_2O(aq)} - 2\Delta\phi_H \tag{22.40}$$

where
- α is the energy level (the *real potential*) of transferring particles
- μ is the chemical potential of them
- $\Delta\phi_H$ is the interfacial potential

As mentioned earlier, for most nonmetallic solids in aqueous solution the interfacial potential, $\Delta\phi_H$, is determined by the acid–base nature of the adsorbed hydroxyl group and increases linearly with decreasing pH of the solution. The corrosion of ionic solids is, hence, dependent on the pH of the solution in which the solids were immersed.

22.6.2 CORROSION RATES AND SURFACE COMPOSITIONS

The anodic cation transfer rate, v_M^+, and the cathodic anion transfer rate, v_O^-, both depend on the interfacial potential, $\Delta\phi_H$, between the ionic solid and the aqueous solution as described in the following equations:

$$v_M^+ = k_M^+ c_{M_s} \exp\left(\frac{2e\beta^+\Delta\phi_H}{kT}\right) \tag{22.41}$$

$$v_O^- = k_O^- c_{H_{aq}^+}^2 c_{O_s} \exp\left(\frac{-2e\beta^-\Delta\phi_H}{kT}\right) \tag{22.42}$$

where
- c_{M_s} is the cation concentration at the solid surface
- c_{O_s} is the anion concentration at the solid surface
- k is the rate constant
- β is the transfer coefficient

These equations suggest that the anodic cation dissolution rate increases with increasing $\Delta\phi_H$, whereas, the cathodic anion dissolution rate decreases with increasing $\Delta\phi_H$. As shown in Figure 22.12, the dissolution rate of a magnetite electrode in acid solution first increases with increasing electrode potential, suggesting that the iron ion transfer is determining the rate. The dissolution rate then turns to decrease with the electrode potential, suggesting that the oxide ion transfer is the rate-determining process [17]. Such potential-dependent dissolution was also observed with FeO, CdO, and NiO in acid solution [17–19].

Because of electrical neutrality, the anodic cation transfer current is always equal to the cathodic anion transfer current: $v_M^+ = v_O^-$. Equations 22.41 and 22.42, then, yield the interfacial potential, $\Delta\phi_H$, as a logarithmic function of the ratio of the anion to the cation in surface concentration:

$$\Delta\phi_H = \frac{kT}{e(\beta^+ + \beta^-)} \ln\left(\frac{k_O^- c_{H_{aq}^+}^2}{k_M^+} \times \frac{c_{O_s}}{c_{M_s}}\right) \tag{22.43}$$

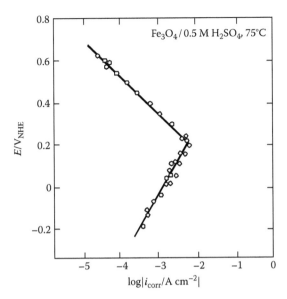

FIGURE 22.12 Corrosion rate i_{corr} as functions of applied electrode potential E for a magnetite Fe_3O_4 electrode in 0.5 kmol m^{-3} sulfuric acid solution at 75°C; interfacial potential $\Delta\phi_H$ is linearly related to applied electrode potential E [17].

where c_{O_s}/c_{M_s} is the ratio of the concentration of the surface anions to the surface cations on the solid. The surface ion ratio is thus an exponential function of the interfacial potential:

$$\frac{c_{M_s}}{c_{O_s}} = \frac{k_O^- c_{H_{aq}^+}^2}{k_M^+} \exp\left(\frac{-e(\beta^+ + \beta^-)\Delta\phi_H}{kT}\right) \tag{22.44}$$

Let us now consider a reference surface of stoichiometric composition where $c_{M_s} = c_{O_s}$. With this stoichiometric surface composition, as shown in Figure 22.13, we obtain the maximum corrosion

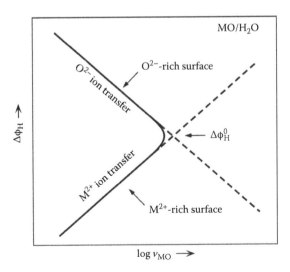

FIGURE 22.13 Schematic corrosion rate versus interfacial potential diagram for the corrosion of ionic solid MO [20]: $\Delta\phi_H$ = interfacial potential of MO, $\Delta\phi_H^0$ = interfacial potential at the surface of stoichiometric composition, and v_{MO} = corrosion rate of MO.

rate of an ionic solid at the interfacial potential of $\Delta\phi_H^0$ in aqueous solution [20]. In the potential range, less positive (more cathodic) than $\Delta\phi_H^0$, where the corrosion rate increases with increasing interfacial potential, the solid surface is enriched with cations, and the corrosion is controlled by the anodic cation transfer reaction. On the contrary, in the potential range more positive (more anodic) than $\Delta\phi_H^0$, where the corrosion rate decreases with increasing interfacial potential, the solid surface is enriched with anions, and the corrosion is controlled by the cathodic anion transfer reaction.

22.6.3 ACIDIC AND BASIC CORROSION OF METAL OXIDES

Normally, the corrosion of metal oxides occurs in both acidic solution and basic solution. In general, as schematically shown in Figure 22.14, the corrosion rate of metal oxides usually increases with decreasing pH in acidic solution; whereas, it increases with increasing pH in basic solution. It is also an accepted fact that the corrosion of metal oxides in basic solution, where OH_{aq}^- ions and basic salts prevail in concentration, apparently differs from that in acidic solution, where prevailing solutes are H_{aq}^+ ions and acidic salts.

As mentioned earlier, the corrosion in an acidic solution occurs through the anodic reaction (Reaction 22.36) and the cathodic reaction (Reaction 22.37) involving H_{aq}^+. In a basic solution, on the other hand, OH_{aq}^- ions will participate in both anodic and cathodic reactions. Most probable processes of metal oxide corrosion in basic solution may be described in terms of metal hydroxide complex ions and hydroxide ions:

$$M_s^{2+} + nOH_{aq}^- \rightarrow M_{aq}^{2+}(OH^-)_n \quad \left(M_s^{2+} \rightarrow M_{aq}^{2+}\right) \quad A_{M^{2+}} \tag{22.45}$$

$$O_s^{2-} + H_2O_{aq} \rightarrow 2OH_{aq}^- \quad \left(O_s^{2-} \rightarrow O_{aq}^{2-}\right) \quad A_{O^{2-}} \tag{22.46}$$

where A denotes again the affinity of the reaction. The overall reaction is then given by

$$MO + nOH^- + H_2O_{aq} \rightarrow M_{aq}^{2+}(OH^-)_n + 2OH_{aq}^- \tag{22.47}$$

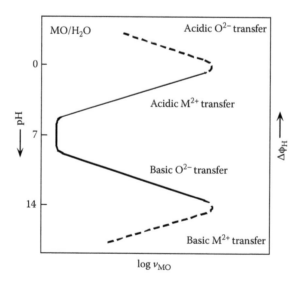

FIGURE 22.14 Schematic diagram for corrosion of ionic bonding metal oxide MO in aqueous solution as a function of solution pH and interfacial potential $\Delta\phi_H$: ν_{MO} = corrosion rate of MO.

We may call these ion transfer processes the basic M^{2+} and O^{2-} transfer processes, in contrast to the acidic M^{2+} and O^{2-} transfer processes shown in Reactions 22.36 and 22.37. The basic oxide ion transfer, for instance, is the transfer of O^{2-} from the metal oxide into the O^{2-}-acceptor level of H_2O molecules, while the acidic oxide ion transfer is the one into the O^{2-}-acceptor level of H^+ ions in acid solution. The basic oxide ion transfer in basic solution and the acidic oxide ion transfer in acidic solution obviously differ from each other, and so do their reaction kinetics.

Figure 22.14 shows schematically the corrosion rate of an ionic solid of metal oxide in aqueous solution over a wide range of the interfacial potential and hence over a wide range of pH of the solution. We see that the corrosion of metal oxides in acid solution is controlled by the acidic metal ion transfer, whose rate increases with decreasing pH; whereas, in basic solution, the rate-determining process is the basic oxide ion transfer, whose rate increases with increasing pH. From the foregoing discussion, it is expected as shown in the figure that in extremely acidic solutions the acidic oxide ion transfer will determine the corrosion rate of MO, whereas the basic metal ion transfer will control the corrosion rate of MO in extremely basic solutions.

22.7 CORROSION OF METALS

22.7.1 CORROSION POTENTIAL

Let us consider the corrosion of metallic iron in acid solution. As mentioned in the foregoing text, metallic corrosion occurs as a coupled reaction of an anodic metal ion transfer and a cathodic electron transfer from the metal into the solution:

$$Fe^{2+}_M \rightarrow Fe^{2+}_{aq} \quad A_M = \alpha_{Fe^{2+}(M)} - \alpha_{Fe^{2+}(aq)} \qquad (22.48a,b)$$

$$2e^-_M \rightarrow 2e^-_{redox} \left(2H^+_{aq} + 2e^-_M \rightarrow H_{2(gas)} \right) \quad A_{redox} = \varepsilon_{F(M)} - \varepsilon_{F(redox)} \qquad (22.49a,b)$$

where
   A denotes the affinity of charge transfer
   ε_F is the electron energy level (Fermi level)
   α is the ion energy level

The cathodic electron transfer is in fact the cathodic reduction of some oxidant such as hydrated protons present in the solution. As we saw in the foregoing section, the energy level of metal ions may be set on the same scale as that of the Fermi level of electrons, that is, on the scale of the electrode potential.

Figure 22.15 schematically shows energy-level diagrams for metallic corrosion consisting of an anodic metal ion transfer and a cathodic redox electron transfer. The corrosion potential, which represents the electronic energy level as well as the ionic energy level of metal ions in a corroding metal electrode, is situated at a level somewhere between the Fermi level, $\varepsilon_{F(redox)}$, of the redox electron and the energy level, $\alpha_{Fe^{2+}(aq)}$, of the hydrated metal ion. At the corrosion potential, as shown in the figure, the anodic metal ion transfer current is equal to the cathodic electron transfer current. The corrosion rate is, in fact, the metal ion transfer rate that we observe at the corrosion potential. It is, then, made clear that the corrosion potential is determined by both the anodic polarization (potential-current) curve of the metal ion transfer and the cathodic polarization curve of the redox electron transfer. Figure 22.16 shows the anodic and cathodic polarization curves of metallic iron in acid solution [21], where the point of intersection of the anodic and cathodic polarization curves describes the state of corrosion.

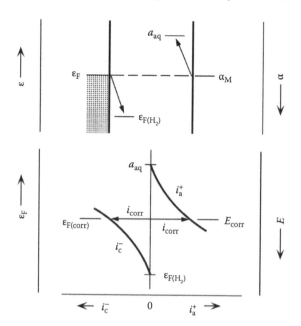

FIGURE 22.15 Schematic energy diagrams and polarization curves of anodic metal dissolution and cathodic proton reduction for metallic corrosion: α = energy level of metal ions in metal and in solution, ε = energy level of electrons, $\varepsilon_{F(H_2)}$ = Fermi level of hydrogen reaction, $\varepsilon_{F(corr)}$ = Fermi level of corroding metal, E_{corr} = corrosion potential, i_a^+ = anodic current, i_c^- = cathodic current, and i_{corr} = corrosion current.

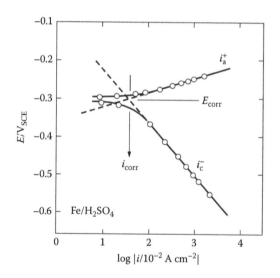

FIGURE 22.16 Polarization curves of metallic iron corroding in acidic sodium sulfate solution at pH 1.7 at 25°C [21].

    The corrosion rate of metals may be controlled by either the anodic reaction or the cathodic reaction. In most cases of metallic corrosion, the cathodic hydrogen reaction controls the corrosion rate in acidic solution, while in neutral solution the cathodic oxygen reaction preferentially controls the corrosion rate of metals. Generally, the presence of corrosion precipitates, such as metal oxides and hydroxides, exerts a considerable influence on the corrosion of metals in neutral solutions. The effects of surface oxides on metallic corrosion will be discussed in the following section.

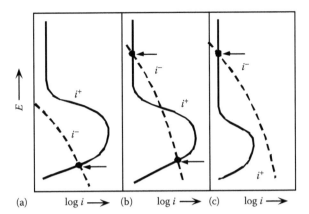

FIGURE 22.17 Metallic passivation schematically illustrated by anodic and cathodic polarization curves of corroding metals: (a) active corrosion, (b) unstable passivity, and (c) stable passivity: i^+ = anodic metal dissolution current and i^- = cathodic oxidant reduction current.

22.7.2 PASSIVATION OF METALS

When the corrosion potential of a metal is made by some means more positive than the passivation potential, the metal will passivate into almost no corrosion because of the formation of a passive oxide film on the metal surface. As shown in Figure 22.17, the passivation of a metal will occur, if the cathodic polarization curve for the redox electron transfer of oxidant reduction goes beyond the anodic polarization curve for the metal ion transfer in the active state of metal dissolution. As far as the anodic polarization curve of metal dissolution exceeds the cathodic polarization curve of oxidant reduction, however, the corrosion potential remains in the active potential range and the metal corrosion progresses in the active state. An unstable passive state will arise if the cathodic polarization curve crosses the anodic polarization curve at two points, one in the passive state and the other in the active sate. In this unstable state, a passivated metal, once its passivity is broken down, can never be repassivated again because of its active dissolution current greater than the cathodic current of oxidant reduction.

Figure 22.18 shows the corrosion rate of metallic nickel in sulfate solution as a function of solution pH, where the cathodic reaction participating in the corrosion is mainly the reduction of gaseous oxygen [22]. We observe that metallic nickel corrodes in the active state in acidic sulfate solution and passivates in basic sulfate solution: the transition from the active corrosion to the passivation occurs discontinuously at the pH around 6.

22.8 CORROSION OF SEMICONDUCTORS

22.8.1 CORROSION POTENTIAL

Let us consider the corrosion of *p-type* germanium in aqueous solution. The anodic reaction of germanium corrosion is the transfer of germanium ions involving holes in the valence band and its overall reaction may be expressed by

$$\text{Ge}_{\text{SC}} + 4\text{h}^+_{\text{SC}} \rightarrow \text{Ge}^{4+}_{\text{aq}} \qquad A_{\text{Ge}} = \varepsilon_{\text{F(Ge}^{4+})} - \varepsilon_{\text{F}} \qquad (22.50\text{a,b})$$

where
  A_{Ge} denotes the reaction affinity for germanium
  $\varepsilon_{\text{F(Ge}^{4+})}$ is the Fermi level of the reaction in equilibrium
  ε_{F} is the Fermi level of the germanium electrode

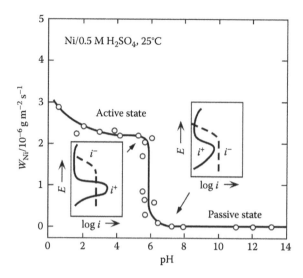

FIGURE 22.18 The corrosion of metallic nickel in aerated 0.5 kmol m^{-3} sodium sulfate solution as a function of pH and inserted sketches of its polarization curves in acidic and basic solutions [22]: W_{Ni} = the corrosion rate of nickel.

This is the anodic hole-emitting dissolution of germanium, and its Fermi level, $\varepsilon_{F(Ge^{4+})}$, stands within the range of the band gap of the germanium electrode. In order for this reaction to occur, $\varepsilon_{F(Ge^{4+})}$ will have to be higher than ε_F [23].

For the cathodic reaction of the corrosion, there are two different charge-transfer processes. One involves holes in the valence band as a cathodic hole-injecting reaction and the other involves electrons in the conduction band as a cathodic electron-emitting reaction:

$$h_{O_2}^+ \rightarrow h_{SC}^+, \ \left(0.5O_{2(gas)} + 2H_{aq}^+ \rightarrow H_2O_{aq} + 2h_{SC}^+\right) \quad A_{O_2} = \varepsilon_F - \varepsilon_{F(O_2)} \quad (22.51a,b)$$

$$e_{SC}^- \rightarrow e_{H_2}^-, \ \left(2H_{aq}^+ + 2e_{SC}^- \rightarrow H_{2(gas)}\right) \quad A_{H_2} = \varepsilon_F - \varepsilon_{F(H_2)} \quad (22.52a,b)$$

where
  $e_{H_2}^-$ is the hydrogen redox electron
  $h_{O_2}^+$ is the oxygen redox hole
  ε_F is the Fermi level of the electrode
  $\varepsilon_{F(H_2)}$ is the Fermi level of the hydrogen reaction
  $\varepsilon_{F(O_2)}$ is the Fermi level of the oxygen reaction
  A is the reaction affinity

Figure 22.19 shows energy diagrams for the corrosion reaction consisting of the anodic hole-emitting germanium dissolution, whose reaction rate increases with increasing hole concentration in the valence band of germanium, and the cathodic hole-injecting oxygen reaction, whose reaction rate increases with increasing oxygen redox holes in the solution. Also shown in the figure is the other corrosion reaction associated with the cathodic electron-emitting hydrogen reaction, whose reaction rate increases with increasing electron concentration in the conduction band of germanium. With a p-type germanium electrode, electrons are minority carriers so that the cathodic electron-emitting hydrogen reaction will hardly occur. As the cathodic polarization increases, however, an

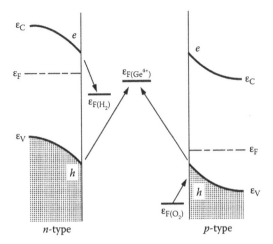

FIGURE 22.19 Schematic electron energy diagrams for the corrosion of n-type and p-type germanium electrodes in aqueous solution: $\varepsilon_{F(Ge^{4+})} =$ Fermi level of germanium dissolution, $\varepsilon_{F(H_2)} =$ Fermi level of hydrogen reaction, and $\varepsilon_{F(O_2)} =$ Fermi level of oxygen reaction.

inversion layer (p-type to n-type) arises and a few electrons participate in the cathodic hydrogen reaction, whose rate is limited though. It is a rough guideline that, if the Fermi level of the cathodic reaction is close to the valence band edge, it is probable that the cathodic hole-injecting reaction will occur; whereas, the cathodic electron-emitting reaction is likely to occur if the Fermi level of the cathodic reaction is close to the conduction band edge.

Figure 22.20 shows schematic polarization curves for the corrosion of a p-type semiconductor, in which the anodic semiconductor dissolution is coupled either with a cathodic hole-injecting oxygen reaction or with a cathodic electron-emitting hydrogen reaction. When the anodic dissolution is coupled with the oxygen reaction, the corrosion potential will be high, approaching the valence band edge potential. When the anodic dissolution is coupled with the hydrogen reaction, by contrast, the

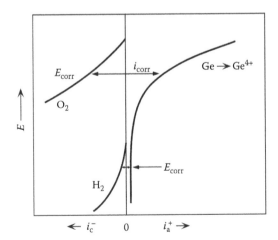

FIGURE 22.20 Schematic polarization curves for the corrosion of p-type germanium in acid solution; the corrosion potential is less positive with the cathodic hydrogen reaction but more positive with the cathodic oxygen reaction.

corrosion potential will be low approaching the conduction band edge potential of the semiconductor electrode. In general, the lower the Fermi level of the cathodic reaction, the more positive (more anodic) the corrosion potential will be and the greater the corrosion rate.

22.8.2 PHOTO-CORROSION

Let us consider a gallium arsenide electrode in acid solution. The anodic dissolution of GaAs in acid solution is a hole-emitting process:

$$GaAs + 2H_2O + 6h^+_{SC} \rightarrow Ga^{3+}_{aq} + HAsO_{2(aq)} + 3H^+_{aq} \quad \varepsilon_{F(Ga^{3+}As^{3+})} \tag{22.53}$$

where $\varepsilon_{F(Ga^{3+}As^{3+})}$ is the Fermi level of the reaction in equilibrium. This dissolution reaction occurs on the p-type electrode but not on the n-type electrode in the dark. Under photoexcitation conditions, however, the n-type electrode is also subject to the hole-emitting dissolution, involving photoexcited holes at the quasi-Fermi level in the space charge layer as referred to Figure 22.6.

In acid solution, the anodic semiconductor dissolution is coupled with the cathodic reduction of hydrogen ions in the solution:

$$2H^+_{aq} + 2e^-_{SC} \rightarrow H_{2(gas)} \quad \varepsilon_{F(H_2)} \tag{22.54}$$

where $\varepsilon_{F(H_2)}$ is the Fermi level of the hydrogen electrode reaction. The corrosion of GaAs proceeds at the corrosion potential, where the rates of the two reactions are balanced with each other giving the corrosion current.

As suggested earlier, the n-type GaAs does not corrode in the dark but does corrode under photoexcitation. Figure 22.21 shows semi-schematically the polarization curves for corrosion of an n-type GaAs electrode in sulfuric acid solution under the dark and photoexcited conditions [5–6]. It is seen that in acid solution the n-type GaAs electrode does not corrode in the dark but does corrode under photoexcitation. The anodic dissolution occurring at the photoexcited n-type electrode is essentially the same as that which will occur at the p-type electrode, except that the potential region

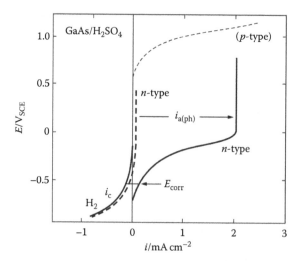

FIGURE 22.21 Polarization curves for the corrosion of a photoexcited n-type GaAs electrode in 0.5 kmol m^{-3} sulfuric acid solution [15]; curves are exaggerated around E_{corr}: solid curve = photoexcited, dashed curve = dark, and H$_2$ = cathodic hydrogen reaction coupled with anodic GaAs dissolution.

where the reaction occurs is much less positive with the photoexcited n-type electrode than that with the p-type electrode as was already shown in Figure 22.10.

In the presence of a ferric–ferrous redox reaction in the solution, the anodic hole-emitting GaAs dissolution of Reaction 22.53 can be coupled with the cathodic reduction of hydrated ferric ions, which injects holes in the valence band as follows:

$$Fe^{3+}_{aq} \rightarrow Fe^{2+}_{aq} + h^+_{SC} \quad \varepsilon_{F(Fe^{2+}/Fe^{3+})} \tag{22.55}$$

where $\varepsilon_{F(Fe^{2+}/Fe^{3+})}$ is the Fermi level of the reaction; $\varepsilon_{F(Fe^{2+}/Fe^{3+})}$ is lower than $\varepsilon_{F(Ga^{3+}As^{3+})}$ in acid solution. The rate of this cathodic reaction is controlled by the diffusion of hydrated ferric ions toward the GaAs electrode in a wide range of electrode potential below the equilibrium potential of the reaction. Injected holes by the cathodic reaction introduce a quasi-Fermi level $_p\varepsilon_F^*$ for holes in the space charge layer of GaAs and immediately go to the anodic hole-emitting GaAs dissolution. Such hole injection is essentially the same as the photoexcited hole creation in the semiconductor. The presence of the hydrated ferric–ferrous redox reaction, therefore, causes the corrosion of both n-type and p-type GaAs electrodes to occur all at the same corrosion rate controlled by ferric ion diffusion, no matter whether they are in the dark or under photoexcitation [5–6,8–11]. The corrosion potential, though, is higher with p-type electrodes than that with n-type electrodes.

It is worth noting that the cathodic electron-emitting hydrogen ion reduction does not occur on the p-type electrode in the dark because of the lack of electrons, but does occur on the photoexcited p-type electrode. This is compared with the fact that, as we saw earlier, the anodic hole-emitting semiconductor dissolution, which normally does not occur on the n-type electrode in the dark, is allowed to occur at the photoexcited n-type electrode. When a p–n heterojunction electrode of GaAs is photoexcited, therefore, accelerated corrosion will occur at the n-part where the photoexcited dissolution is coupled with the photoexcited hydrogen ion reduction at the p-part of the photoexcited heterojunction electrode. Figure 22.22 shows the polarization curves of photoexcited p-type and n-type GaAs electrodes [15]. The corrosion potential is situated somewhere between the flat band potential of the p-type and that of n-type electrodes, and the localized corrosion rate at the n-part is much greater than the corrosion rate of a photoexcited n-type GaAs electrode alone.

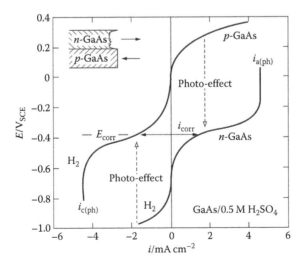

FIGURE 22.22 Polarization curves for localized corrosion of a photoexcited p–n heterojunction electrode of GaAs in 0.5 kmol m^{-3} sulfuric acid solution [15]: $i_{a(ph)}$ = anodic photo-dissolution current and $i_{c(ph)}$ = cathodic photo-hydrogen current.

22.9 PASSIVITY OF METALS AND SEMICONDUCTORS

22.9.1 PASSIVITY OF METALS

Metallic passivity was discovered as far back as 1790, when metallic iron in concentrated nitric acid was found to turn suddenly into the passive state after violent metal dissolution had occurred in the active state [24,25]. It was not until 1960s that we certainly confirmed the presence of an oxide film several nanometers thick on the surface of passivated metals [26]. Passivation was also found to occur with semiconductors in aqueous solution [27]. We may learn the latest overview on the passivity of metals and semiconductors in corrosion literature [11].

Metal passivation may be illustrated most clearly by the anodic polarization curve of metal dissolution in aqueous solution [28] as was already shown for iron in Figure 22.7. The passivation occurs beyond a critical electrode potential, called the *passivation potential*, E_P, which is a function of the solution pH [29]:

$$E_P = E_P^0 - 0.059\text{pH}, \tag{22.56}$$

where 0.059 V $(\text{pH})^{-1}$ is the coefficient at room temperature. In the passive potential range, the passive oxide film grows in thickness with increasing electrode potential at the rate of 1–3 nm V^{-1} equivalent to an electric field 10^6–10^7 V cm^{-1} [29] as shown for iron and titanium in Figure 22.23 [30–32]. Equation 22.56 suggests that the passivation is caused by the formation of an oxide film on the metal. For most metals and alloys, the passive film is less than several nanometers in thickness in the range of electrode potential where water is thermodynamically stable. For some metals such as aluminum and titanium, the oxide film can be made thick by anodic oxidation up to several hundred nanometers.

In the passive state, metal electrodes normally hold extremely small potential-independent dissolution current as shown in Figure 22.7 for metallic iron in acid solution. For some metals such as nickel, however, the passive state changes beyond a certain potential into the *transpassive* state, where the dissolution current, instead of being potential-independent, increases nearly exponentially with

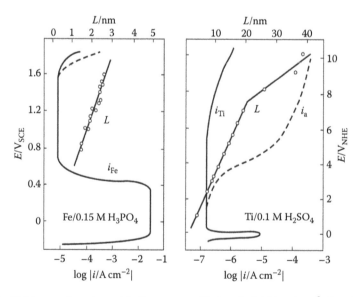

FIGURE 22.23 Thickness of passive oxide films on metallic iron in 0.15 kmol m^{-3} phosphoric acid solution and on metallic titanium in 0.1 kmol m^{-3} sulfuric acid solution as a function of electrode potential [30–32]: L = film thickness, i_{Fe} = iron dissolution current, i_{Ti} = titanium dissolution current, and i_a = anodic total current.

the electrode potential. The passive film is stable as far as it is in the state of the band edge pinning, in which state the interfacial potential, $\Delta\phi_H$, at the film–solution interface stays unchanged and the dissolution rate of the film hence remains constant independent of the electrode potential. This situation of the passive state continues to hold as far as the Fermi level of the passive metal stands within the band gap of the passive oxide film. As we saw in Section 22.5.4, with increasing electrode potential, the Fermi level, ε_F, finally reaches the valence band edge level at the electrode surface and the state of the Fermi-level pinning is effected at the band edge. The interfacial potential, then, turns to be dependent on the electrode potential and the film dissolution rate hence increases with the electrode potential. This situation, which describes the *transpassive* state, remains to hold in a range of potential beyond a critical potential called the *transpassivation potential*, E_{TP}, at which potential the Fermi level of the metal begins to be pinned at the valence band edge of the film. In the transpassive potential range, as a result, the interfacial potential, $\Delta\phi_H$, increases with shifting electrode potential in the positive direction, and hence the transpassive dissolution current increases with the electrode potential.

Figure 22.24 shows the passivation of iron and nickel in acid solution [31]. We see that the potential range of stable passivity is wide with iron; whereas, with nickel the transpassive potential-dependent dissolution occurs beyond the relatively narrow potential range of stable passivity. It is obvious that even in the presence of a surface oxide film the transpassive metal is subject to potential-dependent dissolution, because of quasi-metallization of the oxide film surface due to the Fermi-level pinning at the valence band edge. The iron oxide film, because of its *n*-type character, situates its flat band potential, E_{fb}, at a relatively less positive potential and holds a relatively large potential gap between the flat band potential, E_{fb}, and the valence band edge potential, E_V. This large potential gap will describe a relatively wide potential range for the stable passive film on metallic iron. By contrast, the nickel oxide film, which is of the *p*-type, situates its flat band potential, E_{fb}, at a relatively more positive potential and holds only a small potential gap between E_{fb}, and E_V. This small potential gap will explain a narrow potential range for the stable

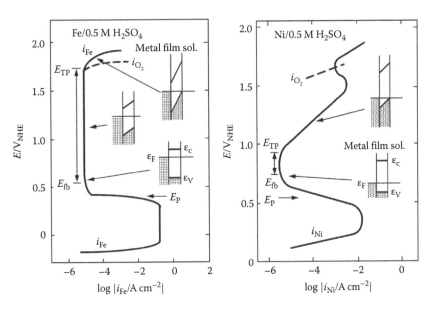

FIGURE 22.24 Anodic polarization curves for passivation and transpassivation of metallic iron and nickel in 0.5 kmol m^{-3} sulfuric acid solution with inserted sketches for electronic energy diagrams of passive films [32]: E_P = passivation potential, E_{TP} = transpassivation potential, E_{fb} = flat band potential, i_{Fe} = anodic dissolution current of metallic iron, i_{Ni} = anodic dissolution current of metallic nickel, and i_{O_2} = anodic oxygen evolution current.

passive film on metallic nickel. Beyond the valence band edge potential, E_V, the passive oxide film on nickel maintains its potential-dependent dissolution because of the surface degeneracy of the electron energy state density in the passive oxide film.

In comparison with the ionic solid corrosion discussed earlier, it is also worth noting that nickel ion transfer from the film into the solution is assumed to control the transpassive nickel dissolution whose rate increases with increasing interfacial potential. We may also see that the rate-determining process changes from the metal ion transfer to the oxide ion transfer near the oxygen evolution potential, beyond which the dissolution rate of the transpassive oxide film decreases with increasing interfacial potential, $\Delta\phi_H$.

22.9.2   PASSIVITY OF SEMICONDUCTORS

The anodic passivation of semiconductors in aqueous solution occurs in much the same way as that of metals and produces a passive oxide film on the semiconductor electrodes. Figure 22.25 shows the anodic dissolution current and the thickness of the passive film as a function of electrode potential for p-type and n-type silicon electrodes in basic sodium hydroxide solution [32,33]. As mentioned earlier, silicon dissolves in the active state as divalent silicon ions and in the passive state a film of quadravalent insoluble silicon dioxide is formed on the silicon electrode. The passive film is in the order of 0.2–1.0 nm thick with an electric field of $10^6 \approx 10^7$ V cm^{-1} in the film within the potential range where water is stable.

22.9.3   PASSIVE FILMS

Normally, passive films are extremely thin, and hence their composition and property are sensitive to the conditions under which they are formed and also to the methods of observation with which

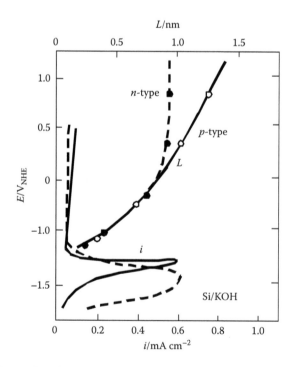

FIGURE 22.25 Thickness of passive oxide films on p-type and n-type semiconductor silicon electrodes in basic 40% potassium hydroxide solution as a function of electrode potential [33,34]: I = anodic dissolution current of silicon, L = thickness of the passive film, $\circ = p$-type silicon, and $\bullet = n$-type silicon.

they are examined. It was reported that the passive film formed on metallic iron in a neutral solution was a two-layered film consisting of an inner layer of Fe_3O_4 and an outer layer of γ-Fe_2O_3 [35]. It was also reported that the equivalent film formed on iron in the neutral solution was a single barrier layer probably of γ-Fe_2O_3 covered with a deposit layer of hydrated ferric oxide, which dissolves out in acidic solution [30,36]. It is worth noticing that a cathodic reduction technique, frequently used for surface film analyses, might introduce a change in the composition of the film.

In the process of film growth, oxide ions migrate toward the metal–oxide interface to react with the solid metal forming an inner part of the passive oxide film; whereas, metal ions migrate toward the oxide–aqua-solution interface to react with water molecules and other hydrated anions in the solution forming an outer anion-incorporating part of the passive film. The anion incorporation in passive oxide films occurs only when migrating metal ions react with anion-containing aqua-solution to form oxide at the oxide–solution interface. The ratio of the thickness of the outer anion-incorporating part to the total film thickness is thus equal to the transport number, τ_M, of the metal ion migration during the film growth. It was found that $\tau_M = 0.7$–0.8 for an anodic oxide film 65 nm thick formed on aluminum in phosphate solution [37]. In most cases, the passive film appears to be amorphous when it is thin, but turns to be crystalline, at least partially, when it grows thicker. As shown in Figure 22.23, the mean electric field in the passive oxide film on titanium decreases discontinuously at 8 V, beyond which the passive film is likely crystallized due probably to internal stresses created in the film during the film growth.

Passive films are either insulators or semiconductors. On metals such as Fe, Ti, Sn, Nb, and W, the passive films are n-type semiconductors with relatively high donor concentrations at 10^{19}–10^{20} cm^{-3}. Some metals such as Ni, Cr, and Cu form the passive film of p-type oxides. The passive films on metallic Al, Ta, and Hf are insulator oxides.

It is a concept that anodic oxide films may be classified into two groups: network (glass) formers, such as Si, Al, Ti, Zr, and Mo; and network modifiers such as Fe, Ni, Co, and Cu [38,39]. The former usually forms a single-layered oxide film and the latter tends to form a multilayered film such as $Co/CoO/Co_2O_3$. As a matter of course, the composition of anodic oxides depends on the electrode potential and higher valence oxides are formed at more positive potentials. Normally, low valence metal oxides appear to be less corrosion-resistive than high valence metal oxides. Furthermore, network forming metal oxides appear to grow by inward oxide ion migration and hence form dehydrated compact films containing no anions other than oxide ions; whereas, network modifying metal oxides appear to grow by outward metal ion migration and hence form anion-containing, less-protective films. There are of course exceptions.

It is worth noting that, as far as they are less than several nanometers thick, the passive films are subject to the quantum mechanical tunneling of electrons. Electron transfer at passive metal electrodes, hence, easily occurs no matter whether the passive film is an insulator or a semiconductor. By contrast, no ionic tunneling is expected to occur across the passive film even if it is extremely thin. The thin passive film is thus a barrier to the ionic transfer but not to the electronic transfer. Redox reactions involving only electron transfer are therefore allowed to occur at passive film-covered metal electrodes just like at metal electrodes with no surface film. It is also noticed, as mentioned earlier, that the interface between the passive film and the solution is equivalent to the interface between the solid metal oxide and the solution, and hence that the interfacial potential is independent of the electrode potential of the passive metal as long as the interface is in the state of band edge pinning.

22.9.4 PASSIVITY BREAKDOWN

In the presence of aggressive anions such as chloride ions in solution, the passive film on metals occasionally breaks down leading the underlying metal into a localized type of corrosion. In general, as shown in Figure 22.26, the chloride-breakdown of passivity occurs beyond a certain critical potential, called the *film-breakdown potential*, E_b. The film-breakdown is then followed either by

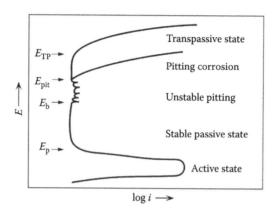

FIGURE 22.26 Schematic polarization curves for anodic metal dissolution, passivation, passivity break-down, pitting corrosion, and transpassivation: E_b = film-breakdown potential and E_{pit} = pitting potential.

the repassivation of the breakdown site around E_b or by pitting corrosion, which occurs at the breakdown site beyond a critical threshold potential called the *pitting potential*, E_{pit}. These specific potentials are obviously dependent on the concentration of film-breaking anions and the pH in the solution. Normally, the film breakdown potential is less positive, the greater the concentration of chloride ions in the solution, and so is the pitting potential. There is a marginal chloride concentration below which passive metals undergo no passivity breakdown in the potential range where water is stable: For reinforcing steel it was 0.004 kmol m^{-3} in a Ca(OH)$_2$ solution at pH 12.6 and 0.4 kmol m^{-3} in a mixed solution of NaOH, KOH, and Ca(OH)$_2$ at pH 13.6 [40].

It was observed for chloride-breakdown of the passive film on metallic iron in neutral borate solution that the amount of chloride ions required for initiating the local passivity breakdown is dependent on the film thickness, film defects, and electric field in the film as well as on the solution pH [41,42]. It was also observed that at the initial stage of the passivity breakdown the passive film locally dissolves and becomes thinner around the breakdown embryo before the underlying metal begins to dissolve in pitting at the passivity breakdown site [42,43]. From these observations, it is likely that the passivity breakdown is not a mechanical rupture of the passive film but a localized mode of dissolution of the passive film accelerated by the adsorption of aggressive anions on the film.

Passivity breakdown appears to occur preferentially at local heterogeneities, such as inclusions, grain boundaries, dislocations, and flaws on the passive metal surface. In the case of stainless steels, the passivity breakdown and pit initiation occur almost exclusively at sites of MnS inclusions, and the pitting potential was observed to decrease linearly with the increasing size of MnS inclusions [44]. With metals containing no apparent defects, however, passivity breakdown is likely to occur in the presence of sufficient concentrations of film breaking ions. It is worth noting that any of the localized phenomena is nondeterministic but somehow stochastic. For stainless steels in chloride solution, the passivity breakdown was found to obey a stochastic distribution [45].

The pitting potential, E_{pit}, at which the film-breakdown site begins to grow into pitting is either more positive (more anodic) or less positive (more cathodic) than the film-breakdown potential, E_b. If E_{pit} is more positive than E_b, the breakdown site of the film will be repassivated, as we see in the case of stainless steels in acid solution [46]. If E_{pit} is less positive than E_b, by contrast, the film breakdown site will immediately grow into a stable pitting site as with iron in acid solution [47].

In corrosion literature, a number of models have been proposed for describing the local breakdown of passive films. These however seem to be still in a stage of controversy. One of the currently prevailing models is an ionic point defect model [48], in which model chloride ions stimulate the injection into the passive film of metal ion vacancies that migrate to and accumulate at the metal–film interface to produce a void, which eventually grows to a size large enough to cause a localized breakdown of the passive film. Another is an interfacial electronic point defect model

[49,50], in which model adsorbed chloride ions introduce an electronic interface state at the film–solution interface. If the interfacial electronic point defects coagulate and become large enough to cause the local Fermi-level pinning at the electronic interface state and thus to cause the local quasi-metallization at the adsorption site on the interface, a localized film dissolution will occur leading to a localized breakdown of the passive film. As mentioned in the foregoing text, the passivity breakdown starts not with a localized mechanical rupture of the passive film but with a localized dissolution of the passive film. There have also been various conventional chemical models for passivity breakdown, in which, for instance, the local thinning of passive films on metallic iron is attributed to the formation of iron chloride complexes that finally dissolve into solution [51].

It is probably worth noting that the passivity breakdown is a process different from the pitting dissolution that follows the passivity breakdown: the former is associated with the passive film itself, whereas the metal that underlies the passive film characterizes the latter.

22.10 LOCALIZED CORROSION

22.10.1 Two Modes of Localized Corrosion

Basically, there are two modes of localized corrosion of metals in the presence of passivity-breakdown salts in aqueous solution. As shown in Figure 22.27a, a metal electrode in aqueous solution is stable in the passive potential range between the passivation potential, E_P, and the pitting potential, E_{pit}. As we will see later, the pitting potential is the one at which a pit embryo (a film breakdown site) reaches its threshold size for stable pitting dissolution. The film breakdown site smaller than the threshold size does not grow into pitting but repassivates into the passive state. For some austenitic stainless steels, the minimum threshold size is about 0.01 mm [46]. Normally, the pitting potential becomes less positive with increasing concentration of aggressive salts such as chloride ions in the solution [52]: for stainless steel it becomes less positive at the rate of 0.09–0.10 V decade^{-1} [53].

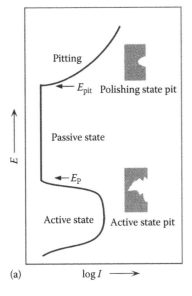
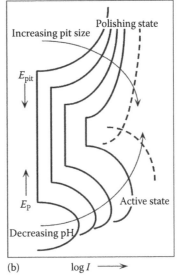

FIGURE 22.27 Schematic polarization diagrams for pitting dissolution of metals: (a) polishing and active modes of metal dissolution and (b) change in dissolution modes: I = metal dissolution current, E_{pit} = pitting potential, and E_P = passivation potential.

In the range of electrode potential more positive (more anodic) than the pitting potential, the pitting corrosion occurs in the presence of chloride ions and the metal dissolution at a pit, initially hemispherical, proceeds through the mode of *electropolishing*, in which concentrated chloride salts in an occluded pit solution will control the pit dissolution. It is likely that the polishing mode of metal dissolution proceeds in the presence of a metal salt layer on the pit surface in the salt-saturated pit solution. It was experimentally found with stainless steels in acid solution [54] that the pit dissolution current density, i_{pit}, is an exponential function of the electrode potential, E (Tafel equation):

$$E = a + b \log i_{pit} \tag{22.57}$$

The Tafel constant was $b = 0.20$ V decade^{-1} for iron electrodes [55] and $b = 0.20$ V decade^{-1} for austenitic stainless steels [54] in acid solution. It is noticed that these Tafel constants are greater than those (0.03–0.1 V) usually observed with general dissolution of metals in acid solution. The other mode of localized corrosion is the *active mode* of corrosion that prevails in the potential range less positive (more cathodic) than the passivation potential, E_P, in which potential range the localized corrosion is mainly controlled by the acidity of the occluded pit solution. In the potential range of active metal dissolution, the anodic dissolution current density is also an exponential function of the electrode potential, except for diffusion-controlled dissolution.

Figure 22.27b schematically shows that the pitting dissolution current of the electropolishing mode increases with increasing pit size. On the other hand, with increasing acidity of the solution, the passivation potential shifts in the more positive direction and the active mode of localized dissolution increases. It is also suggested that the potential range where the electropolishing mode in metal dissolution is stable extends toward less positive potentials as the size of the pit increases. By contrast, the potential range for the active mode of metal dissolution extends toward more positive potentials as the acidity of the solution increases. As a matter of consequence, with increasing pit size and acidity in the occluded solution, the two modes of dissolution become indistinguishable from each other as shown in the figure.

22.10.2 Repassivation of Pits

Once a corrosion pit breaks out, it grows in its size at relatively high potentials. The pit can, however, be repassivated by shifting the electrode potential in the less positive direction below a certain threshold called the *repassivation potential*, E_R, as shown in Figure 22.28: E_R is always less positive (more cathodic) than the pitting potential, E_{pit}, at which the pit initially breaks out. It is in fact a certain critical concentration, $c_{Cl^-}^*$, of chloride salts in the occluded pit solution that determines the stability of pitting corrosion. An electropolishing pit will stay alive if the local chloride concentration, c_{Cl^-}, in the pit is greater than a certain critical chloride concentration of $c_{Cl^-}^*$; whereas, it will repassivate if c_{Cl^-} is less than $c_{Cl^-}^*$. The critical chloride concentration was estimated to be $c_{Cl^-}^* = 1.8$ kmol m^{-3} for austenitic stainless steels in acid solution [54].

Mass transport in the pitting dissolution determines the localized chloride concentration, c_{Cl^-}, which is proportional to the product, $i_{pit} \times r_{pit}$, of the pitting current density, i_{pit}, and the pit radius, r_{pit}. Since i_{pit} is an exponential function of E, the electrode potential will have to be a logarithmic function of the pit radius, r_{pit}, in order for c_{Cl^-} to hold its critical concentration, $c_{Cl^-}^*$. The critical electrode potential, E_R, for pit repassivation, therefore, depends logarithmically on the radius of the pit:

$$E_R = a - b \log r_{pit} \tag{22.58}$$

where r_{pit} is the radius of the pit semispherical in shape. It was in fact found for austenitic stainless steels in acid solution [54] that the repassivation potential, E_R, is a logarithmic function of the pit

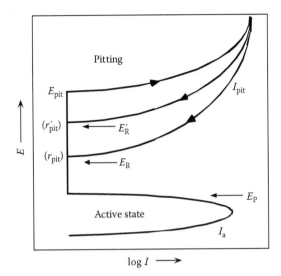

FIGURE 22.28 Schematic polarization diagrams for the repassivation of pitting dissolution of metals: E_R = repassivation potential, r_{pit} = pit radius, I_{pit} = pitting dissolution current, and I_a = metal dissolution current in the active sate.

radius [56]. Furthermore, the critical chloride concentration determines the acidity, $c_{H^+}^*$, of an occluded pit solution, which necessarily becomes much greater than the acidity of the solution bulk. The passivation potential, E_P^*, in the pit solution, consequently, turns to be much more positive than the passivation potential, E_P, in the solution bulk. It is obvious, as a result, that if E_R goes down to potentials lower than E_P^*, no pit repassivation is expected to occur and the electropolishing mode of pitting corrosion will then transform into the active mode of localized corrosion.

From the foregoing discussion on localized corrosion, we see that the repassivation potential, E_R, and the critical passivation potential, E_P^*, in the pit solution play the primary role in determining the stability of pitting corrosion of metals. Under normal conditions, a corrosion pit first breaks out at an electrode potential in the potential range of passivity, where the metal surface remains passive except for pit sites. In this range of electrode potential, the pitting dissolution is of the electropolishing mode. As the corroding pit grows in size, the repassivation potential of the pit shifts in the less positive direction, while the passivation potential, E_P^*, of the pit remains constant in the occlude pit solution, whose concentration is critical of both chloride ions and hydrogen ions. It is obvious that as far as E_R is more positive than E_P^*, the corrosion pit will repassivate in the range of electrode potential between E_R and E_P^*. By contrast, if E_R becomes less positive than E_P^*, the corrosion pit never repassivates and the electropolishing mode of pitting corrosion turns into the active mode of localized corrosion:

$$E_R > E_P^*; \quad \text{pitting at high potentials and repassivation at low potentials} \qquad (22.59)$$

$$E_R < E_P^*; \quad \text{pitting turning into localized corrosion in the active state} \qquad (22.60)$$

Figure 22.29 shows schematic polarization diagrams for the repassivation of pitting corrosion (a) and for the transformation from pitting to active localized corrosion (b). It has been observed with stainless steels that there is a critical pitting temperature below which stable pitting does not occur [57]. Above the critical pitting temperature, the pitting potential decreases

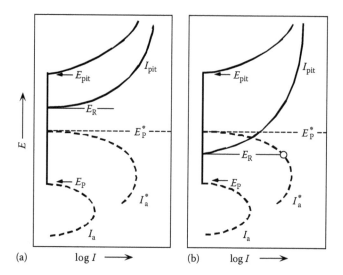

FIGURE 22.29 Schematic polarization diagrams: (a) for the repassivation of pitting dissolution of metals and (b) for transformation from the electropolishing mode of pitting to the active mode of localized dissolution: E_P = passivation potential in the solution bulk, E_P^* = passivation potential in the critical pit solution, E_R = pit repassivation potential, I_{pit} = pitting dissolution current, I_a = anodic metal dissolution current in the active state in the bulk solution, and I_a^* = anodic metal dissolution current in the critical pit solution.

approximately linearly with increasing temperature. Criterions such as critical pitting potentials and temperatures all result from the stability requirement described earlier for pitting in the occluded pit solution.

22.10.3 PROTECTION POTENTIAL OF CREVICE CORROSION

Crevice corrosion is one of the forms of localized metal corrosion, in which the anodic metal dissolution inside a crevice is coupled with a cathodic reaction outside the crevice [58]. This form of localized corrosion occurs only if the structural crevice is thinner than a certain width, for example, 30–40 μm for stainless steels [59], and thus restricted mass transport through the crevice is responsible. For crevice corrosion to occur, a certain induction period of time is required, during which a local cell has formed between the inside and the outside of the crevice. It was also shown that scaling factors in crevice corrosion may be characterized by the aspect ratio, L/a, where L is the depth of crevices and a is the crevice opening. If the aspect ratio is greater than its critical ratio, crevice corrosion will occur, whereas no crevice corrosion will occur if the aspect ratio does not exceed its critical ratio [60].

It is also an accepted fact that the crevice corrosion ceases to grow at potentials less positive than a certain critical potential resulting in crevice protection as shown for austenitic stainless steel in Figure 22.30 [59,61]. The critical potential, E_{crev}, is called *crevice protection potential* or the *critical crevice corrosion potential*. It was found for a cylindrical crevice in austenitic stainless steel that the crevice protection potential shifts in the less positive direction as a logarithmic function of solution chloride concentration [61]:

$$E_{crev} = E_{crev}^0 - \varsigma \ln c_{Cl^-}$$ (22.61)

where
  ς is a constant
  c_{Cl^-} is the chloride concentration in the solution outside the crevice

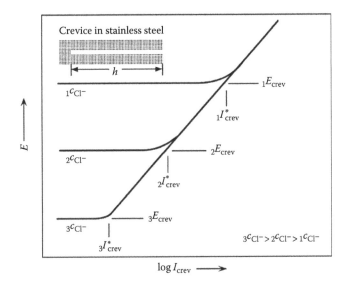

FIGURE 22.30 Schematic polarization curves of a cylindrical crevice in an anode of stainless steel in neutral solutions of three different chloride concentrations [59]: h = crevice depth, I_{crev} = anodic crevice dissolution current, c_{Cl^-} = chloride concentration in the solution bulk, E_{crev} = crevice protection potential, and I_{crev}^* = minimum crevice dissolution current at the critical crevice (protection) potential E_{crev}.

Furthermore, as shown in Figure 22.30, there exists at the crevice protection potential a threshold dissolution current, I_{crev}^*, below which the crevice ceases from corroding. The threshold dissolution current was found to decrease linearly with increasing chloride concentration in the solution bulk [59]:

$$I_{crev}^* = I_{crev}^0 - \xi c_{Cl^-} \qquad (22.62)$$

where ξ is a constant nearly inversely proportional to the depth, h, of the crevice; $\xi \propto 1/h$ [62].

It is in fact the acidification of the occluded crevice solution that triggers the crevice corrosion. The critical acid concentration, $c_{H^+}^*$, for crevice corrosion to occur corresponds to what we call the *passivation–depassivation pH*, beyond which the metal spontaneously passivates. This critical acidity determines the crevice passivation–depassivation potential, E_{crev}^* and hence the crevice protection potential E_{crev}. The electrode potential actually measured consists of the crevice passivation–depassivation potential, E_{crev}^*, and the IR drop, ΔE_{IR}, due to the ion migration through the crevice. Assuming the diffusion current from the crevice bottom to the solution outside, we obtain $\Delta E_{IR} = i_{crev} \times h = $ constant, where i_{crev} is the diffusion-controlled metal dissolution current density at the crevice bottom and h is the crevice depth [62]. Since anodic metal dissolution at the crevice bottom follows a Tafel relation, we obtain E_{crev} as a logarithmic function of the crevice depth:

$$E_{crev} = E_{crev}^* + \Delta E_{IR} = \text{constant} - b' \log h \qquad (22.63)$$

In fact, a linear relationship was found to hold between the crevice protection potential and the logarithm of the crevice depth with $b' = -0.06$ V decade^{-1} for a cylindrical crevice in austenitic stainless steel in chloride solution [62].

Crevice corrosion is, of course, dependent on temperature. No crevice corrosion is found to occur below a certain temperature called the *critical crevice corrosion temperature*, which is similar to the critical pitting temperature mentioned earlier. The critical crevice corrosion temperature has been used as a measure to evaluate metallic materials for the susceptibility to crevice corrosion [63].

22.10.4   Potential-Dimension Diagrams for Localized Corrosion

The stability of localized corrosion including pitting and crevice corrosion may be described in a potential-dimension diagram [62] shown in Figure 22.31. As we saw in the foregoing text, under normal corrosion conditions, a corrosion pit is produced on the passive metal surface at electrode potentials more positive than the pitting potential, E_{pit}, and then the pit grows in its size following a gradual shift of the corrosion potential in the less positive direction. Generally, the corrosion potential is controlled by the supply of cathodic reaction from oxidants. If a sufficient supply of cathodic current is available, a corrosion pit will grow steadily at relatively high potentials. By contrast, if the cathodic current is insufficient, a corroding pit grows relatively slowly following a rather steep shift of the corrosion potential in the less positive direction.

In Figure 22.31, the intersection of the $E_R - \log r_{pit}$ line at the passivation potential, E_P^*, in the critical pit solution stands for a critical pit-radius, r_R^{max}, that decides the stability of localized corrosion. This critical radius, in fact, is the maximum size of repassivable pits corroding in the electropolishing mode. A pit may repassivate if its size does not exceed r_R^{max}. If a corrosion pit grows in its size greater than r_R^{max}, however, the pit never passivates but changes into the active mode of localized corrosion. When the corrosion potential reaches the repassivation potential while the size of a corroding pit remains smaller than its critical size, r_R^{max}, the corrosion pit ceases from growing and repassivates to the passive state. The pit repassivation thus occurs in the region where the electrode potential is less positive than the repassivation potential, E_R, and the pit size is smaller than the critical size, r_R^{max}. On the other hand, if the corrosion potential fails to reach the repassivation potential before a pit grows to its critical size, r_R^{max}, the corrosion pit will be transformed into a localized corrosion site in the active state. The region where the electrode potential is more positive than E_R and the pit size is greater than r_R^{max} is thus for non-repassivable pits leading to localized corrosion in the active state. Such localized corrosion, initiated with pitting in the passive state and then transformed into the active state, is the case that we have frequently encountered in practice.

Furthermore, no crevice corrosion will occur in the region where the electrode potential is less positive than the crevice protection potential, E_{crev}. Crevice corrosion will cease from growing at

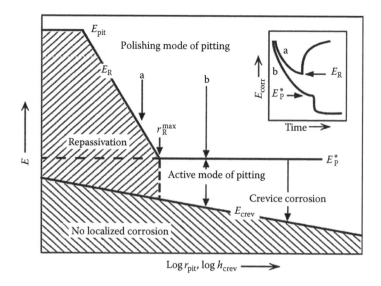

FIGURE 22.31   Schematic potential-dimension diagrams for localized corrosion of stainless steel in aqueous solution [63]: E_{pit} = pitting potential, E_R = pit repassivation potential, E_P^* = passivation potential in the critical pit solution, E_{crev} = crevice protection potential, r_R^{max} = critical pit radius for pit repassivation, a = pit repassivation, and b = transition from the polishing mode to the active mode of localized corrosion.

electrode potentials less positive than E_{crev}, at which potential the crevice solution holds its acidity at the passivation–depassivation pH of the crevice metal. No crevice corrosion is thus expected to occur in the potential region less positive than E_{crev}.

22.11 SURFACE OXIDES IN METALLIC CORROSION

22.11.1 Corrosion Rust Precipitates

Metallic corrosion usually produces corrosion precipitates of metal oxides and hydroxides on the surface of corroding metals in aqueous solution. Corrosion precipitate layers thus formed affect the corrosion of underlying metals. The presence of porous precipitate layers of insoluble rusts such as hydrous metal salts and hydroxides will cause either corrosion acceleration or corrosion inhibition [64–66].

Local cell corrosion of metals involves the transport of hydrated ions in solution, with anions migrating from a cathodic portion to an anodic portion and cations migrating in the reverse direction. In the presence of corrosion precipitates, the transport of hydrated ions occurs through a precipitate layer, and an occluded solution under the layer is hence enriched with either the anions or cations depending on the ion-selective property of the precipitate layer. Rusts of hydrous metal oxides in general are ion selective. Aluminum oxide is anion-selective in sodium chloride solution and cation-selective in sodium hydroxide solution [67]. Hydrous ferric oxide, nickel oxide, and chromium oxide are anion-selective in potassium chloride solution [68–70]. The ion selectivity of hydrous rusts depends on the ionic charge fixed on the inner surface of pores in the rusts: They are anion-selective if the fixed charge is positive and cation-selective if the fixed charge is negative. Basically, the ion selectivity of a hydrous rust layer is evaluated in terms of the transference number, τ, which is the number of moles carried by a migrating ion or molecule for the unit charge flow across the layer: it is $\tau_- = -1$ for the perfectly anion-selective and $\tau_+ = 1$ for the perfectly cation-selective. It was found that for a hydrous ferric oxide layer $\tau_{Cl^-} = -0.94$ and $\tau_{Na^+} = 0.06$ in neutral sodium chloride solution [71]. The osmotic flow of water molecules across the layer is also estimated with its transference number, τ_{H_2O} [72].

Metal hydroxides in general are anion-selective in acid solution and turn to be cation-selective beyond a certain pH, called the point of the *iso-selectivity*, pH_{pis}: it is $pH_{pis} = 10.3$ for ferric oxide and $pH_{pis} = 5.8$ for ferric–ferrous oxide [72]. Adsorption of multivalent ions may also control the ion selectivity of hydrous metal oxides because of its effect on the fixed charge in the oxides. For instance, hydrous ferric oxide, which is anion-selective in neutral sodium chloride solution, turns to be cation-selective by the adsorption of such ions as divalent sulfate ions, divalent molybdate ions, and trivalent phosphate ions [70,73]. It is worth emphasizing that such an ion-selectivity change due to the adsorption of multivalent ions frequently plays a decisive role in the corrosion of metals.

We notice, as a supplement, that the ion-selective nature remains valid only when the concentration of transferring ions is comparable to or less than the concentration of the fixed charge on the inner pore surface in the precipitate layer. This situation will not hold if the concentration of the transferring ions is great compared to the fixed charge concentration and/or if the porosity of the precipitate layer is so great that the fixed charge concentration will be fairly small in comparison. In concentrated solutions or in highly porous rust precipitates, the normal ionic migration and diffusion dominate over the ion-selective transfer.

22.11.2 Anion-Selective Rust Layers

Let us suppose the anodic dissolution of metals occurring under an anion-selective corrosion precipitate layer in chloride solution. The anodic corrosion current carries chloride ions across the anion-selective precipitate layer into an occluded solution under the layer as shown in

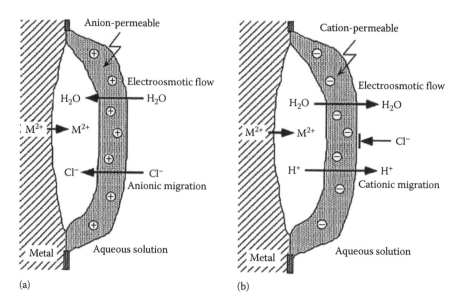

FIGURE 22.32 Anion-selective and cation-selective rust layers on corroding metals in chloride solution: (a) chloride ion condensation and acidification in occluded solution under an anion-selective layer and (b) proton depletion and dehydration in occluded solution under a cation-selective layer.

Figure 22.32a. The accumulation of chloride ions in the occluded solution reaches a stationary level, at which the rate of inward ion migration equals the rate of outward diffusion of the ions. Furthermore, the anodic chloride ion migration is accompanied by an electroosmotic flow of water molecules into the occluded solution. The final chloride concentration thus established in the steady state is determined by the ratio of the chloride ion migration rate to the electroosmotic water flow rate. It is therefore the transference numbers, τ_{Cl^-} and τ_{H_2O}, of chloride ions and water molecules that determine the chloride ion concentration in the occluded solution.

As the local concentration of chloride ions increases, the occluded solution under the anion-selective layer will be acidified and the passive film on metals will break down giving rise eventually to localizing metal corrosion under the layer. It is thus obvious that the presence of an anion-selective rust layer accelerates the corrosion of the underlying metal. In order to prevent such accelerated corrosion under an anion-selective layer, we need to reduce the anion-selective nature of the layer by some way such as the adsorption of multivalent anions on the layer.

22.11.3 CATION-SELECTIVE RUST LAYERS

With a cation-selective rust layer, the anodic ion transport across the rust layer carries mainly hydrated protons migrating outward from an occluded solution into the solution bulk as shown in Figure 22.32b. No accumulation of chloride ions and hydrogen ions is thus expected to occur in the occluded solution. Instead, the outward migration of hydrogen ions leads to the basification of the occluded solution, where the precipitation of corrosion products will then be accelerated.

Furthermore, the electroosmotic outward flow of water molecules, which follows the anodic hydrogen ion transport, counteracts the inward diffusion of water molecules into the occluded solution. The dehydration of the occluded solution will then occur as the corrosion progresses. Since metal dissolution requires water molecules for metal ions to hydrate, the depletion of water molecules will finally result in the deceleration of metal corrosion. The cation-selective rust layer, therefore, will be preventive of the corrosion of underlying metals.

22.11.4 BIPOLAR ION-SELECTIVE RUST LAYERS

Let us consider a composite rust layer consisting of an anion-selective inner layer and a cation-selective outer layer as shown in Figure 22.33a. A bipolar ion-selective layer of this type may be realized when the outer part of an anion-selective layer is made cation-selective by the adsorption of multivalent anions. As with the electronic rectifier of p–n junction semiconductors, a bipolar junction consisting of an anion-selective layer and a cation-selective layer rectifies the ionic current across the bipolar layer. The ion transfer current in the anodic direction is thus suppressed across the bipolar layer as shown in Figure 22.34, where the anodic ion transport current is seen to be restricted across a bipolar ferric hydroxide membrane in sodium chloride solution [70].

Furthermore, it is a consequence of the ionic current rectification that, under anodic polarization conditions, a high electric field arises at the boundary between the anion-selective inner layer and the cation-selective outer layer as shown in Figure 22.33b. Because of this high electric field, the bipolar layer will eventually be dehydrated to change into a compact corrosion-resistive oxide layer. We may therefore assume that the bipolar ion-selective rust layer will inhibit the corrosion and induce the passivation of metals. Pit repassivation in stainless steels containing molybdenum was reasonably described by the formation of a bipolar rust layer consisting of an anion-selective inner layer and a cation-selective outer layer containing MoO_4^{2-} ions [74,75].

22.11.5 REDOX-OXIDE LAYERS

Let us consider a hydrous ferric-ferrous oxide layer, which is sensitive to the reduction–oxidation reaction on the surface of corroding metallic iron:

$$Fe(OH)_3 + H_{aq}^+ + e^- = Fe(OH)_2 + H_2O_{aq} \qquad (22.64)$$

For this redox reaction to occur, electrons and protons will have to migrate through the hydrous oxide layer. In the anodic oxidation, hydrogen ions migrate outward from the layer into the solution and

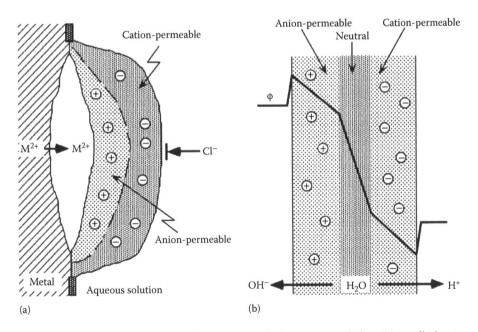

FIGURE 22.33 Bipolar ion-selective rust layers on metals in aqueous solution: (a) anodic ion transport suppressed by the backward bipolar ion selectivity and (b) profile of electrostatic potential across a backward bipolar ion-selective rust layer [65].

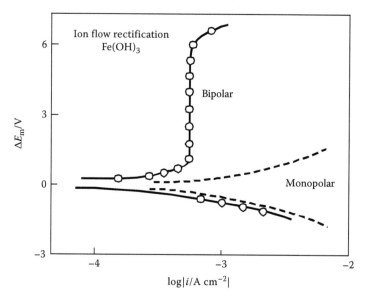

FIGURE 22.34   Polarization curves for a monopolar membrane and for a bipolar membrane of hydrous ferric oxide in sodium chloride solution [69,72].

electrons migrate inward from the layer to the metal. In the cathodic reaction, inversely, hydrogen ions migrate inward into the layer and electrons migrate outward from the metal to the layer. The reaction rate will be controlled either by the electron transfer or by the hydrogen ion transfer.

When the redox reaction is in near equilibrium, the electrode potential of the layer is set around the equilibrium potential, E_{redox}, of the reaction. The electrode potential of the redox reaction of the hydrous oxides is usually more positive than the corrosion potential of metallic iron. The redox oxide in contact with metallic iron, hence, shifts the corrosion potential in the positive (anodic) direction and provides the cathodic reaction for the corrosion:

$$Fe \rightarrow Fe_{aq}^{2+} + 2e^- \quad \text{anodic reaction} \tag{22.65}$$

$$2Fe(OH)_3 + 2H_{aq}^+ + 2e^- \rightarrow 2Fe(OH)_2 + 2H_2O_{aq} \quad \text{cathodic reaction} \tag{22.66}$$

The cathodically reduced oxide will be oxidized again by gaseous oxygen:

$$2Fe(OH)_2 + 2H_2O_{aq} \rightarrow 2Fe(OH)_3 + 2H_{aq}^+ + 2e^- \quad \text{anodic reaction} \tag{22.67}$$

$$(1/2)O_2 + 2H_{aq}^+ + 2e^- \rightarrow H_2O_{aq} \quad \text{cathodic reaction} \tag{22.68}$$

Since the rate of the oxide reduction is usually greater than the rate of the oxygen reduction on metallic iron, the corrosion rate of iron in the presence of the redox oxide is greater than that in the absence of the redox oxide. It is in fact an accepted understanding that the presence of hydrous ferric oxide accelerates the corrosion of mild steels forming hydrous ferrous oxide. The ferrous oxide thus formed is then oxidized to ferric oxide again by atmospheric oxygen.

22.11.6   N-Type Oxides

Let us consider an n-type semiconductor oxide in contact with a corroding metal in aqueous solution. Before contact, the electrode potential of the n-type oxide is set near its flat band potential

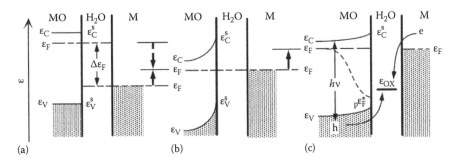

FIGURE 22.35 Electronic energy diagrams for a metal electrode in contact with an *n*-type metal oxide: (a) prior to contact, (b) posterior to contact, and (c) with photoexcitation: $_p\varepsilon_F^*$ = quasi-Fermi level for photoexcited holes in oxide and ε_{OX} = Fermi level of oxygen reaction.

and the electrode potential of the metal is set at its corrosion potential. Figure 22.35a shows electron-level diagrams for the two separate electrodes. The Fermi level, ε_F, is usually higher for *n*-type oxides than for common metals, $\varepsilon_{F(MO)} > \varepsilon_{F(M)}$, and hence the flat band potential of the oxide, for most cases, is less positive than the corrosion potential of metals, $E_{fb} < E_{corr}$. After contact, the two Fermi levels equilibrate with each other introducing a space charge potential in the oxide and shifting the interfacial potential at the metal electrode. Consequently, the oxide potential becomes more positive ($\varepsilon_{F(MO)}$ becomes lower) and the corrosion potential of the metal becomes less positive ($\varepsilon_{F(M)}$ becomes higher) as shown in Figure 22.35b. We thus see that the presence of an *n*-type oxide will reduce the corrosion of metals by shifting the corrosion potential in the less positive direction. The oxides that can make the metal corrosion potential less positive are those whose flat band potential is less positive than the metal corrosion potential.

Furthermore, as we saw in a foregoing section, photoexcitation produces in a semiconductor electrode electron–hole pairs and introduces a photo-potential, which reduces the space charge potential in the semiconductor. With an *n*-type semiconductor in contact with a corroding metal, photoexcitation raises the Fermi level up to the flat band level of the semiconductor, thus shifting the corrosion potential in the less positive direction toward the flat band potential of the *n*-type oxide as shown in Figure 22.35c. Photoexcitation therefore will shift the corrosion potential in the less positive (more cathodic) direction and the corrosion will then be suppressed. With some *n*-type oxides such as titanium oxide, photoexcitation brings the interfacial quasi-Fermi level, $_p\varepsilon_F^*$, down to a level lower than the Fermi level, $\varepsilon_{F(redox)}$, of the oxygen electrode reaction:

$$2H_2O + 4h_{redox}^+ = O_2 + 4H_{aq}^+ \quad \varepsilon_{F(redox)} \tag{22.69}$$

where h_{redox}^+ is the redox hole in the equilibrium of the reaction. As the interfacial quasi-Fermi level, $_p\varepsilon_F^*$, comes down to levels lower than $\varepsilon_{F(redox)}$, the anodic oxygen production will thermodynamically be allowed to occur on the *n*-type oxide, in spite of the fact that the Fermi level, $\varepsilon_{F(MO)} = \varepsilon_{F(M)}$, is much higher than the Fermi level, $\varepsilon_{F(redox)}$, of the oxygen reaction and hence that the electrode potential is much less positive than the oxygen evolution potential:

$$2H_2O + 4h^+ \rightarrow O_2 + 4H_{aq}^+ \quad \text{\textit{n}-type oxide} \tag{22.70}$$

By contrast, on the part of the metal surface, the cathodic oxygen reduction and even the cathodic hydrogen production are thermodynamically allowed to occur at relatively less positive potentials:

$$O_2 + 4H_{aq}^+ + 4e^- \rightarrow 2H_2O \quad \text{metal} \tag{22.71}$$

$$2H_{aq}^+ + 2e^- \rightarrow H_{2(gas)} \quad \text{metal} \tag{22.72}$$

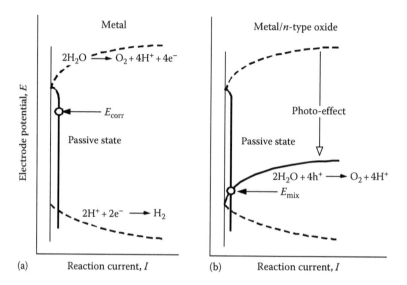

FIGURE 22.36 Schematic polarization diagrams for the corrosion potential of a passive metal electrode (a) alone and (b) in contact with an n-type metal oxide under photoexcitation: $I^*_{a(OX)}$ = anodic oxygen current at photoexcited oxide.

We see, as a consequence, that the anodic oxygen production on the part of n-type oxide may be coupled with the cathodic oxygen reduction and/or with the cathodic hydrogen production on the part of the metal surface. The polarization curves of these reactions are schematically shown for a passive metal in Figure 22.36 and for an active metal in Figure 22.37.

A passive metal, which is subject to pitting corrosion beyond the pitting potential, E_{pit}, in the presence of chloride ions, will be inhibited from pitting corrosion if an n-type oxide makes the electrode potential of the metal less positive than E_{pit}. Furthermore, for a corroding metal in the active

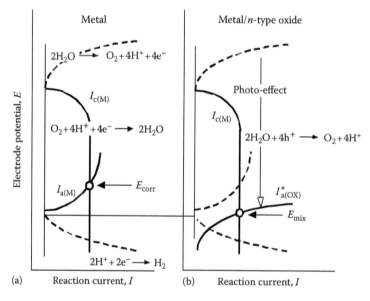

FIGURE 22.37 Schematic polarization diagrams for the corrosion of an active metal electrode (a) alone and (b) in contact with an n-type metal oxide under photoexcitation: $I_{a(M)}$ = anodic metal dissolution current, $I_{c(M)}$ = cathodic current at metal, and $I^*_{a(OX)}$ = anodic oxygen current at photoexcited oxide.

state, the corrosion will be suppressed if the anodic curve of photoexcited oxygen production on an n-type oxide comes out at electrode potentials less positive than the potential region of metal dissolution. The electrode potential will then be set at a mixed electrode potential determined either by a coupled reaction of the anodic photoexcited oxygen production on the oxide part and the cathodic oxygen reduction on the metal part, or by a coupled reaction of the anodic photoexcited oxygen production on the oxide part and the cathodic hydrogen production on the metal part. In both cases of passive metals and active metals, it is evident that n-type oxides will have an inhibitive influence on the corrosion of metals. It was in fact observed that metallic copper and stainless steel were prevented from corrosion when they were brought into contact with n-type titanium oxides [76–78]. The inhibitive effect of course depends on the ratio of surface area of the metal to the oxide; for stainless steel in contact with n-type titanium oxide, the inhibitive effect was observed at the area ratio less than $1/10$ [78].

22.11.7 p-Type Oxides

We consider a p-type semiconductor oxide in contact with a corroding metal in aqueous solution. Normally, as shown in Figure 22.38a, the Fermi level is considerably lower in p-type oxide electrodes than in ordinary metal electrodes, and thus the flat band potential, E_{fb}, of the oxides is fairly more positive than the corrosion potential, E_{corr}, of metals. When the oxide and the metal are brought into contact, the Fermi levels in both the oxide and the metal equilibrate with each other, introducing a space charge potential in the oxide and a change in the interfacial potential at the metal electrode, and consequently the Fermi level of the oxide–metal electrode is lowered (the corrosion potential is made more positive) as shown in Figure 22.38b. Photoexcitation, which reduces the space charge potential in the oxide, makes the Fermi level further lower, and the potential of the oxide–metal electrode approaches the flat band potential of the oxide as shown in Figure 22.38c. As a consequence, in contact with the p-type oxide the metal electrode will make its corrosion potential more positive and hence increase the corrosion of the metals.

As we saw in the foregoing section, pitting corrosion of passive metals occurs beyond the critical pitting potential, E_{pit}. In order to protect passive metals from pitting corrosion, therefore, it is advisable to hold the corrosion potential as far less positive from E_{pit} as possible in the passive potential range. The presence of p-type oxides, however, makes E_{corr} more positive and hence enhances the breakout of pitting corrosion. In the same way, metals corroding in the active state will accelerate their corrosion rates when their electrode potential is made more positive (more anodic) by the presence of p-type oxides.

Under photoexcitation conditions, the p-type oxide receives photo-generated electrons whose quasi-Fermi level, $_n\varepsilon_F^*$, is close to the conduction band edge, ε_C. In some cases, the quasi-Fermi level, $_n\varepsilon_F^*$, is situated higher than the Fermi level, $\varepsilon_{F(H)}$, of the hydrogen electrode reaction:

$$2H_{aq}^+ + 2e^- = H_{2(gas)} \quad \varepsilon_{F(H)} \tag{22.73}$$

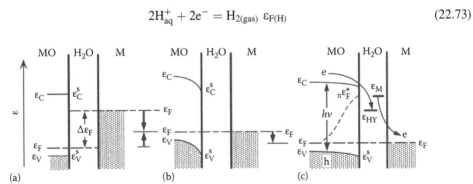

FIGURE 22.38 Electronic energy diagrams for a metal electrode in contact with a p-type semiconducting metal oxide: (a) prior to contact, (b) posterior to contact, and (c) with photoexcitation: $_n\varepsilon_F^* =$ quasi-Fermi level for photoexcited electrons in oxide, $\varepsilon_M =$ Fermi level of metal dissolution reaction, and $\varepsilon_{HY} =$ Fermi level of hydrogen reaction.

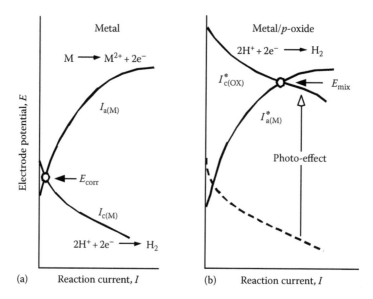

FIGURE 22.39 Schematic polarization diagrams for the corrosion of an active metal: (a) alone and (b) in contact with a p-type metal oxide under photoexcitation: $I^*_{c(OX)}$ = cathodic hydrogen current at photoexcited oxide.

where $\varepsilon_{F(H)}$ is the Fermi level of the reaction. When $_n\varepsilon^*_F$ is higher than $\varepsilon_{F(H)}$, the cathodic hydrogen production is thermodynamically allowed to occur on the part of the p-type oxide, in spite of the fact that the corrosion potential is much more positive than the electrode potential of hydrogen evolution.

Figure 22.39 shows, for a corroding metal in contact with a p-type oxide, schematic polarization curves of the anodic metal dissolution and the cathodic hydrogen reduction that occurs at the photoexcited oxide part. We see, as a consequence, that metals such as copper, which suffer no corrosion in the absence of oxygen, if in contact with a p-type oxide, may become subject to anodic corrosion with the cathodic hydrogen production on the oxide under the photoexcited or radiation-excited conditions.

22.12 ELECTROCATALYSIS IN CORROSION PROCESSES

22.12.1 ADSORPTION ON CORRODING METALS

Electrocatalysis in metallic corrosion may be classified into two groups: Adsorption-induced catalyses and solid precipitate catalyses on the metal surface. In general, the bare surface of metals is "soft acid" in the Lewis acid–base concept and tends to adsorb ions and molecules of "soft base" forming the covalent binding between the metal surface and the adsorbates. The Lewis acidity of the metal surface however may turn gradually to be hard as the electrode potential is made positive, and the bare metal surface will then adsorb species of hard base such as water molecules and hydroxide ions in aqueous solution. Ions and molecules thus adsorbed on the metal surface catalyze or inhibit the corrosion processes. Solid precipitates, on the other hand, are produced by the combination of hydrated cations of hard acid and anions of hard base forming the ionic bonding between the cations and the anions on the metal surface.

The adsorption of molecules and ions on the metal surface affect the processes of metallic corrosion through its anodic metal dissolution and/or cathodic oxidant reduction. We now consider as an example the adsorption of an anion on the bare surface of metals:

$$M + X^- \rightarrow M \cdot X^- \tag{22.74}$$

where

M is the metal surface

X^- is the hydrated anion

$M \cdot X^-$ is the anion adsorbed on the metal surface

As the electrode potential is made positive, the Lewis acidity of the bare metal surface may grow from soft to hard, and hence the preferentially adsorbed anions may change from anions of soft base to anions of hard base.

Anions such as iodide ions, I^-, sulfide ions, S^{2-}, and thiocyanate ions, SCN^-, are the soft base. In contrast, hydroxide ions, OH^-, fluoride ions, F^-, chloride ions, Cl^-, phosphate ions, PO_4^{3-}, sulfate ions, SO_4^{2-}, and chromate ions, CrO_4^{2-}, are the hard base. Bromide ions, Br^-, and sulfite ions, SO_3^{2-}, are situated somewhere between the soft and the hard base.

We see in literature [79,80] that the anodic dissolution of iron in acidic solution is likely to occur through a series of steps involving adsorbed intermediates:

$$Fe + X_{aq}^- \rightarrow (Fe \cdot X^-)_s \tag{22.75}$$

$$(Fe \cdot X^-)_s \rightarrow (FeX)_s + e \tag{22.76}$$

$$(FeX)_s \rightarrow FeX_{aq}^+ + e \tag{22.77}$$

$$FeX_{aq}^+ \rightarrow Fe_{aq}^{2+} + X_{aq}^- \tag{22.78}$$

where

$(Fe \cdot X^-)_s$ is the adsorbed anion

$(FeX)_s$ is the adsorbed intermediate

FeX_{aq}^+ is the hydrated ferrous complex ion in aqueous solution

It is suggested that the anodic dissolution will be inhibited if the adsorbed anion and the reaction intermediate are stable and hardly dissolve in aqueous solution. On the contrary, if the reaction intermediate is relatively unstable and readily dissolves into aqueous solution, the anion will function as an electrocatalyst accelerating the metal dissolution rate. It is now common knowledge that hydroxide ions, OH^-, catalyze the anodic dissolution of metallic iron and nickel in acid solution [81,82]. It is also known that chloride ions inhibit the anodic dissolution of iron in acidic solution [83]. No clear-cut understanding is however seen in literature on why hydroxide ions catalyze but chloride ions inhibit the anodic dissolution of iron, even though the two kinds of anions are in the same group of hard base. We assume that the hardness level in the Lewis base of adsorbed anions will be one of the most effective factors that determine the catalytic activity of the adsorbates. Further clarification on the catalytic characteristics will require a quantum chemical approach to the adsorption of these anions on the metal surface.

It has also been found in literature [84] that iodide ions, I^-; bromide ions, Br^-; and azide ions, N_3^-, inhibit the corrosion of metallic iron in acid solution. On the other hand, the presence of hydrogen sulfide ions, HS^-; thiosulfate ions, $S_2O_3^{2-}$; and thiocyanate ions, SCN^-, accelerate the corrosion of metallic iron in acidic solution [84].

It is apparent that the adsorption of ions and molecules also affects the cathodic reaction of metallic corrosion. The typical cathodic reaction is the cathodic reduction of hydrated protons and atmospheric oxygen molecules. From the standpoint of corrosion science, however, no systematic studies on this topic seem to have been made in the literature. In practice, the cathodic reaction may occur on the bare metal surface and the surface of metal oxides as well, if present on corroding metals. It is likely that for the cathodic reduction of oxygen molecules and hydrated protons the catalytic activity of the bare surface of metals is greater than that of the surface of metal oxides and salts. In the case of metal oxides, the rate of the cathodic reaction is controlled not only by the catalytic activity of the oxide surface but also by the concentration of electrons and holes that participate in the reaction.

Semiconductor corrosion will also be affected by the adsorption of ions and molecules on the semiconductors. We have nevertheless seen in the field of corrosion science almost no reliable research results on the catalytic characteristics of semiconductor corrosion.

22.12.2 PRECIPITATION IN CORROSION PROCESSES

Metal atoms on the metal surface, as mentioned earlier, are soft acid, and hence they combine with anions of soft base on the metal surface. Once these metal surface atoms are ionized, they form metal ions such as iron ions and aluminum ions, and the metal surface turns to be hard acid. The metal ions then combine with anions of hard base such as hydroxide ions, OH^-, oxide ions, O^{2-}, and sulfate ions, SO_4^{2-}, to form insoluble metal oxides and salts of ionic bonding character. The two-dimensional concentration of surface metal ions increases with the electrode potential of the metal, and hence the metal surface gradually becomes harder in the Lewis acidity with increasing electrode potential until it combines with anions of hard base such as oxide ions to form a metal oxide film adhering firmly to the metal surface. The passivation potential of a metal is thus regarded as a threshold potential where the metal surface grows hard enough in the Lewis acidity to combine with a hard base of oxide ions.

It is interesting to see that chromate ions, CrO_4^{2-}, which are strongly hard in the Lewis base and an oxidizing agent, oxidize the iron atoms or ferrous ions on the metallic iron surface and reduce the chromate ions themselves forming a passive film of chromic-ferric oxide, which is extremely thin and highly corrosion resistive:

$$2CrO_4^{2-} + 10H^+ + 6e \rightarrow Cr_2O_3 + 5H_2O \tag{22.79}$$

$$2Fe + 3H_2O \rightarrow Fe_2O_3 + 6H^+ + 6e \quad \text{or} \quad 2Fe^{2+} + 3H_2O \rightarrow Fe_2O_3 + 6H^+ + 2e \tag{22.80}$$

In the same way, molybdate ions, MoO_4^{2-}, and tungstate ions, WO_4^{2-}, can also passivate metallic iron, though only in the presence of oxygen because of their relatively week oxidizing capability.

Some oxidizing agents such as nitrite ions, NO_2^-, which are hard bases to some degree, oxidize metallic iron and ferrous ions directly taking part in the cathodic reaction of metal corrosion and form a passive film of ferric oxide in which no nitrite ions are involved:

$$NO_2^- + 5H_2O + 6e \rightarrow NH_3 + 7OH^- \tag{22.81}$$

$$2Fe \rightarrow 2Fe^{2+} + 4e \quad 2Fe^{2+} + 6OH^- \rightarrow Fe_2O_3 + 3H_2O + 2e \tag{22.82a,b}$$

Since the reaction of nitrite ion reduction sets its redox potential at a relatively high (positive) potential and its reaction rate is great in aqueous solution, metallic iron in neutral solution is readily passivated in the presence of nitrite ions. Nitrite salt is thus an effective passivating agent for metallic iron and steels in aqueous and atmospheric corrosion.

As the hydration of surface metal ions progresses, the hydrated metal ions of hard acid combines with anions of hard base forming a precipitate film of insoluble metal salts and complexes. The precipitate film thus formed, though not firmly adhering to the metal surface, may be capable of inhibiting the anodic metal dissolution. For instance, the corrosion of magnesium alloys can be inhibited by fluoride ions, F^-, which combine with magnesium ions, Mg^{2+}, dissolved from the alloys to form a precipitate film of magnesium fluoride, MgF_2:

$$Mg^{2+} + 2F^- \rightarrow MgF_2 \tag{22.83}$$

The magnesium ion is a hard acid and the fluoride ion is a hard base. The two ions then combine with each other to form a precipitate film of ionic bonding. In the similar way, benzotiazole, BTAH, which dissociates into deprotonated benzotiazole ion, BTA^-, and proton, H^+, inhibits the corrosion of copper in aqueous solution by forming an insoluble precipitate film of cuprous polymer complexes [85]:

$$nCu^+ + nBTA^- \rightarrow [Cu(BTA)]_n \qquad (22.84)$$

Cuprous ion, Cu^+, is a soft acid and BTA^- is a soft base. Benzotiazole also inhibits the corrosion of metallic zinc by forming a precipitate film of polymer complexes of zinc ions:

$$nZn^{2+} + 2nBTA^- \rightarrow [Zn(BTA)_2]_n \qquad (22.85)$$

where zinc ions, Zn^{2+}, are relatively soft acid.

It is worthwhile to mention that both the passivating film and the precipitate film, which are formed in the presence of foreign ions and molecules, usually inhibit not only the anodic metal dissolution but also the cathodic reaction of corroding metals. There are however some inhibitors, which are effective only to one of the anodic and the cathodic reactions of metallic corrosion. In the case of porous precipitate films loosely attached to the metal surface, the anodic metal dissolution may be accelerated at porous sites of the precipitate films. For instance, Zn^{2+} ions, Al^{3+} ions, Co^{2+} ions, and Ce^{3+} ions, which are hard or slightly hard Lewis acid, combine with hydroxide ions of hard base forming a porous precipitate film of metal hydroxide on metallic iron in neutral solution. The porous precipitate film thus formed effectively inhibits the cathodic oxygen reduction, but it may accelerate the anodic dissolution of metallic iron at the porous sites of precipitates [86].

It is interesting to see that bismuth ions, Bi^{3+} or BiO^+, inhibit the corrosion of metallic iron, zinc, and cobalt in perchloric acid solution. The bismuth ions are reduced in the cathodic reaction of metal corrosion forming metallic precipitates of bismuth on the corroding metal surface.

$$Bi^{3-} + 3e \rightarrow Bi \qquad (22.86)$$

Since the overpotential of cathodic proton reduction on metallic bismuth is relatively high, the metallic precipitate of bismuth makes the corrosion inhibited by decreasing the rate of the cathodic proton reduction and hence of the cathodic reaction of metallic corrosion in acid solution [87].

22.12.3 Catalysis in Localized Corrosion

As mentioned in the foregoing text, localized corrosion of metals starts with a local breakdown of passive films and localized metal dissolution then occurs at the breakdown site. The local film breakdown is caused by the adsorption of aggressive anions such as chloride ions, Cl^-, which are hard bases. No reliable information is available about the acid–base characteristic of the passive film surface. It is however expected that the breakdown of passive films will be prevented, if some anions or molecules are firmly adsorbed expelling chloride ions from the film surface. In literature, though, we have seen almost no reliable studies on this subject.

Once the passive film breaks down, the anodic metal dissolution localizes at the breakdown site forming a growing pit, in which ionic condensation will provide an occluded pit solution of high ionic concentration. In such a concentrated ionic solution, the adsorbing type of corrosion inhibitors will be ineffective because of the presence of aggressive anions at high concentration. In contrast, the precipitating type of inhibitors seems to be effective in inhibiting the pitting dissolution of metals. For instance, sodium octylthiopropionate, $NaC_8H_{17}S(CH_2)_2\,CO_2$, whose anions are hard base and easily combine with hydrated ferric ions of hard acid, was found to inhibit the pit initiation in the passivated iron in neutral solution by forming an insoluble precipitate containing sulfur at the passivity breakdown sites [88]. In fact, octylthiopropionate ions make the pitting potential of passivated iron more positive than that in the absence of the ions.

Normally, oxyanions such as CrO_4^{2-}, MoO_4^{2-}, and WO_4^{2-} are hard base and oxidizing agents. Nitrite ions, NO_2^-, whose acid–base level is slightly hard, may also be regarded as a hard acid to an intense degree because the cathodic reduction of nitrite ions produces hydroxide ions of hard acid. As mentioned in the foregoing text, these oxyanions oxidize the metal surface and hydrated metal ions to form a passive oxide film at the pitting sites on the metal surface. In general, the pitting potential of passivated metals appears to be a logarithmic function of the concentrations of aggressive anions and inhibitive anions in a limited range of ionic concentration [89]:

$$E_{pit} = a + b \log\left(\frac{c_i}{c_a}\right) \tag{22.87}$$

where

    c_a is the concentration of aggressive anions
    c_i is the concentration of inhibitive ions
    a and b are constants

22.13 ATMOSPHERIC CORROSION OF STEELS

22.13.1 CHEMISTRY OF ANTICORROSIVE RUST

In a wet atmosphere, iron and steel corrode producing first hydrated ferrous ions at the anodic site and hydrated hydroxide ions, OH_{aq}^-, at the cathodic site where air-oxygen is reduced into OH_{aq}^- ions. The presence of salts such as NaCl is known to enhance corrosion. Hydrated ferrous ions occur in the form of Fe_{aq}^{2+}, $Fe(OH)_{aq}^+$, $Fe(OH)_{2,aq}^0$, and $Fe(OH)_{3,aq}^-$. Gel-like hydroxide of $Fe(OH)_{2,solid}$ precipitates when the concentration of hydrated ferrous ions exceeds its solubility. Thermodynamic calculation gives the proton level to be pH 9.31 and 1.27×10^{-5} mol dm^{-3} of solubility at which the equilibrium is established between gel-like $Fe(OH)_{2,solid}$ and hydrated OH_{aq}^- ions when $Fe(OH)_{2,solid}$ is put in pure water [90].

The hydrated ferrous ions thus formed in corrosion are then oxidized by air-oxygen into a solid precipitate of ferric hydroxide $Fe(OH)_{3,solid}$:

$$\left[Fe_{aq}^{2+} + 2OH_{aq}^-\right] + (1/4)O_2 + (1/2)H_2O \rightarrow Fe(OH)_{3,solid} \tag{22.88}$$

where $\left[Fe_{aq}^{2+} + 2OH_{aq}^-\right]$ is the initial corrosion product. In the presence of $Fe(OH)_{2,solid}$, which keeps the proton level at pH 9.31, the hydrated ferrous ions are readily oxidized into a precipitate of ferric hydroxide, $Fe(OH)_{3,solid}$. With the ionic charge balance

$$\left[H_{aq}^+\right] = \left[OH_{aq}^-\right] + \left[Fe(OH)_{4,aq}^-\right] - 3\left[Fe_{aq}^{3+}\right] - 2\left[FeOH_{aq}^{2+}\right] - \left[Fe(OH)_{2,aq}^+\right] \tag{22.89}$$

established after the completion of the air-oxidation of the concentration of ferrous ions saturated at pH = 9.31, we obtain through equilibrium calculation the solubility of $Fe(OH)_{3,solid}$ at 1.20×10^{-8} mol dm^{-3} and the proton level at pH = 7 [90]. The proton level will thus shift from pH 9.31 to pH 7, when the hydrated ferrous ions oxidize into a precipitate of ferric hydroxide $Fe(OH)_{3,solid}$.

The precipitate of ferric hydroxide then gradually dehydrates into a mass of ferric oxyhydroxide:

$$Fe(OH)_{3,solid} \rightarrow FeOOH_{solid} + H_2O \tag{22.90}$$

where $FeOOH_{solid}$ occurs in the form of γ-$FeOOH_{solid}$; β-$FeOOH_{solid}$, which forms only in the presence of Cl^-_{aq}, α-$FeOOH_{solid}$, which is the most stable, and amorphous $FeOOH_{solid}$, which presumably consists of extremely fine α-$FeOOH_{solid}$ particles. The ferric oxyhydroxide thus formed tends to coagulate into a dense solid aggregate in the pH range from 7 to 9.31, where ferric oxyhydroxide particles have almost no surface charge: the point of zero charge pH for α-$FeOOH_{solid}$ is in the range from 7.5 to 9.38 [91]. Equilibrium calculation shows that the solubility of α-$FeOOH_{solid}$ at pH 9.31 is 1.51×10^{-13} mol dm^{-3}, which is much smaller than that of $FeOOH_{solid}$ [90].

Magnetite may also come out through air-oxidation of the gel-like $Fe(OH)_{2,solid}$ formed in the initial stage of corrosion:

$$3Fe(OH)_{2,solid} + (1/2)O_2 \rightarrow Fe_3O_{4,solid} + 3H_2O \tag{22.91}$$

The gel-like $Fe(OH)_{2,solid}$ dehydrates and partially oxidizes into magnetite without causing any change in the proton level. The solubility of magnetite is 1.10×10^{-13} mol dm^{-3} at pH 9.31, which is close to that of α-$FeOOH_{solid}$ [90].

With steels containing foreign metal, M, as an alloying element, the alloying metal also corrodes forming its hydroxide, which is then incorporated into iron rust in the form of ferrite $MFe_2O_{4,solid}$:

$$M(OH)_{2,solid} + 2Fe(OH)_{2,solid} + (1/2)O_2 \rightarrow MFe_2O_{4,solid} + 2H_2O \tag{22.92}$$

The solubility of transition metal ferrite such as $ZnFe_2O_3$ is found to be as small as that of α-$FeOOH_{solid}$ [90]. Although occupying a minor concentration, transition metal ferrites may play a significant role in developing anticorrosion rust in weathering steels [92].

For atmospheric wet-dry corrosion cycles, we have the well-known Evans model [93], in which the wet corrosion couples the anodic iron dissolution with the cathodic ferric oxyhydroxide reduction into magnetite:

$$Fe \rightarrow Fe^{2+}_{aq} + 2e \tag{22.93}$$

$$8FeOOH + Fe^{2+}_{aq} + 2e \rightarrow 3Fe_3O_4 + 4H_2O \tag{22.94}$$

where Fe_3O_4 is reactive magnetite into which protons can penetrate. The dry stage is the oxidation of reactive magnetite by air into ferric oxyhydroxide:

$$3Fe_3O_4 + (3/4)O_2 + (9/2)H_2O \rightarrow 9FeOOH \tag{22.95}$$

We then obtain the overall corrosion as follows:

$$Fe + (3/4)O_2 + (1/2)H_2O \rightarrow FeOOH \tag{22.96}$$

which suggests the growth of ferric oxyhydroxide during atmospheric steel corrosion.

In general, the atmospheric corrosion of steels develops a multilayered rust structure composed of ferrous hydroxide, ferric hydroxide, magnetite, and ferric oxyhydroxide. As the rusting progresses, less anticorrosive hydroxide rust gradually turns into more anticorrosive oxide rust. The main component of the aged rust is amorphous $FeOOH$ or α-$FeOOH_{solid}$ containing some rate of magnetite. The aged rust usually presents itself in an imbricate pattern, and it is the received wisdom that the smaller the size of the imbricate, the more anticorrosive is the rust layer.

22.13.2   LOCALIZED CORROSION IN WEATHERING STEELS

In the presence of chloride ions, a local breakdown of rust layers makes an anode channel for localized corrosion of underlying steel, and the chloride ions tend to accumulate in the channel as the anodic metal dissolution progresses. Assuming the ferrous chloride concentration at 1 mol dm^{-3} in the anode channel, we obtain the proton level at pH 4.75, where no ferrous hydroxide precipitation is expected to occur because of its solubility greater than 1 mol dm^{-3}. The hydrated ferrous chloride produced by corrosion is then oxidized by air-oxygen in the anode channel:

$$\left[Fe_{aq}^{2+} + 2Cl_{aq}^{-} \right] + (1/4)O_2 + (5/2)H_2O \rightarrow Fe(OH)_{3,solid} + \left[2H_{aq}^{+} + 2Cl_{aq}^{-} \right] \tag{22.97}$$

This oxidation generates hydrochloric acid and makes the anode channel acidified. As the acidification progresses, the rate of the air-oxidation of hydrated ferrous ions steeply decreases and the solubility of ferric hydroxide increases at the same time. The anode channel, as a result, holds a mass of hydrated ferrous and ferric ions, which gradually leaks out of the rust layer.

With the reference reaction of $Fe_{aq}^{2+} + (1/4)O_2 + (1/2)H_2O \rightarrow FeOH_{aq}^{2+}$ at the total iron ion concentration of $[Fe]_T = 1$ mol dm^{-3}, equilibrium calculation gives us the proton level at pH 1.41, the solubility of Fe_{aq}^{3+} at 0.59 mol dm^{-3}, and the solubility of $FeOH_{aq}^{2+}$ at 0.096 mol dm^{-3} after the air-oxidation of $\left[Fe_{aq}^{2+} + 2Cl_{aq}^{-} \right]$ has completed [91]. One-third of the corroded iron at $[Fe]_T = 1$ mol dm^{-3} precipitates as gel-like $Fe(OH)_{3,solid}$ in the anode channel at pH 1.41. Owing to its positive surface charge in acidic water, the gel-like $Fe(OH)_{3,solid}$ does not aggregate but disperses into the acidified anode channel. The gel-like ferric hydroxide remaining in the anode channel, then, gradually dehydrates into β-FeOOH in the presence of chloride ions and comes out as a flowing mass of yellow rust developing no anticorrosive rust. The remaining two-third of corroded iron occurs mainly in the form of soluble complexes of ferric ions and diffuses out of the anode channel to precipitate into a variety of iron rust not in the acidic anode channel but in neutral water outside the aged rust layer.

In the acidified anode channel containing HCl and Fe_{aq}^{3+}, acid corrosion occurs with protons and ferric ions as oxidants:

$$Fe + \left[2H_{aq}^{+} + 2Cl_{aq}^{-} \right] \rightarrow \left[Fe_{aq}^{2+} + 2Cl_{aq}^{-} \right] + H_2 \tag{22.98}$$

$$Fe + \left[2Fe_{aq}^{3+} + 6Cl_{aq}^{-} \right] \rightarrow \left[3Fe_{aq}^{2+} + 6Cl_{aq}^{-} \right] \tag{22.99}$$

These corrosion reactions continue occurring provided the anode channel holds the concentration of hydrochloric acid.

Weathering steels in practice are subject to cyclic wet-dry corrosion in atmospheric air. In the wet period, they corrode locally producing yellow rust that flows through rust crevices out of aged rust layers. In the dry period, the fresh rust as well as aged rust is dehydrated and oxidized by air to form a dense rust deposit. The corrosion rust, as a result, develops into a multilayered structure successively composed of dense and coarse rust layers formed respectively in the dry and wet periods.

22.13.3   CATALYSES FOR ANTICORROSIVE RUST

The key point for atmospheric steel corrosion is to make an initial corrosion product of soluble ferrous ions air-oxidized as soon as possible into an insoluble ferric hydroxide aggregate, which eventually turns into anticorrosive rust. The air-oxidation of hydrated ferrous ions is fast in neutral water but slow in acidic water. Once corrosion-produced ferrous ions are oxidized into gel-like ferric hydroxide in pure water, the solubility equilibrium of both ferrous and ferric hydroxides keeps the water pH around pH 7–9.31, where the air-oxidation rate of hydrated ferrous ions is great. In the presence of chloride ions at the anodic site, however, the acidified anode channel reduces the rate of the air-oxidation of hydrated ferrous ions and hence counteracts the formation of anticorrosive rust embryos.

It comes to mind then, that catalysts might accelerate the air-oxidation of hydrated ferrous ions in acidic water. We notice that steels containing copper and phosphorus are somehow resistive toward atmospheric corrosion. During steel corrosion, alloying copper and phosphorus are oxidized into hydrated cupric ions, Cu_{aq}^{2+}, and hydrated phosphate ions, $PO_{4,aq}^{3-}$, respectively. Both Cu_{aq}^{2+} and $PO_{4,aq}^{3-}$ are catalysts effective for the air-oxidation of Fe_{aq}^{2+} into $Fe(OH)_{3,solid}$ even in acidified water [94]. The role of alloying elements of copper and phosphorus in improving weathering steels could thus be attributed to their catalytic reaction for the air-oxidation of hydrated ferrous ions into insoluble ferric hydroxide in the acidified anode channel. The catalytic mechanism of these two elements, which has not yet been made clear, is waiting for our study from the standpoint of quantum chemistry.

We also see that cupric ions react as a catalyst on rust coagulation and prevent rust crystallization into coarse aggregates [95,96]. The rust then remains amorphous and develops a dense, void-free barrier layer of anticorrosive rust. Transition metal ions such as Ti_{aq}^{4+}, Co_{aq}^{2+}, Cr_{aq}^{3+}, and Ni_{aq}^{2+}, which come from alloying elements in weathering steels, are also found to catalyze the formation of amorphous or poorly crystalline iron rust [97,98]. It is commonly known that the amorphous rust of FeOOH is much more anticorrosive than the coarse crystalline deposits of iron rust.

Lastly, we recall the ion-selective nature of corrosion rust mentioned earlier. We would not however need to re-mention that a layer of iron rust, if cation-selective, makes itself corrosion-protective and that the rust develops localized corrosion if it is anion-selective. It may however be worthwhile to remind that the ion selectivity arises from the surface charge of rust particles: the negative surface charge makes rust cation-selective and the positive surface charge makes rust anion-selective. The factors that determine the surface charge of iron rust are protonation–deprotonation of the surface acid and base hydroxyl groups, specific adsorption of multivalent ions, and nonstoichiometry of the surface composition. The surface charge of iron rust is positive in acidic water and negative in basic water with a border pH at the point of zero charge, which varies with different sorts of rust. It is the accepted understanding that the lower the point of zero charge pH of the rust, the more anticorrosive is the rust. As mentioned earlier, specific adsorption of multivalent anions such as $PO_{4,aq}^{3-}$ and $CrO_{4,aq}^{2-}$, which come from P and Cr in weathering steels, lowers the point of zero charge pH of iron rust and hence improves the corrosion-resisting rust layers.

22.14 CONCLUDING REMARKS

Our comprehensive understanding of materials corrosion fundamentals has advanced considerably over the decades. Modern corrosion science has made it clear that the corrosion process on metals and semiconductors consists of an anodic oxidation and a cathodic reduction both occurring across the material–aqua-solution interface. These reduction–oxidation reactions depend on the interfacial potential and hence on the electrode potential of materials.

The corrosion process also includes acid–base reactions involving solute particles in aqueous solution. As for the acid–base characteristics of corrosion processes, we still have many research issues that remain to be studied, such as the effect of various solutes in aqueous solution on materials corrosion. Novel experimental techniques in the nanometer scale recently developed in electrochemistry and corrosion science will provide, from now on, much information that, we hope, will contribute to elucidating the acid–base characteristics of corrosion processes, particularly the effect of adsorption of solute particles on the corrosion.

In this chapter, we have not discussed materials corrosion under mechanical stresses, such as environmental cracking, stress corrosion cracking, erosion-corrosion, and cavitation-corrosion.

It is accepted knowledge that the mechanical property of materials is subject to the influence of the environment in which the materials stand. The susceptibility of materials to corrosion is also affected by the mechanical stress. These synergetic effects of corrosion and mechanical degradation are of practical importance in the industrialized society and are among the most critical issues for corrosion science.

REFERENCES

1. U. R. Evans, *Metallic Corrosion, Passivation and Protection*, Edward Arnold, London, U.K. (1937).
2. C. Wagner and W. Traud, *Z. Elektrochem.* **44** (1938) 52.
3. W. B. Jensen, *The Lewis Acid–Base Concepts*, pp. 112–336, John Wiley, New York (1980).
4. W. H. Robertson and M. A. Johnson, *Annu. Rev. Phys. Chem.* **54** (2003) 173.
5. T. Watanabe and H. Gerischer, *J. Electroanal. Chem.* **122** (1981) 73.
6. N. Sato, *Electrochemistry at Metal and Semiconductor Electrodes*, pp. 15–117, Elsevier, Amsterdam, New York, Tokyo (1998).
7. S. Trasatti, in *Comprehensive Treatise of Electrochemistry*, Vol. 1, p. 45, J. O'M. Bockris, B. E. Conway, E. Yeager (Eds.), Plenum Press, New York, London, (1980).
8. H. Gerischer, *Electrochim. Acta* **35** (1990) 1677.
9. K. F. Bonhoeffer and K. J. Vetter, *Z. Phys. Chem.* **196** (1950) 127.
10. U. F. Franck and K. Weil, *Z. Elektrochem.* **56** (1952) 814.
11. N. Sato, in *Passivation of Metals and Semiconductors, Corrosion Science*, Vol. 13, pp. 1–19, Pergamon Press, Oxford, U.K. (1989).
12. W. H. Brattain and C. G. Barret, *J. Bell Syst. Technol.* **34** (1961) 129.
13. H. G. G. Philipsen and J. J. Kelly, in *Passivation of Metals and Semiconductors, and Properties of Thin Oxide Layers*, pp. 233–238, P. Marcus and V. Maurice (Eds.), Elsevier, Amsterdam, the Netherlands (2005).
14. R. Memming and J. J. Kelly, in *Proceedings of the International Conference on Photochemical Energy Conversion and Storage*, p. 243, Academic Press, New York (1981).
15. L. Hollan, J. C. Tranchart, and R. Memming, *J. Electrochem. Soc.* **126** (1979) 855.
16. H. Gerischer, *J. Vac. Sci. Technol.* **15** (1978) 1422.
17. H. J. Engell, *Z. Phys. Chem. N. F.* **7** (1956) 158.
18. H. J. Engell, *Z. Elektrochem.* **60** (1956) 905.
19. Y. A. Riga, R. Greef, and E. B. Yeager, *Electrochim. Acta* **13** (1968) 1351.
20. D. A. Vermilyea, *J. Electrochem. Soc.* **113** (1966) 1067.
21. H. Kaesche, *Die Korrosion der Metalle*, 2 Auflage, p.122, Springer-Verlag, New York (1979).
22. G. Okamoto and N. Sato, *J. Jpn. Inst. Met.* **23** (1959) 725.
23. H. Gerischer and J. Wallem-Mattes, *Z. Phys. Chem.* **64** (1969) 187.
24. J. Keir, *Phil. Trans.* **80** (1790) 259.
25. W. Ostward, *Elektrochemie*, pp. 696–697, Johann Ambrosius Barth, Leipzig, Germany (1896).
26. N. Sato, in *Comprehensive Treatise of Electrochemistry*, Vol. 4, pp. 193–245, J. O'M. Bockris, B. E. Conway, E. Yeager, and R. E. White (Eds.), Plenum Publishing Corp., New York, London, U.K. (1981).
27. M. B. Ives, J. L. Luo, and J. R. Rodda (Eds.), in *Passivity of Metals and Semiconductors, Proceedings*, Vol. 99–42, The Electrochemical Society Inc., Pennington, NJ (1999).
28. U. F. Franck, *Z. Naturforsh.* **49** (1949) 378.
29. N. Sato, in *Passivity and Its Breakdown on Iron and Iron Base Alloys*, pp. 1–9, R. Staehle and H. Okada (Eds.), NECA, Houston, TX (1976).
30. N. Sato, K. Kudo, and T. Noda, *Z. Phys. Chem. N. F.* **98** (1975) 271.
31. T. Ohtsuka, M. Masuda, and N. Sato, *J. Electrochem. Soc.* **132** (1985) 787.
32. N. Sato, *Corros. Sci.* **31** (1990) 1.
33. R. L. Smith, *J. Electroanal. Chem.* **238** (1987) 103.
34. R. L. Smith, B. Kloeck, and S. D. Collins, *J. Electrochem. Soc.* **135** (1988) 2001.
35. M. Cohen and M. Nagayama, *J. Electrochem. Soc.* **109** (1962) 781.
36. N. Sato, K. Kudo, and R. Nishimura, *J. Electrochem. Soc.* **123** (1976) 1419.
37. H. Takahashi, F. Fujimoto, H. Konno, and M. Nagayama, *J. Electrochem. Soc.* **131** (1984) 1856.
38. F. P. Fehlner and N. F. Mott, *Oxid. Met.* **2** (1970) 59.
39. T. L. Barr, *J. Phys. Chem.* **82** (1976) 1801.

40. L. Li and A. A. Sagues, in *Passivity of Metals and Semiconductors, Proceedings*, Vol. 99–42, pp. 584–589, M. B. Ives, J. L. Luo, and J. R. Rodda (Eds.), The Electrochemical Society Inc., Pennington, NJ (1999).
41. K. Fushimi and M. Seo, *J. Electrochem. Soc.* **148** (2001) B450.
42. K. Fushimi, K. Azumi, and M. Seo, *J. Electrochem. Soc.* **147** (2000) 552.
43. K. E. Heusler and L. Fischer, *Werkstoffe u. Korrosion* **27** (1976) 551.
44. H. Bohni, St. Matsch, T. Suter, and J. O. Park, in *Passivity and Localized Corrosion, Proceedings*, Vol. 99–27, pp. 483–492, M. Seo, B. MacDougall, H. Takahashi, and R. G. Kelly (Eds.), The Electrochemical Society Inc., Pennington, NJ (1999).
45. T. Shibata, *Corros. Sci.* **31** (1990) 413.
46. Y. Hisamatsu, T. Yoshii, and Y. Matsumura, in *Proceedings of the U. R. Evans Conference on Localized Corrosion*, p. 420, R. W. Staehle, B. F. Brown, J. Kruger, and A. Agrawal (Eds.), NACE, Houston, TX (1974).
47. K. E. Heusler and L. Fischer, *Werkstoff u. Korrosion* **27** (1976) 55.
48. D. D. MacDonald, *J. Electrochem. Soc.* **139** (1992) 3434; *Pure Appl. Chem.* **71** (1999) 951.
49. N. Sato, *J. Electrochem. Soc.* **129** (1982) 255.
50. N. Sato, *Symposium on Passivity and Breakdown, Proceedings*, Vol. 97–26, pp. 1–14, The Electrochemical Society Inc., Pennington, NJ (1997).
51. K. E. Heusler, in *Corrosion and Corrosion Protection, Proceedings*, Vol. 2001–22, pp. 172–180, J. D. Sinclair, R. P. Frankenthal, E. Kalman, and W. Plieth (Eds.), The Electrochemical Society Inc., Pennington, NJ (2001).
52. J. R. Galvele, in *Passivity of Metals*, pp. 285–327, R. P. Frankenthal and J. Kruger (Eds.), The Electrochemical Society Inc., Pennington, NJ (1978).
53. H. J. Laycock and R. C. Newman, *Corros. Sci.* **39** (1997) 1771.
54. Y. Hisamatsu, in *Passivity and Its Breakdown on Iron and Iron base Alloys*, pp. 99–105, R. Staehle and H. Okada (Eds.), NECA, Houston, TX (1976).
55. H. J. Engell and N. D. Strica, *Arch. Eisenhut.* **25** (1959) 1255.
56. N. Sato, *J. Electrochem. Soc.* **129** (1982) 260.
57. H. J. Laycock and R. C. Newman, *Corros. Sci.* **40** (1997) 887.
58. H. W. Pickering, *Corrosion* **29** (1989) 325.
59. S. Tsujikawa, Y. Sono, and Y. Hisamatsu, in *Corrosion Chemistry within Pits*, p. 171, A. Turnbull (Ed.), National Physical Laboratory, Her Majesty's Stationary Office, London, U.K. (1987).
60. B. G. Ateya, M. Abdulsalam, and H. W. Pickering, in *Passivity and Localized Corrosion, Proceedings*, Vol. 99–27, pp. 599–608, M. Seo, B. MacDougall, H. Takahashi, and R. G. Kelly (Eds.), The Electrochemical Society Inc., Pennington, NJ (1999).
61. S. Tsujikawa, *Handbook of Corrosion Technology*, p. 33, Japan Soc. Corr. Eng., Nikkan Kogyo, Tokyo, Japan (1986).
62. N. Sato, *Corros. Sci.* **37** (1995) 1947.
63. U. Steinsmo, T. Rogne, and J. M. Drugli, in *Marine Corrosion of Stainless Steels*, pp. 115–123, D. Feron (Ed.), European Federation of Corrosion Publications No. 33, IOM Communications, London, U.K. (2001).
64. N. Sato, in *Passivity of Metals and Semiconductors, Proceedings*, Vol. 99–42, pp. 281–287, J. B. Ives, J. L. Luo, and J. R. Rodda (Eds.), Electrochemical Society, Inc., Pennington, NJ (2001).
65. N. Sato, *Corros. Sci. Technol.* **31** (2002) 265.
66. N. Sato, *Corrosion* **45** (1989) 354.
67. K. Huber, *Z. Elektrochem.* **59** (1953) 693.
68. M. Suzuki, N. Masuko, and Y. Hisamatsu, *Boshoku Gijutsu (Jpn. Corros. Eng.)* **20** (1977) 319.
69. Y. Yomura, M. Sakashita, and N. Sato, *Boshoku Gijutsu (Jpn. Corros. Eng.)* **28** (1979) 64.
70. M. Sakashita and N. Sato, *Corrosion* **35** (1979) 351.
71. M. Sakashita, Y. Yomura, and N. Sato, *Denki Kagaku (Jpn. Electrochem.)* **45** (1977) 165.
72. M. Sakashita and N. Sato, *J. Electroanal. Chem.* **62** (1975) 127.
73. M. Sakashita and N. Sato, *Corros. Sci.* **17** (1977) 473.
74. C. R. Clayton and Y. C. Lu, *Corros. Sci.* **29** (1989) 881.
75. C. R. Clayton and Y. C. Lu, *J. Electrochem. Soc.* **133** (1986) 2465.
76. J. Yuan and S. Tsujikawa, *J. Electrochem. Soc.* **142** (1995) 3444.
77. J. Yuan, R. Fujisawa, and S. Tsujikawa, *Zairyou-to-Kankyo (Jpn. Corros. Eng.)* **43** (1994) 433.
78. R. Fujisawa and S. Tsujikawa, *Mater. Sci. Forum* **183–189** (1995) 1076.
79. N. Sato, *Electrochemistry at Metal and Semiconductor Electrodes*, pp. 298–323, Elsevier, Amsterdam, New York, Tokyo (1998).

80. E. Mc.Cafferty and N. Hackerman, *J. Electrochem. Soc.* **119** (1972) 999.
81. K. F. Bonhöffer and K. E. Heusler, *Z. Electrochem.* **61** (1957) 122.
82. N. Sato and G. Okamoto, *Electrochem. Soc.* **111** (1964) 897.
83. K. E. Heusler and G. H. Cartledge, *J. Electrochem. Soc.* **108** (1961) 730.
84. K. Aramaki, M. Hagiwara, and H. Nishihara, *J. Electrochem. Soc.* **135** (1988) 1364.
85. K. L. Stewart, J. Zang, S. Li, P. W. Carter, and A. A. Gewirth, *J. Electrochem. Soc.* **154** (2007) D57.
86. K. Aramaki, *Corrosion* **55** (1999) 157.
87. K. Aramaki and H. Nishihara, *Mater. Technol.* **10** (1992) 207.
88. K. Aramaki, *Zairyou-to-Kankyo (Jpn. Corros. Eng.)* **46** (1997) 748.
89. V. S. Sastri, in *Corrosion Inhibitors, Principles and Applications*, p. 567, John Wiley & Sons, Chichester, U.K. (1998).
90. H. Tamura, *Corros. Sci.* **50** (2008) 1872.
91. R. M. Cornell and U. Schwertmann, in *The Iron Oxides*, p. 224, VCH, Weinheim, Germany (1996).
92. H. Tamura and E. Matijević, *J. Colloid Interface Sci.* **90** (1982) 110.
93. U. R. Evans, *Corros. Sci.* **9** (1969) 813.
94. H. Tamura, K. Goto, and M. Nagayama, *J Inorg. Nucl. Chem.* **38** (1974) 113.
95. R. Furuichi, N. Sato, and G. Okamoto, *Chimia* **23** (1969) 455.
96. I. Suzuki, T. Hisamatsu, and N. Masuko, *J. Electrochem. Soc.* **127** (1980) 2210.
97. T. Ishikawa, T. Ueno, A. Yasukawa, K. Kandori, T. Nakayama, and T. Tsubota, *Corrosion* **45** (2003) 1037.
98. T. Ishikawa, A. Maeda, K. Kandori, and A. Tahara, *Corrosion* **62** (2006) 559.

23 Production, Storage, Use, and Delivery of Hydrogen in the Electrochemical Conversion of Energy

C. Fernando Zinola

CONTENTS

Currently, among the most challenging aspects in science and technology are those related to the new energy vectors and efficient methodologies for energy conversion. The success of the use of hydrogen for energy depends on two factors: less expensive devices for the complete conversion technology and more efficient processes for the production of hydrogen and its final conversion to electricity. Therefore, the effective design and implementation of a hydrogen-based energy scheme needs a "complete system" approach. A number of crossover issues will influence the production, storage, delivery, conversion, applications, education, etc., of hydrogen. The most significants are

- Development of national and international legislation, codes, and standards for the use of hydrogen
- Adoption of policies to incorporate the external costs of energy (energy supply security, air quality, and global climate change) to provide a clear signal to the industry and consumers on the benefits of hydrogen energy
- Safety standards and precautions
- Promotion by institutions and governments, and acceptance by consumers by providing the expected performance at a reasonable cost
- Collaborative research and development by research centers
- Technology validation through government–industry partnerships
- Systems analyses to explore various pathways to extensive hydrogen energy use, including full cost accounting for all challenging energy systems
- Easy access to the existing information on new hydrogen technologies without legal or economical barriers

System integration addresses ways in which different parts of a system work together from technical, economic, and social standpoints. In many cases, system optimization may require an approach different from that used for the optimization of a single part. Similarly, system focus makes it easier to identify key technical or market barriers in any part of the system that might impede the development of the whole. Optimization at the system level will require the following:

- Coordination of technology developments between hydrogen producers and end users.
- Fixed, synchronized, and focused research and development agenda in hydrogen storage, production, and use to allow different way(s) by which the economy of hydrogen energy is developed.
- Efficient coordination between supply and demand in the transportation market segments to assure vehicle manufacturers and the end users of assured fuel supply.

23.1 PRODUCTION OF HYDROGEN

23.1.1 Introduction

There are several reasons for reviewing the electrochemical research and industrial applications related to hydrogen production. Firstly, this topic is continuously developing with several new technologies emerging every year. Secondly, the reforming and electrolysis production processes require novel approaches with continuous improvements in the global efficiencies. However, due to the recent applications of hydrogen-enriched fuels, the problem related to intermediate species formation and poisoning needs to be addressed. The poisoning intermediate produced in the course of organic fuels electrooxidation has been identified as CO and not COH as previously believed [1–3]. However, the detected species depends very strongly on the experimental conditions, topography of the surface, nature of electrolyte, etc. Further, it seems that the contamination from carbon and sulfur compounds cannot be avoided.

One of the most important electrochemical industries is the production of molecular hydrogen. However, in current times, the use of hydrogen as a fuel is imperative. Despite its abundance in the universe, hydrogen is not free in our atmosphere, since it reacts fast with other elements. It is found combined with oxygen as water or bound into other molecular compounds. Therefore, it has to be separated from other molecules. It is easy to remove hydrogen from large compounds that are at a higher energy state, such as fossil fuels. This process releases energy, thereby reducing the amount of the required energy. However, the purity of the obtained hydrogen is not as high as required. On the other hand, it takes more energy to extract hydrogen from compounds that are at a lower energy state, such as water, since an external power source needs to be added to the process.

The process of extracting hydrogen from fossil fuels and carbon compounds is termed *reforming*. Today, this is the principal and least expensive method of producing hydrogen. Unfortunately, reforming emits pollutants and consumes nonrenewable fuels. It is very difficult to eliminate or reduce the presence of carbon monoxide, carbon dioxide, and other derivatives.

The process of extracting hydrogen from water is called *electrolysis*. In principle, electrolysis can be entirely nonpolluting and renewable, but it requires the input of large amounts of electrical energy. Consequently, the total environmental impact of acquiring hydrogen through electrolysis is largely dependent on the source of power.

Alternative methods of hydrogen production are thermochemical water decomposition, photo-conversions, photobiological processes, production from biomass, and various industrial processes where it is a by-product.

Hydrogen can be produced on a large scale at dedicated hydrogen production plants or on a small scale at local production facilities. Large-scale production benefits from economies of scale and plants that are located near power and water but suffers from the problems associated with the transportation of hydrogen. Some methods of hydrogen production, such as from coal or biomass can only be undertaken on a large scale. In the case of small-scale production (as in the case of fuel cell applications), the problems related to hydrogen transportation are avoided by using energy that is easily brought to the facility, such as electricity, natural gas, or solar. On the downside, the amount of equipment required proportionate to the hydrogen produced is significantly high compared to large-scale facilities.

At the extreme small scale, fossil fuels could be reformed to hydrogen on board a fuel cell vehicle, but the systems are complex and costly. Currently, the hydrogen manufactured worldwide predominantly originates from fossil fuels, as a by-product in chemical industries or crude oil refining processes. Hydrogen production from renewable energy is not yet feasible on a large scale. Production of hydrogen currently costs from 3 to 15 times more than natural gas and from 1.5 to 9 times more than gasoline, depending on the method employed. When hydrogen is extracted from fossil fuels, the initial production together with the further refining increases the overall cost.

Since the first energy and oil crisis of 1973, there have been a lot of efforts in developing renewable energies. However, in the second energy crisis, there is still the need for a portable fluid fuel to be used in the advanced vehicles of today. One of the most attractive applications is in urban transportation. It seems that besides hydrogen, methanol and ethanol are the most interesting liquid fuels. In this sense, it is important to emphasize that the only proven technology for hydrogen production from nonfossil fuel as a primary energy is water electrolysis. Moreover, because of the intermittent nature of some energy sources, storage of hydrogen fuel is essential. Hydrogen is richer than methanol or ethanol and does not cause contamination of catalyst surfaces upon the fuel cell operation. The problem with the production of hydrogen from coal is that it requires a lot of chemicals for its purification, which makes it more expensive than its production from water electrolysis [4].

The uses of hydrogen in today's industry are quite significant: hydrocracking of petroleum (refining), fats hydrogenation, inorganic and organic synthesis, metallurgical applications, etc.

23.1.2 INDUSTRIAL ELECTROLYSIS

There are two main categories of electrolyzers: the unipolar tank-type cells or the bipolar filter press cells. Two main differences arise from both systems: the electrical connection and the electrolyte used to produce the electrolysis. In the monopolar cell, all faces of the electrode have the same polarity (either positive or negative) and they are linked together in parallel. Likewise, all the counter polarity electrodes are connected in parallel and both types are installed in a gas-sealed tank in a vertical location with large electrode areas. Between each other, we have a diaphragm of asbestos cloth inserted around either the anode or the cathode in order to separate the changes in pH and the evolution of hydrogen on the first electrode and oxygen on the second electrode, respectively. Thus, the unipolar connections develop large currents under low potentials, mainly 2 V. To obtain the desired hydrogen (or oxygen) production rate, a calculated number of modules in series have to be arranged. The main disadvantage in this method is the large electrolyte gap between the electrodes and the low currents obtained with the devices (but long current paths from the terminals of the cell to the electrodes). This is resolved in the case of bipolar cells (Figure 23.1).

In the case of the bipolar cell, all the electrodes except the terminal ones have one side with a positive polarity and the opposite side with a negative polarity. The terminal electrodes that lead the current into and out of the bipolar system are really monopolar. In this case, also, the electrodes are located vertically but are pressed to the electrolyte with a membrane (and sometimes with an electrolyte depending on the type of the electrolyzer). Both the diaphragm and an insulator frame have to be used to separate, firstly, the gases and then the electrodes by pressing them. The use of gaskets is important too to prevent the leakage of gases from the electrolyte. Each electrode is therefore electrically in series with the electrolyte but is insulated from the neighbors. The current

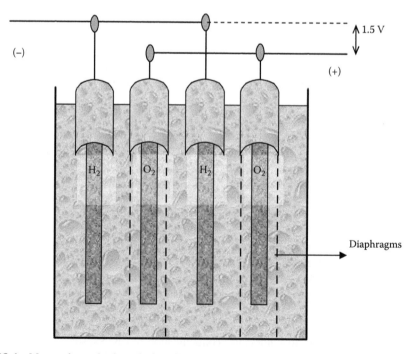

FIGURE 23.1 Monopolar tank electrolysis cell constructed to produce and collect hydrogen and oxygen. Water inlet is depicted inside. The electrolyte is concentrated and aqueous potassium hydroxide solution.

intensity flows in series through all the sandwiches of bipolar plates, diaphragms, etc., from one end to the other. Thus, the overall potential has to be very large since the total value is the accumulation of hundreds of series of connected 2 V units. In the opposite situation, the current is rather low since it offers the lower value. To obtain the desired hydrogen (or oxygen) production rate, a calculated number of modules in series-parallel have to be arranged and matched to produce an output of a suitable rectifier. The main advantage is that the low potential per unit cell involves low power consumption per unit of the produced gas. In this case, very low current paths are observed with narrow interelectrode gap distances. The principal disadvantage is the complexity in the construction: membranes, gaskets, filters, pump, etc.

When applying an external source of electricity to ionize water, it decomposes into its elemental components: hydrogen and oxygen. Electrolysis is often touted as the preferred method of hydrogen production as it is the only process that does not rely on fossil fuels. It also has high product purity and is feasible on small and large scales. Electrolysis can operate over a wide range of electrical energy capacities and takes advantage of abundant electricity at night (Figure 23.2).

At the heart of electrolysis is an electrolyzer. An electrolyzer is a series of cells each with a positive (anode) and a negative (cathode) electrode. The electrodes are immersed in water that has been made electrically conductive by adding hydrogen or hydroxyl ions, usually in the form of highly concentrated alkaline potassium hydroxide.

The anode is typically constructed of nickel and copper and is coated with oxides of metals such as manganese, tungsten, and ruthenium. The anode metals allow quick pairing of atomic oxygen into oxygen pairs at the electrode surface. The cathode is typically made of nickel, which is coated with small quantities of platinum as a catalyst. The catalyst allows the Tafel recombination chemical step into pairs of molecular hydrogen at the electrode surface and increases the rate of hydrogen production. Without the catalyst, atomic hydrogen would build up on the electrode and block the current flow. A gas separator, or diaphragm, is used to prevent the intermixing of the hydrogen and oxygen, although it allows free passage of ions. It is usually made of an asbestos-based material and tends to break apart at above 80°C.

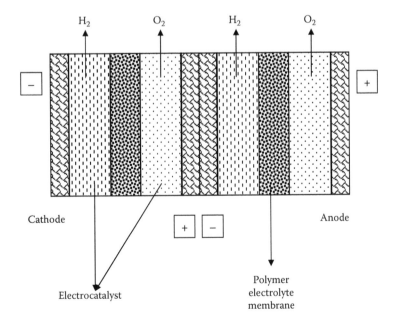

FIGURE 23.2 Bipolar polymer electrolyte electrolysis cell connected in series to enlarge the applied potential. The cathode, anode, current collectors, polymer electrolyte, and electrocatalyst with the evolution of gases are shown.

Due to the new developments [5] in fuel cell technology—the manufacture of carbon supported platinum catalysts and the use of the Nafion® membrane—the cost of bipolar electrolyzers has been reduced a lot, and therefore almost all commercial devices are of this type. In this case, stainless steel or nickel cathodes are used together with nickel anodes in 25%–35% of potassium hydroxide at temperatures between 65°C and 90°C. The hydrogen current density reaches 100–300 mA/cm² at cell potentials of 1.9–2.2 V, denoting a faradaic efficiency of 80% (losses in peripheries). Usually, a pressurized cell is employed to increase their performance and to reduce the size of the bubbles, thus lowering the overpotential associated with the process. This can be done with appropriate membranes and insulators and by using temperatures near 100°C.

The reactions at the cathode are already presented in Chapter 2:

$$H^+ + e^- + Pt(cat) \leftrightarrow [Pt - H]_{ad} \tag{I}$$

$$2[Pt - H]_{ad} \rightarrow 2Pt(cat) + H_2 \tag{II}$$

The highly reactive hydrogen atom is adsorbed on the cathodic metal and it chemically combines with a neighbor-bound hydrogen atom to form a hydrogen molecule that leaves the cathode as a gas after the bubble surpasses the surface tension of the electrode.

On the other hand, the reactions at the anode are

$$OH^- + Pt(cat) \leftrightarrow [Pt - OH]_{ad} + e^- \tag{III}$$

This is the hydroxyl discharge on the surface. After that, the two adsorbates recombine as

$$2[Pt - OH]_{ad} \rightarrow 2Pt(cat) + \tfrac{1}{2}O_2 + H_2O + e^- \tag{IV}$$

The highly reactive oxygen atom or hydroxyl species then bonds to the metal of the anode and recombines as in the Tafel step in the cathode to form an oxygen molecule that leaves the anode as a gas.

The rate of hydrogen production is related to the current density (measured in Amperes per real area). In general, the higher the current density, the higher the source potential required, and the higher the power cost per unit of hydrogen. However, higher potentials decrease the overall size of the electrolyzer and so it results in a lower capital cost. State-of-the-art electrolyzers that have energy efficiencies of 65%–80% and operate at current densities of about 2000 A/m² are reliable. There are many types of electrolyzers, either large devices consuming 180 kW (producing 1500 L/h) or small devices consuming 7 kW (producing 42 L/h). The small industries compress hydrogen gas to approximately 30–100 psig at purities of more than 99.5% [6,7].

For electrolysis, the amount of electrical energy required can be somewhat offset by adding heat energy to the reaction. The minimum amount of potential required to electrodecompose water is more than the thermodynamic value of 1.23 V at 25°C. However, at this potential, the reaction requires heat energy from the outside to proceed since the electrochemical process is in equilibrium. At 1.47 V (and same temperature), no input heat is required. At greater potentials (and same temperature, 25°C), heat is released into the surroundings during water decomposition since the overpotentials are totally compensated by the external potential.

Operating the electrolyzer at lower potentials with added heat is advantageous, as heat energy is usually cheaper than electricity and can be recirculated within the process. Furthermore, the efficiency of the electrolysis increases with increased operating temperature.

When viewed together with fuel cells, hydrogen produced through electrolysis can be seen as a way of storing electrical energy as a gas until it is used. The hydrogen produced by electrolysis is therefore the energy carrier, not the energy source. The energy source is derived from an external power generating plant. In this sense, the process of electrolysis is not very different from charging a battery, which also stores electrical energy.

Viewed as electricity storage medium, hydrogen is competitive with batteries in terms of weight and cost. To be truly clean, the electrical power stored during electrolysis must be derived from nonpolluting, renewable sources. If the power is derived from natural gas or coal, the pollution is not eliminated but only pushed upstream. Moreover, every energy transformation has an associate energy loss. Consequently, fossil fuels may be used with greater efficiency by means other than by driving the electrolysis of hydrogen. Furthermore, the cost of burning fossil fuels to generate electricity for electrolysis is three to five times that of reforming the hydrogen directly from the fossil fuel.

Although a renewable energy source in conjunction with electrolysis would eliminate the dependence on fossil fuels, it does not reduce the number of energy transformations required to produce mechanical work by using hydrogen. If clean, renewable power were available, it could also be used in other ways that require fewer energy transformations, such as direct storage in batteries or compression of air for propulsion. Another problem with electrolysis is the source of water. Water is already a precious substance, and it would be consumed in vast quantities in order to support a large hydrogen-driven economy. Water would also have to be purified prior to use, thereby increasing its cost. However, when hydrogen is used in the fuel cell as an energy converter, water is the main resultant chemical product and therefore can be recirculated again to produce fresh fuel.

23.1.3 Gas Reforming

Reforming is a chemical process in which hydrogen-containing fuels react in the presence of steam, oxygen, or both in a hydrogen-rich gas stream. When applied to solid fuels, the reforming process is called gasification. The resulting hydrogen-rich gas mixture is called the *reformate*. The equipment used to produce the reformate is known as a reformer or fuel processor (Figure 23.3).

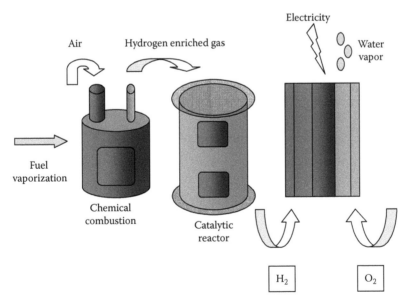

FIGURE 23.3 Simplified scheme of a reforming fuel to be used in a hydrogen/oxygen fuel cell. First, the liquid fuel is vaporized from its liquid form using waste energy from the fuel cell. Second, it is partially oxidized with air to produce carbon monoxide and hydrogen. Third, the sulfur-containing compounds are eliminated to avoid chemical pollution. After that, the purified fuel enters a catalytic reactor with more air to produce carbon dioxide and hydrogen. Lastly, the remaining carbon monoxide is completely oxidized to reduce its concentration to lower than 10 ppm. The fuel now is able to enter the fuel cell engine.

The specific composition of the reformate depends on the source fuel and the process used, but it always contains other compounds such as nitrogen, carbon dioxide, carbon monoxide, and some of the unreacted source fuel. When hydrogen is removed from the reformate, the remaining gas mixture is called *raffinate*.

Reforming a fossil fuel involves the following steps:

1. Feedstock purification (including sulfur removal)
2. Steam reforming or oxidation of feedstock to form hydrogen and carbon oxides
3. Primary purification—conversion of carbon monoxide to carbon dioxide
4. Secondary purification—further removing of carbon monoxide

The steam methane reforming (SMR) process consists of the following two steps [57]:

1. *Reforming of natural gas.* The first step of the SMR process involves methane reacting with steam at 750°C–800°C to produce a synthesis gas (*syngas*), a mixture primarily made up of hydrogen and carbon monoxide.
2. *Shift reaction.* In the second step, known as a water–gas shift (WGS) reaction, the carbon monoxide produced in the first reaction is reacted with steam over a catalyst to form hydrogen and carbon dioxide. This process occurs in two stages: a high temperature shift (HTS) at 350°C and a low temperature shift (LTS) at 190°C–210°C.

Hydrogen produced by the SMR process contains small quantities of carbon monoxide, carbon dioxide, and hydrogen sulfide as impurities. The primary steps for purification include

1. *Feedstock purification.* This process removes poisons, including sulfur and chloride, to increase the life of the downstream steam-reforming process and that of the other catalysts.
2. *Product purification.* In a liquid absorption system, carbon dioxide is removed. The product gas undergoes a methanation step to remove the residual traces of carbon oxides. Recent SMR plants use a pressure swing absorption (PSA) unit instead, producing 99.99% pure hydrogen.

Steam-reforming reactions
Methane:

$$CH_4 + H_2O(+heat) \rightarrow CO + 3H_2 \tag{V}$$

Propane:

$$C_3H_8 + 3H_2O(+heat) \rightarrow 3CO + 7H_2 \tag{VI}$$

Ethanol:

$$C_2H_5OH + H_2O(+heat) \rightarrow 2CO + 4H_2 \tag{VII}$$

Gasoline
(using isooctane and toluene as example compounds from the hundred or more compounds present in gasoline):

$$C_8H_{18} + 8H_2O(+heat) \rightarrow 8CO + 17H_2 \tag{VIII}$$

$$C_7H_8 + 7H_2O(+heat) \rightarrow 7CO + 11H_2 \tag{IX}$$

WGS reaction

$$CO + H_2O \rightarrow CO_2 + H_2(+\text{small amount of heat}) \tag{X}$$

23.1.3.1 Partial Oxidation

In a partial oxidation process, methane and other hydrocarbons in natural gas are combined with a limited amount of oxygen (typically, from air) that is not enough to completely oxidize the hydrocarbons to carbon dioxide and water. With less than the stoichiometric amount of oxygen available for the reaction, the reaction products contain primarily hydrogen and carbon monoxide (and nitrogen, if the reaction is carried out with air rather than pure oxygen) and a relatively small amount of carbon dioxide and other compounds. Subsequently, in the WGS reaction, the carbon monoxide reacts with water to form carbon dioxide and more hydrogen.

The partial oxidation is an exothermic process that gives off more heat. The process is, typically, much faster than steam reforming and requires a smaller reactor vessel. As can be seen from the chemical reactions of the partial oxidation (below), this process initially produces less hydrogen per unit of the input fuel than is obtained by steam reforming of the same fuel.

Partial oxidation reactions
Methane:

$$CH_4 + \tfrac{1}{2}O_2 \rightarrow CO + 2H_2(+\text{heat}) \tag{XI}$$

Propane:

$$C_3H_8 + 1\tfrac{1}{2}O_2 \rightarrow 3CO + 4H_2(+\text{heat}) \tag{XII}$$

Ethanol:

$$C_2H_5OH + \tfrac{1}{2}O_2 \rightarrow 2CO + 3H_2(+\text{heat}) \tag{XIII}$$

Gasoline
(using isooctane and toluene as example compounds from the hundred or more compounds present in gasoline):

$$C_8H_{18} + 4O_2 \rightarrow 8CO + 9H_2(+\text{heat}) \tag{XIV}$$

$$C_7H_8 + 3\tfrac{1}{2}O_2 \rightarrow 7CO + 4H_2(+\text{heat}) \tag{XV}$$

WGS reaction

$$CO + H_2O \rightarrow CO_2 + H_2(+\text{small amount of heat}) \tag{X}$$

High- to ultrahigh-purity hydrogen may be needed for the durable and efficient operation of fuel cells. Impurities, such as those stated above, cause various problems in the current and potential efficiency of the fuel cell, including catalyst poisoning and membrane failure. As such, additional process steps may be required to purify the hydrogen to meet industry quality standards. Additional steps could also be needed if carbon capture and sequestration technologies are developed and utilized as part of this method of hydrogen production.

Natural gas represents one of the most viable pathways for introducing hydrogen as an energy carrier for the hydrogen energy economy because it is among the least expensive feedstocks for producing hydrogen. However, carbon capture and sequestration are needed to eliminate the high level of greenhouse gas emissions associated with natural gas.

The reforming reactions require the input of water and heat. Overall, the reformer thermal efficiency is calculated as the latent heat of vaporization (LHV) of the product hydrogen divided by the LHV of the total input fuel. This thermal efficiency depends on the efficiencies of the individual processes, the effectiveness to which heat can be transferred from one process to another, and the amount of energy that can be recovered through means such as turbochargers. In the end, high-temperature reformer efficiencies are approximately 65% and low-temperature methanol reformers can achieve 70%–75%.

The advantages of reforming fossil fuels are due to the use of existing fuel infrastructures that reduce the requirement of the transportation and storage of hydrogen. Moreover, we do not require large amounts of energy as in electrolysis, and this process is less expensive than other hydrogen production methods.

On the other hand, the disadvantages of reformers are also important since they can have relatively long warm-up times but the electrolysis reaches instantly the steady regime. It is difficult to apply them to vehicle engines because of their irregular demands for power (transient response). Also, they develop complex species in composition that require a chain of expensive devices. The most important disadvantage is that it introduces more losses into the energy conversion process, especially in those that have a small thermal mass. From the environmental point of view, it uses nonrenewable and modified polluting fossil fuels. In this sense, they generate chemical and noise pollutions. The pollution generated by reformers takes three forms:

1. Carbon dioxide emissions
2. Incomplete reactions, leaving carbon monoxide and some of the source fuel in the reformate
3. Production of pollutants through combustion, such as nitrous oxides

Reforming fossil fuels will make sense only if the hydrogen is used directly, as in fuel cell engines. For internal combustion engines, it is always more efficient to use the fossil fuel directly without passing it through a reformer first.

Medium- or large-sized reformers can be installed at fuel cell vehicle-fueling stations. At these scales, the equipment complexity, warm-up time, and transient response are not issues; pollutants can be controlled more effectively; and existing power infrastructures can be used. The facility must store only small amounts of hydrogen and hydrogen transportation is avoided.

Small-sized reformers can be installed in fuel cell vehicles to get rid of the problems connected with fueling, storing, and handling hydrogen directly. In fact, many fuel cell experts think that the true challenge in fuel cell engine design now lies in the development of an efficient, compact, reliable, and highly integrated fuel processor. Other experts think that the use of onboard reformers will never provide a realistic solution due to their size, complexity, and cost. However, the development of new electrocatalysts tolerant to carbon monoxide, sulfur, nitrogen oxides, and anhydrides can shield all these problems.

23.1.4 COAL GASIFICATION

Coal can be transformed into a gaseous mixture of hydrogen, carbon monoxide, carbon dioxide, and other species by applying heat under pressure in the presence of steam and a controlled amount of oxygen (in a unit called a gasifier). The coal is broken apart chemically by the gasifier's heat, steam, and oxygen, setting into motion chemical reactions that produce a *syngas*—a mixture of primarily hydrogen, carbon monoxide, and carbon dioxide. The carbon monoxide reacts (in a separate unit) with water to form carbon dioxide and more hydrogen. Adsorbers or special membranes can separate the hydrogen from this gas stream.

Chemically, coal is a complex and highly variable substance. The carbon and hydrogen in coal may be represented approximately as 0.8 atoms of hydrogen per atom of carbon in bituminous coal. Its gasification reaction may be represented by the (unbalanced) reaction equation:

$$CH_{0.8} + O_2 + H_2O \rightarrow CO + CO_2 + H_2 + \text{other species} \tag{XVI}$$

An advantage of this technology is that carbon dioxide can be separated easily from the *syngas* and captured rather than released into the atmosphere. If carbon dioxide can be successfully sequestered, hydrogen can be produced from coal gasification with near-zero greenhouse gas emissions.

Coal gasification can also be used to produce electricity by routing the *syngas* to a turbine to generate electricity. Coal gasification technology could be used to generate both electricity and hydrogen in one integrated plant operation. Coal gasification technology is most appropriate for large-scale, centralized hydrogen production. This readiness is due to the nature of handling large amounts of coal and the carbon capture and sequestration technologies that must accompany the process.

23.1.5 Biomass Gasification

Many programs and schedules in the United States and Europe consider that biomass gasification could be deployed in the mid-term time frame since it is largely produced from many organic wastes.

Biomass, a renewable organic resource, includes agriculture crop residues, such as corn stove or wheat straw; forest residues; special crops grown specifically for energy use, such as switch grass or willow trees; organic municipal solid waste; and animal wastes. Biomass is converted into a gaseous mixture of hydrogen, carbon monoxide, carbon dioxide, and other compounds by applying heat under pressure in the presence of steam and a controlled amount of oxygen (in a unit called a gasifier). The biomass is broken apart chemically by the gasifier's heat, steam, and oxygen, which sets into motion chemical reactions that produce a *syngas* again. The carbon monoxide then reacts with water to form carbon dioxide and more hydrogen (WGS reaction). Adsorbers or special membranes can separate the hydrogen from this gas stream.

Simplified example reaction

$$C_6H_{12}O_6 + O_2 + H_2O \rightarrow CO + CO_2 + H_2 + \text{other species} \tag{XVII}$$

Note: The above reaction uses glucose as a surrogate for cellulose. Actual biomass has highly variable composition and complexity, with cellulose as one major component.

WGS reaction

$$CO + H_2O \rightarrow CO_2 + H_2(+\text{small amount of heat}) \tag{X}$$

Pyrolysis is the gasification of biomass in the absence of oxygen. In general, biomass does not gasify as easily as coal, and it produces other hydrocarbon compounds in the gas mixture exiting the gasifier; this is especially true when no oxygen is used. As a result, typically, an extra step must be taken to reform these hydrocarbons with a catalyst to yield a clean *syngas*. Then, just as in the gasification process for hydrogen production, a shift reaction step (with steam) converts the carbon monoxide to carbon dioxide. The hydrogen produced is then separated and purified.

Biomass gasification technology is most appropriate for large-scale, centralized hydrogen production, due to the nature of handling large amounts of biomass and the required economy of scale for this type of process. Biomass resources can be converted to ethanol, bio-oils, or other liquid fuels that can be transported at relatively low cost to a refueling station or other points of use and reformed to produce hydrogen. Reforming renewable liquids to hydrogen is very similar to reforming natural gas. The liquid fuel is reacted with steam at high temperatures in the presence of a catalyst to produce a reformate gas composed mostly of hydrogen and carbon monoxide. Additional hydrogen and carbon dioxide are produced after the subsequent reaction of the carbon monoxide (created in the first step) with high-temperature steam in the WGS reaction. Finally, the hydrogen is separated out and purified.

Steam reforming reaction (ethanol)

$$C_2H_5OH + H_2O(+heat) \rightarrow 2CO + 4H_2 \qquad \text{(VII)}$$

Water-gas shift reaction

$$CO + H_2O \rightarrow CO_2 + H_2(+\text{small amount of heat}) \qquad \text{(X)}$$

Biomass-derived liquids, such as ethanol and bio-oils, can be produced at large, central facilities located near the biomass source to take advantage of the economies of scale and reduce the cost of transporting the solid biomass feedstock. The liquids have a high energy density and can be transported with minimal new delivery infrastructure and at relatively low cost to the distributed refueling stations or stationary power sites to be reformed to hydrogen.

23.1.6 THERMOCHEMICAL DECOMPOSITION

High-temperature water splitting (a "thermochemical" process) is a long-term technology in the early stages of development. High-temperature heat (500°C–2000°C) drives a series of chemical reactions that produce hydrogen.

Chemicals used in the process are reused within each cycle, creating a closed loop that consumes only water and produces hydrogen and oxygen. The high-temperature heat needed can be supplied by next-generation nuclear reactors under development (up to about 1000°C) or by using sunlight with solar concentrators (up to about 2000°C). Researchers have identified cycles appropriate to specific temperature ranges and are examining these systems in the laboratory. The more than 200 possible cycles identified have been screened and downselected to about 12 for initial research.

High-temperature water splitting is most suitable for a large-scale and centralized production of hydrogen even though semi-central production from solar-driven cycles might be possible.

23.1.6.1 High-Temperature Water Splitting Using Solar Concentrators

A solar concentrator uses mirrors and a reflective or refractive lens to capture and focus sunlight to produce temperatures up to 2000°C. This high-temperature heat can be used to drive chemical reactions that produce hydrogen.

23.1.6.1.1 Chemical Cycle Example: Zinc/Zinc Oxide Cycle

Zinc oxide powder passes through a reactor heated by a solar concentrator operating at about 1900°C. At this temperature, the zinc oxide dissociates to zinc and oxygen gases. The zinc cools, separates, and reacts with water to form hydrogen gas and solid zinc oxide. The net result is hydrogen and oxygen, produced from water. The hydrogen can be separated and purified. The zinc oxide can be recycled and reused to create more hydrogen through this process:

$$2ZnO + heat \rightarrow 2Zn + O_2 \qquad \text{(XVIII)}$$

$$2Zn + 2H_2O \rightarrow 2ZnO + 2H_2 \qquad \text{(XIX)}$$

23.1.6.2 High-Temperature Water Splitting Using Nuclear Energy

Similar to a solar concentrator, a nuclear reactor produces energy as high-temperature heat that can be used to drive high-temperature thermochemical water splitting cycles. The next-generation nuclear reactors under development could generate temperatures of 800°C–1000°C. We have to emphasize that these temperatures are much lower than those produced by a solar concentrator. A thermochemical process based on nuclear heat would use a different set of chemical reactions to produce hydrogen.

23.1.6.2.1 Chemical Cycle Example: Sulfur–Iodine Cycle

Sulfuric acid, when heated to about 850°C, decomposes to water, oxygen, and sulfur dioxide. The oxygen is removed, the sulfur dioxide and water are cooled, and the sulfur dioxide reacts with water and iodine to form sulfuric acid and hydrogen iodide. The sulfuric acid is separated and removed, and the hydrogen iodide is heated to 300°C, where it breaks down into hydrogen and iodine. The net result is hydrogen and oxygen produced from water. The hydrogen can be separated and purified. The sulfuric acid and iodine are recycled and used to repeat the process:

$$2H_2SO_4 + \text{heat at } 850°C \rightarrow 2H_2O + 2SO_2 + O_2 \tag{XX}$$

$$4H_2O + 2SO_2 + 2I_2 \rightarrow 2H_2SO_4 + 4HI \tag{XXI}$$

$$4HI + \text{heat at } 300°C \rightarrow 2I_2 + 2H_2 \tag{XXII}$$

23.1.7 PHOTOELECTROCHEMICAL WATER SPLITTING

Another promising option for the long term is photoelectrolysis. Here, light shining on a photo-electrochemical cell immersed in water produces bubbles of hydrogen and oxygen. In this process, hydrogen is produced from water using sunlight and specialized semiconductors called photoelectrochemical materials, such as cadmium selenide supported on titanium dioxide. In the photoelectrochemical system, the semiconductor uses light energy to directly dissociate water molecules into hydrogen and oxygen. Different semiconductor materials work at particular wavelengths of light and energies.

Research focuses on finding semiconductors with the correct energies to split water that are also stable when in contact with water. Photoelectrochemical water splitting is in the very early stages of research but offers long-term potential for sustainable hydrogen production with low environmental impact [59].

23.1.8 POSSIBILITIES OF MAKING WATER ELECTROLYSIS AN EFFICIENT METHOD

One of the most important ideas for water electrolysis development is the possibility of using the same technology for fuel cells, but with the reverse reaction. General Electric sought to carry the reverse reaction of the technology device in the Gemini hydrogen–oxygen fuel cell with the Nafion membrane. This possibility offers a new reality. However, the development in the course of spinel-type electrodes such as those of cobalt and nickel, with some inclusions of other metals of low cost, makes the tank-type unipolar alkaline technology still promise.

The precious platinum metal has shown the largest activity for water electrolysis, attaining more than 50% yields. Using electrodes composed of nanometals, it has achieved an efficiency of up to 80% at lower current flow rates (100 mA/cm^2) and approximately 60% efficiency at higher rates (1000 mA/cm^2). Over the next year, it is believed that it will achieve or exceed the target of 75% efficiency at rates beyond 1000 mA/cm^2 through further optimization.

The use of high surface area metallic nanoparticles as liquid and gas diffusion electrodes for water electrolysis by compression and sintering into porous plates seems appropriate [8]. They have an expanded metal surface facing away from the electrolyte for strength and current collection. The nanoparticles have tortuous pathways within them to expose several orders of magnitude of larger surface area to the reacting water and allow the takeoff of the gaseous products. The electrolyte flows through the electrode to sweep away the bubbles as they form. The reference electrode is a zinc wire and the electrolyte is eutectic potassium hydroxide (33% aqueous) since it forms the zinc/zinc oxide system.

A typical electrolysis process diagram is shown in Figure 23.4. Note that different processes will use different parts of equipment. For example, PEM units will not require the potassium hydroxide mixing tank, as no electrolytic solution is needed for these electrolyzers. Another example involves water purification equipment. Water quality requirements differ across electrolyzers. Some units include water purification inside their hydrogen generation unit, while others require an external deionizer or reverse osmosis unit before water is fed to the cell stacks. For systems that do not include a water purifier, one is added in the process flow. A water storage tank may be included to ensure that the process has adequate water in storage, in case the water system is interrupted. Each system has a hydrogen generation unit that integrates the electrolysis stack, gas purification unit and dryer, and heat removal unit. The electrolyte circulation unit is also included in the hydrogen generation unit in alkaline systems. The integrated system is usually enclosed in a container or is installed as a complete package. Oxygen and purified hydrogen are produced from the hydrogen generation unit. If desired, a compressor and hydrogen storage can be added to the system. Although hydrogen storage and compression are included in the process diagram below, for purposes of this analysis, hydrogen storage is not included. It is assumed that as the hydrogen is produced, it is fed directly into a pipeline or truck. In addition, note that there is no oxygen compression and storage.

Deionized or ultrapurified water is essential for water electrolyzers and since it is a closed system device, the impurities such as chlorides, sulfates, and calcium or magnesium have to be fully removed from water to avoid shielding the diaphragms. The hydrogen and oxygen streams have

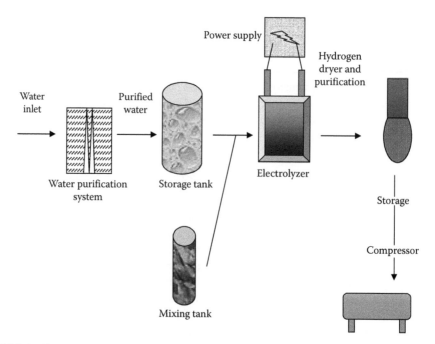

FIGURE 23.4 Simplified process flow diagram for "Hydrogen Global Technology." The extraction, purification, drying, and collection of hydrogen from the electrolyzer are not shown.

to involve very little bubbles; otherwise, they can also shield the diaphragms (bubble overpotentials). Moreover, the gas streams are saturated with water and some amounts of electrolytes. Both have to be eliminated from the stream by successive condensing and retro-osmotic fluxes and then returned to the electrolytic cell.

23.1.8.1 Alkaline Water Electrolyzers

The basics of electrochemistry involved in alkaline electrolysis have been already explained in other reviews and books [3,9,10], but in the case of rather old technologies [11,12]. From all the new concepts, only two have been developed on a commercial level. One of the scopes is the problem of the ohmic drop from the diaphragm and the morphology and composition of the cathode electrode.

Brown Boveri & Cie and Electrolyzer Corporation achieved a main progress in the mid-1980s with respect to the performance of the alkaline monopolar cells. The first company was able to enlarge the efficiency from 2 to 3.4 kA/m^2 with 500 m^3/h [13], on the other hand the second company that was able to achieve only 1.74 V reached 134 mA/cm^2 at 87% efficiency with nickel–boride materials and $NiCo_2O_4$ anodes (prepared from the freeze-dry technique) [14].

In the case of the diaphragm improvements, the Nørsk Hydro Technology produces good operation results with higher loads using highly active electrocatalyst coatings. In this sense, porosities on the order of 5 μm are often used at 1.7 V of applied potentials at 80°C with 3600 A in 25% potassium hydroxide electrolytes. This new diaphragm design achieved one order of magnitude of lower resistances [15] than the former asbestos. Further improvements in order to achieve higher efficiencies have been proposed but working at larger temperatures such as 125°C. The use of nickel/cobalt spinels seems to be not so effective for the counterreaction (oxygen evolution reaction), but it is dependent on the surface roughness factor. These results were extensively studied at the end of the 1980s and beginning of the 1990s [16–19].

In summary, by operating novel electrocatalysts and new materials for the diaphragms at high temperatures, we are able to reduce the power consumption in the hydrogen electrolytic production. This is the main task for achieving the hydrogen fuel economics nowadays. The value of 5.5 kWh of consumption per normal m^3 of hydrogen in the late 1970s has been diminished to 4.2 kWh at present.

Approximately, about 60 kWh will be needed to produce the energy content of a gallon (3.79 L) of gasoline in the form of gaseous hydrogen. In the United States, we can assume electricity costs of 4¢/kWh; however, in other countries these values can increase up to US$ 1/kWh. For the best situation (the first one in the United States), the electricity costs alone would be US$ 20.00/MMBtu, which is equivalent to gasoline costing US$ 2.40 per gal. With conventional electrolyzers that are not in high-volume production, the installed capital costs are in the range of US$ 600/kW, which adds an additional US$ 4.00/MMBtu. This increases the production cost of gaseous hydrogen to about US$ 24.00/MMBtu, which is equivalent to gasoline costing US$ 3.00 per gal. If hydrogen needs to be liquefied, an additional US$ 4.00/MMBtus would be added that would make the production cost of liquid hydrogen about US$ 28.00/MMBtu, which is equivalent to gasoline costing US$ 3.38 per gal [23,24].

In the worst case, that is, for the countries with total thermal conversions of energy, the cost of gaseous hydrogen raises to about US$ 224.00/MMBtu, which is impossible to handle. However, these values are expected since the thermal costs for the countries that do not own oil or carbon will be extremely large. All the other magnitudes reported by the ministries or secretaries of energy in powerful countries are real; however, we believe that new and advanced technologies have been developed with more efficient devices that are fully protected by patents and models that are not in the public domain. Thus, even lower values of cost have to be expected.

23.1.8.2 PEM Electrolyzers

The new advances in the development of novel technologies in PEM electrolyzers are based on the filtration-type configuration that has been conceived in the late 1970s. Before 1975, the PEM electrolyzers were made of three main components: the membrane electrode assembly (MEA), insulator sheet, and a screen-gasket assembly [20]. A flat silicone rubber gasket was put to seal the device; otherwise, the water production on the anode will overflow in the system and the leakage of both gases would be inevitable. A bipolar current collector is placed between the two compartments by a screen package that is in contact with the common metallic separator sheet. In those years, the metallic parts were constructed using titanium and niobium alloys to avoid corrosion problems. The compression of the cell was attained with spring tie rods and adequate gaskets were provided to ensure the cell is sealed. Some of these parts were substituted by carbon supported catalysts and graphite plates [21,22].

The most interesting innovation in these electrolyzers is the use of a very thin membrane that acts as a diaphragm and catalyst support. Dupont's Nafion is the most reliable and efficient proton-conducting polymer with uses in the industry. Nafion is the copolymer of tetrafluoroethylene and perfluor vinyl ether of 4-methyl-3,6-dioxa-7-octene-1-sulfonyl fluoride that is prepared by multiple side chains by patented methods [39]. The long-lasting conditions of work, $\sim$60,000 h, and the high functionality of the pendant sulfonyl fluoride groups (converted to sulfonic superacids) make it the right choice for any industrial aqueous electrochemical process. However, the copolymer's acid capacity is related to the number of comonomers used in the copolymerization (from 0.67 to 1.25 meq/g). It is well known that the final process involves the conversion from SO_2F to SO_3K with aqueous potassium hydroxide in DMSO, followed by acid exchange with nitric acid to SO_3H [40]. The hydration of the membrane produces the weakening of the SO_3-H group, and thus H^+ ions are able to move. The complex micro-phase morphology depicts different phases, hydrated regions (that have to be as thick as possible), tetrafluoroethylene hydrophobic zones, and diluted and strong acidic hydrophilic parts. The latter exhibits large conductivities as 0.1 S/cm under a complete hydration. As pointed out in Ref. [41] for a thickness of 175 μm, the resistivity is 10 Ω cm, so for 0.15 V, the current density can be as high as 0.75 A/cm^2. Also, the large oxygen and hydrogen diffusion coefficient and solubility are from 20 to 30 times over that obtained in a phosphoric acid fuel cell (PAFC). Mechanical properties are also attractive, since elongation of up to 200% is possible with a tensile strength coefficient of up to 40 MPa/cm^2 for 50% relative humidity at 23°C. Nafion permeability for both oxygen and hydrogen is very low, only 1%, which implies 1–10 mA/cm^2 of current loss. Nafion membranes have very few limitations, and some of these have been overcome. For example, proton mobility only occurs under complete hydration, but the incoming extra water with gases has been substituted with a new Nafion 112 (50 μm thick). However, at temperatures well above 80°C, the membrane does not work very well, so the problem of the incoming carbon monoxide (when using hydrogen form the reformate) can be a problem at low temperatures. Dow Chemical and Asian Chemical Co. have made some improvements with new membranes achieving a higher ratio between SO_3H and CF_2 groups [42]. The lower molecular weight produces a lower resistivity and then higher currents. In the case of Nafion 115 ($\sim$100 μm), it is possible to reach 1 A/cm^2 at 0.75 V. The problem is that these thinner devices produce more problems in the gas or methanol crossover by reducing the open-circuit potential. Many research groups such as General Motors, Plug Power, and DuPont have supplied new solutions and developed new membranes with polymer dispersions and coversheets under nitrogen atmospheres to protect from the environment [43–45].

23.1.9 Novel Aspects of the Electrolytic Production of Hydrogen

There are several problems associated with hydrogen production; however, some of them are really a side product associated inconvenience, such as oxygen formation in the course of electrolysis. The use of new anode depolarizers seems to be one of the solutions to avoid the large overpotentials associated with the anodic process. All these anodes are patent protected, but some of them are not new material, but only a product of a side process with lower activation energy. The Westinghouse thermo(electro)chemical hybrid process for hydrogen production is an example [25] with 45% of efficiency with operation potentials as low as 0.6 V. In this case, the anodic reaction is produced from the electrooxidation of sulfurous acid to sulfuric acid instead of oxygen production, since the reversible standard potential is only 0.17 V instead of the large oxygen/water acid electrochemical reaction. The problem here is the large corrosion by sulfuric acid and contamination by the adsorbed sulfur that can be observed after long operation time. The stability of the diaphragms is a problem since sulfurous acid tends to migrate to the cathode depolarizing the surface with the formation of elemental sulfur, that is, acting like a blocking adsorbate. However, the cost of the reactive has to be considered as well in the evaluation of the global economics of the entire process since it is used under saturation conditions. The other problem is of avoiding high operation temperatures since both reactant and product usually decompose at temperatures above 80°C. Another proposed depolarizer is the carbonaceous coal formation from cheap products such as cellulose, where the electric energy input can be reduced by 30%–50% [26].

The reduction of the activation energy in the electrochemical oxidation of water is important in relatively high-temperature operation conditions. Thus, working at 1000°C, the energy requirement considering the conversion of water to hydrogen and oxygen is 20% lower compared to working at room temperature. Also, it is interesting to know that the anodic overpotentials for the oxygen evolution reaction are largely diminished at high temperatures due to the smaller diameters of the bubbles and the higher diffusion coefficients involved in the mass transfer processes. A relationship between parameters has to be established for taking into account the changes in the electrode potentials when working at large temperatures. Hence, a relation between enthalpy changes in the reaction with respect to the molar electric charge involved in the process, that is, a thermoneutral potential, E_{tn}, below the decomposition electrolysis potential value (far above 1.23 V, that is, nearly 1.5 V), where the cell absorbs heat from the ambient and acts as a refrigerator, is defined. At potentials larger than this value, the emission of heat is important and must be removed for an isothermal operation, since E_{tn} is defined as

$$E_{tn} = \frac{\Delta H^\circ}{nF} + \frac{1}{nF} \int_{T_o}^{T} \bar{c}_p(H_2O)dT \tag{23.1}$$

where $\bar{c}_p(H_2O)$ is the mean heat capacity of water between the feed water temperature, T_o, and the operating cell temperature, T. According to [27,28], the expression above is

$$E_{tn} = 1.5475 - 3.23121 \times 10^{-4}T + 4.820 \times 10^{-7}T^2 - 4.820 \times 10^{-10}T^3 \tag{23.2}$$

This value has to be larger than 1.5 V since some heat is required to feed water at the operating temperature and some more to evaporate water carried off out of the cell with the formed gases. More heat losses usually occur in the cell and the peripheral devices and need to be compensated.

Considering the above concepts and the previous knowledge obtained from the solid oxide fuel cells (SOFC), the idea of conducting vapor water electrolysis was well thought-out as a promising possibility in the 1980s. However, the first intent was conducted by means of a carbon dioxide and water reduction to oxygen, carbon, and hydrogen [29]. The catalysts employed then are employed even today, Pt–ZrO_2 (7:3) with ZrO_2: Y_2O_3 as the solid electrolyte. The latter is currently used to make YVO_4-europium and Y_2O_3 europium–phosphorus that give the red color in *TV* picture cathodic tubes. Yttrium oxide is also used to make yttrium iron garnets that are very effective microwave filters. The first report about the use of these devices was by the GE Company [30] and also by Brown Boveri & CIE [31]. The latter designed the same electrolyte as GE but a nickel cathode and a tin-doped In_2O_3 on a $LaMnO_3$/$LaNiO_3$ perovskite was used. On the other hand, it is important to state that the ohmic drop due to large anodic to cathodic gaps and solid electrolytes has been overcome by Westinghouse using thin layer cells [32]. The use of 20 mm multiple interphases of cathode/electrolyte/anode connections that are chemically vapor deposited on a porous substrate cylindrical tube was a great improvement. The multilayer composition of the total cell is nowadays largely used since it avoids ohmic drops and mass transfer problems.

In consideration of the photoelectrochemical processes that have been fully developed in the late 1970s, the pioneering work by Fujishima and Honda [33] leads to a new possibility of making a TiO_2 substrate as an anode and a platinum cathode with a high-yield hydrogen and oxygen production system. The excitation by photons of energies given by a large intensity value must be able to exceed the energy gap, E_{gap}, to separate holes and electrons. The holes go to the anodes, while the cathode is fed with the electrons. The quantum efficiency is only 2% for pure TiO_2 single crystals for $E_{gap} = 2$ eV. These efficiencies can be increased to 10% by using $SrTiO_3$ as anodes with 1.8 eV of band gap [34].

However, it seems that the main advantage in the electrolytic production of hydrogen is the cleanness of the process, but the main problem is the high cost of the electrodes and diaphragms. For the various technologies that were presented above, the cathode, anode, and electrolytic separators require further advances. In the case of the main cathodes that substitute platinum-type catalysts, the use of nickel alloys seems to be the right option. Thus, the value of the exchange current density, j_o, of the process in nickel–carbon-supported catalysts is 2×10^{-4} A/cm^2 at 80°C and 4×10^{-2} A/cm^2 at 264°C with a symmetry factor, $\beta = 1/2$ in 50% potassium hydroxide [27]. However, the problem of using nickel-based spinels for anodes seems to be largely solved. The use of porous carbon surfaces produces a blockage of the high current density points due to bubble blocking. Sintered nickel with anodized NiOOH mixed oxides with large area $NiCo_2O_4$ RuO_2 spinels gave the best results [35]. The value of j_o is 2.5×10^{-6} A/cm^2 in 30% of potassium hydroxide at 80°C.

The use of woven asbestos in alkaline electrolyzers gave one of the best performances for temperatures below 100°C. However, the increase in the performance of both anodes and cathodes at higher temperatures leads to the search for novel membranes to be used in electrolyzers. One of the most attractive membranes is Nafion in 20% of sodium hydroxide at 120°C–160°C, but it works better in acid media [22,36–38].

23.1.10 New Projects for the Production of Clean and Pure Hydrogen

Center for Environmental Research and Technology (CE-CERT) collaborated with The Electrolyzer Corporation (TEC) on the design, construction, and analysis of the next-generation photovoltaic hydrogen production and vehicle refueling station at the Xerox El Segundo facility. The Xerox Solar Hydrogen Project is similar to the experimental Solar Hydrogen Research Facility (SHRF) at UC Riverside, but it provided the opportunity to investigate a larger-scale production facility with line-focusing photovoltaic panels, and the operation of a hydrogen compressor on solar power. The facility first produced hydrogen in December 1995. CE-CERT assisted in selecting the data acquisition system, hydrogen compressor, inverter/converter system, post compression dryer, and water purification system [46].

Schmidt designed the first industrial bipolar electrolyzer around 1900. However, Oerlikon commercialized the modified version, the Oerlikon–Schmidt electrolyzer. In 1967, Brown Boveri joined the Maschinenfabrik Oerlikon, preserving its name. Brown Boveri & Cie was founded in Baden, Switzerland in 1891 by Charles Eugene Lancelot Brown and Walter Boveri who worked at the Maschinenfabrik Oerlikon. In 1970, BBC took over the Maschinenfabrik Oerlikon. In 1988, it merged with ASEA to form ABB. Some novel incursions about the use of more efficient anodes and cathodes have been advanced in the literature. However, some of them are patented in [47]. A porous graphite plate is coated on one face with a layer of TiO_2 doped with a mixture of RuO_2 and IrO_2. The uncoated surface of the porous graphite plate has to be grooved.

On February 5, 1897, the Metallurgische Gesellschaft was founded under Metallgesellschaft AG, who now operates under the name of Gea Group AG. Its middle letters (lurgi) were used as the cable address for the newly founded company. The word Lurgi became popular in all countries and a synonym for top-quality technological performance. Lurgi Gmbh Technology put its first pressure electrolyzer to produce hydrogen using the Zdansky–Lonza cell [48]. The main difference is that it works at high pressure, that is, 450 psig, reducing costs and the size of the entire equipment, leading to a lower electrodecomposition potential value. The rest of the peripheries are similar to the Brown Boveri & CIE technology. The old technology used asbestos diaphragms with nickel-plated steel plates as electric contacts allowing hydrogen and oxygen bubbles to rise behind the electrodes to avoid undesired overpotentials. The largest units that had a current capacity of 6000 A of 350 unit cell modules were able to produce 800 m^3 of hydrogen per hour. The current efficiency is near 98.75% at a cell potential for the electrodecomposition of 1.86 V in 25% of potassium hydroxide at 90°C.

Oronzio De Nora established De Nora in 1923. De Nora is recognized worldwide as a leading supplier of technologies for the production of chlorine, caustic soda, and derivatives, as well as the largest worldwide supplier of noble metal-coated electrodes for the chlor-alkali industry and for the electrochemical industry in general. Energy saving and environmental protection are the group's distinguishing technologies. De Nora S.P.A. technology has built the most attractive water electrolyzer of the filter press type that consists of a number of bipolar plates installed in India with a capacity of 30,000 m^3 of hydrogen per hour. Similarly to the other systems, the anode and cathode have a nickel-based steel substrate that is activated by the sulfide method [49]. The use of activation processes induced by chemicals has been replaced by electric inductions to avoid large periods under strong sulfide-containing solutions. These processes also produce the formation of porous surfaces that are able to facilitate the withdrawal of gases by avoiding large overpotentials. In this technology, a double woven asbestos separator is also used. The construction of these devices can reach currents as high as 12,000 A with 40–100 unit cells.

De Nora is the manufacturer of 100% of the De Nora Tech electrolyzer anode and cathode elements and coatings and is the only company approved as the coating supplier by original equipment suppliers. Three recoating techniques best fit the needs of individual plants:

1. *Direct*. Takes advantage of present coating activity and loading to minimize cost and structure wear
2. *Full*. Complete coating removal and reapplication techniques required when present coating has reached its end life
3. *Remesh*. Enables the recovery of structure value when the mesh has become thin

All recoating options include a full leak check and back pan nickel replacement. Spares can be made available on a lease basis to minimize capital investment while eliminating production losses. Anodes are inspected and, upon customer approval, repaired by using De Nora's proprietary techniques with different anode coatings available for any potential or oxygen evolution requirement. Each coating has different characteristics in resistance to shutdowns and iron, hydrogen overpotential, and price that can be tailored to specific plant needs.

The problem of ohmic drops by diaphragms has been studied for a long time. A laboratory scale diaphragm-less water electrolyzer was developed for hydrogen production at large pressures of up to 140 kPa by electrolysis in an alkaline solution. Porous electrodes with a nickel catalyst and a copper cover layer serve as cathodes, whereas nickel sheets are used as anodes. Modular construction of the electrolyzer permits simple combination of its cells into larger units. Thus, up to 20 cells with disk-shaped electrodes of 7 cm in diameter were connected in series and provided with electrolyte manifolds, automatic pressure, and electrolyte level control devices. The dimensions of the electrolyte manifolds were optimized based on the calculations of parasitic currents [50].

23.1.11 Energy Losses within a Hydrogen Economy

The energy losses in the optimized and efficient electric grids for hydrogen-based energy are less than those from other sources. For modest distribution distances, about 90% of the generated electricity is accessible as free electrical energy. Energy losses at all important stages of the hydrogen economy have recently been analyzed elsewhere [51].

The problem of hydrogen production is only one part of the so-called hydrogen economy, since it involves more stages than the two obvious conversion processes of the electrolyzer and the fuel cell. Even before water is converted into hydrogen and oxygen, high-voltage AC external power has to be transformed to low-voltage DC current, and water has to be filtered, demineralized, distilled, pumped, and stored. The production of hydrogen either from reforming or by electrolysis involves significant energy losses. The generated hydrogen requires that it is packed by the compression, liquefaction, or electrochemical process (metal hydrides) to make it transportable. Then, the synthetic chemical energy carrier has to be transported to the final user by road, rail, ship, or even pipelines. Hydrogen may also be lost by leakages. Because of the physical properties of hydrogen, all these stages require much more energy than is needed for the distribution of liquid fuels to consumers. Finally, the reconversion of hydrogen to DC electricity with fuel cells and the subsequent DC/AC conversions are associated with heavy energy losses. However, some devices such as computers, mobile phones, or calculators do not require a DC/AC conversion. The main losses reflect the physics of hydrogen; only consumers with efficient hydrogen fuel cells can recover a small fraction of the original renewable electricity.

The energy losses or the parasitic energy consumption is presented in percent of the higher heating value (HHV) potential of the delivered hydrogen. It is calculated from the ratio of the enthalpy changes and the charge per mol of the final electrochemical process. The heat of formation of liquid water or the energy released when water is formed in the chemical reaction is 39 kWh/kg of hydrogen, known as the HHV of hydrogen. On the other hand, the heat of formation of steam is 33 kWh/kg of hydrogen and is the lower heating value (LHV) of hydrogen.

The heat losses associated with the electrolysis system can be evaluated comparing the HHV potential with the operational cell potential. The ratio between these two parameters, HHV potential and operation cell potential, multiplied by the electric charge is the energy efficiency in the process. The latter can be improved by working at larger temperatures since the thermodynamic reversible value is lower under these conditions [27]. In the case of energy losses during the electrolyzer operation, the internal ohmic losses are proportional to the current intensities. For current densities between 1 and 2 A/cm^2, power losses in the range of 30% are observed. Higher efficiencies can be obtained for lower current densities and lower potential outputs.

Because of its low specific gravity and weight, the compression of hydrogen requires 8 times more energy than the compression of natural gas and 15 times more than the compression of air. Multistage compressors with intercoolers are required to compact hydrogen close to the ideal isothermal limit. Depending on the final pressure, the compression energy needed amounts to 8%–15% of the energy contained in the fuel itself. On the other hand, when we desire to liquefy hydrogen, much more energy is required. The Linde–Hampsen thermodynamic cycle is usually applied. Thus, low temperatures of about 20 K are reached by multistage counterflow expansion precooled by using liquid nitrogen to temperatures below its Joule–Thomson (JT) coefficient. Hydrogen liquefaction will always remain an energy-intensive process. A 40 ton hydrogen carrier implies about 350 kg of hydrogen gas at 200 bar (3500 psi) or 3500 kg of liquid hydrogen at cryogenic 20 K. It takes 22 tube trailers (200 bar) or 4.5 liquid hydrogen trucks to transport the energy contained in a single gasoline tanker of the same gross weight. Pipeline transport of hydrogen requires much more energy than for natural gas. The gas flow gradually decreases as hydrogen is consumed to energize the pumps. Consequently, the lines are exponentially curved due to evaporations. These rather disturbing numbers can be overcome by the storage of hydrogen as metal hydrides. The storage of hydrogen as metal hydrides was first thought to be more expensive since the energy required for the process was much higher than in the other methods. However, the metal hydrides are long-life devices in comparison with the other storage containers.

The use of hydrogen requires the local availability of electricity and water. To serve 1000 vehicles per day, continuous electric power of at least 30 MW is needed at a water consumption

TABLE 23.1
Energy Losses in the Production and Storage of Hydrogen

Method or Technology	Energy Loss (%)
Hydrogen by electrolysis	25
Compression to 200 bar	8
Compression to 800 bar	13
Liquefaction in small plants	50
Liquefaction in large plants	30
Chemical hydrides	60
200 km road delivery (diesel) at 200 bar	13
200 km road delivery (diesel) as liquid	3
2000 km pipeline	20
Onsite generation (electrolysis)	50
Transfer from 100 to 700 bar tank	8
Reconversion to electricity by fuel cells	50

Source: Bossel, U., *European Fuel Cell Forum International Conference,
Intelec '05*, Berlin, Germany, September 18–22, 2005.

of 110 m^3 per day. There are many sites where these conditions cannot be met. For the ongoing hydrogen debate, the following results are presented for a selection of representative operational parameters.

The initial publication of these numbers has caused irritation among hydrogen promoters. However, this analysis has been vetted by renowned institutions and has been proven good. The only difference is that the values in Table 23.1 compare a future hydrogen economy with a future electron economy while all critics compare the future hydrogen economy with the present fossil fuel situation. In this study, the reduction of greenhouse gas emissions cannot be addressed, as carbon is not involved in either of the two likely alternatives. The tabulated numbers have been used for the illustration of the loss cascade presentation for the energy transport by electrons and hydrogen.

Bossel [52] compares the use of electrons with the use of hydrogen as energy carriers. Only about 25% of the original energy becomes useful for the final use of hydrogen. Only 20%–25% of the source energy can be functional when hydrogen is employed in a transport medium. This is certainly in conflict with the efficiency order of a sustainable energy device. However, the present use of hydrocarbons is more inefficient. Bossel proposes the comparison of the electricity (pure electronic conduction) option to the hydrogen economy to make a reliable association between them. Any comparison reveals that high efficiency cannot be obtained with a chemical energy carrier, even with hydrogen. For a sustainable energy future, hydrogen has to compete with electrons. Hydrogen can never compete with its own energy source, that is, electricity. The comparison of electricity with the use of hydrogen as fuel is more complex and difficult to explain. In a gas-tight system, the hydrogen flow remains unchanged between the electrolyzer and the fuel cell, unless some hydrogen is consumed for gas transport or for energizing compressors in the pipeline systems. Between the connections from the electrolyzer to the fuel cell, parasitic energy and, in particular, electricity must be supplied to power storage plants, compressors, pumps, fuel delivery trucks, etc. Depending on the chosen distribution technology, the parasitic power requirements may amount to 50% or more of the energy content in HHV of the hydrogen delivered to the fuel cells at the end of the line. For this, the electric power grid has to be extended to provide energy for the delivery of hydrogen to the consumers.

It seems that the main problem in the hydrogen production is the energy loss in all the electrolyzers, connectors, peripherals, and the like; however, the large overpotentials involved in the oxygen electroreduction reaction have to be specially considered. The formation of molecular oxygen implies working under a voltammetric region where the platinum oxide formation is the determinant. Therefore, the anodic water discharge produces $[Pt–OH]_{ads}$ and $[Pt–O]_{ads}$ as explained in Chapter 2. At potentials larger than 1.6 V, the surface is completely covered by oxygen so two $Pt = O$ adjacent species, by a Tafel chemical step, produce molecular oxygen and free platinum sites. The latter are really fully covered by the adsorbed water at the platinum surface and are immediately discharged to new $Pt = O$ species [53]. One of the challenges in this technology is to find non-platinum-containing electrocatalysts for the reaction. However, it seems that platinum alloys are the best substrates for this anodic reaction. However, thermally treated macrocyclic compounds, such as metalloporphyrins of cobalt and nickel [54] are currently used in low ionic force media. The problem of working in perfluorosulfonic acid electrolytes is the large corrosion rate and a huge decrease in the faradaic efficiencies. With respect to the counterreaction, which is very similar in the operation mechanism, the use of $NiCoO_2$, NiO, and Co_2O_3 in a spinel configuration produces larger activities, compared with other chalcogenides [55].

23.1.11.1 Technical and Economic Overview of the Electrolytic Hydrogen Production Systems

One of the most interesting reports about economical analysis is that of Ivy [56] that relies on five companies' electrolysis devices: *Stuart IMET*, *Teledyne HM and EC*, *Proton HOGEN*, *Norsk Hydro HPE* and *Atmospheric*, and *Avalence Hydrofiller*. At the end of 2003, the largest electrolyzer unit sold produced only 380,000 kg hydrogen/year. There are two limitations for electrolyzers of this size; if we have to fuel approximately 1900 cars, 2.3 MW of electricity will be required. This electricity demand would likely prevent the purchase of cheaper industrial electricity in the actual scenario, thus raising the price of hydrogen. For example, a 500,000 kg/day hydrogen generation plant using nuclear power and electrolysis would require 500 of the largest electrolyzer devices available today. In this situation, from 10 to 100 times the size of the actual electrolyzers could be utilized.

An initial cost analysis was completed in [56] to determine the effects of the electricity price on hydrogen costs. For each electrolyzer, the specific system energy requirement was used to determine how much electricity is needed to produce hydrogen. At current electrolyzer efficiencies, in order to produce hydrogen at lower values than US$ 3.00/kg, electricity costs must be lower than 4¢/kWh. In a developing country without oil, such as Uruguay, the electricity costs from thermal devices are nearly US$ 1/kWh, so the inferences have to be very different.

For an ideal system operating at 100% efficiency, electricity costs must be less than 7.5 ¢/kWh to produce hydrogen at a price lower than US$ 3.00/kg. This analysis demonstrates that regardless of any additional cost elements, electricity costs will be a major price contributor. This is only possible in the United States or Europe.

The detailed economic analyses were based on three distinct systems for which cost and economic data were available. These data may or may not be representative of the costs and systems within each category. These three systems represent a small neighborhood (~20 kg/day), a small public space (~100 kg/day), and a public space size (~1000 kg/day). In this analysis, the hydrogen selling prices for the United States or Europe were US$ 19.0/kg of H_2 for the small neighborhood size, US$ 8.0/kg of H_2 for the small public space size, and US$ 4.0/kg of H_2 for the open space size.

For the forecourt case, electricity represents 58% of the cost of the hydrogen, and the capital costs represent only 32%. For the small forecourt case, the electricity contribution drops to 35% while the capital costs become the major cost factor at 55%. In the neighborhood case, the capital costs increase to 73%, but electricity costs are significant at 17%. This analysis demonstrates that for all systems, the electricity price is a contributor to the hydrogen price, but for small-sized electrolyzers, capital costs are more significant.

The available electrolysis systems were categorized into five different ranges: home, small neighborhood, neighborhood, small public space, and forecourt. The term forecourt refers to a refueling station. The number of cars served and the hydrogen production rate for each size are as follows:

- Home (single domestic) size will provide the fuel needs of 1–5 cars with a hydrogen production rate of 200–1000 kg/year.
- Small neighborhood size will serve the fuel needs of 5–50 cars with a production rate of 1,000–10,000 kg/year.
- Neighborhood size will supply the fuel needs of 50–150 cars with a production rate of 10,000–30,000 kg/year.
- Small public space size, which could be a single hydrogen pump at an existing station, will supply 150–500 cars with a hydrogen production rate of 30,000–100,000 kg/year.
- A full hydrogen forecourt will serve more then 500 cars per year with a production rate of greater than 100,000 kg hydrogen/year (Table 23.2).

23.1.11.2 Results of Energy and Cost Analysis

Table 23.3 details the energy required for hydrogen production by each manufacturer's largest hydrogen generation system. Note that only Stuart and Nørsk Hydro supply the necessary energy for the electrolyzer. Stuart also provided the energy requirement of the entire system, while Nørsk Hydro's system energy need was evaluated by using the power needs of the system and the hydrogen generation rate. Proton provides energy requirement data based on the entire hydrogen production system. Only Nørsk Hydro's system energy requirements include compression, one water-injected screw compressor followed by a reciprocating compressor to bring the gas to 33 bar.

The conversion efficiency of water to hydrogen is shown in Table 23.4. Overall, the conversion efficiency is high, ranging from 80% to 95%.

The energy efficiency is defined as the HHV of hydrogen divided by the energy consumed by the electrolysis system per kilogram of hydrogen produced (Table 23.5). The energy efficiency ranges from 56% to 73%. Proton's PEM process has the lowest efficiency at 56% and both Stuart's and

TABLE 23.2
Hydrogen Station Size

Manufacturer Model	Hydrogen Production Rate (kg/Year)	Number of Fuelling Cars	Category of Station Size
Avalence Hydrofiller 15	315	1.6	Category I
Proton HOGEN 40	789	3.9	Category I
Avalence Hydrofiller 50	1,182	6	Category II
Teledyne HM-100	4,410	22	Category II
Teledyne HM-200	8,820	44	Category II
Norsk HPE 12	9,450	47	Category II
Nørsk HPE 16	12,600	63	Category III
Nørsk HPE 30	23,622	118	Category III
Teledyne EC-600	26,457	132	Category III
Nørsk HPE 40	31,494	157	Category IV
Nørsk HPE 60	47,241	236	Category IV
Stuart IMET 1000 3 stacks	35,433	177	Category IV
Stuart IMET 1000 4 stacks	70,863	354	Category IV
Nørsk atmospheric type 5020 4 kA *DC*	118,104	591	Category V
Nørsk atmospheric type 5040 5.15 kA *DC*	381,864	1909	Category V

TABLE 23.3
Energy Balance

Manufacturer Model	Energy Required System (kWh/N-m³)	Energy Required Electrolyzer (kWh/N-m³)	Hydrogen Production (N-m³/h)	System Power Requirement (kW)
Stuart IMET 1000 3 stacks	4.8	4.2	60	288
Nørsk atmospheric type 5040 5.15 kA *DC*	4.8	4.3	485	2330
Proton: HOGEN 380	6.3	—	10	63

TABLE 23.4
Conversion Efficiency

Manufacturer Model	Reactants (kg/h)	Hydrogen Production (kg/h)	Oxygen Production (kg/h)	Conversion Efficiency (%)
Stuart IMET 1000 3	60	5.4	43	80
Nørsk atmospheric type 5040 5.15 kA *DC*	485	434	347	80
Proton: HOGEN 380	8.4	0.9	7.1	95

TABLE 23.5
Energy Efficiency

Manufacturer Model	Energy Required (kWh/kg)	HHV for Hydrogen Production (kWh/kg)	System Efficiency (%)	Hydrogen Production Pressure (psig)
Stuart IMET 1000 3	53.4	39	73	360
Nørsk atmospheric type 5040 5.15 kA *DC*	53.5	39	73	435
Proton: HOGEN 380	70.1	39	56	200
Avalence Hydrofiller 175	60.5	39	64	10,000

Nørsk Hydro's bipolar alkaline efficiencies have the highest (73%). However, this percentage includes compression of the hydrogen gas to 6000 psi. These efficiencies would decrease if additional compression up to 6000 psig were included. Only Avalence's energy requirement of 60.5 kWh/kg includes hydrogen pressures reaching up to the 6000 psig range.

The efficiency of the electrolysis process—the theoretical amount of energy needed—39 kWh/kg of hydrogen, needs to be divided by the actual amount of energy used by the electrolysis unit to create hydrogen. Note that in this study, the energy requirement of the entire electrolysis system is used to calculate the efficiency, not just the efficiency of the electrolyzer. For example, the electrolyzer alone for the Stuart IMET 1000 requires 46.8 kWh/kg that corresponds to 83% efficiency when you divide the HHV of hydrogen by the electrolyzer power requirement. However, when you include the rectifier and auxiliaries, the energy requirement becomes 53.5 kWh/kg or 73% efficient for the system. An initial and approximate analysis was performed in [56] to find the effects of electricity price on hydrogen costs. For each electrolyzer, the specific system energy requirement is used to determine how much electricity is needed to produce hydrogen; no capital, operating, or maintenance costs are included in the calculation. The system energy requirement used is the lowest energy requirement reported among all manufacturers.

As mentioned above, for the electrolyzer to be efficient and to produce hydrogen at $3.00/kg, electricity costs must be between 4 and 5.5¢/kWh. In order to produce hydrogen for less than $3.00/kg with a system that is 100% efficient, electricity prices must be less than 7.5¢/kWh. The U.S. Department of Energy reports the year 2002 industrial, commercial, and residential electricity prices at 4.83, 7.89, and 8.45¢/kWh, respectively. Therefore, if only electricity costs were incurred, electrolyzers could fabricate hydrogen for $3.00/kg at less than the industrial electricity prices. An ideal system could produce hydrogen for $3.00/kg at slightly lower than commercial prices.

Example: Conversion efficiency of 1 kg hydrogen to electricity

The conversion efficiency of 1 kg hydrogen to electricity is an important example. Figure 23.5 summarizes the literature findings on energy input, output, and efficiencies in gaseous, liquid, and

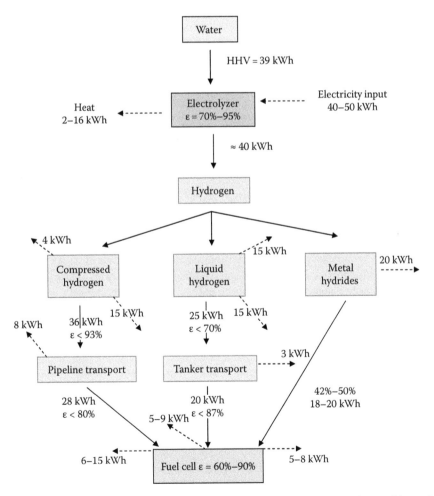

FIGURE 23.5 Conversion efficiency of 1 kg of hydrogen to electricity. The energy losses either as heat or as total energy is indicated in gray-colored writing with dashed arrows. More details are given in the text.

solid hydrogen cradle-to-grave systems for 1 kg of hydrogen assuming the 39 kWh HHV found in the literature. Energy and efficiency ranges reflect the varying results found in the case studies. The electrolyzer and fuel cell efficiencies (ε) are taken directly from the case studies. The energy losses associated with these are obtained by calculations involving input energy and the given efficiencies. Storage and delivery figures are explicitly listed in the case studies as kWh/kg of hydrogen. Values from the literature are shown on the scheme and are used along with the input energy to calculate efficiencies for the intermediate steps.

The overall efficiencies at the end of each pathway reflect the lowest possible (most E_{in}, least E_{out}) and highest possible (least E_{in}, most E_{out}) efficiency values.

For the hydrogen transport as compressed gas in cylinders, $\varepsilon = 25\%$–56% with a total power of electricity between 13 and 23 kWh.

On the other hand, for liquid hydrogen, the overall efficiency from the fuel cell conversion is $\varepsilon = 16\%$–47% with a total power of electricity between 9 and 19 kWh.

Finally, for the transport of hydrogen as a metal hydride, $\varepsilon = 14\%$–35% with a total power of electricity between 8 and 14 kWh.

23.1.11.3   Technical and Economic Overview of the Reforming Hydrogen Production Systems

Nowadays, the compromised energy supply system for electricity involves mostly thermal conversion from gasoline, diesel fuel, and natural gas that serve as energy carriers. These carriers are made by the conversion of primary energy sources, such as coal, petroleum, underground methane, and nuclear energy, into an energy form that is easily transported and delivered in a usable form to industrial, commercial, residential, and transportation end users. The sustainable energy supply system of the future features electricity and hydrogen as the dominant energy carriers. Hydrogen would be produced from a very diverse base of primary energy feedstock, using the resources and processes that are most economical or consciously preferred. Methods to produce hydrogen from natural gas are well developed and account for over 95% of all hydrogen produced in the United States and 48% globally. It is anticipated that hydrogen from natural gas can serve as a foundation to the U.S. transition to a hydrogen energy economy.

Mahler AGS has more than 30 years of experience and expertise in the design and manufacturing of hydrogen generation plants. The HYDROFORM-C system is based on steam reforming of natural gas, LPG, or naphtha. Mahler AGS also offers efficient processes for hydrogen generation by means of methanol reforming within a range of 100–4000 N-m^3/h of hydrogen and purities of up to 99.999+ as % volume [58]. The purification of raw hydrogen is carried out by means of the HYDROSWING® system that provides a separate purification process step. As one of the internationally leading suppliers of these technologies, Mahler AGS develops systems that are exactly tailored to the customer's needs and that can be easily integrated into already existing processes. These processes offer customers maximum quality and security, as well as the capability to efficiently meet hydrogen requirements from 100 to 10000 N-m^3/h at purities of up to 99.999+ as % volume.

23.2   DELIVERY OF HYDROGEN

A key part of the overall hydrogen energy infrastructure is the selection of the delivery system that is able to move the required hydrogen from the production region to an end-use device. Delivery system requirements essentially differ with the production method and end-use application.

Hydrogen is currently transported by pipeline or by road via cylinders, tube trailers, metallic hydrides, and cryogenic tankers. For the latter, a small amount shipped by rail or barge is certainly required. At present, hydrogen is produced in a limited number of plants and is used for making chemicals or upgrading fuels.

Due to the energy intensive nature and the cost associated with hydrogen distribution by high-pressure cylinders and tube trailers, this method of distribution has a range limited to approximately 200 km. The use of metallic hydrides seems to be one of the most promising modes of delivering. However, it is more expensive than hydrogen compression in cylinders.

For longer distances of up to 1500 km, hydrogen is usually transported as a liquid in super-insulated, cryogenic, over-the-road tankers, railcars, or barges and is then vaporized for use at the customer site. This is also an energy intensive and costly process. Pipelines that are owned by hydrogen producers are limited to small areas where large hydrogen refineries and chemical plants are concentrated. A large pipeline system dedicated to the transport of large volumes of hydrogen does not exist yet. It is likely that hydrogen production, transportation, and storage will use both decentralized and centralized approaches. Developing the infrastructure necessary to produce, store, and deliver the large quantities of hydrogen necessary for the future hydrogen economy is one of the major challenges of our century.

Hydrogen delivery methods cost more than conventional fuel delivery because of the nature and properties of the fuel. The high price of hydrogen delivery methods could lead to the use of conventional fuels and associated delivery infrastructure up to the point of use and small-scale conversion systems to produce hydrogen onsite. However, cost-effective means do not currently exist to generate hydrogen in small-scale systems.

Full lifecycle costing has not been applied to the delivery of alternative fuels. Any strategy to select appropriate delivery systems should involve full lifecycle valuation of the options. Lifecycle cost analyses should compare gaseous and liquid hydrogen delivery and hydrogen carrier media such as metal and chemical hydrides, methanol, and ammonia. Multiple delivery infrastructures may be essential, which could add to the price of transitioning to a hydrogen economy.

However, the most important point that is somewhat forgotten nowadays is the increase in research and development on new and efficient delivery systems. Economics and administration (bureaucracy) dominate the "third world countries" and war and power dominate the budget in the "first world countries." The main impediments that do not allow scientific advances to lead in both kinds of countries arise from these speculations and politics that really conduct the world. Improvements are needed in areas such as hydrogen detectors; odorization; materials selection for pipelines, seals, and valves; and transportation containers for hydrogen. Technology validation should address research and development needs for fueling components such as high-pressure, breakaway hoses, hydrogen sensors, compressors, onsite hydrogen generation systems, and robotic fuelers.

23.3 STORAGE OF HYDROGEN

Storage issues intersect the production, transport, delivery, and final applications of hydrogen as an energy vector. Mobile applications are driving the development of safe, space-efficient, and cost-effective hydrogen storage systems. Considerably, other applications will benefit substantially from all technological advances made for these on the means of transportation storage systems.

The main characteristic that places hydrogen in the best position from other energy carriers, except pure electricity, is the larger capacity to be stored for further use in another place and time. In the "old days," storage research has been mostly focused on compressed gas, cryogenic hydrogen, and metal hydrides, but in current times a growing number of alternative methods including nanocarbon novel materials, chemical hydrides, and glass microspheres are being tested. Compressed gas is the most mature storage technology, but compression adds inefficiencies to the hydrogen lifecycle and requires stronger, more expensive materials for tank construction such as titanium instead of aluminum. Extensive materials research is being conducted to improve compressed gas storage technology. Advancements have already been made in carbon fiber wrapped tanks, which are lighter and safer than traditional steel tanks.

Hydrogen is difficult to compress since it is a very small molecule, so positive displacement compressors are used. Hydrogen compressors are expensive, because of the materials and its size. The compression process is energy intensive. For example, for an Inlet–Outlet (psig) ratio of 300:1000, the compression energy is 0.6–0.7 kWh/kg with an adiabatic efficiency of 75%. The main problem of hydrogen compression is the hydrogen embrittlement of metals. At elevated pressure and temperature, hydrogen can permeate carbon steel, resulting in decarburization. To avoid this problem, alloy steels containing chromium and molybdenum have been suggested for compressor materials. The methods to decrease energy requirement mostly involve two points: new mechanical concepts such as a linear or a guided rotor compressors and nonconventional compression approaches, such as electrically driven membranes or hydride compressors.

Liquid hydrogen takes up less storage volume than gas but requires expensive cryogenic containers. Furthermore, the liquefaction of hydrogen is an energy-intensive process and results in large evaporative losses—about one-third of the energy content of the hydrogen is lost in the process. Cryogenic hydrogen is denser than compressed gaseous hydrogen, thus involving a less storage volume. Energy and economic costs associated with cryogenic hydrogen storage are higher than compressed gas storage costs. Between 10% and 30% of the fuel of hydrogen is required for liquefaction, and tanks must be extra insulated to maintain cryogenic temperatures near −250°C. The simplest liquefaction process is the Linde or JT expansion cycle (Figure 23.6). The following are the steps involved in the process. The gas is compressed at ambient pressure and later is cooled in a heat exchanger. It is passed through a throttle valve (isenthalpic *JT* expansion), producing some liquid while it is removed and the cool gas is returned to the compressor *via* a heat exchanger of the second step.

In a temperature vs. entropy diagram (Rankine diagram), we are able to see the four steps to liquefy a gas in the case of the *Linde cycle*. Figure 23.7 shows the starting gas *g* at experiments under isobaric conditions, a heating in a counterflow heat exchanger in (1). Secondly, it is compressed isothermally to (2) and then from (2) to (3) it is isobarically cooled in a counterflow

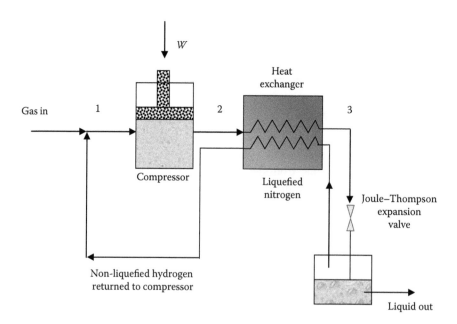

FIGURE 23.6 The simplest JT liquefier cycle: (1) gas compression with a mechanical work *W*, (2) cooling in the heat exchanger, and (3) liquefying in the isenthalpic expansion valve to produce some liquid. In the case of hydrogen precooling with liquid nitrogen is needed.

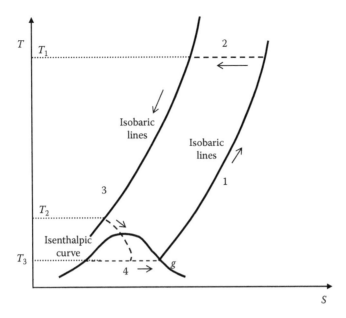

FIGURE 23.7 Temperature vs. entropy diagram for the Linde cycle of liquefaction. Details are given in the text.

heat exchanger. Then, it is isenthalpically expanded to (4), where it is finally expanded again isothermically to g. The fraction liquefied is the difference between (4) and g.

The Linde cycle works for gases, such as nitrogen, that cool upon expansion at room temperature. In the case of hydrogen, it warms upon expansion at room temperature. For hydrogen gas, the cooling upon expansion requires a temperature below its pressure-dependent inversion temperature T_{J-T}, where internal interactions allow the gas to do work when it is expanded.

To reach the inversion temperature, precooling of the hydrogen gas to 78 K ($-319°F$) is done before it reaches the first expansion valve using liquid nitrogen. The nitrogen gas may be recovered and recycled in a continuous refrigeration loop [62,63]. The ideal work of liquefaction for nitrogen is only 0.207 kWh/kg, whereas for hydrogen, the ideal work of liquefaction is 3.228 kWh/kg.

The JT coefficient, μ_{JT}, is defined as the first derivation of the temperature T with respect to the pressure p at constant enthalpy H:

$$\mu_{JT} = \left(\frac{\partial T}{\partial p}\right)_H \tag{23.3a}$$

or equivalently

$$\mu_{JT} = -\frac{1}{c_p}\left(\frac{\partial H}{\partial p}\right)_T \tag{23.3b}$$

where c_p is the isobaric heat capacity. Positive values of μ_{JT} imply a cooling of the fluid as it exceeds an adiabatic throttle.

The inversion curve for gases gives information about the T_{J-T}. In Figure 23.8, it is possible to see that the maximum temperature value at which hydrogen liquefaction is initiated, 202 K at 0 atm. However, since the expansion process has to begin at higher pressures, it is usually started below 100 K [64]. The JT integral inversion curves are those connecting all state points at $\mu_{JT} = 0$.

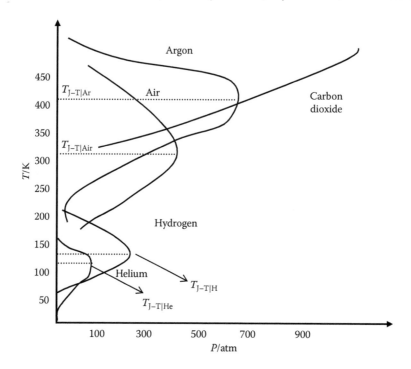

FIGURE 23.8 Inversion of the JT temperature vs. pressure curves for different gases.

The experimental determination of a fluid's T_{J-T} and its inversion curve is difficult since very precise measurements of volumetric or caloric properties at conditions up to 5 times its critical temperature and 12 times its critical pressure. Experimental inversion curve data are therefore scarce, often unreliable [65], and mostly available only for pure fluids.

For a relatively small number of pure fluids, Wagner and Pruß have developed reference equations of state based on thousands of experimental data points on different thermodynamic properties [66]. These reference equations yield precise results in the fluid region over a large temperature and pressure range including the JT inversion curve.

An alternative to the precooled Linde process is to pass some of the high-pressure gas through an expansion engine, which then is sent to a heat exchanger to cool the remaining gas. The theoretical process referred to as *ideal liquefaction* uses a reversible expansion process to reduce the energy required for liquefaction. It consists of an isothermal compressor followed by an isenthalpic expansion to cool the gas and produce a liquid. In practice, an expansion engine can be used only to cool the gas stream and not to condense it because excessive liquid formation in the expansion engine would damage the turbine blades. This is called the *Claude cycle*, which is depicted in Figure 23.9. In this cycle the isenthalpic branch is substituted by different consecutives isentropic branches.

The Claude's cycle is a nonideal isentropic plus an isenthalpic expansion. A preheated gas is firstly isothermally compressed from p_1 to p_2 followed by a fast isobaric cooling through 3 from $T_1 \approx 300$ K. There, an expansion engine with another precooled gas helps the hydrogen gas to step to point 4 and then to $T_2 \approx 80$ K. An expansion valve helps the cooled gas to go isenthalpically to point 5 at an even lower temperature $T_3 \approx 30$ K and p_3 being lower than p_2. An isentropic expansion finally produces the liquefied gas at the initial T_1.

There are several ways of producing liquefied hydrogen. Table 23.6 seeks to classify the most widespread and the most promising cryogenic processes for hydrogen liquefaction. The most common processes are based on thermodynamic gas cycles with compression, heat exchangers, and gas expansion as the main individual subprocesses. These processes will be referred to as gas compression

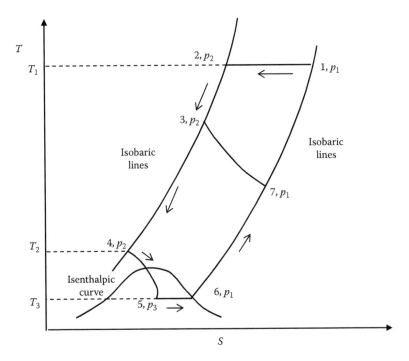

FIGURE 23.9 Temperature vs. entropy diagram for the Claude cycle of liquefaction. Details are given in the text.

TABLE 23.6
The Most Common Liquefaction Processes
for Hydrogen Storage

Thermodynamic Gas Cycles	
Recuperative Cycle	**Regenerative Cycle**
Brayton	Stirling
Claude	Gifford–McMahon
Collins	Vuilleumier
—	Pulse tube

cycles, which seem to be the most realistic processes for producing liquefied hydrogen in considerable amounts for the near future.

Hydrogen molecules exist in two forms, para and ortho (Figure 23.10), depending on the electron configurations. At the hydrogen's boiling point of 20 K (−423°F), the equilibrium concentration is almost all *para*-hydrogen. However, at room temperature or higher, the equilibrium concentration is 25% *para*-hydrogen and 75% *ortho*-hydrogen. Uncatalyzed conversion from *ortho*- to *para*-hydrogen proceeds very slowly. Moreover, the *ortho*- to *para*-hydrogen conversion releases a significant amount of heat (527 kJ/kg). The heat of conversion from normal to para is 0.146 kWhth/kg, a process that can vaporize some hydrogen.

If *ortho*-hydrogen remains after liquefaction, the heat of transformation described previously will be released slowly as the conversion proceeds. Long-term storage of hydrogen requires that the hydrogen be converted from its ortho form to its para form to minimize boil-off losses. This can be accomplished using a number of catalysts including activated carbon, platinized asbestos, ferric

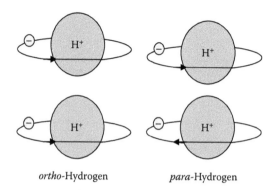

ortho-Hydrogen   *para*-Hydrogen

FIGURE 23.10   The two forms of molecular hydrogen: *ortho*-hydrogen and *para*-hydrogen.

oxide, rare earth metals, uranium compounds, chromic oxide, and nickel compounds. Activated charcoal is used most commonly, but ferric oxide is a cheap material that can be widely used.

Solid storage in metal hydrides is not yet feasible, in spite of the various materials developed by electrochemists after the cold fusion era. However, advanced research suggests that lithium metal and lanthanum hydrides will be the most used in the future hydrogen economy. Using the concept of temperature change, hydrogen is adsorbed within the interstices of metal hydride lattices. The resulting granules can be stored much safely than compressed gas or cryogenic liquid hydrogen. Hydrogen is released from the metal hydrides by applying a little amount of heat, after which it works spontaneously. The high costs currently associated with adsorption make metal hydride storage still impractical, but economic feasibility will increase as technological advances occur.

Metal hydrides have the possibility for reversible onboard hydrogen storage and discharge at low temperatures and pressures. The optimum operating window for a polymer electrolyte fuel cell (PEFC) vehicular application is in the range of 1–10 atm and 25°C–120°C. This is based on using the waste heat from the fuel cell to "release" the hydrogen from the media. Waste heat less than 80°C is available, but as high-temperature membranes are developed, there is a potential for waste heat.

A simple metal hydride such as $LaNi_5H_6$ that incorporates hydrogen into its crystal structure can work in this range, but its gravimetric capacity is too low ($\sim$1.3 wt.%). Thus, its cost is too elevated for any applications. Complex metal hydrides such as alanate (AlH_4^-) materials have the potential for higher gravimetric hydrogen capacities in the operational window than the simple metal hydrides. Alanates can store and release hydrogen reversibly when catalyzed with the titanium dopants, according to the following two-step displacement reaction for sodium alanate:

$$NaAlH_4 \leftrightarrow 1/3Na_3AlH_6 + 2/3Al + H_2 \qquad\qquad (XXIII)$$

$$Na_3AlH_6 \leftrightarrow 3NaH + Al + 3/2H_2 \qquad\qquad (XXIV)$$

At 1 atm pressure, the first reaction becomes thermodynamically favorable at temperatures above 33°C and can release 3.7 wt.% hydrogen, whereas reaction (XXIV) occurs above 110°C and can release 1.8 wt.% hydrogen. The amount of hydrogen that a material can free, rather than only the amount the material can hold, is the key parameter used to determine (net) gravimetric and volumetric capacities.

The maximum material gravimetric capacity of 5.5 wt.% hydrogen for sodium alanate is below the system target of 6 wt.%. Thus far, 4 wt.% reversible hydrogen content has been experimentally demonstrated with alanate materials. In addition, hydrogen release kinetics is too slow for vehicular applications [67]. Furthermore, the packing density of these powders is low (e.g., roughly 50%), and the system-level volumetric capacity is a challenge. Although sodium alanates will not meet the midterm targets, it is envisioned that their sustained study will lead to a fundamental understanding

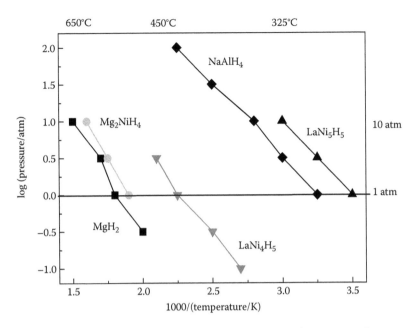

FIGURE 23.11 Hydrogen metal hydrides characterized in a log (pressure/atm) vs. 1000/(temperature/K) plot.

that can be applied to the design and development of improved types of complex metal hydrides (Figure 23.11).

A new complex hydride system based on lithium amide has been developed. For this system, the following reversible displacement reaction takes place at 285°C and 1 atm:

$$Li_2NH + H_2 \leftrightarrow LiNH_2 + LiH \qquad (XXV)$$

In this reaction, 6.5 wt.% hydrogen can be reversibly stored with potential storage up to 10 wt.%. However, the current operating temperature is outside of the vehicular operating window. On the other hand, the temperature of this reaction can be lowered to 220°C with magnesium substitution.

One of the major issues with complex metal hydride materials, due to the reaction enthalpies involved, is thermal management during refueling. Depending on the amount of hydrogen stored and the required refueling times, they must be handled during the recharging of onboard vehicular systems with metal hydrides. Reversibility of these and the new materials must be demonstrated for over a thousand cycles for reliability.

23.4 USE OF HYDROGEN

23.4.1 INTRODUCTION TO FUEL CELLS

The electrochemical conversion of hydrogen and oxygen (or air) into electricity and water in a fuel cell is the most important feature of a prospective hydrogen economy (Figure 23.12). As explained in Chapter 2, it works as a reverse electrolyzer: diatomic hydrogen is broken into electron and proton components at the anode via a Volmer–Tafel–Heyrovský mechanism. Consequently, the electrons flow through an external circuit (electronic conductor) to be consumed as electricity, and simultaneously hydrogen ions (protons) diffuse and migrate through the electrolyte to the cathode (Figures 23.13 and 23.14). There, they are combined with gaseous oxygen in a triphasic element to produce water *via* the Damjanović, Wróblowa, or Bagostkii scheme. Fuel cells are categorized by low- or high-temperature operations and are classified by the type of electrolyte that they contain. Examples of low-temperature fuel cells comprise PAFC and proton exchange membrane fuel cells

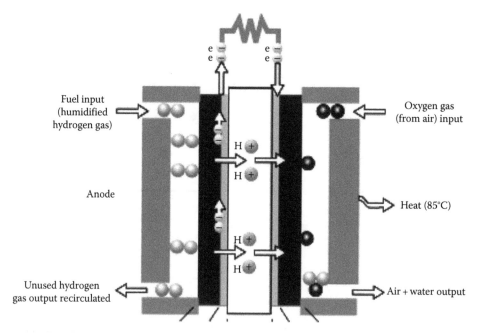

FIGURE 23.12 Hydrogen/oxygen fuel cell diagram with the proton (inner) and electron (outer) transfer mechanisms.

FIGURE 23.13 Bipolar plates made of conducting graphite, sinterized and engraved (machined) with a 1 mm of width and depth.

(PEMFC); high-temperature models include molten carbonate fuel cell (MCFC) and SOFC. The wide range of power outputs available make fuel cells suitable for a variety of applications.

Relatively high fuel cell efficiencies are coupled with high material costs. Research and development efforts will continue to focus on optimization until a feasible model is developed [68,69].

The electrochemical processes that occur at each electrode are really electrocatalytic reactions. At the anode, hydrogen gas or the selected fuel must diffuse through tortuous pathways until a platinum particle is encountered. Platinum electrocatalyzes the dissociation of the hydrogen molecule into two hydrogen atoms (H) bonded to two neighboring platinum atoms. Only then can each H atom release an electron to form a hydrogen ion (H^+). The current flows in the circuit as these H^+ ions are conducted through the membrane to the cathode while the electrons go from the anode to the outer circuit and then to the cathode. The reaction of one oxygen molecule at the cathode is a 4-electron reduction process (see Section 23.1.2), which occurs in a multistep sequence. Expensive platinum-based

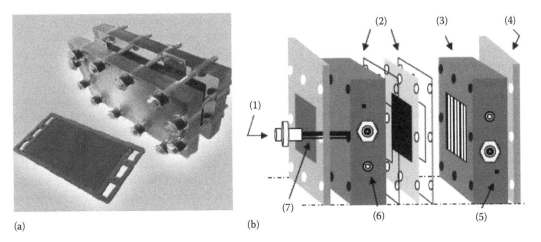

(a) (b)

FIGURE 23.14 (a) Bipolar plates and a three-unit cell stack pressed between two metal collector plates. (b) Scheme of the pressed graphite plate with metal collectors. (1) create uniformed force distribution, (2) capable of applying torque equally to the bolts, (3) adaptable to varied flow-field plate design, (4) gold-plated copper collector plate, (5) reach thermocouple at center of fuel cell test, (6) measure exact cell potential without ohmic drop, and (7) silicone rubber heater provides highest watt density and maintains uniform cell temperature.

catalysts seem to be the only ones capable of generating high rates of oxygen reduction at the relatively low temperatures ($\sim$80°C) at which the PEM fuel cells operate. There is still uncertainty regarding the mechanism of this complex process. The performance of the polymer electrolyte membrane fuel cells is limited primarily by the slow rate of the oxygen reduction reaction that is more than 100 times slower than the hydrogen oxidation reaction.

Impurities often present in the hydrogen fuel feed stream bind to the platinum catalyst at the anode, preventing hydrogen electrooxidation by blocking platinum sites. Alternative catalysts that can oxidize hydrogen while remaining unaffected by impurities, such as carbon monoxide or methane that are often found in the purification process, need to improve their performance in the cell. The rate of the oxygen reduction process at the air electrode is quite low, even by the best platinum catalysts that have been developed until now, resulting in significant performance loss. Alternative catalysts that promote a high rate of oxygen reduction are needed to enhance the fuel cell's performance further [70].

23.4.2 Comparison between the Thermodynamic Engine and the Fuel Cell

Generally, we relate Carnot's law only to the amount of external energy that can be extracted from thermal energy systems. The same law, however, does apply to all internal energy systems whether nuclear, chemical, or thermal. The amount of external energy that can be obtained from all types of internal energy is called the *Carnot ratio*. In the case of fuel cells, it is about 92% under normal conditions. This is much higher than for a gas turbine with a mean temperature of 1000 K and a Carnot ratio of 72%. It is based on a defined ambient temperature of the surroundings and is only related to the absolute temperature scale [71].

Heat engines such as gas turbines are considered inferior to fuel cells because they must convert the high chemical temperature of the chemical energy into low thermal temperature of thermal energy first. A gas turbine cannot operate at the temperature of the chemical energy without melting. When the temperature is reduced, the Carnot ratio is reduced. A large amount of the Helmholtz energy that was available at the higher temperature is lost, since it is transformed into useless bound energy. The fuel cell also loses energy but only because of the resistance and inefficiencies during the ion and electron flow during the production of electricity. Thus, too many types of fuel cells can run efficiently at low temperatures while at the same time converting very high temperature energy.

Present highly advanced gas turbines do not achieve a mean temperature of more than 877°C. In spite of this, gas turbines can be highly efficient if they are in large sizes and produce little pollution. The latest are 60% efficient in converting fuel to electricity. In the future, ceramic gas turbines could reach 70% efficiency. This would result in a higher efficiency than what the fuel cell can achieve by itself [72].

The Clausius energy efficiency of fuel cells can be compared when they operate on hydrocarbon fuel. The fuel cell process is divided into six subsystems. In each subsystem, there are inefficiencies involved that reduce the Clausius energy that is left in the system. In all cases, the electricity that is extracted is still considered to be part of the Clausius energy of the system. It appears that the SOFC system will have an efficiency that is 1.4 times more than that of the others. This is because of the lower reformer and air pressurization losses. The SOFC can reform the fuel inside the stack and utilize some of the stack waste thermal energy. The rest cannot since they operate at a lower temperature. The SOFC does not need to operate at higher than the ambient air pressure. It only uses a low-pressure blower to drive air through the cell.

23.4.3 TYPES OF FUEL CELLS

A variety of fuel cells can be found commercially, but they are in different stages of development. They can be classified by uses, depending on the combination of the type of fuel and oxidant, whether the fuel is processed outside (external reforming) or inside (internal reforming) the fuel cell, the type of electrolyte, the temperature of operation, whether the reactants are fed to the cell by internal or external manifolds, etc. The most ordinary categorization of fuel cells is by the kind of electrolyte employed in the cells:

1. PEFC, ~80°C
2. Alkaline fuel cell (AFC), ~100°C
3. PAFC, ~200°C
4. MCFC, ~650°C
5. Intermediate temperature solid oxide fuel cell (ITSOFC), ~800°C
6. Tubular solid oxide fuel cell (TSOFC) ~1000°C.

In addition, the operating temperature of these devices are listed in this category. The operating temperature and useful life of a fuel cell dictate the electrochemical, electrical, and thermomechanical properties of materials used in the cell components, that is, electrodes, electrolyte, interconnect, current collectors, etc. Aqueous electrolytes are limited to temperatures of about 150°C or lower because of their high water vapor pressure and/or rapid degradation at higher temperatures. The operating temperature also plays an important role in dictating the type of fuel that can be used in a fuel cell. The low-temperature fuel cells with aqueous electrolytes are restricted to hydrogen as a fuel. In high-temperature fuel cells, carbon monoxide and even methane can be used because of the inherently fast electrode kinetics and the lesser need for large catalytic activity at elevated temperatures. However, the higher temperature cells can favor the conversion of carbon monoxide and methane to hydrogen, as explained above (as hydrogen equivalent actual fuels).

A brief description of various electrolyte cells of interest follows, but a detailed description of these fuel cells may be found in [60,61].

23.4.3.1 Polymer Electrolyte Fuel Cell

The electrolyte in this fuel cell is an ion exchange membrane (fluorinated sulfonic acid polymer or other similar polymer) that is an excellent proton conductor. The only liquid in this fuel cell is water; thus, corrosion problems are minimal. Water management in the membrane is critical for efficient performance; the fuel cell must operate under conditions where the by-product water does not evaporate faster than it is produced because the membrane must be hydrated. Because of the limitation on the operating [61] temperature imposed by the polymer, usually less than 120°C, and because of problems with water balance, a hydrogen-enriched gas with minimal or no carbon monoxide (a poison at low temperature) is used. Higher catalyst loading (platinum type metals in most cases) than that used in PAFCs is required for both the anode and the cathode.

23.4.3.2 Alkaline Fuel Cell

The electrolyte in this fuel cell is concentrated (85 wt.%) potassium hydroxide in fuel cells operated at high temperature (~250°C) or less concentrated (35–50 wt.%) potassium hydroxide for lower temperature (<120°C) operation. The electrolyte is retained in a matrix (usually asbestos) and a wide range of electrocatalysts can be used, for example, nickel, silver, metal oxides, spinels, and noble metals. The fuel supply is limited to nonreactive constituents except for hydrogen. Carbon monoxide is a poison and the produced carbon dioxide (in the case of having carbon monoxide) will react with the potassium hydroxide to form solid potassium carbonate, thus altering the electrolyte. Even the small amount of carbon dioxide in air must be considered as a problem in the alkaline cell.

23.4.3.3 Phosphoric Acid Fuel Cell

Phosphoric acid concentrated to 100% is used as an electrolyte in this cell that operates between 150°C and 220°C. At lower temperatures, phosphoric acid is a poor ionic conductor and carbon monoxide poisoning of the platinum electrocatalyst in the anode becomes severe. The relative stability of concentrated phosphoric acid is high compared to other common acids. Consequently, the PAFC is capable of operating at the high end of the acid temperature range (100°C–220°C). In addition, the use of concentrated acid (100%) minimizes the water vapor pressure, so water management in the cell is not difficult. The matrix universally used to retain the acid is silicon carbide and the electrocatalyst in both the anode and the cathode is platinum.

23.4.3.4 Molten Carbonate Fuel Cell

The electrolyte in this fuel cell is usually a combination of alkali carbonates, which is retained in a ceramic matrix of $LiAlO_2$. The fuel cell operates between 600°C and 700°C where the alkali carbonates form a highly conductive molten salt, with carbonate ions providing ionic conduction. At the high operating temperatures in the MCFCs, nickel (anode) and nickel oxide (cathode) are adequate to promote the reaction. Noble metals are not required.

23.4.3.5 Intermediate Temperature Solid Oxide Fuel Cell

The electrolyte and electrode materials in this fuel cell are the same as those used in the TSOFC. The ITSOFC operates at a lower temperature, however, typically between 600°C and 800°C. For this reason, thin film technology is being developed to promote ionic conduction. Alternative electrolyte materials are also being analyzed.

23.4.3.6 Tubular Solid Oxide Fuel Cell

The electrolyte in this fuel cell is a solid, nonporous metal oxide, usually Y_2O_3-stabilized ZrO_2. The cell operates at 1000°C where ionic conduction by oxygen ions takes place. Typically, the anode is $Co-ZrO_2$ or $Ni-ZrO_2$ cermet and the cathode is strontium-doped $LaMnO_3$.

In low-temperature fuel cells (PEFC, AFC, and PAFC), protons or hydroxyl ions are the major charge carriers in the electrolyte, whereas in the high-temperature fuel cells (MCFC, ITSOFC, and TSOFC), carbonate ions and oxygen ions are the charge carriers.

Another important difference in defining a certain type of cell is the method used to produce hydrogen for the anode and the overall cell reaction. Hydrogen can be reformed from natural gas and steam in the presence of a catalyst starting at a temperature of ~760°C. The reaction is endothermic. MCFC, ITSOFC, and TSOFC operating temperatures are high enough for the reforming reactions to occur within the cell, a process referred to as internal reforming. The reforming reaction is motivated by the decrease in hydrogen as the cell produces energy. This internal reforming can be helpful to the system's efficiency because there is an effective transfer of heat from the exothermic cell reaction to ensure the endothermic reforming reaction. A reforming catalyst is required adjacent to the anode gas chamber for the reaction to occur. The cost of an external reformer is neglected and system efficiency is improved, but at the expense of a more complex cell configuration and increased maintenance. This provides developers of high-temperature cells with a choice of an external reforming or internal reforming approach. The internal reforming of a MCFC is limited to operation at ambient pressure, whereas external reforming MCFC can operate at pressures up to 3 atm. The slow rate of the reforming reaction makes internal reforming impractical in the lower temperature cells. Instead, a separate external reformer is used.

The combination of anode, membrane, and cathode is referred as an MEA. Its evolution in the polymer electrolyte membrane fuel cells has been conceived through several creations. The original membrane-electrode assemblies were constructed with $\sim$4 mg of platinum per cm^2 of membrane area. Current technology varies with the manufacturer, but the total platinum loading has decreased from the original to about $0.5\,mg/cm^2$. Laboratory research now uses platinum loadings of $0.1\,mg/cm^2$. This corresponds to an improvement in fuel cell performance from about $0.5\,A/mg$ of platinum to $15\,A/mg$. The thickness of the membrane in a MEA can vary with the type of the membrane. The thickness of the catalyst layers depends upon how much platinum is used in each electrode. For catalyst layers containing about 0.2 mg Pt/cm^2, the thickness of the catalyst layer is close to $10\,\mu m$, less than half the thickness of a sheet of paper. This MEA with a total thickness of about $200\,\mu m$ can generate more than half an ampere of current for every square centimeter of device at an electrode potential between the cathode and the anode of 0.7 V, but only when encased in well-engineered components.

The polymer electrolyte membrane is a plastic-solid, organic polymer, poly-perfluoromethanesul-fonic acid. A typical membrane material, such as Nafion, consists of three regions: the hydrophobic Teflon-like, fluorocarbon backbone, made of hundreds of repeating $-CF_2-CF-CF_2-$ units in length; the hydrophobic side chains, $-O-CF_2-CF-O-CF_2-CF_2-$, that connect the molecular backbone to the third region, where the ion clusters consisting of sulfonic acid ions; and $SO_3^-H^+$ defining the hydrophilic side. The negative SO_3^- ions are permanently attached to the side chain and therefore cannot move. However, when absorbing water, it can hydrate the membrane. This process makes the protons mobile inside the membrane. Ion migration occurs by those carriers, bonded to water molecules, from the SO_3^- site out of the membrane to another SO_3^- site within it. Because of this mechanism, the solid hydrated electrolyte is an excellent conductor of only the hydrogen ions.

The temperature over which water is a liquid border affects the performance of the PEM fuel cells. The membrane must hold water so that the protons can carry the charge within the membrane. Operating PEM fuel cells at temperatures exceeding 100°C is possible under pressurized conditions that are required to keep the water in a liquid state, but this decreases the cell life. At present, polymer electrolyte membranes cost about $\$100/m^2$. However, they are going to drastically decrease as the consumer demand for these fuel cells increases in the near future.

The "hardware" of the fuel cell—*diffusers*, *flow fields*, and *current collectors*—is designed to optimize the current density that can be obtained from a MEA. The so-called backing layers, one next to the anode and the other next to the cathode, are usually made of a porous carbon paper or carbon cloth, typically, of 100–300 μm thick. The backing layers have to be made of a material, such as carbon, that can conduct the electrons out of the anode and into the cathode. The porous nature of the backing material ensures an effective diffusion of each reactant gas to the catalyst on the MEA. The porous structure of the backing layers allows the gas to spread out as it diffuses so that when it accesses the backing, the gas will be in contact with the whole area of the catalyst/membrane device.

The *backing layers* also assist in water management during the operation of the fuel cell; too little or too much water can cause the cell to cease operation. The correct backing material allows the right amount of water vapor to reach the MEA in order to keep the membrane humidified, without overflowing; that is, the liquid water produced at the cathode that leaves the cell does not spread out. The backing layers are often wet proofed with Teflon to ensure that at least some, and hopefully most, of the pores in the carbon cloth do not become clogged with water, which would prevent quick gas diffusion necessary for a good reaction rate at the electrodes.

Pressed against the outer surface of each backing layer is a piece of material, called a *plate* (being bipolar or not) that often serves the dual role of flow field and current collector. In a single fuel cell, these two plates are the last of the components that make up the cell. The plates are made of a lightweight, strong, gas-impermeable, electron-conducting material. Graphite or metals are regularly used although other composed plates are now being developed. The first task of each plate is to provide a gas "flow field." The side of the plate next to the backing layer contains channels machined into the plate. The channels are used to carry the reactant gas from the point at which it enters the fuel cell to the point at which the gas goes away. The pattern of the flow field in the plate as well as the width and depth of the channels have a large impact on the effectiveness of the even distribution of the reactant gases across the active area of the MEA. The design of the flow field also affects water supply to the membrane and water removal from the cathode. The second purpose of the plate is of a current collector. Electrons produced by the oxidation of hydrogen must be conducted through the anode, through the backing layer and through the plate, before they can exit the cell, travel through an external circuit, and reenter the cell at the cathode plate (see Figure 23.13). We have prepared various plates of sinterized graphite for subsequent machining to be used under a proper fluidodynamic profile using a three-dimensional milling device. The engraving was carefully conducted with a precise automatic computer controlled system (Figure 23.14a).

With the addition of the flow fields and current collectors, the PEM fuel cell is now complete. Only a load-containing external circuit, such as an electric motor, is required for the electric current to flow, the power having been generated by passing hydrogen and air on either side of what looks like a piece of food wrap that is painted black.

Since fuel cells operate at less than 100% efficiency, the potential output of one cell is ~ 0.7–0.5 V. As most applications require much higher potentials than this, the required value is obtained by connecting individual fuel cells in series to form a fuel cell "*stack*." If fuel cells were simply lined up next to each other, the anode and cathode current collectors would be side by side. To decrease the overall volume and weight of the stack, instead of two current collectors, only one plate is used with a flow field cut into each side of the plate. This type of plate, called a "*bipolar plate*," separates one cell from the next, with this single plate helping to carry hydrogen gas on one side and oxygen or air on the other. It is important that the bipolar plate be made of gas-impermeable material. Otherwise, the two gases would intermix, leading to direct oxidation of fuel. Without separation of the gases, the electrons will be conducted directly from the hydrogen to the oxygen and are essentially wasted as they cannot be routed through an external circuit to do useful electrical work. The *bipolar plate* must also be electronically conductive because the electrons produced at the anode on one side of the bipolar plate are conducted through the plate where they enter the cathode on the other side of the bipolar plate. Two end plates, one at each end of the complete cell stacks, are linked through an external circuit. The area of a single fuel cell can vary from a few to a thousand square centimeters. A stack can consist of from a few to a hundred or more cells connected in series using the *bipolar plates* (Figure 23.14b). For applications that require large amounts of power, many stacks can be used in series or in parallel combinations.

23.4.4 Remarkable Thoughts on Fuel Cells

The ratio of hydrogen to carbon in the commonly used fuels has increased over the years along with our increased use of natural gas. This is very good news for the environment and humankind. When we burn fossil fuels such as oil or natural gas, the two primary products are carbon dioxide and water. The higher the percentage of hydrogen in a fuel, the less carbon dioxide and the more water that are produced. However, the production of carbon monoxide is larger than expected. Therefore, the use of these fuels in industrial and vehicular applications seems to be problematic. The most likely solution is the use of hydrogen fuel cells.

Fuel cells as a technology is actually quite old. It has not been widely used yet because of speculative reasons and political problems. Present material science may soon make them a reality particularly in specialized applications. The SOFC appears to be the most promising technology for small electric power plants of over 1 kW. The DAFC appears to be the most promising battery replacement option for portable applications such as cellular phones and laptop computers. It is clear that at this moment, fuel cells will be practical for transportation applications such as automobiles and buses, but hybrid vehicles are likely to be more popular in the future in order to avoid unexpected problems. It is unclear whether hydrogen fuel will be widely used or whether a mostly "electric economy" will continue to exist.

Fuel cells are being proposed to replace Otto or diesel engines because they are reliable, simple, and quieter, less polluting, and economical. The internal combustion Otto or diesel cycle engine has been used in automobiles for about 100 years now. A reasonably simple and reliable mechanical device nowadays has a lifespan of up to 400,000 km in automobiles and over 1,000,000 km in larger applications such as buses and trucks.

Fuel cells have the potential to be considerably quieter than Otto or diesel cycle power plants. This would especially reduce the noise in the quiet neighborhoods or highways. At speeds higher than 50 km/h, however, there is still the problem of road noise. Fuel cells produce electricity. This is not the desired form of energy for transportation. The electricity must be converted into mechanical power using an electric motor. The Otto or diesel cycle produces the required mechanical power directly. This gives them an advantage compared to fuel cell–powered automobiles.

Presently, Otto and diesel cycle engines seem to be able to comply with extremely stringent pollution regulations. They are inexpensive to produce under a reasonable fuel economy and are readily available for liquid fuels. Fuel cell vehicles have a much greater chance of being accepted in the future when fuel prices are higher and liquid fossil fuels are in short supply. Fuel cell vehicles will then be competing with electric vehicles that will be cheaper to operate but have problems with recharging. Finally, we are able to say that fuel cells will be commercially viable when the cost of the technology will be affordable by any person who can afford an automobile as of today. This fact will finally silence all the critics.

The Vision 21 Program [73] is promoting research and innovation to maintain cost-competitive options for using a diverse mix of fossil fuels for power generation. A key element of this program is the effective removal of current environmental concerns and impediments associated with producing electricity and transportation of fuels from fossil fuels, including natural gas, petroleum, and coal.

The point is that we still have this technology and some reserves, so we are going to use all of them. Based on current conditions, the outlook for future energy supplies and conversion technologies indicates a growing reliance on natural gas as the economic fuel of choice for the generation of electric power. Although relatively low cost, abundant supplies of natural gas are projected to be available for the near future, we also identify as in Ref. [73] that the future is inherently uncertain and long-term projections of price inclinations and fuel mixtures could easily be wrong.

We must be able to hold a range of scenarios with regard to each possible application. Advanced fossil-fuel technologies, especially coal-based ones, must be able to meet increasingly stringent environmental requirements for "critical" air pollutants (sulfur dioxide, oxides of nitrogen), as well as other environmental issues (such as liquid and solid waste), and still remain cost competitive with other fossil fuels, especially natural gas [73].

Some First World countries are embarked upon an aggressive, well-financed, and deeply coordinated effort to switch from petroleum-based economies to hydrogen-based economies. A hydrogen economy is considered as the final task, coupled with major vehicle efficiency improvements and biofuels as intermediate steps to shift the world away from oil derivatives. This seems to be the new energy revolution since the vapor machine. The first nation to undergo the conversion to a hydrogen economy is Iceland where geothermal energy is used for producing the electrolysis-based hydrogen [74]. In spite of all the problems that we have addressed here, the hydrogen economy is attractive and will soon replace the less efficient and polluting conventional energy convertors.

REFERENCES

1. A. Capon and R. Parsons, *J. Electroanal. Chem.* **45** (1973) 205.
2. S. Wilhelm, T. Iwasita, and W. Vielstich, *J. Electroanal. Chem.* **238** (1987) 383.
3. J. B. Goodenough, A. Hamnett, B. J. Kennedy, and S.A. Weeks, *Electrochim. Acta* **32** (1987) 1233.
4. B. V. Tilak, P. W. Lu, J. E. Colman, and S. Srinivasan, Electrolytic production of hydrogen, in *Comprehensive Treatise of Electrochemistry, Vol. 2: Electrochemical Processing* (J. O.'M. Bockris, B. E. Conway, E. Yeager, and R. E. White, Eds.), Plenum Press, New York, London, 1981.
5. J. H. Russel, *Proceedings of the Symposium on Industrial Water Electrolysis*, Vol. 78–4 (S. Srinivasan, F. J. Salzano, and A. R. Landgrebe, Eds.), Electrochemical Society, Princeton, NJ, 1978, p. 77.
6. Proton Energy Systems, Inc., Hydrogen Technology Group, 10 Technology Drive, Wallingford, CT. www.protonenergy.com
7. National Renewable Energy Laboratory, 1617 Cole Boulevard, Golden, Colorado, www.nrel.gov. U.S. Department of Energy Office of Energy Efficiency and Renewable Energy by Midwest Research Institute.
8. Highly efficient hydrogen generation via water electrolysis using nanometal electrodes, September 2006, Energy Research Laboratory, QuantumSphere Inc., Santa Ana, CA.
9. H. Wuellenweber and J. Mueller, *Proceedings of the Symposium on Industrial Water Electrolysis*, Vol. 78–4 (S. Srinivasan, F. J. Salzano, and A. R. Landgrebe, Eds.), Electrochemical Society, Princeton, NJ, 1978, p. 1.
10. M. S. Casper, Ed., *Hydrogen Manufacture by Electrolysis, Thermal Decomposition and Unusual Techniques*, Noyes Data Corporation, Park Ridge, NJ, 1978.
11. P. W. T. Lu and S. Srinivasan, *J. Appl. Electrochem.* **9** (1979) 269.
12. R. L. Le Roy, Hydrogen in Canada's energy future, in *Hydrogen in Metals*, The Metallurgical Society of CIM, Montreal, Canada, 1978, pp. 1–10, Annual Volume.
13. M. J. Braun, *Proceedings of the Symposium on Industrial Water Electrolysis,* Vol. 78–4 (S. Srinivasan, F. J. Salzano, A. R. Landgrebe, Eds.), Electrochemical Society, Princeton, NJ, 1978, pp. 16–23.
14. R. L. Le Roy and A. K. Stuart, *Proceedings of the Symposium on Industrial Water Electrolysis*, Vol. 78–4 (S. Srinivasan, F. J. Salzano, and A. R. Landgrebe, Eds.), Electrochemical Society, Princeton, NJ, 1978, pp. 117–127.
15. K. Christiansen and T. Grundt, *Proceedings of the Symposium on Industrial Water Electrolysis*, Vol. 78–4 (S. Srinivasan, F. J. Salzano, and A. R. Landgrebe, Eds.), Electrochemical Society, Princeton, NJ, 1978, pp. 24–38.
16. W. E. Triaca, T. Kessler, J. C. Canullo, and A. J. Arvia, *J. Electrochem. Soc.* **134** (1987) 1165.
17. T. de los Ríos, D. Lardizabal Gutiérrez, V. Collins Martínez, and A. López Ortiz, *Int. J. Chem. Reactor Eng.* **3** (2005) A33.
18. A. Gaisch and T. Kessler, *Revista Información Tecnológica del Chile* **7** (1996) 125.
19. C. A. Marozzi and A. C. Chialvo, *Electrochim. Acta* **45** (2000) 2111.
20. L. J. Nuttall, A. P. Fickett, and W. A. Titterington, Hydrogen generation by solid polymer electrolyte technology, American Chemical Society Division of Fuel Chemistry Meeting, Chicago, IL, 1973.
21. D. J. Vaughn, *Du Pont Innovat.* **4** (1973) 10.
22. P. W. T. Lu and S. Srinivasan, *J. Appl. Electrochem.* **9** (1979) 269.
23. T. Kato, M. Kubota, N. Kobayashi, and Y. Suzuoki, Effective utilization of by-product oxygen of electrolysis hydrogen production, Nagoya University, Furo-cho, Chikusa-ku, Nagoya, Japan, 2007.
24. B. E. Eliassen, Norsk hydro electrolyzers electrolyzer hydrogen plant 485 Nm^3/h hydrogen. Norsk Hydro Electrolyzers, AS, Aug. 8, 2002.
25. P. W. Lu and R. L. Ammon, Extended Abstracts of the Fall 1979 Meeting of the Electrochemical Society, Electrochemical Society, Princeton, NJ, October 1979.
26. R. W. Coughlin and M. Farooque, *Nature* **279** (1979) 301.
27. M. H. Miles, G. Kissel, P. W. Lu, and S. Srinivasan, *J. Electrochem. Soc.* **123** (1976) 332.
28. R. L. Leroy, C. T. Bowen, and D. J. Leroy, *J. Electrochem. Soc.* **127** (1980) 1954.

29. L. Elikan, J. P. Morris, and C. K. Wu, Development of a solid electrolyte system for oxygen reclamation, report prepared under Contract NASI-8896, Westinghouse Electric Corporation for NASA, 1971.
30. H. S. Spacil and C. S. Tedmon, Electrochemical dissociation of water vapour in solid oxide electrolyte cells, thermodynamics and materials. Series connected multiple cell stacks, in *Extended Abstracts of Spring Meeting of the Electrochemical Society*, Vol. 68–1, Electrochemical Society, Princeton, NJ, 1968, p. 194.
31. F. J. Rohr, *Proceedings of the Workshop of High Temperature Solid Oxide Fuel Cells*, May 5–6, 1977, Brookhaven National Laboratory, Upton, New York, 1977, p. 122.
32. A. P. Isenberg, *Proceedings of the Symposium in Electrode Materials and Processes for Energy Conversion and Storage*, Vol. 77–6, Spring, 1977, Meeting of the Electrochemical Society, Philadelphia (J. D. Mc. Intyre, S. Srinivasan, and F. G. Wills, Eds.), Electrochemical Society, Princeton, NJ, 1977, p. 682.
33. A. Fujishima and K. Honda, *Nature* **238** (1972) 37.
34. J. O'M. Bockris and K. Uosaki, *Int. J. Hydrogen Energy* **2** (1977) 123.
35. M. H. Miles, Y. H. Huang, and S. Srinivasan, *J. Electrochem. Soc.* **125** (1978) 1931.
36. L. Giuffre, P. M. Spaziante, and A. Nidola, *Proceedings of the Symposium on Industrial Water Electrolysis*, Vol. 78–4 (S. Srinivasan, F. J. Salzano, and A. R. Landgrebe, Eds.), Electrochemical Society, Princeton, NJ, 1978, p. 132.
37. S. Srinivasan and F. J. Salzano, *Int. J. Hydrogen Energy* **2** (1977) 53.
38. P. J. Moran and G. E. Stoner, *Proceedings of the Symposium on Industrial Water Electrolysis*, Vol. 78–4 (S. Srinivasan, F. J. Salzano, and A. R. Landgrebe, Eds.), Electrochemical Society, Princeton, NJ, 1978, p. 169.
39. J. Kiefer, H. Brack, J. Huslage, F. Buchi, A. Tsakada, F. Geiger, and G. Scherer *Proceedings of the European Fuel Cell Forum, Portable Fuel Cells Conference*, Lucerne, Switzerland, 1999, pp. 227–235.
40. R. A. Smith, Co-extruded multilayer cation exchange membranes, U.S. Patent 4,437,952, 1984.
41. K. Dhathathreyan and N. Rajalakshmi, Polymer electrolyte membrane fuel cell, in *Recent Trends in Fuel Cell Science and Technology* (S. Basu, Ed.), Springer Verlag, Anamaya, India, 2007, pp. 40–115.
42. S. Cleghhorn, J. Kolde, and W. Liu, Catalyst coated composite membranes, in *Handbook of Fuel Cells: Fundamentals, Technology and Applications*, Vol. 3, Part 3 (W. Vielstich, A. Lamm, and H. Gasteiger, Eds.), John Wiley & Sons, Chichester, U.K., 2003, pp. 566–575.
43. T. L. Yu, H-L. Lin, K-S- Shen, L-N. Huang, Y-C. Chang, G.-B. Jung, and J. C. Huang, *J. Polym. Res.* **11** (2004) 217.
44. J. Kohler, K.-A. Starz, S. Witthal, and M. Diehl, Process for producing a membrane electrode assembly for fuel cells, U.S. Patent 2002/0064593 A1, 2002.
45. C. Preischel, P. Hedrick, and A. Hahn, Continuous method for manufacturing a laminated electrolyte and electrode assembly, U.S. Patent 6,291,091 B1, 2001.
46. J. M. Norbeck and M.N. McClanahan, Clean air now solar hydrogen project, University of California, Riverside, CA, CE-CERT, 1995–1996.
47. Electrode for the electrolysis of water, U.S. Patent 4,348,268, BBC Aktiengesellschaft Brown, Boveri & CIE, May 18, 1983.
48. H. Wullenweber and J. Mueller, *Proceedings of the Symposium on Industrial Water Electrolysis*, Vol. 78–4 (S. Srinivasan, F. J. Salzano, and A. R. Landgrebe, Eds.), Electrochemical Society, Princeton, NJ, 1978, pp. 1–15.
49. Water electrolysis—A simple method to produce high purity hydrogen, Commercial Brochure from Panchlor Chemicals Ltd., Milan, Italy (new De Nora).
50. F. P. Dousek and K. Micka, *J. Appl. Electrochem.* **23** (1993) 241.
51. J. Rifkin, *The Hydrogen Economy*, Penguin Putman, New York, 2002.
52. U. Bossel, *European Fuel Cell Forum International Conference, Intelec '05*, Berlin, Germany, September 18–22, 2005.
53. A. Damjanovic, M. A. Genshaw, and J. O'M. Bockris, *J. Chem. Phys.* **45** (1966) 4057.
54. R. Holze, I. Vogel, and W. Vielstich, *J. Electroanal. Chem.* **210** (2001) 83.
55. E. Brosha, J-H.Choi, J. Davey, F. Garzón, C. Hamon, B. Piela, J. Ramsey, F. Uribe, and P. Zelenay, Non-precious metal catalysts, in *Hydrogen, Fuel Cells and Infrastructure Technology Program*, 2005 Annual Review, Washington DC, May 23–27, 2005.
56. J. Ivy, Summary of electrolytic hydrogen production milestone completion report, September 2004. National Renewable Energy Laboratory, Golden, Colorado. www.nrel.gov, Operated for the U.S. Department of Energy Office of Energy Efficiency and Renewable Energy by Midwest Research Institute, Battelle Contract No. DE-AC36-99-GO10337.

57. Committee on alternatives and strategies for future hydrogen production and use, National Research Council. The hydrogen economy: Opportunities, costs, barriers, and R&D needs. The National Academy Press, Washington, DC, 2004.
58. Mahler AGS—Advanced Gas Systems, Engineering, Mahler AGS GmbH, Stuttgart, Germany, mahler-ags. com
59. Hydrogen Production, U.S. Department of Energy, Washington, DC, Energy efficiency and renewable energy, 2008 Annual Report.
60. M. Glass, Fuel cell codes and standards summit III summary, April 5–7, 1999, Pacific Northwest Labs, Richland, WA.
61. ASME, New York, Object and scope for the proposed code on fuel cell power systems, August 2000, http://www.asme.org
62. R. Drnevich, Hydrogen delivery: Liquefaction and compression, in Strategic Initiatives for Hydrogen Delivery Workshop, Praxair, Tonawnda, NY, May 7, 2003.
63. T. M. Flynn, *Cryogenic Engineering*, Dekker, New York, 1997, p. 284.
64. P. Häussinger et al., Hydrogen, in *Ullmann's Encyclopedia of Industrial Chemistry*, 5th edn., Vol. A-13, VCH Verlagsgesellschaft, Weinheim, Germany, 1989.
65. M. J. Hiza, A. J. Kidnay, and R. C. Miller, *Equilibrium Properties of Fluid Mixtures*, IFI/Plenum, New York, 1975.
66. W. Wagner and A. Pruβ, The IAPWS formulation 1995 for the thermodynamic properties of ordinary water substance for general and scientific use. *J. Phys. Chem. Ref. Data* **31** (2002) 387–535.
67. L. Tse, Hydrogen and fuel cell photo gallery, 2002 (Hannover Fair), Vision Engineer.
68. M. Porter, Towards a new conception of the environment-competitiveness relationship. Environmental Protection Agency Clean Air Marketplace, Washington, DC, September 1993.
69. A. J. Appleby and F.R. Foulkes, *Fuel Cell Handbook*, Van Norstand Reinhold, New York, 1989.
70. R.F. Buswell, J. V. Clause, R. Cohen, C. Louie, and D.S. Watkins, Hydrocarbon fueled solid polymer fuel cell electric power generation system, Ballard U.S. Patent 5,360,679, 1994.
71. K. Kordesch and G. Simader, *Fuel Cells and Their Applications*, VCH Press, New York, 1996.
72. U. Stimming et al., *Proceedings of the Fifth International Symposium on Solid Oxide Fuel Cells*, Vol. 97–40, The Electrochemical Society, Princeton, NJ, 1997, p. 69.
73. Vision 21: Fossil fuel options for the future committee on R&D opportunities for advanced fossil-fueled energy complexes, Board on Energy and Environmental Systems, National Research Council, Washington, DC, 2000.
74. D. Sjoding and E. Hamernyik, Overview of hydrogen and fuel cells in Washington State, Washington State University, Olympia, WA, September 2008.

Index

633